Paul Donaldson

Progress in Cell Research

Volume 3

Gap Junctions

Progress in Cell Research

Volume 3

Gap Junctions

Editors

J.E. Hall

Department of Physiology and Biophysics
University of California
Irvine, California
USA

G.A. Zampighi

Department of Anatomy and Cell Biology
UCLA Medical Centre
University of California
Los Angeles, California
USA

R.M. Davis

Department of Anatomy and Cell Biology
UCLA Medical Centre
University of California
Los Angeles, California
USA

1993

ELSEVIER
AMSTERDAM — LONDON — NEW YORK — TOKYO

Elsevier Science Publishers B.V.
P.O. Box 211
1000 AE Amsterdam
The Netherlands

Library of Congress Cataloging-in-Publication Data

Gap junctions / editors, J.E. Hall, G.A. Zampighi, R.M. Davis.
p. cm. – (Progress in cell research, ISSN 09248315 ; v. 3)
Includes bibliographical references and index.
ISBN 0-444-89871-9 (alk. paper)
1. Gap junctions (Cell biology) I. Hall, J.E. (James E.)
II. Zampighi, G.A. (Guido A.) III. Davis, R.M. (Ruth M.) IV. Series.
[DNLM: 1. Intercellular Junctions–congresses. W1 PR6671L v. 3
1993 / QH 603.C4 G2113 1993]
QH603.C4G36 1993
574.87'6–dc20
DNLM/DLC
for Library of Congress 92-48435
CIP

ISBN 0-444-89871-9 (volume)
ISSN 0924-8315 (series)

This book is printed on acid-free paper

Printed in The Netherlands

Dedicated to
J.D. Robertson and J.-P. Revel

Prologue

Although specialized regions of cell-to-cell contact were known to cytologists of the last century, notably Malpighi, modern understanding of these contacts (called "junctions") is based on studies performed in the early sixties. A combination of electron microscopy and electrophysiology showed that a variety of tissues from different animal species have their cells connected by junctions having characteristic structures and performing distinct functions. This is beautifully illustrated in epithelia where three different junctions, the tight junction, the gap junction and the desmosome are arranged sequentially between the cells to form a junctional complex. Evidence accumulated showing that the tight junctions control the permeability of the extracellular clefts between cells (the "paracellular" pathway), gap junctions the permeability between the cytoplasm of adjacent cells (the "cell-to-cell" pathway) and desmosomes cell adhesion.

In contrast to tight junctions, found only in epithelia, gap junctions are present in nearly all tissues, regardless of their embryonic origin. These include (but are not limited to) the central nervous system of both vertebrates and invertebrates, cardiac and smooth muscle fibers and, with the exception of mature skeletal muscle and blood, all non-excitable tissues of the body.

Because they are ubiquitous, gap junctions have long been of great interest to scientists from many different disciplines. The 1991 International Meeting on Gap Junctions which took place at Asilomar, California, collected together 157 scientists from 12 countries and almost as many scientific disciplines. This book is based on the proceedings of that meeting. The organization of the book is based on the organization of talks presented at the conference, but in a few cases we felt the reader would be better served by slight rearrangement. A few papers, presented as abstracts, not talks, at the meeting, were included in the book because we felt they had special value. The seven Parts of the book progress (for the most part) from general topics (The Connexin Family, Structure, Hemichannels, Biophysics) to specific ones (Role of Gap Junctions in Various Tissues, Regulation and Biochemistry, Cancer).

The connexin family

Biochemistry and molecular biology have provided the amino acid sequences of the proteins composing gap junctions and have defined the connexin family. The six chapters in Part I describe the connexin family from several points of view. Chapter 1 details differences in the properties of connexins expressed in oocytes and provides a good example of the emerging partnership between electrophysiology and molecular biology. Chapter 2 shows how PCR can be exploited to find new connexins. Chapter 3 addresses the question of how different connexins interact with each other to form (or not to form) hybrids. Chapter 4 describes the expression of Cx43 (α_1) and its co-localization with tight junction proteins and other epitopes. Chapter 5 describes the mouse connexin family and compares various mouse sequences with each other and other connexin proteins. From these chapters the reader can gain an up-to-date idea of what is now known of the homologies, sequences and topology of the members of the connexin family.

Structure

Effective detergent solubilization of membrane proteins revolutionized membrane biochemistry and permitted isolation of highly enriched fractions of gap junctions plaques from liver hepatocyte and heart myocytes. Because the connexons are ordered in hexagonal lattices, their structure can be reconstructed using electron microscopy and low-angle x-ray diffraction of partially oriented pellets. Both methods showed that each channel is composed of two hexamers, one per cell, joined through their external domains. The 12-subunit assembly surrounds a narrow cavity (almost certainly the pore) that spans both plasma membranes and the intervening extracellular gap. A model derived by comparing three-dimensional maps of isolated junctions in two different conformations proposes that the channels gate from the open to the closed conformation by tilting and sliding of the essentially rigid subunits. The four chapters in Part II describe recent work on the structure of gap junctions. Chapter 6 describes the first application of atomic force microscopy to the study of gap junctions and offers the prospect of not only imaging at high resolution, but of determining the magnitudes of the forces which hold junctions together. Chapter 7 provides a detailed structural study of cardiac gap junctions. Chapter 8 describes the use of the baculovirus system to over-produce gap junction protein and the first attempts to grow crystals of β_1. Chapter 9 describes the assembly of gap junctions in vitro and the study of their structure using transmission and STEM electron microscopy. Part II thus summarizes our present knowledge of the structure of gap junctions and suggests that new techniques will soon yield both higher-resolution structures and measurements of the actual forces which hold junctions together.

Hemichannels

We are only beginning to probe the assembly of gap junctions, but already a number of surprises have turned up. It used to be a matter of dogma amongst gap-junction scientists that the large conductance of gap-junction channels meant that functional hemichannels connecting the cytoplasm to extracellular fluid would kill the cell. It appears this notion will have to be re-examined. In Part III, three papers describe evidence that functional hemichannels may arise under a variety of circumstances. Chapter 10 discusses the possibility that ATP-induced pores in macrophages may be hemichannels. Chapter 11 describes the formation of hemichannels formed in oocytes injected with Cx46, and Chapter 12 presents compelling evidence that both hemichannels and true gap junctions exist in retinal horizontal cells.

Biophysics

Methods for studying the functional properties of the channel have advanced impressively. Very early on it was shown that gap-junction channels allowed the diffusion of low-molecular-weight uncharged molecules as well as ions. Double-patch clamping, now routine, has provided values for the single channel conductance of a number of different junctions in a variety of systems. Two-cell voltage clamping has also been successfully applied in the oocyte expression system. The five chapters of Part IV describe the biophysics of gap-junction channels in a variety of interesting systems.* Chapter 13 describes single-channel and macroscopic conductances of gap junctions in developing chick heart. Chapter 14 provides a detailed dissection of the single-channel conductance levels, addressing particularly the question of multiple conductance levels. Chapter 15 discusses the voltage dependence of gap junctions in hepatocytes and of both homotypic and heterotypic channels of Cx26 and Cx32 expressed in oocytes. Chapter 16 expands our knowledge of agents affecting the gating of gap junctions and presents the case that lipophilic compounds close gap junctions by disordering the hydrocarbon region of the membrane near gap-junction proteins. Chapter 17 explores heterogeneous channel behavior in gap junctions of the earthworm and demonstrates that these junctions very likely contain heterogeneous populations of channels. Chapter 18 shows both the promise and the problems of studying gap junctions in transfected cells. The promise is that wonderful data can be obtained on the properties of the transfected connexin and the problem is that the expression system may be modifying the transfected protein so that its properties in the transfected cell differ from those it would have in its normal environment (cf. Chapter 1). Nevertheless, the chapters in Part IV show the biophysics of gap-junction channels is progressing at an exciting pace.

Role of gap junctions in various tissues and organisms

Although knowledge of the biochemical, biophysical and, to some extent, structural properties of gap junctions has grown rapidly in recent years, understanding precisely what gap junctions do is a much tougher problem. Part V emphasizes one reason this problem is so difficult: gap junctions apparently perform a variety of quite different functions ranging from the conduction of the cardiac action potential (perhaps their

* Considerable editorial agony was expended over the possibility that some of the papers in Part IV should perhaps have been in Part I and vice versa.

best-understood role) to synaptic plasticity and pattern formation in development. The diversity of topics covered in Part V is very great. Nevertheless all the chapters are united by the attempt to understand what role gap junctions actually play in a particular system.

Nervous system

Part V begins with the role of gap junctions in the nervous system. Chapter 19 describes the alteration of gap-junction conductance as means of altering synaptic strength in a mixed synapse. Chapter 20 demonstrates that gap junctions are more widely distributed in the CNS than expected and suggests that steroids may regulate their function. Taken together, Chapters 19 and 20 indicate the roles of gap junctions in the central nervous system may be much larger than previously suspected.

Lens and heart

A diverse set of junctional proteins is expressed in the lens, and it is still not quite clear if all of them are gap-junction proteins. Chapters 21, 22, and 23 describe the reconstitution of various junctional proteins from the lens into planar lipid bilayers and liposomes, and each chapter takes a different position on the question of whether MIP26 forms communicating channels. Chapter 24 describes the expression of multiple connexin proteins in both heart and lens, and Chapter 25 describes the expression and organization of various junctional proteins in the lens. Both studies make heavy use of immunocytochemistry.

Chapter 26 could well have been included in Part IV as its thorough discussion of the voltage dependence of cardiac gap-junction channels is heavily biophysical. Chapter 27 describes reduction of the coupling between cardiac myocytes in Chargas' disease and concludes that this reduction likely contributes to the conduction disorders associated with this infection.

Pancreas and salivary gland

Gap junctions are morphologically and physiologically prominent in secreting glands. Chapters 28 and 29 describe the extent and modulation of gap-junctional coupling in pancreas and salivary gland. Both studies elucidate mechanisms of both modulation of junctional conductance and the role of gap junctions in the process of coordinated glandular secretion.

Vascular smooth muscle: Corpus Cavernosum

Chapter 30 describes the role of gap junctions in maintaining vascular tone in smooth muscle. The data presented support the conclusion that gap junctions play an important role in the α_1-adrenergic contractions in human vascular smooth muscle, and suggest gap junctions and smooth muscle will be the object of many future studies.

Development

It has been clear for some time that gap junctions play a role in development, almost certainly in pattern formation in the early embryo, and perhaps in the continuing differentiation of stem cells in the maintenance of adult tissues. Chapter 31 describes the possible role of gap junctions in mesoderm induction in the embryo of the mollusc *Patella vulgata*. Chapter 32 describes the coupling pattern in human epidermis, the effects of retinoic acid on coupling, and the distribution of connexin proteins. Chapter 33 recounts the patterns of expression of α_1 and β_2 gap-junction proteins during the development of rat skin and hair.

Insect

Chapters 34 and 35 both describe studies on the gap junctions of insects, and both are sufficiently biophysical in nature that they could easily have been included in Part IV. (More editorial agony.) Nevertheless, the gap junctions of arthropods, represented here by insects, constitute a special class of gap junctions, distinguished (at least at the time of this writing) by their singular resistance to isolation or cloning. It will be a great day for the field when the first sequence of an arthropod gap junction becomes available.

Regulation and biochemistry

Part V deals with the expanding subject of regulation and biochemistry of gap junctions, and we can be confident that the papers in this section are heralds of many exciting results to come. Chapter 36 describes an investigation into the cell biology of gap junctions: intracellular transport, assembly into junctions and phosphorylation. Chapter 37 covers similar ground in a different cell system and identifies different forms of gap junctions found at different stages of processing. Chapter 38 addresses the role of phosphorylation in regulating junctional communication in MDCK cells, and Chapter 39 investigates the same subject in cardiac myocytes. Chapter 40 presents data suggesting that the external domains of gap-junction proteins are required for the assembly of functional cell-to-cell channels. Chapter 41 describes regulatory elements in the rat Cx32 gene, including the evidence for multiple transcription-initiation sites and the localization of the native promoter.

Cancer

Part VII contains four papers which describe the accumulating evidence that at least some forms of cancer involve loss of gap-junction communication or regulation. There is also evidence hinting that restoring junctional communication may restrict tumor growth in some cases. Chapter 42 describes the up-regulation of junctional communication by both carotenoids and retinoids and correlates these with effects on growth control and cancer prevention. Chapter 43 describes losses of cell-to-cell communication between preneoplastic foci and surrounding cells and investigates alterations in cell-to-cell communication and gap-junction expression during various stages of human and animal carcinogenesis. Chapter 44 describes how transfection of a gap-junction gene into a communication-deficient tumor cell line retards tumor growth *in vivo*. Chapter 45 investigates the role of TPA and growth factors in the suppression of gap-junctional communication.

Summary: nomenclature, a personal note and acknowledgements

Even the not-so-discerning reader will have noticed the studied editorial avoidance of the dread nomenclature issue. Both current nomenclatures have clear advantages and equally clear disadvantages. Cynics frequently point out that the reason semantic problems take so much time to sort out is because there is so little at stake. Everyone agrees that the family of gap junction-forming proteins is called the connexin family. Most investigators identify the gap junction-forming proteins as connexin XX, where the equis represent the M_r of the protein calculated from its DNA sequence. The principal liver hepatocyte connexin is thus Cx32 and the heart myocyte protein Cx43. A newer nomenclature, based on analysis and comparison of different connexin sequences, distinguishes two subfamilies called α and β. Each connexin is denoted α or β on the basis of its sequence and a subscript is attached corresponding to the temporal order of cloning. Thus, Cx32 and Cx43 are β_1 and α_1, respectively. The Human Genome Project accepts the alpha-beta nomenclature. People interested in gap junctions will likely have to be conversant with both nomenclatures for some time to come.

We are all too aware that this book did not edit itself, but we did have an extraordinary amount of help. The organizing committee of the meeting, Dave Spray, Gerhard Dahl, Bruce Nicholson, Barbara Yancey, and *ex officio*, Guido Zampighi and Jim Hall, produced a program which represented a complex field extraordinarily well. The investigators, especially the younger ones attending perhaps their first international meeting, added considerable zest to the proceedings and some abstracts submitted as posters were so interesting they were promoted to papers in this volume. All of the contributors have been forbearing in the extreme. Those few who were unlucky enough to have been victims of editorial slings and arrows ("I'm terribly sorry, but I've lost the disk of your manuscript, could you send me another one. I need it yesterday." ... "We'll have to cut eight of your nine figures. Which one would you like to keep?") were unfailingly cheerful and helpful. In fact interacting with contributors has been one of the most positive aspects of the whole task. Our thanks to all of you.

Without the agencies and individuals who supported the meeting, this volume would not exist. Larry Wheeler of Allergan gave us some seed money to start the whole thing going. Eve Barak and the National Science Foundation (NSF 910034), Loré Anne McNichol of the National Eye Institute and the National Institutes of Health (NIH EY09132), and Igor Vodyanoy and the Office of Naval Research (ONR task 441k911) provided funds for supporting travel and other conference expenses. And, of course, the Asilomar Conference Center staff were essential in helping the meeting to run smoothly in a marvelous natural setting. Again, our thanks to all of you.

August 1992

Jim Hall
Guido Zampighi
Ruth Davis

Contributors

M. Arneson, *Department of Veterinary Biology, University of Minnesota, 1988 Fitch Avenue, Room 295m, St Paul, MN 55108, USA*

M.F. Arnsdorf, *Section of Cardiology, Department of Medicine, University of Chicago, Chicago, IL 60637, USA*

S. Bai, *Marion Bessin Liver Research Center, Department of Medicine, Albert Einstein College of Medicine, 1300 Morris Park Avenue, Bronx, NY 10461, USA*

T.A. Bargiello, *Department of Neuroscience, Albert Einstein College of Medicine, Bronx, NY, USA*

L. Barrio, *Department of Neuroscience, Albert Einstein College of Medicine, 1300 Morris Park Avenue, Bronx, NY, USA*

M.V.L. Bennett, *Department of Neuroscience, Albert Einstein College of Medicine, 1300 Morris Park Avenue, Bronx, NY, USA*

K. Berg, *Department of Pharmacology, SUNY Health Science Center, 750 E. Adams Street, Syracuse, NY 13210, USA*

V.M. Berthoud, *Department of Neuroscience, Albert Einstein College of Medicine, 1300 Morris Park Ave., Bronx, NY 10461, USA*

J.S. Bertram, *Molecular Oncology Unit, Cancer Research Center of Hawaii, University of Hawaii, Honolulu, HI 96813, USA*

E.C. Beyer, *Departments of Pediatrics, Medicine and Cell Biology, Washington University School of Medicine, St. Louis, MO 63110, USA*

J.-P. Briand, *Institut de Biologie Moléculaire et Cellulaire, CNRS, Laboratoire d'Immunochimie, 6708, Strasbourg Cédex, France*

P.R. Brink, *Department of Physiology and Biophysics School of Medicine, HSC, SUNY at Stony Brook, Stony Brook, NY 11794, USA*

R. Bruzzone, *Department. of Anatomy and Cellular Biology, Harvard Medical School, 220 Longwood Avenue, Boston, MA 02115, USA*

F.F. Bukauskas, *Kaunas Medical Academy, Mickeviciaus 9, 233000 Kaunas, Lithuania*

R.D. Burk, *Marion Bessin Liver Research Center, Departments of Pediatrics and Microbiology and Immunology, Albert Einstein College of Medicine, 1300 Morris Park Avenue, Bronx, NY 10461, USA*

J.M. Burt, *Department of Physiology, University of Arizona, Tucson, AZ 85724, USA*

A.C. Campos De Carvalho, *Department of Neuroscience, Albert Einstein College of Medicine, 1300 Morris Park Avenue, Bronx, NY 10461, USA*

F.L. Cao, *Department of Biological Sciences, SUNY Buffalo, Buffalo, NY 14260, USA*

S. Caveney, *Department of Zoology, University of Western Ontario, London, Ontario, Canada N6A 5B7*

M. Chanson, *Department of Morphology, University of Geneva, Medical School, 1211 Geneva, Switzerland*

Y.-H. Chen, *Department of Anatomy and Cell Biology, Emory University Health Science Center, Atlanta, GA 30322, USA*

G.J. Christ, *Department of Urology, Albert Einstein College of Medicine, 1300 Morris Park Avenue, Bronx, NY 10461, USA*

D. Churchill, *Department of Zoology, University of Western Ontario, London, Ontario, Canada N6A 5B7*

A.J. Czernik, *Laboratory of Molecular and Cellular Neuroscience, The Rockefeller University, New York, NY 10021-6399, USA*

E. Dahl, *Institut für Genetik, Abteilung Molekulargenetik, Römerstrasse 164, 53 Bonn 1, Germany*

G. Dahl, *Department of Physiology and Biophysics, University of Miami, School of Medicine, Miami, FL 33101, USA*

P. Damen, *Department of Experimental Zoology, University of Utrecht, 3584 CH Utrecht, The Netherlands*

R.L. DeHaan, *Department of Anatomy and Cell Biology, Emory University Health Science Center, Atlanta, GA 30322, USA*

R. Dermietzel, *Department of Neuroscience, Albert Einstein College of Medicine, Bronx, NY 10461, USA*

S.H. DeVries, *Department of Neurobiology, Stanford University, Fairchild D-238, Stanford, CA 94305, USA*

W.J.A.G. Dictus, *Department of Experimental Zoology, University of Utrecht, 3584 CH Utrecht, The Netherlands*

P. Donaldson, *Department of Cellular and Molecular Biology, University of Auckland, Auckland, New Zealand*

B. Drake, *Imaging Services, Box 9981, Truckee, CA 95737, USA*

E. Dupont, *Department of Pediatrics/Human Development, Michigan State University, East Lansing, MI 48824, USA*

L. Ebihara, *Department of Pharmacology, Columbia University, 630 W. 168th Street, New York, NY 10032, USA*

B. Eghbali, *Department of Neuroscience, Albert Einstein College of Medicine, 1410 Pelham Parkway South, Bronx, NY 10461, USA*

G.R. Ehring, *Department of Physiology and Biophysics, UC Irvine, Irvine, CA 92717, USA*

A. El Aoumari, *Laboratoire de Biologie de la Différenciation Cellulaire, UA CNRS 179, Faculté des Sciences de Luminy, Université d'Aix-Marseille II, 13288, Marseille Cédex 9, France*

A. Engel, *M.E. Müller Institute for High-Resolution Electron Microscopy at the Biocenter, University of Basel, CH-4056 Basel, Switzerland*

D.S. Faber, *Department of Anatomy and Neurobiology, Medical College of Pennsylvania, 3300 Henry Avenue, Philadelphia, PA 19129, USA*

G. Feldmann, *Laboratoire de Biologie Cellulaire, INSERM Unité 327, Faculté de Médecine Xavier Bichat, Université Paris VII, 75018, Paris, France*

R.S. Fisher, *Departments of Psychiatry and Biobehavioral Sciences, Anatomy and Cell Biology, Mental Retardation Research Center and Brain Research Institute, School of Medicine, University of California at Los Angeles, Los Angeles, CA 90024, USA*

G.I. Fishman, *Department of Medicine, Albert Einstein College of Medicine, Bronx, NY 10461, USA*

D.J. Fitzgerald, *Public and Environmental Health Division, South Australia Health Commission, Adelaide SA, Australia*

C. Fourtner, *Department of Biological Sciences, SUNY Buffalo, Buffalo, NY 14260, USA*

C. Fromaget, *Laboratoire de Biologie de la Différenciation Cellulaire, UA CNRS 179, Faculté des Sciences de Luminy, Université d'Aix-Marseille II, 13288, Marseille Cédex 9, France*

N.B. Gilula, *Department of Cell Biology, The Scripps Research Institute, 10666 N. Torrey Pines Road, La Jolla, CA 92037, USA*

C.M. Gondré, *Department of Urology, Albert Einstein College of Medicine, 1300 Morris Park Avenue, Bronx, NY 10461, USA*

D.A. Goodenough, *Department of Anatomy and Cellular Biology, Harvard Medical School, Boston, MA 02115, USA*

D. Gros, *Laboratoire de Biologie de la Différenciation Cellulaire, UA CNRS 179, Faculté des Sciences de Luminy, Université d'Aix-Marseille II, 13288, Marseille Cédex 9, France*

J.-A. Haefliger, *Department. of Anatomy and Cellular Biology, Harvard Medical School, 220 Longwood Avenue, Boston, MA 02115, USA*

J.E. Hall, *Department of Physiology and Biophysics, UC Irvine, Irvine, CA 92717, USA*

P. Hammernick, *Department of Biological Sciences, SUNY Buffalo, Buffalo, NY 14260, USA*

H. Hennemann, *Institut für Genetik, Abteilung Molekulargenetik, Römerstrasse 164, 53 Bonn 1, Germany*

E.L. Hertzberg, *Department of Neuroscience, Albert Einstein College of Medicine, 1300 Morris Park Ave., Bronx, NY 10461, USA*

C. Hirono, *Department of Physiology, Hiroshima University School of Dentistry, 2-3 Kasumi 1 Chome, Minami-ku, Hiroshima 734, Japan*

K.K. Hirschi, *Department of Physiology, University of Arizona, Tucson, AZ 85724, USA*

J.H. Hoh, *Division of Biology, California Institute of Technology, Pasadena, CA 91125, USA*

M.Z. Hossain, *Molecular Oncology Unit, Cancer Research Center of Hawaii, University of Hawaii, Honolulu, HI 96813, USA*

T. Jarry, *Laboratoire de Biologie de la Différenciation Cellulaire, UA CNRS 179, Faculté des Sciences de Luminy, Université d'Aix-Marseille II, 13288, Marseille Cédex 9, France*

L. Jarvis, *Department of Veterinary Biology, University of Minnesota, 1988 Fitch Avenue, Room 295m, St Paul, MN 55108, USA*

S.A. John, *Division of Biology, California Institute of Technology, Pasadena, CA 91125, USA*

R.G. Johnson, *Department of Genetics and Cell Biology, University of Minnesota, St. Paul, MN 55108, USA*

W.M.F. Jongen, *Agrotechnological Research Institute ATO, Wageningen, The Netherlands*

H.J. Jongsma, *Department of Physiology, University of Amsterdam, Meibergdreef 15, 1105 AZ Amsterdam, The Netherlands*

S. Jungbluth, *Institut für Genetik, Abteilung Molekulargenetik, Römerstrasse 164, 53 Bonn 1, Germany*

Y. Kanno, *Department of Physiology, Hiroshima University School of Dentistry, 2-3 Kasumi 1 Chome, Minami-ku, Hiroshima 734, Japan*

H.L. Kanter, *Department of Medicine, Washington University School of Medicine, St. Louis, MO 63110, USA*

C. Kempf, *Department of Biochemistry and Central Laboratory, Blood Transfusion Service, Swiss Red Cross, University of Berne, Freistr. 3, 3012 Berne, Switzerland*

J.A. Kessler, *Departments of Neuroscience and Neurology, Albert Einstein College of Medicine, 1410 Pelham Parkway South, Bronx, NY 10461, USA*

J. Kistler, *Department of Cellular and Molecular Biology, Centre of Gene Technology, University of Auckland, New Zealand*

G.F. Klier, *Department of Cell Biology, The Scripps Research Institute, 10666 N. Torrey Pines Road, La Jolla, CA 92037, USA*

V. Krutovskikh, *Unit of Multistage Carcinogenesis, International Agency for Research on Cancer, 150, cours Albert Thomas, 69372 Lyon Cédex 08, France*

N.M. Kumar, *Department of Cell Biology, The Scripps Research Institute, 10666 N. Torrey Pines Road, La Jolla, CA 92037, USA*

D.W. Laird, *Division of Biology, California Institute of Technology, Pasadena, CA 91125, USA*

R. Lal, *Section of Cardiology, Department of Medicine, University of Chicago, Chicago, IL 60637, USA*

P.D. Lampe, *Department of Genetics and Cell Biology, University of Minnesota, St. Paul, MN 55108, USA*

D.M. Larson, *Mallory Institute of Pathology, Boston University School of Medicine, Boston, MA 02118, USA*

M.L.S. Ledbetter, *Department of Biology, College of the Holy Cross, Worcester, MA 01610, USA*

E. Levine, *Department of Physiology and Biophysics, University of Miami, School of Medicine, Miami, FL 33101, USA*

C.F. Louis, *Departments of Biochemistry and Veterinary Biology, University of Minnesota, 1988 Fitch Avenue, Room 295m, St. Paul, MN 55108, USA*

B.V. Madhukar, *Department of Pediatrics/Human Development, Michigan State University, East Lansing, MI 48824, USA*

K. Manivannan, *Department of Physiology and Biophysics School of Medicine, HSC, SUNY at Stony Brook, Stony Brook, NY 11794, USA*

B. Maro, *Institut Jacques Monod, UMR CNRS 3, Université Paris VII, 75005 Paris, France*

E. Masgrau, *Clinic of Dermatology, University of Geneva, Medical School, 1211 Geneva, Switzerland*

K.D. Massey, *Department of Physiology, University of Arizona, Tucson, AZ 85724, USA*

R.T. Mathias, *Department of Physiology and Biophysics School of Medicine, HSC, SUNY at Stony Brook, Stony Brook, NY 11794, USA*

M. Maurice, *Laboratoire de Biologie Cellulaire, INSERM Unité 327, Faculté de Médecine Xavier Bichat, Université Paris VII, 75018, Paris, France*

P. Meda, *Clinic of Dermatology, University of Geneva, Medical School, 1211 Geneva, Switzerland*

A. Melman, *Department of Urology, Albert Einstein College of Medicine, 1300 Morris Park Avenue, Bronx, NY 10461, USA*

M. Mesnil, *Unit of Multistage Carcinogenesis, International Agency for Research on Cancer, 150, cours Albert Thomas, 69372 Lyon Cédex 08, France*

R.A. Meyer, *Department of Genetics and Cell Biology, University of Minnesota, St. Paul, MN 55108, USA*

P.E Micevych, *Department of Anatomy and Cell Biology, Brain Research Institute and Laboratory of Neuroendocrinology, School of Medicine, University of California at Los Angeles, Los Angeles, CA 90024, USA*

B.N. Minnich, *Department of Physiology, University of Arizona, Tucson, AZ 85724, USA*

L.K. Moore, *Department of Physiology, University of Arizona, Tucson, AZ 85724, USA*

A.P. Moreno, *Department of Neuroscience, Albert Einstein College of Medicine, 1300 Morris Park Avenue, Bronx, NY 10461, USA*

L.S. Musil, *Department of Anatomy and Cellular Biology, Harvard Medical School, Boston, MA 02115, USA*

A.C. Nairn, *Laboratory of Molecular and Cellular Neuroscience, The Rockefeller University, New York, NY 10021-6399, USA*

B.J. Nicholson, *Department of Biological Sciences, SUNY Buffalo, Buffalo, NY 14260, USA*

M. Ovadia, *Department of Physiology, University of Arizona, Tucson, AZ 85724, USA*

M. Oyamada, *Department of Pathology, Sapporo Medical College, Sapporo 060, Japan*

D.L. Paul, *Department. of Anatomy and Cellular Biology, Harvard Medical School, 220 Longwood Avenue, Boston, MA 02115, USA*

C. Peracchia, *Department of Physiology, University of Rochester, School of Medicine and Dentistry, 601 Elmwood Ave., Rochester, NY 14642, USA*

K.L. Puranam, *Howard Hughes Medical Institute, Duke University Medical Center, Durham, NC 27710, USA*

C. Rabadan-Diehl, *Department of Biochemistry, University of Miami, School of Medicine, Miami, FL 33101, USA*

S.V. Ramanan, *Department of Physiology and Biophysics School of Medicine, HSC, SUNY at Stony Brook, Stony Brook, NY 11794, USA*

K.E. Reed, *Departments of Pediatrics, Washington University School of Medicine, St. Louis, MO 63110, USA*

L.M. Reid, *Department of Molecular Pharmacology, Albert Einstein College of Medicine, 1410 Pelham Parkway South, Bronx, NY 10461, USA*

J.-P. Revel, *Division of Biology, California Institute of Technology, Pasadena, CA 91125, USA*

B. Risek, *Department of Cell Biology, The Scripps Research Institute, 10666 N. Torrey Pines Road, La Jolla, CA 92037, USA*

M.B. Rook, *Department of Physiology, University of Amsterdam, Meibergdreef 15, 1105 AZ Amsterdam, The Netherlands*

C. Roy, *Department of Neuroscience, Albert Einstein College of Medicine, 1300 Morris Park Avenue, Bronx, NY 10461, USA*

J.B. Rubin, *Department of Neuroscience, Albert Einstein College of Medicine, Bronx, NY, USA*

D.M. Rup, *Departments of Pediatrics, Washington University School of Medicine, St. Louis, MO 63110, USA*

J.C. Sáez, *Department of Neuroscience, Albert Einstein College of Medicine, 1300 Morris Park Ave., Bronx, NY 10461, USA*

J.E. Saffitz, *Departments of Medicine and Pathology, Washington University School of Medicine, St. Louis, MO 63110, USA*

D. Salomon, *Clinic of Dermatology, University of Geneva, Medical School, 1211 Geneva, Switzerland*

Y. Sasaki, *Department of Physiology, Hiroshima University School of Dentistry, 2-3 Kasumi 1 Chome, Minami-ku, Hiroshima 734, Japan*

J.-H. Saurat, *Clinic of Dermatology, University of Geneva, Medical School, 1211 Geneva, Switzerland*

E.A. Schwartz, *Committee on Neurobiology, The University of Chicago, 947 E. 58th Street, Chicago, IL 60637, USA*

L. Shen, *Department of Physiology, University of Rochester, School of Medicine and Dentistry, 601 Elmwood Ave., Rochester, NY 14642, USA*

Y. Shiba, *Department of Physiology, Hiroshima University School of Dentistry, 2-3 Kasumi 1 Chome, Minami-ku, Hiroshima 734, Japan*

D.C. Spray, *Department of Neuroscience, Albert Einstein College of Medicine, 1300 Morris Park Ave., Bronx, NY 10461, USA*

K.A. Stauffer, *Medical Research Council, Laboratory of Molecular Biology, Hills Road, Cambridge, UK*

T.H. Steinberg, *Washington University School of Medicine, St. Louis, MO, USA*

E. Steiner, *Department of Pharmacology, Columbia University, 630 W. 168th Street, New York, NY 10032, USA*

T. Suchyna, *Department of Biological Sciences, SUNY Buffalo, Buffalo, NY 14260, USA*

B.R. Takens-Kwak, *Department of Physiology, University of Amsterdam, Meibergdreef 15, 1105 AZ Amsterdam, The Netherlands*

H.B. Tanowitz, *Departments of Pathology and Medicine, Albert Einstein College of Medicine, Bronx, NY 10461, USA*

E.M. TenBroek, *Department of Veterinary Biology, University of Minnesota, 1988 Fitch Avenue, Room 295m, St Paul, MN 55108, USA*

J.E. Trosko, *Department of Pediatrics/Human Development, Michigan State University, East Lansing, MI 48824, USA*

N. Unwin, *Medical Research Council, Laboratory of Molecular Biology, Hills Road, Cambridge, UK and Department of Cell Biology, The Scripps Research Institute, 10666 N. Torrey Pines Road, La Jolla, CA 92037, USA*

R.D. Veenstra, *Department of Pharmacology, SUNY Health Science Center, 750 E. Adams Street, Syracuse, NY 13210, USA*

V.K. Verselis, *Department of Neuroscience, Albert Einstein College of Medicine, Bronx, NY, USA*

S. Vischer, *Clinic of Dermatology, University of Geneva, Medical School, 1211 Geneva, Switzerland*

H.-Z. Wang, *Division of Hematology, Washington University School of Medicine, St. Louis, MO 63110, USA*

R. Weingart, *Department of Physiology, University of Berne, Bühlplatz 5, 3012 Berne, Switzerland*

R. Werner, *Department of Biochemistry and Molecular Biology, University of Miami, School of Medicine, Miami, FL 33101, USA*

E.M. Westphale, *Department of Pediatrics, Washington University School of Medicine, St. Louis, MO 63110, USA*

R. Wilders, *Department of Physiology, University of Amsterdam, Meibergdreef 15, 1105 AZ Amsterdam, The Netherlands*

K. Willecke, *Institut für Genetik, Abteilung Molekulargenetik, Römerstrasse 164, 53 Bonn 1, Germany*

M. Wittner, *Department of Pathology, Albert Einstein College of Medicine, Bronx, NY 10461, USA*

L.R. Wolszon, *Biology Department, Columbia University, 1003 Fairchild, New York, NY 10025, USA*

L.X. Xu, *Department of Biological Sciences, SUNY Buffalo, Buffalo, NY 14260, USA*

H. Yamasaki, *Unit of Multistage Carcinogenesis, International Agency for Research on Cancer, 150, cours Albert Thomas, 69372 Lyon Cédex 08, France*

X.-D. Yang, *Neurobiology Laboratory, State University of New York, 313 Cary Hall, Buffalo, NY 14214, USA*

M. Yeager, *Departments of Cell Biology and Molecular Biology, The Scripps Research Institute and Division of Cardiovascular Diseases Scripps Clinic and Research Foundation, La Jolla, CA 92037, USA*

G.A. Zampighi, *Department of Anatomy and Cell Biology, UCLA, School of Medicine, Los Angeles, CA 92204, USA*

L.-X. Zhang, *Molecular Oncology Unit, Cancer Research Center of Hawaii, University of Hawaii, Honolulu, HI 96813, USA*

Contents

Part IV. Biophysics

Part V. Role of Gap Junctions in Various Tissues and Organisms

Nervous system

Ocular lens and heart

Part VII. Cancer

Part I. The Connexin Protein Family

J.E. Hall, G.A. Zampighi and R.M. Davis (Eds.)
Progress in Cell Research, Vol. 3

CHAPTER 1

Divergent properties of different connexins expressed in *Xenopus* oocytes

B.J. NICHOLSON[a], T. SUCHYNA[a], L.X. XU[a], P. HAMMERNICK[a], F.L. CAO[a], C. FOURTNER[a], L. BARRIO[b] and M.V.L. BENNETT[b]

[a]*Department of Biological Sciences, SUNY Buffalo, Buffalo, NY 14260 and* [b]*Department of Neuroscience, Albert Einstein College of Medicine, 1300 Morris Park Avenue, Bronx, NY, USA*

Introduction

To analyze and compare the properties of intercellular channels composed of individual members of the growing connexin family, we have utilized a modification of the paired *Xenopus* oocyte expression system first described by Dahl et al. (1987). In our case, we have taken advantage of the lack of new RNA synthesis in oocytes to irreversibly eliminate endogenous background coupling by injection of antisense oligonucleotides to the principal endogenous connexin (Xe Cx38, Ebihara et al., 1989). The only other known endogenous message for connexins in oocytes, encoding Xe Cx43 (Gimlich et al., 1990), was demonstrated not to contribute to oocyte coupling. Injection of exogenous connexin cRNAs into oocytes induced the production of a protein which preferentially accumulated at the appositional surface of oocytes and formed morphologically identifiable gap junctions. Both dye and electrical coupling demonstrated these junctions were functional. Analyses of the voltage-dependent gating of gap-junctional channels comprised of Cx32, 26, 37 and 40, using dual intracellular voltage clamps, revealed each junction type to have unique characteristics. Cx37, in particular, shows a rapid response to transjunctional voltage which may have physiological significance under normal conditions. A model, based on an earlier proposal by Unwin (1989), which could account for the variability in the kinetics of channel gating is discussed. Evidence is also presented for at least two possible modes of cooperativity between gap-junctional hemichannels.

As the family of connexins has expanded, there has been a growing need to determine how these proteins affect the properties of the intercellular channels which they form. Although intercellular coupling has been studied by electrophysiology, dye transfer and metabolic coupling in a number of systems, the nature of the connexin expressed is, for the most part, unknown. In the cases where some biochemical characterization exists, frequently more than one connexin is present (e.g. hepatocytes, Spray et al., 1986; myocytes, Veenstra, 1990). Thus, in order to definitively establish the properties of homomeric channels comprised of a particular connexin, an exogenous expression system is necessary. The principle problem in developing such a system has been the presence of a background of endogenous junctions, since intercellular coupling of eukaryotic cells seems ubiquitous. Some apparently uncoupled cell lines have been identified, but detailed analysis has almost invariably led to the unmasking of a suppressed expression (e.g. by cAMP or by kinase treatment of coupling-deficient cells, Azarnia et al., 1981 and Wiener and Loewenstein, 1983) or detection of low levels of endogenous coupling (skHepl cells, Moreno et al., this volume). There is even evidence in the *Xenopus* oocyte system that expression of an exogenous connexin (Cx43) can induce the expression of the endogenous Cx38 (Swenson et al., 1989; Werner et al., 1989). To date, studies on cells which would be expected to lack junctions (e.g. yeast or protozoa) have not been successful, principally because of difficulties in obtaining appropriate expression of the protein in such foreign environments (Revel, personal communication).

An alternative approach is to actively eliminate the background of endogenous coupling, thus the advantage of the *Xenopus* oocyte expression system. While these oocytes do contain endogenous connexins, presumably used in vivo to couple with follicle cells, the connexins have been well characterized by Ebihara et al. (1989) and Gimlich et al. (1990), who identified messages for Xe Cx38 and, at lower levels, Xe Cx43, in unfertilized oocytes. However, Dash et al. (1987) have described the unique application of antisense oligonucleotides directed against different portions of both exogenous and endogenous messengers in oocytes as a means of irreversibly inhibiting expression of protein. Endogenous RNaseH activity apparently cleaves the RNA where it hybridizes with the DNA oligonucleotides, leaving it susceptible to further degradation. Since no new RNA synthesis occurs until gastrulation, this represents, for our purposes, a permanent block. Thus, barring an extended lifetime of the junctional proteins, this procedure should irreversibly eliminate endogenous coupling (Table I). Antisense oligonucleotides to Xe Cx38 (nucleotides –5 to 25 or 327 to 353 of the coding region, Ebihara et al., 1989) cause substantially reduced intercellular conductance compared to uninjected or H_2O-injected oocytes. However, antisense Xe Cx43 oligonucleotides (nucleotides –5 to 25 or 320 to 345 of the coding region, Gimlich et al., 1990), either alone, or co-injected with Xe Cx38 antisense, cause no significant decrement in conductance. This suggests that the message for Xe Cx43 is masked or in some other way rendered inactive in oocytes so that no functional protein is produced. This unique means of specifically eliminating "endogenous" coupling and the proven efficiency of *Xenopus* oocytes as an expression system for ion channels have made this system ideal for analyzing the gating properties of different members of the connexin family. A complication was the tendency of some connexins (notably Cx43 and Cx37) to induce the functional expression of Xe Cx38 in the adjacent oocyte (reported previously for Cx43 by Swenson et al., 1989 and Werner et al., 1989). This was observed even when the antisense Xe Cx38 oligonucleotide was co-injected with the exogenous connexin cRNA. The induction of endogenous channels could be consistently eliminated by pre-injection of the antisense oligonucleotide 7 days prior to cRNA injection. The simplest conclusion consistent with these findings is that a pool of endogenous connexins persists which can only be recruited efficiently to form functional channels by the presence of high levels of a compatible connexin (e.g. Cx43 or 37) in the juxtaposed oocyte. The degree to which endogenous coupling could be induced varied significantly between oocyte batches. Thus, in the Cx37 analyses presented below, we have only used populations showing minimal or no endogenous background, even after pairing of injected and uninjected oocytes. It should also be noted that the voltage-dependent responses of the Cx37 channels (Willecke et al., 1991) were readily distinguishable from those of Xe Cx38 or Xe Cx38:Cx37 hybrids.

TABLE I Conductances of paired *Xenopus* oocytes

Oocytes injected with	g_j (nS) (mean ± S.D.)	No. of oocyte pairs
H_2O	53 ± 65	24
Antisense oligo to:		
Xe Cx43	130 ± 60	4
Xe Cx38	2.3 ± 1.6	21
Xe Cx38+Xe Cx4 3	1.8 ± 0.5	4
Antisense Xe Cx38 oligo+:		
Cx32	1330 ± 1150	5
Cx26	336 ± 330	11
Cx37	950 ± 830	13
Cx40	2700 ± 2600	6

Expression of connexins in oocytes

We have consistently been able to functionally express capped, in vitro transcripts of several cDNA and genomic connexin clones, inserted in the pGEM7Zf vector, in the paired *Xenopus* oocyte system: the only exceptions to date being Cx31 and Cx31.1 (see Willecke et al., this volume). In some instances where efficiency was low, expression was improved by insertion of the connexin-coding sequence into the BstXI restriction site separating the 5' and 3' untranslated sequences of *Xenopus* β-globin in the SP64T vector of Kreig and Melton (1984). Oocytes were stripped of their follicle cell layer by collagenase treatment and injected with both antisense oligonucleotide to Xe Cx38 and the cRNA of interest (in the case of Cx43, RNA injection followed antisense injection by 7 days). After 24 h, the vitelline membranes were stripped manually, in some cases with the aid of transient hypertonic shock, and the oocytes paired for 24–48 h before recording. While conductances vary significantly, oocyte pairs expressing exogenous connexins typically have average junctional conductances of 0.3–5 μS, compared to ~50 nS for H_2O-injected oocytes and < 2 nS for Xe Cx38 antisense-oligonucleotide-injected oocytes.

Expression of the connexin proteins can be monitored in a variety of ways, as illustrated in Fig. 1. By co-injecting ^{35}S-Met with Cx43 cRNA, the protein can be immunoprecipitated from ~10 pooled oocytes, using an antibody to a portion of the Cx43 C-terminus (residues 302–319) (Fig. 1A). This same antibody can also be used to localize the majority of Cx43 to the appositional surface of paired oocytes (a heterotypic pairing of Cx43 and Cx32 injected oocytes is shown in Fig. 1B). Less intense labeling of non-junctional membranes immediately adjacent to the apposed domain and punctate labeling in the cytoplasm were also detected in oocytes injected with Cx43, but not with other connexins (Fig. 1B). Electron microscopic analysis of thin sections of the appositional surface of oocytes injected with exogenous connexin cRNAs directly demonstrates the formation of gap junction-like structures between short regions of juxtaposed membrane or, more commonly, between microvillar projections (Fig. 1C). Consistent with these observations, oocyte pairs injected with Cx43, Cx32 or other connexin cRNAs are coupled electrically (e.g. see Fig. 2; Swenson et al., 1989 and Werner et al., 1989) and pass dyes such as Lucifer Yellow (Fig. 1D).

Our experiments with Cx43 (e.g. see Figs. 1A and B) augment recent reports that have correlated the phosphorylation of Cx43, associated with an increase in apparent molecular weight on SDS gels (M_r), with its functional expression (Musil et al., 1990). While it is clear Cx43 is functionally expressed in *Xenopus* oocytes (Swenson et al., 1989, 1990; Werner et al., 1989), im-

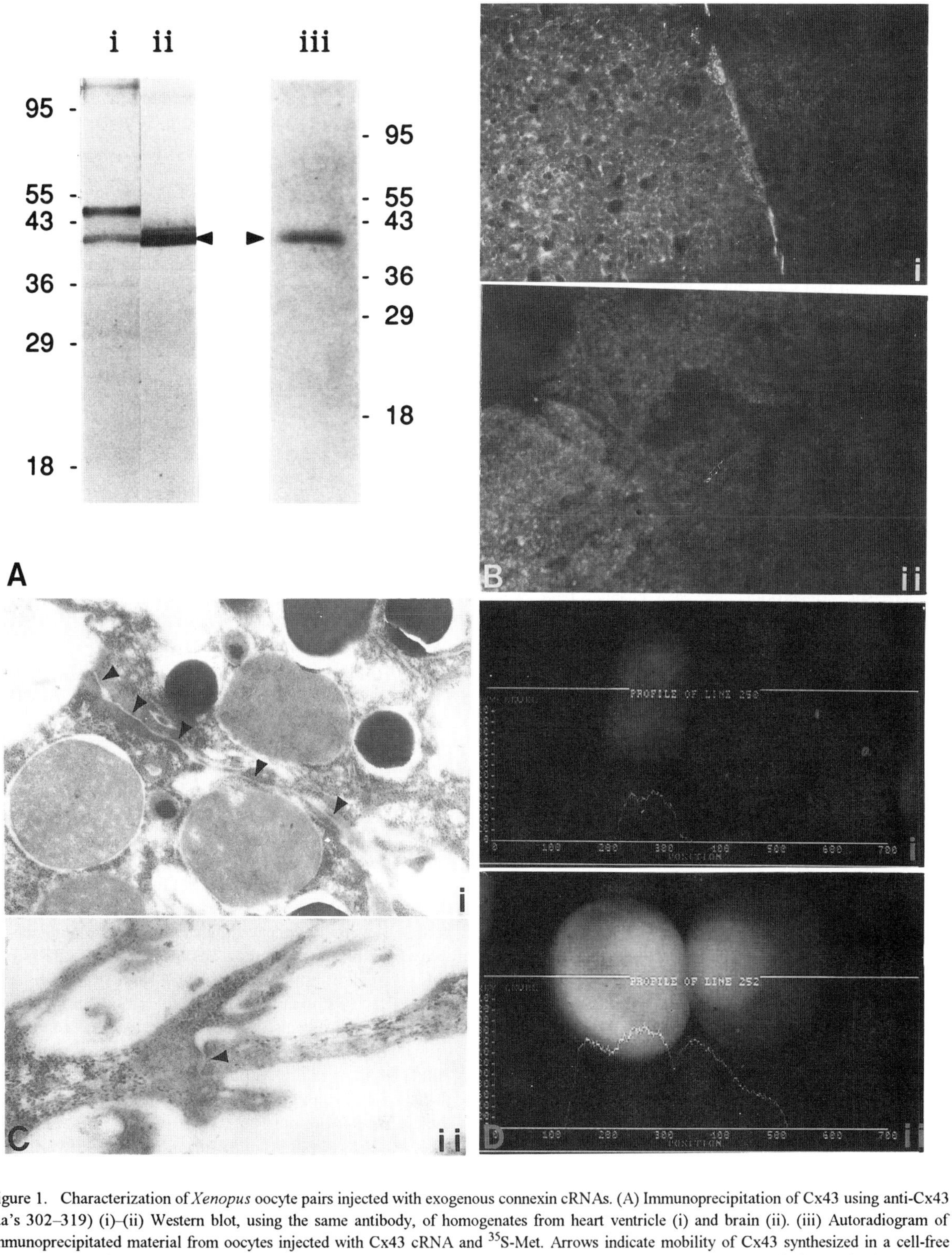

Figure 1. Characterization of *Xenopus* oocyte pairs injected with exogenous connexin cRNAs. (A) Immunoprecipitation of Cx43 using anti-Cx43 (aa's 302–319) (i)–(ii) Western blot, using the same antibody, of homogenates from heart ventricle (i) and brain (ii). (iii) Autoradiogram of immunoprecipitated material from oocytes injected with Cx43 cRNA and ^{35}S-Met. Arrows indicate mobility of Cx43 synthesized in a cell-free system, representing the non-phosphorylated form of the protein. Mobilities of molecular-weight markers are indicated in kDa's. (B) Immunofluorescent staining of paired oocytes injected with Cx43 [left oocyte in (i); both oocytes in (ii)] or Cx32 [right oocyte in (i)] using anti-Cx43 (aa's 302–319) (i) or pre-immune serum (ii)]. (C) Thin section of the interface between an oocyte pair injected with Cx32. A close apposition of membranes characteristic of gap junctions (arrows) is evident both over larger areas (i, ×14,400) and between microvilli (ii, ×22,000). (D) Dye coupling with Lucifer Yellow (injected into the left oocyte) between oocytes injected with either antisense Xe Cx38 (i) or cRNA for Cx32 (ii). Data were analyzed on a Quantex fluorescent-imaging system which superimposes a profile of fluorescence density determined along the equator of the paired oocytes. Images obtained 12 h after dye injection.

munoprecipitation indicates that it is present exclusively in the form with the lowest M_r, corresponding to that of in-vitro-synthesized, or phosphatase-treated Cx43 (Fig. 1A; Kadle et al., 1991). This supports the observations of Swenson et al. (1990), although these authors show that this lowest M_r form can contain some phosphoserine. Functional coupling by this form of Cx43, along with its localization to the appositional surface of oocytes (Fig. 1B and Swenson et al., 1989), indicates that higher-order phosphorylation is not required for functional assembly of junctions, at least in oocytes. This is consistent with the demonstration that Cx43 can be present predominantly in the lowest M_r form in various tissues [Fig. 1A(ii)] and cells, such as the astroglia of the CNS, which are known to be well coupled (Kadle et al., 1991, and submitted).

Different connexins — different fingerprints: an analysis of the voltage dependence of homotypic junctions

We have now expressed a number of different connexins in the paired *Xenopus* oocyte system, and each has displayed quite different properties. In order to establish a characteristic "fingerprint" for each channel type, we have analyzed in detail their gating in response to applied transjunctional voltages (V_j). It should be noted that, with one exception (discussed below), these responses to voltage are so slow that they are unlikely to play a physiological role other than as a means of uncoupling healthy cells from unhealthy ones.

The voltage dependence was determined by independently clamping each oocyte to –40 mV (i.e. within +10 mV of the normal resting potential) and then stepping one cell to a different holding potential for 30 s. The current required to clamp the second oocyte to its original potential represents a direct measure of transjunctional current (Spray et al., and Harris et al., 1981). Transjunctional voltages were generated by both hyper- (left side of plots in Fig. 2) and depolarizations (right side of plots in Fig. 2) of either oocyte of the pair. Initial and steady-state currents were determined by fitting each current trace to an exponential decay and extrapolating to 0 and infinity respectively. Typical intercellular conductances (g_j) of an oocyte pair ranged from means of 0.34 µS for Cx26 to 2.7 µS for Cx40. In all cases, antisense oligonucleotides to Xe Cx38 were co-injected with the cRNA for the appropriate connexin. Conductances plotted in Fig. 2 are normalized to the initial conductance interpolated at 0 mV. In each case, the drop in g_j with increasing or decreasing V_j was fit to a Boltzmann equation (or, in the case of the steady-state response of Cx37, to two nested Boltzmanns) of the form

$$g_j = F(V)$$

$$= [g_j(max) - g_j(min)]/[1 + exp[A(V - V_0]] + g_j(min)$$

(Spray et al., 1981). V_0 is a measure of voltage sensitivity and equals the transjunctional voltage at which one sees half-maximal conductance. A is a measure of cooperativity. The equivalent gating charge which moves across the membrane during channel gating, n, is given by the relationship:

$$n = AkT/q$$

(q = the elementary charge; T = temperature; k = Boltzmann constant). For Cx37, steady state responses were fit by the sum of two Boltzmann distributions.

Other than Cx43 (Swenson et al., 1989; Werner et al., 1989, Fishmann et al., 1990), the channels least sensitive to V_j were composed of Cx26 with a V_0 of +89 mV and n of 4. Since oocytes were initially clamped at –40 mV, hyperpolarizing pulses producing transmembrane potentials in excess of –130 mV were necessary to generate the plot at negative V_j's (Fig. 2A). Given the instability of membranes at these voltages, more reliable data at these V_j's could be produced by clamping the oocytes to a resting potential of –20 mV (dotted trace, Fig. 2A). The slight asymmetry in the steady-state conductance [g_j(ss)] curve and the slope of the initial conductance [g_j(i)] plot both demonstrate that these junctions also show a small gating response to the transmembrane or inside-outside voltage ($V_{i\text{-}o}$). Cx32 channels (Fig. 2B) proved, in our hands, somewhat more sensitive to V_j (V_0 = +57 mV), but less cooperative in their closure (n = 2). As in all other connexins tested, with the exception of Cx26, no sensitivity to $V_{i\text{-}o}$ was detected. The comparisons of these connexins is considered in more detail in Barrio et al. (1991).

It should be noted that the voltage gating of gap junctions composed of Cx32 has not been a constant property when results from different laboratories are compared. Swenson et al. (1989) and Werner et al. (1989) both report a lack of voltage sensitivity of Cx32-induced coupling between oocytes, although Werner et al. note induction of voltage sensitivity under acidic conditions. Voltage sensitivity of hepatocyte coupling has also proven rather variable (compare Spray et al., 1986 and Riverdin and Weingart, 1988 with Moreno et al., 1991), although in these cases the parameters measured reflect the combined contributions of both Cx32 and Cx26. Some of this variability is likely due to variations in the intercellular conductance and consequent changes in access resistance of the channels (Jongsma et al., 1991, also see discussion in Moreno et al., 1991). Another potential source of variation would be the presence of modulatory elements in the cytoplasms of different cells. Though the properties of exogenously expressed connexins may not always reflect properties seen in vivo, expressing them in a constant environment as done here (i.e. Stage VI oocytes) allows them to be compared under comparable conditions, so that any differences observed can be attributed to differences in the connexins themselves.

Cx40 channels (Fig. 2D) are even more sensitive to V_j (V_0 = +34 mV) than those of Cx32 and 26 and display a

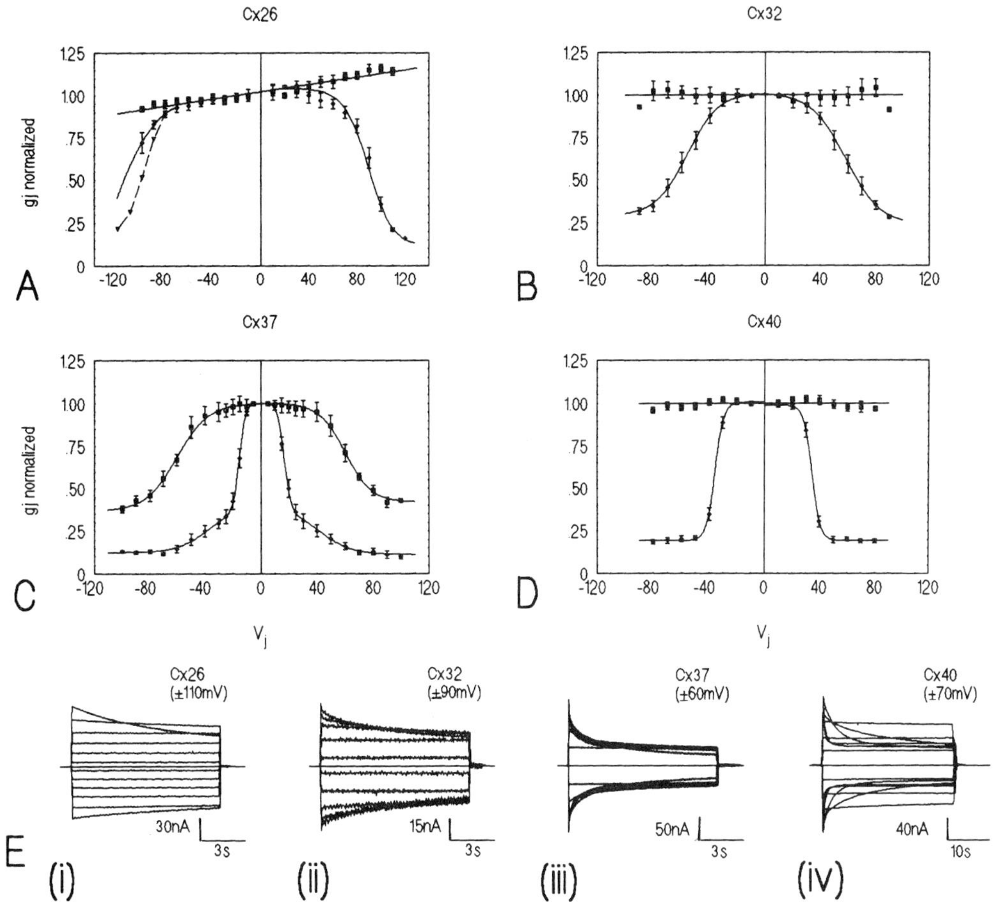

Figure 2. Voltage-dependent gating of different connexins expressed in *Xenopus* oocytes (A–D) Plots of junctional conductance (g_j), normalized to the initial conductance interpolated at 0 mV, vs. transjunctional voltage (V_j) for oocyte pairs injected with the connexins indicated. Two traces are shown, one of g_j at the beginning of each 30 s voltage pulse [g_j(i)] and one of g_j at the end of the 30 s voltage pulse [g_j(ss)]. Positive values of V_j were obtained by depolarizing either oocyte, and negative values by similar hyperpolarizations. All curves are linear, or best fits to the Boltzmann equation given in the text, except for the g_j(i) curve for Cx40, and the g_j(ss) curve for Cx37. The first contains insufficient data and the second is fit by two summed Boltzmann equations. The dotted curve for g_j(ss) at negative V_j's in Cx26-expressing oocytes was obtained from oocytes clamped at −20 mV. All other experiments were done with oocytes clamped at −40 mV. (e) Sample current traces from the clamped oocyte in response to 10 or 30 s voltage pulses, in 5 or 10 mV steps over the range indicated, applied to the adjacent cell. Note the faster decay evident in Cx37 and 40 traces, and the two-phased nature of the current decay in Cx37 traces at the higher voltages.

high level of cooperativity (n = 8). Again, no sensitivity to V_{i-o} is detected. The time constants for current decay in response to V_j (Fig. 2E) are significantly smaller for Cx40 channels than for either Cx32 or 26 channels (~580 ms at $V_j = V_0 + 15$ mV compared to 5.6 and 6.3 s respectively, Table II). In all junctions examined, the time constants decrease with increasing V_j. Given these kinetics, the limited time resolution of our voltage clamps (< 20 ms) may account for the slight deviation from ohmic behavior seen in plots of g_j(i) at high positive and negative V_j's in the Cx40 plot, and to a lesser extent in the plot of Cx32 channels (see Fig. 2B&D). Junctions composed of Cx37 display the most complex behavior of connexins studied to date (Fig. 2C; and Willecke et al., 1991). Steady-state conductance reflects the usual, relatively slow response to V_j, but this is very sensitive (V_0 = +16 mV) and cooperative (n = 11) with a time constant of current decay comparable to, but slower than, that of Cx40 (~950 ms at $V_j = V_0 + 15$ mV). Furthermore, this slow voltage gating has a second, far less sensitive component evident from the complex shape of the plot of g_j(ss) vs. V_j which could be fit by the sum of two Boltzmann equations. This second

TABLE II Boltzmann parameters of voltage gating of different connexin channels expressed in oocytes[1]

Connexin	V_0 (mV)[1]	A (mV^{-1}) (n)[1]	g_{min}[1]	τ^2 (s)
Rat Cx32[3]	±57	0.09 (2)	0.23	5.6
Rat Cx26	±89	0.14 (4)	0.10	6.3
Mouse Cx37[4]				
(slow1)	±16	0.44 (11)		0.95
(slow2)	±40	0.09 (2)	0.12	ND
(fast)	±60	0.09 (2)	0.40	< 0.002
Mouse Cx40	±34	0.32 (8)	0.20	0.58

[1] See text for definitions of V_0, A, n and g_{min}.
[2] τ = time constant of current decay, taken at $V_j = V_0+15$ mV.
[3] Note that data given here correlate well with Barrio et al. (1991) but differ from Swenson et al. (1989) and Werner et al. (1989) in oocytes and Moreno et al. (1991) in transfected cells, see text for discussion.
[4] Reported in Willecke et al. (1991).

gate was also indicated by the observation that current traces at higher V_j's did not fit a simple exponential decay (Fig. 2E). However, most surprising was the Boltzmann-like decay of $g_j(i)$ with V_j (V_0 = +60 mV; n = 2). This indicates that a rapid gating response to V_j occurs in a time frame which is not resolvable by our voltage clamps (response time in cell 2 of < 20 ms). This response is unique among the connexins studied to date and could represent one of the few cases where regulation of intercellular coupling by voltage may have physiological relevance (other than the intriguing cases of rectifying junctions in invertebrates). Certainly, the response is sufficiently rapid that it could be triggered during impulse propagation within an excitable tissue. This could lead to modulation, or even interruption, in the synchronized activity of coupled neural or muscle networks expressing this protein[1].

Connexin structure and channel gating

By correlating the variety of gating properties described here with the deduced sequences of each connexin (Paul, 1986; Zhang and Nicholson 1989; Willecke et al., Chapter 5; Willecke et al., 1991), can we learn anything about the possible mechanisms of the channel gate? The nature of the voltage sensor remains an enigma; it must in some way respond to a transjunctional voltage, moving charged residues within the field and closing the channel. In more rapidly gating, voltage-sensitive Na^+, K^+ and Ca^{2+} channels, this is thought to be effected by a highly basic, (putative) transmembrane S4 helix (Stühmer et al., 1989; Papazian et al., 1991). No such sequence is evident in the connexins. It is also difficult to propose any structural model which could account for the gating charges of 11 or 8 inferred from the cooperativity constants (A) of the Boltzmann equations describing the voltage sensitivity of Cx37 and Cx40 channels, respectively. However, contrary to the simple modeling of channels as electrical conductors and switches, this constant (A or its analog, n) could also reflect cooperative interactions between protein subunits or channels, that is, the closing of one channel may greatly facilitate the closing of others in the vicinity. Clearly, the closely packed arrays of channels within a gap junction make this model appealing. In this way, cooperativity in the Boltzmann equation would reflect both a limited movement of charges across the membrane within a single channel in response to V_j, as well as cooperative interactions between channels. The frequent association of the voltage gating of several connexins (Table II) with an n of 2 suggests this may be the basic "gating charge", with all higher values of n reflecting differing degrees of cooperativity among the 6 "nearest-neighbor" channels in different connexins. A similar suggestion involving cooperativity between subunits of a single channel has been made recently to explain effects of conservative mutations of hydrophobic residues on the cooperativity of K^+-channel gating (McCormack et al., 1991).

Other than sensitivity to V_j, a major difference in the connexins studied here is their gating kinetics. Aside from the sub-20 ms response of Cx37 channels, the "slow" responses of both Cx40 and Cx37 channels are 5–10 times faster than those of Cx32 and 26 (Fig. 2E, Table II). Analysis of the strictly conserved residues among cloned connexins reveals that, in both Cx37 and 40, two of the three conserved phenylalanines in the putative third, amphipathic transmembrane helix (M3) are replaced by smaller, hydrophobic residues (Fig. 3). In Unwin's (1989) earlier model, these phenylalanines may be exposed to the aqueous pore through a tilting of subunits (Fig. 3). This presumably causes a collapse of the pore under hydrophobic interactions. In this model, the substitution of smaller residues for the phenylalanines would certainly be expected to affect the kinetics of closure, either by destabilizing the open state or reducing steric hindrance

[1] As discussed by Willecke et al. in an earlier Chapter 5, Cx37 is expressed at low levels in both brain and heart. We have also detected a highly homologous transcript of the same size in high levels in the brains of lower vertebrates, notably fish (Kadle, Yox, and Nicholson unpublished results).

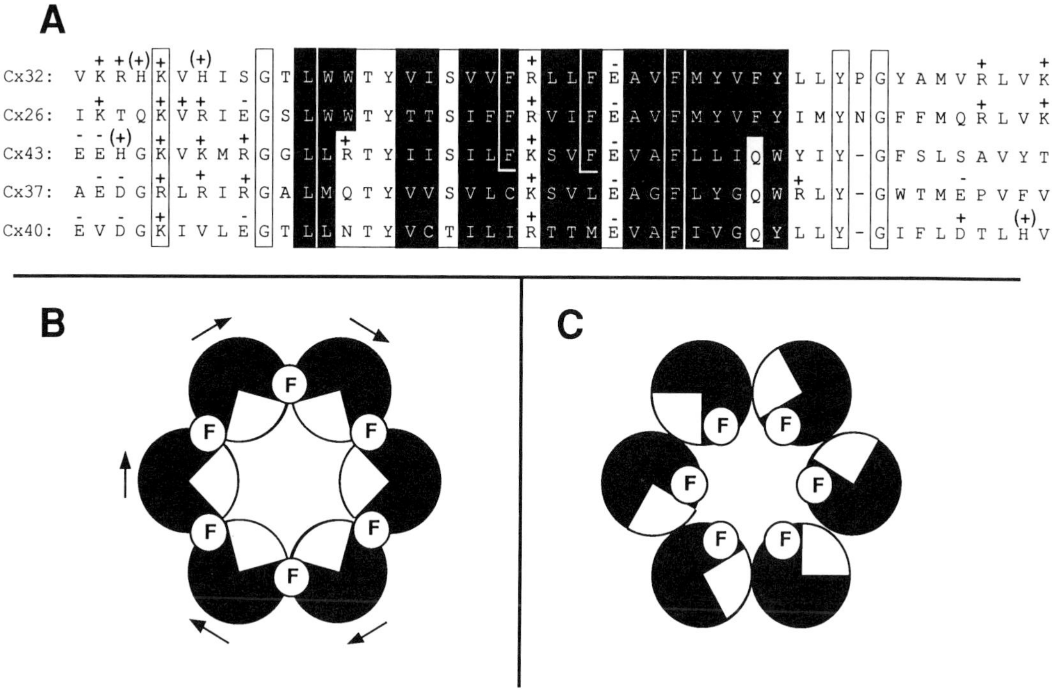

Figure 3. Comparison of different connexin sequences in reference to a possible model of channel gating. (A) Aligned third transmembrane domains (M3) of five connexins. The hydrophobic domain is shaded. Residues which are highly conserved in all known connexins are boxed and include hydrophilic amino acids within the hydrophobic region spaced at every 3rd or 4th position; and three similarly spaced phenylalanines, two of which are located adjacent to the two charged residues. Cx37 and -40 represent exceptions to the latter pattern: two of these phenylalanines are substituted with smaller, non-polar side chains. (B) and (C) The M3 regions in (A) can be modeled as amphipathic helices (shown as filled circles). When each is tilted 15° away from the perpendicular to the lipid bilayer, and arranged as a closely packed hexameric coil, a central "pore" lined by polar side chains (clear area of helix) is formed (B). In this model of the junctional channel, a tilting of the helices consistent with that found by Zampighi and Unwin (1980), combined with a rotation of each helix by a one-quarter turn, results in the conserved phenylalanines (Fs) being exposed to this central cavity (C). This could lead to an exclusion of H_2O and a collapse of the "pore" by rearrangement of the helices. Models shown are schematic representations derived from molecular modeling of the Cx26 M3 regions using SYBYL programs on an Evans and Sutherland PS390 Workstation (courtesy of Dr. David Langs, Buffalo Medical Foundation). They are based on a scheme originally proposed by Unwin (1989).

during the proposed helix sliding. Site-directed mutagenesis of these connexins should be particularly useful in testing this hypothesis.

Unexpected properties of heterotypic junctions

Gap junctions have long been modeled as homopolymeric structures, but the demonstrated coexistence of different connexins within a given cell, or junctional structure (Nicholson et al., 1987; Traub et al., 1989) has raised the possibility of interactions between connexins. Heteromeric interactions between different connexins within a cell have proven difficult to assess in the absence, to date, of procedures for isolating intact channels (but see Stauffer et al., 1991, and Chapter 8). However, heterotypic intercellular channels between rat Cx43 and *Xenopus* Cx38 expressed in different oocytes have been demonstrated by Swenson et al., 1989 and Werner et al., 1989. But it is not known what interactions of physiological significance take place between connexins of the same species expressed in neighboring cells.

To address this question, we analyzed paired oocytes: one was injected with rat Cx32 and the other with rat Cx26. These connexins co-localize in hepatic gap junctions (Nicholson et al., 1987; Traub et al., 1989). We also did analogous experiments with Cx37- and Cx40-injected oocytes, both of which are expressed predominantly in lung (Willecke et al., Chapter 5). The broken traces in Fig. 4 indicate the voltage-gated properties of the respective homotypic channels (as shown in Fig. 2). It is readily evident that the actual responses (indicated by the solid traces) differ significantly from either "parental" form, in contrast to the results reported previously for the interspecies combination of rat Cx43 with *Xenopus* Cx38. In particular, heterotypic Cx32:Cx26 channels (Fig. 4A and B) show a rapid, rectifying response to V_j [i.e.

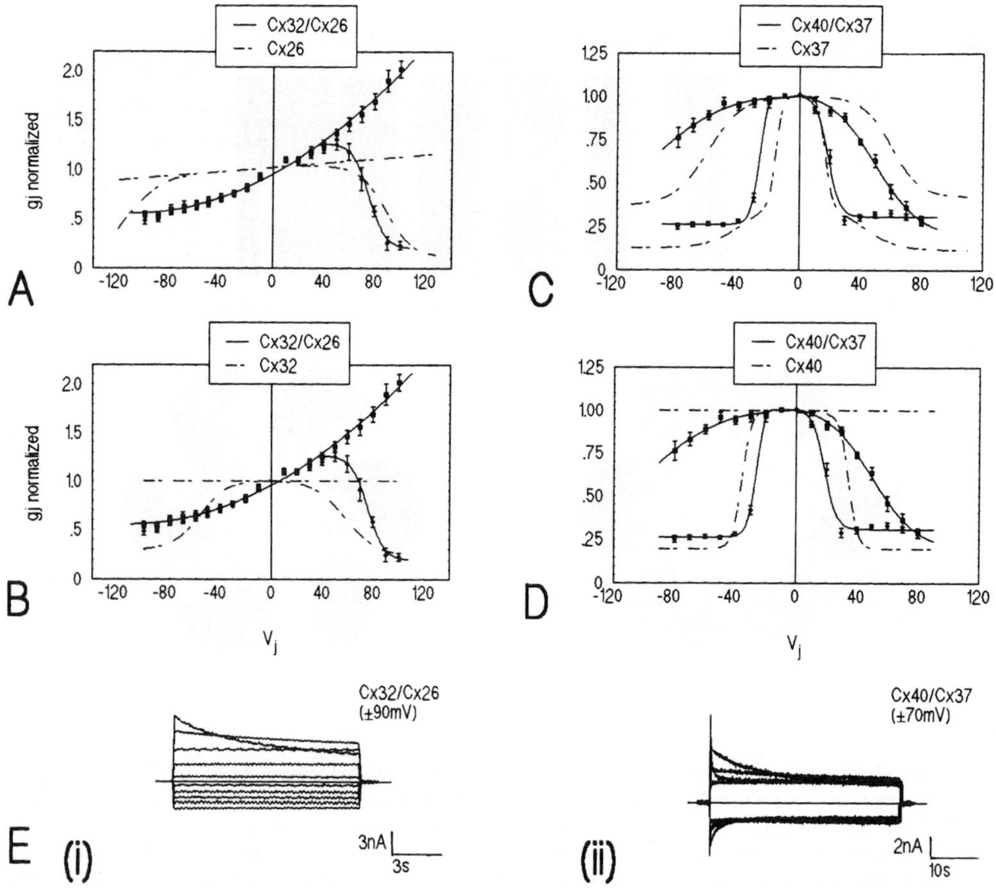

Figure 4. Voltage-dependent gating of heterotypic gap junctions (A) and (B) Paired oocytes, one expressing Cx26 and the other Cx32, show voltage dependent coupling characteristics (solid traces, see Fig. 2 for general description) that are quite distinct from that of either "parental" form (dotted traces (A), Cx26; (B), Cx32). The Cx26 injected cell is defined as the positive pole and data were obtained by both hyperpolerization of one cell and depolarization of the other. Thus, all responses shown represent sensitivity to V_j and not V_{i-o}. (C) and (D) Analogous traces of Cx37:40 hybrids, with the Cx37-expressing oocyte defined as the positive pole. Actual gating behavior (solid traces) also differs from that of either "parental" channel [dotted traces, Cx37 in (C) and Cx40 in (D)], but less so than for Cx26:32. (E) Current traces of the heterotypic channels, analogous to those shown in Fig. 2. Current pulses above the basal current represent responses to positive voltage pulses in the Cx26 (i) or Cx37 (ii) cell, or negative voltage pulses in the opposite cell over the range indicated.

conductance is higher when the Cx26-expressing oocyte is positive with respect to the Cx32-expressing oocyte; also evident in Fig. 4E(i)}, and a slow closure in response to V_j which is now asymmetric (i.e. only recurring when the Cx26-expressing oocyte is > 60 mV positive with respect to the Cx32 oocyte). This latter response can be fit by a Boltzmann equation with a V_0 (+75 mV) intermediate between that of Cx26 and Cx32, and an n of 4.

Deviations from the predicted responses of the Cx37:Cx40 heterotypic channels are less extreme (Fig. 4C&D), but still evident. Previous analyses of amphibian and squid blastomeres (Bennett et al., 1988) and the heterotypic channels of Cx43 and Xe Cx38 (Swenson et al., 1989; Werner et al., 1989), indicate that gating of individual hemichannels occurs at the depolarizing (positive) side of a transjunctional voltage gradient (except for the isolated cases of squid axons and Cx46 hemichannels). Our results are consistent with these observations. Thus, if we accept this premise, we could conclude that the second, less sensitive component of the slow response of Cx37 hemichannels is no longer observed, while the rapid changes in conductance with V_j (reflected in g_j(i) plot) for both hemichannels are notably accentuated. Other minor changes in the V_0 and A values of the Boltzmann equations describing the slow gating response are also apparent in the heterotypic channels.

A question which naturally arises from these studies is whether such heterotypic channels could form in situ, based on both accessibility of the two channel types to each other and their relative affinity. At least in the case of Cx26 and 32, the answer now seems to be yes. It has already been shown that Cx32 and 26 co-localize to the same gap-junctional structures (Nicholson et al., 1987; Traub et al., 1989). By expressing one connexin in oocyte #1 (either Cx26 or Cx32) and mixing both connexins in varying ratios in oocyte #2, the properties of the coupling from their pairing (Fig. 5) indicate that formation of heterotypic channels competes favorably with homotypic channel formation. That is to say, bulk coupling properties reflect those expected from a mixture of heterotypic and homotypic channels. Indeed, based on levels of message injected (protein levels were not determined), there is an indication that Cx26 hemichannels may show a preference towards interactions with hemichannels composed of Cx32 rather than with the homotypic form (Fig. 5 compare A–C with D–F). However, aside from questions of translational efficiency, it should be pointed out that this analysis cannot take into account the possible formation of

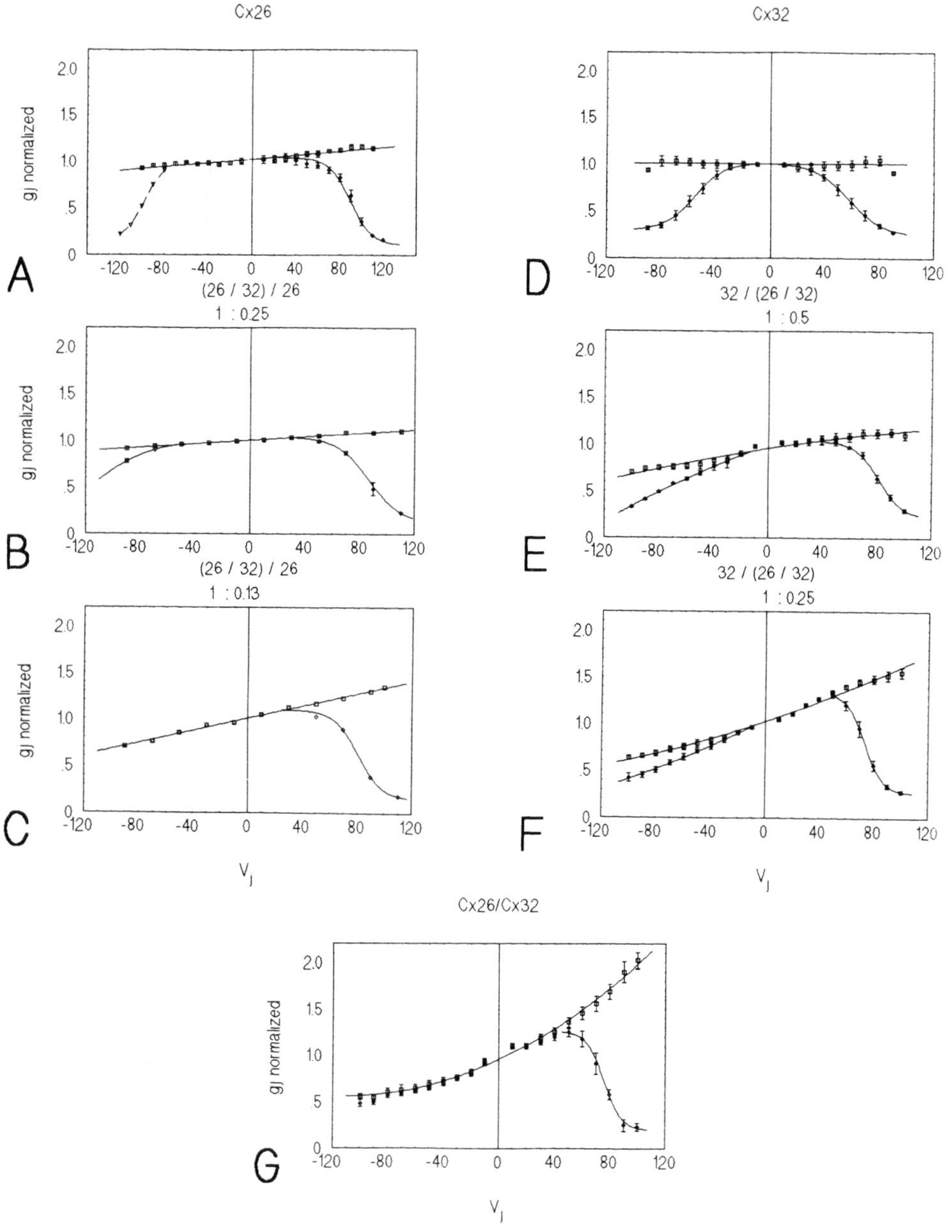

Figure 5. Competition experiments for homo- or heterotypic junctional channels between Cx32 and Cx26 In (A)–(C), equal amounts of Cx26 cRNA are injected into each oocyte of a pair. However, in (B) and (C), increasing amounts of Cx32 cRNA (levels relative to Cx26 RNA are indicated on each trace) were also co-injected in one oocyte of the pair (defined here as the negative pole for V_j). In (D)–(F), Cx32 cRNA was injected into both oocytes of a pair, with one (defined as the positive pole for V_j) receiving successively decreasing amounts. In (E) and (F), Cx26 cRNA was co-injected into the oocyte receiving less Cx32, to produce the relative RNA levels indicated. (G) shows the plot of paired oocytes, one expressing Cx32 (defined as negative V_j pole) and one expressing Cx26 (defined as positive V_j pole). V_j vs. g_j plots derived from these oocytes show that heterologous competition between connexins results in graded changes in the gating properties of the intercellular channels from those characteristic of the homotypic Cx26 (A) or Cx32 (D) channels to those of the pure heterotypic form (G). It is evident that a slight excess of Cx26 over Cx32 RNA will result in a predominance of heterotypic channels with Cx32, while a considerable excess of Cx26 over Cx32 is required to obtain predominantly homotypic Cx26 channels.

heteromeric hemichannels in the co-injected cell. This could add an additional dimension of complexity to the interpretation of these data, which cannot be evaluated until the properties of these putative heteromeric forms are defined.

The modulation of voltage dependence of heterotypic junctions compared to the homomeric "parents" provides direct evidence for allosteric interactions between interacting hemichannels which can substantially affect channel properties. So far, only voltage gating has been analyzed. Effects on channel permeability, especially in light of the rectifying behavior of the Cx32/26 heterotypic channels, will no doubt prove to be a major focus of future studies. Indeed, two recent observations highlight these possibilities. Neveu et al. reported in an abstract at this meeting that islands of pre-neoplastic cells expressing high levels of Cx26 develop in rat liver lobules expressing predominantly Cx32 during the early stages of hepatic tumors (see also Chapter 43). The data presented here raise the possibility of a rectifying flow of molecules, and hence signals, between healthy and pre-neoplastic tissue, at least in the presence of a maintained voltage gradient. Mehta and Loewenstein (1991) recently examined the complex responses of homo- and heterotypic coupling between normal and transformed cells in response to retinoid treatment. Their model proposes multiple channel types (both hetero- and homotypic) with differential affinities between each. Clearly, should multiple connexins be identified in these cells, studies analogous to those presented here for Cx32 and 26 can be used to test many aspects of this model. Thus, there is every indication at the current time that many connexins may be designed to form both homo- and heterotypic contacts, thus contributing further to the variability in properties and regulation of the direct intercellular exchanges mediated by gap junctions.

Conclusions

By essentially eliminating endogenous coupling in the oocyte system by co- or pre-injection of antisense oligonucleotides, we have been able to compare the channel properties of four connexins and heteroptypic combinations of those connexins which are typically co-expressed. Different apparent gating charges are associated with each connexin, yet there is no significant variation in the transmembrane distribution of charged residues. This suggests that the cooperativity constant of the Boltzmann equation describing the voltage gating of gap junctions (from which the apparent gating charge is calculated) is also influenced by cooperative interactions between channels during closure. This can vary by a factor of five between different connexins if a constant gating charge of 2 is assumed. A second mode of interaction was directly demonstrated between hemichannels comprised of different connexins, both of which are co-expressed in the same tissue (Willecke et al., Chapter 5) or gap-junctional structures (Nicholson et al., 1987) in vivo. Such heterotypic junctions display voltage-dependent gating characteristics which are significantly modified from the properties of either "parental" hemichannel. These allosteric interactions between connexins expressed in different cells represent another dimension for diversification of properties and regulation of gap-junctional channels. The great variation in voltage-gated characteristics reported here complements other studies which have linked different single-channel conductances with specific connexins. Both results indicate that a major goal will be to determine if these channels and their hybrids show different permeabilities or even selectivities, for larger metabolites. The present analysis also suggests that gap-junctional channel properties may be significantly influenced by allosteric and cooperative interactions with neighboring channels. New approaches will be needed to model the currents in gap junctions which operate in "community settings". Clearly the diversity in the connexin family of proteins is likely to be only part of the story of diversity in the properties of intercellular channels.

Note added in proof

Recent observations by Rubin, Versailis, Bargiello and Bennett using both chimeric and point mutagenized variants of Cx32 and Cx26 have demonstrated that Cx32 hemichannels respond to V_j of opposite polarity compared to Cx26 (hyperpolarizing for Cx32 and depolarizing for Cx26). This explains the marked asymmetry in the slow response to voltage of Cx32/Cx26 heterotypic channels. The additional presence of a fast response to V_j (opposite in polarity to the slow response) in Cx32 channels, but not in Cx46 channels, could also account for the fast rectification of heterotypic channels.

Acknowledgements

Much of this work was made possible through the isolation of additional connexin clones in the laboratory of Dr. Klaus Willecke (U. Bonn, Germany). We would also like to thank Drs. Chris Loretz and Stan Halvorsen for help in analysis of electrophysiological and dye-transfer data, and Ms. Feng Gao for excellent technical assistance. The aid of Linda Mack, Xochitl Nicholson, Jim Stamos and Alan Siegal in the preparation of the manuscript is also gratefully acknowledged. The work was supported by PHS grants from NIH [CA48049 (BN) and HL37109 (PI-J-P Revel)] and a PEW Scholars award in the Biomedical Sciences to B.N. A NATO travel grant facilitated interactions with Prof. Willecke's laboratory.

References

Azarnia, R., Dahl, G. and Loewenstein, W.R. (1981) Cell junctions and cyclic AMP III. Promotion of junctional membrane permeability and junctional membrane particles in a junction deficient cell type. J. Membr. Biol. 63, 133–146.

Barrio, L.C., Suchyna, T., Bargiello, T., Xu, L.X., Roginski, R.S., Bennett, M.V.L. and Nicholson, B.J. (1991) Gap junctions formed by connexins 26 and 32 alone and in combination are differently affected by applied voltage. Proc. Natl. Acad. Sci. USA 88, 8410–8414.

Bennett, M.V.L., Verselis, V., White, R.L. and Spray, D.C. (1988) Gap-junctional conductance, gating. In: E.L. Hertzberg and R.G. Johnson (Eds.), Gap Junctions, Modern Cell Biology, vol. 7, Alan R. Liss, New York, pp. 287–304.

Dahl, G., Miller, T., Paul, D., Voellmy, R. and Werner, R. (1987) Expression of functional cell-cell channels from cloned rat liver gap junction complementary DNA. Science 236, 1290–1293.

Dash, P., Lotan, I., Knapp, M., Kandel, E. and Goelet, P. (1987) Selective elimination of mRNAs *in vivo*: complementary oligonucleotides promote RNA degradation by RNaseH-like activity. Biochemistry 84, 7896–7900.

Ebihara, L., Beyer, E.C., Swenson, K.I., Paul, D.L. and Goodenough, D.A. (1989) Cloning and expression of a *Xenopus* embryonic gap-junction protein. Science 243, 1194–1195.

Fishman, G.I., Spray, D.C. and Leinwand, L.A. (1990) Molecular characterisation and functional expression of the human cardiac gap-junctional channel. J. Cell Biol. 111, 589–597.

Gimlich, R.L., Kumar, N.M. and Gilula, N.B. (1990) Differential regulation of the level of three gap junction mRNAs in *Xenopus* embryos. J. Cell Biol. 110, 597–605.

Harris, A.L., Spray, D.C. and Bennett, M.V.L. (1981) Kinetic properties of a voltage-dependent junctional conductance. J. Gen. Physiol. 77, 95–117.

Jongsma, H.J., Welders, R., van Ginnekan, A.C.G. and Rook, M.B. (1991) In: C. Peracchia (Ed.), Biophysics of Gap-Junction Channels, CRC Press, Boca Raton, FL, pp. 163–172.

Kadle, R., Zhang, J.T. and Nicholson, B.J. (1991) Tissue-specific distribution of differentially phosphorylated forms of Cx43. Mol. Cell. Biol. 11, 363–369.

Kreig, B.E. and Melton, D.A. (1984) Functional messenger RNAs are produced by SP6 *in vivo* transcription of cloned cDNA. Nucleic Acids Res. 12, 7957–7970.

McCormick, K., Tanouye, M.A., Iverson, L.E., Lin, J.-W., Ramaswami, M., McCormack, T., Campanelli, J.T., Mathew, M.K. and Rudy, R. (1991) A role for the hydrophobic residues in the voltage-dependent gating of *Shaker* K^+ channels. Proc. Natl. Acad. Sci. USA 88, 2931–2935.

Mehta, P.P. and Loewenstein, W.R. (1991) Differential regulation of communication by retinoic acid in homologous and heterologous junctions between normal and transformed cells. J. Cell. Biol. 113, 371–379.

Moreno, A.P., Campos de Carvalho, A.C., Verselis, V., Eghbali, B. and Spray, D.C. (1991) Voltage-dependent gap-junction channels formed by connexin 32, the major gap junction protein of rat liver. Biophys. J. 59, 920–925.

Musil, L.S., Beyer, E.C. and Goodenough, D.A. (1990) Expression of gap-junctional protein connexin 43 in embryonic chick lens: molecular cloning, ultrastructural localization and post-translational phosphorylation. J. Membr. Biol. 116, 163–175.

Nicholson, B.J., Dermietzel, R., Teplow, D., Traub, O., Willecke, K. and Revel, J.-P. (1987) Two homologous components of hepatic gap junctions. Nature 329, 732–734.

Papazian, D.M., Timpe, L.C., Jan, Y.N. and Jan, L.Y. (1991) Alteration of voltage dependence of *Shaker* potassium channel by mutagenesis in the S4 sequence. Nature 349, 305–310.

Paul, D.L. (1986) Molecular cloning of cDNA for rat liver gap-junction protein. J. Cell Biol. 103, 123–134.

Riverdin, E.C. and Weingart, R. (1988) Electrical properties of the gap-junctional membrane studied in rat liver cell pairs. Am. J. Physiol. 254, 226–234.

Spray, D.C., Harris, A.L. and Bennett, M.V.L. (1981) Equilibrium properties of a voltage-dependent junctional conductance. J. Gen. Physiol. 77, 75–94.

Spray, D.C., Ginzberg, R.D., Morales, E.A., Gatmaitan, Z. and Arias, I.M. (1986) Electrophysiological properties of gap junctions between dissociated pairs of rat hepatocytes. J. Cell Biol. 103, 135–144.

Stauffer, K.A., Kumar, N.M., Gilula, N.B. and Unwin, N. (1991) Isolation and purification of gap-junctional channels. J. Cell Biol. 115, 141–150.

Stuhmer, W., Conti, F., Suzuki, H., Wang, X., Noda, M., Yahagi, N., Kubo, H. and Numa, S. (1989) Structural parts involved in activation and inactivation of the sodium channel. Nature 339, 597–603.

Swenson, K.I., Jordan, J.R., Beyer, E.C. and Paul, D.L. (1989) Formation of gap junctions by expression of connexins in *Xenopus* oocyte pairs. Cell 57, 145–155.

Swenson, K.I., Piwnica-Worms, H., McNamee, H. and Paul, D. (1990) Tyrosine phosphorylation of the gap-junction protein connexin 43 is required for the pp 60v-*src*-induced inhibition of communication. Cell Regulation 1, 989–1002.

Traub, O., Look, J., Dermietzel, R., Brummer, F., Hulser, D. and Willecke, K. (1989) Comparative characterization of the 21-kD and 26-kD gap-junction proteins in murine liver and cultured hepatocytes. J. Cell Biol. 108, 1039–1051.

Unwin, P.N.T. and Zampighi, G. (1980) Stucture of the junction between communicating cells. Nature 283, 545–549.

Unwin, N. (1989) The structure of ion channels in membranes of excitable cells. Neuron 3, 665–676.

Veenstra, R.L. (1990) Voltage-dependent gating of gap-junction channels in embryonic chick ventricular cell pairs. Am. J. Physiol. 258, C662–672.

Werner, R., Levine, E., Rabadam-Diehl, C. and Dahl, G. (1989) Formation of hybrid cell-cell channels. Proc. Natl. Acad. Sci. USA 86, 5380–5384.

Wiener, E.C. and Loewenstein, W.R. (1983) Correction of cell-cell communication defect by introduction of a protein kinase into mutant cells. Nature 305, 433–435.

Willecke, K., Heynkes, R., Dahl, E., Stutenkemper, R., Hennemann, M., Jungbluth, S., Suchyna, T. and Nicholson, B.J. (1991) Mouse connexin 37, cloning and functional expression of a gap-junction gene highly expressed in lung. J. Cell Biol. 114, 1049–1057.

Zhang, J.T. and Nicholson, B.J. (1989) Sequence and tissue distribution of a second protein of hepatic gap junctions, Cx26, as deduced from its cDNA. J. Cell Biol. 109, 3391–3401.

J.E. Hall, G.A. Zampighi and R.M. Davis (Eds.)
Progress in Cell Research, Vol. 3

CHAPTER 2

Identification of novel connexins by reduced-stringency hybridization and PCR amplification using degenerate primers

DAVID L. PAUL, ROBERTO BRUZZONE and JACQUES-ANTOINE HAEFLIGER

Department of Anatomy and Cellular Biology, Harvard Medical School, 220 Longwood Avenue, Boston, MA 02115, USA

Introduction

Gap junctions are collections of membrane channels which allow for the movement of ions and small metabolites between cells. It is now clear that there are multiple gap-junction structural proteins. These comprise a family of closely related proteins termed connexins. DNAs encoding five rat connexins have been cloned (Kumar and Gilula, 1986; Paul, 1986; Beyer et al., 1987; Zhang and Nicholson; 1989; Hoh et al., 1991; Paul et al., 1991). These connexins have been designated rat connexin26 (Cx26), Cx31, Cx32, Cx43 and Cx46, using a system of nomenclature based on species of origin and molecular mass predicted by cDNA analysis (Beyer et al., 1987). In addition to sharing a significant amount of primary sequence similarity (35–65%), hydropathy analysis predicts that all connexins share a similar topological orientation in the plasma membrane. Each connexin contains four transmembrane, two extracellular and three cytoplasmic regions, and both N- and C-termini face the cytoplasm. Inspection of connexin primary sequences reveals that certain domains exhibit characteristically high or low similarity. The domains with greatest similarity correspond to the predicted extracellular regions. Predicted transmembrane segments and the very short cytoplasmic region at the N-terminus are also well conserved. In contrast, the central loop and the C-terminal cytoplasmic domains are highly variable in both sequence and size.

The existence of additional rat connexin genes is suggested by several observations. For example, connexins have been described in chicken (Chick Cx45, Chick Cx42; Beyer 1990) and in *Xenopus* (*Xenopus* Cx38; Ebihara et al., 1988) which were not clearly the homologs of identified rat connexins. In addition, partial N-terminal sequence indicates that the lens fiber protein MP70 may be a connexin (Kistler et al., 1988), but DNA encoding MP70 has not been isolated.

Here we report the isolation of genomic DNA fragments encoding four new members of the rat connexin family, Cx31.1, Cx33, Cx37 and Cx40. We describe the use of two methodologies, reduced-stringency hybridization and polymerase chain reaction (PCR) amplification with degenerate oligonucleotides, to identify these genes. The new connexins exhibit all of the conserved structural features of the connexin family, including highly similar extracellular and transmembrane domains, but divergent central and C-terminal cytoplasmic domains. Based on sequence similarity, rat Cx40 may be a homolog of chicken Cx42. Cx37, Cx33 and Cx31.1 do not appear to be the homologs of connexins identified in other organisms.

Cloning strategies

We have utilized two general strategies to identify new connexin genes. The first is reduced-stringency hybridization with probes consisting of full length cDNAs. This screening procedure relies on the significant overall homology between previously identified members of the connexin family. The second strategy involves PCR amplification using degenerate oligonucleotide primers corresponding to domains which exhibit the highest similarity among previously characterized connexins. Thus, the PCR approach depends on the high conservation of specific domains in all connexin family members. For both low-stringency hybridization and PCR amplification, genomic DNAs, rather than cDNAs, are used for templates, principally because all connexin sequences are expected to be equally represented. The presence of introns within coding regions could complicate the analysis of PCR products from genomic templates. However, this was considered unlikely because no introns within the coding region had been identified in any previously characterized connexin gene.

Hybridization screening

A hybridization screen of a rat genomic library in Charon 4A was performed under conditions of reduced stringency.

A fragment of Cx46 cDNA containing the complete protein coding sequence (Paul et al., 1991) was labeled by random primer extension and used to probe the library. Cx46 cDNA was chosen as probe because a primary goal of our studies was to clone the lens gap-junction protein MP70, which is thought to be closely related to Cx46 (Kistler et al., 1988). Hybridization was performed conventionally (Maniatis et al., 1982), but washing steps used only relatively high salt concentrations (2× SSC). To distinguish clones containing the homologous connexin (Cx46) from those containing novel connexins, filters with positive clones were re-washed using normal high-stringency conditions. Clones whose hybridization signals were significantly reduced in the re-wash were selected for further analysis. 180,000 plaques were initially plated which yielded 10 positive clones. The hybridization signal from one of these was eliminated by high-stringency washing. Further analysis showed that a 6 kb EcoRI fragment of this lambda isolate encoded a connexin with a calculated molecular mass of 40,237 (Cx40).

PCR cloning with degenerate primers

Degenerate oligonucleotide primers were constructed based on an analysis of the similarity between previously characterized connexin DNAs. An alignment of selected amino acid and nucleotide sequences from rat Cx26, Cx32, Cx43, and Cx46 is displayed in Fig. 1. Three consensus degenerate oligonucleotides corresponding to these regions (Fig. 1) were used for PCR amplification. Two of the oligonucleotides (24–36-fold degeneracy) corresponded to separate regions of the first extracellular domain (Fig. 1; primers 1 and 3) while the remaining oligonucleotide (54-fold degenerate) was complementary to the sequences encoding the second extracellular domain

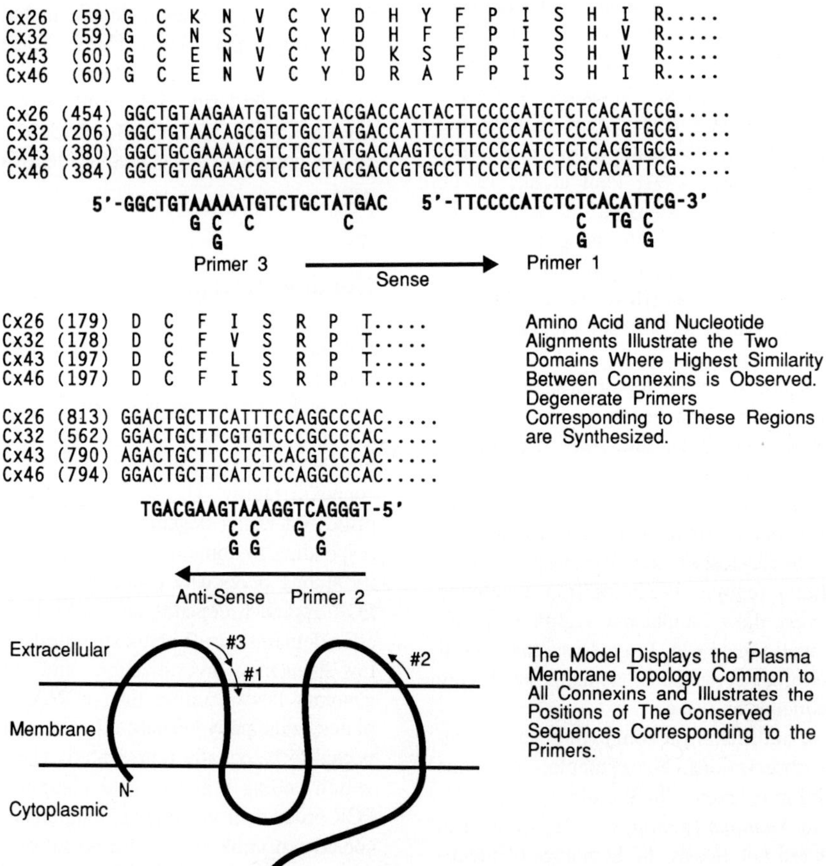

Figure 1. The construction of degenerate oligonucleotides for PCR amplification of connexins. An alignment of amino-acid and nucleotide sequences from two regions of Cx26, Cx32, Cx43 and Cx46 is displayed. The number of the starting amino-acid residue and nucleotide base is given in parentheses . Three consensus degenerate oligonucleotides (primers 1, 2 and 3) corresponding to these regions were used for PCR amplification. The approximate locations of the protein sequence encoded by the primer sequences are indicated in a topological mode for a generic connexin.

(Fig. 1; primer 2). The relatively low degeneracy reflects an attempt to minimize the number of individual sequences in the primers. Thus, not all of the possible sequence variation revealed in the connexin sequence alignment is represented in the primers. The approximate locations of the amino acids encoded by the primer sequences are indicated in a topological model for a generic connexin in Fig. 1. The amplified region corresponded to the second and third transmembrane domains, which are separated by a short cytoplasmic segment. Since the amplification products expected from known connexins range from 351 to 433 bp, depending on primer sets, only bands between 300 and 500 bp were subcloned. Primers 1 and 2 amplified one new connexin (Cx37) while primers 2 and 3 amplified two (Cx33 and Cx31.1). As expected, many of the amplification products corresponded to the connexins on which the primer sequences were based. PCR reactions contained 25–30 cycles followed by 10–15 min of extension at 72°C. Each cycle consisted of 94°C for 1 min; 50–65°C for 2 min; 72°C for 3 min. Annealing temperatures and $MgCl_2$ concentrations were empirically adjusted to optimize the production of bands of the appropriate size. $MgCl_2$ levels were relatively critical (2.0 mM) while annealing temperatures in the range of 50–65°C gave approximately similar patterns of products.

Inverse PCR cloning

The initial PCR amplification produced DNAs containing only a fragment of the new connexin coding sequences. It was necessary to isolate additional clones encompassing the complete coding sequence of the new connexins. A straightforward approach was to screen appropriate cDNA libraries (see Fig. 4 for patterns of expression). This was attempted for Cx33 and for Cx37. However, only one partial Cx37 cDNA, and no cDNAs encoding Cx33, were obtained. Therefore, inverse PCR (Ochman et al., 1988) was used to obtain additional coding sequences for Cx37, Cx33 and Cx31.1, directly from genomic DNA. This procedure is outlined schematically in Fig. 2. Inverse PCR relies on amplification using primers directing synthesis outwards from, rather than towards, each other. Normally, this would not result in geometric amplification. However, if the template for amplification is circular, the reaction will efficiently yield a linear product containing the uncharacterized sequence flanking each end of the known region. To produce a circular template of the appropriate size for amplification (0.75–3.0 kb), it is necessary to obtain restriction digest information from genomic Southern blots. Genomic DNA is then digested with a restriction enzyme cutting outside of the region corresponding to the

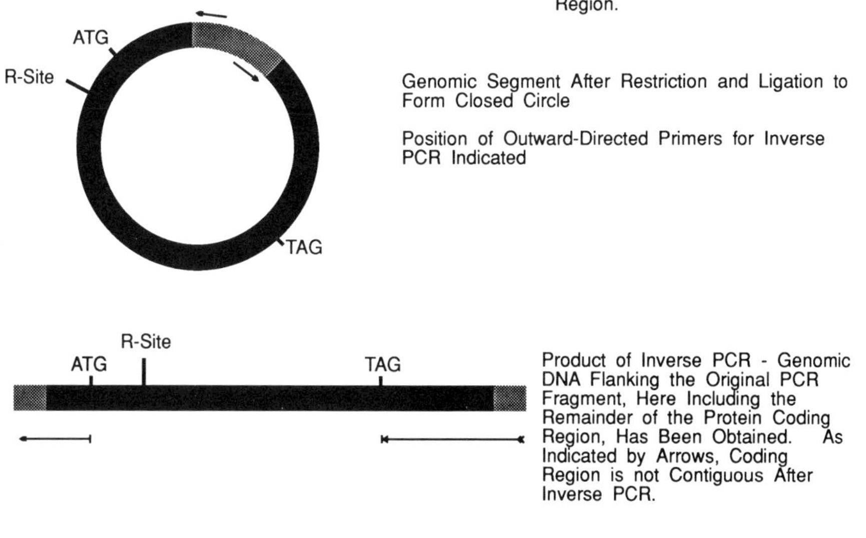

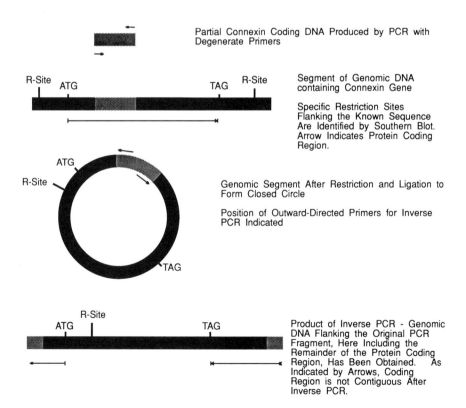

Figure 2. The use of inverse PCR to obtain genomic regions flanking a known DNA sequence.

original DNA fragment produced from amplification with degenerate primers. The genomic DNA is ligated at low concentration, which promotes the formation of closed circles, and amplified using outward-directed primers. It is important to note that DNA sequences of the product of inverse PCR are not contiguous in the order in which they appear in the genome. For future studies involving functional expression, we needed to obtain the complete native coding region. This was produced by another amplification of genomic DNA with primers corresponding to the start and stop codons.

Relationship between connexin family members

Hydropathy analysis of all four new connexins predicts a similar membrane topology, as schematically represented in Fig. 1. All connexins display very high similarity in their extracellular domains, high similarity in their transmembrane domains and N-termini, but relatively little similarity in the other major cytoplasmic regions. In addition, all connexins show an absolute conservation of three cysteines in each extracellular region. Rat Cx33 is much more closely related to rat Cx43 than any connexin identified thus far. Similarly, rat Cx31.1 is more related to rat Cx31 than any other connexin.

A summary of the relationship between nine cloned rat connexins based on amino acid sequence is presented in Fig. 3. This dendrogram was produced by PILEUP (UWGCG ver 7.0; Deveraux et al., 1984). Connexins were differentiated into two broad classes by the computer analysis. It is important to emphasize that this dendrogram does not necessarily indicate evolutionary or phylogenetic relationships between connexins. It is intended only to graphically summarize the primary sequence relationships. A previous analysis of connexin similarity (Bennett et al., 1991) suggested that Cx26 and Cx32 were similar enough to constitute a class, which was designated Group I. Similarly, Cx43 and Cx46 constituted a second class, Group II. Group I and II may correspond to the designations β and α, respectively, introduced by Risek et al. (1990). According to Bennett et al. (1991), the distinction between groups rested on the comparison of the second and fourth membrane-spanning regions and part of the second extracellular domain. Intra-group comparisons in these regions display greater similarities than inter-group comparisons (data not shown). In accordance with these findings, Cx40, Cx37, and Cx33 appear to be most related to Cx43 and Cx46, or Group II connexins. Cx31.1 is most related to Cx31, a Group I connexin.

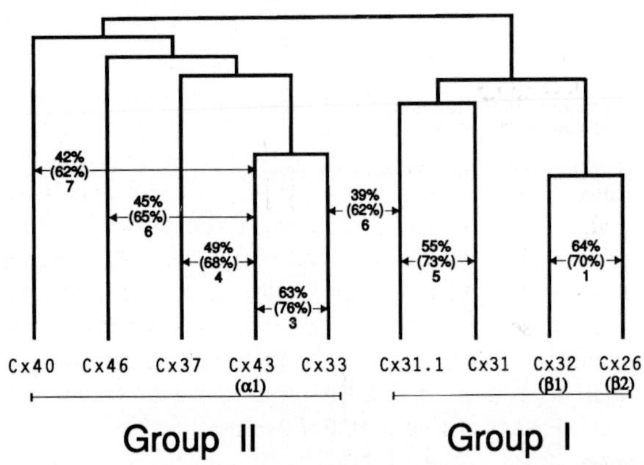

Figure 3. A dendrogram of the relationships between cloned rat connexins based on primary sequence. Connexins may be differentiated into two broad classes, Group I and Group II. Values indicate % identity (top values), percent similarity (parentheses) and number of gaps introduced to obtain the best alignment (bottom value). Within each class, some pairs of connexins are very highly related (e.g. Cx43/Cx33), while others are more distantly related (e.g. Cx43/Cx40).

The numerical values in Fig. 3 were produced by GAP (UWGCG ver 7.0) and indicate the percent identity (top value), percent similarity (in parentheses) and the number of gaps introduced to obtain the best alignment (bottom value). For this analysis, similarity indicates an evolutionary relationship between amino acids based on nucleotide changes, rather than conservative substitution (Lipman and Pearson, 1985). These data indicate that, even between groups, all connexins are strongly related. Intra-class relationships are generally, but not always, higher. Comparison of the new rat connexin sequences to those reported in other species reveals a very close relationship between rat Cx40 and chicken Cx42 (Beyer, 1990), 69% identity (or 81% similarity) with almost complete overlap. Therefore, Cx40 may be the rat homolog of chicken Cx42.

Expression patterns of connexins

The pattern of expression of the four novel connexins was analyzed by Northern blotting total RNA from 12 different adult rat organs; brain, heart, ovary, testis, kidney, spleen, liver, stomach, lung, skin, lens and pancreas. Figure 4 summarizes the results of this analysis. Cx40 mRNA is most abundant in lung, but readily detectable in heart, ovary, and kidney. Cx37 mRNA also predominates in lung, but lower levels are observed in all organs tested except pancreas and lens. In contrast to the relatively broad distribution of Cx40 and Cx37 mRNAs, Cx33 and Cx31.1 mRNAs are observed only in testis and in skin, respectively. For purposes of comparison, the patterns of expression of five previously cloned connexins, analyzed by Northern blot, are also included in this figure.

Discussion

In order to identify new connexins, we have screened genomic DNA by two methodologies: reduced-stringency hybridization and PCR amplification with degenerate primers. We have identified four novel connexins, designated Cx40, Cx37, Cx33 and Cx31.1 Other studies presented in

	Brain	Heart	Ov	Test	Kid	Spl	Liver	Stom	Lung	Skin	Lens	Pan
Cx26	NT	-	NT	-	++	-	+++	+	+/-	++	-	NT
Cx31	-	-	-	-	-	-	-	-	-	+++	-	NT
Cx31.1	-	-	-	-	-	-	-	-	-	+++	-	-
Cx32	+	-	-	-	++	-	+++	+	NT	++	-	+++
Cx33	-	-	-	++	-	-	-	-	-	-	-	-
Cx37	+	++	++	++	+	+/-	+/-	+/-	+++	+	-	-
Cx40	+/-	++	++	-	++	-	-	-	+++	-	-	-
Cx43	+	+++	+++	++	++	+/-	-	+/-	NT	++	+	-
Cx46	-	+	-	-	+/-	-	-	-	-	-	+++	-

Figure 4. The expression pattern of nine cloned connexins analyzed by Northern blot. Of the new connexins, Cx40 mRNA is most abundant in lung, but readily detectable in heart, ovary, and kidney. Cx37 mRNA is also most abundant in lung and present in other organs at low levels, but is not completely coincident with Cx40. In contrast to the relatively broad distribution of Cx40 and Cx37 mRNA, Cx33 and Cx31.1 mRNAs were observed only in testis and skin respectively.

this volume (Chapter 5, Chapter 24) report additional novel connexins and it seems likely that more connexin genes remain to be identified. For example, none of the new connexins reported in this volume can be conclusively identified as the rat homolog of *Xenopus* Cx38. Our failure to isolate more connexin genes is likely the result of the moderate to high-stringency conditions chosen for PCR primer annealing and library hybridization.

Northern blot hybridization suggests that each connexin gene has its own distinct pattern of expression. Most organs and many cell types have been shown to express multiple connexins. Certain combinations of connexins are very often co-expressed in a given organ or cell type. For example, Cx26 is co-expressed with Cx32 in many epithelia (Zhang and Nicholson, 1989). Similarly, transcripts for Cx31 (Hoh et al., 1991) and Cx31.1, which both display a very restricted distribution, are found together in skin. Cx40 and 37 are broadly expressed in many adult organs and tissues, and are particularly abundant in lung. However, the relative levels of transcripts for these two connexins are different in many organs. Thus, their distribution overlaps but is not completely coincident. Cx33 has the most limited expression of any connexin, detected only in testes. Clearly, our analysis of the expression of the new connexins is subject to several limitations. Negative results may only reflect the limited sensitivity of the technique. In addition, these studies indicate the presence of mRNAs, not proteins. It will be important to demonstrate that the corresponding proteins are synthesized in vivo and that they can direct the establishment of intercellular communication.

At the present time, the significance of connexin diversity is not clear. However, there are two likely consequences of variations in connexin primary structure. First, the changes may result in channels that are regulated differently in similar intracellular environments. We have reported one example of a regulatory difference between Cx32 and Cx43 (Swenson et al., 1990). *V-src* expression totally abolished communication based on Cx43 but had little effect on communication based on Cx32. *V-src* expression inhibited Cx43 communication by inducing tyrosine phosphorylation of residue 265. This residue is located in a highly variable, cytoplasmic domain. In contrast, *v-src* induced no tyrosine phosphorylation of Cx32, which is completely dissimilar to Cx43 in this domain.

A second consequence of connexin diversity may be to limit, rather than facilitate, communication between cells. For example, cells expressing *Xenopus* Cx38 established high levels of communication with neighbors expressing rat Cx43 but not rat Cx32 (Swenson et al., 1989; Werner et al., 1989). In contrast, high levels of communication were observed with neighbors expressing the same connexin. Thus, at least theoretically, expression of alternate connexins could be used to permit coupling to certain physically adjacent cells and to prohibit coupling with others. Qualitatively similar patterns of coupling, termed "communication compartments" (Lo and Gilula, 1979), have been reported in both vertebrate and invertebrate organisms (Warner and Lawrence, 1982; Blennerhasset and Caveney, 1984; Weir and Lo, 1984; Kam et al., 1986; Kalimi and Lo, 1988; Serras et al., 1989) as well as in cell-culture studies (Mesnil et al., 1987). The molecular mechanisms underlying compartmentalization have not been established but, in some cases, could result from the intrinsic properties of the connexins expressed.

Acknowledgements

We are grateful to Daniel A. Goodenough for invaluable technical support and helpful discussions. This work was supported by NIH GM37751 to DLP, by Swiss National Science Foundation grants #823A-28370 to JAH, and #83.627.0.88 to RB.

References

Bennett, M.V.L., Barrio, L.C., Gargiello, T.A., Spray, D.C., Hertzberg, E. and Sáez, J.C. (1991) Gap Junctions: new tools, new answers, new questions. Neuron 6, 305–320.

Beyer, E.C., Paul, D.L. and Goodenough, D.A. (1987) A protein from rat heart homologous to a gap-junction protein from liver. J. Cell Biol. 105, 2621–2629.

Beyer, E.C. (1990) Molecular cloning and developmental expression of two chick embryo gap-junction proteins. J. Biol. Chem. 265, 14439–14443.

Blennerhasset, M.G. and Caveney, S. (1984). Separation of developmental compartments by a cell type with reduced junctional permeability. Nature (London) 309, 361–364.

Devereux, J., Haeberli, P. and Smithies, O. (1984) A comprehensive set of sequence analysis programs for the VAX. Nucleic Acids Res. 12, 387–394

Ebihara, L., Beyer, E.C., Swenson, K.I., Paul, D.L. and Goodenough, D.A. (1989). Cloning and expression of a *Xenopus* embryonic gap-junction protein. Science, 243, 1194–1195.

Hoh, J.H., John, S. and Revel, J.-P. (1991) Molecular cloning and characterization of a new member of the gap-junction gene family, connexin-31. J. Biol. Chem. 226, 6524–6531.

Kalimi, G.H. and Lo, C.W. (1988) Communication compartments in the gastrulating mouse embryo. J. Cell Biol. 107, 241–255.

Kam, E., Melville, L. and Pitts, J.D. (1986) Patterns of junctional communication in skin. J. Invest. Dermatol. 87, 748–753.

Kistler, J., Christie, D. and Bullivant, S. (1988) Homologies between gap-junction proteins in lens, heart and liver. Nature 331, 721–723.

Kumar, N. and Gilula, N.B. (1986) Cloning and characterization of human and rat liver cDNAs coding for a gap-junction protein. J. Cell Biol. 103, 767–776.

Lipman, D.J. and Pearson, W.R. (1985) Rapid and sensitive protein similarity searches. Science 227, 1435

Lo, C.W. and Gilula, N.B. (1979) Gap junctional communication in the postimplantation mouse embryo. Cell 18, 411–422.

Maniatis, T., Fritsch, E.F. and Sambrook, J. (1982) Molecular Cloning: A Laboratory Manual. Cold Spring Harbor Press. Cold Spring Harbor, NY, pp. 320–329.

Mesnil, M., Fraslin, J.M., Piccoli, C., Yamasaki, H. and Guguen-Guillouzo, C. (1987) Cell contact but not junctional communication (dye coupling) with biliary epithelial cells is required for hepatocytes to maintain differentiated functions. Exp. Cell Res. 173, 524–533.

Ochman, H., Gerber, A.S. and Hartl, D.L. (1988) Genetic applications of an inverse polymerase chain reaction. Genetics 120, 621–623.

Paul, D.L. (1986). Molecular cloning of cDNA for rat liver gap-junction protein. J. Cell Biol. 103, 123–134.

Paul, D.L., Ebihara, L., Takemoto, L.J., Swenson, K.I., Goodenough, D.A. (1991) Connexin46, a novel lens gap-junction protein, induces voltage-gated currents in nonjunctional plasma membrane of *Xenopus* oocytes. J. Cell Biol. 115, 1077–1089.

Risek, B., Guthrie, S., Kumar, N.M. and Gilula, N.B. (1990) Modulation of gap-junction transcript and protein expression during pregnancy in rat. J. Cell Biol. 110, 269–282.

Serras, F, Damen, P, Dictus, W.J.A.G, Notenboom, R.G.E. and Van den Biggelaar, J.A.M. (1989) Communication compartments in the ectoderm of embryos of *Patella vulgata*. Wilhelm Roux Arch. Dev. Biol. 198, 191–200.

Swenson, K.I., Jordan, J.R., Beyer, E.C. and Paul, D.L. (1989). Formation of gap-junctions by expression of connexins in *Xenopus* oocyte pairs. Cell 57, 145–155.

Swenson, K.I., Piwnica-Worms, H., McNamee, H. and Paul, D. (1990). Tyrosine phosphorylation of the gap-junction protein connexin 43 is required for the pp 60v-*src*-induced inhibition of communication. Cell Reg. 1, 989–1002.

Warner, A.E. and Lawrence, P.A. (1982) Permeability of gap junctions at the segmental border in insect epidermis. Cell 28, 243–252.

Weir, M.P. and Lo, C.W. (1984) Gap junctional communication compartments in the Drosophila wing disk. Dev. Biol. 102, 130–146.

Werner, R., Levine, E., Rabadam-Diehl, C. and Dahl, G. (1989). Formation of hybrid cell-cell channels. Proc. Natl. Acad. Sci. USA 86, 5380–5384.

Zhang, J.T. and Nicholson, B.J. (1989). Sequence and tissue distribution of a second protein of hepatic gap junctions, Cx26, as deduced from its cDNA. J. Cell Biol. 109, 3391–3401.

J.E. Hall, G.A. Zampighi and R.M. Davis (Eds.)
Progress in Cell Research, Vol. 3

CHAPTER 3

Affinities between connexins

RUDOLF WERNER, CRISTINA RABADAN-DIEHL, ERIC LEVINE and GERHARD DAHL

Departments of Biochemistry and Molecular Biology and Physiology and Biophysics, University of Miami, School of Medicine, Miami, FL, USA

Introduction

Connexins, the protein components of cell–cell channels, are members of a distinct family of membrane proteins (Beyer, 1991). The reason for the multiplicity of cell–cell channels is unknown. Their function could be so basic for survival of the organism that during evolution many connexins that evolved were retained with their basic structural elements conserved. For example, the extracellular loops of all connexins are highly conserved. The transmembrane domains that must include the channel-lining sequences are homologous. However, the observation that there is little species, but pronounced tissue specificity among connexins suggests that the diversity of the connexins may reflect tissue-specific functions. Obviously, the situation is more complicated because of the finding that some tissues co-express several types of connexin.

In the present study, we investigate how hemichannels, located in the membranes of apposing cells, interact. This includes the interaction between identical hemichannels as well as between hemichannels of different kinds. We focus on three connexins: Cx32, normally expressed in rat liver, Cx43 expressed in rat heart, and Cx38, the endogenous *Xenopus* oocyte connexin.

Channel formation

The *Xenopus* oocyte expression system is ideal for studying the formation of cell–cell channels (Dahl et al., 1987). The translation in oocytes of microinjected mRNA into connexin proteins and their subsequent incorporation into functional cell–cell channels in the membranes of paired oocytes can be assayed by measuring the junctional conductance in a double voltage-clamp setup. In the symmetric configuration, both oocytes of a pair are injected with the same type of mRNA; in the asymmetric configuration, the injected mRNAs differ.

For an open channel to form, the hemichannels first have to bind to each other (docking); in addition, both hemichannels must open to provide a hydrophilic path between the two cells. Conductance measurements thus do not allow discrimination between docking and opening. For reasons of simplicity we will use the term "affinity" of hemichannels when interpreting data from conductance measurements, with the understanding that this term includes all steps in the channel-formation process.

Oocytes that have been injected with connexin mRNA accumulate a pool of channel precursors from which cell–cell channels can rapidly form upon pairing (Fig. 1). A fraction of this pool is most likely located within the cell membrane (Dahl et al., 1991, 1992), as indicated by surface labeling and immunohistochemistry. These precursors probably are in the hexameric hemichannel form. This conclusion is based on analogy with all other known multimeric membrane proteins which are assembled before they reach the plasma membrane. Furthermore, the observed concentration dependence of the rate of channel formation suggests that channels form from preassembled units (Dahl et al., 1992).

If, as suggested, hemichannels exist in the membrane, they must be closed because the membrane conductance in oocytes containing a large pool of channel precursors is indistinguishable from precursor-free oocytes (Werner et al., 1989). It appears that hemichannels never open to the extracellular milieu and that channel formation is a leak-proof process. Within the limits of resolution of the detection technique, no increase in the non-junctional membrane conductance can be observed, even at the highest rates of channel formation, up to 40 channel openings per second (Fig. 1).

Affinities between hemichannels

Levels of junctional conductance due to the presence of endogenous channels in oocyte pairs are nonexistent or very low, while conductances due to channels made from Cx32 or Cx43 in the same oocytes can reach more than 100 μS within 24 h after pairing (Table I). When hybrid pairs are made between oocytes injected with Cx32 and uninjected oocytes, no significant junctional conductance

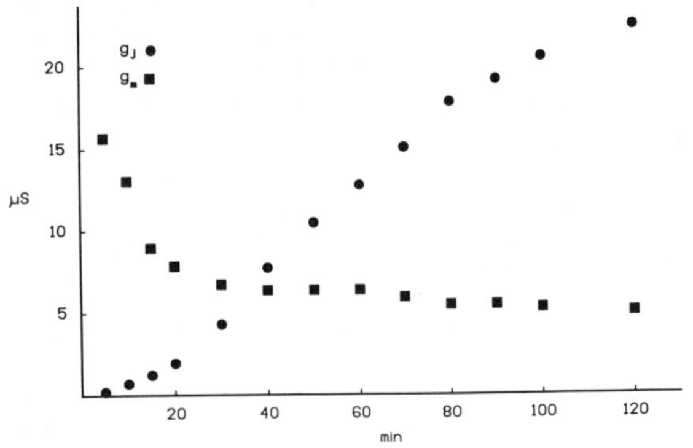

Figure 1. Time course of channel formation in a single oocyte pair. Junctional conductance g_j (●) and membrane conductance g_m (■) are expressed in μS and plotted as a function of time after pairing (time point zero). Oocytes were injected with Cx32 mRNA 24 h prior to pairing. Oocytes were stripped from their vitelline layer manually, incubated in OR2 medium containing 10 μg/ml of Glycine Max (Levine et al., 1991) for 20 min, washed, and then paired. Electrode impalement for dual voltage-clamp experiments occurred within 5 min after pairing. The initial decline in membrane conductance is attributable to damage caused by vitelline removal, Glycine Max effects, and impalement. Note that at the time of maximal rate of channel formation, membrane conductance remains constant. (The input resistance of unstripped oocytes is > 1 MΩ.)

above endogenous levels is observed. On the other hand, hybrid channels between Cx43 and endogenous hemichannels form readily and can be identified by their typical gating properties (rectification, Werner et al., 1989; Swenson et al., 1989). The levels of these hybrid channels vary significantly between different batches of oocytes; corresponding data with pure Cx43 channels or pure endogenous channels do not show such a variation. However, a clear trend can be seen: the levels of hybrid channels are low whenever the endogenous channels cannot be detected.

To get a better understanding of this phenomenon, the hybrid pairing experiments were repeated with Cx38, which probably represents the endogenous oocyte channels (Ebihara et al., 1989). When oocytes are injected with synthesized Cx38 mRNA, the observed junctional conductance was less than 5% of that observed with oocytes that were both injected with similar levels of Cx43 mRNA. On the other hand, when only one oocyte was expressing Cx43 while the other oocyte expressed Cx38 hemichannels, the junctional conductance reached levels comparable to those observed when Cx43 was expressed in both oocytes. This suggests that hemichannels made from Cx38 have a poor affinity for each other, yet pair much more readily with hemichannels made from Cx43. Thus, it would appear that Cx38 is designed to form hybrid channels with other connexins, such as Cx43, rather than pure Cx38 channels. In order to determine the relative affinities of Cx38 and Cx43 more precisely, oocytes were injected with mixtures of Cx38 and Cx43 mRNA at ratios of 1:1 and 9:1. The presence of Cx38-containing channels was then determined by perfusion with 50% CO_2 which leaves only the pure Cx43 channels open (Werner et al., 1991). When mixtures containing equal amounts of Cx43 and Cx38 mRNA were injected into both oocytes, most of the channels formed consisted of pure Cx43 channels: the junctional conductance in such pairs was not reduced by 50% CO_2. Only when the mRNA ratio (Cx43/Cx38) in the mixture was 1:9 did the majority of the channels (95%) contain both Cx43 and Cx38. 5% of the junctional conductance remained in the presence of 50% CO_2, indicating that these channels were pure Cx43 channels. The CO_2-sensitive portion of the channels consists of hybrid channels oriented in either direction and/or of hetero-oligomeric hemichannels (43/38).

The conclusions from these experiments are summarized in Fig. 2, which compares the affinity properties of three connexins: Cx32, Cx43, and Cx38. With regard to its affinity properties, Cx43 seems to assume a middle position between Cx38 and Cx32: it forms hybrid channels with both of the other types of connexin. Cx43–Cx38 hybrid channels are asymmetrically voltage-gated; they rectify. There appears to be no affinity between Cx32 and Cx38, since oocyte pairs expressing Cx32 hemichannels in one cell and Cx38 hemichannels in the other show little, if any, junctional conductance. The low level of conductance that is sometimes observed is symmetrically sensitive to junctional voltage and thus must be due to Cx38 channels.

TABLE I Results from a typical experiment involving oocytes with endogenous channel activity; all oocytes are from the same ovary. Junctional conductances in oocyte pairs injected with Cx32 or Cx43 mRNA, measured 24 h after pairing. All conductances are expressed in μS ± SEM, and the number of cell pairs analyzed is indicated in parentheses. (In this particular experiment, the concentration of Cx43 mRNA was lower than that of Cx32 mRNA.)

Connexin	Conductance [μS]	n
Endogenous	0.10 ± 0.02	10
Cx32	120.28 ± 3.27	9
Cx43	85.50 ± 1.82	9

	cx32	cx43	cx38
cx32	100%	100%	0%
cx43		100%	~50%
cx38			5%

Figure 2. Relative affinities of connexin hemichannels. The numbers indicate junctional conductance expressed as percentage of the conductance observed with wild-type Cx32 channels. Vertical labels refer to the mRNA injected into one oocyte of a pair, horizontal labels to the mRNA injected into the other oocyte.

E1 domains:

```
S F I C N T Q Q P G C E N V C Y D H F F P I S H V R L W
A   V   D   L         N S         K S A   L Y N I   F
D   R       E         T           A A               Y
K   M                             Q Y
E   T                             N
    H
```

E2 domains:

```
L V K C E R F P C P H T V D C F V S R P T E K T V F T
V Y T   D A D     K N R I E   Y L A     M     N I   I
I F V   K Q A         P   N     I                   L
  L I   N V K         K
    Q   S   W         Q
        Q   Y
            I
            S
            V
```

Figure 3. Amino-acid variability in the E1 and E2 domains of presently known members of the connexin family.

Amino acids that determine hemichannel affinity

In order to determine the reasons for the difference in affinities of the three connexins, the contribution of individual amino acids located in the extracellular domains of the protein, and therefore likely to contribute to the specificity of the docking process, was studied. A large number of amino acids in the extracellular loops of all known connexins are identical. These amino acids must be involved in basic channel functions and cannot be responsible for specificity. For example, each of the two extracellular domains contains three cysteines located at the same positions in all known connexins, with a minor exception in Cx31 where the position of one cysteine is shifted by one residue. These cysteines are critical for the channel-formation process. Changing any one of them to serine results in absolute loss of function (Dahl et al., 1991, 1992).

The variations in the extracellular loops of connexin that nature has allowed are shown in Fig. 3. Some of these variations are probably due to the lack of functional constraints. Others, however, must be involved in determining affinities of hemichannel interactions.

In order to study the amino acids involved in hemichannel affinity, four locations were chosen for site-directed mutagenesis, two each in the extracellular loops E1 and E2. The analysis was restricted to the three connexins for which affinities had been determined before. The strategy was to use Cx32 as a backbone and to change any one of the four amino acids into those found at the corresponding position in Cx43. The positions of the four mutants, I52R, N61E, K167T, and N175H are shown in Fig. 4. (Mutant nomenclature: first letter = original amino acid; number = position in Cx32; second letter = new amino acid; multiple mutations are separated by a hyphen.).

The mutants were tested in two ways: under stringent conditions, where both oocytes in a pair expressed the mutant connexin, and under relaxed conditions, where the oocyte expressing mutant connexin was paired with an oocyte expressing wild-type Cx32.

All of the mutants exhibited rather severe loss in channel-forming ability. This was surprising because Cx32 and Cx43 are known to form hybrid channels with high efficiency. Thus, every one of the four amino acids chosen must be involved in determining the specificity of hemichannel interactions.

When tested under stringent conditions, none of the mutants tested, with the exception of N61E and I52R, was capable of forming functional channels within 3 h after pairing the oocytes. (I52R exhibited less than 5% of the channel-forming ability of wild-type Cx32, and N61E 40%.) At 24 h after pairing, however, every mutant formed functional channels with activity levels ranging from less than 5% of the wild-type activity to 100% wild-type activity. Under relaxed conditions, channel-forming ability was much higher. The K167T, and N61E mutants reached wild-type levels even within 3 h after pairing. The other mutants still exhibited lower levels of activity.

These results suggest that all mutants can still form functional channels, but that the channel-forming process now takes considerably more time. It is likely that the process of cell–cell channel formation involves conformational changes in the connexin molecule. In the mutants, the conformation required for binding may not be as likely to occur as in the wild type. Evidence for conformational changes in the connexin molecule during channel formation was first suggested by experiments demonstrating disulfide exchange during this process (Dahl et al., 1991).

Pairing with wild-type Cx32-expressing oocytes was considerably faster in most cases, suggesting that the wild-type hemichannel conformation can induce the mutant hemichannel to assume a conformation compatible for pairing ("induced fit").

Despite their differences in amino-acid sequences, Cx32 and Cx43 hemichannels bind to each other equally well in symmetric and asymmetric configurations. This means that their extracellular loops must have compatible structures. This structure appears to be disrupted by a single amino-acid change from one type of connexin to the other. Restoration of the original structure, therefore, must involve at least one additional change that rescues the lost

```
           Comparison of E1 domains:
Cx32: D E K S S F[I]C N T L Q P G C[N]S V C Y D
Cx43: D E Q S A F[R]C N T Q Q P G C[E]N V C Y D
Cx38: D E Q S D F I C N T Q Q P G C T N V C Y D

           Comparison of E2 domains:
Cx32: G Y A M V R L V[K]C E A F P C P[N]T V D C F
Cx43: G F S L S A V Y[T]C K R D P C P[H]Q V D C F
Cx38: G F V M S P I F V C E R I P C K H K V E C F
```

Figure 4. Comparison of extracellular domains in Cx32, Cx43, and Cx38. The mutated amino acids are indicated in boxes.

function. Therefore, double mutants were constructed in which one position (K167T) was kept constant and an additional substitution was made in 3 different positions.

In general, double mutants were found to be less functional than single mutants, but some exhibited drastically altered pairing properties. For example, the K167T-I52R mutant became capable of forming hybrid channels with Cx38, the endogenous oocyte connexin (Ebihara et al., 1989). The resulting channels were asymmetrically voltage gated and thus rectifying. This is in striking contrast to wild-type Cx32, which does not pair with Cx38. Another mutant, K167T-N175H, in which both mutations are located within the same extracellular loop (E2) and which showed the lowest channel-forming ability both under stringent and relaxed conditions, exhibited now almost wild-type levels of conductance when paired with Cx43.

Conclusions

The poor channel-forming ability of Cx38 by itself and its effectiveness in forming hybrid channels with Cx43 suggest that this protein may be designed to form hybrid rather than symmetrical channels. The existence of rectifying neuronal cell–cell channels (electrical synapses, Furshpan and Potter, 1959) suggests that connexins with a tendency to form hybrid channels may be widely distributed. In the ovary, it appears that the oocytes' Cx38 has a natural hybrid partner in the follicle cells' Cx43 (Gimlich et al., 1990). Later in the developing embryo, the situation is unclear. Considering the poor channel-forming ability of Cx38, the possibility that another partner exists, either in the form of Cx43 or a yet unidentified connexin, merits investigation.

It is tempting to speculate that the conserved amino acids located in the extracellular loops of connexins are involved in the basic mechanism of channel formation, including stabilization of protein folding by disulfide bonds and sealing of the extracellular channel segments to the extracellular milieu by hydrophobic amino acids that are present in surprising abundance in these otherwise hydrophilic domains.

The variable amino acids, on the other hand, may be separated into two groups, one providing specificity for hemichannel–hemichannel interaction, the other serving the function of spacers. The data presented here indicate that the four amino acids tested, and probably more, are responsible for specificity.

No structural data are presently available on the conformation of the extracellular loops. It is also not known which types of bond hold the hemichannels together. It is interesting to note, however, that the two extracellular loops together contain sequences resembling those found in "Zn-finger" motifs. Zn fingers are known for mediating protein–DNA interactions. Recently, however, Zn fingers were also implicated in protein–protein interaction (Froehner, 1991). This observation may provide a lead for further studies on hemichannel interaction.

Acknowledgements

This work was supported by NIH grant GM40583 (R.W.), NSF grant DCB-8911238 (G.D.) and a grant from the Florida Division of the American Heart Association (R.W.).

References

Dahl, G., Miller, T., Paul, D.L., Voellmy, R. and Werner, R. (1987) Expression of functional cell-cell channels from cloned rat liver gap-junction complementary DNA. Science 236, 1290–1293.

Dahl, G., Levine, E., Rabadan-Diehl, C. and Werner, R. (1991) Cell-cell channel formation involves disulfide exchange. Eur. J. Biochem. 197, 141–144.

Dahl, G., Werner, R., Levine, E. and Rabadan-Diehl, C. (1992) Mutational analysis of gap-junction formation. Biophys. J. 62, 187–195.

Ebihara, L., Beyer, E.C., Swenson, K.I. and Paul, D.L. (1989) Cloning and expression of a *Xenopus* embryonic gap-junction protein. Science 243, 1194–1195.

Froehner, S.C. (1991) The submembrane machinery for nicotinic acetylcholine receptor clustering. Cell 114, 1–5.

Furshpan, E.J. and Potter, D.D. (1959) Transmission at the giant motor synapses of the crayfish. J. Physiol. 145, 289–325.

Gimlich, R.L., Kumar, N.M. and Gilula, N.B. (1990) Differential regulation of the levels of three gap-junction mRNAs in *Xenopus* embryos. J. Cell Biol. 110, 597–605.

Levine, E., Werner, R. and Dahl, G. (1991) Cell-cell channel formation and lectins. Am. J. Physiol. 261, C1025–C1032.

Swenson, K.I., Jordan, J.R., Beyer, E.C. and Paul, D.L. (1989) Formation of gap junctions by expression of connexins in *Xenopus* oocyte pairs. Cell 57, 145–155.

Werner, R., Levine, E., Rabadan-Diehl, C. and Dahl, G. (1989) Formation of hybrid cell-cell channels. Proc. Natl. Acad. Sci. USA 86, 5380–5384.

Werner, R., Levine, E., Rabadan-Diehl, C. and Dahl, G. (1991) Gating properties of connexin32 cell-cell channels and their mutants expressed in *Xenopus* oocytes. Proc. R. Soc. London Ser. B 243, 5–11.

J.E. Hall, G.A. Zampighi and R.M. Davis (Eds.)
Progress in Cell Research, Vol. 3

CHAPTER 4

Expression of Cx43 in rat and mouse liver

CATHERINE FROMAGET[a], ABDELHAKIM EL AOUMARI[a], THÉRÈSE JARRY[a], JEAN-PAUL BRIAND[b], MICHÈLE MAURICE[c], GÉRARD FELDMANN[c], BERNARD MARO[d] and DANIEL GROS[a]

[a]*Laboratoire de Biologie de la Différenciation Cellulaire, UA CNRS 179, Faculté des Sciences de Luminy, Université d'Aix-Marseille II, 13288, Marseille Cédex 9,* [b]*Institut de Biologie Moléculaire et Cellulaire, CNRS, Laboratoire d'Immunochimie, 6708, Strasbourg Cédex,* [c]*Laboratoire de Biologie Cellulaire, INSERM Unité 327, Faculté de Médecine Xavier Bichat, Université Paris VII, 75018, Paris and* [d]*Institut Jacques Monod, UMR CNRS 3, Université Paris VII, 75005 Paris, France*

Introduction

Previous studies on the expression of Cx43 and the abundance of its mRNA suggested that connexin 43 (Cx43), the cardiac major gap-junctional protein, is developmentally regulated during the ontogenesis of mouse heart (Fromaget et al., 1990). Complementary immunofluorescence investigations were undertaken at embryonic stages (Fromaget et al., 1992) using whole-embryo frozen sections and antipeptide antibodies directed to residues 314–322 of rat Cx43 (El Aoumari et al., 1990). In addition to the myocardium of mouse embryos, immunofluorescence staining was seen in the lens, brain, skin epithelium, digestive system, lungs, etc., and, unexpectedly, in the liver (from 14 days post-coitum). This prompted us to further examine immunostained frozen sections of adult mouse and rat liver for the presence of Cx43 mRNA.

Results

Immunofluorescent staining of adult rat and mouse liver sections indicate that the epitopes recognized by the anti-Cx43 site-directed antibodies are mainly localized in the hepatic parenchyma (Figs. 1A and 1B). A few small immunoreactive sites are also seen in the portal areas (Fig. 1A), where they are associated with the vascular endothelium, and in the Glisson's capsule, which envelops the liver (not shown). In the hepatic parenchyma, the staining is visible either as pairs of dots, a single continuous line, or as continuous parallel lines resembling railroad tracks (Figs. 1A and 1B; Figs. 2A, B and C, green labels). This pattern of staining is reminiscent of the distribution in the liver of ZO-1 and cingulin, two tight-junction-associated proteins (Stevenson et al., 1986 and 1989; Anderson et al., 1988; Citi et al., 1988 and 1989). To determine the precise localization of the labels, the following antibodies, for which the antigenic sites have been well characterized in hepatic tissue, were used:

(i) Hen polyclonal antibodies directed to residues 266–282 of rat connexin 32 (Cx32), one of the connexins of the gap junctions localized in the lateral domain of hepatocyte plasma membrane (Paul, 1986; Stevenson et al., 1986; Dermietzel et al., 1987; Traub et al., 1989). Figure 1C shows the distribution of Cx32 in an adult rat liver frozen section.

(ii) Mouse monoclonal antibody B10 recognizes an antigen of M_r 125–130 kDa associated with the membrane which delimits the bile canaliculi, i.e. with the hepatocyte apical membrane (Maurice et al., 1985 and 1988). See Fig. 1D.

(iii) Rat monoclonal antibody R26.4C targets ZO-1 (Stevenson et al. 1986). See Fig. 1E.

Omitting the incubation step with the primary antibodies or replacing the primary antibodies by preimmune fractions produced control sections which showed no labeling.

Double indirect-immunofluorescence experiments, in which anti-Cx43 antibodies were used in association with an antibody described above, were analyzed by confocal laser scanning microscopy (see Fig. 2). We found that the sites recognized by anti-Cx43 antibodies in rat and mouse are always associated, but not co-localized, with the B10 antigen. Thus, the epitopes recognized by the anti-Cx43 antibodies are expressed by the hepatocytes and are localized very near the apical pole but they do not co-localize with Cx32. They do co-localize, however, with the tight-junction-associated protein, ZO-1

Northern blots of poly(A)$^+$-rich RNA (10 μg) extracted from adult rat and mouse liver and of total RNA (50 μg) from rat liver, were probed with the EcoRI/AaTI restriction fragment (1.03 kb) excised from the cDNA clone G1

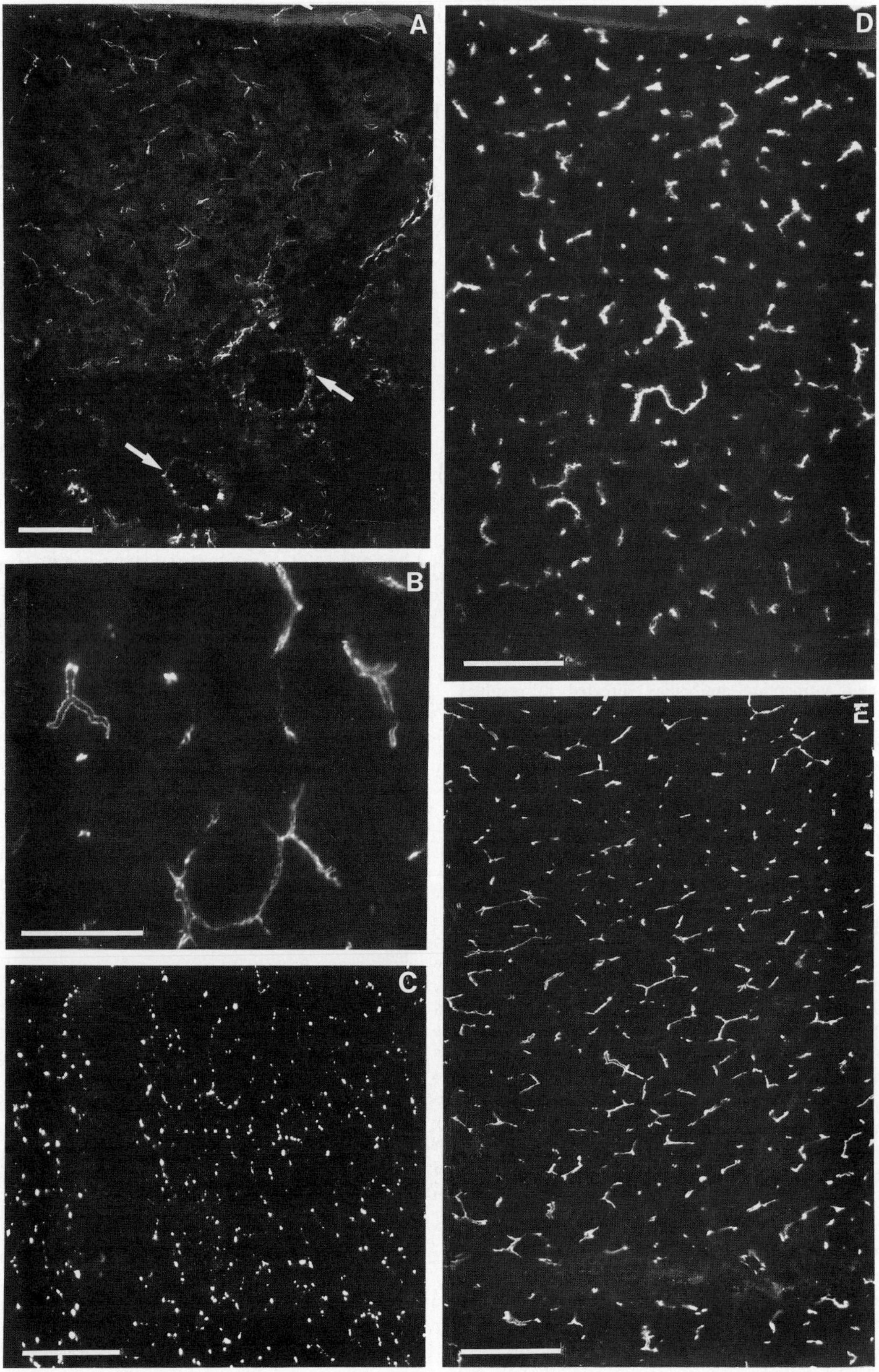
A
B
C
D
E

coding for the rat heart Cx43. This cDNA, kindly provided by Dr. E. Beyer, contains about 92% of the coding sequence of rat Cx43 and a long 3'-untranslated sequence (Beyer et al., 1987). The EcoRI-AaTI restriction fragment contains only the coding region. Figure 3A shows a 3-week exposure of one blot. A single hybridization signal of 3.0 kb, i.e. of the size expected for the Cx43 mRNA, is seen in the liver poly(A)$^+$-rich RNA samples (Fig. 3A, lanes b and c). This message is also seen in the liver total RNA sample (Fig. 3A, lane d) as well as in the control heart total RNA sample (Fig. 3A, lane a). Using the same probe, Cx43 mRNA was sought in mouse liver at various developmental stages, from 14 days post-coitum to adult. A 3.0 kb signal is detected at all the stages investigated (Fig. 3B), and the intensity of the labelings indicates that the abundance of the Cx43 mRNA decreases progressively as development proceeds (compare lanes a and f in Fig. 3B).

Rat liver plasma membrane sheets were prepared according to Hubbard et al. (1983) (Fig. 4) and immunoblotted as described by Dupont et al. (1988). The replicas were probed with affinity-purified rabbit antipeptide antibodies directed to residues 55–66 (which are localized in the extracellular junctional space) or residues 314–322 of rat Cx43 (El Aoumari et al., 1991). Both antibodies label a single protein of M_r 43 kDa, corresponding to Cx43 (Fig. 4A). Plasma membrane fractions were exposed to alkaline media (pH 11) as described by Hubbard and Ma (1983) (see Fig. 4B). The pellets from these fractions, probed with antibodies directed to residues 55–66 or 314–322 of rat Cx43, were shown to contain a 41-kDa protein (Fig. 4B), representing a dephosphorylated form of Cx43 (Crow et al., 1990; Kadle et al., 1991; Laird et al., 1991). It is likely that dephosphorylation occurred during exposure of the plasma membranes to the alkaline media (compare lane c, Fig. 4A, and lane b, Fig. 4B). The 41-kDa protein was not detected in the supernatants of pellets from the plasma membrane exposed to alkaline media, but α-actin (M_r 43 kDa) was. In control experiments, performed either by omitting the incubation step with antipeptide antibodies or by replacing the antipeptide antibodies with an affinity-purified pre-immune serum fraction, no labeling was seen.

Discussion

Northern blots of poly(A)$^+$-rich RNA (Fig. 3A) demonstrate the presence of Cx43 mRNA in rat and mouse liver. Figure 3B also suggests that the expression of Cx43 mRNA is developmentally regulated during mouse liver ontogenesis. Willecke and Traub, in their 1990 review on the molecular biology of connexins, noted the presence of very small amounts of Cx43 mRNA in rat and mouse liver. Moreover, Oyamada et al. (1990) have shown that Cx43 mRNA is abundantly expressed in primary human hepatocellular carcinomas and hypothesized that the large amounts of Cx43 mRNA in hepatic carcinomas could be explained by the stem-cell concept of tumor-cell origin (Pierce et al., 1978). Thus, it is now quite certain that normal liver expresses low levels of Cx43 mRNA which may be overexpressed in some pathological situations.

Immunoblotting using two antipeptide antibodies directed to two distinct domains of rat Cx43 was used to characterize this junctional protein in plasma membranes isolated from rat liver. The results indicate that Cx43 is indeed present both in the untreated fractions of liver plasma membrane (Fig. 4A) and in the pellets from the fractions exposed to alkaline media (pH 11) (Fig. 4B). Exposure to alkaline media results in loss of cytoskeletal proteins associated with these membranes: actin and tonofilament components (Hubbard and Ma, 1983). The re-

Figure 1. Indirect immunofluorescent staining of adult mouse and rat liver frozen sections. Tissue samples were frozen in liquid nitrogen. Frozen sectioning was performed at −20°C using a Reichert–Jung cryostat. The sections were incubated with the saturation solution (PBS plus 1% bovine serum albumin) for 30 min at room temperature, then with the primary antibodies (diluted in the saturation solution) overnight at 4°C. After washing, the sections were incubated with the appropriate fluorescent secondary antibodies (diluted in the saturation solution) for 1 h at room temperature. The preparations were examined with a fluorescent Zeiss photomicroscope. (A). Rat liver. Distribution of antigenic sites recognized by anti-Cx43 antibodies. Primary antibodies: affinity-purified rabbit polyclonal directed to residues 314–322 of rat Cx43 (see El Aoumari et al., 1990, for characterization); 6 μg/ml. Secondary antibodies: fluorescein isothiocyanate (FITC)-labeled goat anti-rabbit antibodies (Sigma); dilution, 1: 100. The immunoreactivity is seen in the portal areas where the reactive sites are associated with the endothelial cells of the vessels (white arrows), and in the hepatic lobules where the labelings have a very characteristic pattern shown at high magnification in (B). Bar: 50 μm. (B). Mouse liver. Distribution of antigenic sites recognized by anti-Cx43 antibodies (see above) in the hepatic lobules. The staining is visible in cross-sections as pairs of dots on either side of the bile canaliculi, as single continuous lines, and as two continuous parallel lines, running along the hepatocyte surfaces. Bar: 50 μm. (C). Distribution of antigenic sites in rat liver recognized by affinity-purified hen polyclonal antibody directed to residues 266–282 of rat Cx32 (Paul, 1986); and visualized with tetraethyl rhodamine isothiocyanate (TRITC)-labeled goat anti-hen antibodies. An IgY fraction was extracted from egg yolks (Polson et al., 1980) of hens immunized with the peptide SPGTGAGLAEKSDRCSAY coupled to keyhole-limpet hemocyanin. The antipeptide antibodies were purified from the IgY fraction by affinity chromatography (Dupont et al., 1988). Cx32 is distributed between hepatocytes as described in previous investigations (Traub et al., 1989). Bar: 50 μm. (D) Distribution of antigenic sites in rat liver recognized by the mouse monoclonal antibody B10 (Maurice et al., 1985 and 1988); and visualized with TRITC-labeled goat anti-mouse antibodies. The bile canaliculi are visible in cross-section as single dots, and in longitudinal section as continuous thick lines. Bar: 50 μ m. (E) Mouse liver. Distribution of ZO-1. Primary antibody: rat monoclonal antibody R26.4C (Developmental Studies Hybridoma Bank; see Stevenson et al., 1986, for characterization); dilution, 1:2. Secondary antibodies: TRITC-labeled goat anti-rat antibodies (Jackson Immunoresearch Laboratories); dilution, 1:500. Note that the distribution of ZO-1 seems similar, if not identical, to that of the sites recognized by the anti-Cx43 antibodies. Bar: 50 μm.

sults shown in Fig. 4B thus eliminate the possibility of a non-specific absorption of the antibodies by the cytoskeletal proteins. The question that remains to be answered at this stage is which cell type(s) express Cx43?

Immunoreactive sites to antibodies directed to residues 314–322 of rat Cx43 were shown to be associated with the thin connective tissue which envelops the liver. In the portal blood vessels the labeling seems to be associated with the endothelial cells. This observation extends the recent results of Larson et al. (1990) who demonstrated that microvascular and macrovascular endothelial cells are dye coupled and express Cx43 mRNA. Immuofluorescent staining was also seen in the hepatic parenchyma. Analyses by confocal laser scanning microscopy of double immunofluorescence experiments have shown that the epitopes recognized by antibodies 314–322 co-localize

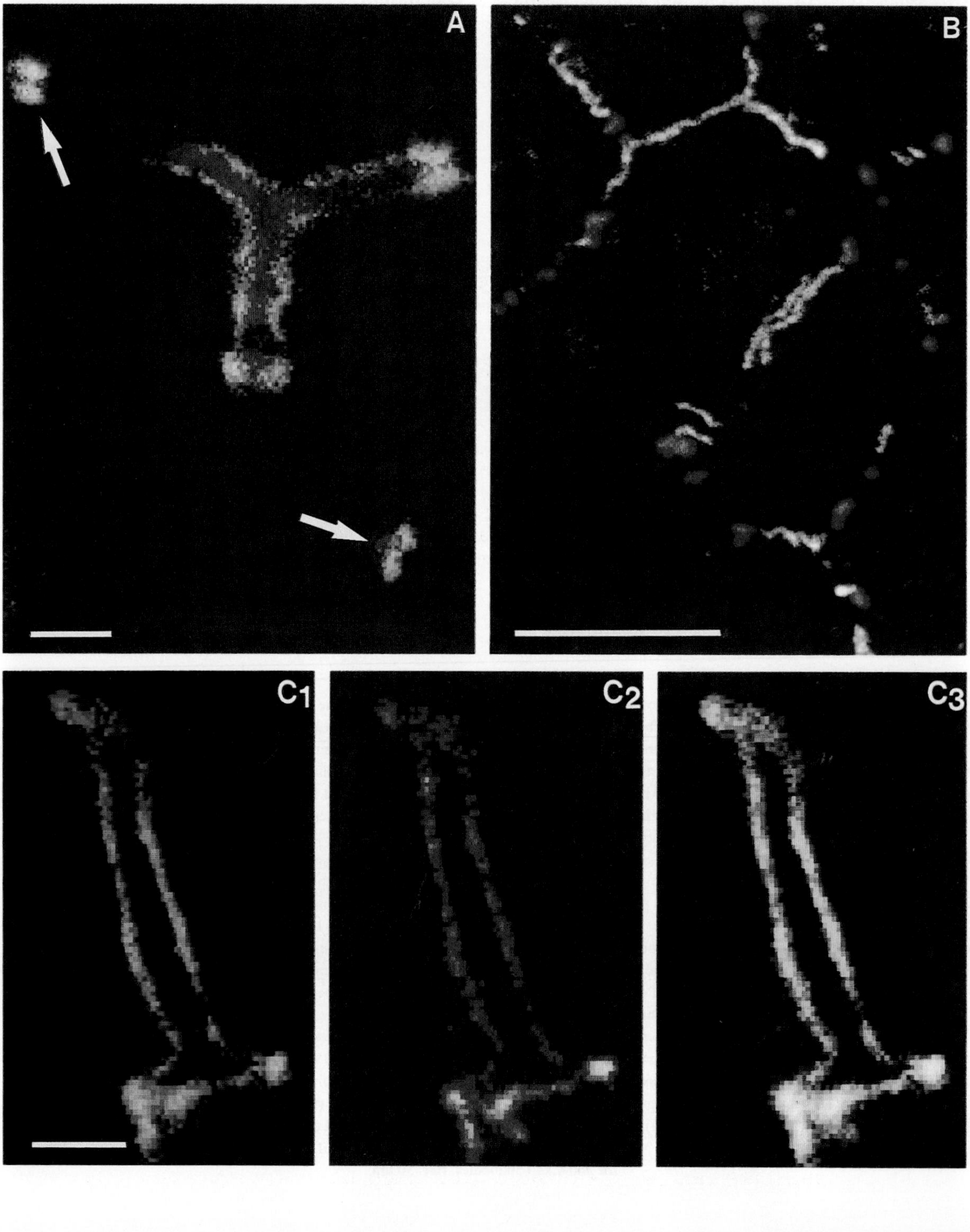

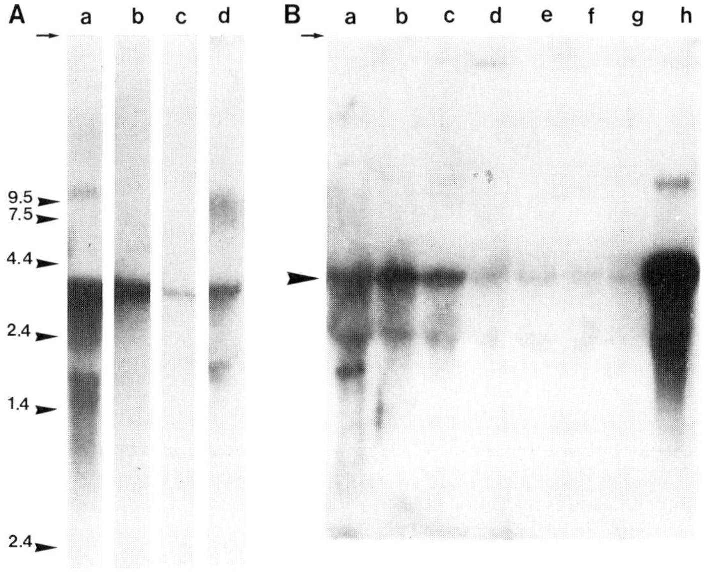

Figure 3. Analysis by Northern blot of liver and heart RNAs. Total cellular RNA from rat heart and from rat and mouse liver was extracted according to the method of Chirgwin et al. (1979). Poly(A)$^+$-rich RNA was purified on an oligo(dT) cellulose column (Aviv and Leder, 1972). The RNA samples were quantified by measuring their absorbance at 260 nm. The samples were fractioned by electrophoresis, capillary blotted onto nylon membranes and probed using the EcoRI/AaTI restriction fragment (1.03 kb; positions 293–1323) of the cDNA clone G1 (2.5 kb) coding for the rat heart Cx43 (Beyer et al., 1987). The probe was phosphate labeled with [α-^{32}P] d-CTP by random hexanucleotide-primed synthesis (Feinberg and Vogelstein, 1983). (A) The electrophoresis wells were loaded with 10 μg of total heart RNA (lane a), 5 μg of rat liver poly(A)$^+$-rich RNA (lane b), 5 μg of mouse liver poly(A)$^+$-rich RNA (lane c) and 50 μg of rat liver total RNA (lane d). A 3.0 kb message corresponding to Cx43 mRNA was detected in the heart RNA sample and in liver RNA samples. The arrowheads on the left indicate the position of RNA size markers; an arrow shows the top of the Northern blot. (B) Lanes a and b correspond to total RNA extracted from 16 and 19 day post-coitum embryonic mouse liver; lanes c, d and e, from 1, 2 and 3 week newborn mouse liver, and lanes f and g from adult mouse and rat liver. Lane h corresponds to total RNA extracted from adult rat heart. All the electrophoresis wells were loaded with 50 μg of RNA except in lane h (10 μg). The arrowhead on the left indicates the position of the 3.0 kb message detected in all the samples; the arrow shows the top of the Northern blot. Note that the intensity of the 3.0 kb hybridization signal decreases progressively as development proceeds.

with ZO-1 and are thus present in the hepatocyte junctional complex. This observation, reported here for the first time, raises a number of questions. Using a single antibody specific for Cx43, one cannot claim with certainty that the epitopes recognized in the hepatocytes really belong to the Cx43. Antibodies directed against other Cx43 domains should be used in order to determine cross-reactivity between Cx43 and other hepatocyte membrane proteins. Small gap junctions, scattered between hepatocyte tight junction strands, have been described by Montesano et al. (1975), Yee and Revel (1978) and Robeneck et al. (1979). If these junctions in fact contain Cx43, one would expect to find labeling in dot form rather than the linear labeling characteristic of ZO-1. Linear labeling could, however, be explained by the presence of aligned connexons linking the gap junctions. Experiments

Figure 2. Analysis of double immunofluorescence experiments by confocal laser scanning microscopy. Adult rat (A and B) and mouse (C) liver frozen sections were incubated with rabbit anti-Cx43 antibodies mixed with one or other of the following antibodies: mouse monoclonal B10 (A) or hen anti-Cx32 polyclonal (B) or rat anti-ZO-1 monoclonal (C). The primary antibodies were revealed with FITC-labeled goat anti-rabbit antibodies mixed with one or other of the following antibodies: TRITC-labeled goat anti-mouse antibodies or TRITC-labeled goat anti-rat antibodies. The conditions of use of the primary and secondary antibodies are given in the legend of Fig. 1. The double immunofluorescence experiments were analyzed by confocal laser scanning microscopy using a BioRad MRC-600 system mounted on an Optiphot II Nikon microscope equipped with a 60× objective. An argon ion laser adjusted to 488 nm and a helium–neon ion laser adjusted to 543 nm were used for excitation of fluorescein and rhodamine, respectively. The field depth was adjusted to about 0.6 μm and the double fluorescence images were acquired in two passes, rhodamine first, fluorescein second. The emitted signals were filtered and digitalized by "photon counting" and each section was scanned 40–80 times. Color pictures from screen images were taken on Kodak Ektachrome 160 film with an Olympus OM-1 35 mm camera. The immunoreactive sites recognized by the anti-Cx43 antibodies appear in green on the photographs. (A) Cross-sections (arrows) and longitudinal sections of bile canaliculi between hepatocytes. The canalicular antigenic sites (red) revealed with the mouse monoclonal B10 are closely associated, but not colocalized, with the epitopes recognized by the anti-Cx43 antibodies (green). Bar: 5 μm. (B) Distribution of Cx32 (red) and anti-Cx43 antibodies immunoreactive sites (green) between hepatocytes. No colocalization. Bar: 5 μm. (C) Longitudinal section of a bile canaliculus. C1 and C2 show the distribution of anti-Cx43 antibodies immunoreactive sites (green) and ZO-1 (red), respectively. C3 is a merged picture of the signals shown in C1 and C3 which demonstrates their colocalization. Bar: 5 μm.

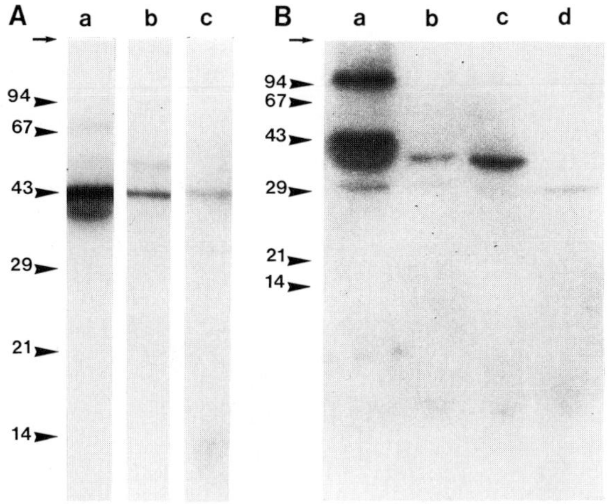

Figure 4. Immunoblotting of rat liver plasma membranes. Whole rat-heart fractions and rat liver plasma-membrane sheets were prepared as described by Dupont et al. (1988) and Hubbard et al. (1983), respectively. Replicas were probed with affinity-purified rabbit antipeptide antibodies directed either to residues 55–66 (El Aoumari et al., 1991) or to residues 314–322 (El Aoumari et al., 1990). Antipeptide antibodies were visualized using protein A labeled with ^{125}I by means of Iodo-Gen. (A) Immunoreplicas of whole rat-heart fractions (lane a) and rat liver plasma membrane (lanes b and c) probed with antibodies directed to residues 314–322 (lanes a and b) and residues 55–66 (lane c) of rat Cx43. The major band, intensively labeled (lane a) corresponds to Cx43 (M_r 43 kDa). The band migrating at 43 kDa in lanes b and c indicates that Cx43 is present in the rat plasma membrane. Size markers (in kDa) are indicated on the left of lane a. The arrowhead shows the top of immunoreplicas. (B) Figure 4B shows immunoreplicas of whole rat-heart fraction (lane a); rat liver plasma membrane (lane b); pellet of plasma membrane after alkaline treatment (lane c); supernatant of plasma membrane after alkaline treatment (lane d). Rat liver plasma-membrane sheets were exposed to an alkaline treatment (pH 11) as described by Hubbard and Ma (1983). The replicas were probed with antipeptide antibodies directed to residues 55–66 or 314–322 of rat Cx43. Identical results were obtained in both cases. In the heart, Cx43 is detected as a large band (M_r 40–43 kDa), including its phosphorylated and dephosphorylated forms. The 100 kDa band also seen in heart membrane probably corresponds to aggregation products. A 41-kDa band, probably representing a dephosphorylated form of Cx43, is present in the untreated plasma-membrane fraction (lane b) and in the pellet of a twin fraction exposed to pH 11 (lane c). The supernatant does not contain this band (lane d).

are in progress to ascertain the presence of Cx43 in hepatocytes. If Cx43 is in fact expressed in hepatocytes, it remains to be determined which structures located in the junction complex actually contain this connexin.

Acknowledgements

We are indebted to Dr. E.C. Beyer (St. Louis University) for providing us with connexin 43 cDNA clones. We thank Dr. J. Davoust and Mr. F. Nicolas for their help and constructive criticisms during the initial investigations using confocal laser scanning microscopy. We wish to thank Mr. G. Géraud and Mr. M. Berthoumieux for excellent technical assistance. This work was supported by INSERM (grant 88-5009), DRET (grant 90-055) and the "Fondation pour la Recherche Médicale".

References

Anderson, J.M., Stevenson, B.R., Jesaitis, L.A., Goodenough, D.A. and Mooseker, M.S. (1988) Characterization of ZO-1, a protein component of the tight junction from mouse liver and Madin-Darby canine kidney cells. J. Cell. Biol. 106, 1141–1149.

Aviv, H. and Leder P. (1972) Purification of biologically active globin messenger RNA by chromatography on oligothymidic acid-cellulose. Proc. Natl. Acad. Sci. USA 69, 1408–1412.

Beyer, E.C., Paul, D.L. and Goodenough, D.A. (1987) Cx43: A protein from rat heart homologous to a gap-junction protein from liver. J. Cell Biol. 105, 2621–2629.

Chirgwin, J.M., Przybyla, A.E., Mac Donald, R.J. and Rutter, W.J. (1979) Isolation of biologically active ribonucleic acid from sources enriched by ribonuclease. Biochemistry 18, 5294–5299.

Citi, S., Sabanay H., Jakes, R., Geiger, B. and Kendrich-Jones, J. (1988) Cingulin, a new peripheral component of tight junctions. Nature 333, 272–275.

Citi, S., Sabanay, H., Kendrich-Jones, J. and Geiger, B. (1989) Cingulin: characterization and localization. J. Cell. Sci. 93, 107–122.

Crow, D.S., Beyer, E.C., Paul, D.L., Kobe, S.S. and Lau, A.F. (1990) Phosphorylation of connexin 43 gap-junction protein in uninfected and Rous sarcoma virus-transformed mammalian fibroblasts. Mol. Cell. Biol. 10, 1754–1768.

Dermietzel, R., Yancey, B., Janssen-Timmen, U., Traub, O., Willicke, K. and Revel, J.P. (1987) Simultaneous light and electron microscopic observation of immunolabeled liver 27 kDa gap-junction protein on ultra-thin cryosections. J. Histochem. Cytochem. 35, 387–392.

Dupont, E., El Aoumari, A., Roustiau-Sévère, S., Briand, J.P. and Gros, D. (1988) Immunological characterization of rat cardiac gap junctions: presence of common antigenic determinants in heart of other vertebrate species and in various organs. J. Membr. Biol. 104, 119–128.

El Aoumari, A., Fromaget, C., Dupont, E., Reggio, H., Durbec, P., Briand, J.P., Rook, M.B., Boller, K. and Gros, D. (1990) Conservation of a cytoplasmic carboxy-terminal domain of connexin 43, a gap-junctional protein, in mammal heart and brain. J. Membr. Biol. 115, 229–240.

El Aoumari, A., Dupont, E., Fromaget, C., Jarry, T., Briand, J.P., Kreitmann, B. and Gros, D. (1991) Immunolocalization of an extracellular domain of connexin 43 in rat heart gap junctions.

Eur. J. Cell Biol. 56, 391–400.

Feinberg, A.P. and Vogelstein, B. (1983) A technique for radiolabeling DNA restriction endonuclease fragments of high specific activity. Anal. Biochem. 132, 6–13.

Fromaget, C., El Aoumari, A., Dupont, E., Briand, J.P. and Gros, D. (1990) Changes in the expression of Cx43, a cardiac gap-junctional protein, during mouse heart development. J. Mol. Cell. Cardiol. 22, 1245–1258.

Fromaget, C., El Aoumari, A. and Gros, D. (1992) Distribution pattern of connexin 43, a gap junctional protein, during the differentiation of mouse heart myocytes. Differentiation 51, 9–20.

Hubbard, A.L., Wall, D.A. and Ma, A. (1983) Isolation of rat hepatocyte plasma membranes. I. Presence of the three major domains. J. Cell. Biol. 96, 217–229.

Hubbard, A.L. and Ma, A. (1983) Isolation of rat hepatocyte plasma membrane. II. Identification of membrane-associated cytoskeletal proteins. J. Cell. Biol. 96, 230–239.

Kadle, R., Zhang, T.J. and Nicholson, B.J. (1991) Tissue specific distribution of differentially phosphorylated forms of Cx43. Mol. Cell. Biol. 11, 363–369.

Laird, D.W., Puranam, K.L. and Revel, S.P. (1991) Turnover and phosphorylation dynamics of connexin 43 gap-junction protein in cultured cardiac myocytes. Biochem. J. 273, 67–72.

Larson, D.M., Haudenschild, C.C. and Beyer, J.C. (1990) Gap-junction messenger RNA expression by vascular wall cells. Circulation Res. 66, 1074–1080.

Maurice, M., Durand-Schneider, A.M., Garbarz, M. and Feldmann, G. (1985) Characterization of rat hepatocyte plasma membrane domains by monoclonal antibodies. Eur. J. Cell. Biol. 39, 122–129.

Maurice, M., Rogier, E., Cassio, D. and Feldmann, G. (1988) Formation of plasma membrane domains in rat hepatocytes and hepatoma cell lines in culture. J. Cell Sci. 90, 79–92.

Montesano, R., Friend, D.S., Perrelet, A. and Orci, L. (1975) *In vivo* assembly of tight junctions in fetal rat liver. J. Cell. Biol. 67, 310–319.

Oyamada, M., Krutovskikh, V.A., Mesnil, M., Partensky, C., Berger, F. and Yamasaki, H. (1990) Aberrant expression of gap-junction gene in primary human hepatocellular carcinomas: increased expression of cardiac-type gap-junction gene connexin 43. Mol. Carcinogen. 3, 273–278.

Paul, D.L. (1986) Molecular cloning of cDNA for rat liver gap-junction protein. J. Cell. Biol. 103, 123–134.

Pierce, G.B., Shikes, R. and Finks, L.M. (1978) Cancer: a problem of developmental biology. Prentice Hall, Englewood Cliffs, New Jersey.

Polson, A., von Wechmar, B.M. and van Regenmortel, M.H.V. (1980) Isolation of viral IgY antibodies from yolks of immunized hens. Immunol. Comm. 9, 475–493.

Robenek, H., Themann, H. and Ott, K. (1979) Carbon tetrachloride induced proliferation of tight junctions in the rat liver as revealed by freeze fracturing. Eur. J. Cell. Biol. 20, 62–70.

Stevenson, B.R., Siliciano, J.D., Mooseker, M.S. and Goodenough, D.A. (1986) Identification of ZO-1: a high molecular weight polypeptide associated with the tight junction (zonula occludens) in a variety of epithelia. J. Cell. Biol. 103, 755–766.

Stevenson, B.R., Heintzelman, M.B., Anderson, J.M., Citi, S. and Mooseker, M.S. (1989) ZO-1 and cingulin: tight junctions proteins with distinct identities and locations. Am. J. Physiol. 257, C621–C628.

Traub, O., Look, J., Dermietzel, R., Brummer, F., Hulser, D. and Willecke, K. (1989) Comparative characterization of the 21 kDa and 26 kDa gap-junction proteins in murine liver and cultured hepatocytes. J. Cell. Biol. 108, 1039–1051.

Willecke, K. and Traub, O. (1990) Molecular biology of mammalian gap junctions. In: W.C. de Mello (Ed.) Cell Intercommunication. CRC Press, Boca Raton, FL, pp. 21–36.

Yee, A.G. and Revel, J.P. (1978) Loss and reappearance of gap junctions in regenerating liver. J. Cell. Biol. 78, 554–564.

J.E. Hall, G.A. Zampighi and R.M. Davis (Eds.)
Progress in Cell Research, Vol. 3

CHAPTER 5

The mouse connexin gene family

KLAUS WILLECKE, HANJO HENNEMANN, EDGAR DAHL and STEFAN JUNGBLUTH

Institut für Genetik, Abteilung Molekulargenetik, Römerstrasse 164, 53 Bonn 1, Germany

Introduction

Morphological and biochemical analysis of gap junctions led originally to the idea that gap junctions and their protein components, now called connexins, are similar in different tissues. After the first connexin cDNA coding for rat and human Cx32 had been cloned (Heynkes et al., 1986; Kumar and Gilula, 1986 ; Paul, 1986), these probes were used in several laboratories under low stringency of hybridization to search for new members of the connexin gene family. Rat Cx43 (Beyer et al., 1987), Cx46 (Beyer et al., 1988) and Cx26 cDNAs (Zhang and Nicholson, 1989) and, most recently, the rat genomic clone of Cx31 (Hoh et al., 1991) have been described. Similarly in *Xenopus*, Cx30 (Gimlich et al., 1988), Cx43 (Gimlich et al., 1990), and Cx38 cDNAs (Ebihara et al., 1989; Gimlich et al., 1990) have been isolated, and chick Cx43 (Musil et al., 1990), Cx45 and Cx42 cDNAs (Beyer, 1990) have been characterized. All these connexin sequences fit the general topological scheme that has been experimentally worked out with peptide-specific antibodies and limited proteolytic degradation of membrane-embedded rat Cx32 (Hertzberg et al., 1988; Milks et al., 1988) and Cx43 (Beyer et al., 1989; Yancey et al., 1989). The basic features of this connexin topology are four transmembrane regions; the third one shows the characteristics of an amphiphilic helix, presumably lining the channel pore. Two extracellular loops are highly conserved in their amino-acid sequence, each containing three cysteine residues that may participate in intramolecular S–S bridges and be required for "docking" of two connexin hemichannels (Dahl et al., 1991). The main differences between connexins are located in the cytoplasmic C-terminal region which can vary from 18 amino acids in Cx26 to 189 amino acids in Cx46.

We chose to investigate the molecular genetics of mouse connexins in some detail since mouse development is the best understood among mammals and there are several hypotheses concerning the role of gap junctions in development and morphogenesis which need verification (cf. Guthrie and Gilula, 1989). Here we present our results concerning the characterization of different mouse connexins. Due to the high degree of conservation of connexin genes, the results obtained for mouse connexins are likely to be very similar to the analogous genes in the rat or human genome.

Summarized characterization of mouse connexins

Table I lists the different mouse genomic and cDNA clones characterized so far. A similar summary including references for rat, human, and chick connexins has recently been published (Willecke et al., 1991a). The nomenclature for connexins (Beyer et al., 1990) uses the abbreviated molecular mass (in kDa; deduced from coding regions) as suffix to the name connexin. Previously chosen names of published connexins needed to be avoided, so that two new connexins mentioned below (Cx31.1 and Cx30.3) were designated, including the first digit of the corresponding molecular mass after the decimal point. An alternative nomenclature for connexins has been suggested (Risek et al., 1990) using Greek letters. This is useful for indicating two connexin subclasses, the α-class, exemplified by Cx43 or gap-junction protein α_1, and the β class, of which Cx32 or gap-junction protein β_1 is the prototype. There are some distinct structural differences between the α and β connexin classes. For example, an insertion of at least 21 amino acids is seen in the central cytoplasmic loop of α-type connexins. Furthermore, there are two consecutive tryptophan residues in the putative third transmembrane region of all β-type connexins. The possible functional consequences of these and other differences (Willecke et al., 1991a) between the two connexin classes remain obscure so far.

Connexin32 and -26

Genomic clones of mouse Cx32 and -26 have been characterized in this laboratory. Mouse and rat Cx32 have essentially the same amino-acid sequence although there are

TABLE I Mouse connexin genes

Abbreviation of connexin	Genomic clone	cDNA clone	Total number of amino acids	M_r (kDa)	Size of of mRNA (kb)	References
Cx26	Yes	No	226	26.414	2.5	1, 8
Cx31	No	Yes	270	30.905	2.3; 1.9	2
Cx30.3	Yes	No	266	30.391	3.2; 1.8	3
Cx31.1	Yes	Yes	271	31.198	1.6	3
Cx32	Yes	Yes	283	32.007	1.6	1, 8
Cx37	Yes	Yes	333	37.600	1.7	4
Cx40	Yes	No	358	40.418	3.5	5
Cx43	Yes	Yes	382	43.009	3.0	5, 6, 8
Cx45	No	Yes	396	45.671	2.3	2
Cx46	Yes	No	Not yet sequenced		3.0	7

(1) Hennemann et al., 1992a; (2) Hennemann et al., 1992b; (3) Hennemann 1992c; (4) Willecke et al., 1991b; (5) Hennemann et al., 1992d; (6) Beyer and Steinberg, 1991; (7) Dahl and Willecke, unpublished results; (8) Nishi et al., 1991.

silent base exchanges in the nucleotide coding region. Mouse Cx26 shows two amino-acid exchanges [at position 68: His (mouse)/Tyr (rat) and position 215: Val (mouse)/-Ile (rat)] compared to its rat analogue (Hennemann et al., 1992a). Both Cx32 and Cx26 genes share similar genomic structures, i.e. the coding regions are uninterrupted by introns but there are introns of 6.1 kb and 3.8 kb, respectively, upstream of the coding sequence. Cx26 and Cx32 transcripts have one strong and at least one weak start site, as shown by primer extension and S1 nuclease protection experiments (Hennemann et al., 1992a). The promoter region and the genomic structure of the rat Cx32 gene (Miller et al., 1988) are very similar to those of its mouse analogue. We have sequenced 600 bp of the mouse Cx26- and 680 bp of the Cx32-promoter regions, upstream of the corresponding transcription start points at the first exon. These promoter/enhancer sequences, including putative binding sites for several transcription factors, are distinct for each gene, although possible binding sites for the transcription factor NF κ B have been found in the control regions of both connexin genes (Hennemann et al., 1992a). Previously it has been shown that Cx32 and Cx26 are co-expressed in several rat and mouse tissues (Zhang and Nicholson, 1989). In leptomeningeal cells of the brain, however, Cx26 is co-expressed with Cx43 without expression of Cx32 (Dermietzel et al., 1989). Recently we have studied Cx32 and Cx26 expression in immortalized mouse embryonic hepatocytes. By shifting from serum-free to serum-containing culture conditions, the hepatocyte-specific phenotype of these cells could be reversibly altered. Expression of Cx32 and Cx26 correlated well with the hepatocyte-specific phenotype. On the other hand, expression of Cx43 mRNA was very low in differentiated hepatocytes but increased in dedifferentiated derivatives (Stutenkemper et al., 1992). Furthermore, Cx32 and Cx26 expression appear to be under different hormonal control in murine hepatocytes and dedifferentiated liver-derived cell lines.

Connexin43 and -46

Mouse Cx43 has been isolated as a genomic clone and a partial cDNA sequence from mouse brain (Hennemann et al., 1992d). Recently the coding sequence of mouse Cx43 has been amplified from genomic DNA by the polymerase chain reaction (Beyer and Steinberg, 1991). One amino-acid exchange (99.7% amino-acid identity, 95% nucleotide identity) at position 341 [Ser (mouse)/Asp (rat)] has been identified by comparison with rat Cx43 cDNA. We have sequenced 150 bp upstream and 900 bp downstream of the mouse Cx43-coding region. We found a possible splice-acceptor site in consensus context at position −10 or −13 and a possible lariat consensus motif at position -41 upstream of the translation start codon, suggesting the end of an intron. Another intron may exist in the 3' non-coding region of mouse Cx43. The sequence information concerning Cx43 is deposited in the EMBL gene data bank under accession number (X62836). The mouse genes for Cx32, Cx26, and Cx43 have been assigned to mouse chromosomes X, 14, and 10, respectively (Hsieh et al., 1991). These assignments correspond to syntenic regions on human chromosomes X, 13, and 6, respectively (Willecke et al., 1990a; Fishman et al., 1991; Hsieh et al., 1991).

Genomic mouse Cx46 DNA has been isolated as an insert of recombinant phages (Dahl, E. and Willecke, K., unpublished), but it has not been characterized in detail. The Cx46 gene has been located in the mouse genome on chromosome 14 or 5 (Hsieh et al., 1991) and in the human genome on chromosome 13 (Willecke et al., 1990a; Hsieh et al., 1991).

Connexin37 and -40

Mouse Cx37 cDNA and its genomic clone (Willecke et al., 1991b) were isolated by low-stringency hybridization using rat Cx32 and Cx26 cDNA, respectively. Although the Cx37 amino-acid sequence shows greater identity with *Xenopus* Cx38 than with mouse Cx32 or Cx43, its expression pattern with regard to cell type and stage of development appears to be different from *Xenopus* Cx38 (Willecke et al., 1991b). Cx37 is a member of the α-class of connexins. Its transcript expression pattern in different tissues shows that the Cx37 mRNA is at least 10-fold more abundant in lung than in any other tissue tested (cf. Table II). It shares these characteristics with another new mouse connexin, Cx40, which we isolated as a genomic clone. Mouse Cx40 appears to be analogous to chick Cx42, based on an overall amino-acid identity of 67% and nucleotide identity of 86%. Similar to Cx37, Cx40 is also a member of the α-subgroup of the connexin gene family. In embryonic kidney, skin, brain, and liver the level of Cx40 mRNA is 2–6 fold higher than in the corresponding adult tissues (Table II). Since the tissue distribution of Cx40 mRNA is similar to that of Cx37, it is possible that both connexins are expressed in the same cells. Cx37 and Cx40 cRNAs have been expressed functionally in *Xenopus* oocytes forming cell-to-cell channels with a pronounced voltage dependency. Hemichannels of Cx37 in *Xenopus* oocytes can form functional cell-to-cell channels with hemichannels built up of Cx40 subunits (Hennemann et al., 1992d). It is not known, however, whether heteromeric channels composed of Cx37 and Cx40 subunits in the same hemichannel can be formed and function as gap-junction channels.

Connexin31 and -45

To characterize connexin genes expressed in early mouse development, we screened a cDNA library from mouse embryonic carcinoma F9 cells under low stringency of hybridization. We isolated cDNAs coding for mouse Cx31, Cx45, Cx43, Cx32 and Cx31.1 (see below). Mouse Cx31 cDNA is the analogue of the rat Cx31 genomic clone (Hoh et al., 1991), from which it differs by eight amino-acid residues. Mouse Cx45 cDNA is possibly analogous to chick Cx42 cDNA (84% amino-acid and 75% nucleotide identity within the coding sequence). Cx31 falls into the β -class, Cx45 is a member of the α-subgroup of connexins. Although Cx31 and Cx45 cDNAs have both been isolated from F9 cells, the expression pattern of these mRNAs in different tissues is not the same. Whereas Cx31 transcripts were found only in adult and embryonic skin (Table II) or testis (Hennemann et al., 1992b), Cx45 mRNA was detected at high abundancy in embryonic kidney, skin and brain, as well as in adult lung (cf. Table II). Hoh et al. (1991) have also found rat Cx31 transcripts in skin, placenta, and Harderian gland. The expression of Cx31 mRNA in skin is likely due to its expression in keratinocytes, since we have detected strong expression of Cx31 mRNA in the mouse keratinocyte cell lines Hel37 and Hel30 but not in mouse embryonic fibroblasts (Hennemann et al., 1992b). Mouse Cx45 mRNA is much more abundant in embryonic than in adult tissues (cf. Table II: embryonic lung has not yet been tested) and thus it fulfills an as yet unknown function in embryonic development. There are basic levels of at least seven different connexin transcripts expressed in mouse F9 embryonic carcinoma cells, as we also found Cx37 mRNA (Willecke et al., 1991b) and Cx26 mRNA (Willecke et al., 1990b) expressed in these cells after induction with cyclic AMP and retinoic acid. Fully differentiated cells appear to express fewer connexins. So far we have not measured the relative abundance of the different connexin mRNAs in mouse F9 cells.

Connexin31.1 and -30.3

Using low-stringency hybridization with the rat Cx26 cDNA probe, we isolated one mouse genomic clone coding for two new connexin sequences, Cx31.1 and Cx30.3

Table II Expression of mouse connexin transcripts in different tissues

Connexin	Embryonic tissues				Adult tissues							
	Kidney	Liver	Skin	Brain	Kidney	Liver	Skin	Brain	Heart	Intestine	Spleen	Lung
Cx31.1	0	0	42	0	0	0	100	0	0	0	0	0
Cx30.3	0	0	50	0	0	0	100	0	0	0	0	0
Cx31	0	0	32	0	0	0	100	0	0	0	0	0
Cx37	3	1	3	4	<1	2	1	<1	1	1	3	100
Cx40	10	1	6	1	6	0	1	<1	3	<1	<1	100
Cx45	77	0	68	80	1	0	0	5	11	5	0	100

Data are presented as % of connexin mRNA level relative to the tissue of maximal expression (set to 100%). All data were obtained by densitometric evaluation of the results of Northern blot hybridizations using total RNA and standardization to the level of cytochrome C oxidase mRNA in the same sample. Note that only the estimated mRNA levels of one type of connexin should be compared. The data do not allow comparison of the relative transcript levels of different connexins. Zero means that the corresponding transcript was not detected by Northern blot hybridization.

which are only 3.3 kb apart from each other. Cx31.1 shows 70% amino-acid identity to Cx30.3 but only 52%, 53%, and 54% identity to Cx26, Cx32, and Cx43, respectively. Both Cx31.1 and Cx30.3 are members of the β-subgroup of connexins. We have isolated a cDNA clone from a library of mouse embryonic carcinoma F9 cells that shows complete nucleotide identity to the presumptive coding region of the mouse Cx31.1 genomic clone. The putative Cx31.1 transcripts are about 2.5-fold more abundant in adult than in embryonic skin. The Cx30.3 genomic sequence is also likely to be expressed, since it hybridizes in total RNA from embryonic and adult skin under full stringency to two transcripts of 3.2 and 1.8 kb (Table I). In order to avoid cross-hybridization with other connexins, we used DNA probes of the Cx31.1 and Cx30.3 sequences for hybridization that code for the C-terminal regions which differ between connexins. Cx30.3 mRNA has been functionally expressed after microinjection in *Xenopus* oocytes (Hennemann et al., 1992c).

Discussion

It is now clear that gap junctions in different animal tissues must be composed of different connexin subunits. While the putative extracellular domains of connexin molecules show high homology, they vary in the cytoplasmic central loop and particularly in the cytoplasmic C-terminal region. This means, in functional terms, that most connexin molecules in gap-junction hemichannels can probably interact and dock with corresponding hemichannels in contacting cells. The docking process is likely to depend on expression of cell-adhesion molecules (Musil et al., 1990b; Jongen et al., 1991). According to the topological model of connexins, the third transmembrane regions of these molecules line the aqueous pore of the channel. All amino-acid sequences of the third putative transmembrane domain of different connexins show two charged residues of opposite polarity (Willecke et al., 1991a) which fits the gating model proposed by Unwin (1989) (cf. Bennett et al., 1991). Cx43, Cx45, and Cx46 exhibit in their putative third transmembrane region at least one more charged amino-acid residue (Willecke et al., 1991a). The details of the gating characteristics will be better understood when site-directed mutagenesis of different connexin sequences is performed, followed by functional expression in *Xenopus* oocytes and/or mammalian cells.

At present one can speculate that the different C-terminal regions of connexins are recognized by different modifying enzymes (kinases, phosphatases, Ca-binding proteins, etc.) that lead to alterations affecting the stability and/or function of the different connexins. It may be advantageous for different cell types to regulate gap-junctional communication according to their own metabolic needs by modification of existing channels. In addition, it seems possible that gap-junction channels built up of different connexins may have different permeabilities for ions or metabolites. The unit conductances measured for gap-junction channels in rat hepatocytes and cardiac myocytes are different (Spray et al., 1986; Burt and Spray, 1988). These conductances can now be studied in coupling-defective mammalian cells after restoration of coupling by transfection with DNA coding for different connexins (Eghbali et al., 1990; Fishman et al., 1990). It may turn out that the different connexin genes have evolved in response to the biological need for homeostasis.

Acknowledgments

We would like to acknowledge the fruitful collaborations and stimulating discussions with Dr. Bruce Nicholson and Tom Suchyna (Dept. of Biological Sciences, State Univrsity of New York at Buffalo, USA). The collaboration between the laboratories in Bonn and Buffalo has been supported by NATO travelling stipends. Work in the Bonn laboratory has been financed by grants from the Deutsche Forschungsgemeinschaft (SFB 284, project C1) and the Fonds der Chemischen Industrie.

References

Bennett, M.V.L., Barrio, L.C., Gargiello, T.A., Spray, D.C., Hertzberg, E. and Sáez, J.C. (1991) Gap Junctions: new tools, new answers, new questions. Neuron 6 305-320.

Beyer, E.C., Paul, D.L. and Goodenough, D.A. (1987) A protein from rat heart homologous to a gap junction protein from liver. J. Cell Biol. 105, 2621–2629.

Beyer, E.C., Goodenough, D.A. and Paul, D.L. (1988) The connexins, a family of related gap-junction proteins. In: E.L. Hertzberg and R.G. Johnson (Eds.), Gap Junctions, Alan Liss, New York, pp. 165–175.

Beyer, E.C., Kistler, J., Paul, D.L. and Goodenough, D.A. (1989) Antisera directed against connexin43 peptides react with a 43-kDa protein localized to gap junctions in myometrium and other tissues. J. Cell Biol. 108, 595–605.

Beyer, E.C. (1990) Molecular cloning and developmental expression of two chick embryo gap-junction proteins. J. Biol. Chem. 265, 14439–14443.

Beyer, E.C., Paul, D.L. and Goodenough, D.A. (1990) Connexin family of gap-junction proteins. J. Membr. Biol. 116, 187–194.

Beyer, E.C. and Steinberg, T.H. (1991) Evidence that the gap-junction protein connexin43 is the ATP-induced pore of mouse macrophages. J. Biol. Chem. 266, 7971–7974.

Burt, J.M. and Spray, D.C. (1988) Single-channel events and gating behaviour of the cardiac gap-junction channel. Proc. Natl. Sci. USA 85, 3431–3434.

Dahl, G., Levine, E., Rabadan-Diehl, C. and Werner, R. (1991) Cell/cell channel formation involves disulfide exchange. Eur. J. Biochem. 197, 141–144.

Dermietzel, R., Traub, O., Hwang, T.K., Beyer, E.C., Bennett, M.V.L., Spray, D.C. and Willecke, K. (1989) Differential expression of three gap-junction proteins in developing and mature brain tissues. Proc. Natl. Acad. Sci. USA 86, 10148–10152.

Ebihara, L., Beyer, E.C., Swenson, K.I., Paul, D.L. and

Goodenough, D.A. (1989) Cloning and expression of a *Xenopus* embryonic gap-junction protein. Science 243, 1194–1195.

Eghbali, B., Kessler, J.A. and Spray, D.C. (1990) Expression of gap-junction channels in communication-incompetent cells after stable transfection with cDNA encoding connexin32. Proc. Natl. Acad. Sci. USA 87, 1328–1331.

Fishman, G.I., Spray, D.C. and Leinwand, L.A. (1990) Molecular characterization and functional expression of the human cardiac gap-junction channel. J. Cell Biol. 111, 589–597.

Fishman, G.I., Eddy, R.L., Shows, T.B., Rosenthal, L. and Leinwand, L.A. (1991) The human connexin gene family of gap-junction proteins: distinct chromosomal locations but similar structures. Genomics 10, 250–256.

Gimlich, R.L., Kumar, N.M. and Gilula, N.B. (1988) Sequence and developmental expression of mRNA coding for a gap-junction protein in *Xenopus*. J. Cell Biol. 107, 1065–1073.

Gimlich, R.L., Kumar, N.M. and Gilula, N.B. (1990) Differential regulation of the level of three gap-junction mRNAs in *Xenopus* embryos. J. Cell Biol. 110, 597–605.

Guthrie, S.C. and Gilula, N.B. (1989) Gap-junctional communication and development. Trends Neural Sci. 12, 12–16.

Hennemann, H., Kozjek, G., Dahl, E., Nicholson, B.J. and Willecke, K. (1992a) Molecular cloning of mouse connexins26 and -32: similar genomic organization but distinct promotor sequences of two gap junction genes. Eur. J. Cell Biol. 58, 81–89.

Hennemann, H., Schwarz, H.J. and Willecke, K. (1992b) Characterization of gap junction genes expressed in F9 embryonic carcinoma cells: molecular cloning of mouse connexin31 and -45 cDNAs. Eur. J. Cell Biol. 57, 51–58.

Hennemann, H., Dahl, E., White, J.B., Schwarz, H.-J., Chang, S., Lalley, P.A., Nicholson, B.J. and Willecke, K. (1992c) Two gap junction genes, connexin31.1 and -30.3 are closely linked on mouse chromosome 4 and preferentially expressed in skin. J. Biol. Chem. 267, 17225–17233.

Hennemann, H., Suchyna, T., Jungbluth, S., Dahl, E., Schwarz, H.J., Nicholson, B.J. and Willecke, K. (1992d) Molecular cloning and functional expression of mouse connexin40, a second gap junction gene preferentially expressed in lung. J. Cell Biol. 117, 1299–1310.

Hertzberg, E.L., Disher, R.M., Tiller, A.A., Zhou, Y. and Cook, R.G. (1988) Topology of the M_r 27,000 liver gap-junction protein. J. Biol. Chem. 263, 19105–19111.

Heynkes, R., Kozjek, G., Traub, O. and Willecke, K. (1986) Identification of a rat liver cDNA and mRNA coding for the 29 kDa gap-junction protein. FEBS Lett. 205, 56–60.

Hoh, J.H., John, S. and Revel, J.-P. (1991) Molecular cloning and characterization of a new member of the gap-junction gene family, connexin-31. J. Biol. Chem. 226, 6524–6531.

Hsieh, C.-L., Kumar, N.M., Gilula, N.B. and Francke, U. (1991) Distribution of genes for gap-junction membrane channel proteins on human and mouse chromosomes. Somat. Cell. Mol. Genet. 17, 191–200.

Jongen, W.M., Fitzgerald, D.J., Asamoto, M., Piccoli, C., Slaga, T.J., Gros, D., Takeichi, M. and Yamasaki, H. (1991) Regulation of connexin43-mediated gap-junctional intercellular communication by Ca^{2+} in mouse epidermal cells is controlled by E-cadherin. J. Cell Biol. 114, 545–555.

Kumar, N. and Gilula, N.B. (1986) Cloning and characterization of human and rat liver cDNAs coding for a gap-junction protein. J. Cell Biol. 103, 767–776.

Milks, L.C., Kumar, N.M., Houghton, R., Unwin, N. and Gilula, N.B. (1988) Topology of the 32-kDa liver gap-junction protein determined by site-directed antibody localizations. EMBO J. 7, 2967–2975.

Miller, T., Dahl, G. and Werner, R. (1988) Structure of a gap-junction gene: connexin32. Biosci. Rep. 8, 455–464.

Musil, L.S., Beyer, E.C. and Goodenough, D.A. (1990a) Expression of the gap-junction protein connexin43 in embryonic chick lens: Molecular cloning, ultrastructural localization and posttranslational phosphorylation. J. Membr. Biol. 116, 163–175.

Musil, L.S., Cunningham, B.A., Edelman, G.M. and Goodenough, D.A. (1990b) Differential phosphorylation of the gap-junction protein connexin43 in junctional communication-competent and -deficient cell lines. J. Cell Biol. 11, 2077–2088.

Nishi, M., Kumar, N.M. and Gilula, N.B. (1991) Developmental regulation of gap junction gene expression during mouse embryonic development, Dev. Biol. 146, 117–130.

Paul, D. (1986) Molecular cloning of cDNA for rat liver gap-junction proteins. J. Cell Biol. 103, 123–134.

Risek, B., Guthrie, S., Kumar, N.M. and Gilula, N.B. (1990) Modulation of gap-junction transcript and protein expression during pregnancy in rat. J. Cell Biol. 110, 269–282.

Spray, D.C., Sáez, J.C., Brosius, D., Bennett, M.V.L. and Hertzberg, E.L. (1986) Isolated liver gap junctions: Gating of transjunctional currents is similar to that in intact pairs of rat hepatocytes. Proc. Natl. Acad. Sci. USA 83, 5494–5497.

Stutenkemper, R., Geisse, S., Look, J., Traub, O., Nicholson, B.J. and Willecke, K. (1992) The hepatocyte specific phenotype of murine liver cells correlates with high expression of connexin32 and -26, but very low expression of connexin43. Exp. Cell Res. 201, 43–54.

Unwin, N. (1989) The structure of ion channels in excitable cells. Neuron 3, 665–676.

Willecke, K., Hennemann, H., Herbers, K., Heynkes, R., Kozjek, G., Look, J., Stutenkemper, R., Traub, O., Winterhager, E. and Nicholson, B.J. (1990a) Molecular heterogeneity of gap junctions in different mammalian tissues. In: A.W. Robards, W.J. Lucas, J.D. Pitts, H.J. Jongsma and D.C. Spray (Eds.), Parallels in Cell-to-Cell Junctions in Plants and Animals, NATO ASI Series, Springer-Verlag, Berlin, London, pp. 21–34.

Willecke, K., Jungbluth, S., Dahl, E., Hennemann, H., Heynkes, R. and Grzeschik, K.-H. (1990b) Six genes of the human connexin gene family coding for gap-junctional proteins are assigned to four different human chromosomes. Eur. J. Cell Biol. 53, 275–280.

Willecke, K., Hennemann, H., Dahl, E., Jungbluth, S. and Heynkes, R. (1991a) The diversity of connexin genes encoding gap-junctional proteins. Eur. J. Cell Biol., 56, 1–7.

Willecke, K., Heynkes, R., Dahl, E., Stutenkemper, R., Hennemann, H., Jungbluth, S., Suchyna, T. and Nicholson, B.J. (1991b) Mouse connexin37: cloning and functional expression of a gap-junction gene highly expressed in lung. J. Cell Biol. 114, 1049–1057.

Yancey, S.B., John, S.A., Ratneshwar, L., Austin, B.J. and Revel, J.-P. (1989) The 43-kDa polypeptide of heart gap junctions. Immunolocalization, topology and functional domains. J. Cell Biol. 108, 2241–2254.

Zhang, J.T. and Nicholson, B.J. (1989) Sequence and tissue distribution of a second protein of hepatic gap junctions, cx26, as deduced from its cDNA. J. Cell Biol. 109, 3391–3401.

Part II. Structure

J.E. Hall, G.A. Zampighi and R.M. Davis (Eds.)
Progress in Cell Research, Vol. 3

CHAPTER 6

Atomic force microscopy of gap junctions

JAN H. HOH[a], RATNESHWAR LAL[b], SCOTT A. JOHN[a], BARNEY DRAKE[c], JEAN-PAUL REVEL[a] and MORTON F. ARNSDORF[b]

[a]Division of Biology, California Institute of Technology, Pasadena, CA 91125, USA, [b]Section of Cardiology, Department of Medicine, University of Chicago, Chicago, IL 60637, USA and [c]Imaging Services, Box 9981, Truckee, CA 95737, USA

Introduction

The atomic force microscope (AFM) is a powerful new tool for high-resolution imaging, and can operate under highly physiological conditions. The AFM belongs to a new class of microscopes often referred to as scanning probe microscopes (SPMs), and is a direct descendant of the first SPM, the scanning tunneling microscope (STM) for which Binnig and Rohrer received the 1986 Nobel Prize in physics (Binnig and Rohrer, 1982). Suitably they shared the prize with Ernst Ruska, who 50 years earlier built the first electron microscope, which has been the centerpiece of structural biology for decades (Knoll and Ruska, 1932). The SPMs all operate by passing a probe over a surface, and collecting information about the surface with a high degree of spatial resolution. Many different probes have been developed, including ones for tunneling current (Binnig et al., 1986), magnetic properties (Martin and Wickramasinghe, 1987), ion conductance (Hansma et al., 1988), capacitance (Matey and Blanc, 1984), electrochemical properties (Bard et al., 1990), and surface topography (Binnig et al., 1986). The resolution of these microscopes varies greatly from about 0.1 nm in the x–y plane for the STM to greater than 100 nm for the electrochemical probe.

The probe in the AFM is a cantilever with a sharp tip, which makes direct physical contact with the sample and returns information about surface topography. Initially these cantilevers consisted of shards of diamond glued to gold foil or tungsten wire (Alexander et al., 1989), but more recently cantilevers microfabricated from Si_3N_4 have become available (Albrecht et al., 1990) (Fig. 1A and B). The image in the AFM is acquired by passing the sample beneath the tip and measuring deflections in the cantilever (Fig. 1C). In the constant-force mode of operation, the signal from the cantilever deflection is used in a feedback circuit to raise or lower the sample as it is scanned to maintain a constant deflection, and thereby maintain constant force applied by the tip to the sample. The deflection of the cantilever is usually measured by the movement of a laser beam reflected off the back of the cantilever (Meyer and Amer, 1988; Alexander et al., 1989), though several other detection systems have been described (for examples see Binnig et al., 1986; Martin et al., 1987; Erlandsson et al., 1988).

The AFM has previously been used to image several biological molecules and structures (for a review, see Engel, 1991) including protein molecules (Drake et al., 1989; Weisenhorn et al., 1990), isolated membranes (Butt et al., 1990; Worcester et al., 1990), and whole cells (Gould et al., 1990; Butt et al., 1991). We here describe initial efforts to apply this new imaging technology to the problem of gap-junction structure (see also Hoh et al., 1991b), which has been of long-standing interest to us (Revel and Karnovsky, 1967; Goodenough and Revel, 1970; Johnson et al., 1974; Finbow et al., 1980; Nicholson et al., 1985; Yancey et al., 1989; Laird and Revel, 1990; Hoh et al., 1991a; John and Revel, 1991). Details of gap-junction structure are extensively discussed in other chapters of this volume.

Experimental methods and results

AFM of isolated liver gap-junction plaques

Isolated native[1] gap junctions from rat liver were adsorbed to a glass coverslip in phosphate buffered saline (PBS), and imaged in PBS with a Nanoscope II Atomic Force Microscope (Digital Instruments, Santa Barbara, CA). For these experiments, we routinely used commercially available 100 μm V-shaped cantilevers (Digital Instruments). The undersides of these cantilevers, including the tip, were coated with chromium to prevent warping from the gold coating. It should be noted that because of improved manufacturing, chromium coating is no longer performed.

[1] The word native is used here mainly to distinguish trypsinized and glutaraldehyde fixed gap junctions from ones not treated.

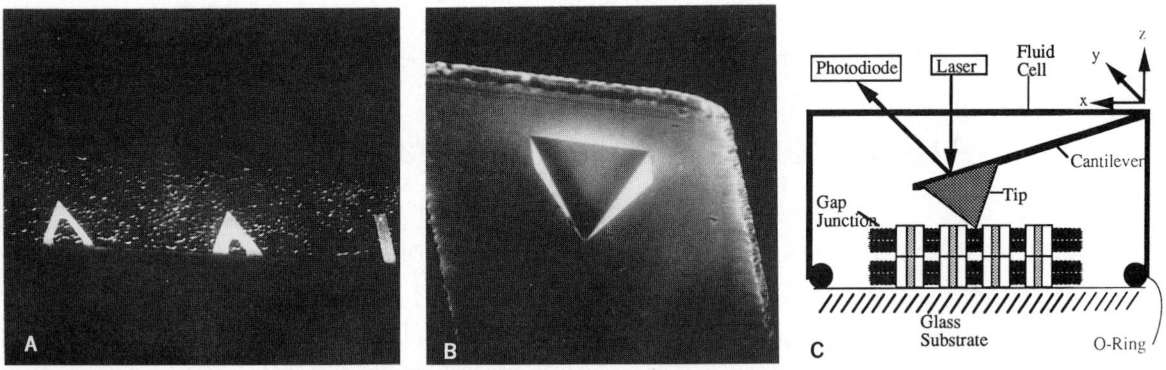

Figure 1. (A) Scanning electron micrograph of the edge of a glass chip on which the Si_3N_4 cantilevers (Park Instruments) are mounted. The different shapes and dimensions of the cantilevers give them different properties such as spring constants. The outside cantilevers (only one shown) are often referred to as springboard cantilevers, while the ones in the middle are V shaped. (B) Higher magnification of the tip on a springboard cantilever. The tips are roughly 2.7 µm on the side and 3.8 µm tall. (C) Schematic of the atomic force microscope fluid cell, cantilever, and a gap-junction plaque. The glass fluid cell filled with PBS is sealed against the coverslip with an o-ring, and encloses the sample and cantilever. The coverslip is mounted on a piezoelectric translator that scans the sample in a raster pattern (x and y) below the tip. The translator maintains the force constant by raising or lowering (z) the sample in response to deflections in the cantilever monitored by a laser reflected off the cantilever onto a segmented photodiode. The drawing is not to scale; the gap junction is not intended as a model.

At low magnification in the AFM, the plaques appear similar in size and distribution to ones seen by negative staining in the electron microscope (Fig. 2). Single plaques are 0.5–1 µm in diameter, and an average of 14.4 nm thick. The surface is generally smooth, though sometimes bumps several nm high and 50–100 nm wide are seen. The source of these bumps is not clear.

High-magnification images of the cytoplasmic surface of the plaque appear smooth, and no connexon substructure is visible. There are several possible explanations for this lack of substructure. For example, the protein protruding on the cytoplasmic face may organize in a very smooth layer on the surface, as suggested by deep-etch experiments of Hirokawa and Heuser (1982). Brownian motion may cause movement in the protein that cannot be resolved, given the relatively long image-acquisition time (usually 10 s or more), or the tip itself may move the protein as it scans the surface.

Trypsinized and glutaraldehyde-fixed gap junctions were also examined. These treatments did not result in any noticeable changes in morphology. However, glutaraldehyde treatment produced a measurable loss of adhesion

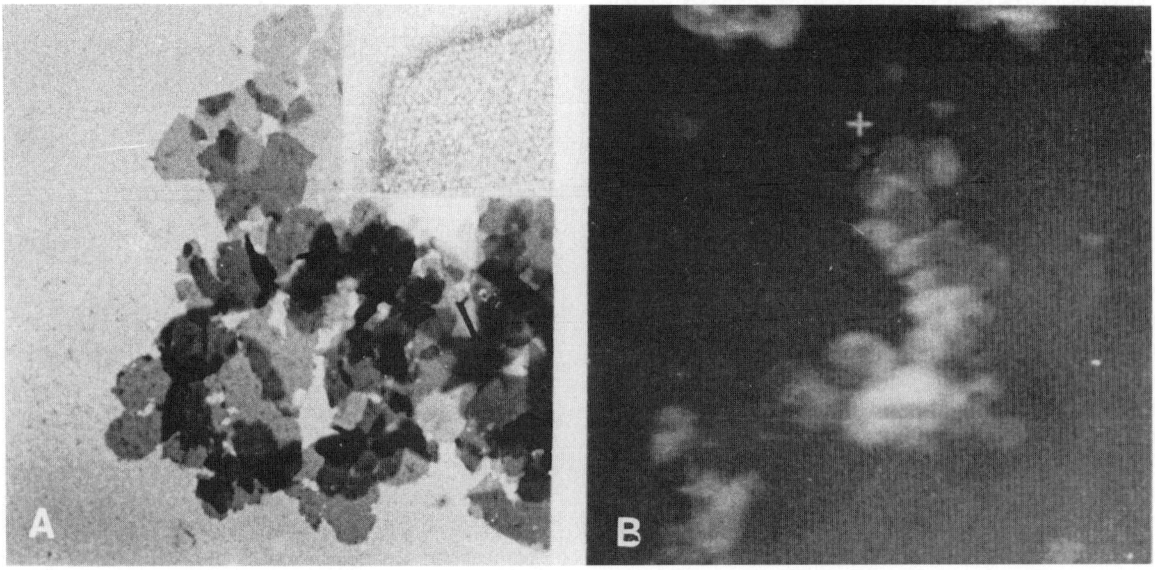

Figure 2. (A) Low-magnification transmission electron micrograph (TEM) of isolated gap junction stained with uranyl acetate. The inset shows individual connexons at high magnification. Rat liver gap junctions were isolated as previously described (Hoh, 1991). The major protein component in these plaques is connexin-32 (Cx32) as determined by gel electrophoresis. (B) Low magnification (18 µm × 18 µm) AFM image of isolated rat liver gap junctions.

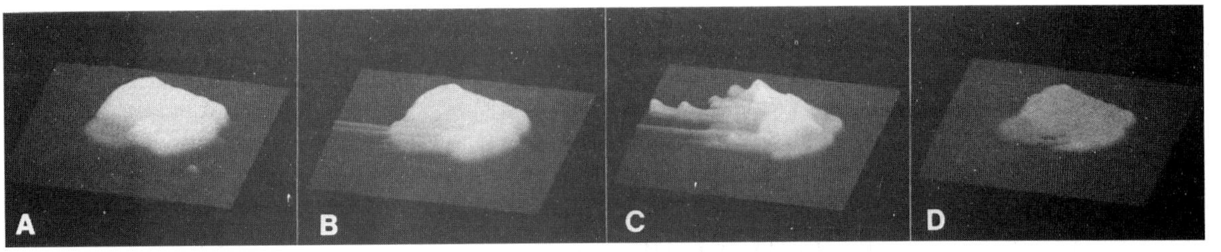

Figure 3. Force dissection of an isolated liver gap-junction plaque. As an increasing force from about 1 to 10 nN is applied (A to D), the thickness of the plaque is reduced from an average of 14.4 nm to 6.7 nm. The remaining membrane could be removed at very high forces. The field is approximately 1.5 μm × 1.5 μm.

between the sample and substrate, resulting in membranes being easily scraped off the surface. This observation suggests that the glutaraldehyde in some way modulates the interaction between the glass and the membrane. Since glutaraldehyde reacts primarily with amines, and glass is negatively charged at neutral pH, the loss of adhesion may result from the disruption of electrostatic interactions between the membrane and the glass.

Force dissection

In an experiment initially designed to examine the compressibility of the gap-junction membrane, isolated rat liver gap junctions were imaged in a series of increasing and decreasing forces. The force applied to the tip was determined from force curves obtained within 30 s of acquiring the image, essentially as described by Weisenhorn et al. (1989). It should be noted that the absolute force applied to the sample depends on a variety of factors such as the z calibration of the piezo scanner, the spring constant of the cantilever, bowing of the cantilever, bond formation between the tip and sample and the contact angle with the sample. Therefore, the values reported here are likely to be accurate relative to each other, but are a simplification (only the vertical applied force is given) and, therefore, an estimate of the real forces applied to the sample.

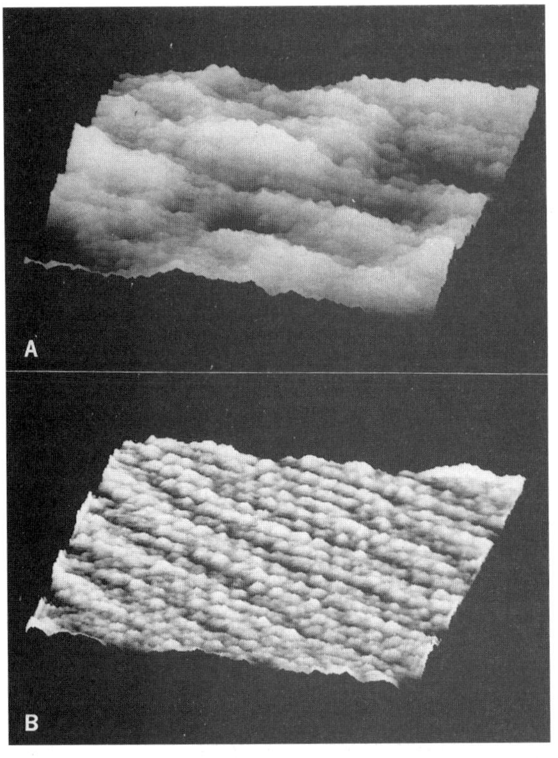

Figure 4. (A) Molecular resolution of connexons on the extracellular surface of an isolated liver gap junction exposed by force dissection. This image was flattened and plane fit. (B) A filtered image in which noise in the Fourier transform was removed. The individual connexons spaced at 9.1 nm, center-to-center, are clearly visible.

At low forces (< 2 nN), liver gap-junction plaques could be scanned for at least tens of min without any change in morphology. However, when the force applied to the tip was gradually increased in successive scans, the top membrane became distorted and was eventually removed completely, exposing the extracellular domains of the molecule, in a process we call "force dissection" (Fig. 3). This demonstrates the powerful micromanipulation capabilities of the AFM, and the controlled access to the extracellular domains of gap junctions will prove to be a valuable tool in future investigations.

Glutaraldehyde fixation did not affect our ability to dissect gap junctions with the AFM tip. This suggests that the glutaraldehyde did not form cross-links between gap-junction proteins across the gap. Using force dissection, further investigations with other cross-linking reagents will provide information about the nature of the interaction between connexons across the gap.

Molecular resolution

As previously noted, we did not see substructures that resembled individual connexons on the cytoplasmic surface of liver gap-junction plaques. However, in gap junctions that had been fixed and trypsinized, the extracellular surface revealed by force dissection exhibited a clearly hexagonal pattern. The subunits in this pattern were spaced 9.1 nm center-to-center, remarkably similar to the spacing reported by electron microscopy (Revel and Karnovsky, 1967) and appeared to protrude 0.3 to 0.5 nm from the surface. This protrusion is not enough to account for the

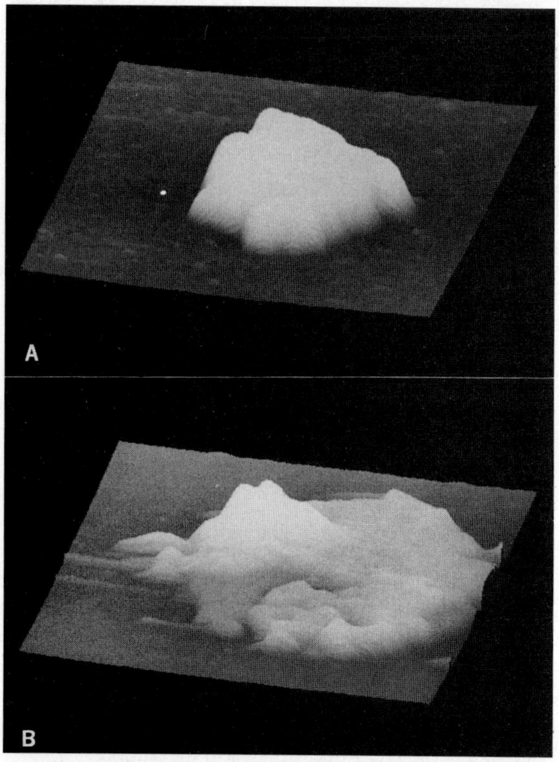

Figure 5. (A) An AFM image of an isolated heart gap junction on glass. The membrane has an apparent thickness of 23–24 nm. Gap junctions were isolated as previously described (Kensler and Goodenough, 1980; Manjunath et al., 1982; Laird and Revel, 1990). The major protein in these plaques is connexin-43 (Cx43) as determined by gel electrophoresis. The size of the field is 2 μm× 2 μm. (B) A trypsinized heart gap junction shown in the same height scale. After trypsin treatment, and removal of the cytoplasmic domains of Cx43, the membrane appears 14–15 nm thick. This membrane has been damaged by the tip and some area of single membrane are clearly visible. The size of the field is 1.5 μm × 1.5 μm.

~2 nm gap between membranes in the gap junctions. The reason for this is not yet clear, though it should be noted that some height information is lost due to the response characteristics of the z piezo. This information can, in principle, still be recovered by monitoring actual cantilever deflections, though the commercial instrument we used does not have this capability.

Molecular resolution was not obtained on unfixed samples. This suggests that glutaraldehyde stabilized the structure of the membrane or modified the surface chemistry in such a way that imaging was possible. It is likely that lower force and improved imaging conditions will allow attainment of molecular resolution on unfixed samples.

Heart gap-junction plaques

Preliminary images of isolated rat heart gap junctions were also obtained in PBS under conditions identical to those for liver gap junctions. These plaques did not adhere well to the glass, and appeared "softer" than the liver gap junctions. In general, these plaques were smaller than the liver gap junctions, but substantially thicker, roughly 22–23 nm compared to 14 nm for liver gap junctions. Images of the heart gap-junction plaques showed rounded edges, and some had lumps that appeared to be large-scale deformations of the membranes. We attribute this appearance to the substantially larger cytoplasmic domain of the major heart gap-junction protein (Cx43) in these preparations.

Trypsinization of the heart gap junctions resulted in a marked decrease in thickness to 14–15 nm, similar to liver gap junctions. This suggests that the cytoplasmic domains of the heart gap junction contributed roughly 8 nm to the thickness of the native plaque observed by AFM. In addition, trypsinization improved the adhesion of the plaque to the glass coverslip.

Conclusions

We have demonstrated that the atomic force microscope can image isolated gap-junction plaques from liver and heart and, under some circumstances, resolve individual connexons. Further, the microscope can be used to micromanipulate the membrane, removing one bilayer, thereby providing experimental access to the extracellular domains. These results provide a basis for further investigations into gap-junction structure and function using this powerful new technology.

Acknowledgements

This work was supported by grants NIH grants HL37109 and BRSG-RR07003 to J.P.R., and R37 HL21788 to M.F.A.. We thank Dr. Dale Laird for the isolated rat heart gap junctions used here. Portions of the work described presented at the meeting and described here have appeared elsewhere (Hoh et al., 1991b). Figures 2 and 3 are reproduced with the permission of the American Association for the Advancement of Science.

References

Albrecht, T.R., Akamine, S., Carver, T.E. and Quate, C.F. (1990) Microfabrication of cantilever styli for the atomic force microscope. J. Vac. Sci. Technol. A 8, 3386–3396.

Alexander, S., Hellemans, L., Marti, O., Schneir, J., Elings, V., Hansma, P.K., Longmire, M. and Gurley, J. (1989) An atomic-resolution atomic-force microscope implemented using an optical lever. J. Appl. Phys. 65, 164–167.

Bard, A.J., Denuault, G., Lee, C., Mandler, D. and Wipf, D.O. (1990) Scanning electrochemical microscopy: a new technique for the characterization and modification of surfaces. Acc. Chem. Res. 23, 357–363.

Binnig, G. and Rohrer, H. (1982) Surface studies by scanning tunneling microscopy. Phys. Rev. Lett. 49, 57.

Binnig, G., Quate, C.F. and Gerber, C. (1986) Atomic force microscope. Phys. Rev. Lett. 56, 930–933.

Butt, H.-J., Downing, K.H. and Hansma, P.K. (1990) Imaging the membrane protein bacteriorhodopsin with an atomic force microscope. Biophys. J. 58, 1473–1480.

Butt, H.-J., Wolf, E.K., Gould, S.A.C. and Hansma, P.K. (1991) Imaging cells with the atomic force microscope. J. Struct. Biol. 105, 54–61.

Drake, B., Prater, C.B., Weissenhorn, A.L., Gould, S.A.C., Albrecht, T.R., Quate, C.F., Cannell, D.S., Hansma, H.G. and Hansma, P.K. (1989) Imaging crystals, polymers, and processes in water with the atomic force microscope. Science 243, 1586–1589.

Engel, A. (1991) Biological applications of scanning probe microscopes. Annu. Rev. Biophys. Biophys. Chem. 20, 79–108.

Erlandson, R., McClelland, G.M., Mate, C.M. and Chiang, S. (1988) Atomic force microscopy using optical interferometry. J. Vac. Sci. Technol. A 6, 266–270.

Finbow, M.E., Yancey, S.B., Johnson, R.G. and Revel, J. (1980) Independent lines of evidence suggesting a major gap-junctional protein with a molecular weight of 26,000. Proc. Natl. Acad. Sci. USA 77, 970–974.

Goodenough, D.A. and Revel, J.-P. (1970) A fine structural analysis of intercellular junctions in the mouse liver. J. Cell Biol. 45, 272–290.

Gould, S.A.C., Drake, B., Prater, C.B., Weissenhorn, A.L., Manne, S., Hansma, H.G., Hansma, P.K., Massie, J., Longmire, M., Elings, V., Northern, B.D., Mukergee, B., Peterson, C.M., Stoeckenius, W., Albrecht, T.R. and Quate, C.F. (1990) From atoms to integrated circuit chips, blood cells, and bacteria with the atomic force microscope. J. Vac. Sci. Technol. A 8, 369–373.

Hansma, P.K., Drake, B., Marti, O., Gould, S.A.C. and Prater, C.B. (1988) The scanning ion-conductance microscope. Science 243, 641–643.

Hoh, J.H. (1991) Studies on the structure and molecular diversity of the gap junction. Thesis. California Institute of Technology.

Hoh, J.H., John, S.A. and Revel, J.-P. (1991a) Molecular cloning and characterization of a new member of the gap-junction gene family, connexin-31. J. Biol. Chem. 266, 6524–6531.

Hoh, J.H., Lal, R., John, S.A., Revel, J.P. and Arnsdorf, M.F. (1991b) Atomic force microscopy and dissection of gap junctions. Science 253, 1405–1408.

John, S.A. and Revel, J.-P. (1991) Connexon integrity is maintained by non-covalent bonds: Intramolecular disufide bonds link the extracellular domains in rat connexin-43. Biochem. Biophys. Res. Commun. 178, 1312–1318.

Johnson, R., Hammer, M., Sheridan, J. and Revel, J.-P. (1974) Gap-junction formation between reaggregated Novikoff hepatoma cells. Proc. Natl. Acad. Sci. USA 71(11), 4536–4540.

Kensler, R.W. and Goodenough, D.A. (1980) Isolation of mouse myocardial gap junctions. J. Cell Biol. 86, 755–764.

Knoll, M. and Ruska, E. (1932) Beitrag zur geometrischen elektronoptik. Ann. Phys. 5, 607–661.

Laird, D.W. and Revel, J.-P. (1990) Biochemical and immunochemical analysis of the arrangement of connexin43 in rat heart gap-junction membranes. J. Cell Sci. 97, 109–117.

Manjunath, C.K., Goings, G.E. and Page, E. (1982) Protein composition of gap junctions from rabbit hearts. Biochem. J. 205, 189–194.

Martin, Y., Williams, C.C. and Wickramasinghe, H.K. (1987) Atomic force microscope force mapping and profiling on a sub-100 Å scale. J. Appl. Phys. 61, 4723–4729.

Martin, Y., Rugar, D. and Wickramasinghe, H.K. (1988) High resolution magnetic imaging of domains in TBFE by force microscopy. Appl. Phys. Lett. 52, 244–246.

Matey, J.R. and Blanc, J. (1985) Scanning capacitance microscopy. J. Appl. Phys. 57, 1437–1444.

Meyer, G. and Amer, N.M. (1988) Novel optical approach to atomic force microscopy. Appl. Phys. Lett. 53, 1045–1047.

Nicholson, B.J., Gros, D.B., Kent, S.B., Hood, L.E. and Revel, J.-P. (1985) The M_r 28,000 gap-junction proteins from rat heart and liver are different but related. J. Biol. Chem. 260(11), 6514–6517.

Revel, J.-P. and Karnovsky, M. (1967) Hexagonal array of subunits in intercellular junctions of the mouse heart and liver. J. Cell Biol. 33, C7–C12.

Weisenhorn, A.L., Drake, B., Prater, C.B., Gould, S.A.C., Hansma, P.K., Ohnerorge, F., Egger, M., Heyn, S.-P. and Gaub, H.E. (1990) Immobilized proteins in buffer imaged at molecular resolution by atomic force microscopy. Biophys. J. 58, 1251–1258.

Worcester, D.L., Kim, H.S., Miller, R.G. and Bryant, P.J. (1990) Imaging bacteriorhodopsin lattices in purple membranes with atomic force microscopy. J. Vac. Sci. Technol. A 8(1), 403–405.

Yancey, S.B., John, S.A., Lal, R., Austin, B.J. and Revel, J.-P. (1989) The 43-kD polypeptide of heart gap junctions: immunolocalization (I), topology (II), and functional domains (III). J. Cell Biol. 108: 2241–2254.

J.E. Hall, G.A. Zampighi and R.M. Davis (Eds.)
Progress in Cell Research, Vol. 3

CHAPTER 7

Structure and design of cardiac gap-junction membrane channels

MARK YEAGER

Departments of Cell Biology and Molecular Biology, The Scripps Research Institute and Division of Cardiovascular Diseases, Scripps Clinic and Research Foundation, La Jolla, CA 92037, USA

Introduction

Gap junctions are specialized plasma membrane regions that allow intercellular communication between cells, thereby coordinating their metabolic and electrical activities (Loewenstein, 1981). Gap-junction proteins (termed connexins) form a multigene family of integral membrane proteins (Beyer et al., 1990; Kumar, 1991) which have been classified based on characteristic features of their amino-acid sequences (Gimlich et al., 1990; Risek et al., 1990). The α class is defined by the 43-kDa (α_1) connexin found in mammalian myocardium (Beyer et al., 1987)[1]. The β class of gap-junction proteins was originally described in mammalian liver, with molecular weights of 32 kDa (β_1) (Paul, 1986; Kumar and Gilula, 1986) and 26 kDa (β_2) (Zhang and Nicholson, 1989). Connexins (Cx) have also been classified according to their molecular weights (Beyer et al., 1990): Cx43 = α_1; Cx32 = β_1 and Cx26 = β_2. Comparison of the amino-acid sequences for the heart α_1 (Beyer et al., 1987) and liver β_1 (Paul, 1986; Kumar and Gilula, 1986) connexins reveals considerable homology in the four hydrophobic domains as well as the extracellular loop regions (Fig. 1). A notable difference is that α_1 connexin (43 kDa) is substantially larger than β_1 (32 kDa) connexin, primarily due to a larger carboxy tail (Manjunath et al., 1987).

Gap junctions in the heart play an important functional role by electrically coupling cells thereby organizing the pattern of current flow to allow coordinated muscle contraction (Barr et al., 1965; De Mello, 1982). Cardiac gap junctions are therefore intimately involved in normal conduction as well as the genesis of potentially lethal arrhythmias (Spach, 1983). Cardiac electrophysiology has provided tremendous insight into the mechanisms of cardiac conduction (Josephson and Seides, 1979) and the basis for intercellular current flow via gap junctions (reviewed by Spray and Burt, 1990). Fundamental to a complete understanding of basic mechanisms of cardiac conduction and arrhythmias is detailed knowledge about the structure of cardiac gap junctions. We are using complementary immunochemical and biophysical techniques as an integrated strategy to delineate the structure and design of cardiac gap-junction membrane channels. Table I summarizes the levels of structure that we are examining, the experimental approach and approximate resolution of the technique, and the major results obtained thus far. Contributions by other investigators are also acknowledged. This report is organized from the lowest level of structural resolution provided by light microscopy to experiments at the molecular level provided by electron microscopy and circular dichroism spectroscopy. The methods and results for the immunolabeling and electron microscopic studies have been summarized from the detailed presentation by Yeager and Gilula (1992).

Tissue distribution of gap-junction proteins: immunofluorescence microscopy

Several laboratories, including our own, have generated antibodies to defined sequences in rat cardiac gap-junction protein (Fig. 1). Immunohistochemical localization of cardiac gap junctions has been examined using phase-contrast and indirect-immunofluorescence microscopy (Fig. 2). Ultrastructural studies have shown that gap junctions in the heart reside in the intercalated disks (McNutt and Weinstein, 1970). In longitudinal cryo-sections of rat heart muscle cells, the intercalated disks appear as thick, high-contrast lines which can be distinguished from the thin striations of the muscle fibrils (see top portion of phase-contrast image in Fig. 2). The corresponding indirect-immunofluorescence micrograph displays fluorescent

[1]*Abbreviations:* 43 kDa rat cardiac gap-junction protein (α_1 = Cx43), 32 kDa rat liver gap-junction protein (β_1 = Cx32); 26 kDa rat liver gap-junction protein (β_2 = Cx26), bovine serum albumin (BSA), keyhole limpet hemocyanin (KLH); phenylmethylsulfonyl fluoride (PMSF), phosphate buffered saline (PBS), phosphotungstic acid (PTA), Tris-buffered saline (TBS), circular dichroism (CD).

labeling that corresponds to the location of intercalated disks. When the muscle cells are sectioned obliquely (bottom portion of micrograph in Fig. 2), the intercalated disks are not readily identified in the phase image, and the corresponding area in the fluorescence image displays a "speckled" pattern. The immunofluorescence images provide a description of the normal patterns for the tissue distribution of gap junctions in heart muscle with more detail than is achieved with conventional ultrastructure analysis. The specific gap-junction labeling by the peptide antibodies demonstrates that the antibodies should also be useful reagents for mapping the distribution of gap junctions in myocardium known by electrophysiologic testing to have abnormal conduction properties.

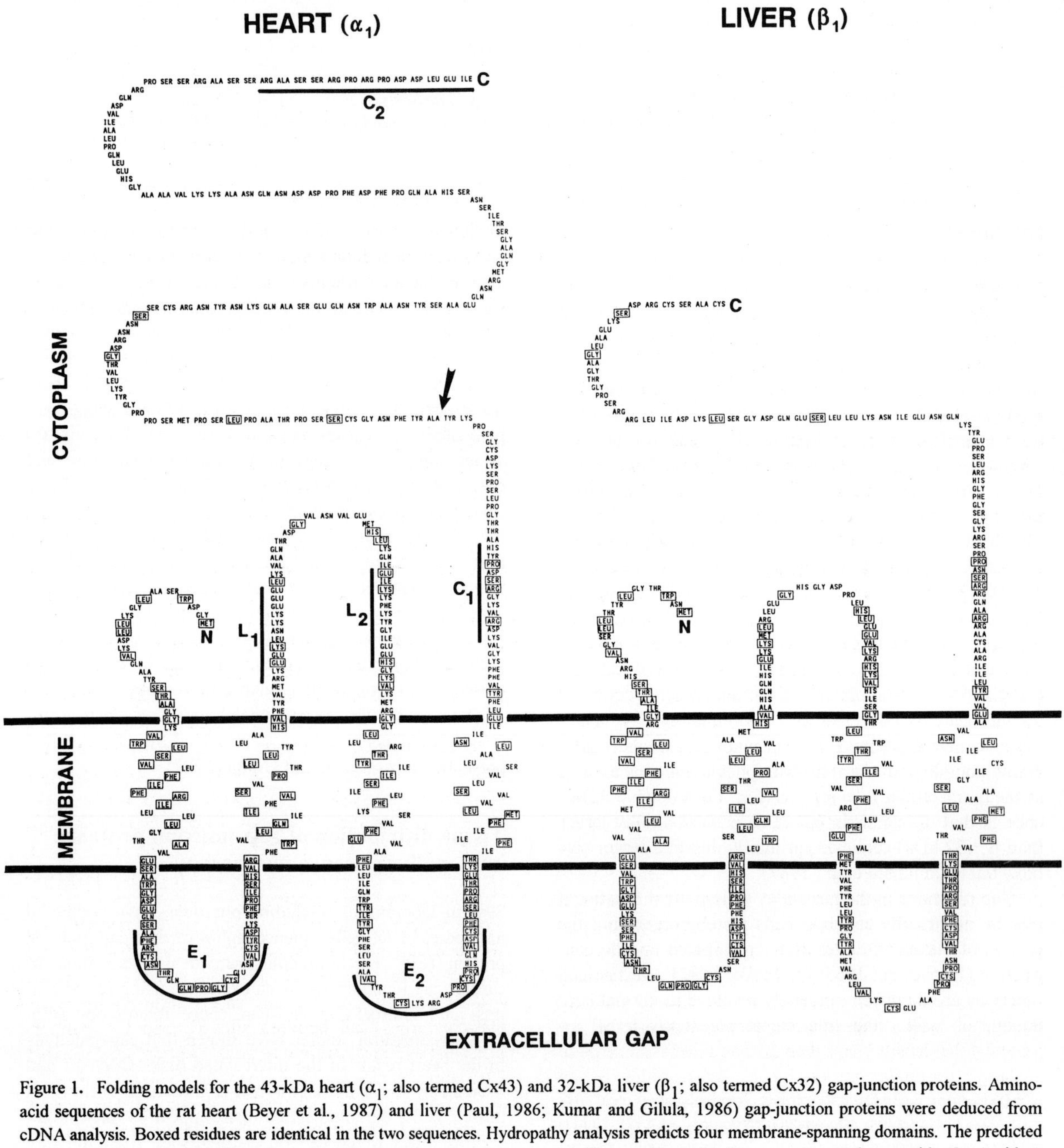

Figure 1. Folding models for the 43-kDa heart (α_1; also termed Cx43) and 32-kDa liver (β_1; also termed Cx32) gap-junction proteins. Amino-acid sequences of the rat heart (Beyer et al., 1987) and liver (Paul, 1986; Kumar and Gilula, 1986) gap-junction proteins were deduced from cDNA analysis. Boxed residues are identical in the two sequences. Hydropathy analysis predicts four membrane-spanning domains. The predicted locations of the extracellular, membrane and cytoplasmic regions are indicated. Site-specific antibodies were generated to peptides L_1 (residues 101–112) and L_2 (residues 131–142) in the cytoplasmic "loop" region, C_1 (residues 237–248) and C_2 (residues 370–382) in the carboxy-terminal domain, and E_1 (residues 51–65) and E_2 (residues 184–198) in the extracellular loops of α_1 connexin (Houghten, 1985; Yeager and Gilula, 1992). The arrow indicates the approximate site of cleavage by an endogenous protease in rat heart which generates a "loop" ~30 kDa membrane-bound fragment.

TABLE I Structure analysis of cardiac gap-junction membrane channels

Level of structure	Experimental approach	Resolution	Results
Tissue architecture	Immunofluorescence microscopy	0.1– 10 μm	Cardiac gap junctions reside in the intercalated disks (1,2,3)
4° structure and shape	Electron microscopy and image analysis	5–100 Å	The ion channels are formed by a hexameric cluster of α_1 subunits (1)
3° structure and membrane topology	Thin-section electron microscopy • Labeling with peptide antibodies • Selective protease cleavage	50–1000 Å	α_1 connexin has 3 accessible cytoplasmic domains (1,2,3,4) • ~13 kDa carboxyterminal domain • ~ 5 kDa cytoplasmic "loop" • Amino-terminal tail
2° structure and α-helical content	Circular dichroism spectroscopy	5–15 Å	The 4 hydrophobic domains of α_1 connexin may be folded as transmembrane α-helices (5)

(1) Yeager and Gilula, 1992; (2) Yancey et al., 1989; (3) Beyer et al., 1989; (4) Laird and Revel, 1990; (5) Yeager, 1991.

Quaternary structure: electron microscopy and image analysis

Gap junctions containing native 43 kDa α_1 connexin have been isolated from rat hearts using cell-fractionation techniques (Manjunath and Page, 1986). When the protease inhibitor PMSF is absent from the isolation buffer, an endogenous enzyme cleaves the native protein to ~30 kDa (indicated by the arrow in Fig. 1) (Manjunath et al., 1985). We have used electron microscopic image analysis to examine the molecular structure of cardiac gap junctions containing native and protease-cleaved α_1 connexin. The lack of two-dimensional crystallinity in isolated cardiac gap junctions has hindered the application of electron crystallography to the structure analysis of these ion channels. Therefore, we have developed a sequential detergent-dialysis method using deoxycholate and dodecylmaltoside to induce crystallization (Yeager, 1987 and 1989) (Fig. 3). These detergents presumably extract membrane lipids to effectively concentrate the protein molecules in the membrane plane and thereby facilitate crystallization. The density maps computed by Fourier transformation have a nominal resolution of 16 Å (Fig. 4), and demonstrate that cardiac gap-junction channels are formed by a hexameric cluster of α_1 polypeptides. Previous work using membrane-impermeable proteases as probes of protein topology has not demonstrated that cleavage of the protein does not change the quaternary structure of the channel. If quaternary structure is altered by cleavage, the pattern of peptides generated by cleavage might not reflect the true membrane topology of the protein. The density maps demonstrate that the hexameric quaternary arrangement of the α_1 polypeptides is not significantly perturbed by protease cleavage of α_1 connexin from M_r 43 kDa to ~30 kDa (Fig. 4). Therefore, the ~13 kDa carboxy-terminal domain is not involved in forming the transmembrane channel. The similar hexameric architecture of cardiac and liver gap-junction connexons indicates conservation in the molecular design of gap-junction channels formed by α and β connexins. This can be understood on the basis of the amino-acid sequences, since protease cleavage generates a truncated α_1 connexin that more closely resembles β_1 connexin (Fig. 1).

Tertiary structure: immunolabeling with site-specific peptide antibodies

The tertiary folding of the α_1 polypeptide (Fig. 1) has been defined using (1) peptide antibodies directed to several sites in the protein sequence and (2) cleavage by an endogenous protease in heart tissue (Yeager and Gilula, 1990 and 1992). Four antibodies were directed to sites in cytoplasmic domains, and two antibodies were directed to the two extracellular-loop domains. Isolated gap junctions could not be labeled by the extracellular loop antibodies (E_1 and E_2) using thin-section immunogold electron microscopy. This is consistent with the known narrowness of the extracellular gap region that would exclude macromolecular antibody probes. Immunoelectron microscopy (Fig. 5) demonstrates that gap junctions containing native 43 kDa α_1 connexin are heavily decorated on their cytoplasmic surfaces by the peptide antibodies directed to cytoplasmic domains (L_1, L_2, C_1 and C_2). The protease-cleaved gap junctions containing ~30 kDa α_1 connexin are also labeled by L_1, L_2 and C_1 antibodies. However, the C_2 antibodies directed against the extreme carboxy-tail do not bind to cleaved gap junctions. This observation proves that the site of protease cleavage resides in the cytoplasmic carboxy-tail between sites C_1 and C_2. Furthermore, the ~13 kDa carboxy-terminal peptide(s) is released after protease cleavage and does not remain attached to the ~30 kDa membrane-bound fragment via noncovalent interactions. (There may in fact be several cleavage products which sum to ~13 kDa.) The arrow in the folding model of α_1 connexin (Fig. 1) indicates the approximate location between C_1 and C_2 that would generate the ~13 kDa carboxy-terminal fragment(s) and the ~30 kDa membrane-bound, amino-terminal fragment with the epitopes for antibodies L_1, L_2, C_1, E_1 and E_2. The site of cleavage has been further delineated using antibodies to residues 252–271 (Beyer et al., 1989). Since these antibodies label both the native and cleaved α_1 polypeptides, the site of protease cleavage must reside between residues 252–271 (Yeager and Gilula, 1992). The amino termini of α_1 and β_1 connexin are homologous (Fig. 1), and antibodies to the amino-terminal sequence of β_1 connexin can bind to α_1 connexin (Zervos et al., 1985). Yancey et al. (1989) have also shown that the amino terminus of α_1 connexin is accessible on the cytoplasmic surface for labeling by peptide antibodies. Thus, the amino-terminal

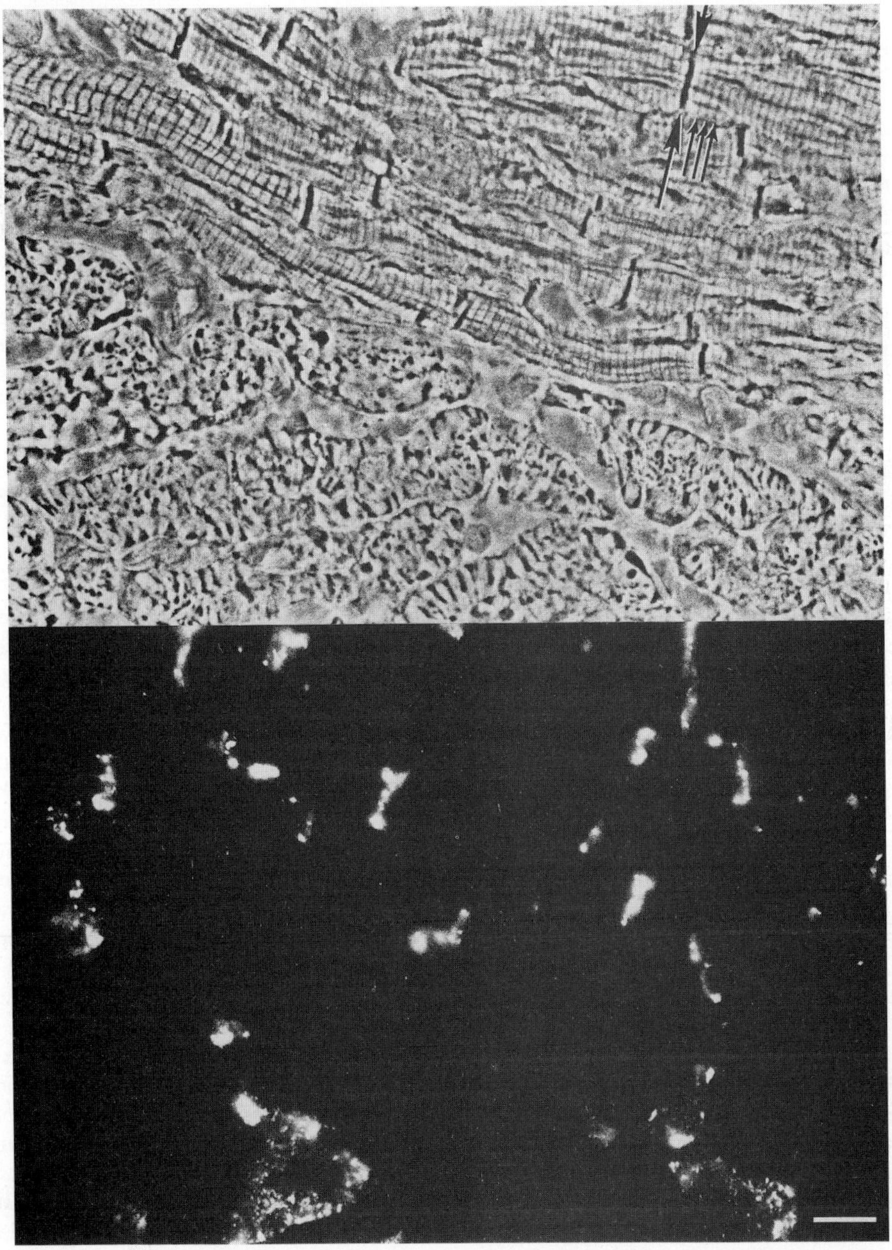

Figure 2. Phase-contrast micrograph (top) and indirect-immunofluorescence micrograph (bottom) for C_2 site-specific peptide antibodies directed against the carboxy-terminal peptide in α_1 connexin (see Fig. 1). In longitudinal sections (top portion of micrograph), the high-contrast linear striations (large arrow) correspond to the intercalated disks, which are readily distinguished from the thin striations of the sarcomeres (small arrows). The correlated fluorescence and phase-contrast images show binding of the antibodies to the intercalated disks in which the gap junctions reside. Cryofixation was performed by rapidly plunging freshly dissected rat heart into liquid nitrogen. Four- to five-μm thick sections were air dried, washed with PBS, exposed to affinity-purified antibody solutions, washed again with PBS and then incubated with fluorescein-conjugated goat anti-rabbit immunoglobulins (Cappel, West Chester, PA), followed by a final PBS wash. Tissue sections were examined using a Zeiss Axiophot microscope. Scale bar = 25 μm.

domain, carboxy-terminal domain and the "loop" domain define three polypeptide regions of α_1 connexin that are accessible on the cytoplasmic side of the membrane. Our results confirm and extend previous topological analyses of α_1 connexin using immunolabeling techniques and protease cleavage (Beyer et al., 1989; Yancey et al., 1989; Laird and Revel, 1990) which provide compelling evidence favoring the folding model in Fig. 1.

Secondary structure: circular dichroism spectroscopy

Circular dichroism (CD) spectroscopy measures the difference in absorption of left and right circularly polarized light. From 190–240 nm, CD spectra of proteins are dominated by the $n \rightarrow \pi^*$ and $\pi \rightarrow \pi^*$ electronic transitions of

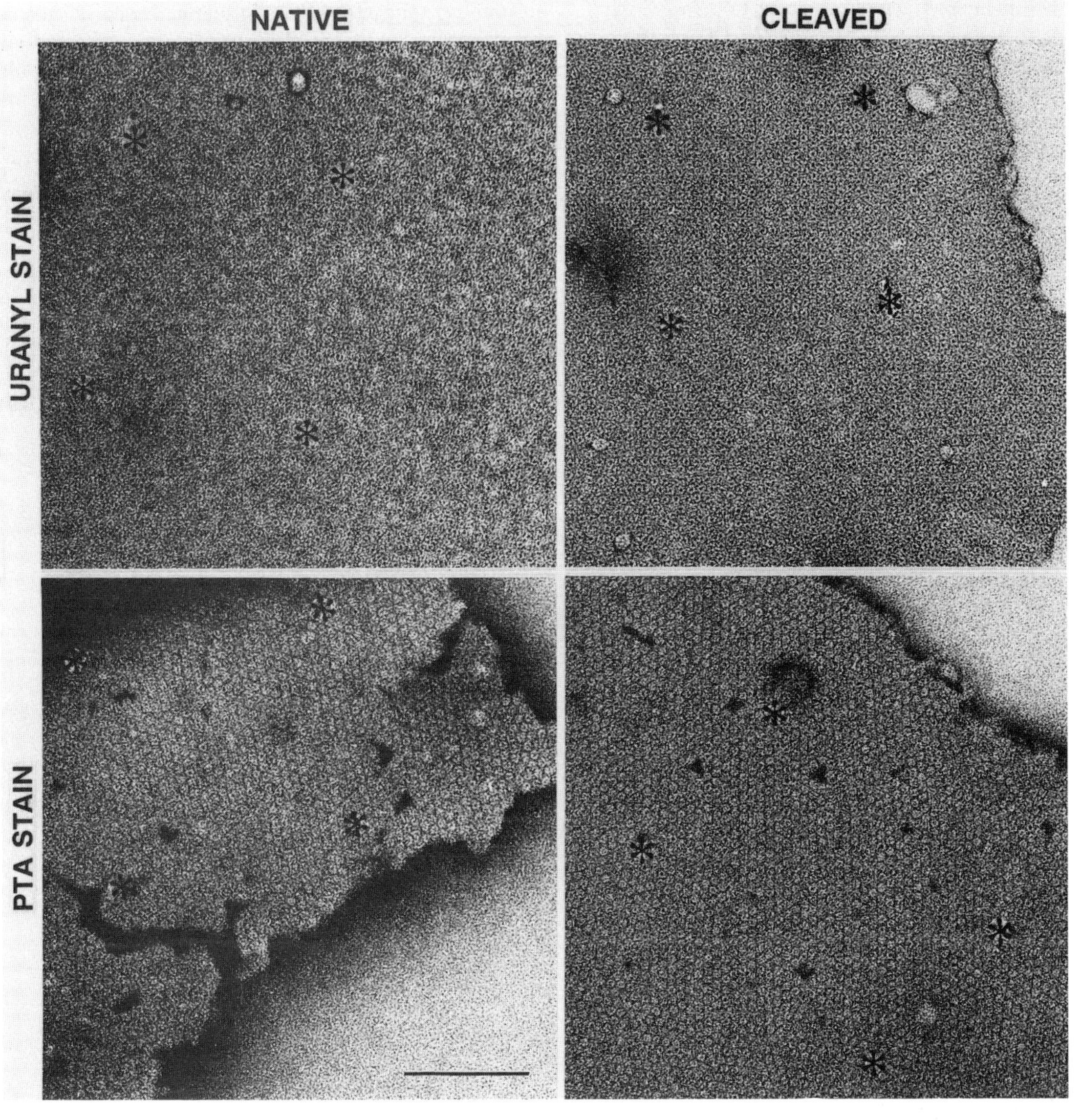

Figure 3. Images of native and protease-cleaved two-dimensional crystalline arrays of cardiac gap junctions stained with either uranyl acetate or phosphotungstic acid (PTA). The hexagonal packing of the oligomers was induced by exposure to deoxycholate and dodecyl-β-D-dodecylmaltoside. Asterisks (*) define the boxed areas judged most crystalline by optical diffraction. Rat ventricular gap junctions containing native α_1 connexin (M_r = 43 kDa) were isolated by the method of Manjunath and Page (1986). Cleaved α_1 gap junctions ($M_r \approx$ 30 kDa) were isolated by following the same protocol as for intact junctions, except the isolation buffer contained 1.0 mM bicarbonate and lacked PMSF (Manjunath et al., 1985). Crystals suitable for electron-image analysis were obtained by sequential ~24 h dialysis of the 35/49% gap junction sucrose gradient fraction against 0.5% sodium deoxycholate (Sigma Chemical Co., St. Louis, MO) and then 0.05% dodecyl-β-D-maltoside (Cal Biochem Corp., La Jolla, CA) in a buffer containing 5 mM-HEPES (pH 8), 0.1 mM $CaCl_2$, 0.5 mM $MgCl_2$ and 0.02% sodium azide. Specimens were examined on copper-coated, glow-discharged grids using a Philips EM400T electron microscope. Scale bar = 1000 Å.

the peptide backbone which are sensitive to molecular geometry. CD spectroscopy is therefore useful for determining protein secondary structure and monitoring conformational changes (Johnson, 1988). The secondary structure of cardiac gap junctions has been examined using CD spectroscopy in order to assess whether the hydrophobic domains may be folded as transmembrane α-helices. Gap junctions containing protease-cleaved α_1 connexin were examined in order to focus on the structure of the hydrophobic domains. The channels have been maintained in their native lipid-membrane environment to preserve the native protein conformation. The corrected CD spectrum recorded from a suspension of cardiac gap junctions (Fig. 6A) displays considerable similarity to the spectrum for a polypeptide with an α-helical conformation (Fig. 6B). A linear least-squares fit of the experimental spectrum (Fig. 6A) to polypeptides with known conformations (Fig. 6B) suggests that there indeed is sufficient α-helical content in the cleaved protein so that the four hydrophobic domains of α_1 connexin (Fig. 1) may have an α-helical conformation (Yeager, 1991).

Implications for the structure and function of cardiac gap-junction membrane channels

Based on (1) the dimensions of the connexon provided by electron crystallography, (2) the folding based on immunolabeling and protease cleavage, (3) the predicted α-helical structure of the transmembrane domains, and (4) the known amino-acid sequence, a working model is proposed in which the transmembrane portion of cardiac gap-junction channels is formed by a hexamer of α_1 connexin subunits, with each subunit, in turn, having a

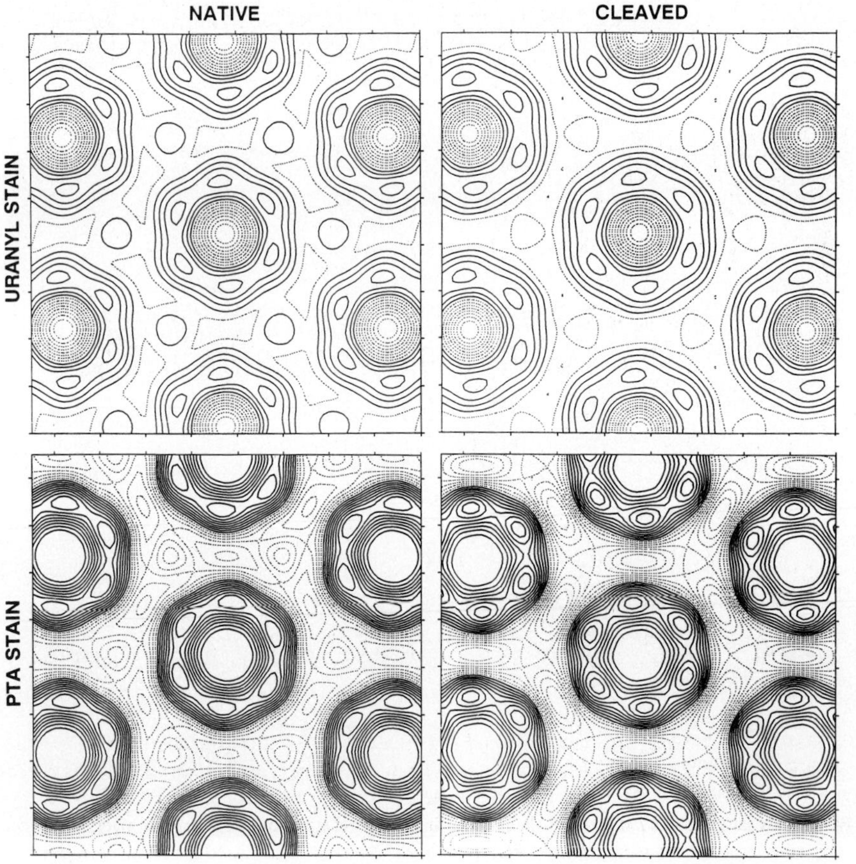

Figure 4. Two-dimensional projection maps of native and protease-cleaved α_1 gap junctions stained with uranyl acetate or PTA. The maps display connexin subunits (continuous contours) arranged in hexameric clusters defining a central channel. The hexagonal channel structure is not detectably altered by protease cleavage. Uranyl acetate penetrates the ion channel (circular dashed contours), whereas PTA is comparatively excluded from the channel and defines the outer boundary of the hexamers. The extraction and refinement of the crystal-lattice parameters and sinc-weighted amplitudes from the digitized images were performed as described (Amos et al., 1982). Lattice distortions were corrected using the procedures of Henderson et al. (1986). The two-dimensional maps were calculated by Fourier transformation using the amplitudes and phases corrected for lattice distortions. (Separation between fiducial marks = 20 Å.)

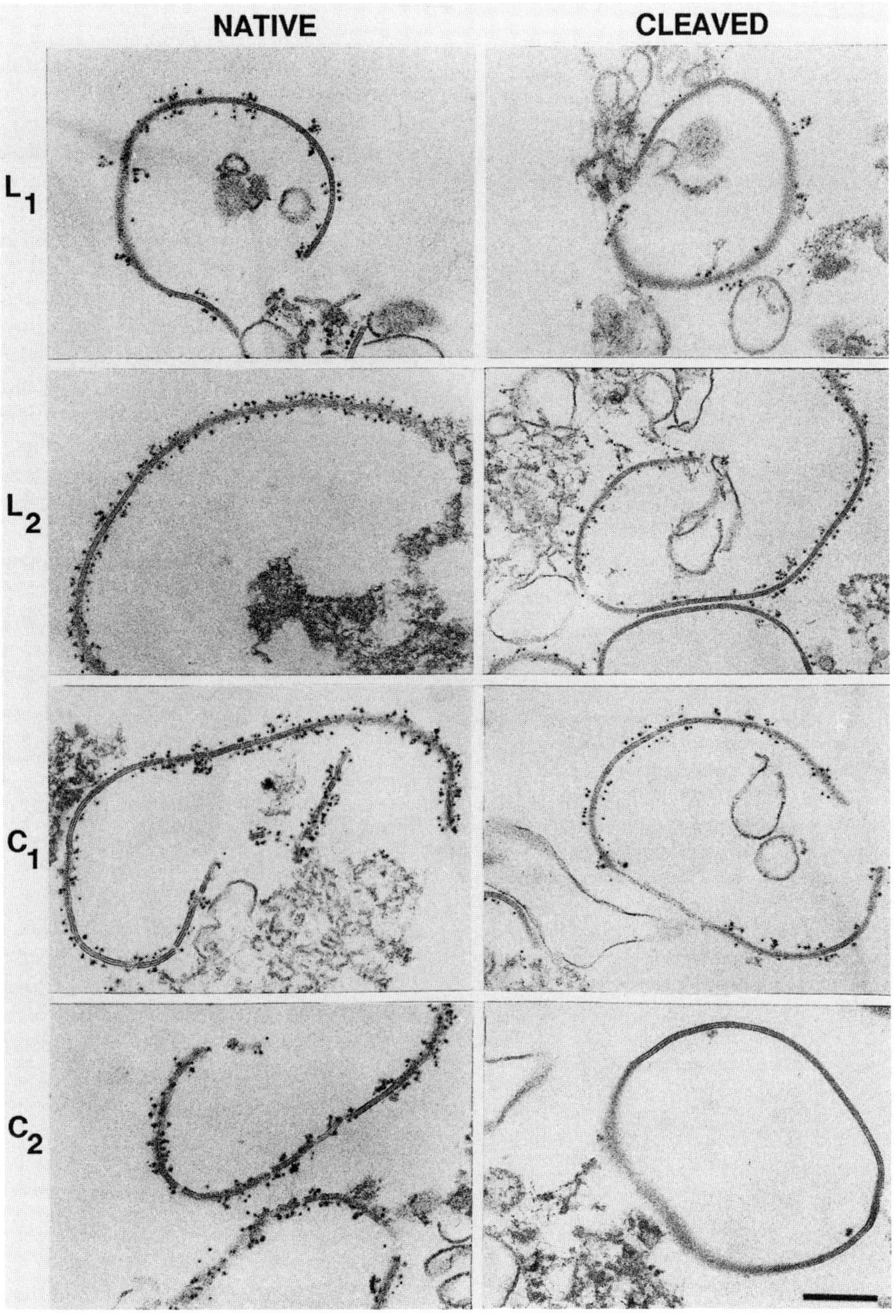

Figure 5. Immunogold labeling of native (left) and protease-cleaved (right) cardiac gap junctions by site-specific peptide antibodies L_1, L_2, C_1 and C_2 directed to cytoplasmic domains in α_1 connexin. All native and protease-cleaved gap junctions are decorated by the antibodies, except for cleaved junctions incubated with C_2 antibodies. Thus, the carboxy-tail epitope for the C_2 antibodies is released after protease cleavage (see model in Fig. 1). Specificity of antibody labeling was indicated by the minimal labeling of contaminating amorphous material and non-junctional single-membrane structures. Sonicated gap-junction suspensions in PBS were blocked with 3% BSA, and then sequentially incubated with primary-peptide antibodies and then anti-rabbit immunoglobulins labeled with 5-nm gold particles (Janssen Life Sciences Products, Olen, Belgium). Fixation, dehydration, embedding, and sectioning were performed as described (Yeager and Gilula, 1992). Specimens were post-stained with lead citrate and uranyl acetate and then examined in a Philips CM12 electron microscope. Scale bar = 0.2 μm.

bundle of four transmembrane α-helices (Yeager, 1991). Area estimates per polypeptide based on the projection maps in Fig. 4 are consistent with four transmembrane, roughly parallel α-helices as has been suggested for β_1 connexin (Milks et al., 1988; Tibbits et al., 1990). However, confirmation of this model will require a three-dimensional structure analysis at higher resolution.

The L_1 and L_2 peptides are close to the membrane surface in the cytoplasmic "loop" flanked by the second and third membrane-spanning domains. Since peptide antibod-

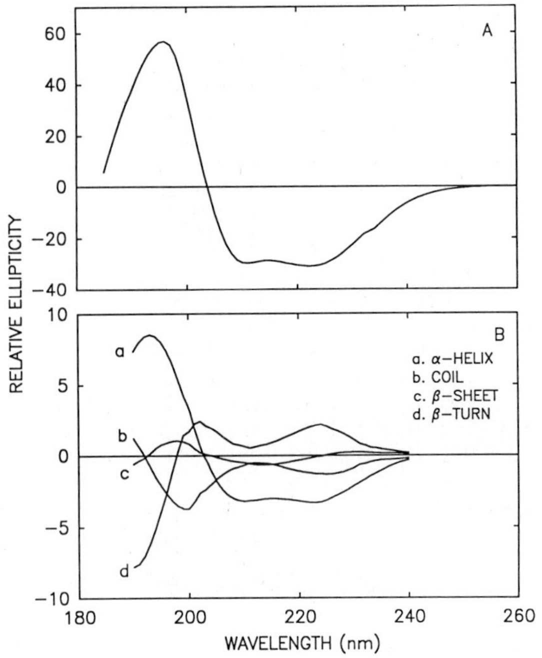

Figure 6. (A) CD spectra of cardiac gap junctions containing cleaved ~30 kDa α_1 connexin (Yeager, 1991). (B) CD spectra for polypeptides in α-helical (a), random coil (b), β-sheet (c) and β-turn (d) conformations (Chang et al., 1978). Note the resemblance between the spectrum recorded from cardiac gap junctions and that for a polypeptide in an α-helical conformation. In order to minimize light-scattering effects when recording CD spectra from membrane suspensions, isolated gap junctions were sonicated for 60 s and then sedimented to remove larger membrane sheets. To correct for any instrumental drift, multiple sample scans from the membrane sample and the dialysis buffer were recorded over several hours using an AVIV spectropolarimeter. The averaged background spectra were then subtracted from the averaged sample spectra, and a smoothing function was applied to the experimental spectrum.

ies L_1 and L_2 both bind to native and cleaved α_1 connexin, the distance between these sites allows one to estimate the size of the cytoplasmic amino-acid "loop" as ~5.1 kDa. The cytoplasmic loop is also accessible for labeling by antibodies to residues 119–142 (Beyer et al., 1989) as well as different polyclonal antibodies generated to isolated gap junctions (Yancey et al. 1989). Thus, virtually the entire polypeptide in the cytoplasmic amino-acid loop domain appears to be readily accessible. In addition, the small amino-terminal and large carboxy-terminal cytoplasmic domains do not shield the loop from the bulk aqueous environment. Even though the sequence of the cytoplasmic amino-acid loop domain is not hydrophobic, it could still function as a molecular "gate" if it sterically penetrates and blocks a hydrophilic membrane channel. Such a model has been proposed for the inactivation of *Drosophila Shaker* K^+ channels by a cytoplasmic amino-terminal domain (Hoshi et al.,, 1990; Zagotta et al., 1990). The α_1 gap junctions were isolated under conditions that would presumably switch the channels to a closed conformation (Unwin and Zampighi, 1980). If the isolated channels are indeed closed, then the accessibility of virtually the entire cytoplasmic loop of α_1 connexin to macromolecular probes argues against a model in which this domain functions as a molecular gate that enters and sterically blocks the transmembrane channel.

The hexameric structure of cardiac gap-junction ion channels supports the concept that ion channels are formed by membrane protein oligomers (Unwin, 1989), as has been found for hexameric gap junctions formed by β connexins (Unwin and Zampighi, 1980; Unwin and Ennis, 1984, Baker et al., 1985) and the acetylcholine receptor channel, which is formed by a pentamer of homologous α_2 βγδ subunits (Toyoshima and Unwin, 1988; Mitra et al., 1989). The similar quaternary design of gap junctions containing α and β connexins indicates conservation in the ular architecture of gap-junction channels. However, the lack of sequence homology and difference in size of the cytoplasmic loop and carboxy-terminal domains of α_1 and β_1 connexins suggests divergence. Such diversity may confer unique functional properties for different connexin proteins. For example, the unique structure of the cytoplasmic domains of α_1 connexin may be related to a regulatory role in mediating cardiac conduction and arrhythmias.

Acknowledgments

I gratefully acknowledge Dr. Richard Houghten for preparing the synthetic peptides and a collaboration with Dr. Norton B. Gilula in generating the peptide antibodies. M.Y. has been supported by a NIH Clinical Investigator Award, a National Center American Heart Association Grant-In-Aid, and a grant from the Gustavus and Louise Pfeiffer Research Foundation.

References

Amos, L.A., Henderson, R. and Unwin, P.N.T. (1982) Three-dimensional structure determination by electron microscopy of two-dimensional crystals. Prog. Biophys. Mol. Biol. 39, 183–231.

Baker, T.S., Sosinsky, G.E., Casper, D.L.D., Gall, C. and Goodenough, D.A. (1985) Gap junction structures VII. Analysis of connexon images obtained with cationic and anionic negative stains. J. Mol. Biol. 184, 81–98.

Barr, L., Dewey, M.M. and Berger W. (1965) Propagation of action potentials and the structure of the nexus in cardiac muscle. J. Gen. Physiol. 48, 797–823.

Beyer, E.C., Kistler, J., Paul, D.L. and Goodenough, D.A. (1989) Antisera directed against connexin43 peptides react with a 43-kD protein localized to gap junctions in myocardium and other tissues. J. Cell Biol. 108, 595–605.

Beyer, E.C., Paul, D.L. and Goodenough, D.A. (1987) Connexin 43: a protein from rat heart homologous to a gap junction protein from liver. J. Cell Biol. 105, 2621–2629.

Beyer, E.C., Paul, D.L. and Goodenough, D.A. (1990) Connexin family of gap junction proteins. J. Membr. Biol. 116, 187–194.

Chang, C.T., Wu, C.-S.C. and Yang, J.T. (1978) Circular dichroic analysis of protein conformation: inclusion of the β-turns. Anal. Biochem. 91, 13–31.

DeMello, W.C. (1982) Intercellular communication in cardiac muscle. Circ. Res. 51, 1–9.

Gimlich, R.L., Kumar, N.M. and Gilula, N.B. (1990) Differential regulation of the levels of three gap junction mRNAs in *Xenopus* embryos. J. Cell Biol. 110, 597–605.

Henderson, R., Baldwin, J.M., Downing, K.H., Lepault, J. and Zemlin, F. (1986) Structure of purple membrane from *Halobacterium halobium*: recording, measurement and evaluation of electron micrographs at 3.5 Å resolution. Ultramicroscopy 19, 147–178.

Hoshi, T., Zagotta, W.N. and Aldrich, R.W. (1990) Biophysical and molecular mechanisms of *Shaker* potassium channel inactivation. Science 250, 533–538.

Houghten, R.A. (1985) General method for the rapid solid-phase synthesis of large numbers of peptides: specificity of antigen-antibody interaction at the level of individual amino acids. Proc. Natl. Acad. Sci. USA 82, 5131–5135.

Johnson, W.C. (1988) Secondary structure of proteins through circular dichroism spectroscopy. Annu. Rev. Biophys. Biophys. Chem. 17, 145–166.

Josephson, M.E. and Seides, S.F. (1979) Clinical Cardiac Electrophysiology. Lea and Febiger, Philadelphia.

Kumar, N.M. (1991) Gap junctions: a multigene family. Adv. Struct. Biol. 1, 209–247.

Kumar, N.M. and Gilula, N.B. (1986) Cloning and characterization of human and rat liver cDNAs coding for a gap junction protein. J. Cell Biol. 103, 767–776.

Laird, D.W. and Revel, J.-P. (1990) Biochemical and immunochemical analysis of the arrangement of connexin43 in rat heart gap junction membranes. J. Cell Sci. 97, 109–117 (1990).

Loewenstein, W.R. (1981) Junctional intercellular communication: the cell-to-cell membrane channel. Physiol. Rev. 61, 829–913.

Manjunath, C.K., Goings, G.E. and Page, E. (1985) Proteolysis of cardiac gap junctions during their isolation from rat hearts. J. Membr. Biol. 85, 159–168.

Manjunath, C.K., Nicholson, B.J., Teplow, D., Hood, L., Page, E. and Revel J.-P. (1987) The cardiac gap junction protein (M_r 47,000) has a tissue-specific cytoplasmic domain of M_r 17,000 at its carboxy-terminus. Biochem. Biophys. Res. Commun. 142, 228–234.

Manjunath, C.K. and Page, E. (1986) Rat heart gap junctions as disulfide-bonded connexon multimers: their depolymerization and solubilization in deoxycholate. J. Membr. Biol. 90, 43–57.

McNutt, N.S. and Weinstein, R.S. (1970) The ultrastructure of the nexus — a correlated thin-section and freeze-cleavage study. J. Cell Biol. 47, 666—688.

Milks, L.C., Kumar, N.M., Houghton, R., Unwin, N. and Gilula, N. B. (1988) Topology of the 32-kD liver gap junction protein determined by site-directed antibody localizations. EMBO J. 7, 2967–2975.

Mitra, A.K., McCarthy, M.P. and Stroud, R.M. (1989) Three-dimensional structure of the nicotinic acetylcholine receptor and location of the major associated 43-kD cytoskeletal protein, determined at 22 Å by low dose electron microscopy and x-ray diffraction to 12.5 Å. J. Cell Biol. 109, 755–774.

Paul, D.L. (1986) Molecular cloning of cDNA for rat liver gap junction protein. J. Cell Biol. 103, 123–134.

Risek, B., Guthrie, S., Kumar, N. and Gilula, N.B. (1990) Modulation of gap junction transcript and protein expression during pregnancy in the rat. J. Cell Biol. 110, 269–282.

Spach, M.S. (1983) The role of cell-to-cell coupling in cardiac conduction disturbances. Adv. Exp. Med. Biol. 161, 61–77.

Spray, D.C. and Burt, J.M. (1990) Structure-activity relations of the cardiac gap junction channel. Am. J. Physiol. 258, C195–205.

Tibbits, T.T., Caspar, D.L.D., Philips, W.C. and Goodenough, D. A. (1990) Diffraction diagnosis of protein folding in gap junction connexons. Biophys. J. 57, 1025–1036.

Toyoshima, C. and Unwin, N. (1988) Ion channel of acetylcholine receptor reconstructed from images of postsynaptic membranes. Nature 336, 247–250.

Unwin, N. (1989) The structure of ion channels in membranes of excitable cells. Neuron 3, 665–676.

Unwin, P.N.T. and Ennis, P.D. (1984) Two configurations of a channel-forming membrane protein. Nature 307, 609–613.

Unwin, P.N.T. and Zampighi, G. (1980) Structure of the junction between communicating cells. Nature 283, 545–549.

Yancey, S.B. (I), John, S.A. (II), Lal, R. (III), Austin, B.J. and Revel, J.-P. (1989) The 43-kD polypeptide of heart gap junctions: immunolocalization (I), topology (II), and functional domains (III). J. Cell Biol. 108, 2241–2254.

Yeager, M. (1987) Cardiac gap junctions: crystallization and electron microscopic image analysis. J. Mol. Cell. Cardiol. 19, S.54.

Yeager, M. (1989) Hexameric structure of an ion channel in the heart: the cardiac gap junction. Circulation 80, II-399.

Yeager, M. and Gilula, N.B. (1990) Membrane topology and quaternary structure of the cardiac gap junction ion channel. J. Am. Coll. Cardiol. 15, 2A.

Yeager, M. and Gilula, N.B. (1992) Membrane topology and quaternary structure of cardiac gap junction ion channels. J. Mol. Biol. 223, 929–948.

Zagotta, W.N., Hoshi, T. and Aldrich, R.W. (1990) Restoration of inactivation in mutants of *Shaker* potassium channels by a peptide derived from ShB. Science 250, 568–571.

Zervos, A.S., Hope, J. and Evans, W.H. (1985) Preparation of a gap junction fraction from uteri of pregnant rats: the 28-kD polypeptides of uterus, liver and heart gap junctions are homologous. J. Cell Biol. 101, 1363–1370.

Zhang, J.-T. and Nicholson, B.J. (1989) Sequence and tissue distribution of a second protein of hepatic gap junctions, Cx26, as deduced from its cDNA. J. Cell Biol. 109, 3391–3401.

J.E. Hall, G.A. Zampighi and R.M. Davis (Eds.)
Progress in Cell Research, Vol. 3

CHAPTER 8

Biochemistry of gap-junction channels

KATHRIN A. STAUFFER[a], NALIN M. KUMAR[b], NORTON B. GILULA[b] and NIGEL UNWIN[a,b]

[a]*Medical Research Council, Laboratory of Molecular Biology, Hills Road, Cambridge, UK and* [b]*Department of Cell Biology, The Scripps Research Institute, 10666 N. Torrey Pines Road, La Jolla, CA, USA*

Gap junctions are known to be formed by proteins called connexins. Several connexins have been characterized which are closely related to each other, and this finding has led to the concept of a gene family of connexins. The connexin family is rapidly expanding, with new gene products being identified all the time. Compared to the speed at which our knowledge of amino-acid sequences increases, progress in the field of connexon structure and biochemistry has been very slow. One reason for this lag is the difficulty of obtaining large amounts of protein, for although connexins are ubiquitous, they are hardly a major component in any known tissue. It was therefore a big step forward when a baculovirus vector was constructed and isolated which contained a cDNA for the human β_1 gap-junction protein (Cx32). We can now express β_1 connexin in insect cells which overproduce it to the level of several milligrams per liter of cultured cells (Stauffer et al., 1991). The isolation of gap-junction plaques proceeds by the simple technique of extracting broken cells with sodium hydroxide, and fractionating the resulting insoluble material on a sucrose step gradient, a procedure very similar to that commonly used for the isolation of rat liver gap junctions (Hertzberg, 1984). In this way, we can typically obtain 2–3 mg of connexin of about 80% purity in half a day's work. Negatively stained preparations of these plaques look indistinguishable from rat liver gap junctions, and in SDS-PAGE the only major difference seems to be the absence of the β_2, or Cx26, band in the recombinant material.

For many biochemical experiments, it is necessary to solubilize membrane proteins with detergents and to remove the native lipids, without denaturing the proteins. In the case of connexin, this means developing a set of conditions which do not disrupt the oligomeric structure of the connexon, because the structure to be investigated is that of the oligomeric gap-junctional channel and not that of a connexin monomer. We tried a wide range of different detergents at various pH values and salt concentrations, assessing the solubilization by electron microscopy; we had previously found that successful solubilization conditions produced single connexons in negative stain (see Fig. 1). Using this assay, we established that optimal conditions for the solubilization of gap junctions at protein concentrations above about 0.1 mg/ml are 1.5–2 M NaCl, pH 10, 10 mM EDTA, at least 50 mM DTT, and 5% dodecyl maltoside. Once the protein was solubilized, the pH as well as the concentrations of salt, reducing agent and detergent could be lowered considerably.

We have been able to purify connexons by hydroxylapatite chromatography followed by anion exchange and gel filtration on Superose 6. The work was done using recombinant-derived β_1 connexin; as a control we have performed the same experiments on isolated rat liver gap junctions and found the results to be the same, except for the contamination of the β_1 connexin with β_2 connexin which is not removed by this procedure.

The final product of this purification procedure was again examined by electron microscopy. It is a homogeneous population of particles of about 80 Å diameter. An accumulation of stain in the centre of each particle suggests the position of the channel. Apparently, the connexons interact with the carbon support film rather specifically, and therefore orient uniformly with the channel axis perpendicular to the carbon film, as in Fig. 1, left panel. Occasionally, populations of connexons are found which are deposited on the film in a more random fashion, as shown in Fig. 1 on the right. These views allowed us to confirm that the species purified is indeed a single connexon, or hemichannel, rather than the whole cell-to-cell channel. They also show that the accumulation of stain is confined to one end of the channel, which has been shown earlier to be the extracellular side (Unwin and Zampighi, 1980). Some images were taken of connexons embedded in amorphous ice, and these allowed us to compute rotational power spectra which confirm that the particle has sixfold symmetry.

Initial attempts to grow crystals of β_1 connexin have produced regular aggregates: at first in one dimension to form filaments (Fig. 2, left) and, under favorable conditions, in two dimensions to form sheets (Fig. 2, right). The filaments apparently form by association of the hydrophilic domains of the channels. Extracellular domains interact with each other, presumably to form a physiological "gap" junction, and the cytoplasmic domains bind to each

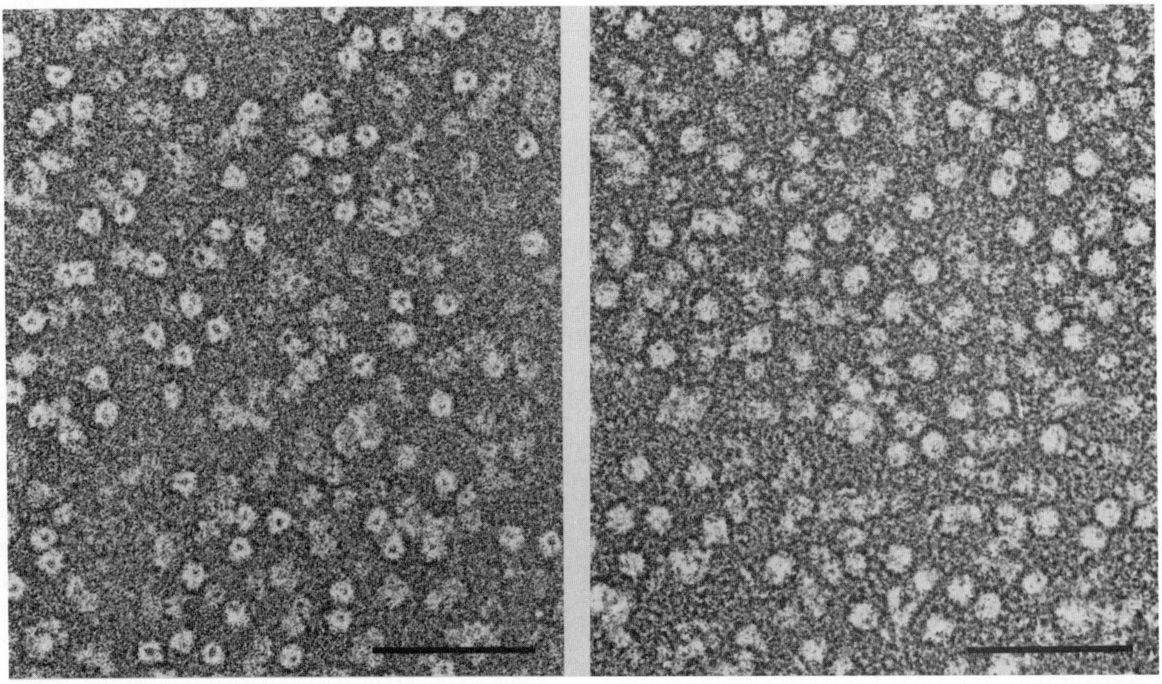

Figure 1. Electron micrographs of solubilized and purified connexons in negative stain. The left panel shows connexons oriented with the channel axis roughly perpendicular to the support film, resulting in a population of uniform particles of about 80 Å diameter with an accumulation of stain in the centre. On the right panel, the orientation is more random, showing tilted as well as some side views of connexons, and thus illustrates that the isolated particles are single connexons or hemichannels. The scale bars represent 50 nm.

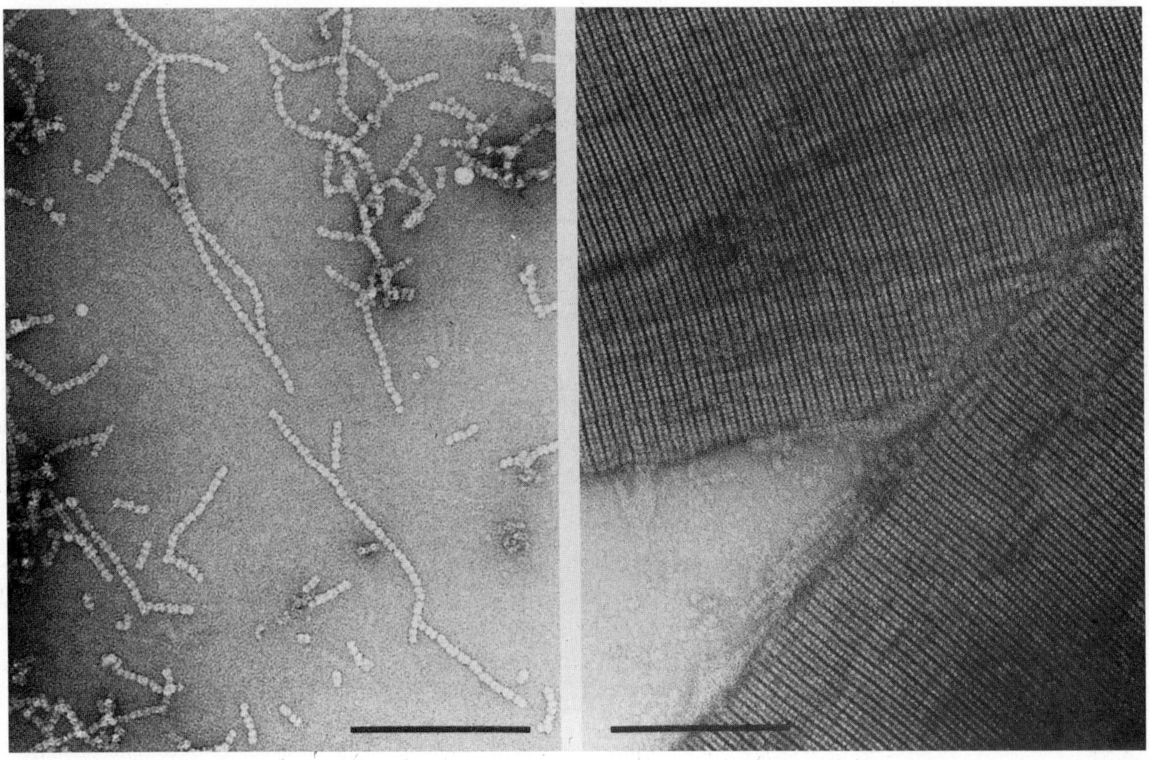

Figure 2. Aggregated connexons in negative stain. On the left, the connexons have formed filaments by interaction of their hydrophilic domains. The axis of the channel runs along these filaments. As shown on the right, filaments can associate sideways to form sheets. The scale bars represent 0.2 mm.

other, resulting in a filament of connexons in alternating orientations which therefore has no overall polarity. The filament width is about 80 Å, a value which agrees with the diameter of a connexon. Sheets are formed by sideways interaction of filaments. The diffraction pattern of such a sheet shows that the order in the direction perpendicular to the filament axis (in the plane of the membrane) is weaker than the order along the filament axis (or perpendicular to the plane of the membrane). This supports the assumption that sideways interactions between connexons are mainly hydrophobic in nature, and therefore not very specific. At present, we are trying to increase our understanding of the biochemistry of gap-junction channels in the hope that such work will help to eventually solve their structure.

References

Hertzberg, E.L. (1984) A detergent-independent procedure for the isolation of gap junctions from rat liver. J. Biol. Chem. 159, 9936–9943.

Stauffer, K.A., Kumar, N.M., Gilula, N.B. and Unwin, N. (1991) Isolation and purification of gap-junction channels. J. Cell Biol. 115, 141–150.

Unwin, P.N.T. and Zampighi, G. (1980) Structure of the junction between communicating cells. Nature 283, 545–550.

J.E. Hall, G.A. Zampighi and R.M. Davis (Eds.)
Progress in Cell Research, Vol. 3

CHAPTER 9

In-vitro assembly of lens gap junctions

ANDREAS ENGEL[a], PAUL D. LAMPE[b] and JOERG KISTLER[c]

[a]*M.E. Müller Institute for High-Resolution Electron Microscopy at the Biocenter, University of Basel, CH-4056 Basel, Switzerland,* [b]*Department of Genetics and Cell Biology, University of Minnesota, St. Paul, MN 55108, USA and* [c]*Department of Cellular and Molecular Biology, Centre of Gene Technology, University of Auckland, New Zealand*

Summary

Gap junction-like structures were assembled in vitro from octyl-β-D-glucopyranoside solubilized components of lens fiber cell membranes. Individual connexons, short double-membrane structures, and amorphous material were found in the solubilized mixture by negative-stain electron microscopy. During the removal of the detergent by dialysis, the connexons associated to form single- and double-layered, two-dimensional hexagonal arrays (unit cell size a = b = 8.5 nm). The size of these arrays was dependent on the lipid-to-protein ratio and the presence of Mg^{2+} ions. Electron micrographs and image processing revealed that lens connexons consisted of six subunits surrounding a stain-filled channel. Upon further detergent treatment, in-vitro-assembled gap junctions were insoluble and formed 3-dimensional stacks while other components were solubilized. SDS-PAGE, immunocytochemistry and mass data from scanning transmission electron microscopy strongly suggest that a 38-kDa polypeptide, which is a processed form of the lens specific gap-junction protein MP70, is a major component of the arrays. The in-vitro assembly of gap junctions opens new avenues for the structural analysis of gap junctions and for the study of the intermolecular interactions between connexons during junctional assembly.

Introduction

Many different membrane proteins have been reconstituted into two-dimensional (2-D) crystals in the presence of phospholipids. These include cytochrome reductase (Wingfield et al., 1979; Hovmöller et al., 1983), complex I from mitochondria (Weiss and Leonard, 1991) and bacterial porins (Dorset et al., 1983; Jap, 1988), light-harvesting complex (Li, 1985), the photosystem I reaction center (Ford et al., 1990) and the reaction center of *Chloroflexus aurantiacus* (Barnakov et al., 1990). In some cases, endogenous lipids did not dissociate from the membrane protein during purification, allowing 2-D crystals to be grown without addition of exogenous lipids, e.g. light-harvesting complex II (Kühlbrandt, 1988), reaction center from *Rhodopseudomonas viridis* (Miller and Jacob, 1983), the photosystem II reaction center (Dekker et al., 1990) and Ca^{2+}-ATPase (Stokes and Green, 1990).

In spite of this, general rules for such crystallization experiments are not available, because processes involved in the assembly of regular 2-D membrane protein arrays are not understood in depth. The assembly of 2-D crystals from solubilized membrane proteins in the presence of phospholipids is entropy driven (Israelachvili et al., 1980; Sharp, 1991), and depends critically on protein-detergent, lipid-detergent, lipid-protein, and protein-protein interactions. The formation of regular structures in vivo is of acute biological interest. We are only beginning to gain some insight into the processes which occur when two cells initiate gap-junction assembly (Johnson et al., 1989). In-vitro assembly could be extremely useful for dissecting this process.

In this paper we report on the in-vitro assembly of gap-junction structures from detergent-solubilized lens membrane proteins. Solubilization of ocular-lens fiber-cell membrane preparations produced, among other structures, connexon-like, hexameric channel complexes detectable by negative-staining electron microscopy. By removing the detergent in the presence of exogenous lipid and Mg^{2+} ions, these connexons could be induced to assemble into 2-D crystalline, junctional arrays. Studies on such samples revealed structures showing a hexameric projection similar to structures of gap junctions isolated from liver and heart (Unwin and Zampighi, 1980; Yeager and Gilula, 1992). The in-vitro-assembled lens gap-junction arrays were induced to stack in a regular fashion by a further solubilization step, effectively forming type I crystals (Deisenhofer and Michel, 1989). SDS-PAGE, cytochemistry, and mass measurements with the scanning transmission electron microscope (STEM) strongly suggested that the main protein component of these arrays was MP38, an in-vivo-processed form of the gap-junction protein MP70 (Gruijters et al., 1987; Kistler et al., 1988; Kistler et al., 1990a). The present study, in combination with an earlier report (Kistler et al., 1990b), shows that the fundamental features of the lens fiber connexon structures are similar to those in hepatocytes or myocardial cells.

Material and methods

Preparation of lens fiber cell membranes; solubilization and reconstitution

Lens fiber cell plasma membranes were isolated from bovine or sheep lenses as described (Goodenough, 1979; Kistler et al., 1990a). Membranes were treated with V8 protease to cleave MP70 to MP38, and subsequently solubilized in 10 mM HEPES, pH 7.0, 20 mM $MgCl_2$, 0.005% NaN_3, 2% octyl-β-D-glucopyranoside for 15 min at room temperature. Insoluble material was removed by centrifugation. The supernatant was dialyzed against buffer without the detergent at room temperature or at 37°C. In some experiments, phospholipase A_2 was added to remove excess lipids, or, alternatively, lipids (dimyristoyl phosphatidylcholine) were added. In-vitro-assembled junctional structures were further enriched by mixing 100λ aliquots of reconstituted membranes with equal volumes of buffer containing 4% octyl-β-D-gluco-

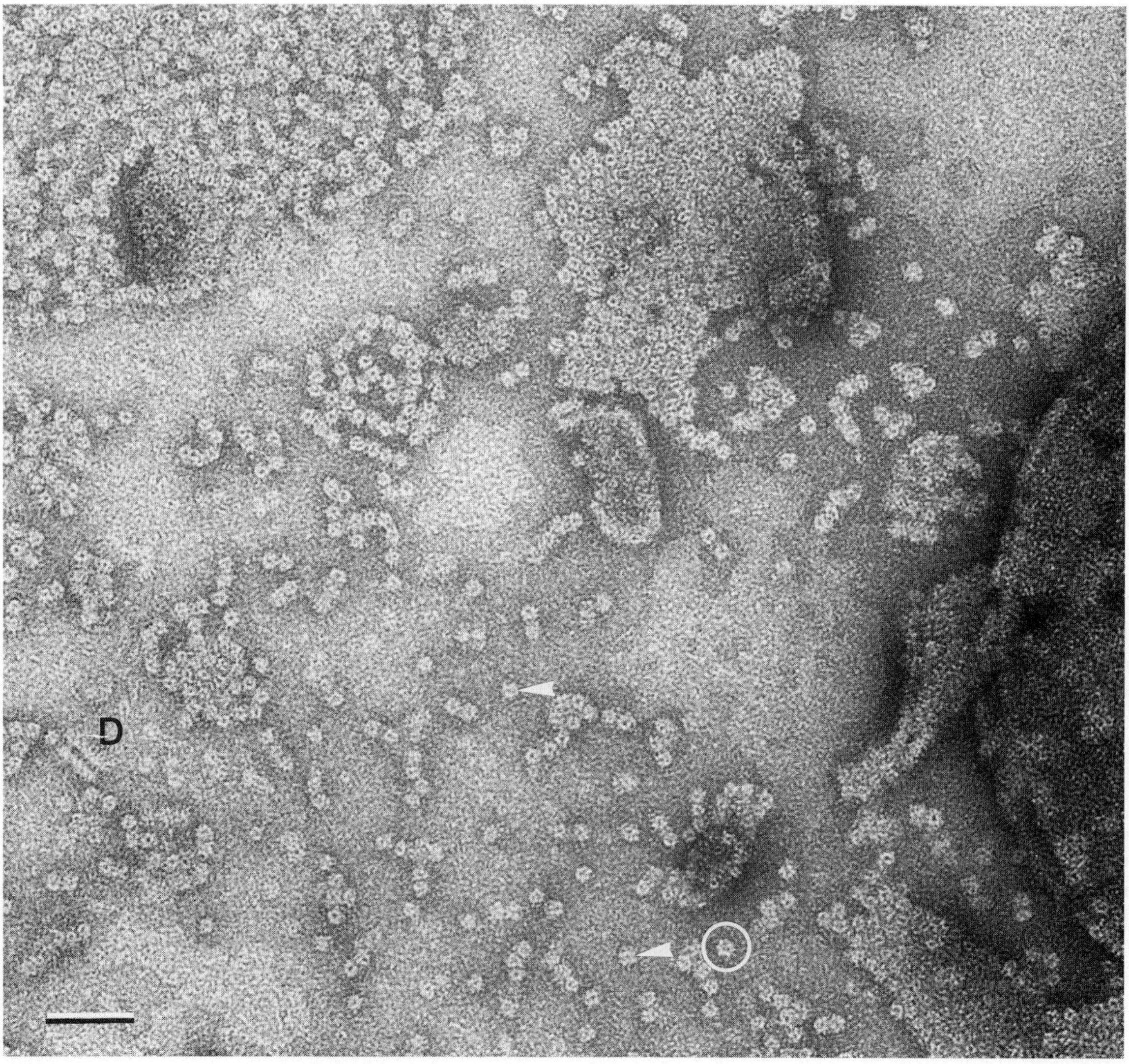

Figure 1. Individual and assembling connexon structures after partial dialysis of a crude lens fiber cell membrane extract. Short double-membrane structures are marked by D. Arrows point to connexons viewed side-on that exhibit a distinct central line. Circle encloses a connexon head-on view where the six-fold symmetry and a stain-filled central pit are evident. Bar, 50 nm.

pyranoside. The insoluble material containing the junctional arrays was separated from solubilized protein by centrifugation and analyzed by negative staining and SDS-PAGE.

Electron microscopy, digital image processing and STEM

In-vitro-assembled material was adsorbed to electron microscope grids covered with carbon-coated Formvar films made hydrophilic by glow discharge, negatively stained with 0.75% uranyl formate (pH 4.2) and air dried. Images were recorded on Kodak SO-163 film at a nominal magnification of 50,000× using an Hitachi H-7000 transmission electron microscope operated at 100 kV. Micrographs were evaluated for focus, astigmatism and drift, as well as for crystallinity of the assembled membranes by optical diffraction. Suitable areas were digitized using an Eikonix 850 CCD camera and stored as single pixels separated by 0.63 nm.

Averaged projections of gap junctions with ordered connexon arrays were calculated taking advantage of the SEMPER image-processing system (Saxton et al., 1979). Correlation averaging (Saxton and Baumeister, 1982) allowed the elimination of residual lattice disorder by locating the unit cells precisely, using a reference that included approximately 2 connexons. After least-squares fitting of a hexagonal lattice, unit cells far from the ideal lattice point were eliminated. A single unit cell was finally obtained from the correlation average calculated in this manner, its phase origin determined and a sixfold symmetrization applied.

For mass measurements with the STEM, membranes were adsorbed to glow-discharged grids covered with a thin carbon film over a fenestrated plastic film. The grids were then washed extensively in double-distilled water, blotted, rapidly frozen and freeze dried within the STEM (Engel and Meyer, 1980). Tobacco mosaic virus particles served as an internal mass standard. Elastic annular dark-field images were acquired in digital format using an HB-5 STEM operated at 80 kV at either 100,000× (pixel size

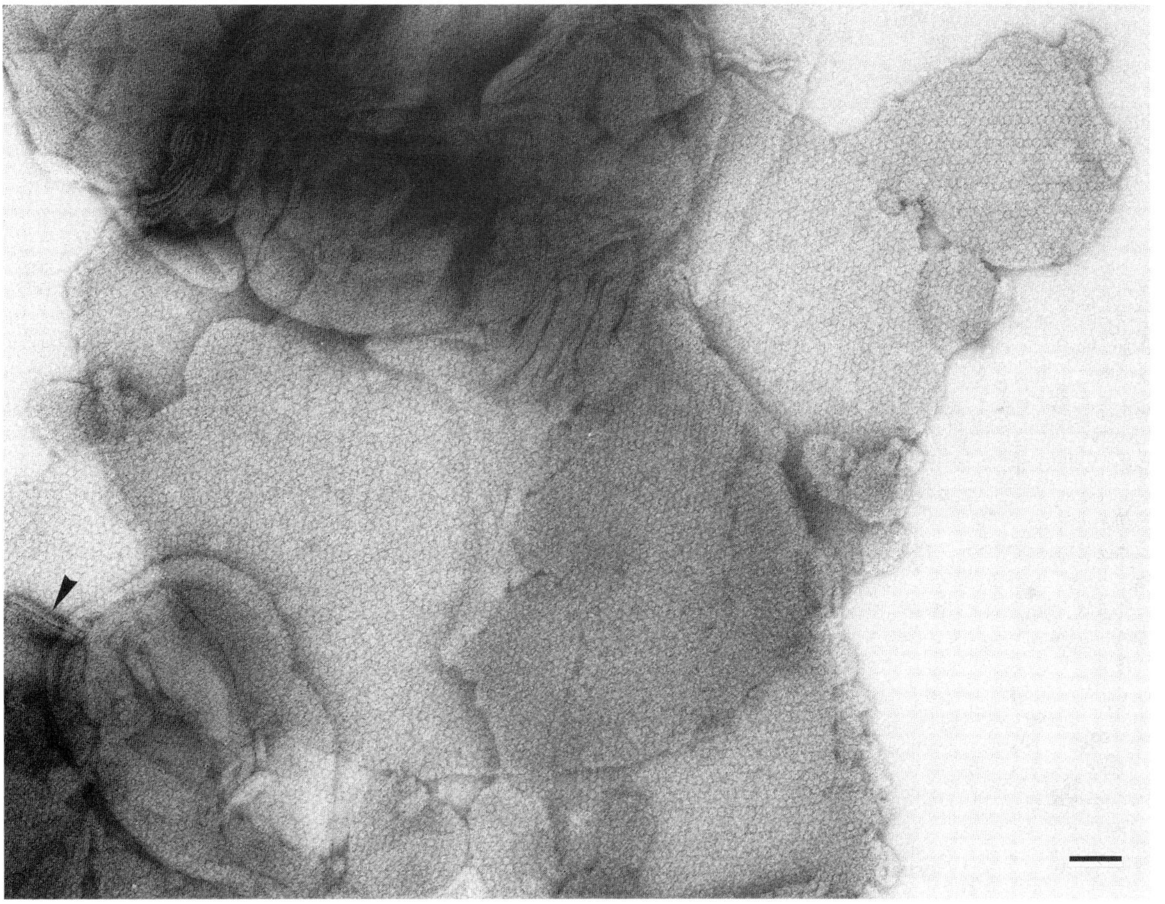

Figure 2. Large hexagonal arrays formed after extensive dialysis of a crude lens fiber cell membrane extract. Arrow points to the double membrane of a rolled-up edge. Bar, 50 nm.

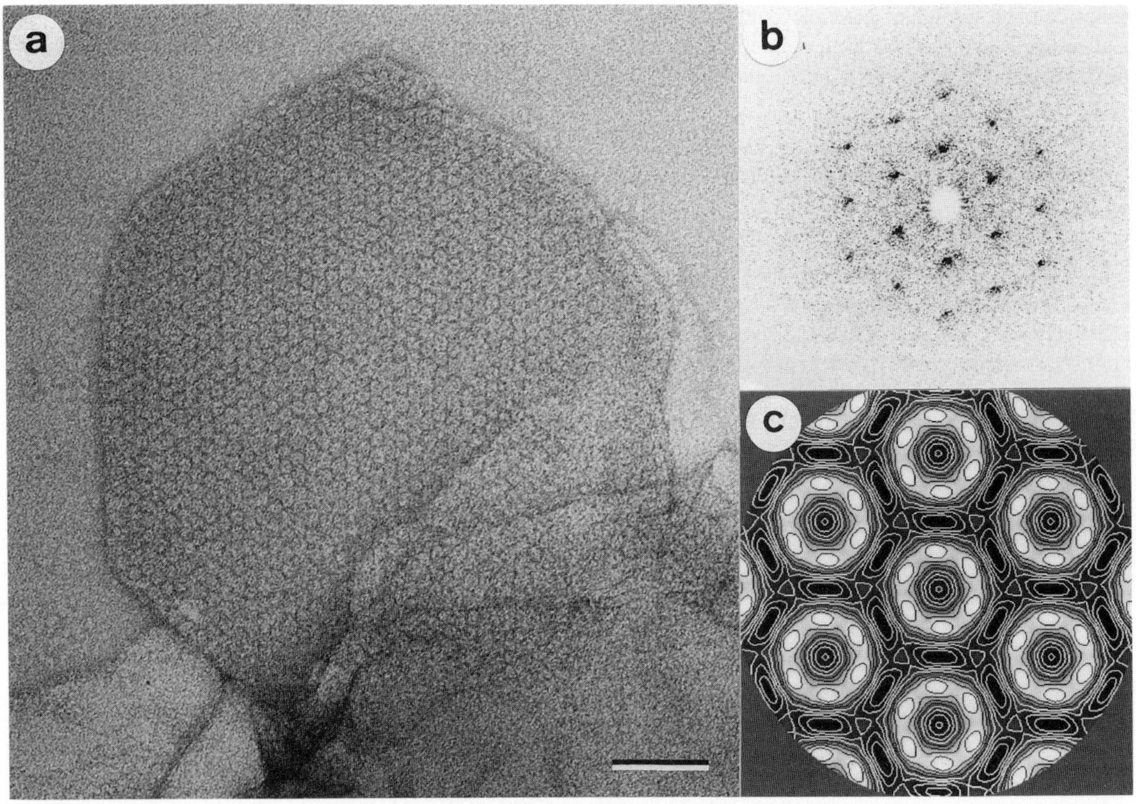

Figure 3. Ordered hexagonal arrays of connexons. (a) An ordered gap-junction plaque which contains approximately 700 channels. Bar, 50 nm. (b) Small displacements of the channels from their ideal lattice position limit the resolution, as documented by the lack of high-order spots in the optical diffraction pattern. (c) The characteristic connexon structure is revealed after correlation averaging (see text) and sixfold symmetrization.

1.87 nm) or 200,000× (pixel size 0.93 nm) nominal magnification (Engel, 1978). Recording doses varied between 100 and 800 e^{-}/nm^{2}. Processing of digital STEM images for mass determination was carried out as described (Engel and Reichelt, 1988). As the hexagonal protein lattices themselves were not visualized under these conditions, membranes selected for mass measurements had indicators such as straight borders, corners with a characteristic 120° angle, or straight cracks. Mass data were compiled in histograms to which Gaussian profiles could be fitted by a Marquardt algorithm (Bevington, 1969).

Results

Lens connexons assemble into gap junctions in vitro

A heterogeneous mixture of proteins and lipids obtained by solubilization of junctional and non-junctional lens fiber membranes was used for the in-vitro assembly of gap junctions. By negative-stain electron microscopy this material showed a variety of globular particles and confirmed the absence of gap-junction arrays. After partial removal of the detergent by dialysis, individual connexon-like particles (Stauffer et al., 1991) and short double-membrane structures (Kistler and Bullivant, 1988) were observed (Fig. 1). Abundant connexon particles were about 9 nm in diameter and had a central pore, seen as a stain-filled pit in head-on views or a central line in side-on views. Connexons associated with each other to form lacy structures or small aggregates, reminiscent of gap junctions observed in the membranes of developing lens fiber cells (Benedetti et al., 1976). After extensive dialysis, hexagonal connexon arrays formed that were as large as 1 μm in diameter and had unit cell dimensions a = b = 8.5 nm (Fig. 2). Well-ordered arrays were often limited to smaller domains within the assembled structures. Many arrays appeared to be single-layered, although definitive determination was not possible with negatively stained preparations. Others were clearly double-layered, as demonstrated by rolled-up edges (arrow in Fig. 2). These edge-on views showed a double-membrane profile 15 nm wide and revealed regularly spaced, stain-excluding particles, as previously described for edge-on views of isolated liver gap junctions (Sosinsky et al., 1988).

The lens connexon structure

Some double-layered hexagonal lattices appeared to be rather well ordered (Fig. 3a), but generally, they did not diffract to better than 3.7 nm (Fig. 3b). This is a result of small displacements of the connexons from their ideal lattice position, as can be readily seen by viewing the array at a glancing angle along a lattice line. Unit cell displacements can be detected and corrected using correlation-averaging procedures (Saxton and Baumeister, 1982). The averaged connexon structure showed a prominent sixfold rotational symmetry which was imposed on the final map (Fig. 3c). The stain-filled pore had a diameter of 2.4 nm and was surrounded by six stain-excluding subunits forming a ring with a diameter of 6.6 nm. At this level of resolution, in-vitro-assembled lens gap junctions have a structure which is virtually indistinguishable from that of gap junctions isolated from liver (Unwin and Zampighi, 1980; Baker et al., 1983; Unwin and Ennis, 1984; Sikerwar and Unwin, 1988) or heart (Yeager and Gilula, 1992).

In-vitro-assembled lens gap junctions contain MP38 as major protein

Major proteins in the reconstituted material were MP18, MP38, and MIP26 and its cleavage products (Fig. 4b). MP18 had previously been identified both as non-junctional (Voorter et al., 1989) and junctional (Louis et al., 1989). MP38 is the cleavage product of the lens-fiber-specific gap-junction protein MP70 (Kistler and Bullivant, 1987). Immunogold labeling indicated that MP18 and MIP26 were components of the amorphous vesicle fraction and not of the in-vitro-assembled junctional structures (data not shown). No antibodies for MP38 are currently available for immunolabeling. Hence, identification of the major protein(s) of in-vitro-assembled lens gap junction was particularly difficult. Since liver gap junctions are resistant to mild detergent treatment, reconstituted material was resolubilized with 2% octyl-β-D-glucopyranoside and separated by centrifugation into an insoluble and a soluble fraction. MP38 and MP18 were enriched in the insoluble fraction (Fig. 4c). MIP26 and its cleavage products partitioned predominantly into the soluble phase, consistent with the localization in the non-junctional membranes (Fig. 4d). This treatment with octyl-β-D-glucopyranoside also induced stacking of the in-vitro-assembled lens gap junctions (Fig. 5). Side views of stacked junctions revealed alternating continuous and regularly interrupted layers of stain (Fig. 5a,b), similar to edge-on views of in-vitro-assembled lens gap junctions (Fig. 2) and gap junctions isolated from liver (Sosinsky et al., 1988). Interrupted stain layers are likely to represent the extracellular "gap" between reconstituted junctional membranes. Furthermore, diffraction analysis of side views of stacks revealed some crystallinity in a direction perpendicular to the plane of the membranes (Fig. 5b) which indicates that these stacks are type I crystals of the lens connexons (Deisenhofer and Michel, 1989).

STEM mass measurements

STEM elastic dark-field images of unstained proteins provide an unique approach for determining the mass of structures identified in the micrograph based on their electron-scattering power. To prevent electron beam-induced mass loss, these images must be recorded at low doses (Engel, 1978), resulting in a significant statistical noise. Hence, in-vitro-assembled lens gap junctions could not be distinguished on-line from amorphous vesicles in freeze-dried preparations. Straight edges, edges subtending an angle of 120°, and straight cracks, however, helped to identify regular arrays (Fig. 6a).

The periodic pattern of the hexagonally packed connexons became visible at higher magnification when the electron quantum noise was eliminated by low-pass filtering (Fig. 6b). Sometimes hexagonal patches similar to that shown in Fig. 3 could be recognized (Fig. 6, insert). Mass-per-area (MPA) analysis (Engel and Reichelt, 1988) on clean homogeneous membranes yield-

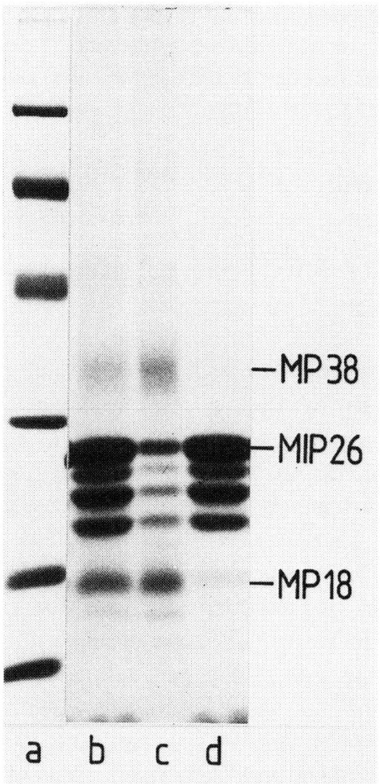

Figure 4. SDS-PAGE (10% acrylamide) of proteins in reconstituted preparations. (a) Molecular weight markers from the top: 97 kDa, 66 kDa, 45 kDa, 31 kDa, 21 kDa, and 14 kDa. (b) Untreated, reconstituted membranes. (c) Insoluble fraction after octyl-β-D-glucopyranoside treatment of reconstituted membranes was enriched in MP38 and MP18. (d) Octyl-β-D-glucopyranoside soluble fraction with enrichment for MIP26.

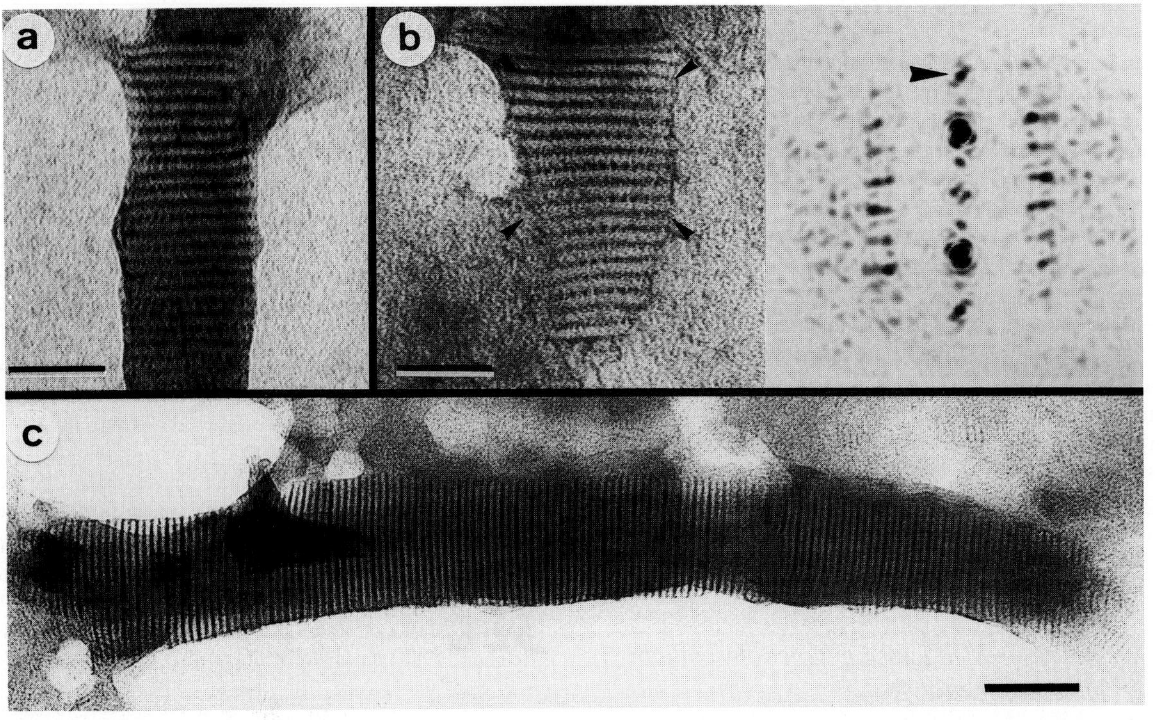

Figure 5. In vitro-assembled gap junctions stack when treated with octyl-β-D-glucopyranoside. (a) Edge-on view of stacked junctions that form a 3-D crystal reveals an alternating staining pattern between layers. Bar, 50 nm. (b) As documented by the diffraction pattern (right), the stacking of gap junctions is coordinated in axial and lateral directions, thus yielding type I crystals (see text). The stacking order is indicated by (0,4) diffraction spots at $(4\ nm)^{-1}$ (arrow in diffraction pattern). Four arrows mark the diffracted frame. Bar, 50 nm. (c) Large stack containing approximately 70 gap junctions. Bar, 100 nm.

ed a complex MPA histogram exhibiting five narrow peaks and one rather wide peak (Fig. 6d) at positions 2.7 ± 0.2, 4.3 ± 0.3, 5.3 ± 0.4, 7.1 ± 1.0, 7.9 ± 0.5, and 9.3 ± 0.4 kDa/nm^2, respectively. As one phospholipid molecule (600–700 Da) occupies approximately 0.5 nm^2, the MPA of a lipid bilayer is close to 2.6 kDa/nm^2.

Therefore, the first peak of the histogram was from single "empty" membranes and peaks at 5.3 kDa/nm^2 and close to 7.8 kDa/nm^2 were likely to originate from double and triple bilayers. The second (4.3 ± 0.3 kDa/nm^2) and the sixth peak (9.3 ± 0.4 kDa/nm^2) were both major peaks and related to hexagonal arrays of connexons. Assuming 8.5 nm for the side of the unit cell and 6.6 nm for the connexon diameter, the lipid moiety was found to cover 45% of the surface area and to contribute 75 kDa per unit cell in a single-layered hexagonal array of connexons. Hence, the mass of one subunit (1/6 of the total protein mass per unit cell) was calculated to be 32 ± 3 kDa for single-layered arrays (4.3 kDa/nm^2) and 36 ± 3 kDa for the double layered gap junctions (9.3 kDa/nm^2). Because the connexon packing arrangement is better preserved in double layers than single layers, the 36 ± 3 kDa mass per subunit is likely to be the more realistic value. Thus, STEM mass data corroborate the finding that in-vitro-assembled lens gap junctions are composed of MP38.

Conclusions

We have demonstrated for the first time that regular connexon arrays and double-membrane gap junction-like structures can be assembled in vitro . This is achieved by removal of the detergent from a crude extract of lens fiber cell membranes containing solubilized integral membrane proteins, protein complexes and lipids. The presence of solubilized complexes as connexons or connexon pairs may affect the assembly of single- or double-layered junctional arrays. Some parameters appeared to be critical for effective reconstitution of gap junction-like structures, and need to be investigated in more detail, e.g., the use of octyl-β-D-glucopyranoside was limited, as prolonged solubilization times (less than 1 h) apparently abolished the ability to form arrays. The best results were obtained in the presence of some exogenous lipids; only small double-membrane structures were assembled in the presence of phospholipase A_2 or when Mg^{2+} was absent.

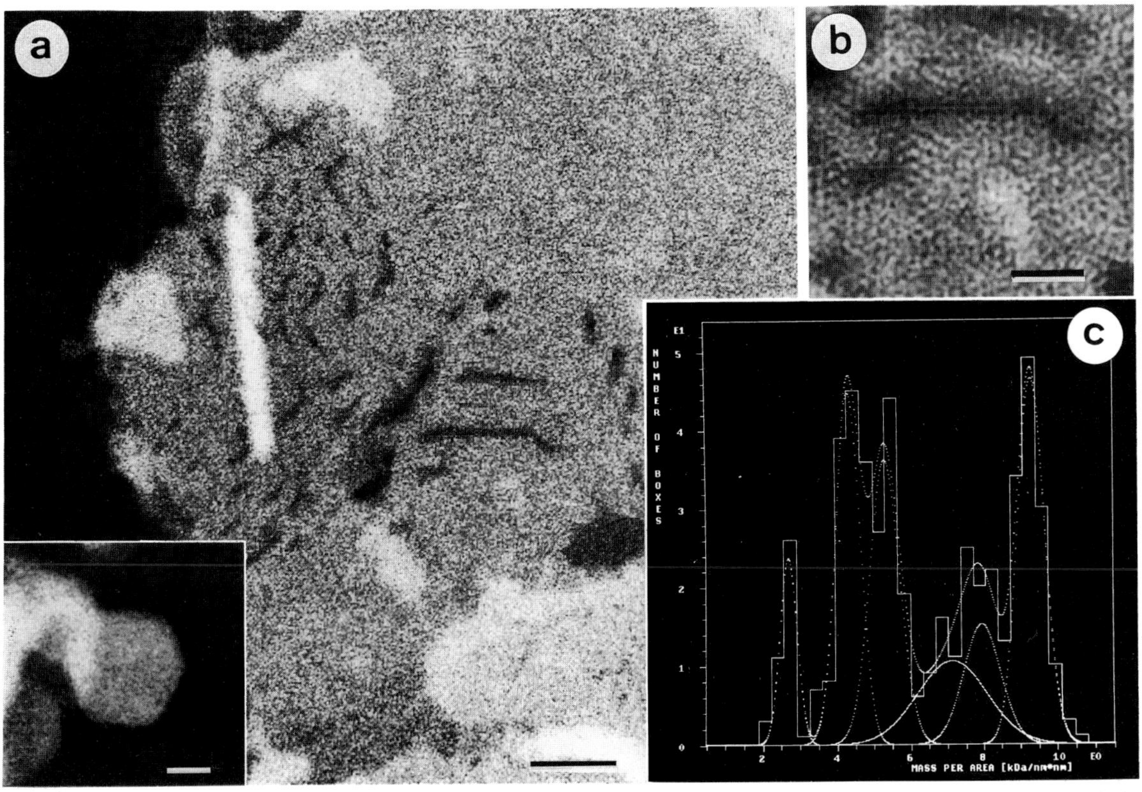

Figure 6. STEM analysis of membrane arrays. (a) STEM image of membrane structure exhibiting straight cracks, sometimes subtending an angle of 120°. Bar, 100 nm. (b) Periodic pattern of hexagonally packed connexons is discernible at higher magnification after elimination of the quantum noise by low-pass filtering. Bar, 50 nm. (c) Hexagonal patch with 120° angles. Bar, 100 nm. (d) Histogram illustrating the results of mass-per-area analysis. The dotted lines represent the six Gaussian profiles fitted to the data (see text).

This simple in-vitro assembly system for lens gap junctions opens a new way for the effective isolation of MP38 and for the study of the molecular interactions which drive gap-junction assembly.

Acknowledgements

We are grateful to K. Stauffer for providing a preprint of her work on recombinant Cx32, to A. Hefti, S. Müller, and J. Bond for excellent collaboration, and to R.G. Johnson for his enthusiastic support. Also we would like to thank Ms. H. Frefel and Ms. M. Zoller for their expert photographic work. This project was supported by grants from the National Institutes of Health, GM37230 and CA28548 (to P.D.L. and R.G.J.), from the Medical Research Council of New Zealand (to J.K.), from the Swiss National Foundation for Scientific Research, 31-25684.88 (to A.E.), the M.E. Müller-Foundation of Switzerland, and the Department of Education of the Kanton Basel-Stadt.

References

Baker, T.S., Caspar, D.L.D., Hollingshead, C.J. and Goodenough, D.A. (1983) Gap-junction structures. IV. Asymmetric features revealed by low-irradiation microscopy. J. Cell Biol. 96, 204–216.

Barnakov, A.N., Demin, V.V., Kuzin, A.P., Zargarov, A.A., Zolotarev, A.S. and Abdulaev, N.G. (1990) Two-dimensional crystallization of reaction centers from *Chloroflexus aurantiacus*. FEBS Lett. 265, 126–128.

Benedetti, E.C., Dunia, I., Bentzel, C.J., Vermorken, A.J.M., Kibbelaar, M. and Bloemendal, H. (1976) A portrait of plasma membrane specializations in eye lens epithelium and fibers. Biochim. Biophys. Acta. 457, 353–384.

Bevington, P.R. (1969) Data Reduction and Error Analysis for the Physical Sciences. McGraw-Hill, New York.

Deisenhofer, M. and Michel, H. (1989) The photosynthetic reaction center from the purple bacterium *Rhodopseudomonas viridis*. Science. 245, 1463–1473.

Dekker, J.P., Betts, S.D., Yocum, C.F. and Boekema, E.J. (1990) Characterization by electron microscopy of isolated particles and two-dimensional crystals of the CP47-D1-D2-cytochrome b-559 complex of photosystem II. Biochemistry 29, 3220–3225.

Dorset, D.L., Engel, A., Hner, M., Massalski, A. and Rosenbusch, J. P. (1983) Two-dimensional crystal packing of matrix porin, a channel forming protein in *Escherichia coli* outer membranes. J. Mol. Biol. 165, 701–710.

Engel, A. (1978) Molecular weight determination by scanning transmission electron microscopy. Ultramicroscopy 3, 273–281.

Engel, A. and Meyer, J. (1980) Preparation of unstained protein structures for mass determination by scanning transmission electron microscopy. J. Ultrastruct. Res. 72, 212–222.

Engel, A. and Reichelt, R. (1988) Processing of quantitative scanning transmission electron micrographs. Scanning Microsc. Suppl. 2, 285–293.

Ford, R.C., Hefti, A. and Engel, A. (1990) Ordered arrays of the photosystem I reaction centre after reconstitution: projections and surface reliefs of the complex at 2 nm resolution. EMBO J. 9, 3067–3075.

Goodenough, D.A. (1979) Lens gap junctions: a structural hypothesis for non-regulated low-resistance intercellular pathways. Invest. Ophthalmol. Visual Sci. 18, 1104–1122.

Gruijters, W.T.M., Kistler, J., Bullivant, S. and Goodenough, D.A. (1987) Immunolocalization of MP70 in lens fiber 16–17nm intercellular junctions. J. Cell. Biol. 104, 565–572.

Hovmöller, S., Slaughter, M., Berriman, J., Karlsson, B., Weiss, H. and Leonard, K. (1983) Structural studies of cytochrome reductase. Improved membrane crystals of the enzyme complex and crystallization of a subcomplex. J. Mol. Biol. 165, 401–406.

Israelachvili, J.N., Marcelja, S. and Horn, R.G. (1980) Physical principles of membrane organization. Q. Rev. Biophys. 13, 121–200.

Jap, B.K. (1988) High-resolution electron diffraction of reconstituted PhoE porin. J. Mol. Biol. 199, 229–231.

Johnson, R.G., Meyer, R.A. and Lampe, P.D. (1989) Gap-junction formation: a self-assembly model involving membrane domains of lipid and protein. In: N. Sperelakis and W.C. Cole (Eds.), Cell Interactions and Gap Junctions, CRC Press, Boca Raton, pp. 159–179

Kistler, J. and Bullivant, S. (1987) Protein processing in lens intercellular junctions: cleavage of MP70 to MP38. Invest. Ophthalmol. Vis. Sci. 28, 1687–1692.

Kistler, J. and Bullivant, S. (1988) Dissociation of lens fibre gap junctions releases MP70. J. Cell Sci. 91, 415–421.

Kistler, J., Christie, D. and Bullivant, S. (1988) Homologies between gap-junction proteins in lens, heart and liver. Nature. 331, 721–723.

Kistler, J., Schaller, J. and Sigrist, H. (1990a) MP38 contains the membrane-embedded domain of the lens fiber gap-junction protein MP70. J. Biol. Chem. 265, 13357–13361.

Kistler, J., Berriman, J., Evans, C.W., Gruijters, W. T., Christie, D., Corin, A. and Bullivant, S. (1990b) Molecular portrait of lens gap-junction protein MP70. J. Struct. Biol. 103, 204–211.

Kühlbrandt, W. (1988) Structure of light-harvesting chlorophyll a/b protein complex from plant photosynthetic membranes at 7 Å resolution in projection. J. Mol. Biol. 202, 849–864.

Li, J. (1985) Light-harvesting chlorophyll *a/b*-protein: three-dimensional structure of a reconstituted membrane lattice in negative stain. Proc. Natl. Acad. Sci. USA 82, 386–390.

Louis, C.F., Hur, K.C., Galvan, A.C., Tenbroek, E.M., Jarvis, L.J., Eccleston, E.D. and Howard, J.B. (1989) Identification of an 18,000-Dalton protein in mammalian lens fiber cell membranes. J. Biol. Chem. 264, 19967–19973.

Miller, K.R. and Jacob, J.S. (1983) Two-dimensional crystals formed from photosynthetic reaction centers. J. Cell Biol. 97, 1266–1270.

Saxton, W.O., Pitt, J.T. and Horner, M. (1979) The SEMPER image processing system. Ultramicroscopy 4, 343–354.

Saxton, W.O. and Baumeister, W. (1982) The correlation averaging of a regularly arranged bacterial cell envelope protein. J. Microsc. 127, 127–138.

Sharp, K.A. (1991) The hydrophobic effect. Curr. Opinion Struct. Biol. 1, 171–174.

Sikerwar, S.S. and Unwin, P.N.T. (1988) Three-dimensional structure of gap junctions in fragmented plasma membranes from rat liver. Biophys. J. 54, 113–119.

Sosinsky, G.E., Jesior, J.C., Caspar, D.L.D. and Goodenough, D.A. (1988) Gap-junction structures. VIII. Membrane cross-sections. Biophys. J. 53, 709–722.

Stauffer, K.A., Kumar, N.M., Gilula, N.B. and Unwin, N. (1991) Isolation and purification of gap-junction channels. J. Cell Biol. 115, 141–150.

Stokes, D.L. and Green, N.M. (1990) Structure of CaATPase: Electron microscopy of frozen-hydrated crystals at 6 Å resolution in projection. J. Mol. Biol. 213, 529–538.

Unwin, P.N.T. and Zampighi, G. (1980) Structure of the junction between communicating cells. Nature 283, 545–549.

Unwin, P.N.T. and Ennis, P.D. (1984) Two configurations of a channel-forming membrane protein. Nature. 307, 545–550.

Voorter, C.E.M., Kistler, J., Gruijters, W.T.M., Mulders, J.W.M., Christie, D. and de Jong, W.W. (1989) Distribution of MP17 in isolated lens fiber membranes. Curr. Eye Res. 8, 697–706.

Weiss, H. and Leonard, K. (1991) Preparation of membrane crystals of mitochondrial NADH:ubiquinone reductase and ubiquinol:cytochrome c reductase and structure analysis by electron microscopy. In: H. Michel (Ed.), Crystallization of Membrane Proteins, CRC Press, Boca Raton, pp. 197–211.

Wingfield, P., Arad, T., Leonard, K. and Weiss, H. (1979) Membrane crystals of ubiquinone:cytochrome c reductase from *Neurospora* mitochondria. Nature 280, 696–697.

Yeager, M. and Gilula, N.B. (1992) Membrane topology and quaternary structure of cardiac gap junction ion channels. J. Mol. Biol. 223, 929–948.

Part III. Hemichannels

J.E. Hall, G.A. Zampighi and R.M. Davis (Eds.)
Progress in Cell Research, Vol. 3

CHAPTER 10

Connexins, gap-junction proteins, and ATP-induced pores in macrophages

ERIC C. BEYER and THOMAS H. STEINBERG

Washington University School of Medicine, St. Louis, MO, USA

Introduction

Gap junctions are structures containing channels that link adjacent cells in most solid tissues of the body. While there may be turnover of their constituent proteins, they are typically considered to be relatively stable structures, linking relatively static cell populations. We have recently been investigating the expression of gap-junction sequences in a migratory cell, the macrophage. If macrophages form gap junctions, they are likely to mediate much more transient interactions. In the studies summarized below, we show that a mouse macrophage cell line expresses the gap-junction protein Cx43.

Mouse macrophages possess ATP-induced plasma membrane pores. The biological function of ATP-induced pores is unknown, but the similarity of their size exclusion to that of gap junctions led us to explore the possibility that the protein which forms ATP-induced pores might be structurally related to a gap-junction protein. We provide evidence below suggesting that Cx43 is involved in ATP-induced permeabilization.

Mouse macrophages express the gap-junction protein Cx43

The existence of gap junctions in macrophages and other cells of the immune system has been suggested by several investigators (Levy et al., 1976; Porvaznik and MacVittie, 1979; Ori et al., 1980), but definitive proof of their existence has been lacking. We determined whether cells of the J774 mouse macrophage-like cell line express any of the known connexin gap-junction proteins. Northern blots of total RNA prepared from J774 cells were hybridized with ^{32}P-labeled cDNA probes for Cx26 (Zhang and Nicholson, 1989), Cx32 (Paul, 1986; Kumar and Gilula, 1986), Cx40 (Kanter et al., 1991), Cx43 (Beyer et al., 1987), and Cx46 (Beyer et al., 1988). Only Cx43 was detected on the RNA blots (Fig. 1). The Cx43 cDNA probe hybridized to a 3.0 kb band in J774 cell RNA that co-migrated with the hybridizing band in mouse heart RNA (Beyer and Steinberg, 1991). Cx43 was also detected in RNA extracted from thioglycollate-elicited mouse peritoneal macrophages and mouse bone marrow macrophages (Fig. 1). Cx43 mRNA was not detected in human peripheral blood monocytes nor in the mouse myeloblast cell line M1 (Steinberg, Robertson, Westphale, and Beyer, unpublished data). These data suggest that expression of Cx43 may be induced during macrophage differentiation.

We also detected Cx43 protein in J774 cells by immunoblot and immunoprecipitation, using several different rabbit polyclonal antibodies directed against peptide sequences from Cx43 (Fig. 2). These antibodies recognized a single protein of approximately 40 kDa in J774 cells that was detected neither by pre-immune sera nor by immune sera incubated with 100 μg/ml of the immunizing peptide. In contrast, these sera recognized two or three bands in heart. The smallest immunoreactive band in heart co-migrated with the band detected in J774 cells; the other bands were approximately 2 and 3 kDa larger. We found that these higher molecular weight Cx43 bands in heart represent phosphorylated forms of the protein, as previously described (Crow et al., 1990; Musil et al., 1990; Laird et al., 1991). After alkaline-phosphatase treatment of solubilized heart samples, only the 40-kDa form of Cx43 which co-migrates with the J774 band was recognized (unpublished data). The J774 band was unaffected by such treatment. Thus, the Cx43 protein present in the macrophages may be the unphosphorylated form.

These experiments demonstrate that mouse macrophages and J774 cells express the gap-junction protein Cx43, but not other known members of this family of proteins. By analogy with the role of gap junctions in other cells, Cx43 may comprise a means of cell-cell communication between macrophages and other cells. Macrophages have many functions and interact with diverse cell types, including other macrophages, lymphocytes, endothelial cells, smooth muscle cells, and cells of solid tissues. It is possible that macrophages form transient, complete gap junctions composed of Cx43 hexamers, with themselves or other cells. We have previously demonstrated that vari-

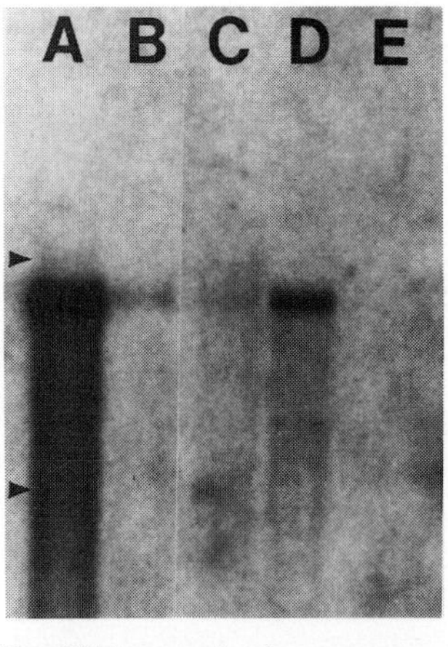

Figure 1. Cx43 mRNA expression by mouse macrophages. Total cellular RNA was isolated from (A) mouse heart, (B) thioglycollate-elicited mouse macrophages, (C) mouse bone marrow macrophages, (D) J774 cells, and (E) ATPR B2 cells, separated on agarose-formaldehyde gels (10 μg/lane), and transferred to nylon membranes. These blots were hybridized with [^{32}P]-labeled probes for rat Cx43 (clone G2). Arrowheads indicate the positions of 18S and 28S rRNA.

ous blood vessel wall cells, including endothelial and smooth muscle cells also express Cx43 (Larson et al., 1990; Moore et al., 1991). Navab et al. (1991) have recently suggested that interaction between human monocytes and co-cultured aortic wall cells increases Cx43 mRNA.

ATP-induced pores in mouse macrophages

Mouse macrophages and J774 cells possess plasma-membrane pores that form in response to extracellular ATP and have a size selectivity similar to that of gap junctions. ATP-induced permeabilization of the plasma membrane appears to be mediated by an ATP-sensitive plasma-membrane pore that opens rapidly (within 40 ms) when the extracellular ATP^{4-} concentration is greater than 100 μM, and closes rapidly upon removal of the ATP (Fig. 3) (Steinberg et al., 1987; Buisman et al., 1988). Other nucleotides are ineffective, and the presence of high concentrations of divalent cations inhibit pore formation by complexing with the ATP (Steinberg and Silverstein, 1987). The pore allows molecules of < 900 Da to enter or leave the cells, and thus causes membrane depolarization and monovalent cation fluxes, in addition to allowing passage of nucleotides, sugar phosphates, and fluorescent dyes.

Disruption of plasma-membrane integrity is highly undesirable; macrophages survive in extracellular ATP for only 30 min (Steinberg and Silverstein, 1987). The lethal effect of prolonged exposure to ATP allowed us to select variant J774 cell lines, ATPR cells, that do not permeabilize in response to extracellular ATP. These cells appear identical to J774 cells by morphologic criteria, ability to ingest IgG-coated particles, ecto-ATPase activity, and calcium mobilization in response to ATP. Thus, ATPR cells appear to differ from J774 cells only by the absence of ATP-induced permeabilization.

ATP pores and Cx43

The biological function of ATP-induced pores in macrophages remains unclear. However, these pores are similar to gap junctions in that both lack ion selectivity and both are permeable to molecules of a similar size range

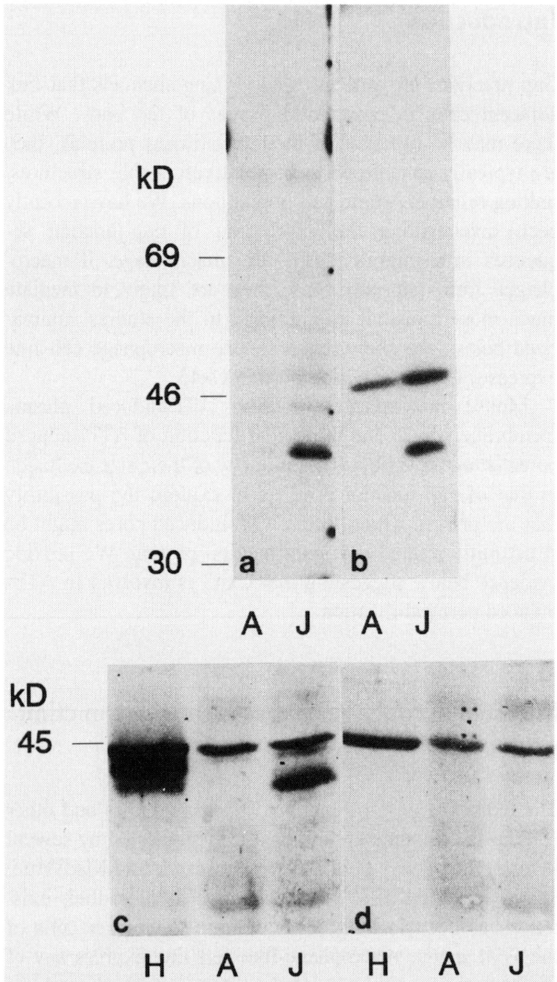

Figure 2. Cx43 protein expression by J774, but not ATPR cells. Western blots of membrane fractions from J774 cells, ATPR B2 cells, and mouse heart were probed with polyclonal anti-Cx43 antibodies (panels a–c). Addition of the immunizing peptide to the anti-Cx43 antibodies abolished antibody binding to the 40-kDa band in the J774 cells and to the 43- and 40-kDa polypeptides in mouse heart (panel d). J, J774 cells; A, ATPR B2 cells; H, mouse heart. Molecular mass standards are indicated. From Beyer and Steinberg, 1991, by permission.

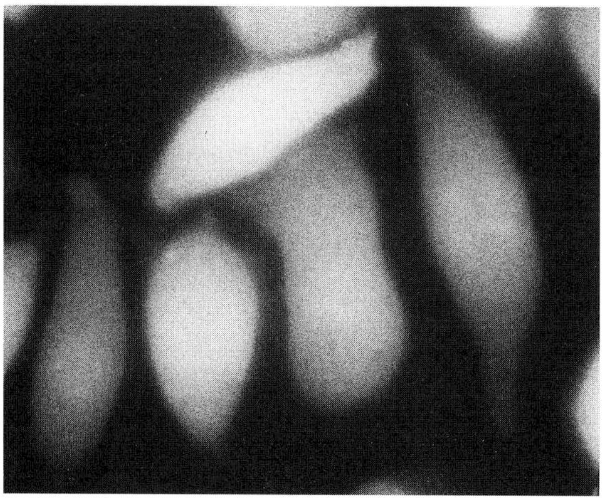

Figure 3. ATP-induced permeabilization of J774 cells. Adherent J774 cells were incubated in medium containing 5 mM ATP and 0.5 mg/ml Lucifer Yellow for 5 min, washed, and viewed by fluorescence microscopy.

(Simpson et al., 1977). To determine whether there was an association between Cx43 and the ATP-induced pores of J774 cells, we attempted to identify Cx43 in the ATPR cell lines (Beyer and Steinberg, 1991). By Northern blot analysis, Cx43 was undetectable in the ATPR lines B2, D5, and G6, under conditions which revealed Cx43 RNA in J774 cells (Fig. 1). We confirmed that J774 lanes and ATPR lanes contained equivalent amounts of RNA by hybridizing blots with an actin probe. In addition, we were unable to detect Cx43 protein by immunoblot or by immunoprecipitation under conditions which revealed immunospecific staining of equivalent amounts of SDS-solubilized J774 cell protein (Fig. 2).

ATPR cells were selected from the parent J774 cell line solely on the basis of their resistance to ATP permeabilization. The contrast between the presence of Cx43 in J774 cells and its absence in ATPR cells provides circumstantial evidence that the ATP-induced pore is Cx43. This suggests that Cx43 may form "half gap junctions", or pores within the plasma membrane of a single cell, in addition to forming channels between two adherent cells. It also suggests a role for ATP in regulation of Cx43 channels. The link between Cx43 and the ATP pore has not yet been directly verified. We are currently attempting to determine whether Cx43 is itself the ATP-induced pore and whether other membrane proteins are involved in this process.

Acknowledgements

We thank Eileen Westphale, Audra Robertson, Wei Li, and Jennifer Frankenfeld for technical assistance and Dr. Diane Rup for helpful discussions. This work was supported by USPHS Grants EY08368 (ECB), HL45466 (ECB), and AI00893 (THS), the American Heart Association (ECB), the McDonnell Foundation (ECB) and the Markey Foundation (THS).

References

Beyer, E.C., Paul, D.L. and Goodenough, D.A. (1987) Connexin43: a protein from rat heart homologous to a gap-junction protein from liver. J. Cell Biol. 105, 2621–2629.

Beyer, E.C., Paul, D.L. and Goodenough, D.A. (1988) The connexins: a family of related gap-junction proteins. In: E.L. Hertzberg and R. Johnson (Eds.) Gap Junctions. Alan R. Liss, New York, pp. 167–175.

Beyer, E.C. and Steinberg, T.H. (1991) Evidence that the gap-junction protein connexin-43 is the ATP-induced pore of mouse macrophages. J. Biol. Chem. 266, 7971–7974.

Buisman, H.P., Steinberg, T.H., Fischbarg, J., Silverstein, S.C., Vogelzang, S.A., Ince, C., Ypey, D.L. and Leijh, P.C.J. (1988) Extracellular ATP induces a large, nonselective conductance in macrophage plasma membranes. Proc. Natl. Acad. Sci. USA 85, 7988–7992.

Crow, D.S., Beyer, E.C., Paul, D.L., Kobe, S.S. and Lau, A.F. (1990) Phosphorylation of connexin43 gap-junction protein in uninfected and RSV-transformed mammalian fibroblasts. Mol. Cell. Biol. 10, 1754–1763.

Kanter, H.L., Saffitz, J.E. and Beyer, E.C. (1991) Canine cardiac myocytes express multiple gap-junction proteins. Clin. Res. 39, 193a.

Kumar, N.M. and Gilula, N.B. (1986) Cloning and characterization of human and rat liver cDNAs coding for a gap-junction protein. J. Cell Biol. 103, 767–776.

Laird, D.W., Paranam, K.L. and Revel, J.-P. (1991) Turnover and phosphorylation dynamics of connexin43 gap-junction protein in cultured cardiac myocytes. Biochem. J. 273, 67–72.

Larson, D.M., Haudenschild, C.C. and Beyer, E.C. (1990) Gap-junction messenger RNA expression by vascular wall cells. Circ. Res. 66, 1074–1080.

Levy, J.A., Weiss, R.M., Dirksen, E.R. and Rosen, M.R. (1976) Possible communication between murine macrophages oriented in linear chains in tissue culture. Exp. Cell Res. 103, 375–385.

Moore, L.K., Beyer, E.C. and Burt, J.M. (1991) Characterization of gap-junction channels in A7r5 vascular smooth muscle cells. Am. J. Physiol. 260, C975–C981.

Musil, L.S., Beyer, E.C. and Goodenough, D.A. (1990) Expression of the gap-junction protein connexin43 in embryonic chick lens: molecular cloning, ultrastructural localization, biosynthesis and post-translational phosphorylation. J. Membr. Biol. 116, 163–175.

Navab, M., Liao, F., Hough, G.P., Ross, L.A., Van Lenten, B.J., Rajvashisth, T.B., Lusis, A.J., Laks, H., Drinkwater, D.C. and Fogelman, A.M. (1991) Interaction of monocytes with cocultures of human aortic wall cells involves interleukins 1 and 6 with marked increases in connexin43 message. J. Clin. Invest. 87, 1763–1772.

Ori, M., Shiba, Y. and Kanno, Y. (1980) Facilitation of cell coupling formation between mouse macrophages with an increase in the external calcium ion concentration. Cell Struct. Funct. 5, 259–263.

Paul, D.L. (1986) Molecular cloning of cDNA for rat liver gap-junction protein. J. Cell Biol. 103, 123–134.

Porvaznik, M. and MacVittie, T.J. (1979) Detection of gap junctions between the progeny of a canine macrophage colony-forming cell *in vitro*. J. Cell Biol. 82, 555–564.

Simpson, I., Rose, B. and Lowenstein, W.R. (1977) Size limit of molecules permeating the junctional channels. Science 195, 294–297.

Steinberg, T.H., Newman, A.S., Swanson, J.A. and Silverstein, S.C. (1987) ATP^{4-} permeabilizes the plasma membrane of mouse macrophages to fluorescent dyes. J. Biol. Chem. 262, 8884–8888.

Steinberg, T.H. and Silverstein, S.C. (1987) Extracellular ATP^{4-} promotes cation fluxes in the J774 mouse macrophage cell line. J. Biol. Chem. 262, 3118–3122.

Zhang, J.T. and Nicholson, B.J. (1989) Sequence and tissue distribution of a second protein of hepatic gap junctions, Cx26, as deduced from its cDNA. J. Cell Biol. 109, 3391–3401.

J.E. Hall, G.A. Zampighi and R.M. Davis (Eds.)
Progress in Cell Research, Vol. 3

CHAPTER 11

Connexin46 forms gap-junctional hemichannels in *Xenopus* oocytes

L. EBIHARA and E. STEINER

Department of Pharmacology, Columbia University, 630 W. 168th Street, New York, NY 10032, USA

Introduction

The steps involved in the assembly of gap-junctional proteins into cell-to-cell channels are poorly understood. Swenson et al. (1989) showed that gap-junctional proteins are diffusely distributed throughout the cytoplasm in unpaired *Xenopus* oocytes injected with mRNA for Cx43 or Cx32. Following pairing, the gap-junctional proteins are drawn toward the region of cell-cell contact. This observation raises the possibility that the channel precursors are inserted into the junctional membrane immediately prior to their assembly into cell-to-cell channels. Alternatively, connexins may be initially inserted into the non-junctional membrane as connexons or hemichannels. Cell-to-cell channels may then form when a connexon from one cell comes randomly into contact with a connexon from a neighboring cell in regions of close membrane apposition.

In this paper, we report that the expression of a lens gap-junctional protein, Cx46, in *Xenopus* oocytes results in the development of a nonselective, time- and voltage-dependent current. Thus Cx46 may form open gap-junctional hemichannels in the non-junctional plasma membrane of *Xenopus* oocytes.

Connexin46 is a gap-junctional protein

The cDNA for Cx46 was isolated from a rat lens cDNA library by low-stringency hybridization, using rat liver cDNA as a probe (Paul et al., 1991). The Cx46 cDNA encodes a protein of 416 amino acids which has a predicted molecular mass of 46 kDa. The amino-acid sequence of Cx46 resembles the sequences of other gap-junctional proteins, indicating that Cx46 is a member of the gap-junctional channel family (see Beyer et al., 1988; Beyer et al., 1990). Furthermore, polyclonal antibodies raised against synthetic peptide segments of Cx46 bind specifically to lens fiber-fiber gap-junctions.

The tissue distribution of Cx46 was investigated by Northern blot analysis using Cx46 cDNA as probe (Paul et al., 1991). Cx46 mRNA was present at high levels in the lens and was also detectable in rat heart and kidney, indicating that Cx46 is not a lens-specific protein.

Expression of Cx46 in *Xenopus* oocytes

Expression of Cx46 in *Xenopus* oocytes results in loss of resting potential and cell lysis (Paul et al., 1991). The loss of resting potential occurred gradually over the first 6–8 h following injection of Cx46 mRNA. By 8 h post-injection, the oocytes showed pigmentation changes and by 16–24 h, the oocytes had lysed. The oocytes could be "rescued" from lysis by addition of 5% Ficoll to the external solution. Cx46 mRNA-injected oocytes incubated in 5% Ficoll were permeable to Lucifer Yellow while oocytes injected with Cx43 mRNA were impermeable to Lucifer Yellow.

In order to determine whether these changes were due to the formation of open connexons in the non-junctional membrane or to some other nonspecific mechanism, we performed voltage-clamp experiments on single oocytes injected with mRNA for Cx46 diluted to 0.05–0.002 μg/μl. Oocytes so injected expressed a large, time- and voltage-dependent current, I_h. This current was present in about 250/260 oocytes injected with Cx46 mRNA. No current of similar amplitude and properties was observed in 95 noninjected oocytes or 44 oocytes injected with mRNA for either Cx43 or Cx32 (Paul et al., 1991; Ebihara and Steiner, submitted). Cx46-injected oocytes also tended to have lower resting potentials and input resistances than those of oocytes injected with mRNA for Cx43 or Cx32. The size of I_h was variable, and depended on both the amount of mRNA injected and the health of the oocytes. Figure 1 shows a family of membrane currents recorded from a Cx46 mRNA-injected oocyte during a series of depolarizing pulses from a holding potential of –40 mV. I_h activated at membrane potentials positive to –20 mV and grew progressively larger as the membrane was made more positive (Fig. 2). Upon repolarization to –40 mV, a large inward tail current was observed due to closure of Cx46 channels which were

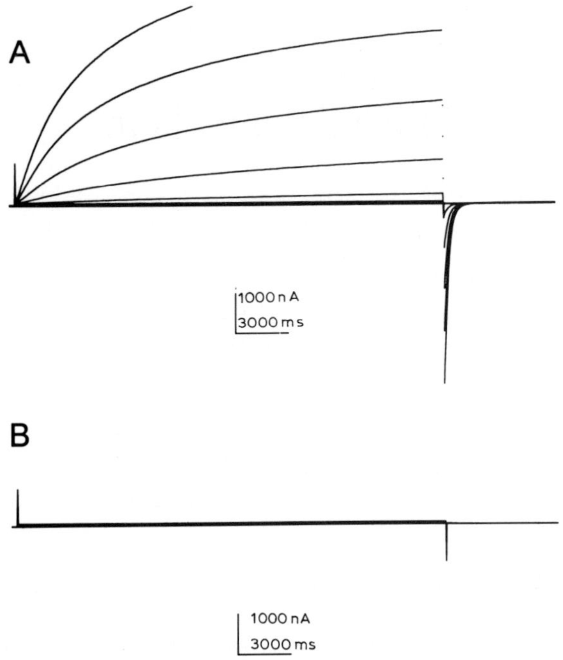

Figure 1. Current traces recorded in response to a series of depolarizing voltage-clamp steps from a holding potential of -40 mV to potentials between −30 and 40 mV. (A) Oocyte injected with mRNA for Cx46. (B) Control oocyte injected with mRNA for Cx43. The oocytes were clamped using a standard 2-microelectrode voltage-clamp technique. The oocytes were bathed in modified Barth's solution containing (in mM): 88 NaCl, 1 KCl, 2.4 $NaHCO_3$, 15 Hepes, 0.3 Ca NO_3, 0.41 $CaCl_2$, 0.82 $MgSO_4$.

activated during the preceding depolarizing pulse. The current reversed polarity between −5 and −15 mV. Furthermore, the reversal potential was not altered by substitution of methanesulfonate for chloride in the external bathing solution, suggesting that I_h is either a

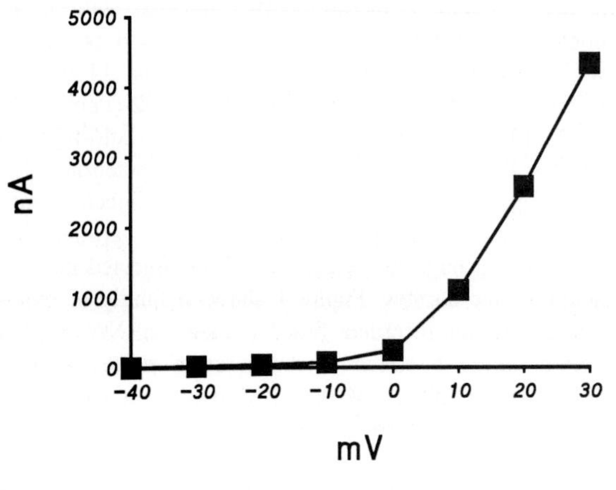

Figure 2. Current-voltage relationship for the same preparation as in Fig. 1A. The amplitude of the current was measured at the end of the pulse and plotted as a function of voltage.

nonselective cation current or a generally nonselective current which cannot discriminate between chloride and methanesulfonate (Ebihara and Steiner, submitted).

Regulation of I_h

It has been previously shown that gap-junctional coupling can be modulated by a number of different agents, including pH, calcium, volatile anesthetics, phosphorylation and the higher alcohols, 1-heptanol and 1-octanol (reviewed by Spray et al., 1985; Spray and Burt, 1990). Therefore, we decided to test the effect(s) of some of these agents on I_h. The effect of external pH was investigated at pH_o 6.8, 7.4 and 7.9. Reducing pH_o from 7.4 to 6.8 caused a marked reduction in the amplitude of I_h (Fig. 3); this effect was fully reversible. Increasing pH_o from 7.4 to 7.9 had no significant effect on I_h.

The effect of changing external calcium was also examined. When calcium was removed from the external bathing solution, the oocyte rapidly depolarized to about −10 mV and became extremely leaky. Voltage-clamp experiments demonstrated that this phenomenon could be attributed to I_h (Ebihara and Steiner, submitted). We also found that the trivalent lanthanide, gadolinium, was an extremely potent blocker of I_h. At 10 μM Gd^{+3}, I_h was completely and irreversibly suppressed. Lower concentrations of Gd^{+3} partially blocked I_h.

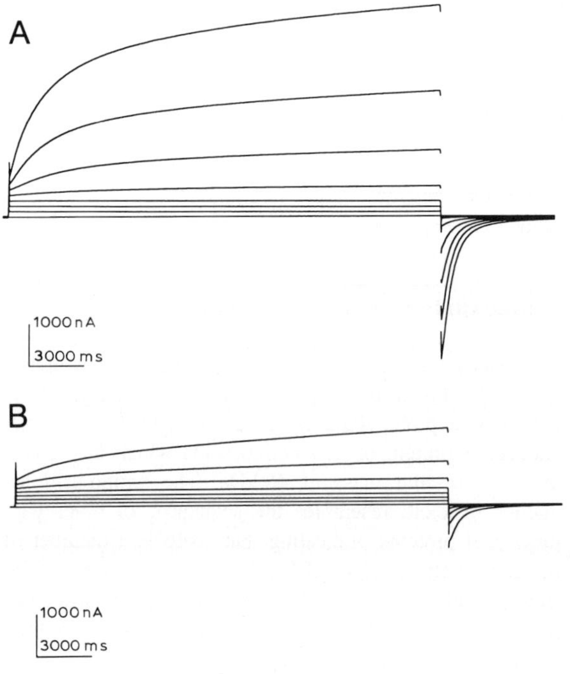

Figure 3. Effect of changing external pH. Families of Cx46 current records recorded at pH_o 7.4 (A) and 6.8 (B). Holding potential −40 mV. The test solution at pH 6.8 was composed of modified Barth's solution buffered with 15 mM Pipes.

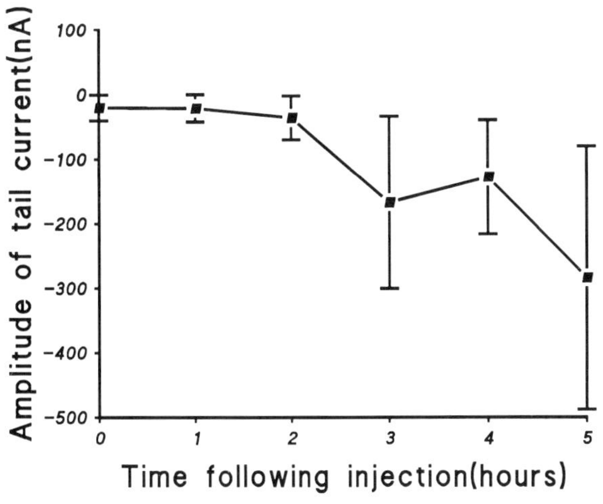

Figure 4. Time course of development of I_h. I_h was determined by measuring the amplitude of the Cx46 tail current elicited in response to application of a 24 s depolarizing pulse to 20 mV from a holding potential of −40 mV. Each point represents the mean value of the tail current recorded in 5 different oocytes.

Time course of development of I_h

Tail currents in oocyte membranes were measured at various times following injection of Cx46 mRNA (Fig. 4.) Each point represents the mean amplitude of the tail current measured in 5 oocytes. I_h was detectable within 3 h following mRNA injection and grew progressively larger with time. The time course of development of I_h was similar to that of gap-junctional protein synthesis (Swenson et al., 1989).

Conclusions

Overexpression of Cx46 in *Xenopus* oocytes results in cellular depolarization and osmotic lysis. Voltage-clamp experiments on oocytes injected with low concentrations of Cx46 mRNA demonstrate that these changes are associated with the development of a large, nonselective transmembrane current which may be due to the presence of open connexons on the surface of the cell. This unusual behavior is not observed when Cx43 or Cx32 is expressed.

It is not clear whether Cx46 forms non-junctional ionic channels in vivo. To date, there have been no reports of a transmembrane current with similar properties in either lens fiber cells or cardiac myocytes. However, a current similar to I_h has been recently reported in catfish horizontal cells (DeVries and Schwartz, Chapter 12). This current is only observed at low external calcium concentrations and can be suppressed by dopamine. In addition, Beyer and Steinberg (1991 and Chapter 10) have reported that the ATP-induced pore in mouse macrophages may be a gap-junctional hemichannel.

If Cx46 forms gap-junctional hemichannels in lens fiber cells or cardiac myocytes, what might the physiological role of hemichannels be? One possibility is that hemichannels represent a necessary precursor stage in the formation of cell-to-cell channels. Under physiological conditions, these channels are closed by external calcium and voltage. Thus, they probably do not contribute to the membrane conductance in the lens or heart. It would be interesting to determine whether the "leakage" current observed in a number of cell types, including cardiac myocytes and *Xenopus* oocytes at low external calcium concentrations, is due to current flowing through open connexons.

Acknowledgements

We thank Dr. D.A. Goodenough for providing us with Cx46 and Marcella Kopal for excellent technical assistance. Supported by NIH grant HL-45377, a Grant-in-Aid from the New York City Affiliate of the American Heart Association, and an Irma Hirschl Career Scientist Award.

References

Beyer, E.C., Goodenough, D.A. and Paul, D. (1988) The connexins: a family of related gap-junction proteins. In: E.L. Hertzberg and R.G. Johnson (Eds.), Gap Junctions, Alan R. Liss, New York, pp. 167–175.

Beyer, E.C., Paul, D.L. and Goodenough, D.A. (1990) Connexin family of gap-junction proteins. J. Membr. Biol. 116, 187–194.

Beyer, E.C., Steinberg, T.H. (1991) Evidence that the gap-junction protein connexin43 is the ATP-induced pore of mouse macrophages. J. Biol. Chem. 266, 7971–7974.

DeVries, S.H. and Schwartz, E.A. (1992) Hemi-gap junction channels in solitary horizontal cells of the catfish retina. J. Physiol. (London) 445, 201–230.

Paul, D.L., Ebihara, L., Takemoto, L.J., Swenson, K.I., and Goodenough, D.A. (1991) Connexin46, a novel lens gap-junction protein, induces voltage-gated currents in non-junctional plasma membrane of *Xenopus* oocytes. J. Cell Biol. 115, 1077–1089.

Spray, D.C. and Bennett, M.V.L. (1985) Physiology and pharmacology of gap junctions. Annu. Rev. Physiol. 47, 281–330.

Spray, D.C. and Burt, J.M. (1990) Structure-activity relations of the cardiac gap-junction channel. Am. J. Physiol. 258, C195–C205.

Swenson, K.I., Jordan, J.R., Beyer, E.C. and Paul, D.L. (1989) Formation of gap junctions by expression of connexins in *Xenopus* oocyte pairs. Cell 57, 145–155.

J.E. Hall, G.A. Zampighi and R.M. Davis (Eds.)
Progress in Cell Research, Vol. 3

CHAPTER 12

Solitary retinal horizontal cells express hemi-gap-junction channels

S.H. DeVRIES* and E.A. SCHWARTZ

Committee on Neurobiology, The University of Chicago, 947 E. 58th Street, Chicago, IL 60637, USA

Introduction

Gap-junction channels are aqueous pores that span the membranes of two adjacent cells. An individual gap-junction channel is built from two half channels; each half channel is contributed by one of the adjacent cells; and, the two halves are linked in tandem. Like channels which only span the membrane of a single cell, the gap-junction channel may be regulated by voltage and second messengers. Voltage control will depend upon the position of voltage-sensing domains within each half of the channel. A voltage-sensing domain may (i) protrude into the plane of the lipid bilayer and sense the membrane voltage, (ii) protrude into the aqueous pore and sense the voltage within this region, or (iii) reside within the wall of the channel and respond to the transjunctional voltage. Of course, intermediate situations may occur, contributing to the complex voltage-dependent behavior of these channels. In addition to regulation by voltage, gap-junction channels may be modulated by cytoplasmic messenger molecules. For example, the gap junctions that interconnect retinal horizontal cells are modulated by three intracellular cascades (Lasater, 1987; DeVries and Schwartz, 1989). The first cascade begins when dopamine binds to receptors on the cell surface and activates an adenylate cyclase. Subsequent activation of a cyclic AMP-dependent protein kinase is followed by the closure of gap-junction channels. A second cascade begins when nitric oxide diffuses across the horizontal cell membrane and activates a soluble guanylate cyclase. An increase in intracellular cyclic GMP concentration probably activates a second kinase and finally also closes gap-junction channels. The third cascade is perhaps the simplest. Small increases in intracellular proton concentration directly close gap junctions. Although three pathways can be identified, the details of each are still uncertain. For example, because the gap-junction channel does not have a face exposed to the extracellular milieu, it has not been possible to use conventional patch-pipette recording techniques to isolate a membrane patch and apply kinases to the cytoplasmic face. Thus, it is not evident whether kinases act directly on the gap-junction protein or whether additional steps intervene.

In order to study voltage control and modulation, it would be advantageous to have a preparation in which hemi-gap-junction channels were active in the membrane of a solitary cell. The voltage dependence of the hemichannel could then be compared to that of the whole channel, and additional inferences drawn about the mechanisms of voltage-dependent gating. In addition, membrane patches containing hemichannels could be isolated and excised, and regulation at the cytoplasmic face studied. We now show that solitary horizontal cells express a novel dopamine-sensitive membrane current which is present only when the extracellular Ca^{2+} concentration is reduced. We compare the modulation, voltage dependence, and permeation of this new conductance to that of the gap junction and conclude that the dopamine-sensitive membrane current is produced by the flow of ions through hemi-gap-junction channels (see DeVries and Schwartz, 1992). Hemi-gap-junction channels that are "cryptic" at physiological Ca^{2+} concentrations may be intermediates utilized in the assembly of whole gap-junction channels. The ability to isolate a part of the gap channel in the membrane of one cell may allow single-channel properties, modulation, and mechanisms of voltage dependence to be studied.

Reducing the extracellular Ca^{2+} concentration reveals a dopamine-sensitive current, I_γ

Dopamine has littie effect on membrane current when solitary horizontal cells are bathed in a physiological saline solution (Lasater and Dowling, 1982; DeVries and Schwartz, 1989). However, when the extracellular Ca^{2+}

*Present address: Department of Neurobiology, Stanford University, Fairchild D-238, Stanford, CA 94305, USA.

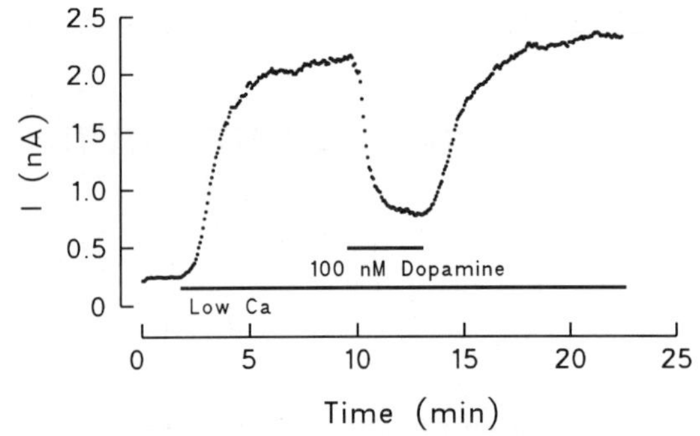

Figure 1. A membrane current is suppressed by dopamine when cells are bathed in a solution which contains Ca^{2+} at a reduced concentration. The membrane voltage of a solitary horizontal cell was maintained at 20 mV and membrane current was measured at 1 s intervals. Each point represents the average of 4 such measurements plotted against time. At the onset of the lower horizontal bar, the superfusate was changed from one that contained 2 mM Ca^{2+} to one in which the free Ca^{2+} concentration was buffered to 10 μM. After the membrane current stabilized at a new, elevated value, 100 nM dopamine was added to the superfusate (upper horizontal bar). Dopamine suppressed the current increase by more than 70%. (Reproduced from DeVries and Schwartz, 1992)

concentration is lowered, a membrane current is produced that can subsequently be suppressed by dopamine. An example is shown in Fig. 1. The membrane voltage of a horizontal cell was maintained at 20 mV with a patch-pipette voltage clamp. After the superfusate was switched from a saline solution that contained 2 mM Ca^{2+} to one that contained 10 μM Ca^{2+} (lower horizontal bar), an outward current developed. The subsequent addition of dopamine (100 nM) suppressed the current by more than 70%. Larger concentrations of dopamine could suppress the current by more than 90%. Thus, horizontal cells possess a dopamine-sensitive membrane current, I_γ which is revealed by lowering the extracellular Ca^{2+} concentration. More complete experiments (DeVries and Schwartz, 1992) demonstrate that I_γ is maximal at Ca^{2+} concentrations less than 10 μM, half maximal at 220 mM, and less than 4% active at a physiological concentrations of approximately 1 mM Ca^{2+}.

I_γ is conducted by hemi-gap-junction channels

In the following experiments, we compare the behavior of I_γ and gap-junction channels and demonstrate that both channels are coordinately modulated by three intracellular cascades, and that both channels are large, non-selective pores that are permeable to Lucifer Yellow CH, and have similar voltage-sensing domains.

Preliminary results indicated that, like gap-junction channels, I_γ channels are suppressed (i) by dopamine acting through an adenylate cyclase/cyclic AMP-dependent protein kinase cascade, (ii) by nitric oxide acting through a guanylate cyclase/cyclic GMP cascade, and (iii) by intracellular acidification. In the first set of experiments, we simultaneously measured membrane current and junctional conductance in a cell pair and show that I_γ and gap-junction channels are coordinately modulated by each of the three cascades (Fig. 2). The experiment was performed as follows: The membrane voltage of each cell was separately controlled with a switching voltage clamp operating through a patch pipette. First, total membrane current was measured when both cells were held at 20 mV. At regular intervals, one cell was briefly stepped to 0 mV, transjunctional current measured, and the junctional conductance calculated. Reducing the extracellular Ca^{2+} concentration produced an increase in membrane current without a significant change in junctional conductance. Thereafter, the cell pair was sequentially exposed to dopamine, acetate, and nitroprusside (nitroprusside in solution produces nitric oxide; the un-ionized form of acetate enters cells and acidifies the cytoplasm). Dopamine and nitroprusside each produced alterations in junctional conductance and membrane current that had similar time courses and amplitudes. Acetate produced alterations which, although comparable, occasionally showed small differences in time course and relative amplitude which may reflect non-uniform changes in intracellular pH. The similarity of the traces indicates that

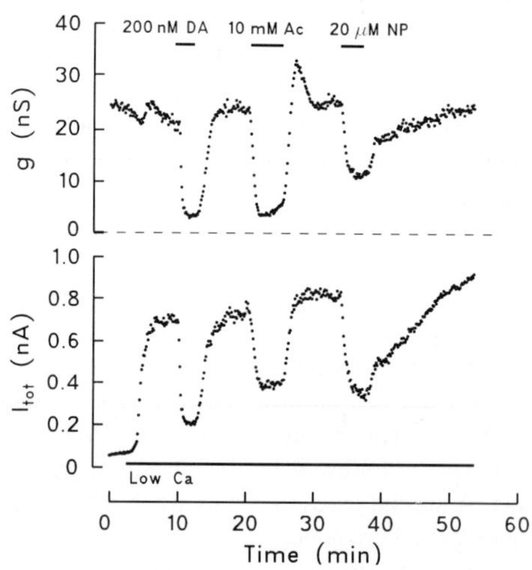

Figure 2. Dopamine, acetate, and nitroprusside produce coordinated changes in both membrane current and junctional conductance. The membrane voltage of each cell in a pair was maintained at 20 mV and the total membrane current of the two cells was measured. Subsequently, one cell was briefly stepped to 0 mV and the resulting junctional current was measured. Junctional conductance (above) and membrane current (below) are plotted against time. At the time indicated by the lower horizontal bar, the cell pair was superfused with a solution containing 100 nM free Ca^{2+}. The upper horizontal bars denote the times during which the pair was superfused with either 200 nM dopamine (DA), 10 mM Na-acetate (Ac), or 20 μM nitroprusside (NP). (Reproduced from DeVries and Schwartz, 1992).

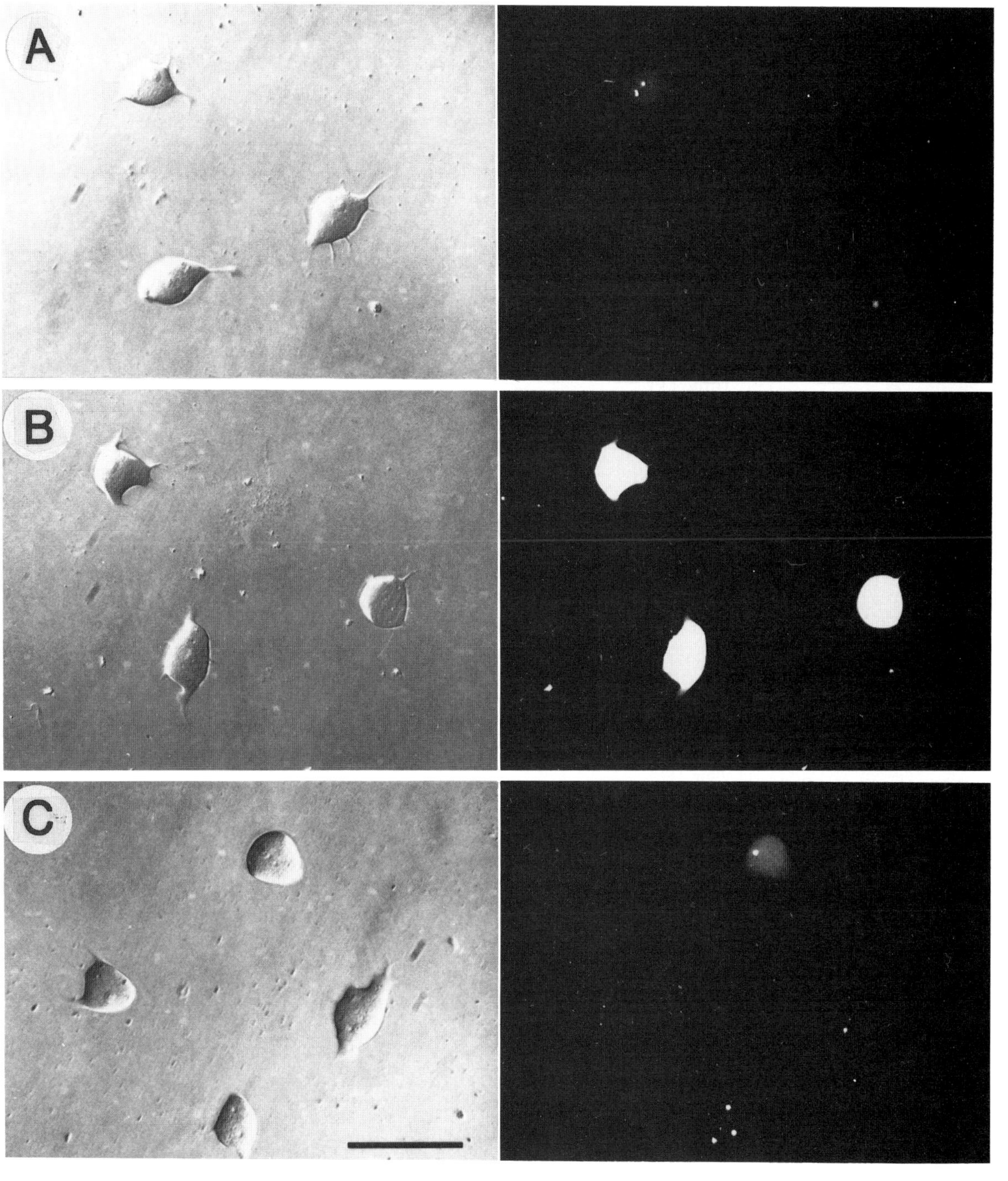

Figure 3. Lucifer Yellow CH permeates I_γ channels. Horizontal cells were incubated for 30 min in three different solutions, each containing 90 mM K-D-aspartate and 10 mM Lucifer Yellow CH. In addition, the solutions contained either (A) 2 mM free Ca^{2+}; (B) 100 nM free Ca^{2+}; (C) 100 nM free-Ca^{2+} plus 5 μM dopamine. Afterwards, the cells were rinsed and photographed. Micrographs on the right were taken with fluorescence optics; micrographs on the left are of the same fields taken with Hoffman interference optics. Lucifer Yellow CH influx (part B) was reduced by either external Ca^{2+} (part A) or by dopamine (part C). (Reproduced from DeVries and Schwartz, 1992)

both junctional and I_γ channels are modulated in concert by the three agents and suggests that both channels have similar regulatory domains at their cytoplasmic faces.

The anionic dye Lucifer Yellow CH (443 Da) permeates gap junctions between teleost horizontal cells in the intact retina (Teranishi et al., 1983), and between paired cells in culture (DeVries and Schwartz, unpublished observations). The second set of experiments tests whether the I_γ channel is also permeable to Lucifer yellow CH (Fig. 3). Horizontal cells were first separated on a Percoll gradient from other retinal cells and debris and then plated into culture dishes. Cells were incubated in

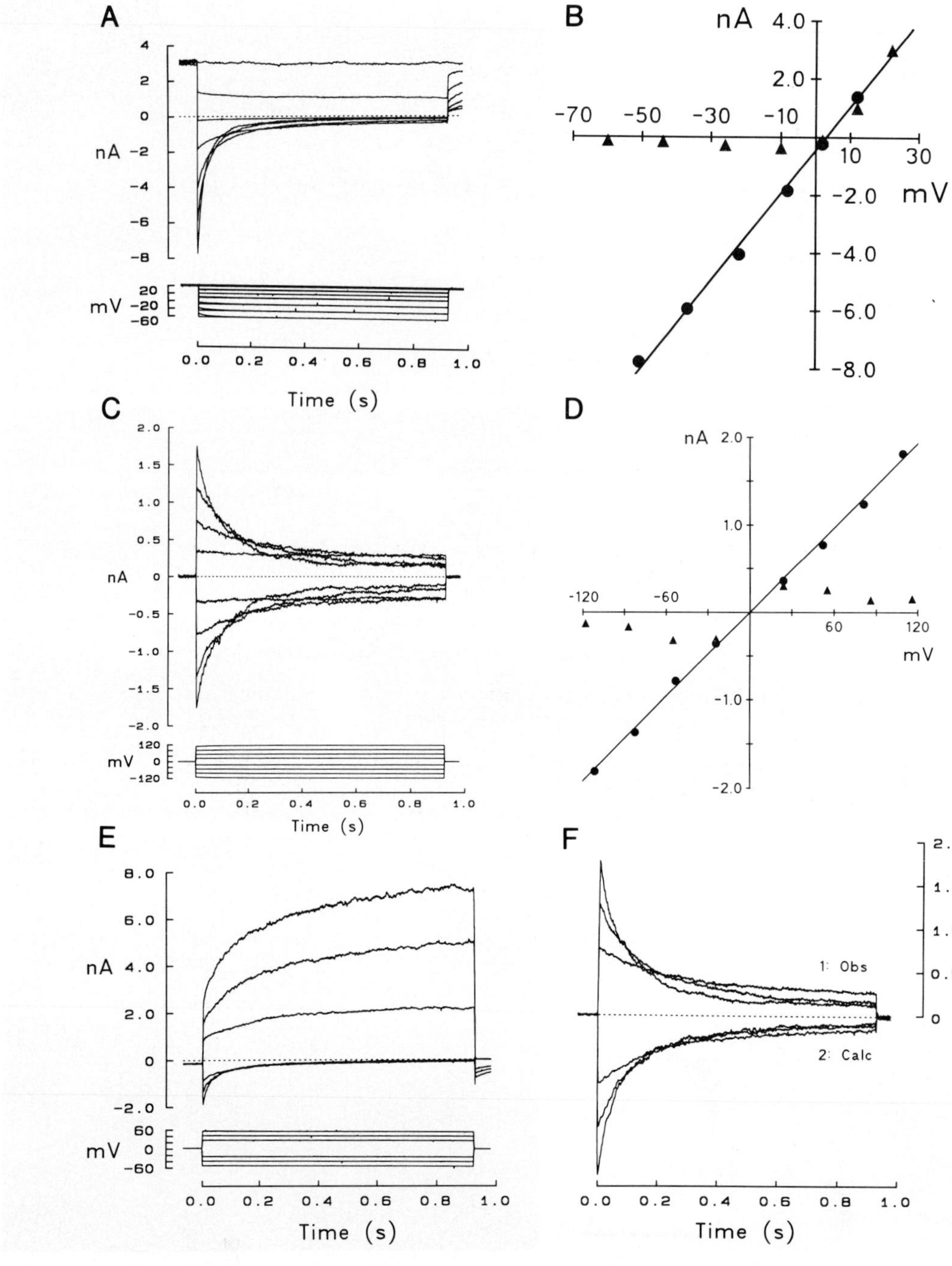

Figure 4. The voltage-dependent behavior of the gap-junction channel can be predicted from the voltage-dependent properties of the I_γ channel. (A) The current-voltage properties of I_γ. The extracellular saline solution contained 10 μM Ca^{2+}. Membrane voltage was stepped from 26 mV to a series of voltages between 14 and -61 mV. Next, 500 nM dopamine was added to the external medium and the series of voltage steps repeated. The difference currents obtained by subtraction are shown. (B) Initial (filled circles) and steady-state (filled triangles) currents from A are plotted against membrane voltage. (C) The current-voltage properties of the gap junction between a pair of horizontal cells. The membrane voltage of each cell was maintained at 20 mV. Then the membrane voltage of one cell was stepped to a series of voltages between –5 and –95 mV. The resulting junctional currents are displayed as upward-going traces. Next, the membrane voltage of the other cell was stepped in a similar fashion and the resulting junctional currents are displayed as the downward going traces. (D) Initial (filled circles) and steady-state (filled triangles) currents from C are plotted against membrane voltage. (E) A solitary cell was bathed in a solution containing a reduced concentration of Ca^{2+} and the dopamine-sensitive current was measured during voltage steps from 0 mV to a series of voltages having equal amplitudes, but opposite polarities (27, 42, 57 mV). (F) Junctional currents were calculated as described in the text using the currents in E, and are displayed as downward-going traces. For comparison, the upward current is the actual junctional current resulting from transjunctional voltage steps of 55, 85, and 115 mV (each trace is the average obtained from a symmetric pair of voltage steps) in C. (Reproduced from DeVries and Schwartz, 1992)

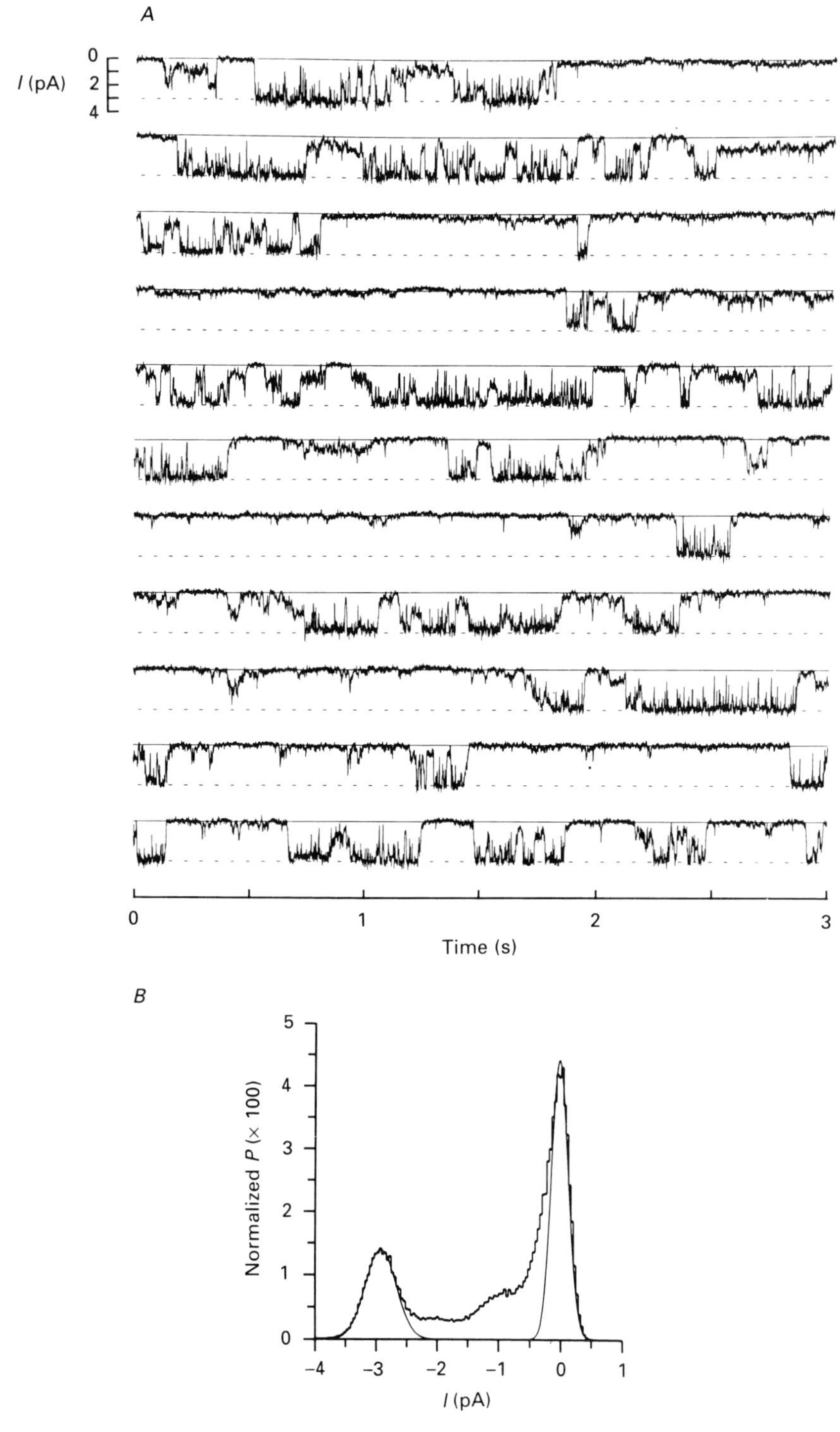

Figure 5. Single-channel activity in an excised patch. (A) After dopamine had suppressed channel activity in a cell-attached patch, the patch was excised. Membrane voltage was continuously maintained at 20 mV and both faces of the membrane patch were bathed in identical solutions containing 70 mM cesium-D-aspartate, 25 mM tetraethylammonium chloride, and 1 μM free Ca^{2+}. Illustrated is a continuous 33-s sequence selected from a longer 83-s record. The dashed line indicates a current amplitude of 3.0 pA. (B) An amplitude histogram obtained from the 83-s record. A maximum open-channel size can be described by a Gaussian curve with a 2.9 pA mean, corresponding to a single-channel conductance of 145 nS if the current reverses at 0 mV in symmetric solutions. (Reproduced from DeVries and Schwartz, 1992)

solutions containing 90 mM K-D-aspartate, 10 mM Lucifer Yellow CH, and either (A) 2 mM free Ca^{2+}, (B) 100 nM free Ca^{2+}, or (C) 100 nM free Ca^{2+} plus 5 µM dopamine. The high concentration of extracellular potassium depolarized the cells to approximately 0 mV, a voltage at which I_γ channels are open (see below). Cells were rinsed to remove the extracellular dye and then photographed using Hoffman interference and epifluorescence optics. Lucifer Yellow CH entered cells that were bathed in a solution containing a reduced Ca^{2+} concentration alone (part B). The addition of either Ca^{2+} (part A) or dopamine (part C) prevented dye influx. Thus, influx of Lucifer Yellow CH occurs through channels that are regulated by Ca^{2+} and dopamine. We conclude that both the I_γ and gap-junction channels have large, non-selective pores.

In the third set of experiments, the voltage-dependent properties of I_γ and the junctional conductance were characterized and compared. The current-voltage properties of I_γ were determined by first bathing solitary cells in a solution containing a reduced Ca^{2+} concentration, and then stepping membrane voltage from a maintained value of 26 mV to a series of more negative potentials. The sequence of steps was then repeated in the presence of dopamine. The difference currents are shown in Fig. 4A. In Fig. 4B, the initial current, measured 2 ms after the onset of a step (circles), and in the steady state, measured at the end of a step (triangles), are plotted against membrane voltage. I_γ is ohmic immediately after a change in voltage and then inactivates to a steady state characterized by a region of negative resistance at potentials more hyperpolarized than –15 mV, and outward rectification at potentials more depolarized than 0 mV.

The current-voltage properties of the junctional conductance were determined as follows: The membrane voltage of each cell of a pair was controlled by a switching voltage clamp operating through a patch pipette. Initially, the voltages of both horizontal cells were maintained at 20 mV. Then, one cell was stepped to a series of voltages between –5 and –95 mV. The junctional current during each step is plotted as an upward-going trace in Fig. 4C. The stimulus regime was reversed and the resulting junctional currents are plotted as downward-going traces in Fig. 4C. Initial (circles) and steady-state (triangles) currents are plotted against junctional voltage in Fig. 4D. The junctional conductance rectifies bidirectionally with regions of negative resistance at voltages exceeding 30 mV.

The distinctive voltage-dependent behavior of the gap-junction conductance can be deduced from that of the I_γ conductance by making two assumptions: First, the channel which conducts I_γ is a hemi-gap-junction channel, and a whole gap-junction channel is composed of two hemichannels in series with their extracellular faces apposed. Second, each half of an intact gap-junction channel senses and responds to 1/2 of the transjunctional voltage. Of course, since the two hemichannels that are apposed to form an intact gap-junction channel are oriented oppositely in the transjunctional voltage field, the voltages sensed by each half will be of opposite polarity. The conductance of the whole gap junction would then be related to that of the I_γ channel by an equation for conductances in series:

$$g_j = (g_1 g_2)/(g_1 + g_2)$$

where g_1 and g_2 are the conductances of the two I_γ channels and g_j is the gap-junction conductance. Note that g_1 and g_2 are time dependent and respond to half the voltage field.

The applicability of this equation for describing the voltage-dependent behavior of the gap junction was tested as follows: The membrane voltage of a solitary cell was maintained at 20 mV, stepped to a prepulse potential of 0 mV for 400 ms, and then to one of a series of symmetrical values (27, 42, and 57 mV). The resulting dopamine-sensitive currents are shown in Fig. 4E. Currents were converted to time-varying conductances using the reversal potential (4 mV); and a symmetrical pair of conductance traces, e.g., for +57 and –57 mV, were used as g_1 and g_2 in the equation. The calculated junctional conductance, g_j, was multiplied by the amplitude of the transjunctional voltage step (e.g. 114 mV) and the resulting currents displayed as the downward-going traces in Fig. 4F. For comparison, the three traces from Fig. 4D for steps of 55, 85, and 115 mV, or approximately double the amplitude of the steps imposed across the solitary cell membrane, are plotted as upward-going currents. The two sets of traces have remarkably similar time courses. The similarity between the two sets of traces indicates that the voltage-dependent behavior of the whole gap junction can be accounted for on the basis of the voltage-dependent behavior of two I_γ channels opposed end-to-end in series.

Thus, the demonstration that gap-junction channels and I_γ channels have (i) similar regulatory domains at their cytoplasmic faces, (ii) similar large, non-selective ion pores, and (iii) similar voltage-sensing domains indicate that the I_γ channel is the hemi-gap-junction channel.

Single-channel properties

Current through I_γ channels could be observed with patch pipettes in the "cell-attached" configuration (Hamill et al., 1981). In a typical experiment, a patch pipette filled with a solution containing 1 µM free-Ca^{2+} was sealed to a cell and the underlying patch of membrane depolarized. Individual patches frequently contained several channels. Next, the cell was bathed in a solution which contained dopamine. If all of the channel activity was suppressed by dopamine, we concluded that the patch contained only I_γ channels and the patch was then excised and bathed in symmetric solutions containing 70 mM cesium-D-aspartate and 25 mM tetraethylammonium chloride. In two patches, only a single channel remained active after excision (Fig. 5A). Amplitude histograms demonstrated a maximum open-channel size of 145 pS (Fig. 5B). Two 145 pS channels in series would form a complete gap-junction channel with a maximum conductance of approximately 70 pS, slightly larger than the 50–60 pS previously reported for the unitary conductance of the gap-junction channel in teleost horizontal cells (McMahon et al., 1989). However, the expected distribution of open states produced by two I_γ channels in series would be quite complex.

Conclusions

The expression of hemi-gap-junction channels by solitary horizontal cells has implications for understanding the mechanisms of gap-junction modulation, voltage dependence, and formation.

Modulation

The sequence that begins when dopamine binds to a membrane receptor includes the activation of adenylate cyclase and cyclic AMP-dependent protein kinase, and concludes with the closure of gap-junction channels. Nitric oxide initiates a separate pathway (DeVries and Schwartz, 1992). Because cyclic GMP itself does not close channels in excised patches, we believe that a cyclic GMP-dependent kinase may be involved (DeVries and Schwartz, 1992). It is not known whether either kinase acts directly on the gap-junction channel or whether additional steps intervene. In other preparations, direct phosphorylation of a connexin can alter cell-cell coupling. For example, Cx43 must be phosphorylated before junctional communication can occur (Musil et al., 1990). However, it is thought that this change in coupling is not due to a change in channel gating, but to a change in stability of a connexin in the membrane. The excised-patch preparation provides, for the first time, the ability to test whether kinases or other intermediaries act directly to change the conductance of gap-junction channels.

Voltage dependence

Gap junctions are produced by a family of proteins. Our results indicate that the gap junctions between horizontal cells respond to transjunctional voltage; each hemichannel operates independently and senses half the transmembrane voltage. In contrast, gap junctions that connect cells in *Drosophila* salivary glands respond both to transmembrane and transjunctional voltage (Verselis et al., 1991); and, the gap junctions that connect amphibian blastomeres behave as though their voltage sensors were within the aqueous pore and the gating of each hemichannel were dependent upon the state of its neighbor (Harris et al, 1981). The structural basis for these differences in voltage dependence is unknown.

Junctional formation

We suggest the following scenario for the formation of complete gap junctions: Hemi-gap-junction channels are continuously synthesized and inserted into the plasma membrane, where they accumulate. These hemi-gap-junction channels are prevented from opening by both the hyperpolarized operating range of horizontal cells (–20 to –80 mV) and physiological concentrations of extracellular Ca^{2+}. Presumably, hemichannels diffuse laterally in the membrane and the chance alignment of channels in adjacent cells allows them to join. The association of two hemichannels into a complete junction removes the control by extracellular Ca^{2+} and, since each horizontal cell has a similar membrane voltage, produces a transjunctional voltage that is relatively small. In this case, each half of a gap junction opens and gates independently. A similar scheme is indicated by the experiments of Musil and Goodenough (in this volume) for Cx43 in a normal rat kidney (NRK) cell line.

Lisa Ebihara (Chapter 11) has described the formation of large, non-selective channels following the expression of Cx46 in *Xenopus* oocytes. Cx46 was isolated from a rat lens library and may form gap junctions between lens fiber cells. The channel formed by Cx46 is constitutively open when expressed in unpaired oocytes bathed in a saline solution containing 0.5 mM Ca^{2+}. Both I_γ and Cx46 channels produce currents that are outwardly rectifying and share similar kinetics of activation and inactivation. Moreover, both channels have a unitary conductance of approximately 140 pS in symmetric bathing solution containing Cs^+ as the predominant cation. Further comparison between Cx46 and the horizontal cell I_γ channel or gap junction must wait until the amino-acid sequence of the horizontal cell channel is known.

Acknowledgements

This work was supported by a National Institutes of Health Research Grant EY-02440 and by a Medical Scientist Training Program Grant GM-07281 (S.H.D.)

References

DeVries, S.H. and Schwartz, E.A. (1989) Modulation of an electrical synapse between solitary pairs of catfish horizontal cells by dopamine and second messengers. J. Physiol. (London) 414, 351–375.

DeVries, S.H. and Schwartz, E.A. (1992) Hemi-gap-junction channels in solitary horizontal cells of the catfish retina. J. Physiol. (London) 445, 201–230.

Hamill, O.P., Marty, A., Neher, E., Sakmann, B. and Sigworth, F.J. (1981) Improved patch-clamp techniques for high-resolution current recording from cells and cell-free membrane patches. Pflügers Arch. 391, 85–100.

Harris, A.L., Spray, D.C. and Bennett, M.V.L. (1981) Kinetic properties of a voltage-dependent junctional conductance. J. Gen.

Physiol. 77, 95–117.

Lasater, E.M. and Dowling, J.E. (1982) Carp horizontal cells in culture respond to L-glutamate and its agonists. Proc. Natl. Acad. Sci. USA 79, 936–940.

Lasater, E. M. (1987) Retinal horizontal cell gap-junctional conductance is modulated by dopamine through a cyclic AMP-dependent protein kinase. Proc. Natl. Acad. Sci. USA 84, 7319–7323.

McMahon, D.G., Knapp, A.G. and Dowling, J.E. (1989) Horizontal cell gap junctions: Single-channel conductance and modulation by dopamine. Proc. Natl. Acad. Sci. USA 86, 7639–7643.

Musil, L.S., Cunningham, B.A., Edelman, G.M. and Goodenough, D.A. (1990) Differential phosphorylation of the gap-junction protein connexin43 in junctional communication-competent and -deficient cell lines. J. Cell Biol. 111, 2077–2088.

Teranish, T., Negishi, K. and Kato, S. (1983) Dopamine modulates S-potential amplitude and dye coupling between external horizontal cells in carp retina. Nature 301, 243–246.

Verselis, V.K., Bennett, M.V.L. and Bargiello, T.A. (1991) A voltage-dependent gap junction in *Drosophila melanogaster*. Biophys. J. 59, 114–126.

Part IV. Biophysics

J.E. Hall, G.A. Zampighi and R.M. Davis (Eds.)
Progress in Cell Research, Vol. 3

CHAPTER 13

Molecular and biophysical properties of the connexins from developing chick heart

RICHARD D. VEENSTRA[a], KATHRIN BERG[a], HONG-ZHAN WANG[b], EILEEN M. WESTPHALE and ERIC C. BEYER

[a]*Department of Pharmacology, SUNY Health Science Center, 750 E. Adams Street, Syracuse, NY 13210 and* [b]*Division of Hematology, Washington University School of Medicine, St. Louis, MO 63110, USA*

Introduction

Electrophysiological evidence indicates the existence of more than one unitary junctional conductance state between embryonic chick ventricular myocytes (40–240 pS, Chen and DeHaan, 1989) Gap-junction channel conductances (γ_j) of 40–80 and 140–160 pS predominate in 7-day embryonic chick heart (Veenstra and DeHaan, 1988). The macroscopic junctional conductance (g_j) of 7-day embryonic chick heart is apparently regulated by transjunctional voltages (V_j) as are other V_j-dependent gap junctions (DeHaan et al., 1989; Veenstra, 1990, 1991a). These findings were confirmed in hearts from 4-day, 14-day, and 18-day embryos.

These studies showed that the half-inactivation voltage (V_o) shifts towards larger V_j values, and the voltage-insensitive g_j component (g_{min}) increases with developmental age (Veenstra, 1991b,c). Multiple conductance states and developmental changes in the regulation of g_j by V_j can be explained by a variety of mechanisms, including functional modulation by post-translational modification (i.e. protein phosphorylation), differential V_j-dependent subconductance states, or the expression of multiple gap-junction proteins (connexins) having different functional electrical properties.

Recently, the protein composition of gap junctions in embryonic chick heart has been deduced by the molecular cloning of three distinct connexins, chick Cx42, 43, and 45 (ChCx42, 43, 45) from a chick embryo cDNA library (Musil et al., 1990; Beyer, 1990). The expression of multiple connexins adds another level of complexity to the regulation of intercellular communication within a given tissue.

In order to determine the functional properties of individual connexins, we have expressed the connexins by stable transfection of the connexin cDNAs into the communication-deficient SK Hep1 human hepatoma cell line (Eghbali et al., 1989; Fishman et al., 1990).

Connexins of developing chick heart

The first chick connexin to be cloned and sequenced had 92% amino-acid identity to rat Cx43 (Musil et al., 1990). Of the 24 amino-acid substitutions found in chick Cx43, only 12 are non-conservative (e.g. nonpolar-polar or acidic-basic amino-acid) substitutions. Hence, chick Cx43 is one of the most highly conserved connexins to have been identified. Other mammalian Cx43 isoforms found in bovine vascular smooth muscle and human heart have only 8 or 9 amino-acid substitutions (Fishman et al., 1990; Lash et al., 1990). Chick Cx43 RNA was detected in Northern blots of 10–12-day embryonic hearts, but was found in greater abundance in 12-day chick lenses (Musil et al., 1990). Sequence-specific rat Cx43 antisera cross-react with ChCx43 and immunocytochemical localization detects ChCx43 in junctional surfaces of chick lens epithelia. Yet, in contrast to mammalian heart, immunocytochemical analysis has failed to detect ChCx43 in embryonic chick or adult chicken heart (El Aoumari et al., 1990). This information suggests that ChCx43 may be a minor component of chick heart gap junctions.

The other two chick connexins, ChCx42 and ChCx45, were identified with the rat Cx43 probe using the same 10-day chick embryo cDNA library under low-stringency conditions (Beyer, 1990). The deduced amino-acid sequences for these two connexins are compared to ChCx43 in Fig. 1A. The amino-acid sequences are displayed according to the topological model for the connexin family of proteins (Beyer et al., 1990). All three chick connexins have three conserved cysteines in each of the two putative extracellular domains; the highest degree of homology is in the transmembrane, extracellular, and amino-terminal domains, which are common properties of the connexins. A unique feature of any of the chick connexins is the presence of a highly charged (3 acidic and 8 basic residues, including one histidine) 23 amino-acid sequence located at the end of the cytoplasmic loop nearest to the second transmembrane domain of ChCx45 (enclosed by the brackets in Fig. 1A).

All three connexins are expressed in embryonic chick heart, as revealed by Northern blot analysis of total RNA from embryonic and adult chicken heart (Fig. 1B). Compared to RNA levels in day-6 embryonic heart, all three connexins are expressed in the highest amounts at day 9 and are present in lower amounts in the adult heart. Quantitative comparisons of the relative amount of each connexin RNA to the total RNA at a given age cannot be made since the specific activity of the connexin-specific cDNA probes and exposure times were not standardized to a common value. While the RNA levels for chick Cx42, -43, and -45 are lower in the adult heart, this difference is most dramatic for ChCx45, which declines by 90% between day-6 embryos and the adult heart. The co-expression of multiple connexins in the same tissue raises some interesting questions about whether the amino-acid sequence variations confer different electrical and regulatory properties on the individual connexins. This would imply that there might be differential regulatory pathways for controlling intercellular communication within a given tissue. Additional experiments are required to determine if all three connexins are expressed in the same cell.

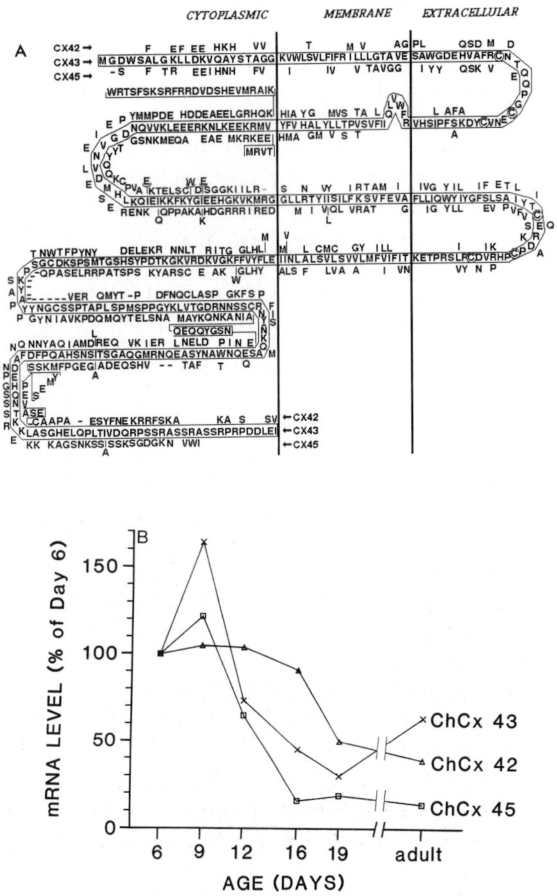

Figure 1. Developmental expression of three connexins in embryonic chick heart. (A) Topological representation of the primary amino-acid sequence for chick Cx43 (ChCx43, center) and the alternative amino-acid residues found in chick Cx42 (ChCx42, left) and Cx45 (ChCx45, right). Conventional single letter abbreviations depicting the identity of each amino-acid residue of ChCx43 are enclosed. Other letter codes indicate the amino acid substitutions and the dashes (–) indicate spaces inserted into the sequence to optimize the alignment of homologous amino acids for chick Cx42 or Cx45 relative to chick Cx43. ChCx45 sequence terminates with the final three typed residues VWI. (B) Relative amounts of RNA for each connexin expressed at different ages. Northern blots of total heart RNA prepared from embryonic and adult chickens were probed with connexin-specific ^{32}P-labeled probes. The intensities of hybridizing bands were quantified by scanning with a laser densitometer. For each connexin, the value at day 6 was arbitrarily assigned the value of 100% and every other time point is expressed as a percentage of the day 6 amount.

Developmental changes in regulation of junctional conductance in embryonic chick heart

Developmental changes in the regulation of junctional conductance were demonstrated by determining the g_j–V_j relationships for four ages of embryonic heart (Veenstra, 1990; 1991b,c). Figure 2 summarizes the responses of steady-state g_j to V_j in 4-, 7-, 14-, and 18-day embryonic chick ventricular myocyte pairs. In all experiments, V_j was varied between −100 and +100 mV by stepping V_1 from a holding potential of −40 mV to a new potential for 2 s. Each V_j pulse was separated by a recovery interval of 5 s ($V_j = 0$ mV $= V_2 - V_1$). Junctional-current measurements were taken at the beginning and end of each pulse, and the instantaneous and steady-state I_j values were plot-

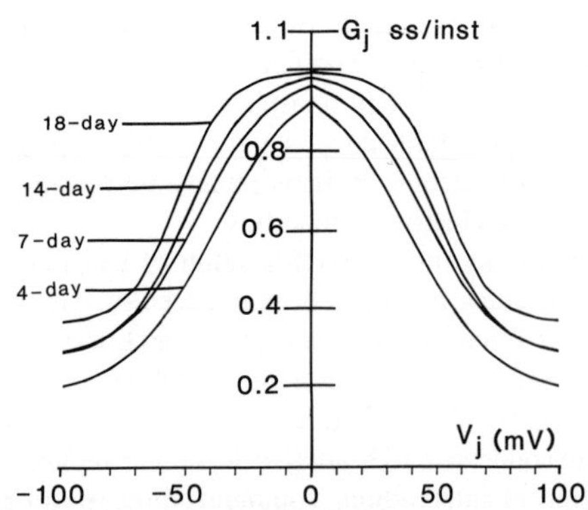

Figure 2. Developmental changes in voltage dependence of embryonic chick heart gap junctions. Boltzmann curves describing the steady-state normalized g_j–V_j relationships for 4-, 7-, 14-, and 18-day embryonic chick heart. An increase in the residual V_j-insensitive g_j (g_{min}) and a shift in the half-inactivation voltage towards larger potentials is evident with increasing age. The area under the 18-day curve is significantly greater than the 4-day curve and is indicative of a developmental loss in V_j-dependent regulation of g_j in embryonic chick heart. Reprinted with permission from Veenstra (1991c).

ted as a function of V_j. Since the instantaneous I_j–V_j relationship was linear for all cell pairs (i.e. instantaneous g_j is constant, $g_j = I_j/V_j$), regardless of age or magnitude of g_j, the steady-state g_j (g_{ss}) was normalized to the instantaneous g_j (g_{inst}) at each V_j and the data pooled according to developmental age. The mean normalized g_{ss}–V_j relationships were then fit with a two-state Boltzmann distribution used to describe the V_j dependence of other embryonic gap junctions (Spray et al., 1981). The equation can be written as follows:

$$g_{ss}/g_{inst} = \{(g_{max} - g_{min})/(1 + \exp[A(V_j - V_o)])\} + g_{min} \quad (1)$$

where g_{max} is the maximum conductance (= 1, normalized to instantaneous g_j), g_{min} is the minimum conductance attained during each experiment, V_o is the voltage where g_j lies midway between g_{max} and g_{min}, and A is a slope factor related to the gating charge of the system. The constant A can be expressed as zq/kT, where z is the number of equivalent electron charges, q acts as the voltage sensor, and kT represent Boltzmann's constant and absolute temperature, respectively. The values for these relevant parameters are listed in Table I according to embryonic age of the heart (Veenstra, 1990, 1991b,c).

TABLE I Developmental differences in V_j-dependence of embryonic chick cardiac gap junctions

Age	g_{max}	g_{min}	N	A	V_o (mV)	z
4-day	1.0	0.18	8	0.059	37	1.50
7-day	1.0	0.28	4	0.068	45	1.72
14-day	1.0	0.28	8	0.081	49	2.06
18-day	1.0	0.36	11	0.105	53	2.67

Although there is an apparent trend in all of the parameters listed in Table I, only the difference in g_{min} and the area under the 4- and 18-day curves were significantly different ($p < 0.05$, ANOVA). This shift in the V_j sensitivity of g_j in embryonic heart towards larger potentials provides the first direct experimental evidence for developmental changes in the regulation of intercellular communication within a given tissue. Previous experimental data also revealed that signal averaging of repetitive junctional current traces containing 160 pS channel activity produces an exponentially decaying I_j characteristic of

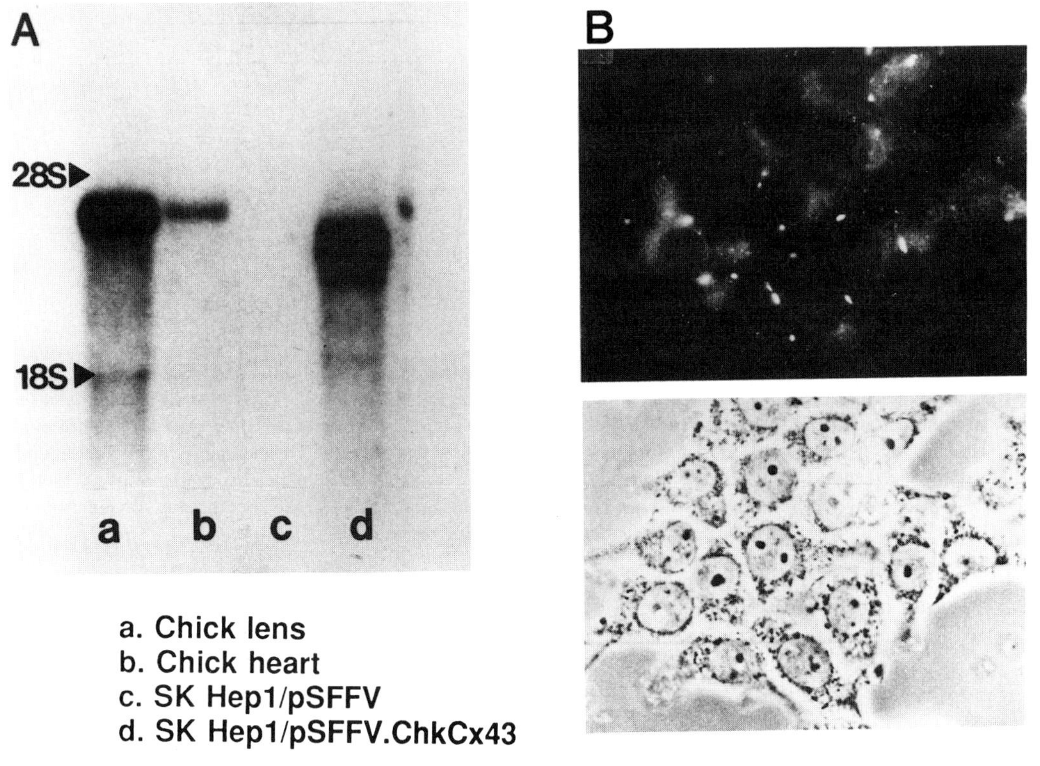

Figure 3. Expression of chick Cx43 mRNA and protein in SK Hep1 cells. (A) Total RNA from whole 12-day chick lens (a), whole 12-day chick heart (b), SK Hep1 cells transfected with the pSFFV-*neo* expression vector without the ChCx43 insert (c), and SK Hep1 cells transfected with the pSFFV-*neo* expression vector containing the ChCx43 cDNA insert. 10 μg of RNA was loaded in each lane and the RNAs were probed at high-stringency with cDNA probes specific for ChCx43. (B) Phase-contrast (upper panel) and immunofluorescent (lower panel) photomicrographs of ChCx43-transfected cells. Antisera against rat Cx43 were used as the primary antibody for immunofluorescent staining of the transfected cells. Punctate staining occurs at intercellular contacts between the transfected cells and presumably corresponds to intercellular gap-junction plaques.

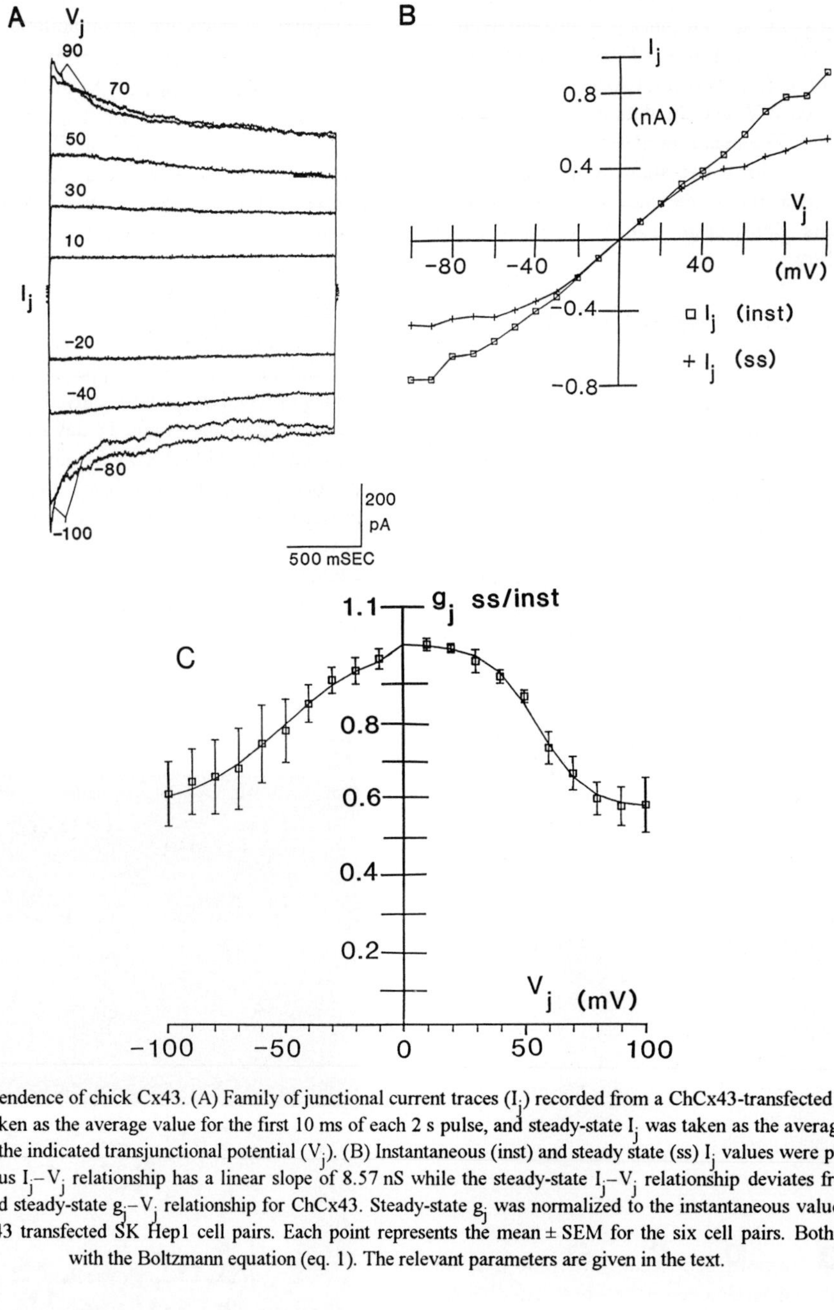

Figure 4. Voltage dependence of chick Cx43. (A) Family of junctional current traces (I_j) recorded from a ChCx43-transfected SK Hep1 cell pair. Instantaneous I_j was taken as the average value for the first 10 ms of each 2 s pulse, and steady-state I_j was taken as the average value for the last 10 ms of each pulse to the indicated transjunctional potential (V_j). (B) Instantaneous (inst) and steady state (ss) I_j values were plotted as a function of V_j. The instantaneous I_j–V_j relationship has a linear slope of 8.57 nS while the steady-state I_j–V_j relationship deviates from linearity above 20 mV. (C) Normalized steady-state g_j–V_j relationship for ChCx43. Steady-state g_j was normalized to the instantaneous value at each V_j for six experiments on ChCx43 transfected SK Hep1 cell pairs. Each point represents the mean ± SEM for the six cell pairs. Both polarities were fit with the Boltzmann equation (eq. 1). The relevant parameters are given in the text.

V_j-dependent gap junctions (Veenstra, 1990, 1991b). Although the V_j-dependent closure of a population of 160 pS gap-junction channels may be involved in the time-dependent decline in I_j observed in embryonic chick heart, this observation alone cannot explain the developmental changes observed in this tissue. Therefore, by determining functional expression of the individual connexins in a communication-deficient cell line we tested the hypothesis that developmental expression of the three chick connexins can account for the observed changes in V_j sensitivity and multiple channel conductances found in embryonic chick heart.

Functional expression of chick Cx43

The first chick connexin to be functionally expressed by stable transfection was ChCx43. The SK Hep1 human hepatoma cell line was selected for transfection using the eukaryotic expression vector pSFFV-*neo*. Control transfections were performed using only the pSFFV-*neo* vector to confer neomycin resistance, while other cells were transfected with the ChCx43 cDNA inserted into the EcoR1 site of the expression vector. The ChCx43 cDNA insert contained the entire coding sequence plus 200 bases of the 5' and 3' noncoding regions. Expression of ChCx43

TABLE II V_j-dependence of embryonic chick Cx43

Polarity	g_{max}	g_{min}	N	A	V_o (mV)	z
Negative	1.0	0.57	6	−0.059	−51	1.32
Positive	1.0	0.57	6	0.101	56	2.57

was examined by Northern blot analysis of total RNA prepared from 12-day embryonic chick lens and heart, in addition to control and ChCx43-transfected cells grown in the presence of 0.5 mg/ml G418 (Fig. 3A). Each lane was loaded with the same amount of RNA and probed by high-stringency hybridization with a ChCx43-specific probe. Chick lens, chick heart, and pSFFV-ChCx43 transfected SK Hep1 cell RNA all hybridized with the probe; no ChCx43 RNA was detected in the control pSFFV-transfected cells. ChCx43 protein was also detected in transfected cells by immunocytochemical localization using rat Cx43(252–271) antisera (Beyer et al., 1989), which cross-reacts with ChCx43 (Musil et al., 1990). Comparison of the phase-contrast (upper panel, Fig. 3B) and immunofluorescent (lower panel, Fig. 3B) photomicrographs reveals a punctate staining pattern only at regions of intercellular contact between the ChCx43-transfected cells.

Junctional currents were determined in pairs of ChCx43-transfected SK Hep1 cells using standard double whole-cell recording procedures as described for embryonic chick ventricular myocyte pairs (Veenstra, 1990, 1991b). From a total of 12 cell pairs, g_j averaged 3.90 ± 0.91 nS (mean ± standard error of the mean). Figure 4A displays a family of junctional currents acquired from one of these ChCx43-transfected cell pairs. At transjunctional voltages of 30 mV, I_j appears to be time independent, while a slow time-dependent current is evident at V_j=40 mV. The time-dependent relaxation of I_j becomes faster and more pronounced at V_j values of 70 and above. If one plots the I_j values measured at the beginning (instantaneous) and end (steady state) of each V_j pulse, the I_j–V_j relationships depicted in Fig. 4B are obtained. The instantaneous I_j–V_j relationship approximates a straight line with a slope of 8.57 nS, which is equal to the g_j value for this cell pair. The steady-state I_j–V_j relationship becomes non-linear beyond 30 mV, but does not have the negative slope region observed in 4- and 14-day chick heart cell pairs (Veenstra, 1991b). The shape of this curve more closely resembles that from 18-day embryonic chick heart cell pairs.

A more definitive answer of how closely the V_j dependence of ChCx43 resembles that of developing chick heart can be obtained by comparing the normalized steady-state g_j–V_j curves, provided that each can be adequately defined by the classical two-state Boltzmann model (equation 1) applied to all previous V_j-dependent gap junctions. The normalized steady-state g_j–V_j curve for ChCx43 is shown in Fig. 4C. The solid lines represent the best fits of the experimental data points generated using equation 1. The best fit for each polarity was determined using the least-squares criterion. The relevant parameters are listed in Table II. The normalized steady-state g_j–V_j curve is approximately symmetrical around 0 mV with the exception of the slope factor, A (and consequently the valence, z). The g_{min} value is 20% higher than that observed in 18-day chick heart and indicates that ChCx43 cannot account for much of the V_j-dependence observed at any age of the developing chick heart. This latest observation is consistent with the findings that ChCx43 is not detected by immunofluorescence and may be a minor component of the embryonic and adult chicken heart.

In three poorly coupled cell pairs, single-channel currents could be resolved (Fig. 5). With the voltage of cell 2 held constant at 0 mV, the voltage of cell 1 was stepped to −100 mV as indicated by the large jump in whole-cell current for cell 1 (I_1). The paired current traces displayed here represent a 6.76 s recording interval at V_j=100 mV. I_j is taken as the change in I_2 during the V_j pulse, as confirmed by the return to the prepulse holding current upon

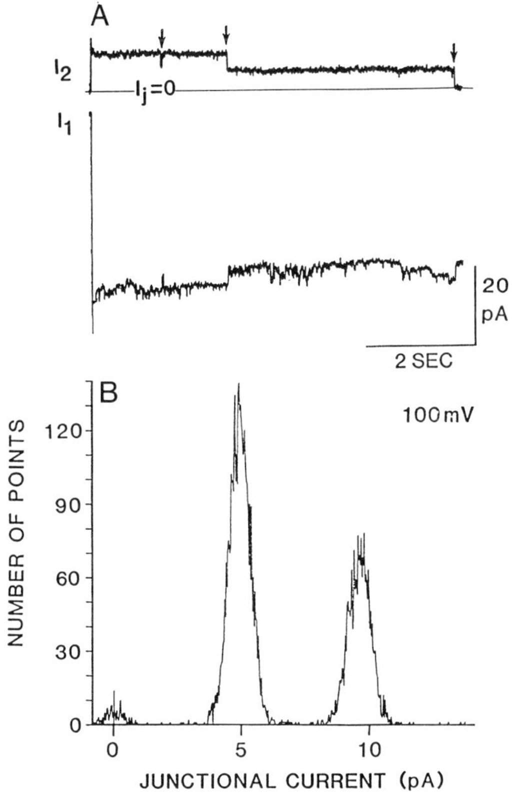

Figure 5. Unitary channel conductance of chick Cx43. (A) Whole-cell currents (I_1 and I_2) recorded from a pair of ChCx43-transfected cells during a step in V_j from 0 to +100 mV. Both records were low-pass filtered at 125 Hz and digitized at 1 kHz. The arrows indicate channel transitions observed during the 6.76 s record. Note that the final channel closure returns to the ground state, indicating that no channels were open at that moment in time and there is no leak in the I_j record. (B) Channel amplitude histogram of the I_j current trace shown in panel A. Each stable current level is evident as a peak in the amplitude histogram. The differences between the means of the three Gaussian peaks are 4.7 and 4.8 pA, which corresponds to a channel conductance of 47–48 pS.

closure of a second gap-junction channel. Figure 5B is the channel amplitude histogram compiled from all 6763 points in the I_2 current record from Fig. 5A. Three peaks which correspond to the ground state ($I_j = 0$) and two open states are evident in the junctional current histogram. The differences between the means of adjacent peaks measure 4.7 and 4.8 pA, corresponding to a γ_j of 47–48 pS. Similar γ_j values of 43–60 pS have been reported for mammalian heart and Cx43 (Burt and Spray, 1988; Rook et al., 1988; Rüdisüli and Weingart, 1989; Fishman et al., 1990).

Preliminary observations with chick Cx42 and -45

Because of low levels of endogenous coupling in SK Hep1 cells (Eghbali et al., 1990), we chose to express the remaining chick connexins in the mouse N2A neuroblastoma cell line by stable transfection of ChCx42 and ChCx45 cDNAs incorporated into the pSFFV-*neo* eukaryotic expression vector. Northern blot analysis did not reveal any known connexins, and no endogenous coupling was detected in > 30 electrophysiological experiments in N2A cells. A ChCx45 clone has been obtained in which electrical coupling was detected in over 55% of the cell pairs examined, with a mean g_j of 4.17 ± 1.32 nS (N = 12). A time-dependent decline of I_j was observed when V_j exceeded 10 mV, and became more pronounced as V_j increased. The instantaneous I_j–V_j relationships were linear over the entire V_j range of 100 mV. Hence, steady-state g_j was normalized to the instantaneous g_j for each experiment and the Boltzmann distribution was determined for ChCx45. The normalized steady-state g_j–V_j curve for 6 cell pairs was symmetrical around 0 mV with a g_{min} of 0.10, a slope factor of 0.70, and a half-inactivation voltage near 30 mV for both polarities of V_j. Channel currents were observed over the 70 to 100 mV range in one poorly coupled cell pair and the I_j–V_j relationship had a linear slope of 30 pS. This new information, combined with the evidence for a developmental decrease in the RNA levels for ChCx45, suggests that the expression of ChCx45 can account for the V_j-dependence observed in the early stages of embryonic chick heart, especially at voltages < 50 mV.

ChCx42 was also expressed in an N2A cell clone and very preliminary results from 3 cell pairs indicate that V_j dependence begins at 20 mV. I_j declined to a minimum value of 35% of instantaneous I_j at 100 mV, with a half-inactivation voltage of 40–50 mV for both polarities. These values are similar to those reported for 14- to 18-day embryonic chick heart (Table I). Channel activity was observed in one of these cell pairs and preliminary current amplitude measurements indicate that ChCx42 forms a gap-junction channel with a γ_j in excess of 100 pS, which could account for the of 140–160 pS channel activity observed in paired embryonic chick heart myocytes. Work is continuing to determine the functional properties of all three chick connexins expressed in mouse N2A cells.

Acknowledgements

We thank Mark Chilton for his excellent technical assistance in maintaining the clonal cell cultures received from Dr. Beyer's laboratory. This work was supported by NIH grants HL45466, EY08368, and HL42220.

References

Beyer, E.C. (1990) Molecular cloning and developmental expression of two chick embryo gap-junction proteins. J. Biol. Chem 265, 14439–14443.

Beyer, E.C., Kistler, J., Paul, D.L. and Goodenough, D.A. (1989) Antisera directed against connexin43 peptides react with a 43-kDa protein localized to gap junctions in myocardium and other tissues. J. Cell. Biol. 108, 595–605.

Beyer, E.C., Paul, D.L. and Goodenough, D.A. (1990) Connexin family of gap-junction proteins. J. Membr. Biol. 116, 187–194.

Burt, J.M. and Spray, D.C. (1988) Single-channel events and gating behavior of the cardiac gap-junction channel. Proc. Natl. Acad. Sci. USA 85, 3431–3434.

Chen, Y.-h. and DeHaan, R.L. (1989) Cardiac gap-junction channels shift to lower conductance states when temperature is reduced. Biophys. J. 55, 152a (abstr.).

DeHaan, R.L., Chen, Y.-h. and Penrod, R.L. (1989) Voltage dependence of junctional conductance in the embryonic heart. In: L. Hondeghem (Ed.), Molecular and Cellular Mechanisms of Antiarrhythmic Agents, Futura, Mount Kisco, NY, pp. 19–43.

Eghbali, B., Kessler, J.A. and Spray, D.C. (1989) Expression of gap-junction channels in communication-incompetent cells after stable transfection with cDNA encoding connexin32. Proc. Natl. Acad. Sci. USA 87, 1328–1331.

El Aoumari, A., Fromaget, C., Dupont, E., Reggio, H., Durbec, P., Briand, J.-P., Bller, K. and Gros, D. (1990) Conservation of a cytoplasmic carboxy-terminal domain of connexin43, a gap-junction protein, in mammalian heart and brain. J. Membr. Biol. 115, 229–240.

Fishman, G.I., Spray, D.C. and Leinwand, L.A. (1990) Molecular characterization and functional expression of the human cardiac gap-junction channel. J. Cell. Biol. 111, 589–598.

Lash, J.A., Critser, E.S. and Pressler, M.L. (1990) Cloning of a gap-junctional protein from vascular smooth muscle and expression in two-cell mouse embryo. J. Biol. Chem. 265, 13113–13117.

Musil, L.S., Beyer, E.C. and Goodenough, D.A. (1990) Expression of the gap-junction protein connexin43 in embryonic chick lens: Molecular cloning, ultrastructural localization, and post-translational phosphorylation. J. Membr. Biol. 116, 163–175.

Rook, M.B., Jongsma, H.J. and van Ginneken, A.C.G. (1988) Properties of single gap-junctional channels between isolated neonatal rat heart cells. Am. J. Physiol. 255, H770–H782.

Rüdisüli, A. and Weingart, R. (1989) Electrical properties of gap-junction channels in guinea-pig ventricular cell pairs revealed by exposure to heptanol. Pflügers Arch. 415, 12–21.

Spray, D.C., Harris, A.L. and Bennett, M.V.L. (1981) Equilibrium properties of a voltage-dependent junctional conductance. J. Gen. Physiol. 77, 77–93.

Veenstra, R.D. (1990) Voltage-dependent gating of gap-junction channels in embryonic chick ventricular cell pairs. Am. J. Physiol. 258, C662–C672, 1990.

Veenstra, R.D. (1991a) Comparative physiology of cardiac gap-

junction channels. In: C. Peracchia (Ed.), Biophysics of Gap Junction Channels, CRC Press, Boca Raton, FL, pp. 131–144.
Veenstra, R.D. (1991b) Developmental changes in regulation of embryonic chick heart gap junctions. J. Membr. Biol. 119, 253–265.
Veenstra, R.D. (1991c) Physiological modulation of cardiac gap-junction channels. J. Cardiovasc. Electrophys. 2, 168–189.
Veenstra, R.D. and DeHaan, R.L. (1988) Cardiac gap-junction channel activity in embryonic chick ventricle cells. Am. J. Physiol. 254, H170–H180.

J.E. Hall, G.A. Zampighi and R.M. Davis (Eds.)
Progress in Cell Research, Vol. 3

CHAPTER 14

Multiple channel conductance states in gap junctions

YAN-HUA CHEN and ROBERT L. DeHAAN

Department of Anatomy and Cell Biology, Emory University Health Science Center, Atlanta, GA 30322, USA

Introduction

Many gap junctions, including those in embryonic chick cardiac cells, have multiple channel conductance states. We propose three models for gap-junction structure: (a) small (γ_{40}) channels that can exhibit cooperativity; (b) large (γ_{240}) channels with evenly spaced substates; or (3) several classes of channels of different sizes. We suggest a gating mechanism that would predict channels with evenly spaced sub-states. Our model differs from the Unwin tilting helix in that it would permit the phenylalanine "gate" associated with each connexin M3 transmembrane domain to twist into the connexon pore independently of the movements of the other five members of the connexin hexamer.

Measurements of unit channel conductance (γ) in diverse preparations reveal that many, perhaps most, of the voltage-gated and ligand-gated membrane channels that allow ions to move between the extracellular milieu and the cytosol exhibit more than one conductance level (reviewed in Fox, 1987; Patlak, 1991). Examples include SR K^+ channels (Fox, 1987; Hill et al., 1990), neuronal glutamate-activated channels (Cull-Candy and Usowicz, 1987; Jahr and Stevens, 1987), the cardiac inward rectifier (Sakmann and Trube, 1984; Kell and DeFelice, 1988), and sodium channels (Patlak, 1988). Investigators studying gap-junction channels have also reported both single- and multiple-conductance states, raising the question of whether gating in gap-junction channels normally entails an abrupt switch from a closed state to a unique open-pore configuration, or if gap-junction channels, under some conditions, can open in a graded or step-wise fashion that results in more than one distinguishable conductance level.

A rapidly accumulating body of evidence indicates that gap junctions are formed from a multi-gene family of closely related proteins, the connexins. DNAs encoding nine rat connexins have been cloned and sequenced. Using a nomenclature based on species of origin and molecular mass predicted by cDNA analysis, these connexins have been designated rat connexin 26 (Cx26), Cx31, Cx31.1, Cx32, Cx33, Cx37, Cx40, Cx43, and Cx46 (Beyer et al. 1990; Haefliger et al., 1992). Homologs of many of these connexins have been identified in gap junctions of other species, e.g. chicken (Beyer, 1990; Musil et al., 1990); *Xenopus* (Gimlich et al., 1988; Ebihara et al., 1989; Gimlich et al., 1990); mouse (Hennemann et al., 1992; this volume, Chapter 5); dog (Kanter et al., 1992); humans (Kumar and Gilula, 1986; Fishman et al., 1990). Five of these connexins are associated with gap junctions in heart tissue: Cx37, Cx40, Cx42, Cx43, and Cx45 (Beyer, 1990; Haefliger et al., 1992; Gourdie et al., 1992; Kanter et al., 1992).

Nucleotide sequences and derived amino-acid sequences of these connexins, coupled with hydropathy analysis suggest that they are pore-forming membrane proteins comprised of four transmembrane α-helices (labeled M1–M4), with two extracellular loops and three cytoplasmic domains. One of the α-helices (M3) is amphiphilic and is thought to line a portion of the wall of the aqueous pore (Milks et al., 1988). Among the connexin family members, the extracellular, transmembrane and N-terminal cytoplasmic domains are all well conserved within and between species, while the central and C-terminal cytoplasmic regions are highly variable in both sequence and length.

Are gap-junction channels two-state devices?

Substantial evidence has been obtained recently supporting the idea that gap-junction channels in differentiated tissues operate as two-state devices. According to this view, any given channel can be either closed in a non-conducting configuration, or fully open to a characteristic conductance level. If this model is correct, substates, or conductance levels smaller than the fully open state would be excluded, and junctions that exhibit different conductance levels would necessarily have distinctive classes of channels. Variation around a mean value of γ might result from small fluctuations in pore size from one opening to the next in a given channel, or from small individual variations among the channels of a given class. Cardiac gap-junction channels, in which Cx43 predominates (Beyer et al., 1990; Yancey et al., 1990; Dermietzel et al., 1990), have been described in these terms (Spray and Burt, 1990). Burt and Spray (1988), using 135 mM K-glutamate as the main charge carrier, reported that neonatal rat heart junctional channels have a single characteristic conductance of 53 pS under control conditions, and of 55–60 pS when the total junctional conductance (G_j) was reduced with cytoplasmic acidification or agents such as halothane or heptanol. Cardiac junctional channel conductance between pairs of adult rat ventricular myocytes bathed in 3 mM heptanol was 43 pS (Rüdisüli and Weingart, 1990). In neonatal hamster heart cells it was 64 pS (Veenstra, 1990b; 120 mM K-glutamate). In cultured striatal astrocytes, whose gap junctions are also composed of Cx43, γ was 56 pS (Giaume et al., 1991; 140 mM KCl). Similarly, transfection of communication-deficient SKHep1 cells with cDNA encoding the human cardiac gap-junction (HCGJ) Cx43 protein endows these cells, again in the presence of halothane, with a 55–60 pS gap-junction channel (Fishman et al., 1990). Preparations in which the major gap-junction protein is Cx32 appear to have a larger unitary conductance, on the order of 120–150 pS (Young et al., 1987; Eghbali et al., 1990). Spray et al. (1991) reported three conductance levels at 20–30 pS, 80–90 pS and 100–120 pS in junctions of WB cells, a clonal cell line derived from rat liver. In these preparations, immunocytochemical evidence of both Cx43 and Cx32, and ultrastructural images showing two distinct particle sizes, led the investigators to infer that the junctions contained two separate classes of junctional channels (~25 pS and ~85 pS), that occasionally sum to the 110 pS level. This interpretation is in agreement with the two-state model, rather than one that envisions a single-channel class with sub-conductance states. But it leaves open the question of why 50–60 pS and 160–170 pS channel states (i.e. sums of two simultaneous 25 pS and 85 pS openings) were never reported. Furthermore, even some of the reports that emphasize a single dominant value of γ illustrate channel events of more than one size in their current records (e.g. Young et al., 1987; Burt and Spray, 1988; Moreno et al., 1991).

The evidence cited above supports the contention that there are different size classes of gap-junction channels, each composed of a given connexin protein and each with a single unique conductance. Other results, however, suggest that γ can have more than one value. In pairs of neonatal rat cardiac myocytes, Rook et al. (1988) reported that γ = 43 pS, but a substate at about half that value was also noted. Communication-deficient N2A cells, stably transfected with chick Cx43, had junctional channels with a main γ of 44 pS but which also had openings to 28 pS and 67 pS (Veenstra et al., 1992). Pairs of A7r5 smooth muscle cells from rat aorta exhibited two conductance levels, 36 and 89 pS (Moore et al., 1991). In these preparations, Cx43 was the only gap-junction sequence that could be detected by Northern blots, immunoblots, or immunochemical staining. Similarly, preparations containing primarily Cx32 may show more than one γ level

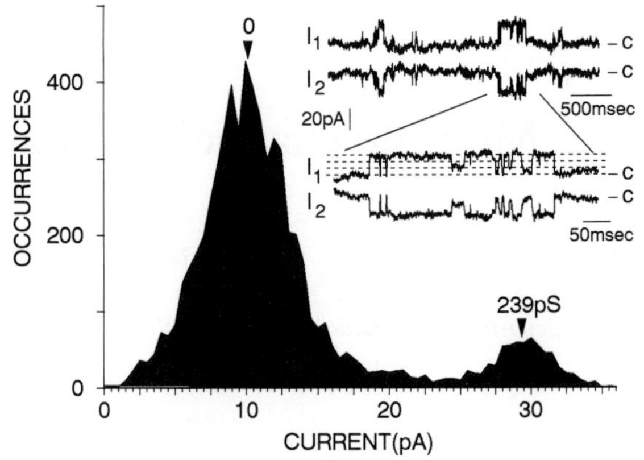

Figure 1. Analysis of steady-state gap-junction channel activity from a typical pair of 7-day chicken ventricle cells, showing records of channel current and an amplitude histogram obtained with dual patch-electrode voltage clamp as described previously (Veenstra and DeHaan, 1988). Each electrode had an access resistance to the cell interior (2–6 MΩ) and a seal resistance (5–50 GΩ) between the cell interior and the external reference electrode. The membrane (input) resistance of each cell was 2–5 GΩ, while R_j, the junctional resistance between the cells, was 0.1–5 GΩ ($G_j = 1/R_j = 0.2$–10 nS). Channel current amplitude (i_j) was measured in two ways, by computer-generated amplitude histograms (MSINCH Program, Goolsby), and by direct measurement of the current transitions from the oscilloscope screen (see Fig. 2). MSINCH was designed to construct current amplitude histograms, and to calculate i_j from the interval between peaks. The amplitude histogram was prepared from a 56-s epoch of gap-junction channel recording, during which V_1 was held constant at −10 mV and $V_2 = -90$ mV ($V_j = 80$ mV). Although 80 pS and 160 pS conductance states can be observed in the records, the amplitude histogram shows no peaks at these levels because there were too few of these events. Mean conductance ($\gamma = i_j/V_j$) from the amplitude histogram is 239 pS. (Inset) A 4-s segment of the record from which the amplitude histogram was prepared. Unitary currents (i_j) through junctional channels were seen as symmetrical signals of equal size and opposite polarity in the two electrodes (Veenstra and DeHaan, 1988). Because V_j was positive, opening channels allowed outward current to flow from cell 1 (shown in I_1 as an upward deflection) across the junction into cell 2 (inward current shown in I_2 as a downward deflection). The lower pair of traces expands a section of channel activity with dotted lines at each 80 pS of conductance. During the 4-s "burst" of activity shown in this record, the open state probability (P_o) was 0.11. Cell pair 040187, IPS#57.

(Somogyi and Kolb, 1988; Kolb and Somogyi, 1990). More convincing results are from Veenstra et al. (1992) who found that junctions in N2A cells expressing only chick Cx42 had a maximal γ of 236 pS with four smaller substates at approximately 40 pS increments.

In a recent study of gap-junction channel activity in embryonic chick cardiac cell pairs (Chen and DeHaan, 1992a,b) channel events that occurred during 56-s epochs of steady-state transjunctional voltage ($V_j = 80$ mV) were grouped in spontaneous bursts, seen as periods of intense activity (Fig. 1), and separated by long periods of rare openings. In these preparations, most junctions showed clear multiple-conductance levels (Fig. 2A and Table I.

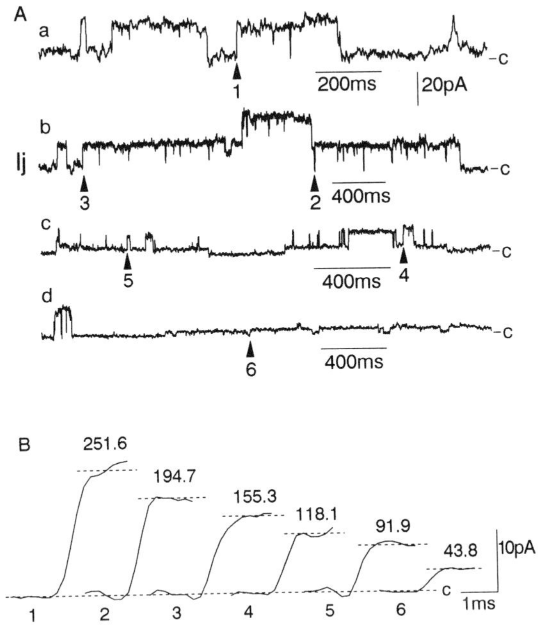

Figure 2. Unitary channel open-close events during steady-state recording at 80 mV V_j ($V_1 = -10$ mV, $V_2 = -90$ mV). (A) Records a–d illustrate typical channel activity, recorded simultaneously in both electrodes and displayed at time scales selected to show individual open-close events. Only I_1 is shown in epochs of 1.5 s (a) or 3 s (b–d). Openings to each of the common conductance levels are labeled (upward arrowheads 1–6). (B) Each current transition labeled in A is displayed at fast sweep speed to demonstrate the direct method of measuring γ from the oscilloscope screen. Transitions between stable i_j levels were selected. A stable i_j level was defined as a period of at least 1 ms before or after a smooth transition, during which current fluctuated no more than 1 pA around the mean. A transition was defined as a monotonically changing sequence of points from one i_j level to the next. Since the values of γ centered around multiples of 40 pS (see Table I), we refer in the text to the different channel sizes by their nominal conductance (γ_{40}, γ_{80}...γ_{240}). Data was filtered at 1 kHz and displayed at 200 µs/point. The current traces shown in this figure were obtained from the following cell pairs: (a) 040187; (b) 121687b; (c,d) 121687a.

The dominant open states were near 80 pS (γ_{80}) and 160 pS (γ_{160}), as reported earlier (Veenstra and DeHaan, 1988; DeHaan, 1988; DeHaan et al., 1989; Veenstra, 1990a), with more than 50% of over one thousand discrete events measured at these two conductance levels (Table I. But the frequencies of openings to the γ_{40} and γ_{120} states were almost as great, and 10% of the events were 240 pS. Although γ_{240} was present as the maximal uninterrupted current transition in all but one of eight experiments (in which only γ_{40} conductance states appeared), small openings were favored in some junctions, while others exhibited a greater frequency of large-channel events.

Our finding of several conductance levels cannot be ascribed to the simultaneous opening of members of a single population of stochastically active γ_{40} channels. Such a model can be ruled out by their observed characteristics and frequencies of the openings (Lauger, 1985; Fox, 1987). If openings like those pictured in Fig. 2A (to levels between 80 pS and 240 pS) resulted from the random simultaneous superposition of two to six γ_{40} channels, the frequency of appearance of the progressively larger conductances should approximate an inverse power series, with the likelihood of appearance of the γ_{240} level becoming vanishingly small. Furthermore, random superposition would only very rarely result in smooth transitions to the larger conductance levels, whereas we commonly observed such transitions (Fig. 2B). The data instead are consistent with three other models: (1) 40 pS channels that can exhibit cooperativity; (2) γ_{240} channels with substates; or (3) channels of different sizes.

Both models 1 and 3 would conform to the concept that gap-junction channels are two-state devices that can be either closed, or fully open to a homogeneous pore size. A version of model 1 is a multiport like that proposed by Hunter and Giebisch (1987) to explain the behavior of K^+-channels in renal tubules. Groups of six individually gated γ_{40} channels sharing a common gate, or connected by a mechanism that caused one to six members of the group to open simultaneously on a random basis, would result in an apparent γ_{240} channel as the maximal conductance with five equally spaced substates. There is little evidence in support of this model, and no such cooperative mechanism is known. The purest preparations examined with high-resolution electron microscopy show no evidence at an ultrastuctural level for any physical connections or common gate-linking groups of six connexons (e.g. Stauffer et al., 1991). Moreover, when channels closed by physiological agents are examined after rapid freezing, the connexons are not even seen in hexagonally packed paracrystalline arrays (Miller and Goodenough, 1985), as is commonly

TABLE I Frequency of occurrence and conductance of embryonic chick cardiac gap junction channel openings[1]

Nominal γ (pS)	γ_{40}	γ_{80}	γ_{120}	γ_{160}	γ_{200}	γ_{240}
Mean γ	42.6	80.7	119.6	157.7	200.4	240.3
Std. dev.	6.8	7.9	8.2	9.0	8.5	19.1
N^2 (total = 1104)	193	321	170	231	78	111
% total	17.5	29.2	15.5	21.0	7.1	10.1

[1] Data from Chen and DeHaan (1992b), pooled from eight 7-day chick embryo ventricular cell pairs in which γ was measured as described in the legend to Fig. 1. G_j of the 8 junctions was 0.24–1.65 nS.
[2] N = number of channel openings during 59.1 min of total recording time.

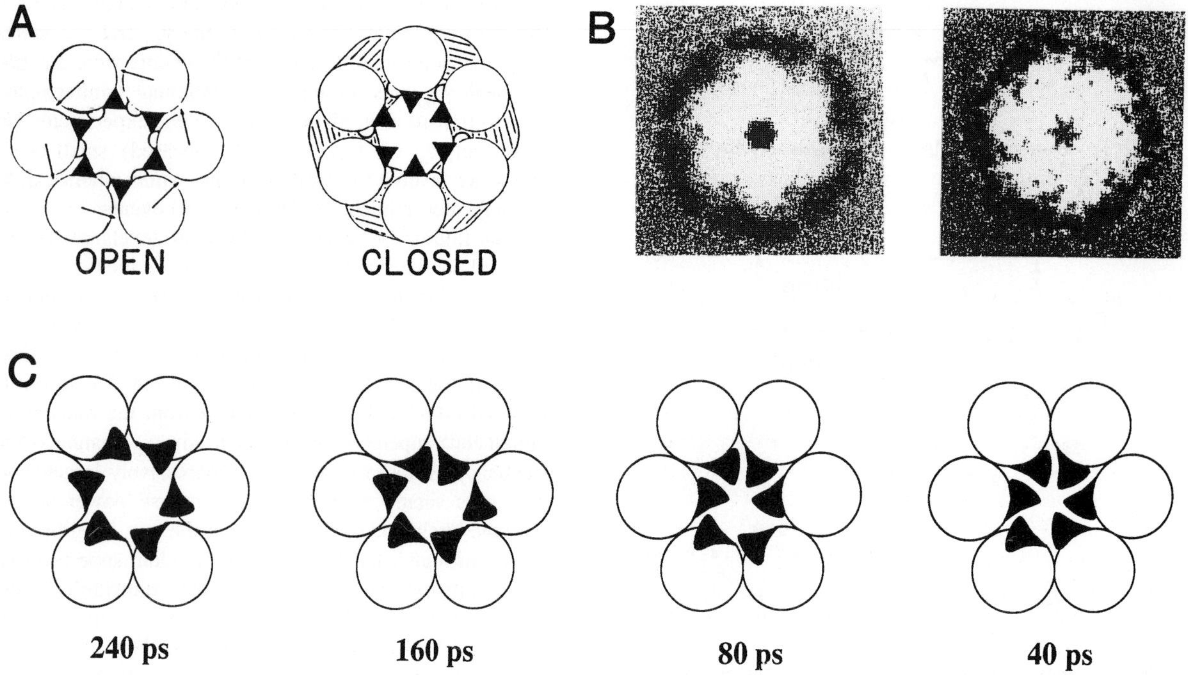

Figure 3. Models of gap-junction gating. (A) Tilt model of connexon closure. The M3 membrane-spanning domains of each of the six connexin subunits (circles) are thought to line the channel pore. In the open state (left), each M3 helix is nearly orthogonal to the plane of the membrane. Here they are viewed end-on from the cytoplasmic side. Closure of the pore (right) occurs when each M3 helix tilts in the direction of the arrow, moving bulky phenylalanine groups into the lumen that were previously tucked out of the way (adapted from Bennett et al. 1991). (B) High resolution electron micrographs of purified rat liver connexons. Sixfold rotationally filtered images of two connexons in the same electron micrographic field (from Stauffer et al, 1991, with permission). (C) Proposed twist model of connexon closure that allows for conductance substates. Fully open state is shown at left. Intermediate conductance levels would result if the phenylalanine "gating" group associated with each M3 helix (shown as black leaflets) were free to rotate individually into the lumen. If the conductance of the fully open pore (γ_{max} = 240 pS) is reduced to zero by occlusion with six gating units, partial occlusion by one leaflet is assumed to yield a 200 pS substate, two would produce 160 pS, and so on.

reported after aldehyde fixation. In the absence of such proximity or regularity, it is difficult to imagine a physical substratum that could mediate cooperativity.

Model 2, in contrast, presupposes a channel that can exist in six different configurations, each with a different conductance. Recent models of closure of ion channels visualize a helical twisting (Catterall, 1988; Millhauser, 1990) or tilting (Unwin, 1989) of the pore-forming subunits.

In the open state, a polar region of an amphipathic transmembrane helix is thought to line the wall of the aqueous pore. The tilting or twisting motion of the subunits could move non-polar groups into the channel, thereby occluding it. Bennett et al. (1991) have elaborated on an Unwin-based model for the connexon (Fig. 3A). Moreover, images that could represent the two states have been seen with high-resolution electron microscopy of purified liver connexons (Stauffer et al., 1991 and Fig 3B). While the authors of this work are careful not to speculate about the functional significance of these images, one reasonable interpretation could be that the connexon in 3B (left) has the hexameric connexin units nearly vertical, and thus has an open pore, while the one in 3B (right) shows the tilted arrangement, with a "bulky" phenylalanine associated with each M3 protruding into the pore, thereby occluding the lumen. Other interpretations might account for the striking disparity in the appearance of the paired micrographs in Fig. 3B, such as differences in biochemical state unrelated to open-closed conditions, or different connexin composition.

An alternative to the Unwin-Bennett model (Fig. 3C) could account for the six equally spaced conductance levels that we have found in embryonic chick junctions. This model differs mainly in that it relaxes the requirement that all of the units tilt in unison. Movements of regions within a helical domain are limited, since the interiors of folded polypeptide chains consist of fairly tightly packed, interlocking side chains that allow only restricted flexibility (Chothia and Lesk, 1985). However, within these limits, we may suppose that each connexin unit is stochastically free to twist, and thereby individually swing the phenylalanine "gate" associated with its M3 helix into the lumen, along the lines suggested by the theory of polymer reptation dynamics (Millhauser, 1990). If the conductance of the fully open pore (γ_{max}) is reduced to zero by occlusion with six M3-associated gating groups, then a parsimonious prediction is that each one would reduce the conductance by one-sixth of γ_{max}. If $\gamma_{max} = 240$ pS, partial occlusion by the gating group from one connexin molecule would yield a 200 pS substate, two would produce 160 pS, and so on (Fig. 3C). Furthermore, different conductance levels could be favored by different environments surrounding the connexons. The conductance of gap-junction channels formed from Cx32 expressed in transfected SKHep1 cells was about 150 pS (Eghbali et al., 1990), whereas conductances of about 150 pS and 80 pS resulted when hepatocyte gap-junction channels, also composed of Cx32, were incorporated into lipid bilayers (Moreno et al., 1991; corrected for $[K^+]$). Direct evidence supporting model 2 comes from recent work of Veenstra et al. (1992). These workers found that in communication-deficient N2A cells, stably transfected with cDNA for chick Cx42, junctional channels opened to $\gamma_{max} = 236$ pS with four substates at 201, 158, 121, and 86 pS, in remarkable confirmation of our data obtained from chick cardiac myocytes (Chen and DeHaan, 1992b, and Fig. 2B).

The third model envisions several different types of channels, each with a characteristic open state value for c. Channels of different sizes (model 3) might result from isohexamers of different connexins (see Veenstra et al., this volume), or from heterohexamers. With immunocytochemical techniques, different connexins can be found in a single gap-junction interface (Traub et al., 1989; Spray et al., 1991). Moreover, at least five connexin mRNAs have been identified in heart tissue: Cx37, Cx40, Cx42, Cx43, and Cx45 (Beyer, 1990; Haefliger et al., 1992). Expression of Cx43 and Cx45 is developmentally regulated, suggesting that the composition of the connexons might change during embryonic development (Beyer, 1990; Fromaget et al., 1990; Fishman et al., 1991; Nishi et al., 1991).

Does γ change under conditions that alter G_j?

One way to distinguish a homogeneous population of small, cooperative channels (model 1) from either of the alternative models that postulate the existence of channels with higher conductance states, would be to measure pore-size directly by diffusion of marker molecules of known dimensions. The effective diameter of junctional channel pores was reported to change under the influence of agents that alter G_j (reviewed in Loewenstein, 1981). For example, Schwarzmann et al. (1981) used a series of linear uncharged fluorescently labeled glycopeptides of progressively increasing molecular diameter as probes in cultured mammalian cells. With junctions maintained under conditions that promoted high conductance, the cut-off size for tracer passage was about 1.6 nm, similar to the size found in pairs of mammalian ventricular cells (Imanaga et al., 1987). Elevation of intracellular calcium caused concomitant decreases in G_j and in cut-off size, suggesting a constriction in mean diameter of the channel pores. In contrast, Zimmerman and Rose (1985) co-injected pairs of tracers of different size and fluorescence wavelength into *Chironomus* salivary gland cells, and measured the rate constant of diffusion for each dye into a neighboring cell. Under conditions that modified overall junctional conductance by more than 7-fold, the permeability ratio for large and small molecules in these preparations remained constant, suggesting that differences in G_j resulted from regulation of the number of open channels at any instant, not the channel pore size or selectivity. Similarly, in pairs of amphibian blastomeres, Verselis et al. (1986) demonstrated that junctional permeability (P_j) and G_j were linearly related over a wide range of G_j values. This result suggests that junctional channels are either open or closed with no intermediate states. Similar dye-diffusion experiments with embryonic heart cells might be instructive.

We reported earlier that 166 ± 51 pS was the predominant junctional channel γ in chick cardiac myocytes (Veenstra and DeHaan, 1988). From independent data sets, Veenstra (1990a, b) found that mean γ was 166 ± 37 pS and we reported more recently a value of 157 ± 9 pS (Table I). Openings at the γ_{160} level were found over the entire range of V_j values tested (−80 mV to 80 mV), in the face of wide disparities of intracellular $[Ca^{2+}]$, and in preparations spanning wide differences in G_j. This finding is corroborated by the evidence that channel current is linear and γ is unaffected by V_j from −80 mV to 80 mV (Veenstra, 1990a; Chen and DeHaan, 1992b). These results suggest that if both γ and N in the relation $G_j = \gamma N P_o$ remained constant with G_j, changes in G_j must be explainable entirely on the basis of P_o (Veenstra and DeHaan, 1988). However, an alternative interpretation is that agents that alter G_j change the frequency of occurrence of the various conductance levels. Evidence supporting this view came from experiments in which large (+80 mV) transjunctional voltage steps were applied to embryonic heart cell pairs. In response, the large openings that were prevalent during the initial phase of current decay of G_j were largely suppressed in favor of 40 pS openings at the end of the pulse (Chen and DeHaan, 1992b). The effects of octanol (Veenstra and DeHaan, 1988) can be interpreted similarly as limiting channels from opening to their largest configurations. Thus agents that reduce G_j in embryonic gap junctions may do so as a result of two effects: reducing P_o; and shifting from large conductance states to small (Chen and DeHaan, 1989), thereby decreasing average pore size. In a dye-diffusion experiment this would result in a retardation of large tracer molecules.

Conclusions

1. In many cell types gap junctions have multiple channel conductance states.

2. Three hypothetical models of channel structure are consistent with the data: (a) small (γ_{40}) channels that can exhibit cooperativity; (b) large (γ_{240}) channels with evenly spaced substates; or (3) several classes of channels of different sizes.

3. We suggest a gating mechanism that would predict channels with evenly spaced substates. It differs from the Unwin tilting helix model in that it would permit the phenylalanine "gate" associated with each connexin M3 transmembrane domain to twist into the connexon pore independently of the movements of the other five members of the connexin hexamer.

References

Bennett, M.V.L., Barrio, L.C., Bargiello, T.A., Spray, D.C., Hertzberg, E. and Sáez, J.C. (1991) Gap junctions: new tools, new answers, new questions. Neuron, 6, 305–320.

Beyer, E.C. (1990) Molecular cloning and developmental expression of two chick embryo gap-junction proteins. J. Biol. Chem. 265, 14439–14443.

Beyer, E.C., Paul, D.L. and Goodenough, D.A. (1990) Connexin family of gap-junction proteins. J. Membr. Biol. 116, 187–194.

Burt, J.M. and Spray, D.C. (1988) Single-channel events and gating behavior of the cardiac gap-junction channel. Proc. Natl. Acad. Sci. 85, 3431–3434.

Catterall, W.A. (1988) Structure and function of voltage-sensitive ion channels. Science. 242, 50–61.

Chen, Y-H. and DeHaan, R.L. (1989) Cardiac gap-junction channels shift to lower conductance states when temperature is reduced. Biophys. J. 55, 152a.

Chen, Y-H. and DeHaan, R.L. (1992a) Multiple channel conductance states in embryonic chick cardiac gap junctions. Biophys. J. 61, abstr.

Chen, Y-H. and DeHaan, R.L. (1992b) Multiple channel conductance states and voltage regulation of embryonic chick cardiac gap junctions. J. Membr. Biol. 127, 95–111.

Chothia, C. and Lesk, A.M. (1985) Helix movements in proteins. Trends Biochem. Sci. 10, 116–118.

Cull-Candy, S. and Usowicz, G.M.M. (1987) Multiple-conductance channels activated by excitatory amino acids in cerebellar neurons. Nature 325, 525–528.

DeHaan, R.L. (1988) Dynamic behavior of cardiac gap-junction channels. In: E. Hertzberg and R. Johnson (Eds.), Gap Junctions, Alan R. Liss, New York, pp. 305–320.

DeHaan, R.L., Chen, Y.-H. and Penrod, R.L. (1989) Voltage dependence of junctional conductance in the embryonic heart. In: L. Hondeghem (Ed.), Molecular and Cellular Mechanisms of Antiarrhythmic Agents, Futura, Mount Kisco, New York, pp. 19–43.

Dermietzel, R., Hwang, T.K. and Spray, D.C. (1990) The gap-junction family, structure, function and chemistry. Anat. Embryol. 182, 517–528.

Ebihara, L., Beyer, E.C., Swenson, K.I., Paul, D.L. and Goodenough, D.A. (1989) Cloning and expression of a *Xenopus* embryonic gap-junction protein. Science 243, 1194–1195.

Eghbali, B., Kessler, J. A. and Spray, D.C. (1990) Expression of gap-junction channels in communication-incompetent cells after stable transfectin with cDNA encoding connexin 32. Proc. Natl. Acad. Sci. USA 87, 1328–1331.

Fishman, G. I., Spray, D.C. and Leinwand, L.A. (1990) Molecular characterization and functional expression of the human cardiac gap-junction channel. J. Cell Biol. 111, 589–598.

Fishman, G. I., Hertzberg, E.L., Spray, D.C. and Leinwand, L.A. (1991) Expression of Connexin43 in the developing rat heart. Circ. Res. 68, 782–787.

Fox, J. A. (1987) Ion channel subconductance states. J. Membr. Biol. 97, 1–8.

Fromaget, C., El Aoumari, A., Dupont, E., Briand, J.P. and Gros, D. (1990) Changes in the expression of Connexin 43, a cardiac gap-junction channel protein, during mouse heart development. J. Mol. Cell Cardiol. 22, 1245–1258.

Giaume, C., Fromaget, C., Aoumari, A.E., Cordier, J., Glowinski, J. and Gros, . (1991) Gap junctions in cultured astrocytes, single-channel currents and characterization of channel-forming protein. Neuron. 6, 133–143.

Gimlich, R. L., Kumar, N.M. and Gilula N.B. (1988) Sequence and developmental expression of mRNA coding for a gap-junction protein in *Xenopus*. J. Cell Biol. 107, 1065–1073.

Gimlich, R. L., Kumar, N.M. and Gilula, N.B. (1990) Differential regulation of the levels of three gap-junction mRNAs in *Xenopus* embryos. J. Cell Biol. 110, 597–605.

Gourdie, R.G., Green, C., Severs, N.J. and Thompson, R.B. (1992) Immunolabeling patterns of gap junction connexins in the developing and mature rat heart. Anat. Embryol. 185, 363–378.

Haefliger, J-A., Bruzzone, R., Jenkins, N.A., Gilbert, D. J., Copeland, N.G. and Paul D.L. (1992) Four novel members of the connexin family of gap-junction proteins: molecular cloning, expression and chromosome mapping. J. Biol. Chem. 267, 2057–2064.

Hennemann, H., Suchyna, T., Lichtenberg-Frate, H., Jungbluth, S., Dahl, E., Schwarz, J., Nicholson, B.J. and Willeke, K. (1992) Molecular cloning and functional expression of mouse connexin40, a second gap junction gene preferentially expressed in lung. J. Cell Biol. 117, 1299–1310.

Hill, Jr., J.A., Coronado, R. and Strauss, H.C. (1990) Open-channel subconductance state of K^+ channels from cardiac sarcoplasmic reticulum. Am. J. Physiol. 258, H159–H164.

Hunter, M. and Giebisch, G. (1987) Multi-barrelled K channels in renal tubules. Nature 327, 522–524.

Imanaga, I., Kameyama, M. and Irisawa, H. (1987) Cell-to-cell diffusion of fluorescent dyes in paired ventricular cells. Am. J. Physiol. 252, H223–H232.

Jahr, C. E. and Stevens, C.F. (1987) Glutamate activated multiple single-channel conductances in hippocampal neurons. Nature 325, 522–523.

Kanter, H.L., Saffitz, J.E. and Beyer, E.C. (1992) Cardiac myocytes express multiple gap junction proteins. Circ. Res. 70, 438–444.

Kell, M.J. and DeFelice, L.J. (1988) Surface charge in cardiac inward-rectifier channels measured from single-channel conductance. J. Membr. Biol. 102, 1–10.

Kolb, H.-A and Somogyi, R. (1990) Characteristics of single channels of pancreatic acinar gap junctions subject to different uncoupling procedures. In: C. Peracchia (Ed.), Biophysics of Gap Junction Channels. CRC Press, Boca Raton. 209–228.

Kumar, N.M. and Gilula, N.B. (1986) Cloning and characterization of human and rat liver cDNAs coding for a gap junction protein. J. Cell Biol. 103, 767–776.

Lauger, P. (1985) Ionic channels with conformational substates. Biophys. J. 47, 581–591.

Loewenstein, W.R. (1981) Junctional intercellular communication: the cell-to-cell membrane channel. Physiol. Rev. 61, 829–913.

Miller, T.M. and Goodenough, D.A. (1985) Gap junction structures after experimental alteration of junctional channel conductance. J. Cell Biol. 101, 1741–1748.

Millhauser, G.L. (1990) Reptation theory of ion channel gating. Biophys. J. 57, 857–864.

Milks, L.C., Kumar, N.M., Houghten, R., Unwin, N. and Gilula, N.B. (1988) Topology of the 32 kDa liver gap-junction protein determined by site-directed antibody localizations. EMBO J. 7, 2967–2975.

Moore, L.K., Beyer, E.C. and Burt, J.M. (1991) Characterization of gap-junction channels in A7r5 vascular smooth muscle cells. Am. J. Physiol. (Cell) 260, C975–C981.

Moreno, A.P, Campos de Carvalho, A.C., Verselis, V., Eghbali, B. and Spray, D.C. (1991) Voltage-dependent gap-junction channels are formed by connexin32, the major gap-junction protein of rat liver. Biophys. J. 59, 920–925.

Musil, L.S., Beyer, E.C. and Goodenough, D.A. (1990) Expression of the gap-junction protein connexin43 in embryonic chick lens: molecular cloning, ultrastructural localization and post-translational phosphorylation. J. Membr. Biol. 116, 163–175.

Nishi, M., Kumar, N.M. and Gilula, N.B. (1991) Developmental regulation of gap-junction gene expression during mouse embryonic development. Dev. Biol. 146, 117–130.

Patlak, J. (1988) Sodium channel subconductance levels measured with a new variance-mean analysis. J. Gen. Physiol. 92, 413–430.

Patlak, J. (1991) Molecular kinetics of voltage-dependent Na^+ channels. Physiol. Rev. 71, 1047–1080.

Rook, M.B., Jongsma, H.J. and van Ginneken, A.C.G. (1988) Properties of single gap-junctional channels between isolated neonatal rat heart cells. Am. J. Physiol. 255, H770–H782.

Rüdisüli, A. and Weingart, R. (1990) Gap junctions in adult ventricular muscle. In: C. Peracchia (Ed.), Biophysics of Gap Junction Channels, CRC Press, Boca Raton, pp. 43–56.

Sakmann, B. and Trube, G. (1984) Voltage-dependent inactivation of inwardly-rectifying single-channel currents in guinea pig ventricle. J. Physiol. (London) 347, 659–683.

Schwarzmann, G., Wiegandt, H., Rose, B., Zimmerman, A., Ben-Haim, D. and Loewenstein, W.R. (1981) Diameter of the cell-to-cell junctional channels as probed with neutral molecules. Science. 213, 551–553.

Somogyi, R. and Kolb, H.-A. (1988) Cell-to-cell channel conductance during loss of gap-junctional coupling in pairs of pancreatic acinar and Chinese hamster ovary cells. Pflügers Arch. 412, 54–65.

Spray, D.C. and Burt, J.M. (1990) Structure-activity relations of the cardiac gap-junction channel. Am. J. Physiol. 258, C195–C205.

Spray, D.C., Chanson, M., Moreno, A.P., Dermietzel, R. and Meda, P. (1991) Distinctive gap-junction channel types connect WB cells, a clonal cell line derived from rat liver. Am. J. Physiol. 260, C513–C527.

Stauffer, K.A., Kumar, N.M., Gilula, N.B. and Unwin, N. (1991) Isolation and purification of gap-junction channels. J. Cell Biol. 115, 141–150.

Traub, O., Look, J., Dermietzel, R., Brummer, F., Hulser, D. and Willecke, K. (1989) Comparative characterization of the 21-kDa and 26-kDa gap-junction proteins in murine liver and cultured hepatocytes. J. Cell Biol. 108, 1039–1051.

Unwin, N. (1989) The structure of ion channels in membranes of excitable cells. Neuron 3, 665–676.

Veenstra, R.D. (1990a) Voltage-dependent gating of gap-junction channels in embryonic chick ventricular cell pairs. Am. J. Physiol. 285, C662–C672.

Veenstra, R.D. (1990b) Comparative physiology of cardiac gap-junction channels. In: C. Peracchia (Ed.), Biophysics of Gap Junction Channels, CRC Press, Boca Raton, pp. 131–144.

Veenstra, R.D. and DeHaan, R.L. (1988) Cardiac gap-junction channel activity in embryonic chick ventricle cells. Am. J. Physiol. 254, H170–H180.

Veenstra, R.D., Wang, H.-Z., Westphale, E.M. and Beyer, E.C. (1992) Multiple connexins confer distinct regulatory and conductance properties of gap junctions in developing heart. Circ. Res., in press.

Verselis, V., White, R.L., Spray, D.C. and Bennett, M.V.L. (1986) Gap-junctional conductance and permeability are linearly related. Science 234, 462–464.

Yancey, S.B. (I), John, S.A. (II), Lal, R. (III), Austin, B.J. and Revel, J.-P. (1989) The 43-kDa polypeptide of heart gap junctions: Immunolocalization (I), Topology (II) and Functional Domains (III). J. Cell Biol. 108, 2241–2254.

Young, J.D.-E., Cohn, Z.A. and Gilula, N.B. (1987) Functional assembly of gap-junction conductance in lipid bilayers: demonstration that the major 27 kDa protein forms the junctional channel. Cell. 48, 733–743.

Zimmerman, A.L. and Rose, B. (1985) Permeability properties of cell-to-cell channels: kinetics of fluorescent tracer diffusion through a cell junction. J. Membr. Biol. 84, 269–283.

J.E. Hall, G.A. Zampighi and R.M. Davis (Eds.)
Progress in Cell Research, Vol. 3

CHAPTER 15

Comparison of voltage dependent properties of gap junctions in hepatocytes and in *Xenopus* oocytes expressing Cx32 and Cx26

VYTAS K. VERSELIS, THADDEUS A. BARGIELLO, JOSHUA B. RUBIN and MICHAEL V.L. BENNETT

Department of Neuroscience, Albert Einstein College of Medicine, Bronx, NY, USA

Introduction

A majority of gap junctions, including some for which cloned DNAs are available, display voltage-dependent junctional conductance, g_j. Voltage gating of gap junctions is presumed to be unique as junctional proteins (connexins) bear no primary sequence homology to the family of voltage-sensitive cation channels (see Catterall, 1988; Bennett et al., 1991). Also, gap junctions, because of their construction and topology, possess an inherent complexity in their gating not seen in other voltage-gated channels. There are two types of voltage stimuli that can cause gap-junction channels to gate: (1) V_j, the transjunctional voltage, is the voltage of one cell relative to its coupled neighbor, and (2) $V_{i\text{-}o}$, the inside/outside voltage, is the voltage of either cell, and the channel lumen, relative to the extracellular space.(see Verselis and Bargiello, 1991; Verselis et al., 1991). The differential sensitivity and kinetics of the changes in g_j caused by V_j and $V_{i\text{-}o}$ suggest that the mechanisms of gating are different and possibly result from the actions of distinct protein domains. Because gap junctions are composed of two halves, or hemichannels, even the simplest junctions, i.e., homotypic gap junctions formed by the union of structurally equivalent hemichannels composed of the same connexin, would appear to possess two complete sets of voltage-gating elements, one contributed by each of the two hemichannels. Since the hemichannels are joined with mid-channel (but not mirror) symmetry parallel to the plane of the membrane, the gating elements, although identical in composition, would be situated in opposite orientation with respect to a transjunctional field. This arrangement of V_j-sensitive gating elements can explain the often-observed symmetric reduction in g_j about $V_j = 0$, as originally proposed for amphibian blastomeres (Harris et al., 1981; Spray et al., 1981). When $V_{i\text{-}o}$ sensitivity is present, with or without V_j sensitivity, hyperpolarization and depolarization of one cell produces an asymmetric g_j/V_j relation about $V_j = 0$. However, the asymmetry is the same for equal polarizations of either cell.

Added complexity arises from the formation of heterotypic junctions, in which hemichannels composed of different connexins are joined, and may also arise from the formation of heteromeric junctions in which the hemichannels themselves are composed of different connexins. Heterotypic junctions displaying sensitivity only to V_j would presumably show asymmetry in their g_j/V_j relations, if their homotypic counterparts differed in their V_j sensitivity. If the hemichannels acted independently, heterotypic g_j/V_j relations could be predicted from the behavior of the homotypic junctions. If each type of hemichannel were closed by the same polarity of V_j, changes in g_j for one polarity of V_j would resemble those of homotypic junctions composed of one type of hemichannel, and changes in g_j for the opposite polarity would resemble those of junctions composed of the other type of hemichannel. Sensitivity to $V_{i\text{-}o}$ would add or subtract depending on whether a given polarization acted to increase or decrease g_j. There is no obvious prediction for properties of junctions formed by heteromeric hemichannels, since many types are possible even when only two different connexins are considered.

The formation of heterotypic junctions was first shown in *Xenopus* oocytes expressing different connexins (Swenson et al., 1989), although inter-tissue and inter-specific junctions had been known (cf. Bennett et al., 1991). The heterotypic junction was composed of rat connexin Cx43 and *Xenopus* Cx38. The issue of whether heterotypic or heteromeric junctions are formed in vivo was raised by studies showing that more than one connexin could be expressed in the same cell type and could even be localized to the same junctional plaque (Nicholson et al. 1987; Traub et al., 1989). Since antibody localization was not resolved to the single-channel level, the issue remains unresolved.

Here we provide evidence that heterotypic channels do form in hepatocytes in vivo based on comparison of junctional properties in isolated hepatocyte pairs and pairs of *Xenopus* oocytes expressing Cx32 and Cx26. We have no evidence for or against heteromeric hemichannels.

Methods

Preparation of hepatocytes

Hepatocyte cell pairs were obtained by incomplete dissociation according to Berry and Friend (1969). Cells were plated in Waymouth's medium, MB 752 (Gibco, Grand Island, NY), supplemented with 10% fetal calf serum at a density of 2–5 × 10^5 cells/petri dish at 37°C and 5% CO_2. For recording, the petri dishes were rinsed with PBS (Gibco) supplemented with HEPES (5 mM). All recordings were done 4–6 h after plating.

Preparation of RNA

RNA was transcribed from linearized plasmids using T7 RNA polymerase (New England Biolabs) in the presence of the cap analogue m7G(5')ppp(5')G (Boehringer-Mannheim) at a 10:1 ratio to added rGTP for 85 min at 37°C under standard reaction conditions, followed by an additional 5 min synthesis in the presence of equimolar rGTP to ensure full-length transcription of initiated capped transcripts. The reaction was terminated by digestion of the DNA template with RNase-free DNase 1 (RQ1 DNase, Promega). RNA was extracted with phenol and chloroform, then ethanol precipitated.

Expression of junctional currents in pairs of Xenopus oocytes

Adult female *Xenopus laevis* frogs were purchased from *Xenopus* I, Ann Arbor, Michigan and maintained at 18°C. Oocytes were defolliculated in 2 mg/ml collagenase Type 1A (Sigma) in calcium-free ND96 medium containing in mM: 96 NaCl, 2 KCl, 2 $MgCl_2$, 2.5 Na Pyruvate, 5 HEPES (pH 7.6). Defolliculated oocytes were placed in ND96 medium containing 1.8 mM $CaCl_2$, allowed to recover overnight and either injected with 50 µl of an aqueous solution containing 0.25 µg/µl of two antisense oligonucleotides *5'-GCT TTA GTA ATT CCC ATC CTG CCA TGT TTC-3'* and *5'- TTC CTA AGG CAC TCC AGT CAC CCA TGC TCA-3'* or co-injected with an aqueous solution containing 0.25 µg/µl of the antisense oligonucleotides and 1 µg/µl cRNA. The antisense oligonucleotides are

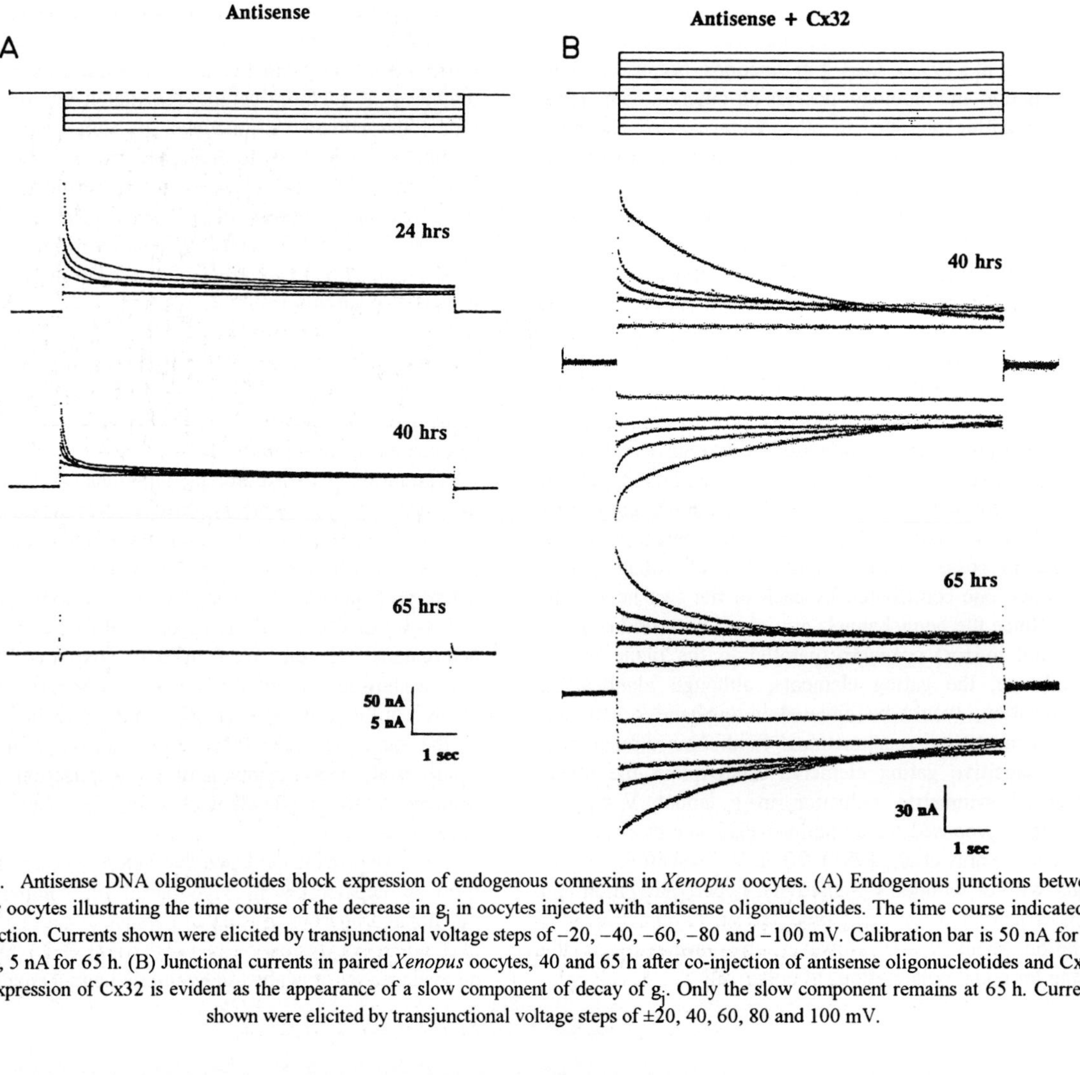

Figure 1. Antisense DNA oligonucleotides block expression of endogenous connexins in *Xenopus* oocytes. (A) Endogenous junctions between *Xenopus* oocytes illustrating the time course of the decrease in g_j in oocytes injected with antisense oligonucleotides. The time course indicated is post-injection. Currents shown were elicited by transjunctional voltage steps of −20, −40, −60, −80 and −100 mV. Calibration bar is 50 nA for 24 and 40 h, 5 nA for 65 h. (B) Junctional currents in paired *Xenopus* oocytes, 40 and 65 h after co-injection of antisense oligonucleotides and Cx32 RNA. Expression of Cx32 is evident as the appearance of a slow component of decay of g_j. Only the slow component remains at 65 h. Currents shown were elicited by transjunctional voltage steps of ±20, 40, 60, 80 and 100 mV.

complementary to endogenous *Xenopus* Cx38 (commencing at nt –5 in the sequence reported by Ebihara et al., 1989) and Cx43 (commencing at nt 190 in the sequence reported by Gimlich et al., 1990) mRNA, respectively. 48 h post-injection, vitelline membranes were manually removed in hypertonic medium containing in mM: 200 K-aspartate, 20 KCl, 1 $MgCl_2$, 10 HEPES, pH 7.6 (Methfessel et al., 1986) and the oocytes were paired in ND96 medium containing calcium.

Measurement of junctional conductance

Junctional conductance was measured with a dual voltage clamp as described by Verselis et al. (1991). Data were collected with a PC-AT microcomputer using pCLAMP software (Axon Instruments Inc., Foster City , CA) and a LABMASTER/TL-1 interface. Initial and steady-state g_j (g_{j0} and $g_{j\infty}$) were obtained by extrapolating exponential fits to step changes in V_j. Steady-state conductance, $g_{j\infty}$,

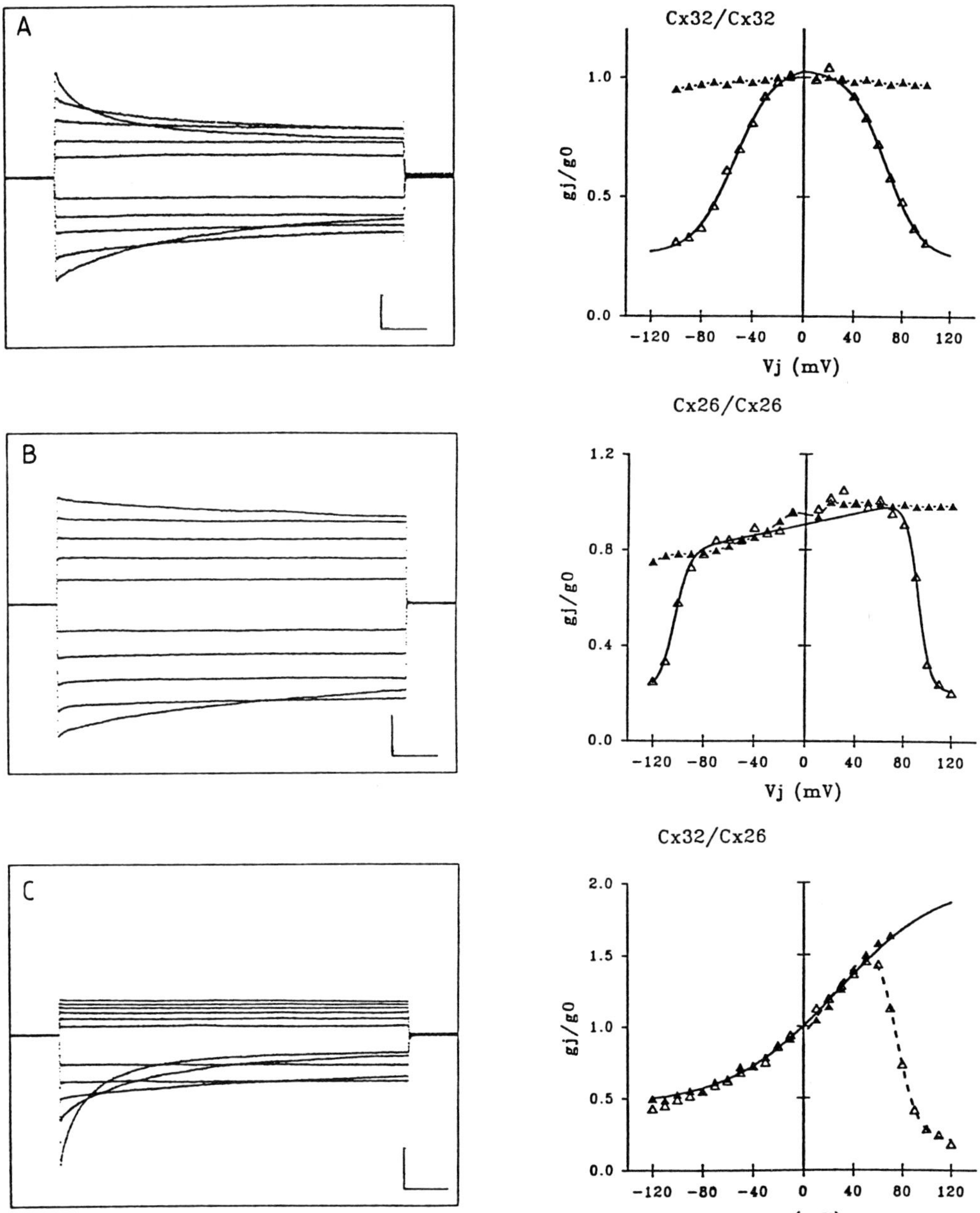

Figure 2. Voltage dependence of g_j in oocytes expressing Cx32 and Cx26. Representative junctional currents and initial and steady-state conductance voltage relations for homotypic Cx32/Cx32 junctions (A) and Cx26/Cx26 junctions (B) and heterotypic Cx32/Cx26 (C) junctions. The junctional currents shown were elicited by ±20, 40, 60, 80 and 100 mV voltage steps applied to either oocyte for the homotypic junctions and the oocyte expressing Cx26 for the heterotypic junctions. Calibration bars are 20, 30 and 30 nA for A–C and the time calibration is 1 s. The conductances are plotted as normalized to g_j at $V_j = 0$; normalized initial g_j is plotted as closed symbols and normalized steady-state g_j as open symbols. The smooth curves are fitted Boltzmann equations (see Methods).

was fit to the Boltzmann relation of the form:

$$g_{j\infty} = \{(g_{jmax} - g_{jmin})/(1 + \exp(A(V_j - V_0)))\} + g_{jmin}$$

where g_{jmax} is the maximum conductance, g_{jmin} is the residual conductance approached at large V_j, V_0 is the voltage at which $g_{j\infty}$ is half maximal $((g_{jmax} + g_{jmin})/2)$, and A is a constant expressing voltage sensitivity. For the heterotypic junction Cx32/Cx26 expressed in *Xenopus* oocytes, the initial g_j was fit to a Boltzmann relation of the same form, but with different parameters. The subsequent slow changes in g_j were fit to the ratio $g_{j\infty}g_{j0}$. For both homotypic and heterotypic junctions in Fig. 2, the relations plotted are normalized to g_j at $V_j = 0$ (g_0).

Results and discussion

Expression of exogenous gap junctions in Xenopus oocytes

Antisense oligonucleotides prevent coupling by endogenous gap-junction channels

Two genes encoding distinct *Xenopus* connexins (Cx38 and *Xen* Cx43) are expressed in unfertilized oocytes (Gimlich et al., 1990, Ebihara et al., 1989). In our hands, the levels of endogenous coupling, although variable, were often high (as much as 10 µS) and precluded the quantitative analysis of exogenously expressed junctional currents. We used the published sequences of these cDNAs (Ebihara et al., 1989; Gimlich et al. 1990) to construct antisense oligonucleotides that blocked the expression of endogenous mRNA coding for these connexins (see Methods, and also Barrio et al., 1991). The time course of the decline in endogenous coupling following injection of our antisense oligonucleotides is shown in Fig. 1A. Electrical coupling between antisense-injected oocytes became detectable 4–6 h after pairing at room temperature (22–25°C). The level of coupling increased rapidly, peaking at about 12 h, after which it steadily declined over the course of 36–48 h. We followed pairs of antisense-injected oocytes for up to 5 days, during which the oocyte pairs remained uncoupled. Control oocyte pairs, either uninjected or water-injected, followed a similar time course in the appearance of coupling (4–6 h), but g_j continued to increase over the course of a few days, often reaching 5–

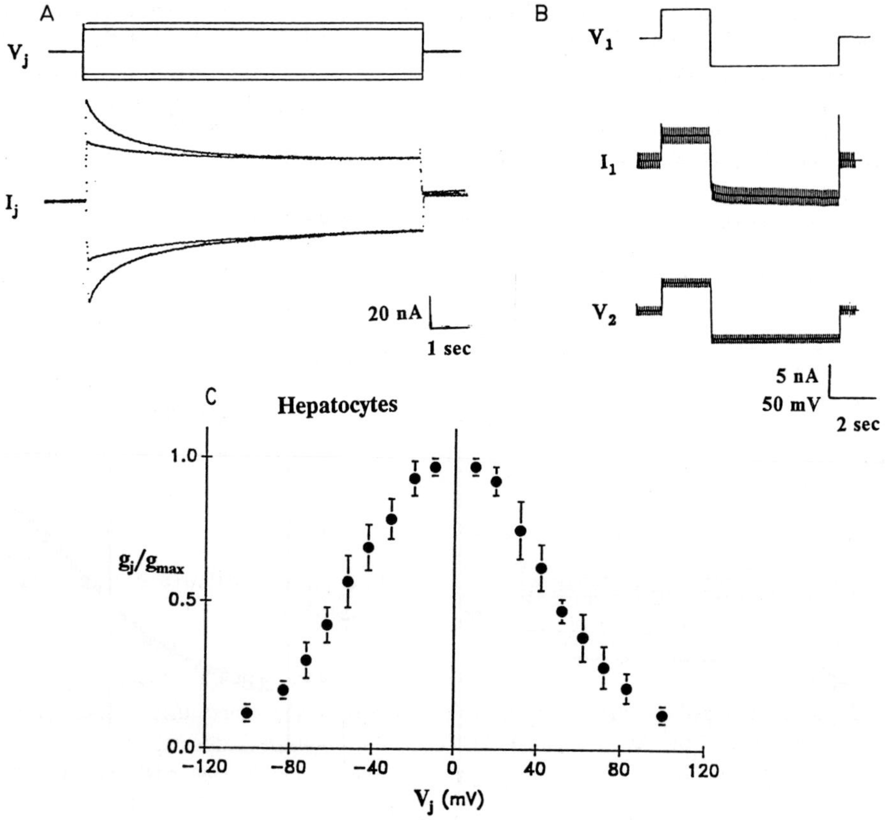

Figure 3. Voltage dependence of g_j in a typical hepatocyte pair. (A) Junctional currents (I_j) for ±50 and ±70 mV transjunctional voltage steps (V_j) are shown. (B) Current and voltage records from the same cell pair demonstrating a lack of V_{i-o} dependence. A long-lasting depolarizing, then hyperpolarizing voltage was applied simultaneously to both cells (V_1, V_2). Small test voltages alternating in polarity were applied to cell 2 to monitor junctional current (the small deflections in I_1). (C) The g_j/V_j relation is symmetrical around $V_j = 0$; g_j was normalized to g_{max} (n = 9, see Methods).

10 μS. The initial development of coupling, even in the presence of antisense nucleotides, indicates the existence of an endogenous pool of connexin protein which turns over in the course of about 48 h.

There appeared to be no deleterious side effects of these antisense oligonucleotides on resting membrane potentials, input resistances, non-junctional membrane I–V curves or expression of exogenous connexins. The characteristics of junctional currents in *Xenopus* oocytes co-injected with the antisense oligonucleotides and rat Cx32 cRNA synthesized in vitro from a full-length cDNA clone (Paul, 1986) are illustrated in Fig. 1B. The time course of the voltage-dependent decline in junctional current 40 h after pairing (top panel) displayed two distinct components, a fast component resembling that of endogenous currents, and a much slower component presumably due to the expression of exogenous Cx32 junctions. After 65 h, only the slower component was evident (bottom panel), coincident with the loss of endogenous coupling in oocytes injected only with antisense oligonucleotides.

Expression of homotypic Cx32 and Cx26 junctions in Xenopus oocytes

Junctional currents (left) and both the initial and steady-state g_j/V_j relations (right) for homotypic junctions composed of Cx32 and Cx26 expressed in oocytes are shown in Fig. 2 (adapted from Rubin et al., 1992). For Cx32 (Fig. 2A), the initial g_j showed a slight decline with transjunctional voltages ranging up to ±100 mV, whereas steady-state g_j declined with transjunctional voltages exceeding ±40–50 mV. These changes in g_j were slow to reach steady state (seconds) and were symmetric about $V_j = 0$. Insensitivity to $V_{i\text{-}o}$ was demonstrated directly by equally and simultaneously altering the potential in both cells (not shown). Although the transitions between states that underlie the changes in g_j are likely to be complex, data were fit to a simple Boltzmann two-state model characterized by transitions between open and closed (non-conducting) states. The Boltzmann constant A, which characterizes the steepness of the g_j/V_j relation, and the constant V_0, which provides a measure of the zero-field intrinsic free energy difference between open and closed states, are indices that can be used to compare the properties of different connexins whether or not the model is correct. Boltzmann parameters for Cx32 were $A = 0.074 \pm 0.019\ \text{mV}^{-1}$, $V_0 = 56.8 \pm 8.1$ mV ($n = 6$) (see also Barrio et al., 1991; Rubin et al., 1992).

Unlike Cx32, junctions formed by Cx26 exhibit a more complex voltage dependence (Fig. 2B). Initial currents displayed a fast rectification that could not be resolved by the voltage clamp (time resolution < 5 ms). This fast rectification of initial currents must depend on $V_{i\text{-}o}$ because the application of hyperpolarizing voltages to either cell (which created equal but opposite V_j's) decreased the initial g_j. Similarly, depolarization of either cell increased the initial g_j. This rapid $V_{i\text{-}o}$ dependence resembled that seen in a small percentage of hepatocytes (see Fig. 4 below). For transjunctional voltages exceeding 60 mV, junctional currents decayed slowly (seconds) for either polarity of V_j. Boltzmann parameters (measured as the ratio $g_{j\infty}/g_{j0}$, see Methods) were $A = 0.160 \pm 0.029\ \text{mV}^{-1}$, $V_0 = 93 \pm 6.1$ mV for hyperpolarizing V_j's and $A = 0.150 \pm 0.027\ \text{mV}^{-1}$, $V_0 = 98.8 \pm 4.5$ mV for depolarizing V_j's and indicate that Cx26 junctions possess a steeper voltage dependence, but require a larger V_j (increased V_0) to reduce g_j compared to Cx32 junctions. The slight asymmetry in V_0 is ascribable to interaction between $V_{i\text{-}o}$-dependent processes and the slow V_j dependence (Barrio et al., 1991).

Expression of heterotypic Cx26/Cx32 junctions in Xenopus oocytes

Heterotypic gap junctions were formed by pairing an oocyte expressing only Cx26 with one expressing only Cx32. The initial and steady-state g_j/V_j relations for these heterotypic junctions are shown in Fig. 2C. The behavior of these channels differed markedly from that of either Cx26 and Cx32. The initial g_j substantially increased on depolarization and decreased on hyperpolarization of the cell expressing Cx26. The same increase and decrease in initial g_j could be achieved by hyperpolarization and depolarization of the cell expressing Cx32, indicating that the change depended only on V_j. The slow decrease in g_j for voltages that make the Cx26 cell more positive gave a steady-state g_j/V_j relation that resembled the relation for homotypic junctions formed by Cx26. Boltzmann parameters for this one polarity of V_j were $A = 0.13 \pm 0.009\ \text{mV}^{-1}$ and $V_0 = 75.7 \pm 3.3$ mV ($n = 3$). We observed little or no slow decrease in g_j for voltages that made the Cx32 side more positive. These results, along with the observation that Cx38 hemichannels are closed by positivity on their cytoplasmic side (see Verselis et al., 1987; Bennett et al., 1988; Swenson et al., 1989; Werner et al., 1989), suggest that the slow gating process closed the Cx26 hemichannels in response to relatively positive voltages, and that the slow gating process was largely ineffective in the Cx32 hemichannels.

Voltage dependence of gap junctions in rat hepatocytes

Voltage dependence of hepatocyte gap junctions has a controversial history that has been resolved by two-electrode voltage-clamp analysis (this paper and see also Moreno et al., 1991), as well as examination of Cx32 in various exogenous expression systems (Barrio et al., 1991; Harris, 1991; Moreno et al., 1991; Rubin et al., 1992). Below we describe properties exhibited in hepatocytes that, based on properties of Cx32 and Cx26 in *Xenopus* oocytes, suggest that both homotypic and heterotypic junctions do form.

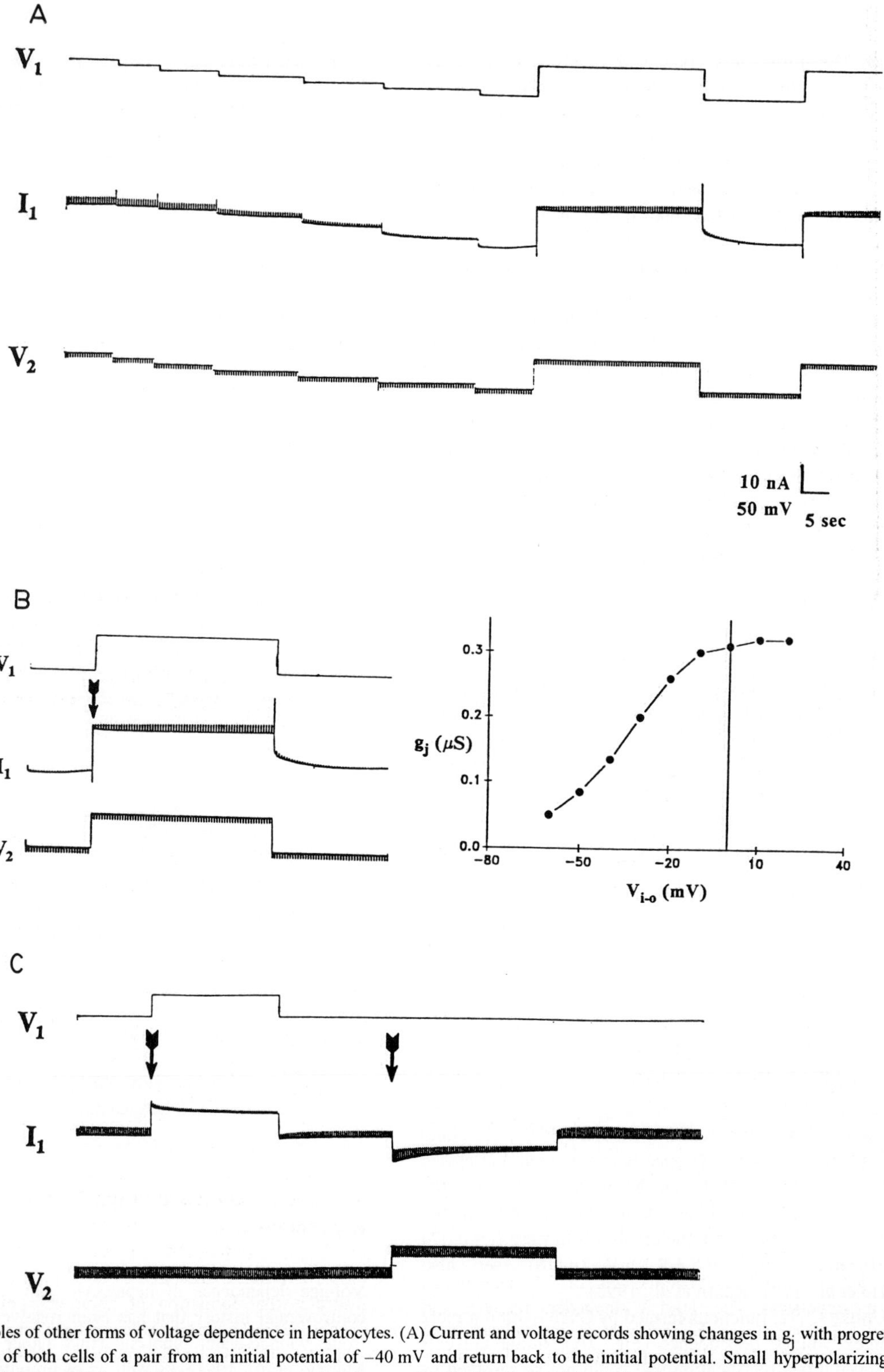

Figure 4. Examples of other forms of voltage dependence in hepatocytes. (A) Current and voltage records showing changes in g_j with progressive hyperpolarization of both cells of a pair from an initial potential of −40 mV and return back to the initial potential. Small hyperpolarizing test voltages were applied to cell 2 to monitor junctional currents (small upward deflections in I_2). (B) A 50-mV depolarizing stimulus was applied simultaneously to both cells, illustrating fast and slow changes in g_j. At the onset of the pulse (arrow), a fast increase in g_j (not resolved by the test pulse interval) was followed by a slow increase to a new steady state. A similar time course was seen upon removal of the stimulus. The steady-state $g_j/V_{i\text{-}o}$ relation for this cell pair was sigmoidal, with g_j increasing and decreasing upon equal depolarization and hyperpolarization, respectively, of both cells. (C) Fast voltage dependence in another hepatocyte pair. Depolarization of cell 1 (V_1) by 50 mV produced a fast (first arrow) and then a slow decrease in g_j. An identical voltage step applied to cell 2 (V_2) produced a fast increase (second arrow) followed by a slow decrease in g_j to below the level prior to the voltage step. After the step, g_j increased with fast and slow phases apparently overshooting the initial level. The time course of the fast changes in g_j was not resolved.

The traces shown in Fig. 3A are the transjunctional voltage, V_j, and the corresponding junctional current, I_j, obtained from a typical pair of hepatocytes in double voltage clamp. Upon application of V_j, I_j declined slowly (seconds) from an initial peak to a new steady state. Shown are I_j's in response to V_j's of ±50 and ±70 mV. Essentially identical changes in I_j were achieved with V_j of either polarity, whether elicited by hyperpolarization or depolarization of either cell. Thus, the g_j/V_j relation was symmetric about $V_j = 0$ (Fig. 3C) as seen when Cx32 was expressed in *Xenopus* oocytes. As indicated by symmetry, there was no dependence on $V_{i\text{-}o}$ determined directly by applying equal voltage steps simultaneously to both cells (Fig. 3B).

In some pairs of rat hepatocytes, g_j displayed complex voltage dependence. Figure 4A and B show records from a cell pair displaying a marked $V_{i\text{-}o}$ dependence. Both cells of this pair were simultaneously depolarized from a holding potential of −40 mV; small test voltages were applied to one cell (cell 2) to monitor g_j. Depolarization produced an increase in g_j consisting of two components: a fast (< 100 ms) increase in g_j which was evident by comparing the brief step currents immediately preceding and following the imposition of the long $V_{i\text{-}o}$ step (arrow in Fig. 4B), and a subsequent slow increase in g_j that took seconds to reach steady state, as well as to recover when returned to the resting potential. The resulting steady-state $g_j/V_{i\text{-}o}$ relation appears sigmoidal, with g_j increasing to a plateau upon equal depolarization, and decreasing upon equal hyperpolarization of both cells (Fig. 4B).

To examine the effects of V_j on this cell pair, large voltages of the same polarity were applied to either cell (Fig. 4C). Smaller (20 mV), brief (100 ms) steps were applied to cell 2 to monitor g_j. Depolarization of cell 1 (first arrow) decreased g_j. Because depolarization of both cells increased g_j (Fig. 4A,B), the decrease in g_j must have resulted from V_j dependence. As with $V_{i\text{-}o}$ dependence, the V_j dependent decrease in g_j occurred first rapidly (< 100 ms), then slowly (seconds). The same depolarization applied to cell 2 (second arrow), which imposed the opposite V_j and the same $V_{i\text{-}o}$, except in cell 2 rather than cell 1, produced a rapid increase in g_j followed by a slow decrease to below the pre-pulse level. The rapid increase in g_j could have resulted from $V_{i\text{-}o}$ dependence, but the slower decrease in g_j, as argued above, must have resulted from V_j dependence. These results indicate that both V_j and $V_{i\text{-}o}$ dependence were present, both appeared to possess fast and slow components, and the V_j dependence was asymmetrical. The rapid V_j-dependent increase and decrease in g_j can be explained qualitatively by the behavior of heterotypic Cx26/Cx32 junctions. If cell 2 expressed more Cx26 than Cx32, and cell 1 predominantly expressed Cx32, heterotypic junctions with Cx26 hemichannels contributed by cell 2 would be expected to cause a rapid V_j-dependent increase in g_j upon depolarization of that cell, and a rapid decrease in g_j upon depolarization of cell 1 (Fig. 4C). Slow reductions in g_j for either polarity of V_j would result from slow V_j dependence contributed by homotypic Cx32 and Cx26 junctions. Fast $V_{i\text{-}o}$ dependence, as observed for Cx26 expressed in oocytes, could contribute to the increase in g_j upon depolarization of either or both cells, but the effect in oocytes is small and cannot fully account for the changes in g_j in hepatocytes described above.

Conclusions

These results and the results of Barrio et al. (1991) concerning connexins expressed in oocytes are in general agreement and provide the following picture of voltage dependence of homotypic and heterotypic channels formed by Cx26 and Cx32: Homotypic channels formed by Cx32 show only V_j dependence. There is no evidence of slow or fast $V_{i\text{-}o}$ dependence. Homotypic junctions formed by Cx26 display slow V_j dependence and a small degree of fast $V_{i\text{-}o}$ dependence. Homotypic junctions do not show the slow $V_{i\text{-}o}$ dependence that is seen in rat hepatocytes. Heterotypic channels formed by the union of Cx26 with Cx32 demonstrate a marked V_j dependence and asymmetry that are not present in either homotypic channel. Slow V_j dependence is present when the Cx26 cell is made relatively positive to the Cx32 side, but is absent when the Cx32 side is made relatively positive to the Cx26 side. There is no evidence for $V_{i\text{-}o}$ dependence in heterotypic channels.

The relative abundance and distribution of Cx32 and Cx26 in rat liver and their ability to form junctions with different parameters of voltage dependence explains much of the complex and variable behavior seen in hepatocyte pairs. Whereas Cx32 is more abundant and expressed in all regions of the rat liver, Cx26 is expressed along with Cx32 in periportal regions (Traub et al., 1989). Therefore, it is likely that, in the rat, most hepatocyte pairs form homotypic junctions composed of Cx32, consistent with observations like those in Fig. 3. In cells pairs that co-expressed Cx26 and Cx32 a more complex situation would exist, as heterotypic channels could be formed from the pairing of hemichannels that were Cx26 or Cx32 homomers or from hemichannels that were Cx26 and Cx32 heteromers. When Cx26 and Cx32 are co-expressed in one cell, there are 14 different possible types of hemichannel. Oocytes expressing both Cx26 and Cx32 form junctions with oocytes expressing Cx32 (Barrio et al., 1991). The properties of these junctions are intermediate between Cx32/Cx26 heterotypic junctions, which indicates that formation of homotypic junctions is not highly favored over formation of junctions containing the two connexins. These observations could be explained by formation of heteromeric hemichannels with intermediate properties or by exclusive formation of homomeric hemichannels of the two types.

If the amounts of Cx32 and of Cx26 expression were the same in each of two hepatocytes, junctions between them would be expected to include Cx32/Cx26 or other asymmetric junctions of either orientation with equal frequencies. Little fast V_j dependence would be observed because the changes in g_j for the two orientations of channels would be opposite in direction. A small component of fast V_{i-o} dependence would be contributed by homotypic Cx26 channels, but would be difficult to detect. Symmetry of the steady-state g_j/V_j relation should also be maintained, except that its shape would be determined by relative contributions of homotypic Cx32 and Cx26 channels and asymmetric channels in parallel. The gradient of Cx26 expression may account for the asymmetry of voltage dependence in some hepatocyte pairs (Fig. 4). From the oocyte data, differential expression of Cx32 and Cx26 in hepatocytes could explain the fast and slow V_j-dependent changes in g_j and fast V_{i-o}-dependent changes in g_j. Slow V_{i-o} dependence like that observed in the hepatocyte pair in Fig. 4 was never seen in oocytes. This phenomenon may be a result of expression in hepatocytes or might be a property of junctions composed of heteromeric hemichannels that was missed in the experiments in oocytes in which Cx26 and Cx32 were co-expressed (Barrio et al., 1991).

Large V_j's are not likely to be generated in well-coupled cells, and the slow V_j-dependent changes in g_j observed in homotypic junctions are not likely to be of physiological significance. The functional significance of the fast V_j-dependent changes in g_j that operate over the entire voltage range may better be assessed with knowledge of the changes that occur in hepatocyte membrane potentials. The slow V_{i-o} dependence that in some cell pairs markedly altered g_j may be relevant to normal function, particularly since the resting potential in hepatocytes (i.e., −30 to −50 mV) resides on the steep part of the g_j/V_{i-o} relation, and glucagon and insulin have marked effects on this potential (Wondergem, 1987).

References

Barrio, L.C., Suchyna, T., Bargiello, T.A., Xu, L.X., Roginski, R., Bennett, M.V.L. and Nicholson, B. (1991) Gap junctions formed by connexins 26 and 32 alone and in combination are differently affected by applied voltage. Proc. Natl. Acad. Sci. USA 88, 8410–8414.

Bennett, M.V.L., Barrio, L.C., Bargiello, T.A., Spray, D.C., Hertzberg, E. and Sáez, J.C. (1991) Gap Junctions: new tools, new answers, new questions. Neuron 6, 305–320.

Bennett, M.V.L., Verselis, V.K., White, R.L. and Spray, D.C. (1988) Gap-junctional conductance: Gating. In: E. Hertzberg and R. Johnson (Eds.), Gap Junctions, Alan R. Liss, New York, pp. 287–304.

Berry, M.N. and Friend, D.S. (1969) High-yield preparation of isolated rat liver parenchymal cells. J. Cell Biol. 43, 506–520.

Catterall, W.A. (1988) Structure and function of voltage-sensitive ion channels. Science 242, 50–61.

Ebihara, L., Beyer, E.C, Swenson, K.I., Paul, D.L. and Goodenough, D.A. (1989) Cloning and expression of a *Xenopus* embryonic gap-junction protein. Science 243, 1194–1195.

Gimlich, R.L., Kumar, N.M. and Gilula, N.B. (1990) Differential regulation of the levels of three gap-junction mRNAs in *Xenopus* embryos. J. Cell Biol. 110, 597–605.

Harris, A.L. Connexin32 forms ion channels in single artificial membranes. In: C. Perrachia (Ed.), Biophysics of Gap Junction Channels, CRC Press, Boca Raton FL, pp. 373–389.

Harris, A., Spray, D.C. and Bennett, M.V.L. (1981) Kinetic properties of a voltage-dependent junctional conductance. J. Gen Physiol. 77, 95–117.

Methfessel, C., Witzemann, V., Takehashi, M., Mishina, M., Numa, S. and Sakmann, B. (1986) Patch-clamp measurements on *Xenopus laevis* oocytes: currents through endogenous channels and implanted acetylcholine receptor and sodium channels. Pflügers Arch. 407, 577–588.

Moreno, A.P., Carvalho, A.C., Verselis, V.K., Eghbali, B. and D.C. Spray (1991) Voltage-dependent gap-junction channels are formed by connexin32, the major gap-junction protein of rat liver. Biophys. J. 59, 920–925.

Nicholson, B.J., Dermietzel, R., Teplow, D., Traub, O., Willecke, K. and Revel, J.P. (1987) Two homologous protein components of hepatic gap junctions. Nature 329, 732–734.

Paul, D.L. (1986) Molecular cloning of cDNA for rat liver gap-junction protein. J. Cell Biol. 103, 123–134.

Rubin J.B., Verselis, V.K., Bennett, M.V.L. and Bargiello, T.A. Molecular analysis of voltage dependence of heterotypic gap junctions formed by connexins 26 and 32. Biophys. J. 62, 183–195.

Spray, D.C., Harris, A.L. and Bennett, M.V.L. (1981) Equilibrium properties of a voltage-dependent junctional conductance. J. Gen. Physiol. 77, 77–83.

Swenson, K.I., Jordan, J.R., Beyer, E.C. and Paul, D.L. (1989) Formation of gap junctions by expression of connexins in *Xenopus* oocyte pairs. Cell 57, 145–155.

Traub, O., Look. J., Dermitzel, R., Brummer, F., Hulser, D. and Willecke, K. (1988) Comparative characterization of the 21-kD and 26-kD gap-junction proteins in murine liver and cultured hepatocytes. J. Cell Biol. 108, 1039–1051.

Verselis, V.K. and Bargiello, T.A. (1991) Dual voltage control in a *Drosophila* gap-junction channel. In: C. Perrachia (Ed.), Biophysics of Gap-Junction Channels, CRC Press, Baton Rouge, LA, pp. 117–129.

Verselis, V.K., Bennett, M.V.L. and Bargiello, T.A. (1991) A voltage dependent gap-junction channel in *Drosophila*. Biophys. J., 59, 114–126.

Verselis, V.K., White, R.L., Spray, D.C. and Bennett, M.V.L. (1987) Induced asymmetry of gating of gap-junction channels in amphibian blastomeres. J. Cell Biol., 105, 309a

Werner, R., Levine, E., Rabadan-Diehl, C. and Dahl, G. (1989) Formation of hybrid cell-cell channels. Proc. Natl. Acad. Sci. USA. 86, 5380–5384.

Wondergem, R. (1987) Effects of glucagon and insulin on cell membrane potential in liver tissue culture. In: N. Kraus-Friedman (Ed.), Hormonal Control of Gluconeogenesis, Vol. II, Signal Transmission, CRC Press, Boca Raton, FL, pp. 105–112.

J.E. Hall, G.A. Zampighi and R.M. Davis (Eds.)
Progress in Cell Research, Vol. 3

CHAPTER 16

Influence of lipophilic compounds on gap-junction channels

JANIS M. BURT, BRIAN N. MINNICH, KENNETH D. MASSEY, MARC OVADIA, LISA K. MOORE and KAREN K. HIRSCHI

Department of Physiology, University of Arizona, Tucson, AZ 85724, USA

Introduction

Gap junctions play a critical role in the normal function of the heart. They constitute the low resistance pathway through which intercellular propagation of the action potential occurs and consequently they provide for coordination of the heart's contractile activity. First-, second- and third-degree heart block, as well as re-entrant arrhythmias are more likely to occur as the conductance of this pathway is reduced from normal levels towards zero (Delmar et al., 1987; Delmar et al., 1989; Jalife et al., 1989; Joyner et al., 1989; Tan and Joyner, 1990). During myocardial ischemia and reperfusion, the incidence of heart block and re-entrant arrhythmias is greatly increased (Janse and Wit, 1989). Numerous changes in cellular composition and function (e.g., increased concentrations of intracellular Ca^{2+} and H^{+}, altered levels of second messenger molecules, and increased phospholipid breakdown) occur during myocardial ischemia (Hearse et al., 1981), many of which appear to influence cardiac gap-junction function (Spray and Burt, 1990). Of particular interest is the apparent link between increased incidence of arrhythmias and increased concentrations of non-esterified fatty acids (NEFA) in the affected tissue (Corr et al., 1981; van Bilsen et al., 1989).

NEFA and other lipophilic substances, especially alkanols and volatile anesthetics, have been demonstrated to alter the activity of a variety of membrane channels and enzymes in a variety of tissues (Mentz et al., 1976; Andreasen and McNamee, 1980; Katz et al., 1982; Philipson and Ward, 1985; Takenaka et al., 1987; Vemuri and Philipson, 1987; Hwang et al., 1990; Burt et al., 1991). Some investigators believe these compounds exert their effects on these channels or enzymes indirectly, by altering the bulk properties of the membrane (typically the membrane is disordered by the compounds) or by altering the activity of other cellular constituents that subsequently influence the channel or enzyme (e.g. kinases, phosphatases, pH). Others believe that the lipophiles exert their effects directly on the channel, either by binding to it or by altering its interaction with the immediately adjacent membrane (annular lipid) (Pringle et al., 1981; Goldstein, 1984; Franks and Lieb, 1986; Gruber, 1988; Akeson and Deamer, 1991). As yet, there is no consensus regarding the mechanism of action of these compounds. However, in at least one case, that of the acetylcholine receptor, it is apparent that those fatty acids which have a disordering effect on the membrane can interact with the channel at the channel-membrane interface and thereby alter the channel's activity (Andreasen and McNamee, 1980).

Intercellular communication between a variety of cell types is reduced by some fatty acids, short chain alkanols and volatile anesthetics (Hauswirth, 1968; Hauswirth, 1969; Wojtczak, 1985; White et al., 1985; Aylsworth et al., 1986; Agrawal and Daniel, 1986; Aylsworth et al., 1987; Terrar et al., 1988; Burt and Spray, 1988a; Burt and Spray, 1989a; Burt and Spray, 1989b; Burt, 1989; Niggli et al., 1989; Rüdisüli and Weingart, 1989; Fluri et al., 1990; Hauswirth, 1990; Burt et al., 1991; Hasler et al., 1991). We have used a structure-activity approach to examine the mechanism(s) of action underlying the uncoupling process (Burt, 1989; Burt, 1991; Burt et al., 1991). Most of our studies have been conducted on neonatal rat heart cells because of the importance of coupling to cardiac function, and because these cells appear to express only one gap junction-channel protein, Cx43. However, recently we have begun to compare the results obtained in neonatal heart cells to those obtained in other tissues in order to address the question of whether there are tissue- or connexin-specific differences in the effects of these lipophiles.

A structurally diverse group of lipophiles that includes fatty acids, fatty alcohols and volatile anesthetics, causes neonatal rat heart cells to uncouple from one another (Table I) (Burt, 1989; Burt and Spray, 1989a; Burt and Spray, 1989b; Fluri et al., 1990; Burt, 1991; Burt et al., 1991). The time course of uncoupling is concentration dependent, and can be quite rapid at higher concentrations (Fig. 1). Typically, coupling is rapidly restored when the lipophile is washed away; washout of the longer-chain species appears to require the presence of bovine serum albumin in the washout solution.

Table I. Effects of lipophiles on coupling between neonatal rat heart cells.[a]

Molecule type	Acyl chain length							
	7	8	9	10	12	14	16	18
Saturated fatty acid				Y	Y	Y	N	N
Saturated fatty alcohol	Y	Y	Y	Y	Y	N	N	N
Unsaturated fatty acid *cis* 9*						Y	Y	Y
Unsaturated fatty acid *trans* 9								N
Unsaturated fatty alcohol *cis* 9								Y

[a] Y indicates cells were uncoupled by the compound; N indicates no effect at the maximum aqueous solubility; blank space indicates the compound was not tested.
[b] C18:*cis* 11, C18:*cis* 6, and arachidonic acid also uncouple.

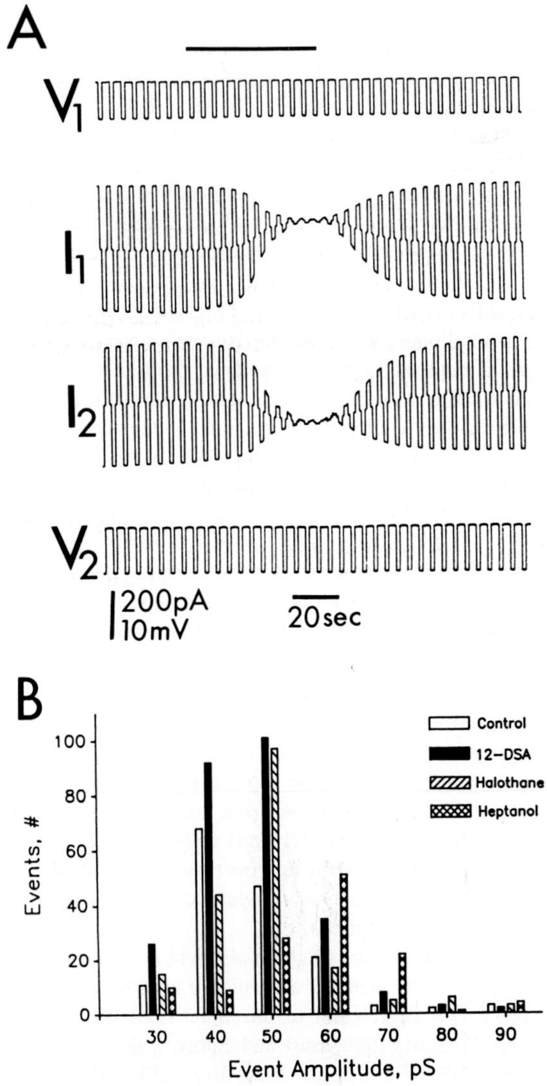

Figure 1. (A) Decanoic acid (2 mM) rapidly and reversibly uncouples heart cells; junctional conductance before and after exposure to the fatty acid is unchanged. (B) The unitary conductances of gap-junction channels between heart cells is constant under control conditions and during uncoupling by doxyl stearic acid, heptanol and halothane. These observations suggest that the observed uncoupling effect reflects a reduction of P_o rather than alteration of N or γ_j. (Part B of this figure was previously published (Burt, 1991))

The reduction in junctional conductance (g_j) produced by these lipophiles could reflect a reduction in the number of gap-junction channels available (N) (Burt and Spray, 1988a), a reduction in the unitary conductance of the channels (γ_j), or a reduction in the probability of the channel being in the open state (P_o). The rapid time course and essentially complete reversibility (Fig. 1A) of the uncoupling-recoupling effect argues against channel removal and re-insertion as the underlying mechanism for the uncoupling process. Comparison of unitary-event amplitudes in the presence and absence of various lipophiles reveals that c_j is unaffected by the lipophiles (Fig. 1B). Thus, it appears that the lipophiles reduce the open-time probability of the channel.

Gap junction-channel activity is influenced by second messengers (Burt and Spray, 1988b; Spray and Sáez, 1988; Spray and Burt, 1990), intracellular acidosis (Spray et al., 1981; Burt and Spray, 1988a; White et al., 1990), intracellular calcium elevation (Burt, 1987; Noma and Tsuboi, 1987; White et al., 1990), and voltage (Spray et al., 1991). In the case of neonatal rat cardiac cells, arachidonic acid and its metabolites appear to be the only second-messenger system that causes complete uncoupling. Arachidonic acid itself causes uncoupling and some of its lipoxygenase metabolites may be more potent uncoupling agents than the parent compound (Burt and Spray, 1989a; Fluri et al., 1990; Massey et al., 1992). However, most of the lipophiles that we have tested that uncouple neonatal rat heart cells do not serve as substrates for the arachidonic acid cascade. Thus, it is unlikely that uncoupling by lipophiles is second-messenger mediated. It is also unlikely that elevated intracellular proton or calcium levels mediate uncoupling by lipophiles (Burt, 1989). In the whole-cell voltage-clamp configuration, the intracellular pH and pCa are well buffered by the 10 mM HEPES and 10 mM EGTA present in the essentially infinite volume (relative to cell volume) of the electrode solution. Finally, the gap junctions in neonatal rat cardiac cells do not exhibit much voltage sensitivity under normal conditions (White et al., 1985; Burt and Spray, 1988a; Spray and Burt, 1990). While it is possible that the uncoupling lipophiles induce voltage sensitivity, this seems unlikely because at the lipophile concentrations that result in partial uncoupling, increased voltage sensitivity is not

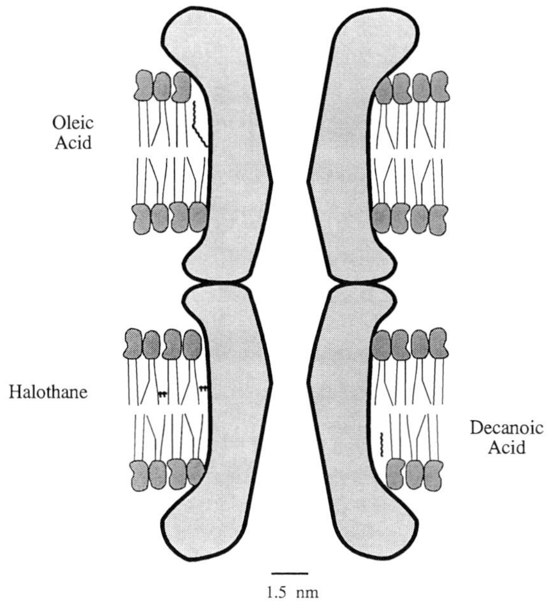

Figure 2. Hypothetical model for interaction of lipophiles with the gap-junction channel. The lipophiles are thought to interact with the channel at its interface with the membrane. By virtue of the disordering effect of lipophiles in this environment, the channel is destabilized and its P_o reduced. The channel, phospholipids and lipophiles are drawn to scale.

observed. Thus, none of the previously identified gating mechanisms seems a likely candidate to mediate uncoupling by lipophiles.

Results and discussion

In view of the above, we concluded that lipophiles directly influence gap-junction channel activity. We proposed (Burt, 1989; Burt, 1991) that lipophiles which have an uncoupling effect disorder the gap-junction channel-membrane interface, causing channel destabilization and consequent reduction in P_o (see Fig. 2 for model). If this model is correct, several predictions follow. First, all of the compounds must act by a common physical mechanism. Second, the activity of some membrane channels and enzymes would be expected to be less sensitive to the effects of the lipophiles than the gap-junction channels. Third, if the uncoupling effect reflects disordering of the channel-membrane interface rather than of the entire membrane, then the effective membrane concentration should be low, on the order of the protein concentration (rather than the lipid concentration) in the membrane. And fourth, the effects of the compounds should be additive, i.e. the dose-response relationship of a lipophile should be shifted in the direction of increased effectiveness when a second lipophile is present. Each of these predictions is borne out by the data, as discussed below.

Physical effect of lipophiles on membrane structure

The lipophiles that cause uncoupling of neonatal rat cardiac cells have one property in common: they all increase disorder in the interiors of biological membranes (C9–C18 region) (Klausner et al., 1980; Pringle et al., 1981; Goldstein, 1984; Gruber, 1988; Gruber and Low, 1988). The thermodynamically favored positions of the uncoupling lipophiles (and some structural analogs) in membranes are illustrated in Fig. 3. All of these compounds have high rotational as well as lateral mobility in membranes. The fixed "kink" of the *cis*-mono-unsaturated chain requires greater membrane volume for rotation, especially towards the interior of the membrane, than straight-chain analogs, and consequently enhances the disorder of the membrane towards its interior. The short-chain compounds and arachidonic acid occupy a disproportionate volume in the exterior (C1–C9) versus the interior (C10–C18) of the membrane, again enhancing disorder in the interior of the membrane. Halothane preferentially dissolves in the interior of the membrane, and thereby enhances disorder in this domain. Thus, it is apparent that all of the lipophilic uncoupling agents, despite their seemingly diverse structures, have a common physical effect on membranes, namely disordering.

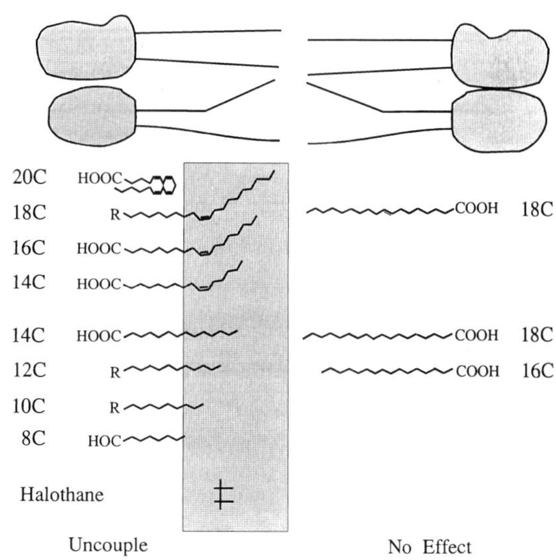

Figure 3. All of the lipophiles which result in uncoupling have a disordering effect on the interior of membranes (shaded region). Disordering results from the vacancy in the inner membrane created by compounds like arachidonic acid and the short-chain saturated fatty acids and alcohols, or from the additional space required for spinning of the kinked chains of the unsaturated fatty acids and alcohols. The positions of the tested compounds relative to the phospholipid bilayer reflect their preferred positions when dissolved in the membrane. Components are drawn to scale.

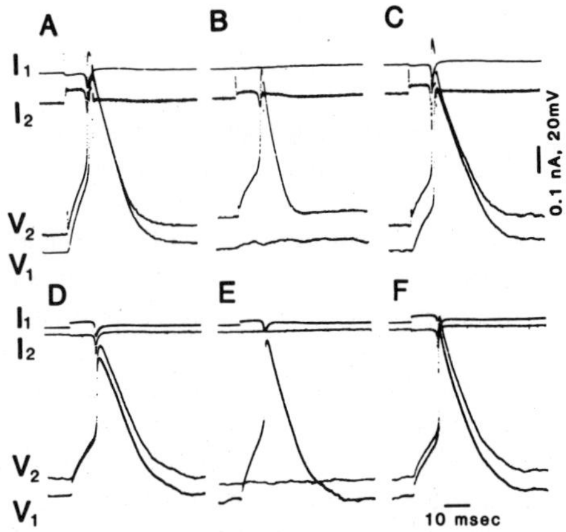

Figure 4. At concentrations of decanoic (A–C) or palmitoleic (D–F) acids sufficient to completely uncouple heart cells, the channels supporting the action potential remain active, suggesting specificity of the lipophiles for the gap-junction channel relative to other channels (from Burt et al., 1991)

Specificity of uncoupling lipophiles for the gap-junction channel

If the lipophiles have a direct effect at the channel membrane interface, then one would expect some specificity for the gap-junction channel compared to other channels in the membrane. This appears to be the case for at least three structurally disparate members of this group of lipophiles, halothane (Burt and Spray, 1989b), decanoic acid, and palmitoleic acid (Burt et al., 1991). As shown in Fig. 4, decanoic and palmitoleic acids completely uncouple neonatal rat myocardial cells without drastically altering the action potentials of the cells. In other words, the channels that support the action potential in the neonatal heart cells appear to be less sensitive to these lipophiles than the gap-junction channels.

Low effective membrane concentration

If the lipophiles cause uncoupling by disordering the channel-membrane interface rather than the bulk membrane, then the membrane concentration at which at least some of these compounds are effective should be low, certainly less than the protein concentration in the membrane (on the order of 1 mol protein/100 mol phospholipid). Incorporation of oleic acid, stearic acid, and oleyl alcohol as a function of time into the lipid fraction of cells is illustrated in Fig. 5. Despite its high concentration in cellular membranes, stearic acid had no effect on coupling. In contrast, at 120 min, when coupling is reduced to very low levels by oleic acid or oleyl alcohol, the lipid fraction of the cell contained approximately 2–3 mol% of the lipophile, i.e. 2–3 mol of oleic acid or oleyl alcohol for every 100 moles of phospholipid. Concentrations of this magnitude have been shown by several investigators to have only minor effects on the bulk properties of membranes, corresponding to the effects produced by a temperature change of only 1 or 2°C (Akeson and Deamer, 1991). The concentration of oleic acid in the plasma membrane (vs. total cellular lipid fraction) could be significantly less. To test this possibility, the percentage of the total incorporated oleic acid washed out by albumin was examined. As mentioned above, the uncoupling effect of the longer chain lipophiles, including oleic acid, is reversed only when albumin is present in the washout solution. Typically, recovery of coupling begins within 2 min. As illustrated in Fig. 6, during this 2-min period only a fraction of the total fatty acid (approximately 20% in this case) incorporated by the cells is washed out. If equilibration of fatty acid between intracellular locations and the plasma membrane is rapid, then this estimate of plasma membrane concentration is also an overestimate. These data suggest that the effective membrane concentration is significantly less than 1 mol%, and support the conclusion that the disordering effect exerted by oleic acid is at the level of the channel rather than on the bulk properties of the membrane.

The stoichiometry for uncoupling by these lipophiles is unknown. Based on our model, one would predict that the effective membrane concentration would be related to the magnitude of the disordering effect produced by the compounds in the region of the channel. In other words, higher membrane concentrations would be required for compounds whose disordering effect is small (e.g. decanoic vs. oleic acid). This could explain the high concentration (Burt et al., 1991) of decanoic acid, approaching 40 mol%, required for uncoupling (recall that despite this high concentration, decanoic acid uncoupled the cells without significantly altering the cell's action potential). It

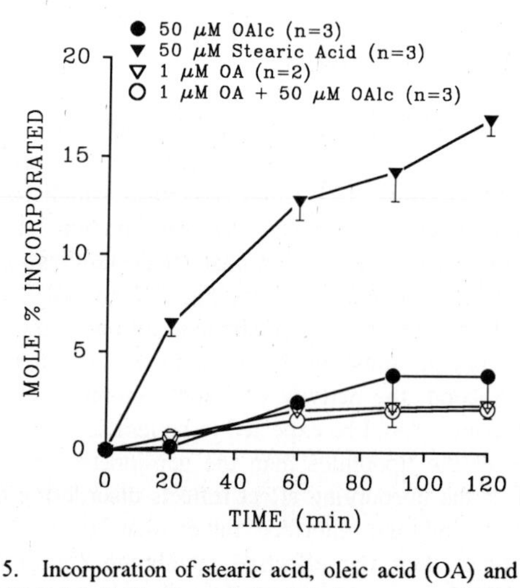

Figure 5. Incorporation of stearic acid, oleic acid (OA) and oleyl alcohol (OAlc) individually, and of OA in the presence of OAlc, into the lipid fraction of neonatal heart cells. OA and OAlc result in essentially complete uncoupling by 120 min of exposure (see Fig. 7) to the indicated aqueous concentrations; the resultant membrane concentration was 2–3 mol%. Despite significantly higher incorporation, stearic acid has no effect on coupling. Note that uptake of OA in the presence vs. absence of OAlc is not different.

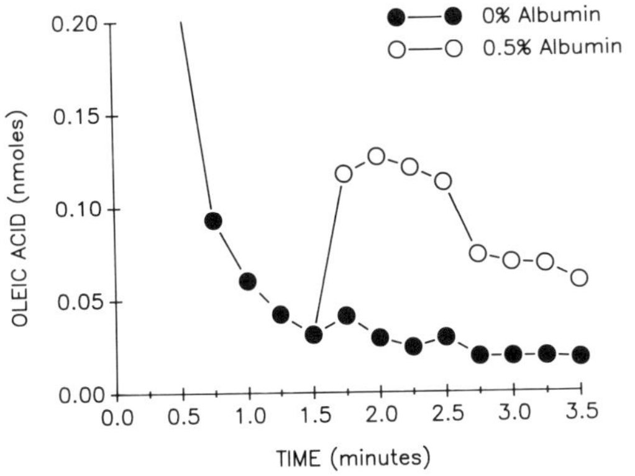

Figure 6. Albumin facilitates the washout of oleic acid, relative to washout by protein-free balanced salts solution, from heart cells. Four dishes (60 mm) of cells were loaded with 25 μM ^{3}H-oleic acid for 20 min, and washed 6 times with 5 ml aliquots of protein-free solution followed by 8 washes with 0.5% albumin (2 dishes, O) or 8 washes with protein-free solution (2 dishes, ●). Points represent the average of the nanomoles of radiolabeled fatty acid in each wash aliquot. In this experiment, approximately 0.65 nmol of OA were washed out by the albumin over the 2-min wash period compared to 0.25 nmol by the protein-free solution. Approximately 2 nmol of labeled fatty acid remained in the cells at the end of the washout period.

is also possible that the effective concentration of some compounds might be reduced relative to others with equal disordering effect because of preferential binding of the compound to the channel proteins. This possibility was thought to explain the apparent insensitivity to oleyl alcohol (Burt et al., 1991). However, the mol% incorporation data presented above for oleyl alcohol and oleic acid indicate that these compounds produce uncoupling at approximately equal membrane concentrations, despite a 50-fold difference in aqueous concentration.

Multiple lipophiles have additive effects

If all of the lipophiles that cause uncoupling act by a common mechanism, then they should have additive effects. In Fig. 7A, it can be seen that 750 μM decanoic acid and 1 mM halothane, which by themselves have no effect on junctional conductance, together significantly reduce coupling between heart cells. Similarly, the time courses for uncoupling by 1 μM oleic acid and 50 μM oleyl alcohol by themselves are similar. When applied simultaneously, uncoupling occurs with a faster time course, consistent with an additive effect of the two compounds. Indeed, the incorporation data presented in Fig. 5 demonstrate that uptake of oleic acid in the presence versus the absence of oleyl alcohol is not different. Consequently, it is likely that when both oleic acid and oleyl alcohol are present, the concentration of uncoupling lipophile in the membrane represents the sum of the concentrations of the individual compounds at the equivalent time point. These data support a common mechanism or site of action for these lipophiles.

Tissue- or connexin-specific differences in sensitivity to uncoupling by lipophiles

The data presented above are consistent with our model in which disordering of the channel-membrane interface and consequent destabilization of the channel is the mechanism underlying reduction of P_o Given the overall sequence and structural similarity of the different connexin proteins and the apparent similarity in lipid composition of mammalian membranes, it seemed likely that these lipophiles would have comparable effects on the conductance of gap-junctions of different connexin composition and tissue origins. To test this possibility, the effects of halothane and oleic acid on coupling between A7r5 cells, a neonatal rat aortic smooth muscle cell line, and between adult rat heart cells were examined.

A7r5 cells exhibit gap-junction channels of two unitary conductances, 36 and 90 pS (Moore et al., 1991). When the cells are exposed to halothane, they uncouple without a shift in the relative frequency of these two channel sizes (Fig. 8). This result is similar to that observed for the neonatal heart cells, and is consistent with a generic effect

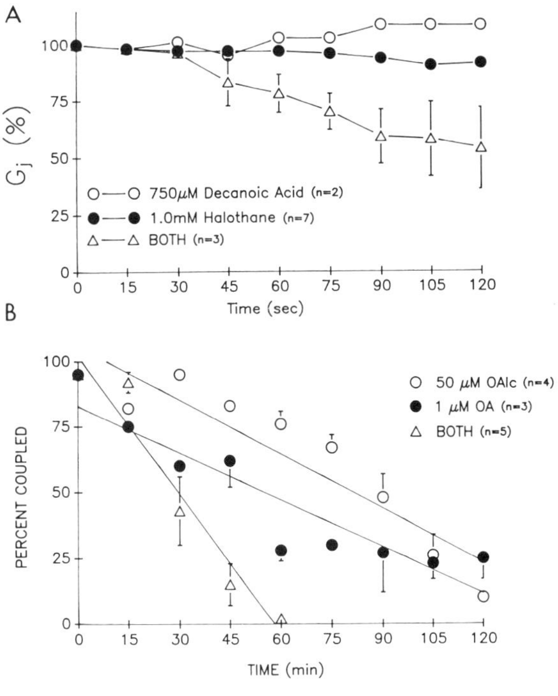

Figure 7. Concentrations of halothane and decanoic acid (A), which by themselves had no effect on coupling, together significantly reduced coupling. Concentrations of OA and OAlc (B), which by themselves reduced coupling to the same extent and with approximately equivalent time courses, together produced uncoupling with a faster time course. The data in Fig. 4 suggest that the presence of two lipophiles does not alter their uptake relative to that when only one is present. Consequently the faster time course most likely reflects an approximate twofold increase in the membrane concentration of the effective lipophiles.

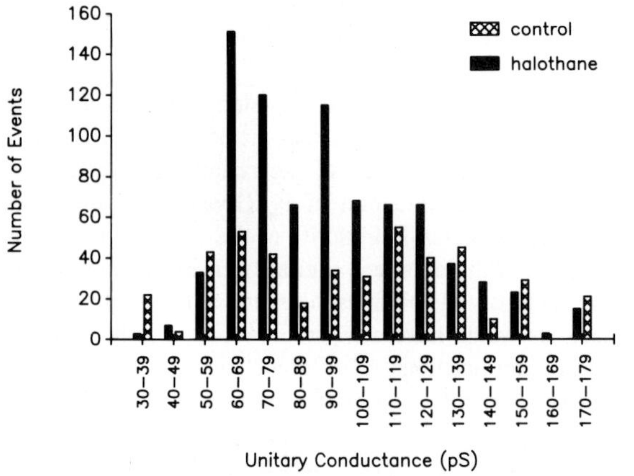

Figure 8. Halothane reversibly uncouples A7r5 cells, revealing single-channel events whose frequency relative to amplitude is not different from that observed in naturally occurring low-conductance pairs (controls). Dual whole-cell voltage-clamp techniques were used to collect the summarized data. The control data were obtained from 6 pairs of cells (153 events); the halothane data from 11 pairs (790 events). The patch electrode contained in mM: 67 CsCl, 67 K-glutamate, 10 HEPES, 10 EGTA, 0.5 $CaCl_2$, 0.3 $MgCl_2$, 10 KCl, 5 glucose, and 5 Na_2ATP. Under these conditions the mean conductances for the observed channels are 75 pS and 130 pS (Moore et al., 1991), vs. 36 and 90 pS measured with a solution in which K-glutamate replaces the CsCl.

of lipophiles on gap-junction channels. However, the sensitivity of A7r5 cells to uncoupling by oleic acid differed significantly from that observed in neonatal heart (Fig. 9) (Hirschi et al., 1991). The A7r5 cells exhibit a dose-dependent decrease in coupling that "saturates" at approximately 50% coupling. In contrast, in neonatal heart cells this decrease "saturates" at 0% coupling. The component of coupling in the A7r5 cells that is affected by oleic acid appears to be more sensitive to oleic acid than that in the neonatal heart cells (compare the effects of low concentrations).

Adult rat heart cells are also relatively resistant to uncoupling by all the lipophiles tested (Ovadia and Burt, 1991). Halothane (4 mM) decreased coupling by only 50% over a 3 min period compared to 100% uncoupling of neonatal heart cells by 2 mM halothane over a 0.5 min period (Burt and Spray, 1989b). Similarly, decanoic (1 mM) and oleic acids (100 μM) also decreased coupling by approximately 50% over an 8 min period compared to 100% uncoupling by 5 μM oleic acid over a similar period (Burt et al., 1991).

The basis for the differential sensitivity to uncoupling by oleic acid of the A7r5 and adult heart cells relative to the neonatal heart cells is uncertain. One possibility is connexin-specific differences. In the A7r5 cells, one of the two channel types might be insensitive to the disordering effects of oleic acid. The 36 pS channel in these cells appears to be composed of Cx40 and the 90 pS channel of Cx43 (Moore et al., 1992). Although the neonatal cells express Cx43, the unitary conductance of the channels, 50 pS, differs from that observed in the A7r5 cells. Whether other properties of the Cx43 channels also differ (e.g. sensitivity to destabilization by oleic acid) between the A7r5 and neonatal heart cells remains to be examined. Cx43 is also expressed in adult heart, although the conductance of the channel, 50 vs. 90 pS, is unknown. The adult cells may express other connexin proteins as well. A second possibility is tissue-specific differences in uptake, subcellular distribution, or metabolism of the fatty acid. The extent of incorporation of oleic acid (at low aqueous concentrations) into A7r5 and neonatal heart cells is the same (data not shown); however, unequal incorporation into the plasma membrane could account for the observed differences in uncoupling.

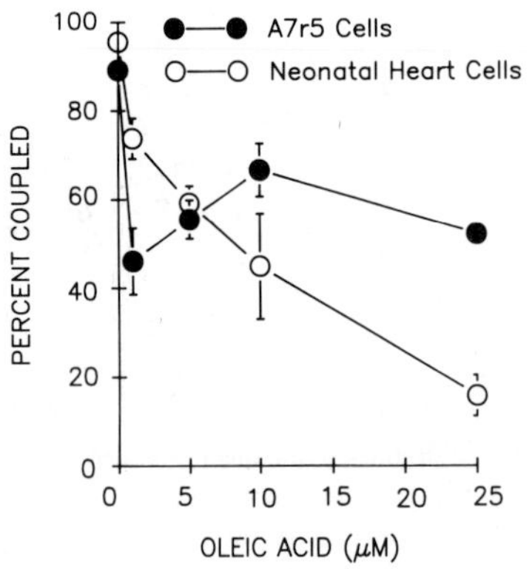

Figure 9. A7r5 cells are more sensitive than neonatal heart cells to uncoupling by oleic acid at low aqueous concentrations, but less sensitive at high aqueous concentrations. Coupling was measured using Lucifer Yellow dye-injection techniques. The cells were scored as coupled if, after 1 min, dye could be detected in any neighbor to the injected cell. Points represent the means ± SEM for five experiments.

Summary

The data presented support a mechanism for the uncoupling of neonatal rat heart cells by lipophiles that involves disordering of the channel-membrane interface with consequent channel destabilization and reduction of channel open-time probability. There appear to be functional differences, that could be connexin or tissue specific, in the sensitivity of coupling to oleic acid between neonatal heart, adult heart and vascular smooth muscle.

Acknowledgements

The authors wish to thank Drs. Eric Beyer and David Spray for their contributions to some of the studies discussed in this article. These studies were supported in part by HL31008, HL39795, the Arizona Disease Control Research Commission, and The American Heart Association, Arizona affiliate.

References

Agrawal, R. and Daniel, E.E. (1986) Control of gap-junction formation in canine trachea by arachidonic acid metabolites. Am. J. Physiol. 250, C495–C505.

Akeson, M. and Deamer, D.W. (1991) Anesthetics and membranes: a critical review. In: R. Aloia (Ed.), Drug and Anesthetic Effects on Membrane Structure and Function, Wiley-Liss, New York, pp. 71–89.

Andreasen, T.J. and McNamee, M.G. (1980) Inhibition of ion permeability control properties of acetylcholine receptor from *Torpedo californica* by long-chain fatty acids. Biochemistry 19, 4719–4726.

Aylsworth, C.F., Trosko, J.E. and Welsh, C.W. (1986) Influence of lipids on gap-junction-mediated intercellular communication between Chinese hamster cells *in vitro*. Cancer Res. 46, 4527–4533.

Aylsworth, C.F., Welxch, C.W., Kabara, J.J. and Trosko, J.E. (1987) Effects of fatty acids on gap-junctional communication: possible role in tumor promotion by dietary fat. Lipids 6, 445–454.

Burt, J.M. (1987) Block of intercellular communication: interaction of intracellular H^+ and Ca^{2+}. Am. J. Physiol. 253, C607–C612.

Burt, J.M. (1989) Uncoupling of cardiac cells by doxyl stearic acids: specificity and mechanism of action. Am. J. Physiol. 256, C913–C924.

Burt, J.M. (1991) Modulation of cardiac gap-junction channel activity by the membrane lipid environment. In: C. Peracchia (Ed.), Biophysics of Gap-Junction Channels, CRC Press, Boca Raton, pp. 75–93.

Burt, J.M., Massey, K.D. and Minnich, B.N. (1991) Uncoupling of cardiac cells by fatty acids: structure-activity relationships. Am. J. Physiol. 260, C439–C448.

Burt, J.M. and Spray, D.C. (1988a) Single-channel events and gating behavior of the cardiac gap-junction channel. Proc. Natl. Acad. Sci. USA 85, 3431–3434.

Burt, J.M. and Spray, D.C. (1988b) Inotropic agents modulate gap-junctional conductance between cardiac myocytes. Am. J. Physiol. 254, H1206–H1210.

Burt, J.M. and Spray, D.C. (1989a) Arachidonic acid uncouples cardiac myocytes. Biophys. J. 55, 217. (Abstract)

Burt, J.M. and Spray, D.C. (1989b) Volatile anesthetics block intercellular communication between neonatal rat myocardial cells. Circ. Res. 65, 829–837.

Corr, P.B., Snyder, D.W., Cain, M.E., Crafford Jr., W.A., Gross, R.W. and Sobel, B.E. (1981) Electrophysiological effects of amphiphiles on canine Purkinje fibers: implications for dysrhythmia secondary to ischemia. Circ. Res. 49, 354–363.

Delmar, M., Michaels, D.C., Johnson, T. and Jalife, J. (1987) Effects of increasing intercellular resistance on transverse and longitudinal propagation in sheep epicardial muscle. Circ. Res. 60, 780–785.

Delmar, M., Glass, L., Michaels, D.C. and Jalife, J. (1989) Ionic basis and analytical solution of the Wenckebach phenomenon in guinea pig ventricular myocytes. Circ. Res. 65, 775–788.

Fluri, G.S., Rüdisüli, A., Willi, M., Rohr, S. and Weingart, R. (1990) Effects of arachidonic acid on the gap junction of neonatal rat heart cells. Pflügers. Arch. 417, 149–156.

Franks, N.P. and Lieb, W.R. (1986) Partitioning of long-chain alcohols into lipid bilayers: implications for mechanisms of general anesthesia. Proc. Natl. Acad. Sci. USA 83, 5116–5120.

Goldstein, D.B. (1984) The effects of drugs on membrane fluidity. Annu. Rev. Pharmacol. Toxicol. 24, 43–64.

Gruber, H.J. (1988) Interaction of amphiphiles with integral membrane proteins. II. A simple, minimal model for the nonspecific interaction of amphiphiles with the anion exchanger of the erythrocyte membrane. Biochim. Biophys. Acta 944, 425–436.

Gruber, H.J. and Low, P.S. (1988) Interaction of amphiphiles with integral membrane proteins. I. Structural destabilization of the anion transport protein of the erythrocyte membrane by fatty acids, fatty alcohols and fatty amines. Biochim. Biophys. Acta 944, 414–424.

Hasler, C.M., Bennink, M.R. and Trosko, J.E. (1991) Inhibition of gap junction-mediated intercellular communication by α-linolenate. Am. J. Physiol. 261, C161–C168.

Hauswirth, O. (1968) The influence of halothane on the electrical properties of cardiac Purkinje fibres. J. Physiol. 201, 42–43.

Hauswirth, O. (1969) Effects of halothane on single atrial, ventricular and Purkinje fibers. Circ. Res. 24, 745–750.

Hauswirth, O. (1990) Halothane on cardiac Purkinje fibres: unmasking of the "true" passive membrane parameters. Second Congress of the Hungarian Pharmacological Society Budapest 13–20.

Hearse, D.J., Braimbridge, M.V. and Jynge, P. (1981) Protection of the Ischemic Myocardium: Cardioplegia, Raven Press, New York, pp. 1–420.

Hirschi, K.K., Minnich, B.N. and Burt, J.M. (1991) Differential effects of oleic acid on coupling between heart and vascular smooth muscle cells. J. Cell Biol. 115, 192a (Abstract).

Hwang, T.-C., Guggino, S.E. and Guggino, W.B. (1990) Direct modulation of secretory chloride channels by arachidonic and other *cis*-unsaturated fatty acids. Proc. Natl. Acad. Sci. 87, 5706–5709.

Jalife, J., Sicouri, S., Delmar, M. and Michaels, D.C. (1989) Electrical uncoupling and impulse propagation in isolated sheep Purkinje fibers. Am. J. Physiol. 257, H179–H189.

Janse, M.J. and Wit, A.L. (1989) Electrophysiological mechanisms of ventricular arrhythmias resulting from myocardial ischemia and infarction. Physiolog. Rev. 69(4), 1049–1169.

Joyner, R.W., Ramza, B.M., Tan, R.C., Matsuda, J. and Do, T.T. (1989) Effects of tissue geometry on initiation of a cardiac action potential. Am. J. Physiol. 256, H391–H403.

Katz, A.M., Hash-Adler, P., Watras, J., Messineo, F.C., Takenaka, H. and Louis, C.F. (1982) Fatty acid effects on calcium influx and efflux in sarcoplasmic reticulum vesicles from rabbit skeletal muscle. Biochim. Biophys. Acta 687, 17–26.

Klausner, R.D., Kleinfeld, A.M., Hoover, R.L. and Karnovsky, M.J. (1980) Lipid domains in membranes: evidence derived from structural perturbations induced by free fatty acids and lifetime heterogeneity analysis. J. Biol. Chem. 255, 1286–1295.

Massey, K.D., Minnich, B.N. and Burt, J.M. (1992) Arachidonic acid and lipoxygenase metabolites uncouple neonatal rat cardiac myocyte pairs. Am. J. Physiol. 263, C494–C501.

Mentz, P.H., Opitz, H. and Forster, W. (1976) Action of unsaturated fatty acids and prostaglandins on parameters of excitation and contraction in the guinea pig myocardium. Acta Biol. Med. Ger. 35, 1165–1166.

Moore, L.K., Beyer, E.C. and Burt, J.M. (1991) Characterization of gap-junction channels in A7r5 vascular smooth muscle cells. Am. J. Physiol. 260, C975–C981.

Moore, L.K., Beyer, E.C. and Burt, J.M. (1992) Selective block of gap-junction channel expression with antisense DNA. Biophys. J. 61, A408 (Abstract).

Niggli, E., Rüdisüli, A., Maurer, P. and Weingart, R. (1989) Effects of general anesthetics on current flow across membranes in guinea pig myocytes. Am. J. Physiol. 256, C273–C281.

Noma, A. and Tsuboi, N. (1987) Dependence of junctional conductance on proton, calcium and magnesium ions in cardiac paired cells of guinea pig. J. Physiol. 382, 193–211.

Ovadia, M. and Burt, J.M. (1991) Developmental modulation of susceptibility to arrhythmogenesis in myocardial ischemia: reduced sensitivity of adult vs. neonatal heart cells to uncoupling by lipophilic substances. Circulation 84, II-324 (Abstract).

Philipson, K.D. and Ward, R. (1985) Effects of fatty acids on Na^+-Ca^{2+} exchange and Ca^{2+} permeability of cardiac sarcolemmal vesicles. J. Biol. Chem. 260, 9666–9671.

Pringle, M.J., Brown, K.B. and Miller, K.W. (1981) Can the lipid theories of anesthesia account for the cutoff in anesthetic potency in homologous series of alcohols? Mol. Pharmacol. 19, 49–55.

Rüdisüli, A. and Weingart, R. (1989) Electrical properties of gap-junction channels in guinea-pig ventricular cell pairs revealed by exposure to heptanol. Pflügers Arch. 415, 12–21.

Spray, D.C., Harris, A.L. and Bennett, M.V.L. (1981) Gap-junctional conductance is a simple and sensitive function of intracellular pH. Science 211, 712–714.

Spray, D.C., Bennett, M.V.L., Campos de Carvalho, A.C., Eghbali, B., Moreno, A.P. and Verselis, V. (1991) Transjunctional voltage dependence of gap-junction channels. In: Peracchia, C. (Ed.), Biophysics of Gap-Junction Channels, CRC Press, Baton Rouge, pp. 97–116.

Spray, D.C. and Burt, J.M. (1990) Structure-activity relations of the cardiac gap-junction channel. Am. J. Physiol. 258, C195–C205.

Spray, D.C. and Sáez, J.C. (1988) Agents that affect gap-junctional conductance: sites of action and specificities. In: M.A. Mehlman (Ed.), Biochemical Regulation of Intercellular Communication, Alan R. Liss, New York, pp. 1–26.

Takenaka, T., Horie, H. and Hori, H. (1987) Effects of fatty acids on membrane current in the squid giant axon. J. Membr. Biol. 95, 113–120.

Tan, R.C. and Joyner, R.W. (1990) Electrotonic influences on action potentials from isolated ventricular cells. Circ. Res. 67, 1071–1081.

Terrar, D.A., Jason, G.G. and Victory, B.A. (1988) Isoflurane depresses membrane currents associated with contraction in myocytes isolated from guinea-pig ventricle. Anesthesiology 69, 742–749.

van Bilsen, M., van der Vusse, G.J., Willemsen, P.H.M., Coumans, W.A., Roemen, T.H.M. and Reneman, R.S. (1989) Lipid alterations in isolated, working rat hearts during ischemia and reperfusion: its relation to myocardial damage. Circ. Res. 64, 304–314.

Vemuri, R. and Philipson, K.D. (1987) Phospholipid composition modulates the Na^+-Ca^{2+} exchange activity of cardiac sarcolemma in reconstituted vesicles. Biochim. Biophys. Acta 937, 258–268.

White, R.L., Spray, D.C., Campos de Carvalho, A.C., Wittenberg, B.A. and Bennett, M.V.L. (1985) Some electrical and pharmacological properties of gap junctions between adult ventricular myocytes. Am. J. Physiol. 249, C447–C455.

White, R.L., Doeller, J.E., Verselis, V.K. and Wittenberg, B.A. (1990) Gap-junctional conductance between pairs of ventricular myocytes is modulated synergistically by H^+ and Ca^{2+}. J. Gen. Physiol. 95, 1061–1075.

Wojtczak, J.A. (1985) Electrical uncoupling induced by general anesthetics: A calcium-independent process? In: M.V.L. Bennett and D.C. Spray (Eds.), Gap Junctions, Cold Spring Harbor Symposium, Cold Spring Harbor, NY, pp. 167–177.

J.E. Hall, G.A. Zampighi and R.M. Davis (Eds.)
Progress in Cell Research, Vol. 3

CHAPTER 17

Evidence for heterogeneous channel behavior in gap junctions

S.V. RAMANAN, K. MANIVANNAN, R.T. MATHIAS and P.R. BRINK

Department of Physiology and Biophysics School of Medicine, HSC, SUNY at Stony Brook, Stony Brook, NY 11794, USA

Introduction

Elucidation of the factors that regulate gap-junction channel patency has been a major focus within the field. For the most part, regulation has been viewed in terms of how "extrinsic" agents could or would affect channel gating. Some regulatory schemes include the complex pathways of second messengers, where the outcome is direct phosphorylation (or dephosphorylation) of protein subunits in gap-junction channels. Covalent alteration would presumably lead to altered gating (Pressler and Hathaway, 1987; Kolb and Somogyi, 1991). Other means of affecting gap junction channel-mediated communication are (1) regulation of the rate of synthesis of various connexin subunit proteins (Cx32–43 etc.), (2) controlling insertion and (3) "activation" of assembled subunits within the membrane. (3) implies that not all assembled channels function. Tuttle et al.(1986) have shown that only a fraction of gap-junction channels are operative at any time between two adjacent coupled cells.

Questions about how cell-to-cell communication could be regulated are difficult to address directly. Macroscopic data cannot distinguish, for example, between the following possibilities: (1) a small population of channels, all with high open probability; (2) a large number of channels with low open probability; (3) some mixture of the two; or (4) cooperative interactions between channels. Data at the single-channel level can provide evidence for different types of channel behavior or heterogeneous channel populations. Channels are classified based on parameters such as selectivity, voltage dependence and sensitivity to blocking agents. These parameters can affect (a) the kinetics of the channel, or grossly, its open probability and (b) the conductance of the channel, and its ability to substate; such substating could affect the type and form of cytoplasmic elements that diffuse from cell to cell. We term voltage-independent gap-junction channels with identical selectivity and blocking characteristics "similar" channels. Such similar channels may however differ in their open probabilities or in their conductances. This paper focuses on our efforts to determine if there are similar channel types within gap junctions and whether substates exist.

Heterogeneous channel types

Multichannel records are fit, in general, by assuming that all the channels in the patch have identical gating characteristics and, by implication, the same open probability. In extended multichannel recordings of earthworm septum, our results do not support this hypothesis. At least two channel types, distinguished by their open probabilities, are found to be necessary to develop a satisfactory model. The theory for fitting the amplitude histograms, i.e., histograms of the observed current through the patch, to a model of a collection of channels with differing probabilities is straightforward (see e.g., Manivannan et al., 1992).

A useful parameter for the analysis of multichannel records may be defined as follows: Let subscripts P_0, P_1 and P_2 be the probability that no channels are open, one channel is open and two channels are open, respectively. Define $D = P_1{}^2/P_0P_2$. Manivannan et al. (1992) present tests that D has to satisfy if all channels in the patch have identical open probabilities. Here we restrict ourselves to the case where only two channels are present in the patch. There are then three possibilities:

1. $D = 4$: Both channels have the same open probability. (See Fig. 1a)
2. $D > 4$: The two channels have differing open probabilities. (Fig 1b)
3. $D < 4$: The two channels cannot be independent. In Fig. 1c, the dotted lines show the best fit with two independent channels.

Note that even if $D > 4$, the channels could be acting non-independently, i.e., *cooperatively*. However, if $D < 4$ then, independent of kinetics, the channels cannot be independent; the condition $D < 4$ is thus a necessary condition for independence.

By varying the concentration of salts in the bath, we are able to determine channel selectivity. Channels, though they may have different open probabilities, have the same selectivity. Note that, in all cases, the solutions contain the same external blocking agents, e.g. Zn, tetraethylammonium, tetrodotoxin. These channels, having the same selectivity and blocking characteristics but different gating behavior, may be considered "similar" channels (Manivannan et al., 1992).

Substates

The 100 pS channel exhibits a variety of substates. Figure 2a,c displays a record where two substates with conductances of 50% and 70% of the maximal conductance are seen. Step transitions from every conductance state to every other conductance state may be seen in the records. This demonstrates that the various conductance levels originate from a single channel, and not a collection of channels; in the latter case, some of the step transitions would be "forbidden".

Both the conductance of the substates and their actual number seem to vary from record to record. We have been unable to correlate the occurrence of the substates with any of the external agents in the bath; and the high variability of the substates precludes any quantitative assessment of the phenomenon at the present time.

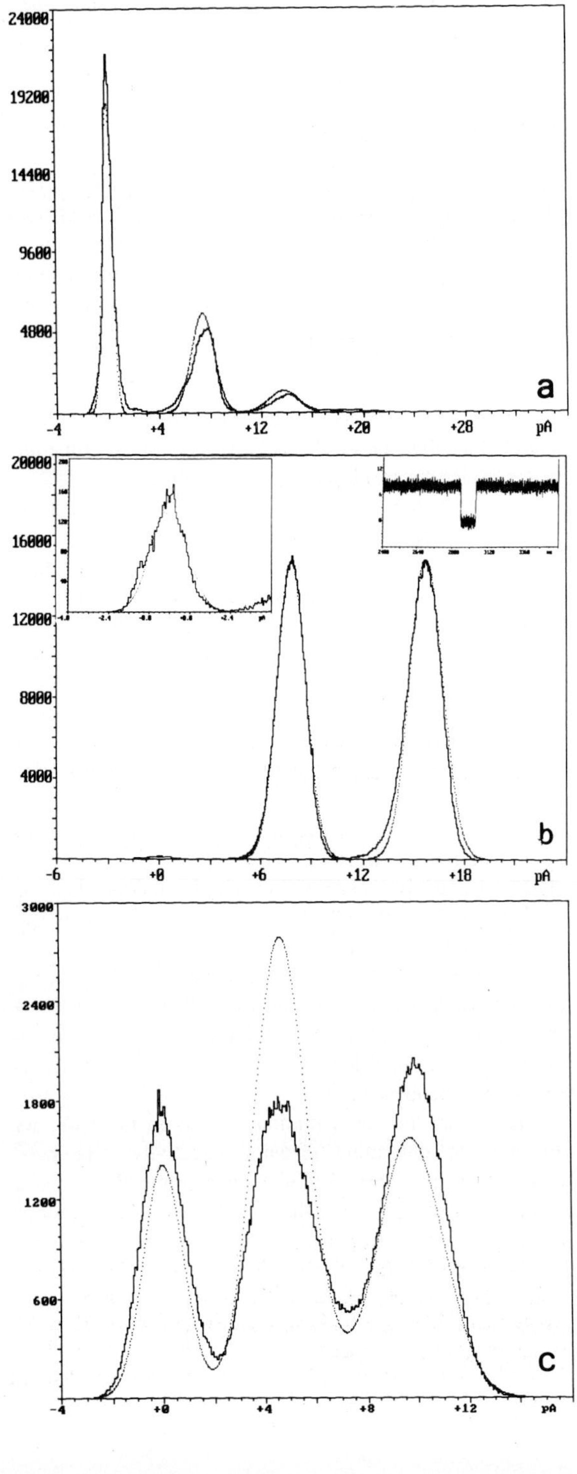

Figure 1. All three panels are amplitude histograms of multichannel data from different patches of the earthworm septum. In all panels, the heavier line represents the amplitude histogram of the data, while the dotted line is the fit to the data. Panel (a) shows a case when all the channels are identical. At least four channels are present in the patch; the holding potential is 77 mV. The fit is made by assuming that there are five channels, all with an open probability of 0.19. Panel (b) shows a case where there are two channels in the patch; the holding potential is 61 mV. D, in this case, was 105, thus two channels with different open probabilities of 0.99 and 0.54 are required to fit the data. The inset shows one of the 34 complete closures in the record. Two channels are again present in the patch in the data represented in panel (c) where the holding potential is 50 mV. The parameter D = 1.4, thus independent channels cannot fit the data. The dotted line shows the best fit of two independent channels to the data; this requires that both channels have an open probability of 0.43. The mismatch of fit to data suggests that there is cooperativity or interactions between the channels. (reprinted with the permission of the Biophysical Journal).

Conductance shifts

In a multichannel record, one expects that the current passed by two open channels should be twice the current passed by one open channel, the current passed by three channels should be thrice the current passed by one open channel, etc. Figure 3a shows the amplitude histogram from a multichannel patch from the earthworm septum where 2 channels were seen to be open simultaneously. The dotted line shows the fit where every channel passes the same current, namely that associated with the first peak, which corresponds to one channel open. It may be seen that as the number of open channels increases, the discrepancy between the expected current and the actual current increases. The net effect is that the current passed when several channels are open is less than a multiple of the current when only one channel is open, a phenomenon dubbed as a "downward" shift. Such downward shifts are found ubiquitously in multichannel records of the 100 pS channel.

Figure 3b shows the opposite case, namely that where the current passed by several channels is greater than a multiple of the current due to only one open channel. Again, the dotted line shows the fit where every channel carries the current associated with the first peak. This phenomenon, the "upward" shift, is quite uncommon compared to the downward shifts illustrated in Fig. 3a and the magnitudes of these upward shifts are also found to be much less than the magnitudes of the downward shifts.

One elementary explanation of these shifts is that they arise from a series resistance, either in the pipette or in the bath. As more channels open, the net current increases, but such an increased current also induces a greater voltage drop across the series resistance. This leaves a lesser potential to be applied across the channels,

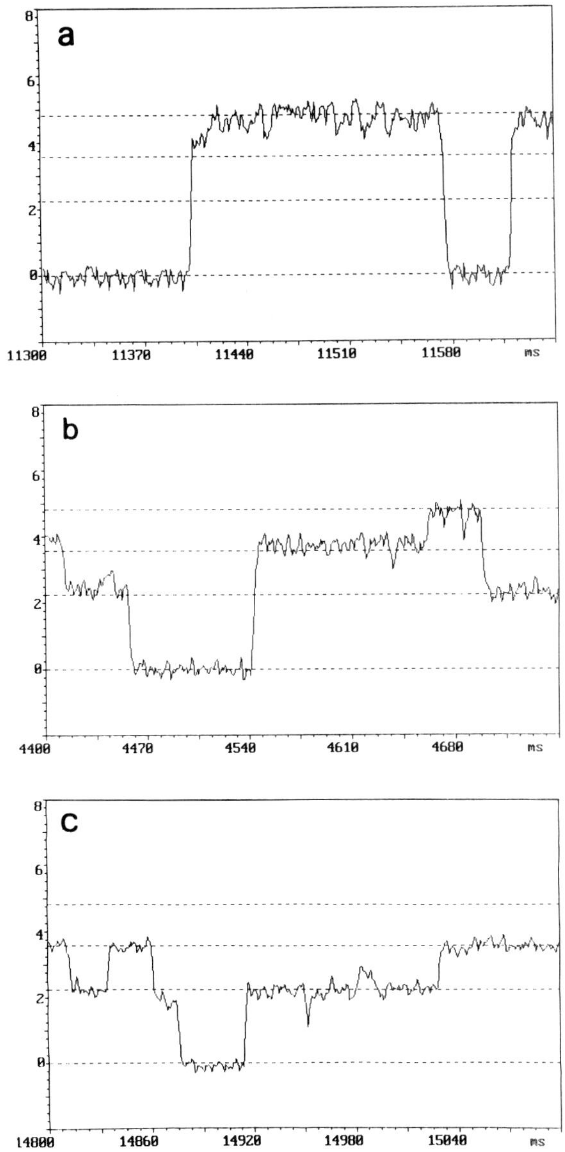

Figure 2. Panels a–c are all from a patch recording (031489) of a septum at a holding potential of 70 mV. Two substates at 50% and 70% of the unitary conductance state are seen. All possible transitions between the four current levels (including zero) can be seen. If the current levels were due to independent channels and not due to substates, some of these transitions would represent a simultaneous double step, and would not be seen.

which in turn reduces the current to less than a multiple of the "unitary" current when only one channel is open. An elementary model of this hypothesis shows that the magnitude of the series resistance must be in the range of 300 MΩ to 3 GΩ, the variability being from different patches. Apart from the fact that such a high series resistance is not a reasonable one, it also fails to account for (1) upward shifts and (2) the phenomenon that the shifts tend to increase relatively with increasing applied voltage.

Another explanation arises from the long-lived substates already shown in the previous section. We model this as follows: Let the net effect of the substates be represented as a variation of the average current passed through the channel with its open probability. Consider the following example of a patch with only two channels. A channel which is open 90% of the time has an average conductance of 105 pS while a channel which is open only 20% of the time has an average conductance of 95 pS. Such variation of the average conductance could occur because the different substates have a different relative occupation probability for the two channels. The average resistance when one channel is open would then be

$$\frac{(1/105)(0.9)(0.8) + (1/95)(0.1)(0.2)}{(0.9)(0.8) + (0.1)(0.2)} = (104.70\ \mathrm{pS})^{-1}$$

while the conductance when both channels are open is $105 \pm 95 = 200$ pS $< 2 \times 104.70$. This gives an example of the downward shift. Exchanging the conductances of the two channels yields a patch which shows an upward shift. Due to the number and variability of the substates noted above, it is very difficult to quantify the argument more precisely on the data. We therefore present the above only as a plausible explanation for the data.

Conclusions

The data presented here clearly indicate that there are heterogeneous populations of channels with regard to open probability. Further, the data support the idea that the channels can interact or behave in a cooperative way. The existence of substates opens up the possibility that gap-junction channels can temporarily affect the rate of small-solute diffusion. If the substates can, in the future, be correlated with effective channel diameter, then the "long-lived" substates would take on new meaning in terms of their role in intercellular communication.

There is anatomical evidence for many of these speculations. Since there are a number of known connexin types within cells such as Cx27, 32 and 43, to mention just a few, the formation of similar channel types might come about as a result of the combination of channels formed solely of one type of connexin, while others might arise as mixtures of two or more connexin types expressed by a cell. The gating and conductive properties might well be different for a channel composed solely of Cx43 as op-

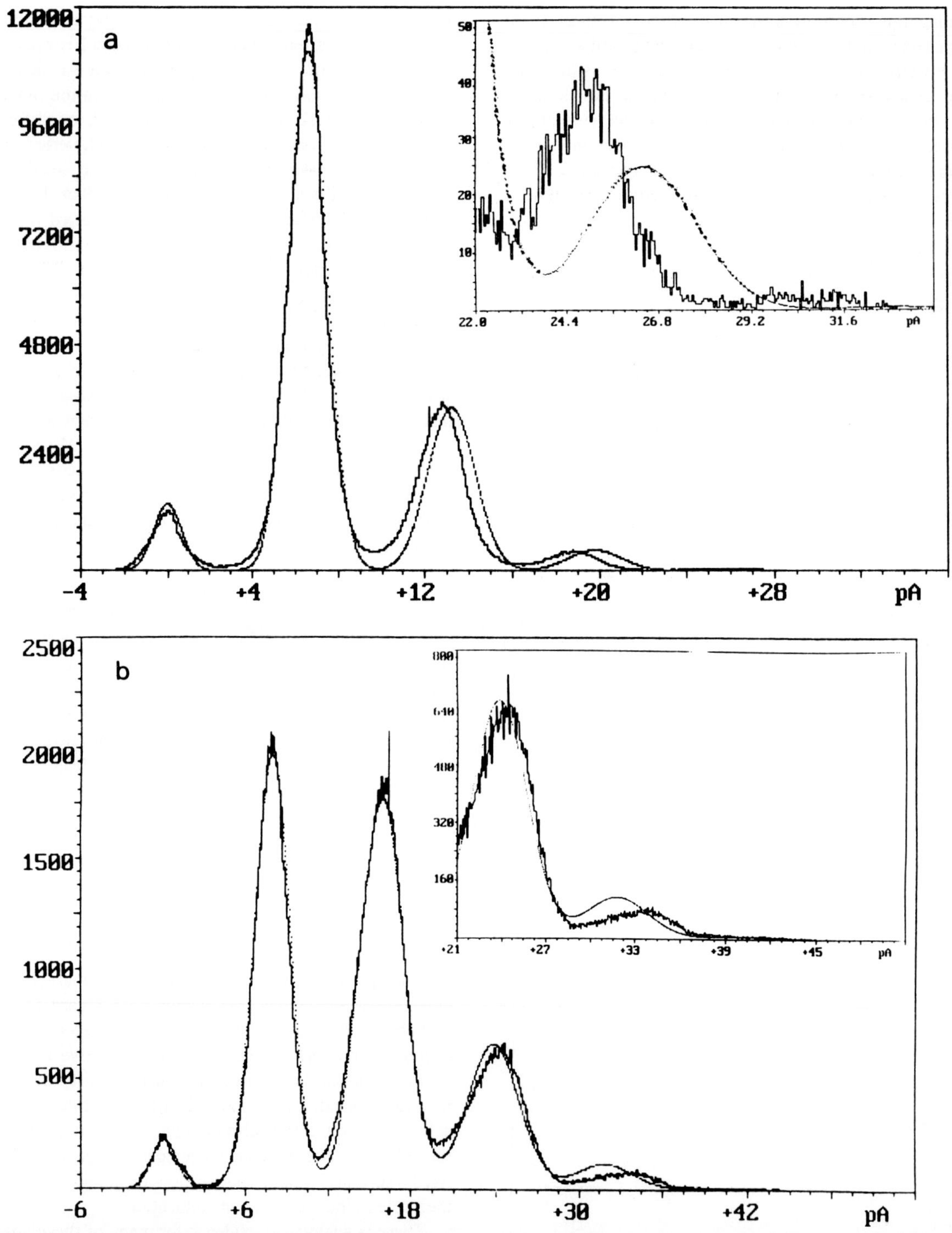

Figure 3. Panel (a) shows four min of a recording of a patch (081790) where four channels are present. The dotted lines are the fit, assuming that all the channels carry the same current, namely that associated with the peak where only one channel is open (6.6 pA). The actual current passed when two or three channels are open is seen to be much less than the expected current, a "downward" shift. The inset shows an enlargement of the smaller peaks. The holding potential is +48 mV; one channel with open probability of 0.911 and four channels with open probability of 0.088 were used to get the fit. The concentration of CsCl in the bath and the pipette are both 170 mM. Panel (b) shows the opposite case in another patch (072590A) at a holding potential of +57 mV, with identical concentration of ions in both sides (170 mM). Here the current when several channels are open is greater than a multiple of the current passed by a single open channel. The dotted line is the fit with one channel of open probability of 0.92 and four channels with open probability of 0.22; in the fit all channels are assumed to pass the same current (7.91 pA). The inset shows an enlargement of some of the smaller peaks.

posed to one containing equal mixes of 43 and 32, for example. The cooperative behavior observed is also consistent with the anatomy. Normally, gap-junction channels sit in loose arrays such that the individual channels are in close proximity to one another. This intimacy is one logical requisite for channel interaction which could lead to cooperative behavior.

The evidence suggests that there are functionally different gap-junction channels within gap-junction plaques. Whether these differences will turn out to be due to the ratios of connexin forms or to the states of phosphorylation, for example, is not known.

Acknowledgement

This work was supported by NIH HL31299.

References

Kolb, H-A. and Somogyi, R. (1991) Characteristics of single channels of pancreatic acinar gap junctions subject to different uncoupling procedures. In: C. Peracchia (Ed.), Biophysics of Gap-Junction Channels, CRC Press, Boca Raton, pp. 209-228.

Manivannan, K., Ramanan, S.V., Mathias, R.T. and Brink, P.R. (1992) Multichannel recordings from membranes containing gap junctions. Biophys. J., in press.

Pressler, M.L. and Hathaway, D.R. (1987) Phosphorylation of purified dog heart junctions by cAMP dependent protein kinase. Circulation, 76 (Suppl IV), 18.

Tuttle, R., Masuko, S. and Nakajima, Y. (1986) Freeze-fracture studies of the large myelinated club ending synapse of the goldfish Mauthner cell: special reference to the quantitative analysis of gap junctions. J. Comp. Neurol., 246, 202.

J.E. Hall, G.A. Zampighi and R.M. Davis (Eds.)
Progress in Cell Research, Vol. 3

CHAPTER 18

Unmasking electrophysiological properties of connexins 32 and 43: transfection of communication-deficient cells with wild type and mutant connexins

A.P. MORENO[a], G.I. FISHMAN[b], B. EGHBALI[a] and D.C. SPRAY[a]

Departments of [a]Neuroscience and [b]Medicine, Albert Einstein College of Medicine, Bronx, NY 10461, USA

Introduction

Direct communication between animal cells in most tissues is primarily achieved through membrane complexes known as gap junctions. These junctions are groups of membrane channels whose homologous protein subunits are termed connexins (for review see Bennett et al., 1991). cDNAs that encode numerous connexins have now been cloned (see Part I, The Connexin Family), and it has become clear that individual connexin isoforms can be expressed in many tissues and that some cells co-express two or more connexins. Co-expression provides important physiological possibilities for cells during normal functions such as development and differentiation. Co-expression of multiple connexins also presents a problem when the characteristics of individual connexins are to be studied. One possible solution is to use potent vectors to transfect communication-deficient cell lines with cDNAs encoding the connexin of interest. In contrast to the expression of connexin cRNAs in oocytes, stable transfection offers the opportunity to record single gap junctional-channel activity and to evaluate gating properties. Other advantages include: (1) rapid exchange of intracellular medium (by internal dialysis) or, alternatively, using nystatin to gain electrical access to the cell without disrupting its regulatory mechanisms. (Horn and Marty, 1988); (2) uniformity of expression over long periods of time; (3) expression of mammalian connexins within a more phylogenetically related milieu, (4) the high surface area to volume ratio, which should enhance resolution in trafficking experiments. A disadvantage is that the time required for stable transfection and selection of stably transfected clones can be several months.

This chapter summarizes our experience in studying the properties of gap-junction channels formed by wild type Cx32, wild type Cx43, and mutated Cx43 tranfected into a communication-deficient cell line, SKHep1. This cell line is not coupled, i.e. does not pass Lucifer Yellow; this simplifies the selection of coupled cells after transfection (Eghbali et al., 1990; Fishman et al., 1990). Unitary conductance as well as voltage dependence of the wild type channels that connexins form are clearly distinguishable. We use site-directed mutagenesis to investigate the effects of the C-terminus on single-channel conductance, and we investigate the voltage dependence and the effect of phosphorylation on conduction properties.

Methods

Transfection

SKHep1 cells were co-transfected with vectors containing connexin cDNA and the neomycin-resistant gene, using the Ca_2PO_4-precipitation method (Graham and van der Eb, 1973). Cells were selected first in neomycin and then colonies where LY transfer was observed were picked. After selection, clones were grown in several dishes and were frozen for subsequent experiments. Freshly dissociated cell pairs from subconfluent cultured dishes were used for electrophysiological recordings.

Electrophysiology

In brief, two heat-polished patch-type electrodes were used, one on each cell. Seals to cell surfaces were achieved with suction and were monitored by applying 20 ms, 4 mV pulses at 10 Hz simultaneously through both electrodes. During suction, the current from the pipette was measured in the Track Mode of the voltage-clamp system (Model 1-C, Axon Instruments, Foster City, CA). After gigaohm seals were formed, the systems were switched to "hold". Whole-cell voltage-clamp was achieved after abrupt suction, and junctional conductance (g_j) was determined by dividing the junctional current by

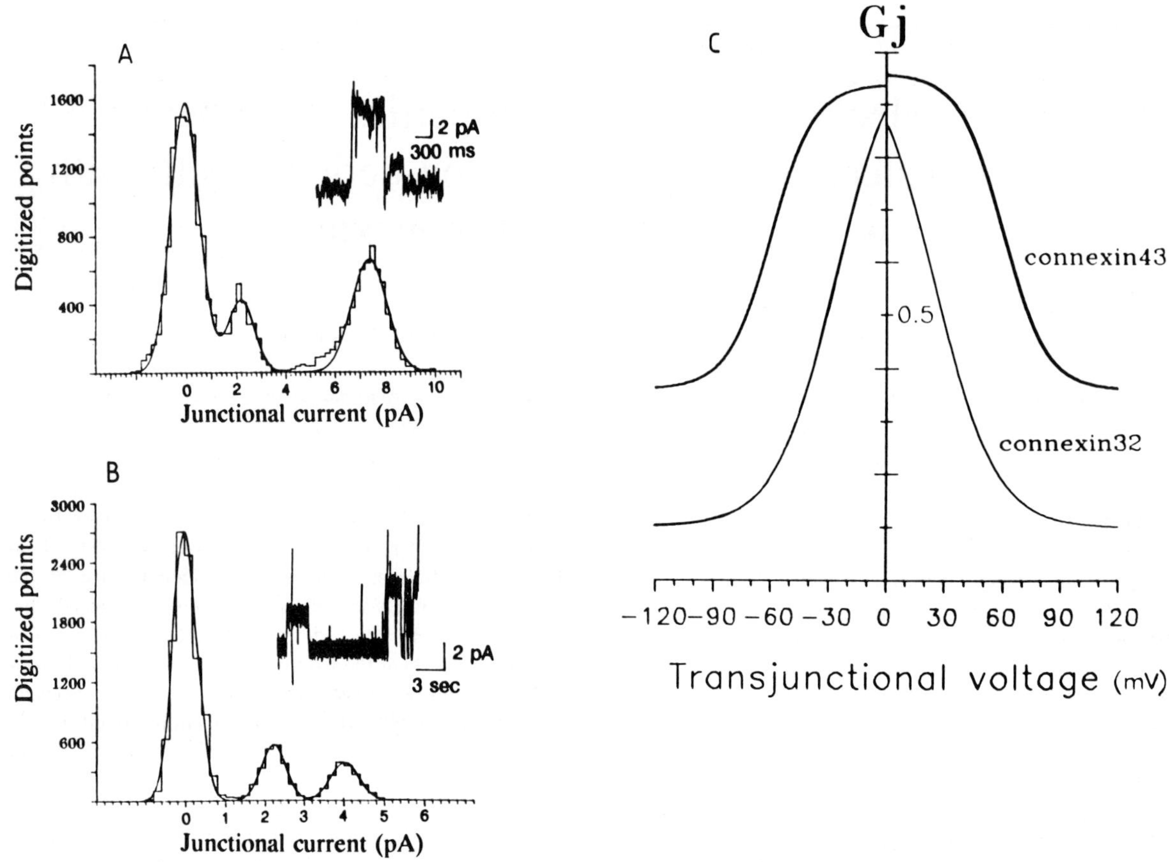

Figure 1. Comparison of unitary conductance and voltage dependence of Cx32 and Cx43 evaluated in stably transfected SKHep1 cells. (A) Dwell-time histogram from a digitized current trace of one cell from a pair stably transfected with rat Cx32 cDNA. The driving force was 60 mV; data were acquired at 1 kHz. The largest current transition corresponds to the exogenous Cx32 channel; the smaller is associated with the endogenous channels. The resolution of the technique allows us to clearly differentiate even the smaller channels from the noise level. (Inset: digitized current trace that corresponds to the histogram shown). (B) Dwell-time histogram as in A, but from a current trace of one cell from a pair transfected with human Cx43 cDNA. Driving force was 40 mV and acquisition frequency 1 kHz. In this case, both of the current transitions correspond to the Cx43 channels; no endogenous channels were observed. (C) Voltage dependence of Cx32 and Cx43. When expressed in SKHep1 cells, channels formed of both Cx32 and Cx43 are sensitive to transjunctional voltage (V_j) of either polarity. This graph shows the best fit of the Boltzmann equation to the dependence of normalized junctional conductance (g_j) on V_j. The parameters to fit this equation are: for Cx32, $V_0 \approx 25$ mV, $g_{min}/g_{max} = 0.1$ and n = 1.75; for Cx43: $V_0 \approx 60$ mV, $g_{min}/g_{max} = 0.37$ and n = 2.65.

the voltage steps applied to each of the cells. To observe single-channel currents through gap junctions, g_j was reduced by exposure to 2 mM halothane (Burt and Spray, 1989) and a transjunctional driving force (V_j, from ±30 to ±60 mV) was established. When g_j was very low, events that were simultaneous and opposite in sign were observed in the current traces of the two-cell current records (Fig. 1). To quantify the voltage dependence of the channels, 10-s V_j steps (20–100 mV) were applied to either cell while the junctional current was recorded. The current at the end of the pulse (steady state) was normalized with respect to the instantaneous current at the beginning of each pulse. This value, divided by the driving force, equals G_∞ (normalized steady-state g_j; Fig. 1B). Plots of G_∞ vs. V_j were compared for different connexins and after various treatments (see below).

Solutions and reagents

Experiments were performed during perfusion of bath solution (in mM): 133 NaCl, 3.6 KCl, 1.0 $CaCl_2$, 0.3 $MgCl_2$, 3 Hepes, pH 7.4. Each heat-polished patch pipette (3–7 MΩ) was filled with internal solution containing (in mM): 135 CsCl, 0.5 $CaCl_2$, 2 Na_2ATP, 3 MgATP, 0 Hepes, 10 EGTA, pH 7.2. Phosphorylating treatments included the membrane-permeant cAMP derivative 8-Br-cAMP (8-bromo 1 mM); the adenyl-cyclase activator forskolin (5 μM), the protein kinase C activator TPA (12-O-tetradecanoylphorbol-13-acetate; 0.1 μM), and the phosphatase inhibitor okadaic acid (300 nM). Dephosphorylating agents included the kinase inhibitor staurosporine (Antibiotic AM-2282, 300 nM) and alkaline phosphatase (AP from *E. coli*), applied intracellularly.

One or both of the patch-cell electrodes were filled with AP in the internal solution to a final concentration of 5–25 μg/ml. To decrease protease activity, the enzyme solution was preheated for 1 min at 100°C.

Differences between Cx32 and Cx43 expressed in the same system

Electrophysiological studies of the parental SKHep1 cells revealed the occasional (< 15% of cell pairs) presence of small (30 pS) endogenous channels (Moreno et al., 1991b). Northern and Western blots and immunostaining for Cx43 and Cx32 were negative. After transfection with cDNA encoding Cx32, Northern and Western blots showed the appearance of mRNA and protein corresponding to Cx32 (Eghbali et al., 1990).

Junctional conductance in these Cx32 transfectants was as high as 20 nS. Unitary conductances of the channels formed by Cx32 are about 130 ps (Fig. 1A). Event histograms of 8 experiments in which at least 100 single-channel events were measured show that the unitary conductance (γ_j) of the newly formed channels was > 4 times larger than that of the endogenous channels.

Cx32 gap-junction channels are closed by transjunctional voltages of either polarity, shown in the G_j–V_j relation in Fig. 1C. Although the voltage dependence of transfectants was steeper than had previously been appreciated (Spray et al., 1986, Riverdin and Weingart, 1988), we have now demonstrated in three preparations (transfectants, hepatocytes and bilayers) that V_0 is ≈25 mV, $g_{min}/g_{max} \approx 0.1$ and n (the number of the equivalent gating charges) ≈ 2. (Moreno et al., 1991a). It is generally believed that the Cx32 channel is formed by two hemichannels connected in series. These gates are activated contingently, so that the closure of one cell's gate is dependent upon the gating status of the apposed hemichannel (Harris et al., 1981; Moreno et al., 1991c).

Human Cx43 has also been characterized in the same parental cell line. LY was used to select for stable transfectants. Immunofluorescence studies (using rat antibodies, courtesy of Dr. E.L. Hertzberg) identified human Cx43 proteins in the appositional membranes of the transfected cells (Fishman et al., 1990).

Junctional conductance of human Cx43 transfectants averaged about 10 nS. Values of unitary conductance were surprising in that there were two discrete populations with γ_j of 60 and 90 pS (Fig. 1B). In Western blots of Cx43, two bands were identified, with M_r of about 43 kDa and 41 kDa. Several bands have been reported in other preparations where Cx43 is detected, including heart (Crow et al., 1990); astrocytes (which show mainly the 41-kDa band; Dermietzel et al., 1991); and cell lines such as SKHep1 transfected with human Cx43 (Fishman et al., 1990), WB (Spray et al., 1991) or NRK (Musil et al., 1990).

Transjunctional voltage also closes junctional channels formed of Cx43, but the sensitivity is much less than that of Cx32 channels, where $V_0 \approx 60$ mV (compare Fig. 1C).

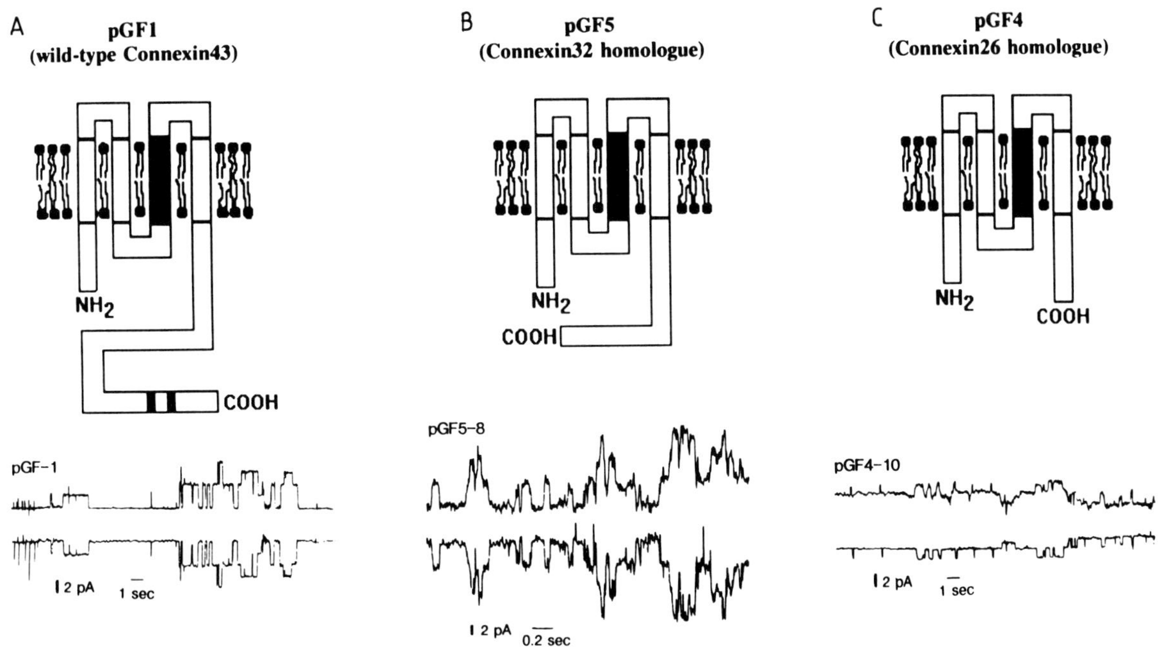

Figure 2. Unitary conductances recorded from wild type and mutant Cx43. Cell clones labeled pGF5 (B) and pGF4 (C) correspond to cells transfected with a mutated cDNA encoding truncated connexins homologous in length to wild type Cx32 and Cx26, respectively. Below each diagrammed molecule is a record of the current traces obtained from cell pairs with the corresponding truncation.

Similar connexins expressed in different tissues

A major concern in expressing any connexin in an exogenous environment was the possibility that the internal milieu might change the connexin's characteristics (e.g. unitary conductance or voltage dependence). We were amazed to discover that all of the freshly dissociated cells from different tissues (in astrocytes: Dermietzel et al., 1991; in leptomeningeal cells: Spray et al., 1991; in corpus cavernosum smooth muscle: Moreno et al., 1990), as well as the SKHep1 cell line transfected with rat Cx43 (Rook et al., 1992), show two conductance levels, one between 55 and 65 pS and the other between 85 and 105 pS. The relative dominance of one or the other conductance value varies among cell types. Voltage-dependent characteristics differed somewhat: V_0 may vary from 50 to 70 mV, g_{min}/g_{max} from 0.25 to 0.5 and, n, while not determined in all instances, is about 2. These results demonstrate that unitary conductance is an intrinsic property to the connexin type, independent of the tissue where expressed.

Cx43 mutants where unitary conductance changes

Because Cx32 and Cx43 are, to a large degree, homologous (differing primarily in length), we have begun to explore which protein domains are responsible for differences in unitary conductance and voltage dependence. First, we investigated the effect of the length of the carboxyl regions. We stably transfected SKHep1 cells with cDNAs encoding truncated proteins with lengths similar to either Cx32 or Cx26 by placing a stop codon in the appropriate position in the Cx43 sequence. Stably transfected clones encoding both truncation mutants allowed LY dye transfer. Their voltage dependence resembled that of wild type human Cx43 (Fishman et al., 1991). Unitary conductances, however, were quite different: γ_j of the Cx32 homologue was ≈160 pS (only slightly higher than that of wild type Cx32), while that of the Cx26 homologue was about 50 pS (slightly lower than that of Cx26 transfectants) (Fig. 2C).

Effects of phosphorylation on human Cx43 channels

In neonatal heart, single-channel recordings indicate γ_j values of mostly 60 pS (Spray and Burt, 1988; Rook et al., 1989); Western blot analyses reveal that most of the protein exists in the phosphorylated (M_r 43–45) forms. In other preparations, two conductive states (60 and 90 pS) are observed; the dephosphorylated isoform of Cx43 is more prominent, as shown by condensation of bands into

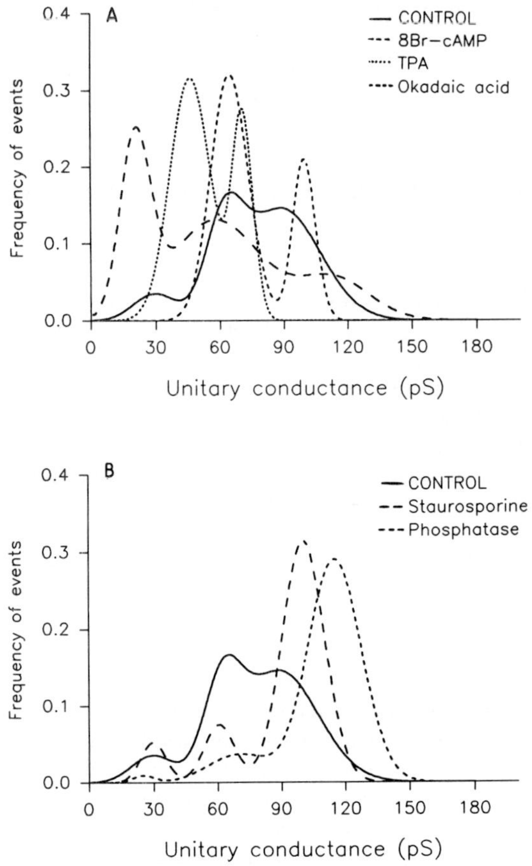

Figure 3. Shifts in frequency distributions of unitary conductances recorded in Cx43-transfected cells after phosphorylating (A) and dephosphorylating treatments (B). (A) Each discontinuous line represents the Gaussian fit to the average of 3 experiments where a phosphorylating treatment (Br-cAMP, TPA or okadaic Acid) was used. The continuous line corresponds to the fit of the average of 14 control experiments. Note that smaller events predominate after each treatment. (B) Discontinuous lines correspond to the Gaussian fit to the average of 3 experiments where dephosphorylating treatments were used. A shift toward larger unitary conductance is clear when compared with the control experiments (continuous curve).

the 41-kDa isoform after alkaline phosphatase treatment (Musil et al., 1990). We designed experiments to determine whether post-translational phosphorylation could account for the different conductance states of Cx43 channels. As shown in Fig. 3A, we found that treatments which activate protein kinases A and C shift the distribution of the γ_j values toward the 60 pS channels, even though no appreciable change is observed in macroscopic junctional conductance. Another agent, staurosporine, which decreases the phosphorylation kinetics of protein kinase C, induces a shift towards the 90 pS channel (Fig. 3B). These shifts were not all-or-none changes, and were reversible. For example, the effects of staurosporine could be reversed with TPA, and cAMP effects reverted towards control after rinsing for a few minutes, or could be shifted toward the higher γ_j level after exposure to staurosporine.

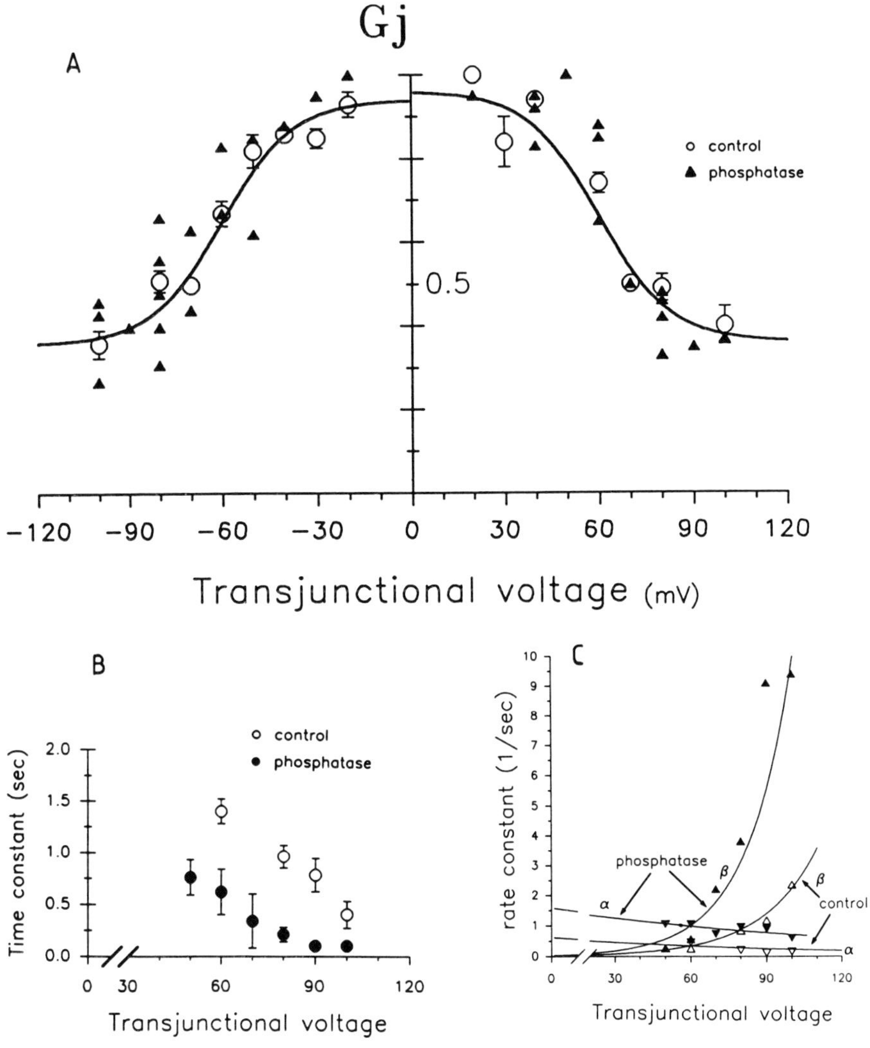

Figure 4. Alkaline-phosphatase treatment does not change the steady-state voltage dependence of Cx43 gap-junction channels to transjunctional voltage but alters the voltage-dependent time (B) and rate (C) constants. (A) Steady-state G_j–V_j relation with a continuous line showing the best fit of control data (open circles) to the Boltzman equation (see Fig. 1C). Superimposed are results from experiments where alkaline phosphatase was applied inside of one cell of a pair (solid triangles). The principal features of the relation are unchanged by phosphatase treatment. (B) This plot shows that the time constant of inactivation decreases for larger V_j pulses (open circles). After phosphatase was applied inside of one of the cells the time constants were substantially reduced at all voltages. (C) Comparison of the voltage-dependent rate constants α and β for transitions between open and closed states before (open symbols) and after phosphatase (filled symbols).

Finally, alkaline phosphatase was introduced directly into the cells through patch pipettes. A final internal concentration of 14 µg/ml of AP was found to 1) abolish g_j when added to both sides of the junction, and 2) to reduce g_j almost by half when applied in one cell. Either method of application led to two major changes in junctional properties. First, the unitary conductance shifted almost completely towards the higher γ_j values (Fig. 3B), and second, the kinetics of voltage dependence gating were increased 3–4 fold (Fig. 4A and B) with no change in the steady-state voltage dependence (Fig. 4A). Thus, the parameters (A_α and A_β) that determine the voltage sensitivity of the rate constants α and β, were increased proportionately so that the rate constants *per se* increased 3–4 fold without changing V_0.

Aims and possibilities

The study of current flow through gap-junction channels is slightly more complicated than through non-junctional channels. With a patch pipette, access to a channel in a non-junctional membrane is direct; a single channel in the patch can be isolated and the composition of the solutions on both sides of the pore can be controlled. Studies of the biophysical properties of gap-junction channels have been limited to whole-cell recording from cell pairs, where generally a large number of channels are expressed and cell shape may not be optimal for space clamp. Molecular biology techniques should soon permit reducing the expression of both endogenous and exogenous connexins, analysis of individual properties of myriad new connexins, as well as determination of the influences of connexin gating properties on physiological functions in particular tissues. Site-directed mutagenesis has made it possible to alter channel proteins. Simply by changing the length of the protein, unitary conductance and gating properties have been modified. The various gating domains that years ago were proposed to regulate junctional coupling between cells are now accessible to genetic dissection.

Conclusions

Cx32 and Cx43 expressed in SKHep1 cells display distinct and characteristic unitary conductances which remain stable, even after numerous passages. The size of the cytoplasmic carboxyl terminus of Cx43 determines the conductance of the channel; if the length of the Cx43 translation product corresponds to that of Cx32, the resulting channels are similar in γ_j to those obtained after transfection with Cx32. Phosphorylating treatments, including stimulation of protein kinases C and A, change the distribution of unitary conductances of human Cx43 toward the lower conductances (60 pS). Dephosphorylating treatments (staurosporine or unilateral application of alkaline phosphatase) shift the distribution to higher values (90 pS). Phosphatase treatment reduces the time constant of voltage-dependent channel closure without changing steady-state voltage dependence, presumably as a result of a reduction in the energy necessary for the channel to transit between the open and shut conformational states, giving rise to closing rates that could be physiologically relevant in excitable cells (Furshpan and Potter, 1957).

References

Bennett, M.V., Barrio, L.C., Bargiello, T.A., Spray, D.C., Hertzberg, E. and Sáez, J.C. (1991) Gap junctions: new tools, new answers, new questions. Neuron. 6, 305–320.

Burt, J.M. and Spray, D.C. (1988) Single-channel events and gating behavior of the cardiac gap-junction channel. Proc. Natl. Acad. Sci. USA 85, 3431–3434.

Burt, J.M. and Spray, D.C. (1989) Volatile anesthetics block intercellular communication between neonatal rat myocardial cells. Circ. Res. 65, 829–837.

Crow, D.S, Beyer E.C., Paul, D.L., Kobe S.S. and Lau, A.F. (1990) Phosphorylation of connexin43 gap-junction protein in uninfected and Rous sarcoma virus-transformed mammalian fibroblasts. Mol. Cell Biol. 10, 1754–1763.

Dermietzel, R., Hertzberg, E.L., Kessler, J.A. and Spray, D.C. (1991) Gap junctions between cultured astrocytes: immunocytochemical, molecular and electrophysiological analysis. J. Neurosci. 11, 1421–1432.

Eghbali, B., Kessler, J.A. and Spray, D.C. (1990) Expression of gap-junction channels in communication-incompetent cells after stable transfection with cDNA encoding connexin 32. Proc. Natl. Acad. Sci. USA 87, 1328–1331.

Fishman, G.I., Spray, D.C. and Leinwand, L.A. (1990) Molecular characterization and functional expression of the human cardiac gap-junction channel. J. Cell Biol. 111, 589–598.

Fishman, G.I., Moreno, A.P., Spray, D.C. and Leinwand, L.A. (1991) Functional analysis of human cardiac gap-junction channel mutants. Proc. Natl. Acad. Sci. USA 88, 3525–3529.

Graham F.L. and van der Eb, A.J. (1973) A new technique for the assay of infectivity of human adenovirus 5 DNA. Virology 52, 456.

Harris A.L., Spray, D.C. and Bennett M.V.L. (1981) Kinetic properties of a voltage-dependent junctional conductance. J. Gen. Physiol. 77, 95–117.

Horn, R. and Marty, A. (1988) Muscarinic activation of ionic currents measured by a new whole-cell recording method. J. Gen. Physiol. 92, 145–160.

Moreno, A.P., Campos de Carvalho, A.C., Christ, G.J., Melman, A., Hertzberg, E.L. and Spray, D.C. (1990) Gap junctions between human corpus cavernosum smooth muscle cells in primary culture: electrophysiological and biochemical characteristics. Int. J. Impotence Res. 2 (Suppl 2), 55–56.

Moreno, A.P., Campos de, Carvalho, A., Verselis, V., Eghbali, B. and Spray, D.C. (1991) Voltage-dependent gap-junction channels are formed by connexin32, the major gap-junction protein of rat liver. Biophys. J. 59, 920–925.

Moreno, A.P., Eghbali, B. and Spray, D.C. (1991) Connexin32 gap junctions in stably transfected cells: Unitary conductance. Biophys. J. 60, 1254–1266.

Moreno, A.P., Eghbali, B. and Spray, D.C. (1991) Connexin32 gap junctions in stably transfected cells: Equilibrium and kinetic properties. Biophys. J. 60, 1267–1277.

Moreno, A.P., Fishman, G.I. and Spray, D.C. (1992) Phosphorylation shifts unitary conductance and modifies voltage-dependent kinetics of human connexin43 gap-junction channels. Biophys. J. 62, 51–53.

Musil, L.S., Cunningham, B.A., Edelman, G.M. and Goodenough, D.A. (1990) Differential phosphorylation of the gap-junction protein connexin43 in junctional communication-competent and -deficient cell lines. J. Cell Biol. 111, 2077–2088.

Riverdin, E.C. and Weingart, R. (1988) Electrical properties of the gap-junction membrane studied in rat liver cell pairs. Am. J. Physiol. 254, C226–C234.

Rook, M.B., Jongsma, H.J. and Jonge, B. (1989) Single-channel currents of homo- and heterologous gap junctions between cardiac fibroblasts and myocytes. Pflügers Arch. 414, 95–98.

Rook, M.B., Moreno, A.P., Fishman, G.I. and Spray, D.C. (1992) Rat and human connexin43 (Cx43) gap-junction channels in stably transfected cells: comparison of unitary conductance and voltage sensitivity. Biophys. J 61, A505 (Abstr).

Spray, D.C., Ginzberg, R.D., Morales, E.A., Gatmaitan, Z. and Arias, I.M. (1986) Electrophysiological properties of gap junctions between dissociated pairs of rat hepatocytes. J. Cell Biol. 103, 135–144.

Spray, D.C., Chanson, M., Moreno, A.P., Dermietzel, R. and Meda, P. (1991) Distinctive gap-junction channel types connect WB cells, a clonal cell line derived from rat liver. Am. J. Physiol. 260, C513–C527.

Spray, D.C., Moreno, A.P., Kessler, J.A. and Dermietzel, R. (1991) Characterization of gap junctions between cultured leptomeningeal cells. Brain Res. 598, 1–14.

Part V. Role of Gap Junctions in Various Tissues and Organisms

Nervous system – Ocular lens and heart – Pancreas and salivary gland – Corpus cavernosum – Development – Insect

J.E. Hall, G.A. Zampighi and R.M. Davis (Eds.)
Progress in Cell Research, Vol. 3

CHAPTER 19

Plasticity of gap junctions at mixed synapses

D.S. FABER*, X.-D. YANG** and L.R. WOLSZON***

Neurobiology Laboratory, State University of New York, 313 Cary Hall, Buffalo, NY 14214, USA

Introduction

Although electrotonic synaptic transmission has been observed at a number of synapses in the nervous systems of vertebrates and invertebrates, relatively little is known about its modifiability. In fact, Bennett (1977) suggested that the most probable adaptive advantage held by chemical excitatory synapses over electrotonic ones is in plasticity, a term used for modifications in synaptic strength produced either by prior activity or by the action of hormones or specific modulators. Thus far, the best-known example of a modification of electrotonic coupling due to a direct effect on gap junctions is the dopamine-induced decoupling of retinal horizontal cells (Piccolino et al., 1982; Lasater and Dowling, 1985). Our focus has been on the properties of mixed synapses, that is, the contacts between afferent fibers and a postsynaptic neuron that have structural specializations consistent with both electrotonic and chemical synaptic transmission. In particular, we have been studying the mixed synapses between large myelinated club endings of eighth nerve afferents and the lateral dendrite of the goldfish Mauthner (M-) cell. This review is concerned with the properties of these connections and their activity-dependent plasticities. As described below, electrotonic synapses can be as modifiable as chemical ones, with similar functional consequences.

The large myelinated club endings are the dominant excitatory input to the M-cell from the eighth nerve. These afferents carry auditory information, with each nerve fiber sending only one axonal branch to the lateral dendrite, as shown with intracellular injections of horseradish peroxidase (Lin and Faber, 1988a). The structural organization of these contacts, first studied by Robertson (1963) and Robertson et al. (1963), indicates mixed synaptic transmission: these contacts contain multiple gap junctions, numerous synaptic vesicles, and at least ten active zones or release sites (Kohno and Noguchi, 1968; Nakajima, 1974; Nakajima and Kohno, 1978; Tuttle et al., 1986). Although electrotonic transmission at these synapses was first demonstrated by Furshpan in 1964, proof that single endings transmitted in both modes was only obtained recently, with simultaneous pre- and postsynaptic intracellular recordings (Lin and Faber, 1988a,b). Their studies suggested that these synapses, including the electrotonic component, were modifiable. First, the junctional conductance (64 nS), calculated from anti- and orthodromic coupling coefficients, corresponds to less than 1000 opened gap-junction channels, while the number of intramembranous particles (the presumed morphological correlates of the channels) is about two orders of magnitude greater (Kohno and Noguchi, 1986; Tuttle et al., 1986). Thus, only a small fraction of the channels are open at any time. Second, the chemically mediated excitatory postsynaptic potential (EPSP) is small, due in part to a somewhat low probability of release of transmitter. Third, the majority of the chemical connections are functionally silent, i.e., a presynaptic action potential does not produce a detectable EPSP (Lin and Faber, 1988b). All of these observations suggested that, minimally, there should be the possibility of increasing the gain of these synapses. We have adopted two convergent approaches for studying this problem: (i) analysis of the mechanisms regulating junctional conductance, and (ii) delineation of physiological conditions for inducing or triggering synaptic modifications.

cAMP regulation of junctional transmission

An advantage of the M-cell system is that it is possible to record intracellularly from an identified region of this neuron, in an intact *in vivo* preparation and, through the same electrode, to inject substances which act specifically on a given second messenger pathway. Thus, recordings are routinely obtained from the lateral dendrite at the site of the eighth nerve input (Fig. 1A). We initially asked whether elevated levels of cAMP in the lateral dendrite

---

* Present address: Department of Anatomy and Neurobiology, Medical College of Pennsylvania, 3300 henry Avenue, Philadelphia, PA 19129, USA.

** Present address: Department of Neuroscience, Roche Institute of Molecular Biology, Nutley, NJ 07110, USA.

*** Present address: Biology Department, Columbia University, 1003 Fairchild, New York, NY 10025, USA.

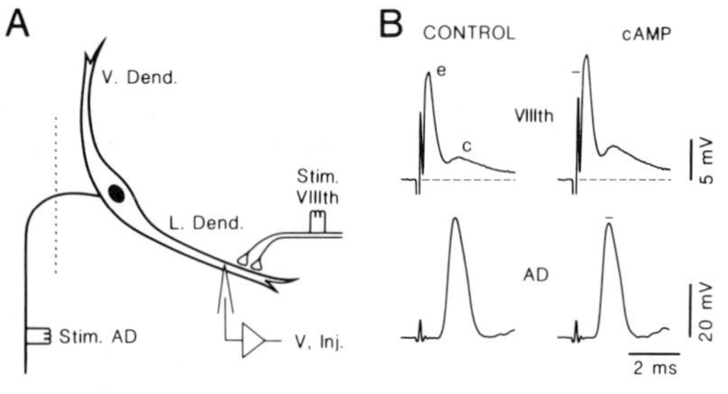

Figure 1. Cyclic AMP enhances electrotonic coupling at the Mauthner cell excitatory synapses. (A) Schematic drawing of the Mauthner (M-) cell, projected onto a horizontal plane. Club endings of ipsilateral eighth (VIIIth) nerve fibers terminate on the distal lateral dendrite (L. Dend). In the experiments described here the recording (V) was from the same site, and the microelectrode could also be used for intracellular injections (Inj.). The M-axon was stimulated in the spinal cord (Stim. AD), to produce a second test response, the AD spike, and to trigger a feedback inhibitory circuit (not shown). V. Dend., ventral dendrite. (B) Upper traces are intradendritic recordings of VIIIth nerve EPSPs before (control) and after (cAMP) injections of cAMP plus aminophylline. Both the electrical (e) and chemical (c) components of the EPSP were enhanced. Horizontal bar represents control amplitude of the electrotonic PSP. Lower traces: In contrast, the antidromically evoked spike was slightly reduced after the injection, suggesting a reduction of the input resistance. In this and subsequent figures, all traces are averages of 10 or more responses.

altered the electrotonic or chemical EPSPs, since cAMP increases the conductance of gap junctions and the number of electrotonic junctions formed between cells in various non-neuronal systems (see Spray and Bennett, 1985). For this purpose, cAMP, the phosphodiesterase inhibitor, aminophylline, or the two together were injected intradendritically while monitoring the amplitudes of the two excitatory synaptic responses (Wolszon and Faber, 1989). As shown in Fig. 1B (upper) both the electrotonic (e) and chemical (c) components were enhanced by the injections. These increases, which were at times as large as 200 to 500% relative to pre-injection values, were typically in the range of 15 to 25% (medians = 16% for electrotonic coupling and 23% for chemical transmission). The latter values probably underestimate the potential for synaptic enhancement, given the large effects seen in some instances and the technical limitations of such experiments.

Increases in both components could be due to a non-specific change in M-cell input resistance. However, two lines of evidence ruled out this possibility. First, the amplitude of the antidromic spike, which propagates passively from the axon hillock through the soma and dendrites and is an indirect measure of M-cell input conductance (see Faber and Korn, 1978), was not altered by cAMP (Fig. lB, lower panel). Second, in 4 experiments the M-cell was penetrated with two microelectrodes, allowing direct measurements of the input, or transfer, resistance; this parameter actually decreased slightly while the synaptic responses were enhanced. Alternatively, there might be only one site of cAMP action, that is, cAMP would diffuse across the gap junctions and increase presynaptic spike amplitude and width, thereby secondarily increasing transmitter release (Lin and Faber, 1988b). However, this is also unlikely, since presynaptic spike width, as inferred from the duration of the electrical EPSP, was not affected by cAMP (Wolszon and Faber, 1989). Also, postsynaptic injections of the catalytic subunit of the cAMP-dependent protein kinase (PKA) produce effects comparable to those of cAMP, and PKA is too large a molecule to diffuse across gap junctions (Pereda et al., 1990).

We have recently begun to search for an appropriate trigger for the cAMP-dependent cascade, since the evidence that aminophylline produces the same effects as cAMP indicates the presence of an endogenous regulatory system. Interestingly, the neurotransmitter, dopamine, which, as mentioned, de-couples retinal horizontal cells, increases electrotonic coupling (and chemical transmission) at the M-cell synapses, apparently by a cAMP-dependent phosphorylation mechanism, and there is a dopaminergic innervation of the lateral dendritic region (Pereda et al., 1991).

Activity-dependent changes in synaptic strength

It is generally assumed that long-term potentiation (LTP) of chemical synapses, produced by repetitive presynaptic activation (Bliss and Lomo, 1973), is a good experimental paradigm for learning and memory (Brown et al., 1988; Gufstafsson and Wigstrom, 1988). Studies of LTP in the vertebrate central nervous system, and of its opposite, long-term depression (LTD), have focused on excitatory connections in hippocampus and neocortex, structures important for global processing and cognitive functions. We have now shown that both phenomena occur at the mixed synapses onto the M-cell (Yang et al., 1990, Yang and Faber, 1991), that not only chemical transmission but also gap junctions are modified, and that, in the case of LTP, the mechanisms underlying its induction are the same as those in higher structures.

The experimental paradigms we have employed are quite straightforward. Extracellular stimulation of the eighth nerve is used to evoke test responses and for the tetanizations, with the two stimulus strengths being independently controlled. The tetani typically consist of bursts of stimuli at 500 Hz applied every 2 s for 1 to 5 min. As shown in Fig. 2A, LTP is produced when a strong tetanus is used, one which produces an orthodromic action potential at the beginning of each burst. Then, both the electrotonic coupling potential and the later monosynaptic chemical EPSP are enhanced. These potentiations, which averaged 25% and 75% for the electrotonic and chemical components, respectively, were apparent immediately af-

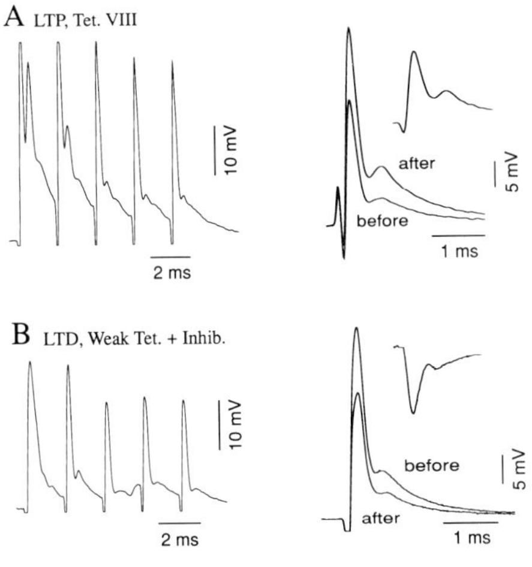

Figure 2. Activity-dependent changes in junctional conductance and chemical transmission at mixed synapses. (A) Long-term potentiation (LTP) of both EPSP components produced by a strong presynaptic tetanus. Left: Intradendritic recordings during a tetanus of 5 stimuli at 500 Hz. The first VIIIth nerve stimulus in the train evoked an orthodromic action potential. The first two coupling potentials are off-scale and truncated. Right: Superimposed records of VIIIth nerve evoked responses 2 min before and 8 min after conditioning with 20 such tetani separated by 2 s intervals. Inset is the computer-calculated difference between the two, showing that both electrotonic and chemical components were potentiated. (B) Pairing a weak tetanus with antidromically evoked feedback inhibition produces long-term depression (LTD). Left panel: example of the intradendritic responses evoked by the pairing. The antidromic spike is superimposed on the first synaptic response, and inhibition is signalled by the shunt of the last three PSPs. Right panel: Averaged VIIIth nerve responses obtained 4 min before and 6 min after the pairing, repeated at 2 s intervals for 4 min. Note that now both components are depressed, as shown by the difference potential in the inset. In both A and B, there were no changes in antidromic spike height or resting membrane potential.

ter terminating the conditioning regime, and they always persisted throughout the recording session (up to 90 min after tetanization). As with LTP of excitatory synapses in the hippocampus, the modifications were specific to the tetanized afferents and did not involve changes in the presynaptic volley. Since, in addition, the M-cell input resistance and the presynaptic spike width were unchanged, and the two synaptic responses did not always change in parallel, we concluded that potentiation of the electrotonic coupling was due to a specific alteration in junctional conductance (Yang et al., 1990).

Figure 2B presents the experimental conditions most effective for producing a persistent LTD of the coupling potential and of the chemically mediated EPSP (Yang and Faber, 1991). LTD is best induced by pairing a weak presynaptic tetanus, one that on its own would generally produce little or no potentiation, with postsynaptic inhibition. In the M-cell system, the latter was evoked by antidromic stimulation of the M-axon in the spinal cord, which triggers a feedback glycinergic inhibition (Korn and Faber, 1990). The presence of this inhibition, which lasts only 10 to 20 ms, is signalled in the left panel of Fig. 2B by the reduced amplitudes (shunts) of the third through fifth composite PSPs in the tetanus, and the pronounced depression is shown to the right. In some cases, the depression was quite large, such that the synaptic potentials were reduced to less than 20% of their control amplitude. As with LTP, this activity-dependent depression lasted for the duration of the recording session and was specific to the tetanized pathway, and the M-cell input resistance and resting membrane potential remained constant. Finally, the postulated role of inhibition was confirmed by blocking inhibition pharmacologically, a procedure which also blocked the induction of LTD (Yang and Faber, 1991). Thus, it appears that at this primary synaptic relay between sensory afferents and an identified reticulospinal neuron, junctional conductance can be persistently enhanced or reduced by activity paradigms typically considered from the perspective of high cortical functions. It will certainly be interesting to learn if similar modifications of gap junctions occur in other structures, including hippocampus and neocortex (Dudek et al., 1983). Also, our results implicating inhibition in the induction of LTD may help to establish conditions for reliably generating depression, a prerequisite for studying its underlying mechanisms.

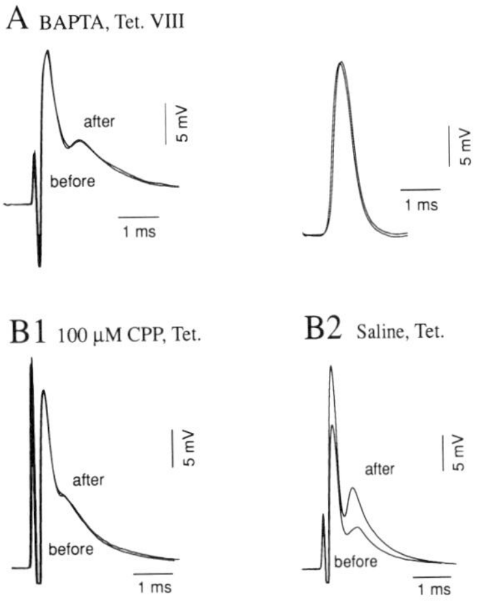

Figure 3. Evidence that LTP of gap junctions requires an increased intracellular Ca^{2+} concentration, via NMDA receptor activation. (A) Superimposed average responses to VIIIth nerve (left) and antidromic (right) stimulation recorded 3 min before and 6 min after tetanization in an experiment where BAPTA (10 mM in 2.5 M KCl) was continuously injected iontophoretically (−5 nA) into the M-cell. Under these conditions, LTP does not occur and M-cell membrane properties remain stable. B1, B2: results from an experiment in which tetanization was first applied while superfusing the brain with 100 μM CPP, an NMDA antagonist (B1, 6 min before and 6 min after tetanization), and then repeated after washing for 40 min in saline (B2, 2 min before and 20 min after tetanization). LTP did not occur in the presence of CPP, but after the wash period the same conditioning protocol produced immediate increases in both the coupling potential and the chemical EPSP.

A great deal is known about the initial steps in the induction of LTP in the hippocampus; in most instances it requires a postsynaptic increase in Ca^{2+} at the level of the tetanized synapses and is mediated via synaptic activation of the NMDA type of glutamate receptor (Collingridge and Bliss, 1987). Since the chemically modulated component of the eighth-nerve-evoked EPSP in the lateral dendrite is most likely mediated by glutamate, with involvement of an NMDA receptor (Wolszon and Faber, 1988), we asked if the same mechanism pertained for LTP. Our results demonstrate that the same scheme applies not only for the chemical EPSPs but also for the potentiation of the junctional conductance. Specifically, two pharmacological experiments were used to block LTP: (i) postsynaptic intracellular injections of the calcium chelator, BAPTA (Fig. 3A), and (ii) superfusion with 100 μM CPP, a specific NMDA receptor antagonist (Fig. 3B1). In both conditions, a strong presynaptic tetanus was then ineffective, while, in the case of the CPP experiments, the same tetanus subsequently produced a typical LTP when applied after washing out the drug (Fig. 3B2). Therefore, although an increased intracellular concentration of calcium generally decouples cells (Loewenstein, 1981), it may also trigger a long-lasting enhancement of gap-junctional conductance, presumably by serving as an intermediate in a metabolic cascade. In view of our results with cAMP and PKA, it is tempting to speculate that LTP expression involves such a phosphorylation mechanism.

Discussion

Activity-dependent modifications of electrotonic coupling at this first-order sensory relay may have significant functional consequences. The M-cell action potential mediates a vital escape reflex (Zottoli, 1977; Eaton et al., 1981, 1991; Faber et al., 1989, 1991), one that can be triggered by sound stimuli. Furthermore, because the preferred auditory frequency of the eighth-nerve afferents that terminate on the lateral dendrite as large myelinated club endings is about 500 Hz (Fay and Olsho, 1979), we chose that as the tetanizing frequency. Consequently, auditory signals, such as those produced by diving birds or other predators, can be expected to have long-term effects on the threshold of this startle response or, since the afferents project to the ipsilateral M-cell but not to its contralateral homologue, on its directionality. Eighth-nerve tetanization also produces an LTP of a disynaptic inhibition of the opposite M-cell (Korn et al., 1991), an effect which would magnify the postulated increase in directionality. Finally, the modifications of electrotonic coupling in this system are functionally significant because the coupling potentials alone may trigger orthodromic spike initiation in the M-cell alone or in combination with the slightly delayed chemical EPSP.

The M-cell system has the advantages that the experiments described here and similar ones can be conducted *in vivo*, and that this neuron is large and identifiable. Thus, it should be possible to identify the connexins at these junctions (Yox et al., 1990) and to isolate single M-cells under various conditions, in order to directly study, at a molecular level, the mechanisms postulated to account for activity-dependent modifications, such as connexin phosphorylation.

Acknowledgments

Supported in part by NIH grants NS 15335 and 21848.

References

Bennett, M.V.L. (1977) Electrical transmission: A functional analysis and comparison to chemical transmission. In: J.M. Brookhart and V.B. Mountcastle (Eds.), Handbook of Physiology, Section 1: The Nervous System. 1, Williams and Wilkins, Baltimore, MD, pp. 357–416.

Bliss, T.V.P. and Lomo, T. (1973) Long-lasting potentiation of synaptic transmission in the dentate area of the anaesthetized rabbit following stimulation of perforant path. J. Physiol. (London) 232, 331–356.

Brown, T.H., Chapman, P.F., Kairis, E.W. and Keenan, C.L. (1988) Long-term synaptic potentiation. Science 242, 724–728.

Collingridge, G.L. and Bliss, T.V.P. (1987) NMDA receptors – their role in long-term potentiation. Trends Neurosci. 10, 288–293.

Dudek, F.E., Andrew, R.D., MacVicar, B.A., Snow, R.W. and Taylor, C.P. (1983) Recent evidence for and possible significance of gap junctions and electrotonic synapses in the mammalian brain. In: H.H. Jasper and N.M. van Gelder (Eds.), Basic Mechanisms of Neuronal Hyperexcitability, Alan R. Liss, New York, pp. 31–73.

Eaton, R.C., DiDomenico, R. and Nissanov, J. (1991) The role of the Mauthner cell in sensorimotor integration by the brain stem escape network. Brain Behav. Evol. 37, 272–285.

Eaton, R.C., Labender, W.A. and Wieland, C.M. (1981) Identification of Mauthner initiated response patterns in goldfish: evidence from simultaneous cinematography and electrophysiology. J. Comp. Physiol. [A] 144, 521–531.

Faber, D.S., Fetcho, J.R. and Korn, H. (1989) Neuronal networks underlying the escape response in goldfish: general implications for motor control. Ann. N.Y. Acad. Sci. 563, 11–33.

Faber, D.S. and Korn, H. (1978) Electrophysiology of the Mauthner cell: Basic properties, synaptic mechanisms, and associated network. In: D.S. Faber and H. Korn (Eds.), Neurobiology of the Mauthner Cell, Raven Press, New York, pp. 47–132.

Faber, D.S., Korn, H. and Lin, J.-W. (1991) Role of medullary networks and postsynaptic membrane properties in regulating Mauthner cell responsiveness to sensory excitation. Brain Behav. Evol. 37, 286–297.

Fay, R.R. and Olsho, L.W. (1979) Discharge patterns of lagenar and saccular neurons of goldfish eighth nerve: Displacement sensitivity and directional characteristics. Comp. Biochem. Physiol. 62A, 377–386.

Furshpan, E.J. (1964) Electrical transmission at an excitatory synapse in a vertebrate brain. Science 144, 878–880.

Gustafsson, B. and Wigstrom, H. (1988) Physiological mechanisms underlying long-term potentiation. Trends. Neurosci. 11, 156–162.

Kohno, K. and Noguchi, N. (1986) Large myelinated club endings on the Mauthner cell in the goldfish. Anat. Embryol. 173, 361–370.

Korn, H. and Faber, D.S. (1990) Quantitative electrophysiological studies and modeling of central glycinergic synapses. In: O.P. Ottersen and J. Storm-Mathisen (Eds.), Glycine Neurotransmission, John Wiley and Sons, New York, pp. 139–170.

Korn, H., Oda, Y. and Faber, D.S. (1992) Long-term potentiation of inhibitory circuits and synapses in the central nervous system. Proc. Natl. Acad. Sci. USA 89, 440–443.

Lasater, E.M. and Dowling, J.E. (1985) Dopamine decreases conductance of the electrical junctions between cultured retinal horizontal cells. Proc. Natl. Acad. Sci. USA 82, 3025–3029.

Lin, J.-W. and Faber, D.S. (1988a) Synaptic transmission mediated by single club endings on the goldfish Mauthner cell. I. Characteristics of electrotonic and chemical postsynaptic potentials. J. Neurosci. 8, 1302–1312.

Lin, J.-W. and Faber, D.S. (1988b) Synaptic transmission mediated by single club endings on the goldfish Mauthner cell, II. Plasticity of excitatory postsynaptic potentials. J. Neurosci., 8, 1313–1325.

Lowenstein, W.R. (198l) Junctional intercellular communication: The cell-to-cell membrane channel. Physiol. Rev. 61, 829–913.

Nakajima, Y. (1974) Fine structure of the synaptic endings on the Mauthner cell of the goldfish. J. Comp. Neurol. 156, 375–402.

Nakajima, Y. and Kohno, K. (1978) Fine structure of the Mauthner cell, synaptic topography and comparative study. In: D.S. Faber and H. Korn (Eds.), Neurobiology of the Mauthner Cell, Raven Press, New York, pp. 133–166.

Pereda, A., Nairn, A., Greengard, P. and Faber, D.S. (1990) cAMP-dependent protein kinase (PKA) injection enhances both electrotonic and chemical excitatory responses at mixed synapses on the Mauthner cell. Soc. Neurosci. Abstr. 16, 674.

Pereda, A., Triller, A., Korn, H. and Faber, D.S. (1991) Dopamine enhances both electrotonic coupling and chemical EPSP at mixed synapses on the Mauthner (M) cell. Soc. Neurosci. Abstr. 17, 583.

Piccolino, M., Neyton, J. and Gerschenfeld, H.M. (1984) Decrease of gap-junction permeability induced by dopamine and cyclic adenosine 3':5'- monophosphate in horizontal cells of turtle retina. J. Neurosci. 4, 247– 248.

Robertson, J.D. (1963) The occurrence of subunit patterns in the unit membranes of club endings in Mauthner cell synapses in goldfish brains. J. Cell Biol. 19, 201–221.

Robertson, J.D., Bodenheimer, T.S. and Stage, D.E. (1963) The ultrastructure of Mauthner cell synapses and nodes in goldfish brains. J. Cell Biol. 19, 159–199.

Spray, D.C. and Bennett, M.V.L. (1985) Physiology and pharmacology of gap junctions. Annu. Rev. Physiol. 47, 281–303.

Tuttle, R., Masuko, S. and Nakajima, Y. (1986) Freeze-fracture study of the large myelinated club ending synapse on the goldfish Mauthner cell: Special reference to the quantitative analysis of gap junctions. J. Comp. Neurol. 246, 202–211.

Wolszon, L.R. and Faber, D.S. (1988) Fast EPSPs evoked in the goldfish Mauthner cell by sensory afferents are due to NMDA receptor activation. Soc. Neurosci. Abstr. 14, 939.

Wolszon, L.R. and Faber, D.S. (1989) The effects of postsynaptic levels of cAMP on excitatory and inhibitory responses of an identified central neuron. J. Neurosci. 9, 784–797.

Yang, X-D., Korn, H. and Faber, D.S. (1990) Long-term potentiation of electrotonic coupling at mixed synapses. Nature 348, 542–545.

Yang, X-D. and Faber, D.S. (1991) Initial synaptic efficacy influences induction and expression of long-term changes in transmission. Proc. Natl. Acad. Sci. USA 88, 4299–4303.

Yox, D.P., Faber, D.S. and Nicholson, B.J. (1990) Gap-junction antibody attenuates electrical coupling between the goldfish Mauthner (M–) cell and its 8th nerve afferents. Soc. Neurosci. Abstr. 16, 185.

Zottoli, S.J. (1977) Correlation of the startle reflex and Mauthner cell auditory responses in unrestrained goldfish. J. Exp. Biol. 66, 243–254.

J.E. Hall, G.A. Zampighi and R.M. Davis (Eds.)
Progress in Cell Research, Vol. 3

CHAPTER 20

Regulation of connexin32 in motor networks of mammalian neurons

ROBIN S. FISHER[a] and PAUL E MICEVYCH[b]

[a]Departments of Psychiatry and Biobehavioral Sciences, Anatomy and Cell Biology, Mental Retardation Research Center and Brain Research Institute, School of Medicine, University of California at Los Angeles, Los Angeles, CA 90024, USA and [b]Department of Anatomy and Cell Biology, Brain Research Institute and Laboratory of Neuroendocrinology, School of Medicine, University of California at Los Angeles, Los Angeles, CA 90024, USA

Introduction

The distributions of gap junctions are well characterized in sensory and integrative networks of the central nervous system of mammals (Bennett et al., 1991). These gap junctions are believed to mediate neuronal dye coupling and electrotonic transmission between cells (Bennett, 1977). They are usually recognized as plaques of aggregated channels ("connexons") between adjacent neurons or glia (Peters et al., 1991). The connexons are formed from two hexameric hemijunctions embedded in external limiting membranes of adjoining cells. The central pores elaborated within the dodecamers provide cytoplasmic continuity, diffusion of small molecules and low-resistance electrical paths between communicating cells (Goodenough and Gilula, 1974). The connexon subunits are a diverse family of homologous proteins (connexins) expressed by different tissue types. The best-known channel protein is the 32-kDa connexin (Cx32) of rat hepatocytes (Paul et al., 1987). The Cx32 protein and mRNA have comparable and distinctive tissue distributions in the lung, stomach, pancreas and brain of mammals (Dermeitzel et al., 1989a,b; Bennett et al., 1991; Micevych and Abelson, 1991).

Several interesting questions have emerged from recent discoveries of the distribution of connexin mRNAs and proteins in the central nervous system (CNS) of mammals. These issues are based, in part, on the complex structure of neuronal tissue, the technical difficulties of channel detection and the uncertain identities of the gap junction-coupled cells. For example, the full extent of neuronal gap junctions is likely to be uncharted because of variations in plaque assembly, size and dispersion. Moreover, hemijunctions and isolated connexins are likely to escape recognition by conventional ultrastructural methods because these forms of the proteins are not part of the crystalline gap-junction plaques. Finally, the mechanisms regulating junctional permeability, connexon assembly and connexin expression are largely unexplored in the mammalian CNS.

New molecular probes and cell culture techniques should contribute to the resolution of these problems. However, the intact CNS has important structural and developmental constraints on cell commitment and differentiation that may not be replicated in dispersed cell models. These tissue characteristics may influence the expression and membrane insertion of connexins as well as the aggregation of separate hemichannels into distinct gap-junctional plaques. Therefore, we investigated the distribution and regulation of Cx32 in the intact CNS. We focused particularly on the localization of Cx32 in mammalian motor systems because very little is known about the organization of gap junctions within these circuits, unlike comparable invertebrate and lower vertebrate motor systems (e.g., Mauthner cells). Our studies on the regulation of Cx32 focus on androgens and catecholamines. Previously, we described gap-junction plaques between motor neurons of the spinal nucleus of the bulbocavernosus (SNB) and dorsolateral nucleus of the rat (Matsumoto et al., 1988; 1989). We demonstrated that mammalian spinal motor neurons, like certain invertebrate motor systems, are coupled by gap junctions. The number and size of SNB gap-junction plaques were reduced between androgen receptor-containing and steroid-sensitive motor neurons following reduction of circulating androgen levels. Moreover, we examined the incidence of dye coupling in the neostriatum and observed alterations of the numbers of coupled cells in response to manipulations of the catecholamine input (Cepeda et al., 1989; Walsh et al., 1989). In order to identify the connexins responsible and to define the various intercellular signals regulating the expression of connexins in the mammalian CNS, we performed a series of new morphological experiments in rats based on *in situ* hybridization of Cx32 mRNA in conjunction with immunohistochemical localizations of the Cx32 protein (Cx32i).

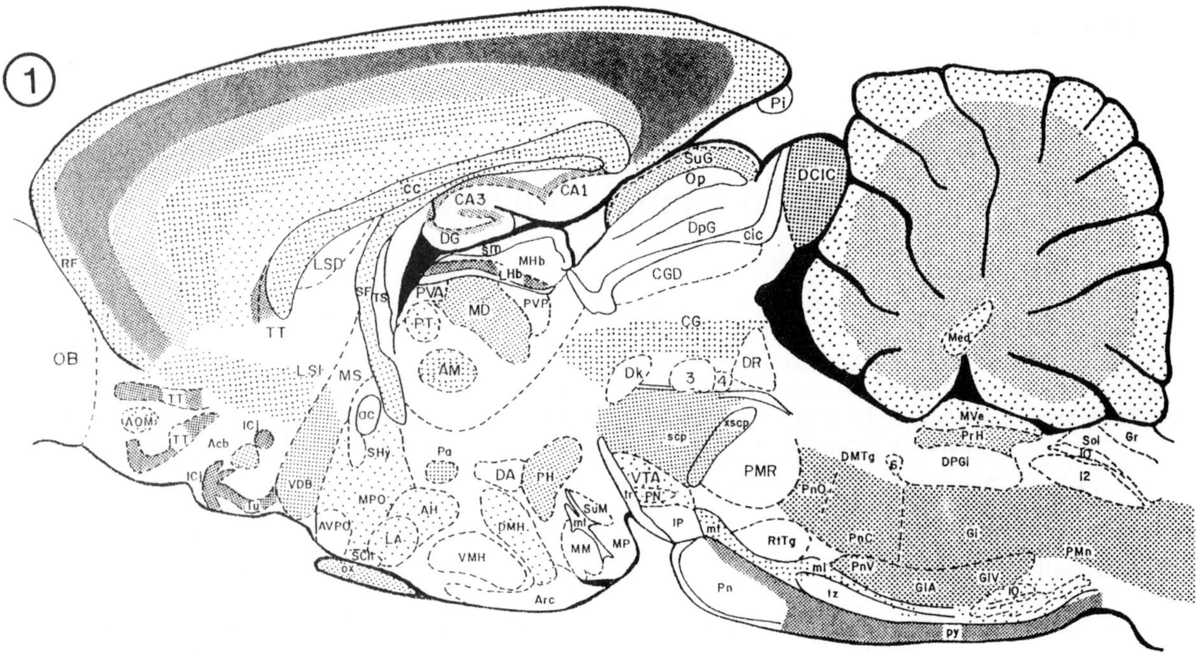

Figure 1. Schematic representation of the distribution of Cx32 mRNA in cytoarchitectonically defined cell groups in a parasagittal section of the rat brain. Shading indicates the relative density of cells containing Cx32 mRNA determined by in situ hybridization as described in Micevych and Abelson (1991). The shading represents a scale with five levels: no labeled cells (no shading), scattered cells, moderate number of labeled cells, high number of labeled cells and very high number of labeled cells (darkest shading). Drawing and abbreviations adapted from Paxinos and Watson (1986).

Our results support three significant and novel conclusions: (1) Gap junctions have a much wider and more dispersed distribution in the mammalian CNS than expected from previous reports. (2) Androgenic steroids direct the modeling of gender phenotypes to control synaptic and electrotonic communications between neurons. (3) The neuropathology of catecholamine depletion in Parkinson's disease is likely to involve defects of synaptic and electrotonic communications between neurons.

The localization of Cx32 in motor networks

Our first aim was to determine the localization and arrangement of Cx32 in the motor networks of adult rats. The experiments were designed to correlate the distributions of Cx32 mRNA and Cx32i in the CNS motor systems. Brains were fixed by aldehyde perfusion, sectioned in coronal or sagittal planes and processed for Cx32 mRNA detection (n = 6) or Cx32i detection (n = 12). Specimens for immunoelectron microscopy and Golgi/gold toning were also selected (Fisher et al., 1987; 1988). These studies used a 1.5 kbp probe coding for Cx32, obtained from Dr. D. Paul (Harvard University) in a pGEM-3Z vector. We transcribed it in the presence of ^{35}S-UTP and used an autoradiographic detection method (Micevych and Abelson, 1991). Control experiments for *in situ* hybridization included pretreatment of tissue sections with RNase prior to hybridization, hybridization with the Cx32 sense sequence and hybridization with a nonsense sequence. A summary of the distribution of Cx32 mRNA is presented in Fig. 1 (derived from Micevych and Abelson, 1991).

The Cx32 protein was detected with a monoclonal antibody, R5.21C, obtained from Dr. D. Goodenough (Harvard Univ.). Antibody-antigen complexes were detected using the avidin-biotin complex (ABC Kit, Vector Laboratories) with hydrogen peroxide as the substrate and diaminobenzidine (DAB) as the chromogen. Control experiments for the immunohistochemistry included substituting the primary antiserum with a non-immune serum and pre-absorbing the R5.21C antibody with acetone-extracted rat liver, heart or striated muscle powders. No staining was detected with the nonimmune serum. The only preabsorption that attenuated specific staining was the liver powder. Long-term detergent treatment (0.1–1.0% Triton-X100) reduced cytoplasmic but not punctate Cx32i labeling patterns associated with cell membranes.

In situ hybridization revealed Cx32 in oligodendroglia and neurons. By contrast, Cx43 mRNA was expressed in astrocytes and ependymal cells. Cx32 had a less uniform anatomical distribution than Cx43 (Micevych and Abelson, 1991). In motor sites (cerebral cortex, neostriatum, cerebellar cortex, brainstem/spinal motor neurons), we observed the specific hybridization of Cx32 mRNA (Figs. 1, 2e, and 5).

The distribution of specific Cx32i matched exactly the regional and cellular patterns of hybridization for Cx32 mRNA in the cerebral cortex (area 4), neostriatum, cerebellar cortex and cranial/spinal motor neuronal pools (Fig. 1). In gray matter, Cx32 mRNA was localized over larger cell profiles which, based on Nissl counterstaining, are likely to be neurons (nuclear diameters > 5 μm, somatic diameters > 15 μm) (Fig. 2e). Cx32i labeling was charac-

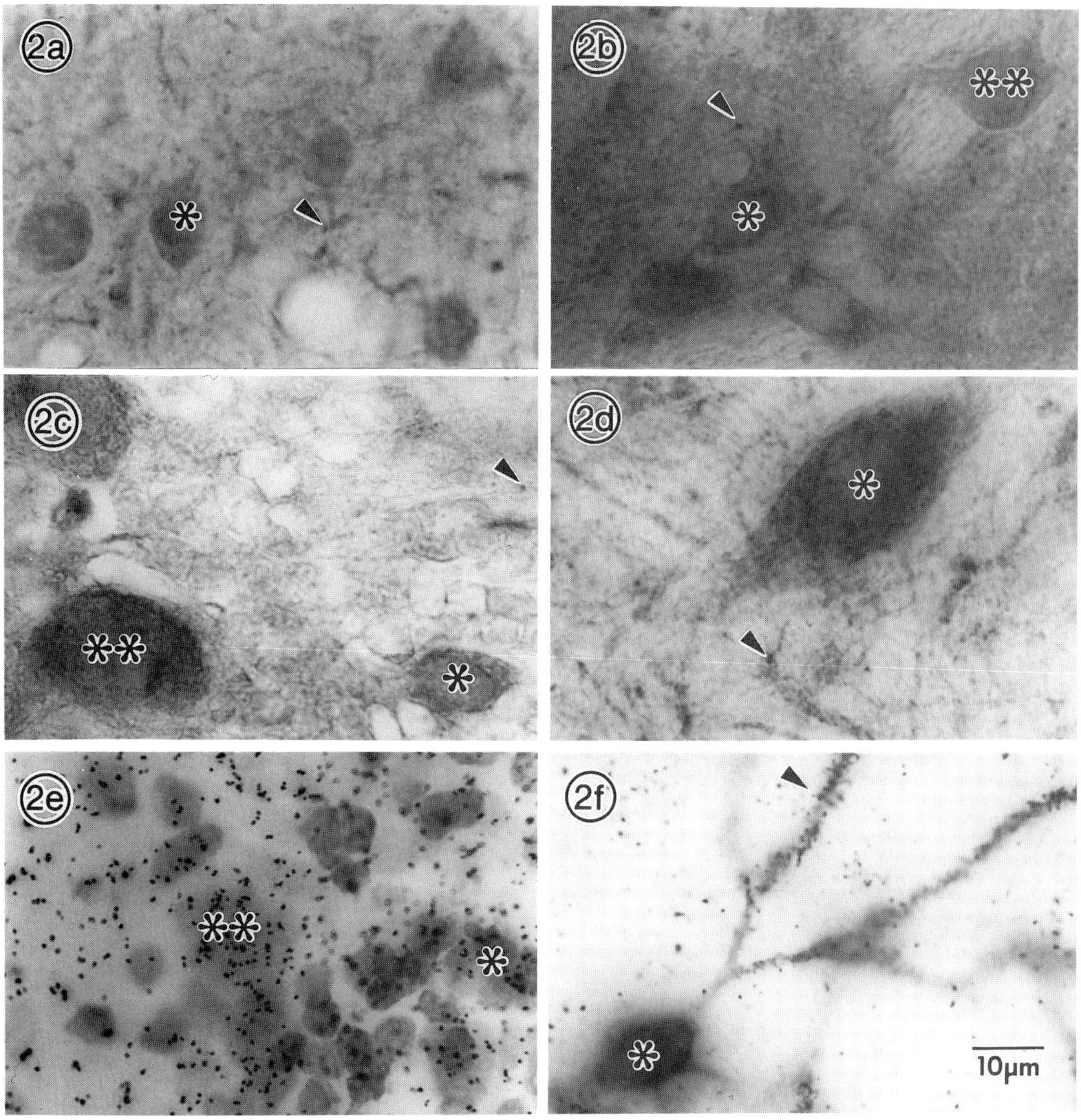

Figure 2. Photomicrographs of Cx32-expressing neurons in central motor circuits of adult rats. Panel 2a: Cx32i neurons (asterisk) and punctae (arrow) in layer 3 of motor neocortex (area 4). Panel 2b: Medium (single asterisk) and large (double asterisks) Cx32i neurons and punctae (arrow) in neostriatum. Panel 2c: Medium (single asterisk) and large (double asterisks) Cx32i neurons and punctae (arrow) in cerebellar cortex. The molecular layer was to the left in the panel, the large immunoreactive (Purkinje) neurons were in the Purkinje cell layer and the medium immunoreactive (Golgi II) neuron was in the granule cell layer. Panel 2d: Cx32i motor neuron (asterisk) and punctae (arrow) in the ventral horn of level C7 of spinal cord. Panel 2e: Bright-field photomicrograph demonstrating the autoradiographic labeling of Cx32 mRNA in cerebellar cortex (Nissl counterstain). The molecular layer was to the left and the granule cell layer was to the right. Grains were accumulated over a larger Purkinje cell (double asterisks) and smaller neurons (single asterisk) in the granule cell layer. Panel 2f: Cx32i medium spiny neuron (asterisk) in the neostriatum counterstained by Golgi impregnation and gold toning. Immunoreactivity was observed in the perikaryal cytoplasm while metallic staining revealed the dendritic arbor and dendritic spines (arrow).

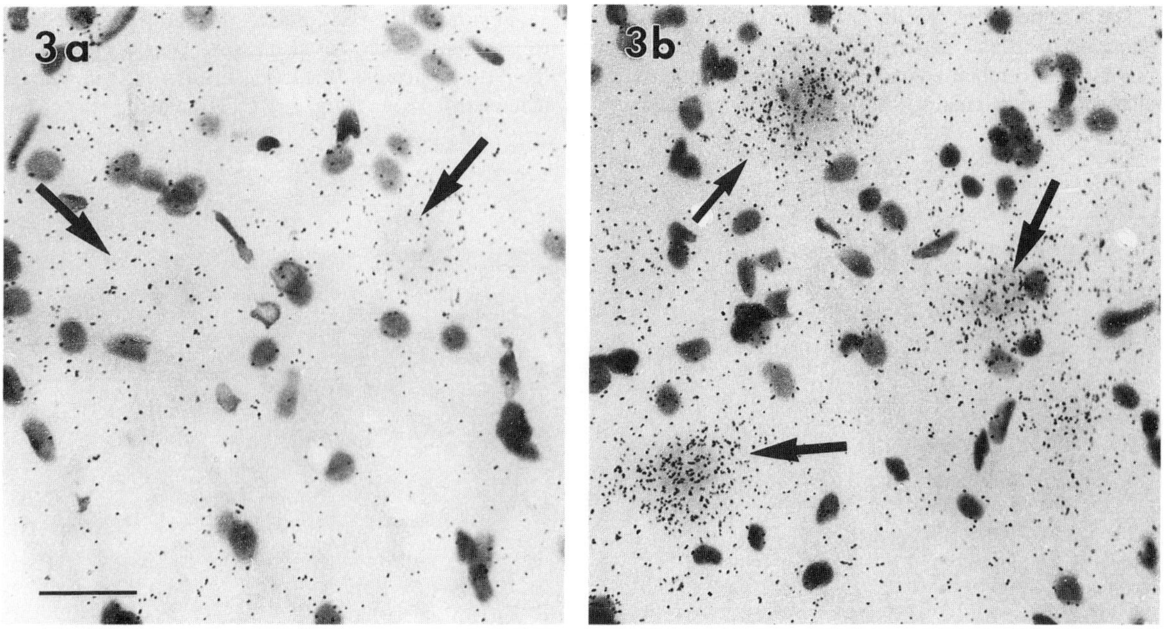

Figure 3. Autoradiograms of sections through the lumbar spinal cord retrodorsolateral nucleus illustrating patterns of hybridization with the Cx32 riboprobe. Panel 3a is from tissues that had the post-hybridization washes terminated at 0.1× SSC. Arrows indicate unlabelled motor neurons. Panel 3b: Cx32 riboprobe hybridization is localized to motor neurons (arrows) in an adjacent section to the tissue in 3a. In panel 3b, the post-hybridization washes were terminated after 2× SSC, a lower stringency condition. Calibration bar = 100 μm. (From Micevych and Abelson (1991) with permission.)

terized by two patterns: (1) diffuse staining in the perikaryal cytoplasm of cells, and (2) punctate staining alongside the cell profiles or scattered through the neurophil (Figs. 2a–d).

The Cx32i cells had the dendritic arbors and dendritic specializations of the diverse morphological phenotypes of neurons characteristic for each of the examined regions. In the cerebral cortex, Cx32i neurons were seen frequently in most layers, but were sparse in layer 1 (Fig. 2a). Double labeling with immunohistochemistry and Golgi/gold toning showed that Cx32i neurons consist of pyramidal and nonpyramidal neurons. In the neostriatum, numerous Cx32i neurons were seen throughout the nucleus (Figs. 2b and 5); they were identified as spiny and aspiny neurons (Fig. 2f). In the cerebellar cortex, Cx32i neurons (in this instance Purkinje, Golgi II, granule and stellate neurons) were most apparent in the Purkinje cell layer but were also present in the granular and molecular layers (Fig. 2c, e).

Preliminary immunoelectron microscopy studies indicate that the Cx32i neurons receive synaptic inputs in the neostriatum and cerebellar cortex (data not shown). The cytoplasmic immunoreactivity was sequestered from the interior of mitochondria and Golgi cisternae, but precipitated onto practically all of the cytoplasmic organelles typically seen in pre-embeddment immunhistochemical preparations of cytosolic proteins. The punctate Cx32i labeling was resolved as immunoreactivity in dendritic and axonal cytoplasm. Occasionally, immunoreactivity was observed in association with interneuronal appositions arranged as plaque or dispersed gap junctions. The observation that distinct interneuronal gap-junction plaques adjacent to Cx32i junctions contain no immunoreactivity suggested that other connexins not recognized by the R5.21C antibody might be expressed by neurons. These results are consistent with the hypothesis that an unidentified neuronal gap junction is revealed by low-stringency hybridization with the Cx32 or Cx43 nucleotide probes.

Although lower-stringency conditions (final posthybridization wash in 2× SSC) were necessary to show Cx32 mRNA in all motor pools (except for the facial motor nucleus), there was definite immunohistochemical labeling in all motor columns examined in the brainstem and spinal cord (Fig. 2d). As previously reported (Micevych and Abelson, 1991), we did not observe specific Cx32 mRNA hybridization in spinal motor neurons. By using a lower-stringency post-hybridization wash, a strong signal was detected in motor neurons of all spinal motor pools (Fig. 3b). A similar experiment under low-stringency conditions demonstrated that the Cx43 probe also hybridized to motor neurons. These results implied that motor neurons might: 1) co-express unusual forms of Cx32 and Cx43 mRNA, 2) express a third, unknown connexin mRNA with considerable sequence homology to Cx32 and Cx43 mRNAs, and/or 3) express a conserved sequence in a non-connexin-encoding mRNA (Beyer et al., 1988; Dermeitzel et al., 1989). Such interpretations are consistent with our preliminary immunohistochemical results demonstrating

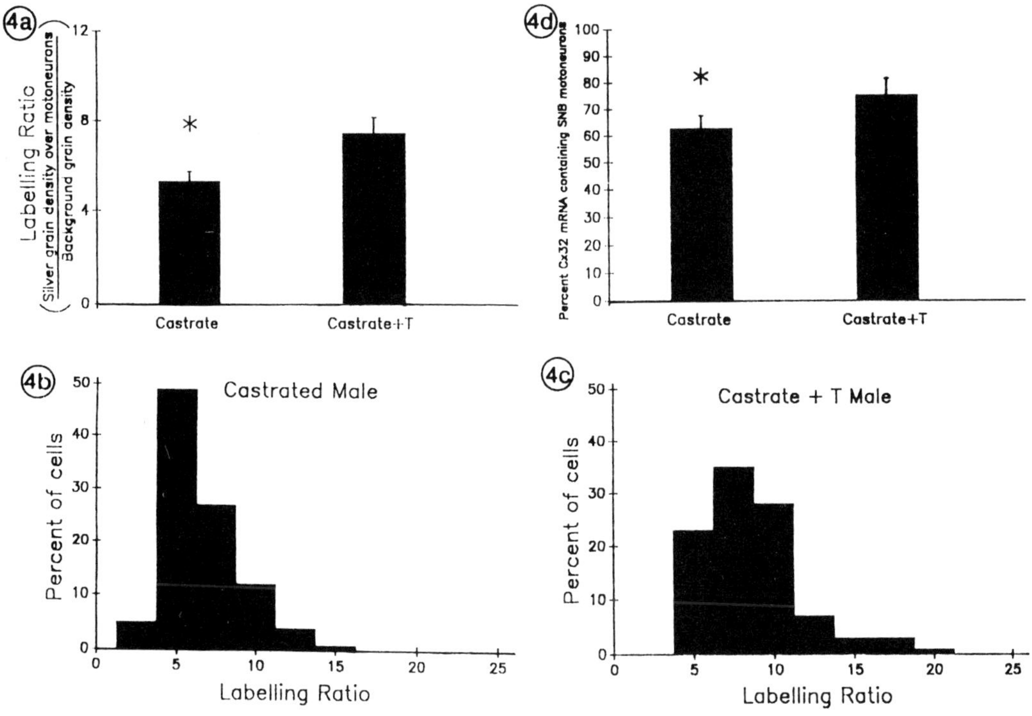

Figure 4. Histograms illustrating the amount and distribution of Cx32 mRNA in motor neurons of the SNB after castration and androgen replacement in adult male rats. Panel 4a: The mean density of labeling is significantly lower in castrated males than in castrated rats treated with testosterone (T) for 4 weeks. Panels 4b and 4c: These histograms demonstrate that after testosterone (T) treatment of castrates, the frequency distribution of the labeling ratio is shifted to the right, indicating the up-regulation of Cx32 copy number (labeling ratio) in Cx32-expressing motor neurons. Panel 4d: This histogram shows the alterations of the labeling ratio in castrated and T treated rats reflected in the proportion of SNB motor neurons judged to contain Cx32 mRNA.

Cx32i motor neurons (Fig. 2d) and Cx43i motor neurons (J.P. Revel, Cx43 antibody). Other investigators (Nagy et al., 1988) report similar positive findings.

Therefore, Cx32 is probably encoded, expressed and accumulated by neurons in mammalian motor networks and myelinating oligodendroglia in adjacent fiber tracts. The distributions of Cx32 mRNA and Cx32i suggest that Cx32 is synthesized in the perikaryal cytoplasm and transported to peripheral sites for membrane insertion and assembly into dispersed gap junctions as well as crystalline plaques of gap junctions. This may account for unexpected observations of Cx32i in cerebellar cortex, a site of few known gap-junction plaques. Finally, Cx32 mRNA and Cx32i are expressed by diverse types of neurons, but may not account for all of the gap junction-forming proteins produced by neurons.

Androgenic regulation of Cx32 in spinal motor neurons

Our second aim was to test the hypothesis that connexin expression is regulated by androgens in SNB motor neurons. As shown by electron microscopy and freeze-fracture replication, gap junctions are present in this androgen-sensitive pool of motor neurons, but decrease in number after androgen elimination by castration. The molecular composition and the pattern of androgenic regulation of SNB neuronal gap junctions are unknown. Eight adult male rats were castrated to eliminate androgens; four of these castrates received no further treatment. The other four castrates were implanted with 4.5 mm Silastic capsules containing testosterone. These capsules produce plasma testosterone levels in the high physiologic range. Four intact male rats served as controls. Spinal cord sections were processed for in situ hybridization as described previously (Micevych and Abelson, 1991). The Cx32 probe was used for in situ hybridization with lowered-stringency post-hybridization conditions in order to detect connexin mRNA (Figs. 3a-b). SNB motor neurons with $> 3\times$ background grain density were analyzed by a computer-assisted grain-counting system (Cue-2, Olympus Corp.).

Low androgen levels after castration produced a consistent, but statistically unreliable, trend toward reductions in the Cx32 mRNA copy number and the frequency of Cx32 mRNA motor neurons. Castrated, testosterone-treated animals have more Cx32 cells, and their density of labeling in SNB motor neurons was significantly higher than in castrates ($p < 0.05$; Figs. 4a–b). These results and our previous ultrastructural analysis of the androgenic effects on gap junctions in the SNB support the conclusion that a steroid-mediated regulation of the expression of Cx32 (or other putative connexins) may occur in motor neurons. Cx32 or a related connexin seems to be part of the gap-junctional connectivity of the SNB motor neurons. The functional impact of the SNB gap junctions may lie in the electrotonic coordination of motor neuron output.

Steroid regulation of gap junctions is also evident in the effect of ecdysone on developing insects (Caveney, 1978). Although androgen up-regulation of Cx32 in SNB motor neurons may involve transcriptional, post-transcriptional or mRNA-stabilization mechanisms (Shapiro et al., 1987), androgen-induced gene expression studies and oocyte injection/coupling experiments suggest that steroids induce the transcription of Cx32 mRNA (Werner et al., 1985; O'Malley et al., 1986; Beato et al., 1987; Dahl et al., 1987). The difference between our ultrastructural results (decreased frequency and size of gap-junction plaques) and the present studies which show a lack of effect after castration suggests that there may be a complex interaction between connexin gene transcription and gap-junction plaque formation. Removal of androgens may decrease the stability of the gap-junction plaque, decrease the rate of plaque formation or attenuate the rate of connexin insertion into membrane. All of these mechanisms would produce a decrease in morphologically identifiable plaques without changing the steady-state levels of connexin mRNA in the SNB motor neurons. However, our data are consistent with an androgen-induced up-regulation of connexin message preceding the increase in plaque size and number from castrate levels to testosterone-treated intact levels.

Dopaminergic regulation of Cx32 in neostriatal neurons

Our third aim was to determine whether connexin expression is regulated by dopaminergic afferents to neostriatal neurons. The nigrostriatal tract is largely undecussated and is the principal source of dopamine for the mammalian neostriatum. Dopamine application diminishes the frequency of dye coupling between retinal neurons in teleost fish (Mangel and Dowling, 1985). In the mammalian neostriatum, dye coupling diminishes with age as dopamine levels increase, and is reinstated by dopamine-depleting lesions of the substantia nigra (Cepada et al., 1989; Walsh et al., 1989). The dopaminergic regulation of neuronal dye coupling may involve changes in the frequency of gap junctions or altered junctional permeability.

Unilateral and nearly complete electrolytic lesions of the substantia nigra were used to destroy the origins of the dopaminergic nigrostriatal pathway in five adult rats. In each case, the neostriatum ipsilateral to the lesion was dopamine depleted while the neostriatum contralateral to the lesion served as a within-case control. Partial unilateral lesions of the substantia nigra (n = 1), unilateral control lesions of the cerebral peduncle (containing striatonigral axons) (n = 1) and sham control lesions (n = 1) were also tested. Rats were sacrificed 14 days after surgery and their brains were fixed for histological procedures. Alternating serial sections were used to detect tyrosine hydroxylase (THi, a catecholamine-synthesizing enzyme indi

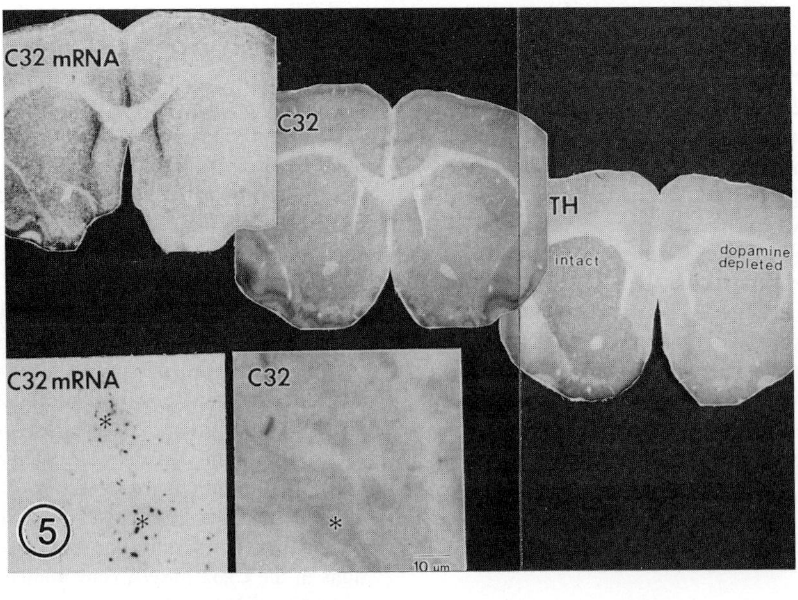

Figure 5. Photomicrographs of alterations of labeling of Cx32 in the neostriatum of an adult rat following unilateral (left) electrolytic destruction of the dopaminergic nigrostriatal tract. Note the different degrees of labeling in the right and left neostriatum in the serial coronal sections of the forebrain. Cx32 mRNA increased in the left neostriatum (left section) and Cx32i increased in the left neostriatum (middle section) when THi decreased in the left neostriatum (right section). Insets show Cx32 mRNA and Cx32i labeling of neurons (asterisks) in the left neostriatum of this case.

cating the status of dopaminergic afferents; immunohistochemical labeling), Cx32i and Cx32 mRNA.

Dopamine depletion, as indicated by the depression of THi levels, was associated consistently and reliably with the increased labeling of Cx32i and Cx32 mRNA in the neostriatum (Fig. 5). The changes occurred in neurons, particularly in the principal medium-sized spiny cells of the neostriatum. In neurons and interfascicular oligodendroglia of the internal capsule, both of these molecular markers were expressed at normal levels in "intact" control neostriata and control cases. The lesion-induced alterations of Cx32 were due to dopaminergic de-afferentation, since partial lesions of the substantia nigra produced a regionalized shift in immunolabeling and no changes of immunolabeling were apparent after axotomy of neostriatal neurons (control lesions of cerebral peduncle).

The evidence indicates that dopamine down-regulates the expression of gap-junction proteins in the mammalian neostriatum (Fisher et al., 1990). The loss of dopaminergic nigrostriatal inputs leads to supranormal accumulations of Cx32 mRNA and Cx32i in association with enhanced dye coupling between neurons. The altered regulation is persistent. Since dopamine modulation occurs through a second-messenger system, the regulation may encompass transcriptional, post-transcriptional and stabilization processes. Short-term changes in dye coupling may be due to altered gap-junction permeability, but long-term changes in dye coupling also involve the levels of gap-junction proteins and, in all likelihood, the frequency of neuronal gap junctions. These results are significant because they suggest that the progressive establishment of electrotonic routes of neuronal communication through gap junctions may contribute to the abnormal control of voluntary movement during the course of nigrostriatal dopaminergic denervation in Parkinson's disease.

Acknowledgements

Expert technical assistance was provided by L. Abelson and J. Miyashiro. We are grateful for the advice of our collaborators G. Zampighi and M. Levine. This research was supported by USPHS grants HD05958 (RSF) and NS21220 (PEM).

References

Beato, M., Arnemann, J., Chalepakis, G., Slater, E. and Willmann, T. (1987) Gene regulation of steroid hormones. J. Steroid Biochem. 27, 9–14.

Bennett, M.V.L., Barrio, L.C., Bargiello, T.A., Spray, D.C., Hertzberg, E. and Sáez, J.C. (1991) Gap junctions: New tools, new answers, new questions. Neuron 6, 305–320.

Beyer, E.C., Kistler, J., Paul, D.L. and Goodenough, D.A. (1989) Antisera directed against connexin 43 peptides react with a 43-kDa protein localized to gap junctions in myocardium and other tissues. J. Cell Biol. 108, 595–605.

Beyer, E.C., Paul, D.L. and Goodenough, D.A. (1987) Connexin 43: A protein from rat heart homologous to a gap-junction protein from liver. J. Cell Biol. 105, 2621–2629.

Caveney, S. (1978) Intercellular communication in insect development is hormonally controlled. Science 199, 192–195.

Cepeda, C., Walsh, J., Hull, C., Howard, S., Buchwald, N. and Levine, M. (1989) Dye coupling in the neostriatum of the rat. I. Modulation by dopamine-depleting lesions. Synapse 4, 229–237.

Dahl, G., Miller, T.M., Paul, D.L., Voellmy, R. and Werner, R. (1987) Expression of functional cell-cell channels cloned from rat liver gap-junction complementary DNA. Science 236, 1290–1293.

Dermietezel, R., Traub, O., Hwang, T.K., Beyer, K., Bennett, M.V. L., Spray, D.C. and Willecke, K. (1989a) Differential expression of three gap-junction proteins in developing and mature brain tissues. Proc. Natl. Acad. Sci. USA 86, 10148–10152.

Dermietzel, R., Völker, M., Hwang, T.K., Berzborn, R.J. and Meyer, H. E. (1989b) A 16-kDa protein co-isolating with gap junctions from brain tissue belonging to the class of proteolipids of the vacuolar H^+-ATPases. FEBS Lett. 253, 1–5.

Fisher, R., Hull, C., Buchwald, N., Adinolfi, A. and Levine, M. (1987) The morphogenesis of glutamic acid decarboxylase in the neostriatum of the cat: Neuronal and ultrastructural localization. Dev. Brain Res. 33, 215–234.

Fisher, R., Buchwald, N., Hull, C. and Levine, M. (1988) GABAergic basal forebrain neurons project to the neocortex: The localization of glutamic acid decarboxylase and choline acetyl-transferase in feline corticopetal neurons. J. Comp. Neurol. 272, 489–502.

Fisher, R., Levine, M., Buchwald, N., Micevych, P. and Zampighi, G. (1990) Regulation of gap-junction proteins in neostriatum: expression of connexin32 after substantia nigra destruction. Soc. Neurosci. Abst. 16, 1230.

Goodenough, D. and Gilula, N. (1974) The splitting of hepatocyte gap junctions and zonulae occludentes with hypertonic disaccharides. J. Cell Biol. 61, 575–590.

Mangel, S. and Dowling, J. (1985) Responsiveness and receptive field size of carp horizontal cells are reduced by prolonged darkness and dopamine. Science 29, 1107–1109.

Matsumoto, A., Arnold, A.P., Zampighi, G.A. and Micevych, P.E. (1988) Androgenic regulation of gap junctions between motor neurons in the rat spinal cord. J. Neurosci. 8, 4177–4183.

Matsumoto, A., Arnold, A.P. and Micevych, P.E. (1989) Gap junctions between lateral spinal motor neurons in the rat. Brain Res. 495, 362–366.

Micevych, P.E. and Abelson, L. (1991) Distribution of mRNA coding for liver and heart gap-junction proteins in the rat central nervous system. J. Comp. Neurol. 305, 96–118.

Nagy, J.I., Yamamoto, T., Shiosaka, S., Dewer, K.M., Whittaker, M.E. and Hertzberg, E.L. (1988) Immunohistochemical localization of gap-junction protein in rat CNS: A preliminary account. In E. Hertzberg and R. Johnson (Eds.), Gap Junctions. Alan R. Liss, New York, pp. 375–389.

O'Malley, B.W., Schrader, W.T. and Tsai, M.J. (1986) Molecular actions of steroid hormones. Adv. Exp. Med. Biol.: Steroid Horm. Res. 196, 1–10.

Paul, D.L. (1986) Molecular cloning of cDNA for rat liver gap-junction protein. J. Cell Biol. 103, 123–134.

Peters, A. , Palay, S.L. and Webster, H. (1991) The Fine Structure of the Nervous System. Oxford University Press, Oxford.

Shapiro, D.J., Blume, J. and Nielsen, D. (1987) Regulation of messenger RNA stability in eukaryotic cells. Bio. Essays 6, 221–226.

Walsh, J., Cepeda, C., Hull, C., Fisher, R., Levine, M. and Buchwald, N. (1989) Dye coupling in the neostriatum of the rat. II. Decreased coupling between neurons during development. Synapse 4, 238–247.

Werner, R., Miller, T., Azarnia, R. and Dahl, G. (1985) Translation and functional expression of cell-cell channel mRNA in *Xenopus* oocytes. J. Membr. Biol. 87, 253–268.

J.E. Hall, G.A. Zampighi and R.M. Davis (Eds.)
Progress in Cell Research, Vol. 3

CHAPTER 21

Channel reconstitution from lens MP70 enriched preparations

PAUL DONALDSON and JOERG KISTLER

Department of Cellular and Molecular Biology, University of Auckland, Auckland, New Zealand

Introduction

The mammalian ocular lens is a functional syncitium. Goodenough et al., (1980) showed that abundant gap junctions provide the pathways for the transport of nutrients from the aqueous humour and the epithelium to the fiber cells situated deeper in the lens. These syncitial properties were also revealed by experiments which demonstrated extensive electrical and dye coupling between lens cells (Cooper et al., 1991). Transmembrane pathways play an important role for lens homeostasis and transparency. In order to identify the components of these low-resistance pathways, lens fiber membranes were isolated and gap junctions partially purified. Immunolocalization studies using such preparations indicated that the major intrinsic polypeptide MIP26 was a principal component of the lens fiber gap junctions (Sas et al., 1985). Structure predictions based on sequence data had the polypeptide chain traversing the lipid bilayer several times, a topology consistent with that of a channel-forming protein (Gorin et al., 1984). Purified MIP26 was reconstituted into planar lipid bilayers and shown to form channels (Zampighi et al., 1985, Ehring et al., 1990). However, in the light of more recent immunolocalization experiments (Paul and Goodenough 1983, Zampighi et al., 1989, Voorter et al., 1989) and of sequence comparisons (Baker and Saier 1990), it now appears certain that the channels formed by MIP26 are distinct from those of the lens fiber gap junctions. Instead, it has been proposed that MIP26 channels play a role in minimizing extracellular space in the lens (Ehring et al., 1990).

In the search for channel proteins which mediate intercellular communication between the lens fiber cells, MP70 is an obvious candidate (Kistler et al., 1988). This membrane protein is a specific lens fiber gap-junction component and is related to the connexin family. Pore structures have been demonstrated in lens fiber gap-junctions after proteolysis-induced crystallization and reveal hexameric connexons with a stain-filled center (Kistler et al., 1990a). However, the ability of MP70 to form channels has yet to be demonstrated, although a closely related protein, Cx46, was recently expressed in *Xenopus* oocytes and shown to form channels (Paul et al., 1991).

In this article, we report on our efforts to purify ovine MP70 in the form of structures which are consistent with connexon pairs. We present single-channel conductance measurements obtained from the reconstitution of MP70-enriched preparations into planar lipid bilayers.

Preparation of MP70 enriched fractions

It has proven to be difficult to obtain a purified lens fiber gap-junction fraction, but by using tissue only from the outer cortex for membrane isolation, gap junctions can be

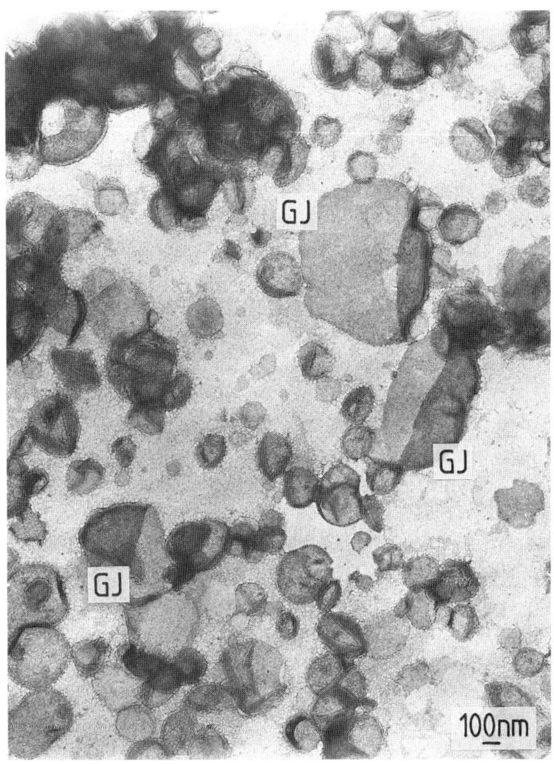

Figure 1. Negatively stained membrane preparation from sheep lens outer cortex. Gap junctions (GJ) are enriched but co-purify with abundant non-junctional membrane vesicles.

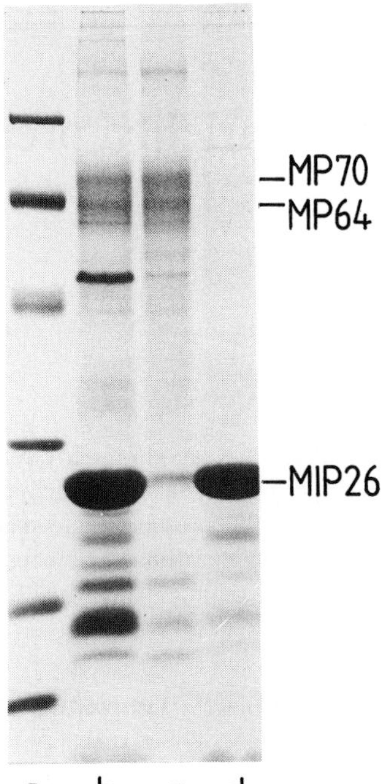

Figure 2. SDS-PAGE of MP70-enriched fraction. (a) Molecular weight markers from top: 97 kDa, 66 kDa, 45 kDa, 31 kDa, 21 kDa, 14 kDa. (b) Urea/alkali-stripped membranes from sheep lens outer cortex. (c) Triton X-100 (0.5%) solubilized proteins, enriched in MP70 and its 64 kDa derivative. (d) Triton X-100 insoluble proteins, mostly MIP26.

significantly enriched. The membranes were washed free of cytoplasmic proteins with buffer and further stripped of adhering proteins using a urea/alkali treatment protocol (Kistler and Bullivant, 1987). When examined by negative-stain electron microscopy, these preparations showed some enrichment for junctional plaques, but great amounts of non-junctional vesicles were always present (Fig. 1). The latter contain MIP26 and MP17/MP18, which strongly dominate the protein profile (Fig. 2b; Voorter et al., 1989).

Considerable enrichment of the gap-junction component MP70 can be obtained using selected detergents. Unlike liver and heart, gap-junction structures from lens fibers dissociate upon detergent treatment (Kistler and Bullivant , 1988). Triton X-100 (0.1–0.5%) solubilizes MP70 together with several minor proteins (Fig. 2c) but leaves most MIP26 membrane-bound (Fig. 2d). Examination of this MP70-enriched fraction by negative-stain electron microscopy reveals abundant "hamburger"-like structures (Fig. 3). The shortest of these are approximately 10 nm long and 15 nm wide and are probably side views of intact connexon pairs. Longer structures most likely represent multiple connexon pairs which have not been fully dissociated by the detergent treatment.

MP70 is solubilized predominantly as a 17S complex and has been further enriched by sucrose gradient centrifugation (Kistler and Bullivant , 1988). However, neither this nor gel filtration methods achieves total separation from minor amounts of MIP26. Efforts are continuing to further purify MP70.

Single-channel recordings from MP70-enriched preparations

To assay for channel activity, MP70-enriched fractions were used for reconstitution into painted bilayers. These were formed from a 3:7 PC/PE mixture in decane (50 mg/ml) following established procedures (Hamilton et al., 1989). Chambers contained 1 M KCl, 10 mM HEPES, pH 7.4 and 1–10 μg detergent-solubilized proteins (volume 15–40 μl) were pipetted directly into one or both chambers. Channel activity was usually detected 10–20 min after addition of protein. The potential across the bilayer was controlled by a computer using an A/D board and commercially available software (pClamp 5.5; Axon Instruments, Foster City, CA). Single-channel data were filtered at 500 Hz using an 8-pole Bessel filter, sampled on line at 0.5 ms/pt and analyzed using the pClamp 5.5 suite of programs.

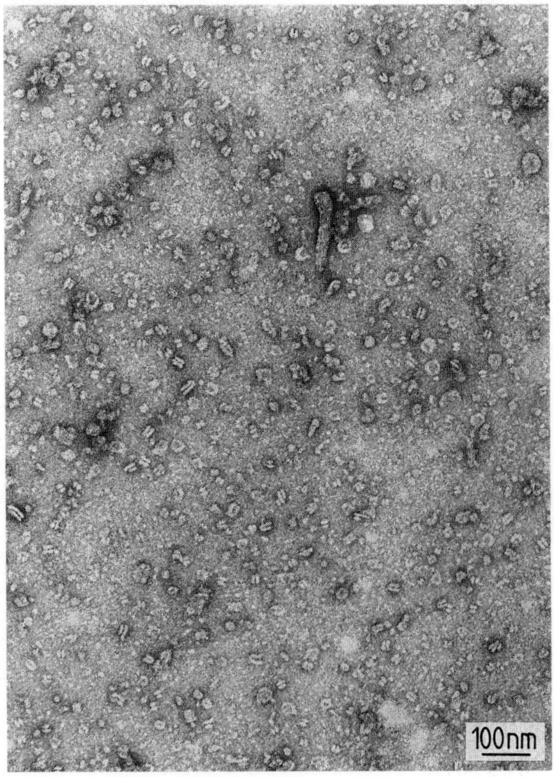

Figure 3. Uranyl acetate-stained preparation of MP70-enriched fraction. "Hamburger"-like structures are abundant and the shortest (approx. 10 nm) are consistent with side views of connexon pairs. Longer structures probably represent incompletely solubilized clusters of connexon pairs.

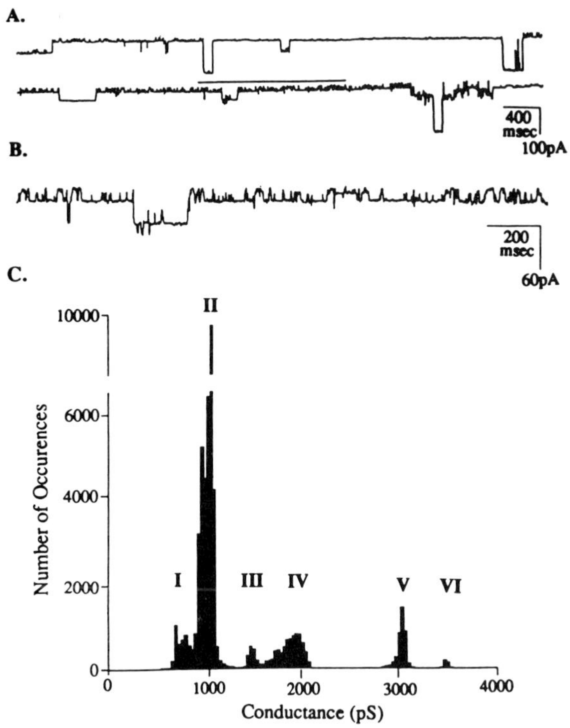

Figure 4. Single-channel recording of Triton X-100 solubilized proteins in Fig. 2c reconstituted into painted bilayers. Symmetrical 1 M KCl solutions. (a) Records at a command potential of −60 mV, downward deflections represent channel openings. (b) Expanded segment of record overlined in (a). (c) Events histogram showing discrete conductance classes. Interpretation: I = closed state, II = 312 pS class activity, III = 620 pS, IV = 312 ± 620 pS, V = 2266 pS, VI = 2266 ± 312 pS. Peak I is offset from zero due to membrane leakage or minimum conductance state of MIP26 channel (Ehring et al., 1990).

In 16 experiments using Triton X-100 solubilized ovine lens membrane proteins for reconstitution, 3 distinct and predominant classes of channel activity were observed. These activities had conductances of 312 ± 24 pS (n = 6), 620 ± 29 pS (n = 9) and 2266 ± 109 pS (n = 11). An example of an experiment in which all 3 channel activities were present is shown in Fig. 4. In other experiments, the different classes of channel activity appeared individually or in different combinations, indicating that the 312 pS and 620 pS conductance classes represent distinct channel activities and not substates of the 2266 pS channel.

Conductance steps of around 2000 pS in 1 M KCl have previously been associated with MIP26 channels (Ehring et al., 1990). We have found that the 2266 pS channel is nonselective for cations but slightly anion selective; has rapid opening and slow closing rates; and exhibits symmetrical voltage dependency such that the open probability of these channels is significantly reduced at command potentials over ±50 mV. Hence, it is likely that the 2266 pS activity is associated with minor amounts of MIP26 in our preparations.

Conductances of 312 pS and 620 pS in 1 M KCl are novel for reconstituted lens fiber proteins. Our single-channel records show that these are not subconductance states of MIP26. Channel activities showed no cation selectivity and exhibit linear current–voltage relationships between −100 and +100 mV. While the characterization of these novel conductances is still in a preliminary stage, the data obtained so far are consistent with the properties of gap-junction channels from tissues other than lens.

Although this is somewhat premature, it is tempting to speculate on the relationship between the 312 pS and 620 pS channels. One possibility is that the 620 pS activity is associated with single connexons and that the 312 pS activity represents lens connexon pairs (i.e., two 620 pS channels in series). Alternatively, the two novel conductance classes could represent distinct lens connexon types, for example, one composed of MP70, the other of the closely related Cx46. In ovine lens, both proteins co-migrate to the same position on SDS gels, and our preparations may indeed contain both polypeptides. At this time, however, it is not possible to reliably identify one or both conductances with lens connexons because the MP70-enriched preparations contain other minor protein components.

Outlook

The further characterization of these novel lens channel activities is a priority. This aim will be expedited by obtaining more highly purified complexes of MP70 or Cx46. Our approach also opens a new avenue to study the functional implications of the age-related processing of lens fiber gap-junction proteins. Deeper in the lens, MP70 is cleaved to a 38-kDa protein. This amino-terminal fragment contains the membrane-embedded and putative channel-forming segments of MP70 (Kistler et al., 1990b). Cleavage also removes the susceptibility of the molecule to phosphorylation with cAMP-dependent protein kinase (Voorter and Kistler, 1989). Comparison of the channel properties of MP70 and MP38 by reconstitution into planar lipid bilayers may shed light on the functional role of this age-dependent processing of the MP70 polypeptide.

Acknowledgements

We thank Dr S. Bullivant for his continued support and stimulating discussions. This work was supported by grants from the Medical Research Council of New Zealand and from the New Zealand Lottery Grants Board.

References

Baker M.E. and Saier M.H. (1990) A common ancestor for bovine lens fiber major intrinsic protein, soybean nodulin-26 protein, and *E. coli* glycerol facilitator. Cell 60, 185–186.

Cooper K., Mathias R.T. and Rae J.L. (1991) The physiology of lens junctions. In: C. Peracchia (Ed.) Biophysics of Gap-Junction Channels, CRC Press, Boca Raton, pp. 57–74.

Ehring G.R., Zampighi G., Horwitz J., Bok D. and Hall J.E. (1990) Properties of channels reconstituted from the major intrinsic protein of the lens fiber membranes. J. Gen. Physiol. 96, 631–664.

Goodenough D.A., Dick J.S.B. and Lyons J.E. (1980) Lens metabolic cooperation: a study of mouse lens transport and permeability visualized with freeze-substitution autoradiography and electron microscopy. J. Cell Biol. 86, 576–589.

Gorin M.B., Yancey S.B., Cline J., Revel J.P. and Horwitz J. (1984) The major intrinsic protein (MIP) of the bovine lens fiber membrane: characterization and structure based on cDNA cloning. Cell 39, 49–59.

Hamilton S.L., Alvarez R.M., Fill M., Hawkes M.J., Brush K.L., Schilling W.P. and Stefani E. (1989) [^{3}H] P N200–110 and [^{3}H] ryanodine binding and reconstitution of ion channel activity with skeletal muscle membranes. Anal. Biochem. 183, 31–41.

Kistler J. and Bullivant S. (1987) Protein processing in lens intercellular junctions: cleavage of MP70 to MP38. Invest. Ophthalmol. Vis. Sci. 28, 1687–1692.

Kistler J., Christie D. and Bullivant S. (1988) Homologies between gap-junction proteins in lens, heart and liver. Nature 331, 721–723.

Kistler J. and Bullivant S. (1988) Dissociation of lens fibre gap junctions releases MP70. J. Cell Sci. 91, 415–421.

Kistler J., Berriman J., Evans C.W., Gruijters W.T.M., Christie D., Corin A. and Bullivant S. (1990a) Molecular portrait of lens gap-junction protein MP70. J. Struct. Biol. 103, 204–211.

Kistler J., Schaller J. and Sigrist H. (1990b) MP38 contains the membrane-embedded domain of the lens fiber gap-junction protein MP70. J. Biol. Chem. 265, 13357–13361.

Paul D.L. and Goodenough D.A. (1983) Preparation, characterization and localization of antisera against bovine MP26, an integral protein from lens fiber plasma membrane. J. Cell Biol. 96, 625–632.

Paul D.L., Ebihara L., Takemoto L.J., Swenson K.I. and Goodenough D.A. (1991) Connexin46, a novel lens gap-junction protein, forms open channels in non-junctional plasma membrane of *Xenopus* oocytes. J. Cell Biol. 115, 1077–1089.

Sas D.F., Sas M.J., Johnson K.R., Menko R.S. and Johnson R.G. (1985) Junctions between lens fiber cells are labeled with a monoclonal antibody shown to be specific for MP26. J. Cell Biol. 100, 216–225.

Voorter C.E.M., Kistler J., Gruijters W.T.M., Mulders J.W.M., Christie D. and de Jong W.W. (1989) Distribution of MP17 in isolated lens fibre membranes. Curr. Eye Res. 8, 697–706.

Voorter C.E.M. and Kistler J. (1989) cAMP-dependent protein kinase phosphorylates gap-junction protein in lens cortex but not in lens nucleus. Biochim. Biophys. Acta 986, 8–10.

Zampighi G.A., Hall J.E. and Kreman M. (1985) Purified lens junction protein forms channels in planar lipid films. Proc. Natl. Acad. Sci. USA 82, 8468–8472.

Zampighi G.A., Hall J.E., Ehring G.R. and Simon S.A. (1989) The structural organization and protein composition of lens fiber junctions. J. Cell Biol. 108, 2255–2275.

J.E. Hall, G.A. Zampighi and R.M. Davis (Eds.)
Progress in Cell Research, Vol. 3

CHAPTER 22

Does MIP play a role in cell–cell communication?

GEORGE R. EHRING[a], GUIDO A. ZAMPIGHI[b] and JAMES E. HALL[a]

[a]*Department of Physiology and Biophysics, UC Irvine, Irvine, CA 92717, USA and*
[b]*Department of Anatomy, UCLA, Los Angeles, CA 92204, USA*

Introduction

The well-coupled lens

The ocular lens consists of two types of cells: (1) a single layer of cuboidal epithelium that covers its anterior surface and performs nearly all active metabolic functions; (2) a mass of fiber cells that forms the bulk of the lens and determines its transparency and refractive power. Transparency requires that the lens remain avascular and that extracellular space be minimal (Bettelheim, 1985; Benedek, 1971). These constraints limit metabolic exchange between the fiber cells and the surrounding vitreous and aqueous humors. Fiber cells also lack the intracellular organelles required for oxidative metabolism, yet they remain viable for the life of the organism. Thus, understanding metabolic exchange between fiber cells and anterior epithelial cells is crucial to understanding the physiology of the lens.

Consider the lens as a sphere

Early electrophysiological experiments indicated that ionic current injected into the lens flowed freely until reaching the lens surface membranes. Duncan (1969) suggested that the results of the current-injection experiments could be explained by treating the lens as a large spherical cell surrounded by a single membrane. According to Duncan's model, this membrane was the only barrier to electrical and metabolic exchange. But analysis of intercellular coupling by diffusion of low molecular weight fluorescent dyes (Scheutze and Goodenough, 1982; Miller and Goodenough, 1986) or radiolabeled metabolites (Goodenough et al., 1980) supported the hypothesis that the lens was composed of discrete cells coupled together by intercellular junctions, a view confirmed by electrical impedance studies (Mathias et al., 1979; Mathias et al., 1981; Rae et al., 1982). The results of these impedance experiments were best fit by modeling the lens as a non-uniform spherical syncitium (Mathias et al., 1981; Mathias and Rae, 1985). Calculations based on this model supported the hypothesis that gap junctions link the metabolic and ionic homeostasis of fiber cells to metabolic exchange between anterior epithelial cells and the aqueous humor.

Guilt by association

Extensive intercellular coupling in the lens was correlated with the presence of large regions of closely apposed fiber-cell plasma membranes reminiscent of gap junctions in other tissues (Dickson and Crock, 1972; Kuwabara, 1975; Benedetti et al., 1976). One protein, Major Intrinsic Protein (MIP), comprises 50–70% of the lens-membrane protein (Bloemendal et al., 1972; Alcala et al., 1975), and 60 to 70% of the membrane area of the lens is junctional (Kuszak et al., 1978; Maisel et al., 1981). Thus, from its initial discovery, MIP was considered likely to be a gap-junction protein. This view was first called into question by the observation that MIP was found in both non-junctional as well as junctional membranes of fiber cells (Kistler et al., 1985; Sas et al., 1985a; Fitzgerald et al., 1985; Paul and Goodenough, 1983). In addition, fiber cell junctions exhibited greater heterogeneity in their structure than gap junctions in other tissues (Zampighi et al., 1982; Simon et al., 1982).

A junction by any other name.

The appearance of gap junctions in the electron microscope is much the same in most tissues. In transverse section, gap junctions exhibit a characteristic septalaminar structure. They are constructed of two closely apposed plasma membranes separated by a 2- to 3-nm-wide extracellular gap and have a total thickness of 15–18 nm. Viewed in the plane of the membrane, gap junctions appear as plaques of intramembrane particles (IMPs), usually arranged randomly or with hexagonal symmetry. Complementary arrangements of particles and pits in the PF (cytoplasmic-facing) and EF (extracellular-facing) membrane faces seen in freeze-fractured junctions support the idea that these particles form coaxially aligned channels spanning both bilayers of the junction (for review see Zampighi, 1987).

Fiber-cell junctions are also formed of closely apposed plasma membranes. However, they differ from the gap junctions of other tissues in their overall thickness, the width of their extracellular gap and the arrangement of the particles within the junctional plaques. Based on their overall thickness, fiber cell junctions have been classified into thin junctions (11–13 nm thickness) and thick junctions (18–20 nm thickness) (Zampighi et al., 1992). Most of the junctions in the lens are of the thin type (Simon et al., 1982; Zampighi et al., 1982). Freeze-fracture replicas show that these junctions are composed of 6–7 nm diameter IMPs arranged in tetragonal arrays. Immunolabeling of the 11–13 nm junctions shows that they contain MIP (Zampighi et al., 1989). The thick junctions contain IMPs with diameters of 7–8 nm arranged randomly within the plaque. The protein composition of the thick junctions is less clear; in the outer cortical regions these junctions are labeled by antibodies to MP 70 (Gruijters et al., 1987), a membrane protein that shows sequence homology to the connexin proteins (Kistler et al., 1988). As shown in Chapter 21 by Donaldson and Kistler, MP 70 forms channels similar to gap junction channels when reconstituted into planar bilayers. Towards the lens nucleus, however, the labeling of the 18–20 nm junctions with antibodies against MP70 is not demonstrable. It has been suggested that in the nuclear region, a proteolytic degradation product of MP 70 forms the nuclear junctions (Kistler and Bullivant, 1987). The protein composition of the 18–20 nm junctions is further complicated by the identification of a second connexin, Cx46, in the same cortical junctions as MP70. Cx46 is also present in junctions closer to the lens nucleus (Paul et al., 1991). The injection of mRNA coding for Cx46 has been reported to result in open hemichannels in single *Xenopus* oocytes (Chapter 11, and Paul et al., 1991) and in junctional channels between pairs of oocytes (Ebihara and Steiner, 1992). Thus, the question of which proteins form what kind of channels in lens fiber-cell junctions remains unresolved.

Since most lens junctions are composed of plaques with tetragonal arrays, thin junctions are often considered to be "the communicating junctions of the lens" (Benedetti et al., 1990). Studies of the structure of these junctions using freeze-fracture electromicroscopy have led to conflicting claims: some authors suggest that both thick and thin junctions consist of orthogonal or rhombic arrays of particles with complementary particles and pits in the PF and EF fracture faces (Peracchia and Peracchia, 1980a,b; Bernardini and Peracchia, 1981), while others describe an asymmetry in the arrangement of particles and pits in tetragonally crystallized junctions (Lo and Harding, 1984; Costello et al., 1989). In these structural studies, the investigators infer the presence of MIP from the crystalline order of the junctions.

In order to determine the protein composition and IMP arrangement of the thin junctions, we used immunocytochemistry and thin-section or label-fracture electron microscopy. Highly enriched preparations of lens junctional membranes were isolated with the junctional-membrane pairs still adherent at their extracellular surfaces. Thus, only their cytoplasmic surfaces were exposed to solution. The junctions were labeled with primary antibodies specific to MIP's cytoplasmic carboxy terminal. Protein A conjugated to electron-dense gold particles was then bound to the antibodies. The gold-labeled junctions were thin-sectioned (Fig. 1A) or affixed to a glass slide, quick frozen, and freeze-fractured for label-fracture experiments (Fig. 1B). Examination of gold-labeled thin-section preparations showed that most of the thin junctions had gold particles only on one side of the junction. This is not expected for gap junctions, where the symmetry of the channel would result in cytoplasmic labeling on both sides of the junction.

These data imply that at least some MIP-containing junctions are not gap junctions. But failure to label both sides of the junctions in these experiments could be the result of an alteration in the antigenic epitope in one half of the junction, for example by proteolytic degradation. Alternatively, MIP and an entirely different protein might combine to form an asymmetric cell-to-cell channel. Label-fracture experiments rule out both of these proposals. In label-fracture experiments, membranes are affixed to a glass slide after antibody-gold labeling. The glass provides a plane of reference for determining the height of the fractured membrane faces, the location of IMPs and the location of the gold particles. Three points are of particular relevance to the argument that follows: First, since junctions can only adhere to the glass by their cytoplasmic surfaces, all PF membrane faces come from membranes adhered to the glass and all EF membrane faces come from the apposing junctional membrane. Second, single membranes can readily be distinguished from junctions, because in junctions the EF and PF faces are separated by small fracture steps. Third, because the membranes are attached to the glass slide before application of the platinum-carbon coating, gold particles on the membrane surface away from the slide will cast shadows, whereas those between the membrane and the slide will not cast shadows. In label-fracture experiments, the antibody-gold labels tetragonal arrays of IMPs in the PF fracture faces and also labels smooth particle-free EF fracture faces. In both cases, the absence of a shadow indicates that the gold particle is on the membrane closest to the glass. As shown in Fig. 1B, the fracture plane sometimes jumps from a labeled PF face to a complementary particle-free EF face, demonstrating that in these junctions molecules of MIP in one membrane face a membrane devoid of IMPs. This result rules out the possibility that MIP is forming an asymmetric junction with some unlabeled protein or with immunologically unreactive MIP. An alternative explanation is that during the isolation of the membranes, the junctions separate, shift out of register and then reform into artifactual junctions. This explanation is favored by some authors (see for example Peracchia and Shen,

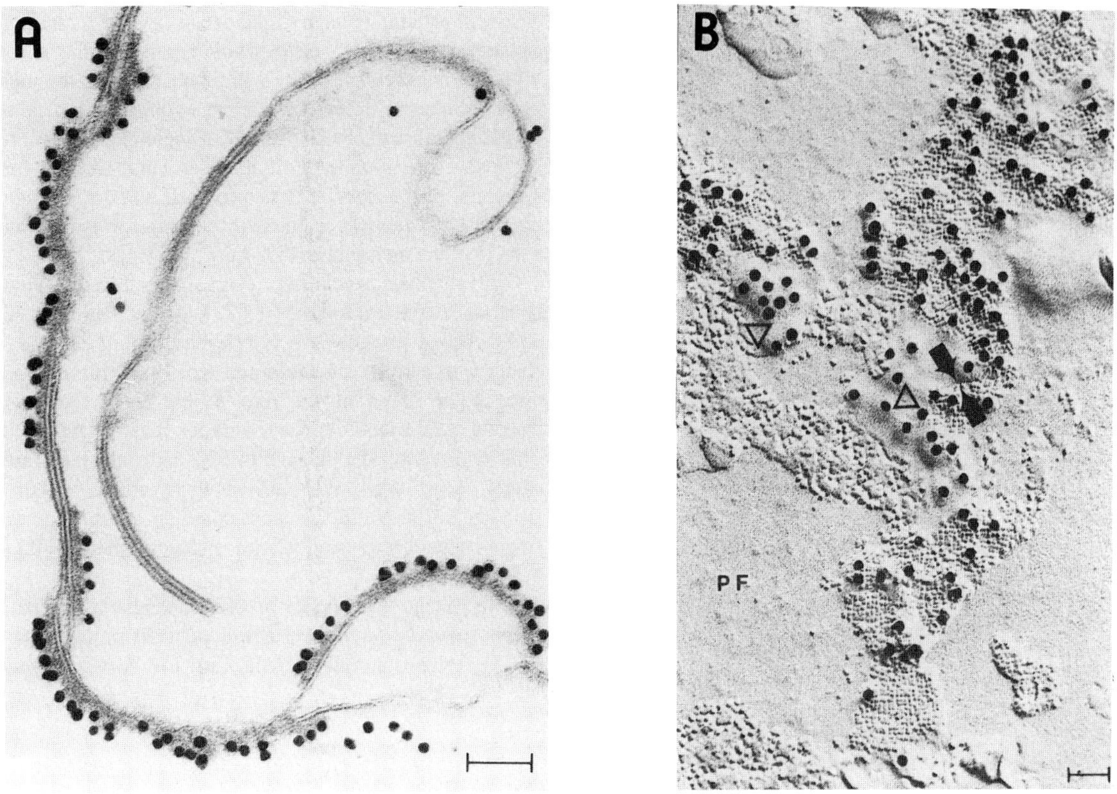

Figure 1. Junctions isolated from calf lenses labeled with a polyclonal antibody to MIP and visualized with protein-A gold. (A) Thin section: Gold particles are present on the convex membrane of the curved thin junctions but absent on the apposing membrane. Note that section of unlabeled thick junctions is interposed between two labeled thin-junction segments. Scale bar 0.1 μm. (B) Label-fracture replica: An extensive PF face of a tetragonal crystal is labeled with the gold particles. In addition, gold particles were present under smooth EF faces (open triangles). The tetragonal arrays were separated from smooth EF faces by fracture faces of small height, indicating the presence of small intervening gaps (black arrows). Scale bar, 62.5 nm. Reprinted with permission from Zampighi et al., 1989.

Chapter 23). We regard it as unlikely for two reasons: First, it is not easy to separate lens junctions. Separation requires much harsher conditions (Sas et al., 1985b) than were used in the label-fracture experiments. Second, in the same preparation where we find asymmetrical labeling of MIP junctions, we also find junctions that label symmetrically with antibodies to the putative gap junction protein, MP70. MIP is more resistant to solubilization than MP70, a fact that seems difficult to reconcile with the hypothesis that MIP junctions can separate and recombine, but MP70 ones do not.

What can MIP do?

All members of the connexin family tested to date form channels when RNA coding for them is injected into oocyte pairs. In contrast, injections of mRNA coding for MIP did not induce coupling between pairs of oocytes, even though the oocytes produced the protein (Swenson et al., 1989). The failure of MIP to form cell-to-cell channels in the oocyte could result from improper co-translational or post-translational processing in the amphibian cells. However, transfection of the cDNA for MIP into the mammalian BHK cell line also failed to induce coupling between cell pairs. In contrast, transfection with Cx32 significantly increased cell-to-cell coupling (Miller, 1992). Nonetheless, these negative results do not rule out the possibility that MIP forms communicating channels, since it is possible that MIP requires lens-specific co-factors to form gap junctions.

There is some evidence that MIP may be involved in intercellular communication (Johnson et al., 1988). In this study, intracellular injections of immune serum containing polyclonal antibodies against MIP reduced dye coupling between lentoid cells, whereas equivalent injections of pre-immune serum did not affect coupling. This result has been taken to imply that MIP channels are mediating the coupling between lens cells (Shen et al., 1991; Benedetti et al., 1990). However, as pointed out in the original report several alternate hypotheses can also explain this observation. For example, mechanical damage to the cell during injection or a secondary effect of altered MIP function could influence junctional coupling. In addition, non-specific serum effects on ion channels formed by

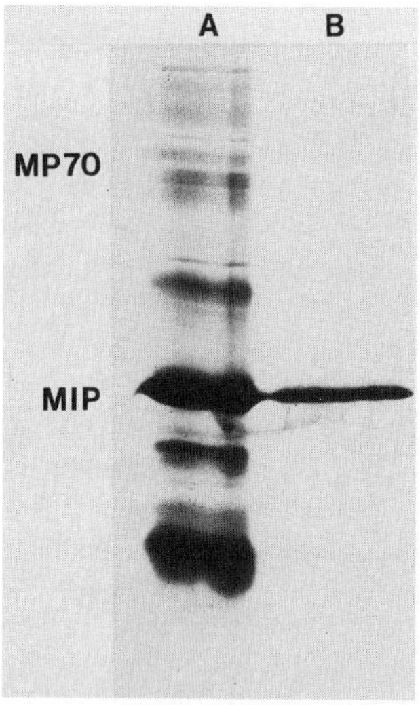

Figure 2. SDS-PAGE of the total protein from isolated calf lens membranes and HPLC-purified MIP. A 12% polyacrylamide gel was loaded with approximately 1 μg of protein and silver stained. Calf lens membrane isolate contained at least 14 protein bands (lane A) with molecular weights in the range of 14 to 96 kDa. In contrast, a fraction that eluted from the HPLC after a retention time of 26 min contained one predominant band with a molecular weight of 26 kDa (lane B). Thus in the HPLC-purified preparation, contaminant proteins could only be present in amounts less than the nanogram resolution of the silver-staining technique. Reprinted with permission from Ehring et al., 1990.

connexins or other proteins are difficult to rule out. Using only one concentration of pre-immune serum is an inadequate control, as the relative concentrations of the active agent can vary dramatically between lots of sera (Ehring et al., 1988). The evidence that MIP forms communicating channels is thus equivocal.

Reconstitution studies have provided the most direct evidence that MIP can form channels. When added to liposomes, preparations containing MIP increase the membrane permeability to molecules < 1 kDa in molecular weight (Girsch and Peracchia, 1985; Gooden et al., 1985a; Nikaido and Rosenberg, 1985; Scaglione and Rintoul, 1989). The changes in permeability can be modulated by calmodulin (Girsch and Peracchia, 1985; Peracchia and Girsch, 1989), pH (Gooden et al., 1985b), and enzymatic proteolysis (Peracchia and Girsch, 1985).

Channel-forming properties of MIP were measured by reconstituting enriched MIP preparations into planar lipid bilayers (Zampighi et al., 1985). We confirmed that channel formation was very likely due to MIP by reconstituting HPLC-purified MIP (Fig. 2). Purification of MIP by gel-permeation HPLC (size separation) or anionic-exchange HPLC (charge separation) results in the reconstitution of identical channels (Ehring et al., 1990; Ehring et al., 1992). The principal characteristics of the reconstituted MIP channel are large single-channel conductance, limited anionic selectivity, symmetrical voltage dependence, rapid opening and slow closing rates, and insensitivity to calcium and hydrogen ion concentration (Ehring et al., 1990). The symmetrical voltage dependence of MIP currents is illustrated in Fig. 3. The currents depend on the magnitude, but not the sign of the membrane voltage; the membrane conductance decreases exponentially during pulses to voltages greater than 20 mV. Single-channel studies show that the symmetry of the I–V curve results from symmetry of the channel structure itself and not just equal numbers of channels oriented in opposite directions with respect to the membrane (Fig. 4).

MIP channels have a fully open single-channel conductance of 360 pS in 100 mM KCl, but they also have several subconductances and a complex gating scheme. If changes in current are measured from an arbitrary baseline, rather than from the zero current level, the single-channel conductance appears smaller than its full value. This is because the reconstituted single channel rarely closes completely over the voltage range +100 to −100 mV. Closure to the most common substate gives a single-channel step conductance of approximately 200 pS in 100 mM KCl.

The voltage dependence of MIP channels is modulated by phosphorylation with cAMP-dependent protein kinase (PKA). Dephosphorylated MIP forms almost voltage-in-

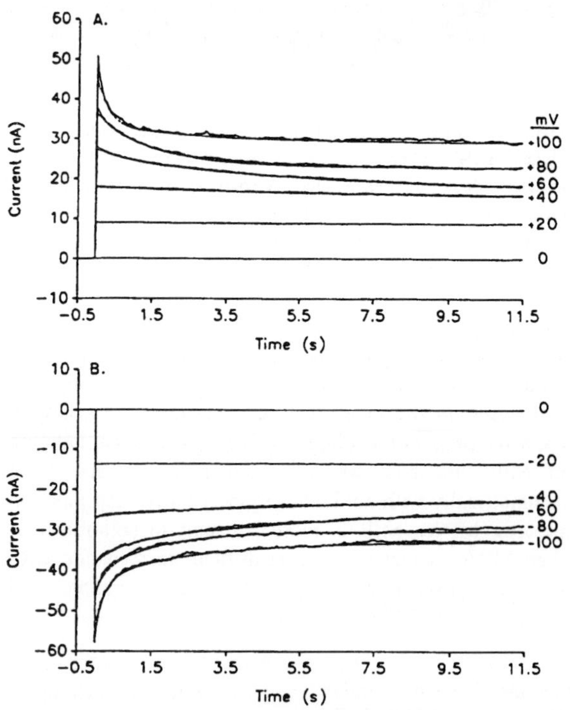

Figure 3. Currents through MIP channels were symmetric both in amplitude and time course during pulses to positive voltages (A) and negative voltages (B). The current measured immediately after the application of the voltage pulse was proportional to the voltage. The conductance of the membrane was maximal near 0 mV. When the voltage pulse was larger than 20 mV, the current decreased with time until it reached steady state. The rate of channel closure also increased with increases in the amplitude of the pulse. Reprinted with permission from Ehring et al., 1990.

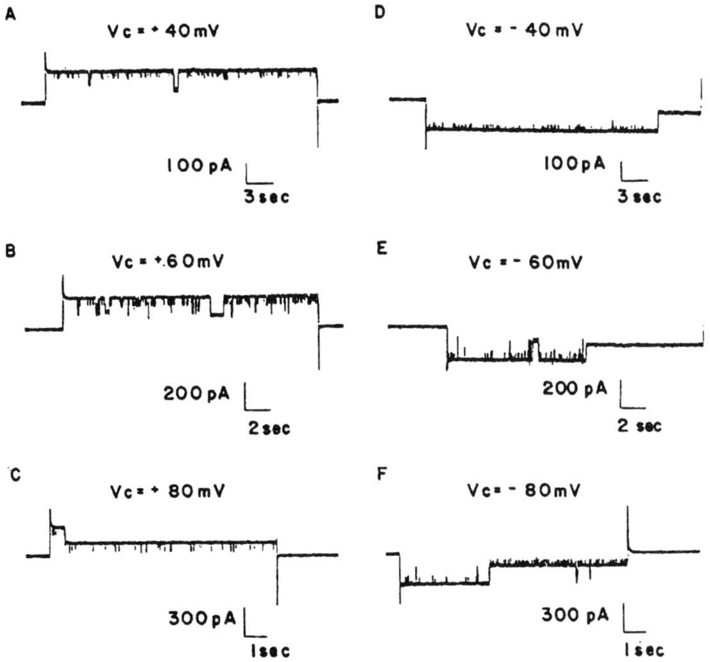

Figure 4. Currents through a single MIP channel in response to voltage pulses in 1 M KCl. In all records, the holding potential (V_h) was 0 mV; the test pulse was marked by brief capacitive currents on the step to V_c and the return to V_h. Bars in each panel indicate the time and voltage scales. The single-channel currents demonstrate the same symmetry, with respect to the sign of V_c, as the multi-channel currents shown in Fig. 3. Reprinted with permission from Ehring et al., 1990.

dependent channels while phosphorylated MIP forms voltage-dependent channels (Ehring et al., 1992). A serine on MIP's cytoplasmic carboxy terminal (Fig. 5) has been identified as the PKA phosphorylation site (Lampe and Johnson, 1991). Proteolytic removal of the carboxy terminal mimics the effects of phosphorylation (Ehring et al., 1991). Single-channel data show that the effect on the channel voltage dependence results from an alteration in channel kinetics rather than a change in single-channel conductance. These results suggest that the carboxy terminal stabilizes the channel in the closed state.

All in the family

The well-defined connexin family

All proteins unequivocally demonstrated to form gap junctions have substantial sequence identity and belong to the connexin family (see Part I). Connexins share a common motif with four transmembrane domains (M1 – M4), one of which (M3) has a stretch of amphiphilic amino acids that could line the hydrophilic pore of the "communicating" channel. The cytoplasmic amino-terminal and the extracellular domains that connect M1 to M2, and M3 to M4 are highly conserved across tissues and species. The carboxyl terminal and the loop connecting M2 to M3 extend into the cytoplasm. These regions show greater sequence diversity and are believed to contain regulatory sites for the channels (see Part I).

MIP and company

The deduced amino acid sequence for MIP is not homologous with that of any connexin (Fig. 5 and Part I). Instead MIP belongs to a new family of membrane proteins found in diverse species distributed amongst several phyla. Hydropathy plots, amino acid analysis, antibody binding, and enzymatic cleavage suggest MIP is a protein with six transmembrane domains (Fig. 5). One of these domains (F) is amphiphilic, consistent with the possibility that an oligomer of MIP could form a hydrophilic pore. Analysis of MIP's amino acid sequence shows a twofold repeat; the first three transmembrane domains have 27% sequence identity to the second three domains (Wistow et al., 1991; Pao et al., 1991). A consensus sequence for MIP-like proteins (GAXΦNPAX[TS]Φ[GA]*) is duplicated in the cytoplasmic loop connecting the transmembrane domains B and C, and in the extracellular loop connecting transmembrane domains E and F (Pao et al., 1991; Wistow et al., 1991).

* Primary sequence represented with the standard one letter amino acid code, Φ indicates any hydrophobic amino acid and pairs of amino acids enclosed in brackets indicate that either could be present.

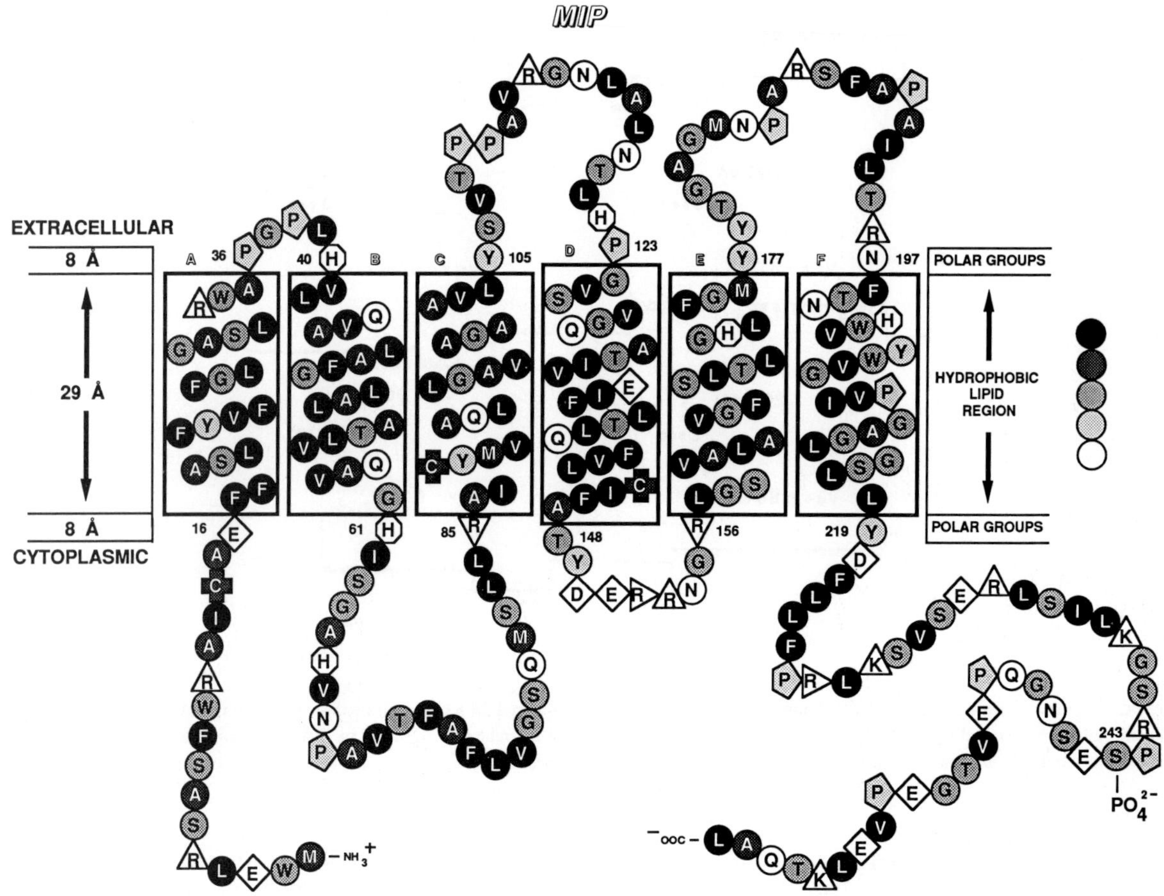

Figure 5. This figure illustrates the model of MIP proposed by Gorin et al. (1984). In this model there are six putative transmembrane domains (A–F) and both the carboxy and amino terminals are on the cytoplasmic side of the membrane. Lampe and Johnson (1990) have shown that the serine at location 243 on the carboxy terminal is phosphorylated by protein kinase A. Reprinted with permission from Ehring et al., 1992.

The MIP family

Nodulin 26

The first indication that MIP might belong to a family of proteins was the cloning of Nodulin 26 from the peribacteriod membrane in soybean nodules (Sandal and Marcker, 1988). Bacteriods are the remnants of the symbiotic nitrogen-fixing bacteria, *Bradyrhizobium japonicum*. The bacteriod, which includes the original inner and outer bacterial membranes and the bacterial enzymes which catalyze the conversion of atmospheric nitrogen to ammonia, is segregated from the plant cytoplasm by the peribacteriod membrane. This membrane, derived from the host cell membrane, replaces the bacterial cell wall and is believed to prevent host cell defences from attacking the bacteriod (Shanmugam et al., 1978). As in the lens-fiber membrane, specific transport molecules and permeases are expressed in the peribacteriod membrane (Fortin et al., 1985). The peribacteriod membrane must protect the bacteriod (which lacks the rigid cell wall of the intact bacterium) against changes in the tonicity of the plant cell cytoplasm. Thus, while no functional studies have been completed on Nodulin 26, its high amino acid sequence identity with MIP and its presence in the peribacteriod membrane have led to the suggestion that it plays a role in membrane transport (Baker and Saier, 1990).

Glycerol facilitator

The glycerol facilitator from the inner plasma membrane of *E. coli* has been proposed as a third member of this putative family of membrane proteins. Based on an alignment of amino acid sequences, Baker and Saier (1990) estimated the probability that MIP is not related to Nodulin 26 or the glycerol facilitator as less than 10^{-20}. The glycerol facilitator transports straight-chain polyols up to approximately 200 Da in molecular weight across the plasma membrane, while restricting access to their analogous cyclic sugars (Heller et al., 1980). High substrate turnover (2×10^5 glycerol molecules per s per facilitator), similar influx and efflux rates, lack of substrate saturation of the transport mechanism and low temperature sensitivity of substrate uptake imply that the glycerol facilitator functions as a membrane channel (Heller et al., 1980; Saier, 1985).

TIP

Most plant cells contain large storage vacuoles whose contents are segregated from the cytoplasm by a membrane called the tonoplast. Concentration gradients of both ions and metabolites between intravacuolar space and the cytoplasm imply the presence of transport mechanisms within the tonoplast. In particular, osmotic equilibrium across the tonoplast membrane must be regulated dynamically, since the tonicity of the vacuole can vary widely relative to that of the cytoplasm (Matile, 1978). Tonoplast Intrinsic Proteins (TIP), a series of integral membrane proteins with an approximate molecular weight of 28 kDa, are the major components of the tonoplast in both germinative (Johnson et al., 1990) and vegetative plant tissue (Hofte et al., 1992). The first TIP protein to be cloned (from seeds of the bean plant *Phaeolus vulgaris)* has 36% sequence identity with MIP (Johnson et al., 1990). A transport function for TIP has been proposed, and this hypothesis is now being tested experimentally in several laboratories. Patch-clamp studies on isolated tonoplasts and reconstitution studies of partially purified tonoplast membrane proteins have revealed a rich mixture of ion channels (Klughammer et al., 1992). Interestingly, one of these exhibited an approximately 200 pS step-size single-channel conductance (Klughammer et al., 1992). Although this conductance step was nearly identical to the step between the minimum and maximum conductance states of reconstituted MIP, it is not possible assign the 200 pS conductance to TIP because of the complex mixture of proteins in the vacuolar membrane preparations. Nevertheless, comparative reconstitution studies of purified TIP and MIP should prove informative.

Turgor-responsive gene products

In the plant cell, the cytoplasm is hypertonic to the extracellular environment. As a result of this osmotic imbalance, water enters the cell and a counterbalancing hydrostatic pressure develops when the cell membrane is forced against a rigid cell wall. This positive turgor pressure is necessary for the mechanical stability of plant tissue (Baker and Hall, 1988). Changes in the net flux of inorganic ions and small organic molecules across the cell membrane and the tonoplast are believed to regulate the tonicity of plant cytoplasm (Baker and Hall, 1988). In the pea plant, *Pisum sativum*, reduction of turgor following dehydration induces an increased expression of several genes. One of these (clone 7a) encodes for a 289 amino acid protein, whose C-terminal 180 amino acids have 33% sequence identity with Nodulin 26 (Guerrero et al., 1990). Based on enhanced transcription of mRNA for clone 7A, Guerrero et al. (1990) have inferred increased protein expression and have suggested that this gene product may be involved in ion transport.

Root-specific gene products

Screening of cDNA libraries constructed from tobacco and *Aradiposis* roots yielded two cDNA clones (TobRB7 and AtRB7), whose deduced amino acid sequences have high sequence similarities to MIP (Yamamoto et al., 1990). Since the primary function of root tissue is the transport of water and nutrients, Conkling and co-workers (1990) proposed that the proteins expressed from these genes function as membrane channels.

Yeast FPS1 gene product

The phenotype of the *fdp1* mutant strain of the yeast *Saccharomyces cerevisiae* includes a growth defect on fermentable sugars as the result of a deficiency in a glucose-induced, *ras*-mediated cyclic AMP signaling pathway. A second yeast gene, *FPS1*, suppresses the growth defect without restoring the signaling pathway. The *FPS1* gene codes for a 669 amino acid protein with an estimated molecular weight of approximately 73 kDa. A 280 amino acid sequence from the middle of the predicted *FPS1* gene product has 30% sequence identity with the entire glycerol facilitator and a high sequence similarity to MIP (Van Aelst et al., 1991). Because *FPS1* complements the *fdp1* mutant without correcting its primary signaling defect and because *FPS1* also alleviates a glucose-induced acidification of *fdp1* cells, Aelst et al., (1991) have postulated that the *FPS1* protein may act as proton pump or channel. Such a role would be consistent with the relatively broad range of transport functions proposed for the MIP family.

Drosophila bib gene product

Big Brain or *bib* is one of seven neurogenic genes in *Drosophila melanogaster* that determine whether embryonic ventral ecotodermal cells differentiate into neuroblasts or epidermoblasts. Mutations at the *bib* gene locus result in all ventral ectodermal cells becoming neuroblasts and in consequent neural hyperplasia, hence the name *Big Brain* (Campos-Ortega, 1991). Cells destined to become epidermal cells, convert to a neurogenic fate following ablation of neighboring neuroblasts (Doe and Goodman, 1985). These observations led to the hypothesis that neuroblasts produce an inhibitory signal that acts on epidermal cells preventing their differentiation into neuroblasts (Campos-Ortega, 1988). It has been suggested that the *bib* gene product is involved in an intracellular signaling pathway, but how the developmental signal is passed between cells is not known. While it is possible that communication occurs via gap-junctional pathways, a cell-surface interaction or an autocrine signal are equally likely. Transplantation of single cells homozygous for the *bib* allele into wild-type embryos gave rise to clones with the wild-type phenotype (Technau and Campos-Ortega, 1987). The mutant cells were capable of responding to signals provided by surrounding wild-type cells. Therefore, the *bib* mutation does not eliminate the possibility of intercellular communication, but may decrease the inhibitory signals produced by the mutant cells. Rao et al.,

(1990) have cloned the *bib* locus complementary and genomic DNAs. The cDNA encodes for a 700 amino acid protein whose amino-terminal half includes a stretch of 224 amino acids that are 40% identical with those of MIP. Hydropathy plots suggest that the *bib* protein, like MIP, has six transmembrane domains with the amino and carboxy terminals oriented towards the cytoplasm, but that the *bib* protein has a long hydrophilic extension of its carboxy terminal. Rao et al., (1990) suggest that the *bib* protein functions as a transmembrane channel or transporter that releases diffusible messenger from presumptive neuroblasts to control the differentiation of epidermoblasts.

CHIP

Recently, Peter Agre and co-workers identified and cloned a homolog of MIP, which is localized to the plasma membrane of erythrocytes and the apical brush border of the proximal convoluted tubules of the kidney (Denker et al., 1988; Smith and Agre, 1991; Agre and Preston, 1992). Based on its 37% sequence identity to MIP, they named this protein CHIP28 for channel-like integral membrane protein (Agre and Preston, 1992). Injection of mRNA encoding for CHIP28 into *Xenopus* oocytes resulted in a 10-fold increase in osmotic water permeability (Preston et al., 1992). As with the physiologic water channel from kidney, CHIP-induced water permeability could be reduced by $HgCl_2$ and restored by β-mercaptoethanol. These changes were observed without corresponding changes in the ionic conductance of the injected oocytes. Thus, Agre et al. (1992) have proposed that CHIP28 is a functional component of the water channel. Unlike MIP, CHIP28 has an N-glycosylated isoform. The glycoslyated and non-glycosylated forms of CHIP can exist together as heterotetramers (Smith and Agre, 1991). It is not known which isoform induces enhanced water permeability of injected oocytes.

Discussion

There is little direct evidence that MIP contributes to the coupling of lens fiber cells. That MIP induces increased solute permeability in liposomes and voltage-dependent channels in lipid bilayers does not argue for or against a role as a gap-junctional channel. Comparisons of the voltage-dependence and single-channel conductances of reconstituted MIP channels with reconstituted gap junction channels, likewise, cannot provide a satisfactory determination of the role of MIP. These parameters have been shown to be quite variable for both types of reconstituted channel. Because it is yet to be determined if MIP forms channels in vivo, even the claim that MIP is an "ancient channel-forming protein" (Agre and Preston, 1992) must be considered with due care.

It is clear that MIP does not belong to the connexin family, but instead to a rapidly growing family of its own. This alone does not rule out that MIP forms cell-to-cell channels. On the other hand, members of the MIP family of proteins exist primarily in single membranes and appear to be involved in transport of small molecules, salts and water, especially under conditions of osmotic stress. This is a plausible role for MIP in the lens, a role that could be complemented by gap junctions connecting fiber cells and providing true cell-to-cell communication. If MIP does form cell-to-cell channels, it would be the first member of its family found to do so, and many members of the family have already been shown to have a particular single-membrane transport function.

Acknowledgements

We would like to thank Drs. Alan Miller, Ana Chepelinsky, and Maarten Chrispeels for helpful discussions while preparing this manuscript; Mary Hawley and Mike Kreman for expert technical help; and the National Eye Institute for a post-doctoral fellowship (to GRE) and grants EY05661 (to JEH) and EY04110 (to GAZ) which made this work possible.

References

Agre, P. and Preston, G.M. (1992) Isolation of the cDNA for erythrocyte integral membrane protein of 28 kilodaltons: Member of an ancient family. Proc. Natl. Acad. Sci. USA 88, 11110–11114.

Alcala, J., Lieska, N. and Maisel, H. (1975) Protein composition of bovine lens cortical fiber cell membranes. Exp. Eye Res. 21, 591–595.

Baker, D.A. and Hall, J.L. (1988) Solute Transport in Plant Cells and Tissues. Longman Scientific and Technical, Essex.

Baker, M.E. and Saier, M.H. (1990) A common ancestor for bovine lens fiber major intrinsic protein, soybean nodulin-26 protein and *E. coli* glycerol facilitator. Cell 60, 185–186.

Benedek, G.B. (1971) Theory of transparency of the eye. Appl. Optics 10, 459–473.

Benedetti, E.L., Dunia, I., Bentzel, C.J., Vermorken, A.J.M., Kibbelaar, M. and Bloemendal, H. (1976) A portrait of plasma membrane specializations in eye lens epithelium fibers. Biochim. Biophys. Acta 457, 353–384.

Benedetti, E.L., Dunia, I., Manenti, S. and Bloemendal, H. (1990) Biochemical and structural properties of the protein constituent of junctional domains in eye lens fiber plasma membrane. In: A.W. Robards (Ed.), Parallels in Cell-to-Cell Junctions in Plants and Animals. Springer-Verlag, Berlin, pp. 35–52.

Bernardini, G. and Peracchia, C. (1981) Gap junction crystallization in lens fibers after an increase in cell calcium. Invest. Ophthalmol. Vis. Sci. 21, 291–299.

Bettelheim, F.A. (1985) Physical basis of lens transparency. In: H. Maisel (Ed.), The Ocular Lens: Structure, Function and Patholoigy. Marcel Dekker, New York, pp. 265–300.

Bloemendal, H., Zweers, A., Vermorken, F., Dunia, I. and Benedetti, E.L. (1972) The plasma membranes of the eye lens fibres. Biochemical and structural characterization. Cell Diff. 1, 91–106.

Campos-Ortega, J.A. (1988) Cellular interactions during early neurogenesis of *Drosophila melanogaster*. Trends. Neurosci. 11, 400–405.

Campos-Ortega, J.A. (1991) Genetic and molecular bases of neurogenesis in *Drosophila melanogaster*. Annu. Rev. Neurosci. 14, 399–420.

Conkling, M.A., Cheng, C.L., Yamamoto, Y.T. and Goodman, H.M. (1990) Isolation of transcriptionally regulated root-specific genes from tobacco. Plant Physiol. 93, 1203–1211.

Costello, M.J., McIntosh, T.J. and Robertson, J.D. (1989) Distribution of gap junctions and square array junctions in the mammalian lens. Invest. Ophthalmol. Vis. Sci. 30, 975–989.

Denker, B.M., Smith, B.L., Kuhajda, F.P. and Agre, P. (1988) Identification, purification, and partial characterization of a novel M_r 28,000 integral membrane protein from erythrocytes and renal tubules. J. Biol. Chem. 263, 15634–15642.

Dickson, D.H. and Crock, G.W. (1972) Interlocking patterns on primate lens fibers. Invest. Ophthalmol. 11, 809–815.

Doe, C.Q. and Goodman, C.S. (1985) Early events in insect neurogenesis. Dev. Biol. 111, 193–205.

Duncan, G. (1969) The site of the ion restricting membranes in the toad lens. Exp. Eye Res. 8, 406–412.

Duncan, G. and Croghan, P.C. (1969) Mechanisms for the regulation of cell volume with particular reference to the lens. Exp. Eye Res. 8, 421–430.

Ebihara, L. and Steiner, E. (1992) Connexin46 forms gap-junctional hemichannels in *Xenopus* oocytes. Biophys. J. 61, A408 (Abstract).

Ehring, G.R., Zampighi, G.A. and Hall, J.E. (1988) Properties of MIP 26 channels reconstituted into planar lipid bilayers. In: R.G. Johnson and E.L. Hertzberg (Eds.), Gap Junctions. Alan R. Liss, New York, pp. 335–346.

Ehring,, G.R., Zampighi, G.A., Horwitz, J., Bok, D. and Hall, J.E. (1990) Properties of channels reconstituted from the major intrinsic protein of lens fiber membranes. J. Gen. Physiol. 96, 631–664.

Ehring, G.R., Lagos, N., Zampighi, G.A. and Hall, J.E. (1992) Phosphorylation modulates the voltage dependence of channels reconstituted from the major intrinsic protein of lens fiber membranes. J. Membr. Biol. 126, 75–88.

Ehring, G.R., Lang, P. and Hall, J.E. (1991) Proteolytic digestion with trypsin reduces the voltage-dependent closing of MIP channels incorporated into planar lipid bilayer membranes (BLM). Biophys. J. 59, 393a (Abstract).

Fitzgerald, P.G., Bok, D. and Horwitz, J. (1985) The distribution of the main intrinsic membrane polypeptide in ocular lens. Curr. Eye Res. 4, 1203–1217.

Fortin, M.G., Zelechowska, M. and Verma, D.P. (1985) Specific targeting of membrane nodulins to the bacteriod-enclosing compartment in soybean nodules. EMBO J. 4, 3041–3046.

Girsch, S.J. and Peracchia, C. (1985) Lens cell-to-cell channel protein: I. Self-assembly into liposomes and permeability regulation by calmodulin. J. Membr. Biol. 83, 217–225.

Gooden, M.M., Rintoul, D.A., Takehana, M. and Takemoto, L.J. (1985a) Major intrinsic polypeptide (MIP26K) from lens membrane: reconstitution into vesicles and inhibition of channel forming activity by peptide antiserum. Biochem. Biophys. Res. Commun. 128, 993–999.

Gooden, M.M., Takemoto, L.J. and Rintoul, D.A. (1985b) Reconstitution of MIP26 from single human lenses into artificial membranes. I. Differences in pH sensitivity of cataractous vs. normal human lens fiber cell proteins. Curr. Eye Res. 4, 1107–1115.

Goodenough, D.A., Dick, J. S.B. and Lyons, J. E. (1980) Lens metabolic cooperation: a study of mouse lens permeability visualized with freeze-substitution autoradiography and electron microscopy. J. Cell Biol. 86, 576–589.

Gruijters, W.T., Kistler, J., Bullivant, S. and Goodenough, D.A. (1987) Immunolocalization of MP70 in lens fiber 16–17-nm intercellular junctions. J. Cell Biol. 565–572.

Guerrero, F.D., Jones, J.T. and Mullet, J.E. (1990) Turgor-responsive gene transcription and RNA levels increase rapidly when pea shoots are wilted. Sequence and expression of three inducible genes. Plant Mol. Biol. 15, 11–26.

Heller, K.B., Lin, E.C. and Wilson, T.H. (1980) Substrate specificity and transport properties of the glycerol facilitator of *Escherichia coli*. J. Bacteriol. 144, 274–278.

Hofte, H., Hubbard, L., Reizer, J., Ludevid, D., Herman, E.M. and Chrispeels, M.J. (1992) Vegetative and seed-specific forms of tonoplast intrinsic protein in the vacuolar membrane of *Arabidopsis thaliana*. Plant Physiol., in press.

Johnson, K.D., Hofte, H. and Chrispeels, M.J. (1990) An intrinsic tonoplast protein of protein storage vacuoles in seeds is structurally related to a bacterial solute transporter (GlpF). The Plant Cell 2, 525–532.

Johnson,, R.G., Klukas, K.A., Tze-Hong, L. and Spray, D.C. (1988) Antibodies to MP28 are localized to lens junctions, alter intercellular permeability and demonstrate increased expression during development. In: E.L. Hertzberg and R.G. Johnson (Eds.), Gap Junctions. Alan R. Liss, New York, pp. 81–98.

Kistler, J., Kirkland, B. and Bullivant, S. (1985) Identification of a 70,000-D protein in lens membrane junctional domains. J. Cell Biol. 101, 28–35.

Kistler, J., Christie, D. and Bullivant, S. (1988) Homologies between gap junction proteins in lens, heart and liver. Nature 721–773.

Kistler, J. and Bullivant, S. (1987) Protein processing in lens intercellular junctions: cleavage of MP70 to MP38. Invest. Ophthalmol. Vis. Sci. 1687–1192.

Klughammer, B., Benz, R., Betz, M., Thume, M. and Dietz, K.-J. (1992) Reconstitution of vacuolar ion channels into planar lipid bilayers. Biochim. Biophys, Acta 1104, 308–316.

Kuszak, J.R., Maisel, H. and Harding, C.V. (1978) Gap junctions of chick lens fiber cells. Exp. Eye Res. 27, 495–498.

Kuwabara, T. (1975) The maturation of the lens cell: a morphologic study. Exp. Eye Res. 20, 427–433.

Lampe, P.D. and Johnson, R.G. (1991) Amino acid sequence of *in vivo* phosphorylation sites in the main intrinsic protein (MIP) of lens membrane. Eur. J. Biochem. 194, 541–547.

Lo, W.K. and Harding, C.V. (1984) Square arrays and their role in ridge formation in human lens fibers. J. Ultrastruct. Res. 228–245.

Maisel, H., Harding, C.V., Alcala, J., Kuszak, J.R. and Bradley, R. (1981) The morphology of the lens. In: H. Bloemendal (Ed.), Molecular and Cellular Biology of the Eye Lens. Wiley, New York.

Mathias, R.T., Rae, J.L. and Eisenberg, R.S. (1979) Electrical properties of structural components of the crystalline lens. Biophys. J. 25, 181–201.

Mathias, R.T., Rae, J.L. and Eisenberg, R.S. (1981) The lens as a nonuniform spherical syncytium. Biophys. J. 34, 61–83.

Mathias, R.T. and Rae, J.L. (1985) Transport properties of the lens. Am. J. Physiol. 249, C181–C190.

Matile, P. (1978) Biochemistry and function of vacuoles. Annu. Rev. Plant Physiol. 29, 193–213.

Miller, A. (1992) Uncovering physiological roles for ion channels and gap junction channels in the lens of the eye. Dissertation,

University of Califorina, Irvine.
Miller, T.M. and Goodenough, D.A. (1986) Evidence for two physiologically distinct gap junctions expressed by the chick lens epithelial cell. J. Cell Biol. 102, 194–199.
Nikaido, H. and Rosenberg, E.Y. (1985) Functional reconstitution of lens gap junction proteins into proteoliposomes. J. Membr. Biol. 85, 87–92.
Pao, G.M., Wu, L.F., Johnson, K.D., Hofte, H., Chrispeels, M.J., Sweet, G., Sandal, N.N. and Saier, Jr., M.H. (1991) Evolution of the MIP family of integral membrane transport proteins. Mol. Microbiol. 5, 33–37.
Paul, D.L., Ebihara, L., Takemoto, L.J., Swenson, K.I. and Goodenough, D.A. (1991) Connexin46, A novel lens gap junction protein, induces voltage-gated currents in non-junctional plasma membrane of *Xenopus* oocytes. J. Cell Biol. 115, 1077–1089.
Paul, D.L. and Goodenough, D.A. (1983) Preparation, characterization, and localization of antisera against bovine MP26, an integral protein from lens fiber plasma membrane. J. Cell Biol. 96, 625–632.
Peracchia, C. and Girsch, S.J. (1985) Is the C-terminal arm of lens gap junction channel protein the channel gate? Biochem. Biophys. Res. Commun. 133, 688–695.
Peracchia, C. and Girsch, S.J. (1989) Calmodulin site at the C-terminus of the putative lens gap junction protein MIP26. Lens. Eye Toxic. Res. 6, 613–621.
Peracchia, C. and Peracchia, L.L. (1980a) Gap junction dynamics: reversible effects of hydrogen ions. J. Cell Biol. 87, 719–727.
Peracchia, C. and Peracchia, L.L. (1980b) Gap junction dynamics: reversible effects of divalent cations. J. Cell Biol. 87, 708–718.
Preston, G.M., Carroll, T.P., Guggino, W.B. and Agre, P. (1992) Appearance of water channels in *Xenopus* oocytes expressing red cell CHIP28 protein. Science 256, 385–387.
Rae, J.L., Mathias, R.T. and Eisenberg, R.S. (1982) Physiological role of the membranes and extracellular space with the ocular lens. Exp. Eye Res. 35, 471–489.
Rao, Y., Yan, L.Y. and Jan, Y.N. (1990) Similarity of the product of *Drosophila* neurogenic gene *big brain* to transmembrane channel protein. Nature 345, 163–167.
Saier, M.H. (1985) Mechanisms and Regulation of Carbohydrate Transport in Bacteria. Academic Press, New York.
Sandal, N.N. and Marcker, K.A. (1988) Soybean nodulin 26 is homologous to the major intrinsic protein of the bovine lens fiber membrane. Nucleic Acids Res. 16, 9347.
Sas, D.F., Sas, M.J., Johnson, K.R., Menko, A.S. and Johnson, R.G. (1985a) Junctions between lens fiber cells are labeled with a monoclonal antibody shown to be specific for MP26. J. Cell Biol. 100, 216–225.
Sas, D.F., Sas, M.J. and Johnson, R.G. (1985b) Lens junctional protein: analyzing MP26 with monoclonal antibodies. Curr. Eye Res. 4, 1171–1182.
Scaglione, B.A. and Rintoul, D.A. (1989) A fluorescence-quenching assay for measuring permeability of reconstituted lens MIP26. Invest. Ophthalmol. Vis. Sci. 30, 961–966.
Scheutze, S.M. and Goodenough, D.A. (1982) Dye transfer between cells of the embryonic chick lens becomes less sensitive to CO_2 treatment with development. J. Cell Biol. 92, 694–705.
Shanmugam, K.T., O'Gara, F., Andersen, K. and Valentine, R.C. (1978) Biological nitrogen fixation. Annu. Rev. Plant Physiol. 29, 263–276.
Shen, L., Shrager, P., Girsch, S.J., Donaldson, P.J. and Peracchia, C. (1991) Channel reconstitution in liposomes and planar bilayers with HPLC-purified MIP26 of bovine lens. J. Membr. Biol. 124, 21–32.
Simon, S.A., Zampighi, G., McIntosh, T.J., Costello, M.J., Tingbeall, H.P. and Robertson, J.D. (1982) The structure of junctions between lens fiber cells. Biosci. Rep. 2, 333–341.
Smith, B.L. and Agre, P. (1991) Erythrocyte M 28,000 transmembrane protein exists as a multisubunit oligomer similar to channel proteins. J. Biol. Chem. 266, 6407–6415.
Swenson, K.I., Jordan, J.R., Beyer, E.C. and Paul, D.L. (1989) Formation of gap junctions by expression of connexins in *Xenopus* oocyte pairs. Cell 57, 145–155.
Technau, G.M. and Campos-Ortega, J.A. (1987) Cell autonomy of expression of neurogenic genes of *Drosophila melanogaster*. Proc. Natl. Acad. Sci. USA 84, 4500–4504.
Van Aelst, L., Hohmann, S., Zimmermann, F.K., Jans, A.W.H. and Thevelein, J.M. (1991) A yeast homologue of the bovine lens fibre MIP gene family complements the growth defect of a *Saccharomyces cerevisiae* mutant on fermentable sugars but not its glucose-induced *ras*-mediated cAMP signaling. EMBO J. 10, 2095–2104.
Yamamoto, Y.T., Cheng, C.L. and Conkling, M.A. (1990) Root-specific genes from tobacco and *Arabidopsis* homologous to an evolutionarily conserved gene family of membrane channel proteins. Nucleic Acids Res. 18, 7449.
Wistow, G., Pisano, M.M. and Chepelinsky, A.B. (1991) Tandem sequence repeats in transmembrane channel proteins. Trends Biochem. Sci. 16, 170–171.
Zampighi, G., Simon, S.A., Robertson, J.D., McIntosh, T.J. and Costello, M.J. (1982) On the structural organization of isolated bovine lens fiber junctions. J. Cell Biol. 93, 175–189.
Zampighi, G.A., Hall, J.E. and Kreman, M. (1985) Purified lens junctional protein forms channels in planar lipid films. Proc. Natl. Acad. Sci. USA 82, 8468–8472.
Zampighi, G.A. (1987) Gap junction structure. In: W.C. DeMello (ed.), Cell-to-Cell Communication. Plenum, New York, pp. 1–28.
Zampighi, G.A., Hall, J.E., Ehring, G.R. and Simon, S.A. (1989) The structural organization and protein composition of lens fiber junctions. J. Cell Biol. 108, 2255–2276.
Zampighi, G.A., Simon, S.A. and Hall, J.E. (1992) The specialized junctions of the lens. Int. Rev. Cytol. 136, 185–225.

J.E. Hall, G.A. Zampighi and R.M. Davis (Eds.)
Progress in Cell Research, Vol. 3

CHAPTER 23

Gap-junction channel reconstitution in artificial bilayers and evidence for calmodulin binding sites in MIP26 and connexins from rat heart, liver and *Xenopus* embryo

CAMILLO PERACCHIA and LILI SHEN

Department of Physiology, University of Rochester, School of Medicine and Dentistry, 601 Elmwood Ave., Rochester, NY 14642, USA

Introduction

This chapter reviews recent data on permeability and gating of in vitro-reconstituted MIP26 and Cx43 channels and presents evidence from spectrofluorimetry and circular dichroism spectroscopy (CD) for the presence of calmodulin (CaM) receptor sites in connexins of rat heart (Cx43), liver (Cx32) and early *Xenopus* embryo (Cx38), and in the lens-fiber protein MIP26. We report permeability and electrical properties of MIP26 channels studied in liposomes, by a spectrophotometric osmotic assay, and in planar bilayers following liposome fusion. We also used a computer algorithm based on helical hydrophobic moment, average hydrophobicity and charge to locate CaM-binding sites in MIP26, Cx43, Cx32 and Cx38 . Four 19-mer peptides: Pep C (MIP26), Cx43-Pep, Cx32-Pep and Cx38-Pep were synthesized and tested for CaM binding and conformational change with Ca^{2+}-CaM. The results obtained suggest a direct interaction between CaM and the peptides and support previous evidence for an involvement of CaM-like proteins in gap-junction function.

Direct exchange of small molecules between cells is mediated by channels that couple electrically and metabolically adjacent cells at gap junctions and that are regulated primarily by $[Ca^{2+}]_i$ and $[H^+]_i$. During the past decade, progress in understanding mechanisms of regulation and modulation of channel permeability has lagged behind our rapidly improving knowledge of channel molecular structure. Uncoupling agents have been identified, but still unclear is whether they act directly on the channel gates or via intermediates. Ca^{2+} and H^+ appear to affect all cells, but their effectiveness varies and there is evidence that H^+ may indirectly uncouple by increasing $[Ca^{2+}]$i (Peracchia, 1990a, b). In turn, Ca^{2+} may work via a CaM-like intermediate. The CaM hypothesis (Peracchia, 1988) is based on the capacity of CaM inhibitors to prevent cell uncoupling (Peracchia et al., 1983; Peracchia, 1984, 1987; Wojtczak, 1985; Peracchia, 1988; Tuganowski et al., 1989; Gandolfi et al., 1990), and of Ca^{2+}-activated CaM to both close junctional channels incorporated into liposomes and interact directly with gap-junction proteins (Peracchia, 1988).

In vitro reconstitution of gap-junction channels

A few years ago, we adapted a simple in vitro approach for studying channel permeability and regulation to gap junctions (Peracchia and Girsch, 1985). It involves incorporation of gap-junction proteins (connexins) into liposomes and monitoring channel function by a spectrophotometric osmotic assay. Recently, we have complemented the liposome approach with channel reconstitution in planar bilayers (Shen et al., 1991).

Reconstitution of HPLC-purified MIP26 channels into liposomes and planar bilayers

The eye lens possesses an extensive system of cell communication that mediates bidirectional transfer of ions and metabolites between deep and superficial fiber cells. Fiber-to-fiber communication is mediated by junctions similar to gap junctions whose protein composition is still unclear. The major intrinsic protein of lens fibers is a 28.2-kDa component that contains six transmembrane chains and displays both carboxyl and amino termini at the cytoplasmic side (Gorin et al., 1984).

MIP26 fractions of different degrees of purity have been successfully incorporated into artificial lipid systems and functional channels have been reconstituted (Girsch and Peracchia, 1985a; Goodin et al., 1985; Nikaido and Rosenberg, 1985; Zampighi et al., 1985; Ehring et al., 1990; Lea and Duncan, 1991; Shen et al., 1991). In

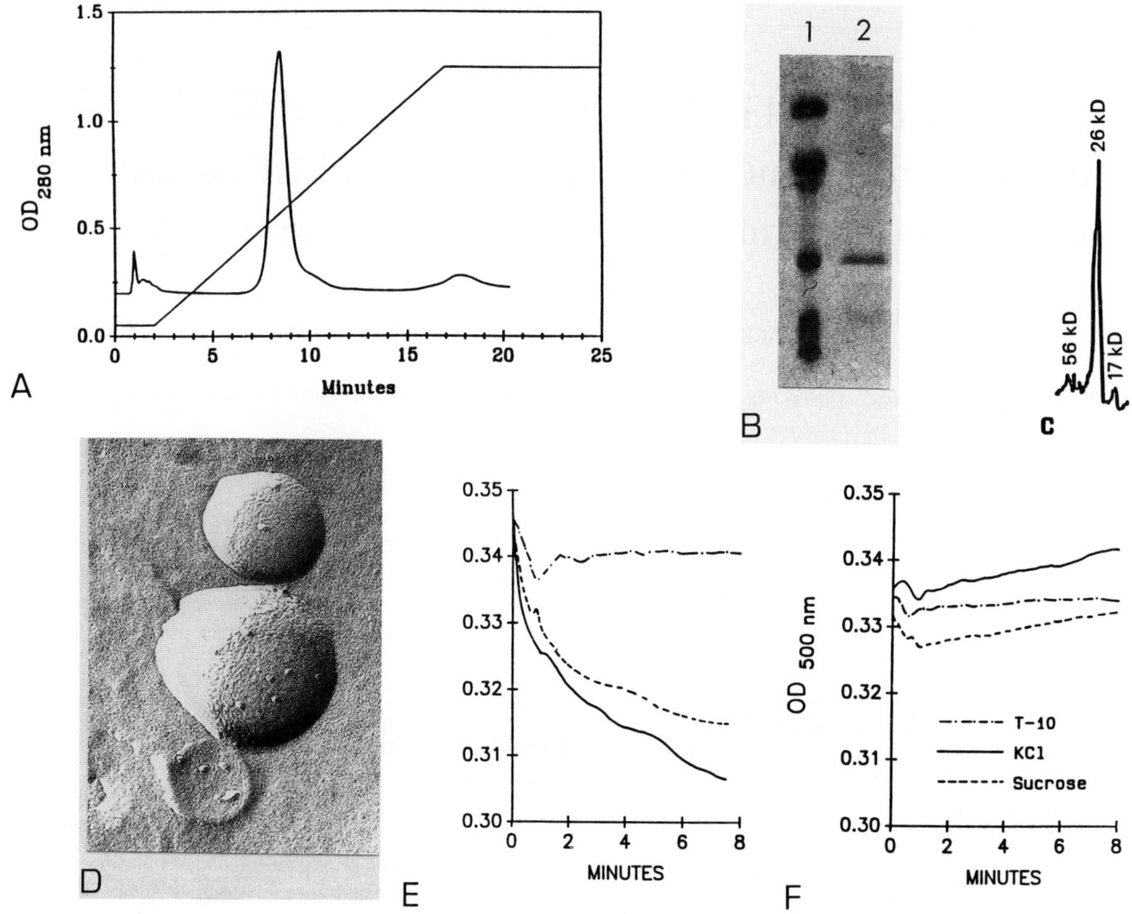

Figure 1. (A) HPLC elution profile of lens fiber membrane components detected by UV absorption (α = 280 nm). The major component (MIP26) eluted as a narrow symmetrical peak. To avoid collecting a contaminant protein closely associated with the right shallow shoulder of the main peak, the main fraction was collected at an OD greater than 0.4. (B) SDS-PAGE of HPLC-purified MIP26 (lane 2); markers (M_r = 66, 45, 34, 24, 18 and 14) are on line 1. (C) Densitometric scan of SDS-PAGE gel stained with Coomassie blue; the HPLC-purified MIP26 monomer was 93 % pure. (D) Freeze fracture replica of MIP26-liposomes. Liposome diameter ranged from 0.1 to 0.8 μm and displayed IMPs and pits on both the convex and concave surfaces. Mag. = 95,300×. (E,F) Traces of optical-density change (light scattering) in suspensions of liposomes reconstituted with (E) and without (F) HPLC-purified MIP26. The liposomes were loaded with a channel impermeant (Dextran T-10, MW = 10,000) and suspended into solutions of channel permeants (KCl or sucrose) slightly hypertonic to T-10. MIP26-liposomes swelled both in KCl and sucrose solutions, as these permeants (and water) diffused through channels into the liposomes, while T-10 remained inside. This resulted in a rapid decrease in optical density (E). In contrast, protein-free liposomes did not swell in any of the permeants (F), as they had no channels.

liposomes the channels passed molecules as large as 1.5 kDa (Girsch and Peracchia, 1985a) and closed with Ca^{2+} only in the presence of CaM (Girsch and Peracchia, 1985a). In planar bilayers (Zampighi et al., 1985; Ehring et al., 1990; Shen et al., 1991) and in vesicles formed at the tip of patch pipettes (Brewer, 1991), voltage-dependent channels of various conductances have been reported.

Recently, we have monitored permeability and electrical properties of channels reconstituted into liposomes and planar bilayers with MIP26 purified from bovine lenses by high-performance liquid chromatography (HPLC) (Fig. 1A–C) (Shen et al., 1991). In freeze-fracture replicas, liposomes reconstituted with MIP26 ranged in diameter from 0.2 to 0.8 μm and contained IMPs on either surface (Fig. 1D). For determining channel competency, liposomes reconstituted with MIP26 (MIP26-liposomes) were loaded with channel-impermeant Dextran T-10. Upon exposure to solutions of gap-junction permeants (sucrose or KCl) slightly hypertonic to T-10, MIP26-liposomes shrank briefly and then swelled, because the initial osmotic gradient caused rapid water efflux, and subsequently the slower influx of channel permeants increased the internal tonicity, driving water back into the liposomes. Swelling was monitored by recording the slow decrease in optical density of the suspension (Fig. 1E) as the enlarging liposomes became more transparent. As a control, MIP26-liposomes were suspended in Dextran T-10 loading solutions (Fig. 1E) and protein-free liposomes were exposed to any of the permeants (Fig. 1F). The data indicate that MIP26 channels are permeable to molecules larger than monovalent ions and confirm previous results with less pure MIP26 (Goodin et al., 1985; Girsch and Peracchia, 1985a; Nikaido and Rosenberg, 1985).

In parallel experiments, MIP26 channels were reconstituted in planar bilayers by adding MIP26-liposomes to both sides of the bilayer chamber. Channel opening and closing events (Fig. 2A) were only seen at transbilayer voltages greater than ±20 mV. At voltages ranging from ±20 to ±60 mV, the channels were preferentially open, and a channel open probability of 50% was at ±60–80 mV. This indicates that the channels are voltage dependent and remain open at voltages lower than ±20 mV. Amplitude histograms representing all points showed two Gaussian distributions that peaked at 11 and 1 pA (Fig. 2B), representing channel open and closed states, respectively (conductance = ~120 pS). The frequency distributions of channel closed and open times were well fitted by single exponential functions with time constants of 1.9 and 0.13 s, respectively (Fig. 2C and D).

To eliminate the possibility that the channels were contaminant K^+-channels, specific blockers (Cs^+ and TEA) and Na^+ substitution of K^+ were tested. Cs^+ did not decrease channel conductance but changed channel kinetics, reducing mean open time from 0.20 to 0.038 s, and mean closed time from 1.5 to 0.38 s. This increased channel "flickering" but did not change open-channel equilibrium probability (~0.1), indicating that Cs^+ only speeds the transition between the two states, possibly via a channel blocking/unblocking mechanism. In contrast, TEA affected neither channel conductance nor kinetics. Channel events were also observed in Na^+ solutions, but channel activity was significantly reduced such that bursts were more rare and short lived.

To our knowledge this is the first example of successful reconstitution of a functional membrane channel after reverse-phase HPLC purification and demonstrates that

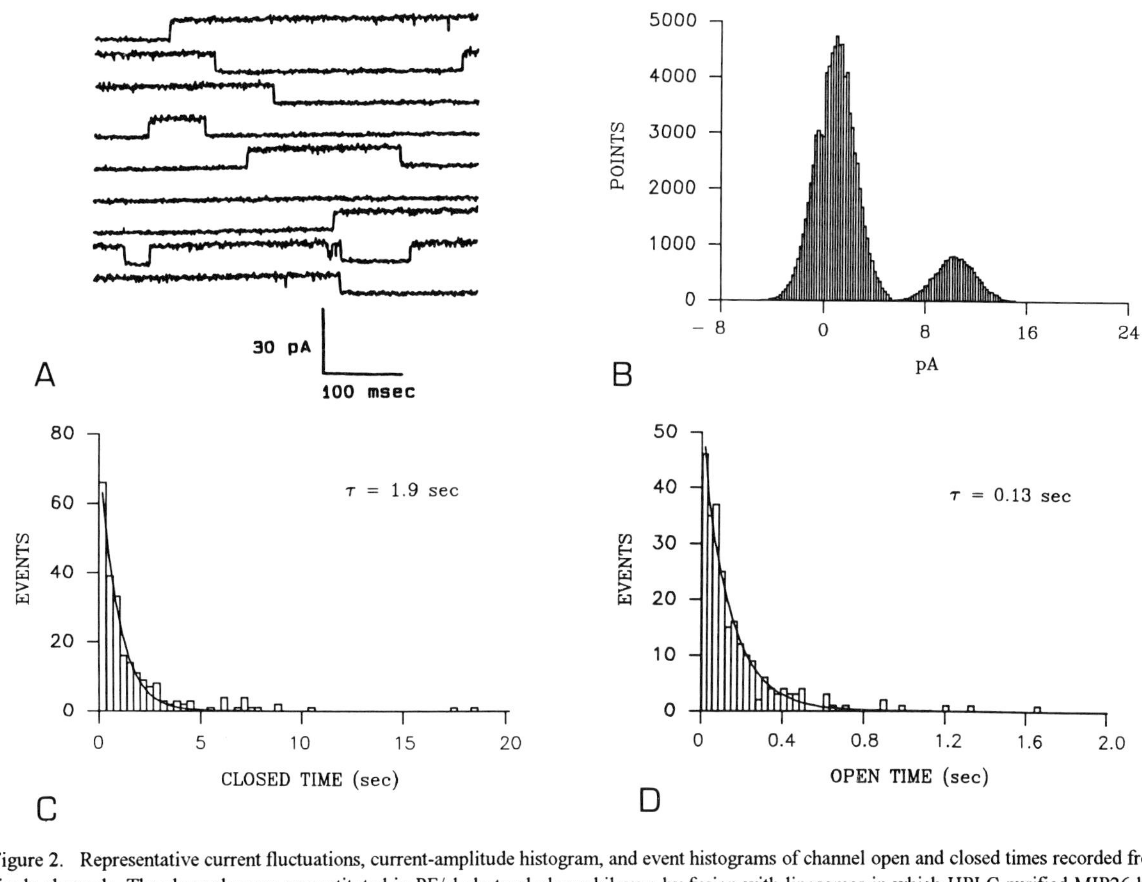

Figure 2. Representative current fluctuations, current-amplitude histogram, and event histograms of channel open and closed times recorded from single channels. The channels were reconstituted in PE/cholesterol planar bilayers by fusion with liposomes in which HPLC-purified MIP26 had been incorporated. The bilayers were subjected to an 80 mV holding potential in symmetrical solutions (100 mM KCl, 1 mM Ca^{2+}, 20 mM HEPES–Tris, pH 6.0). (A) Consecutive current traces filtered at 1.1 kHz and sampled at an acquisition rate of 4.4 kHz (upward deflection indicates channel opening). (B) Single-channel current amplitude histogram (fc = 300 Hz, sampling = 1.2 kHz). The two Gaussian distribution curves correspond to channel closed (1 pA peak) and open (11 pA peak, corresponding to ~120 pS) states. Judging from the relative areas under the curve, the channel was preferentially closed at 80 mV. (C,D) Event histograms of channel closed (C) and open (D) times. Histogram analysis was set for a threshold of 50% (i.e. only events with amplitudes greater than 50% of mean channel amplitude were analyzed) and an acquisition rate of 1.2 kHz. The frequency histograms are fitted by single-exponential functions with time constants (τ) of 0.13 and 1.9 s for open and closed events, respectively.

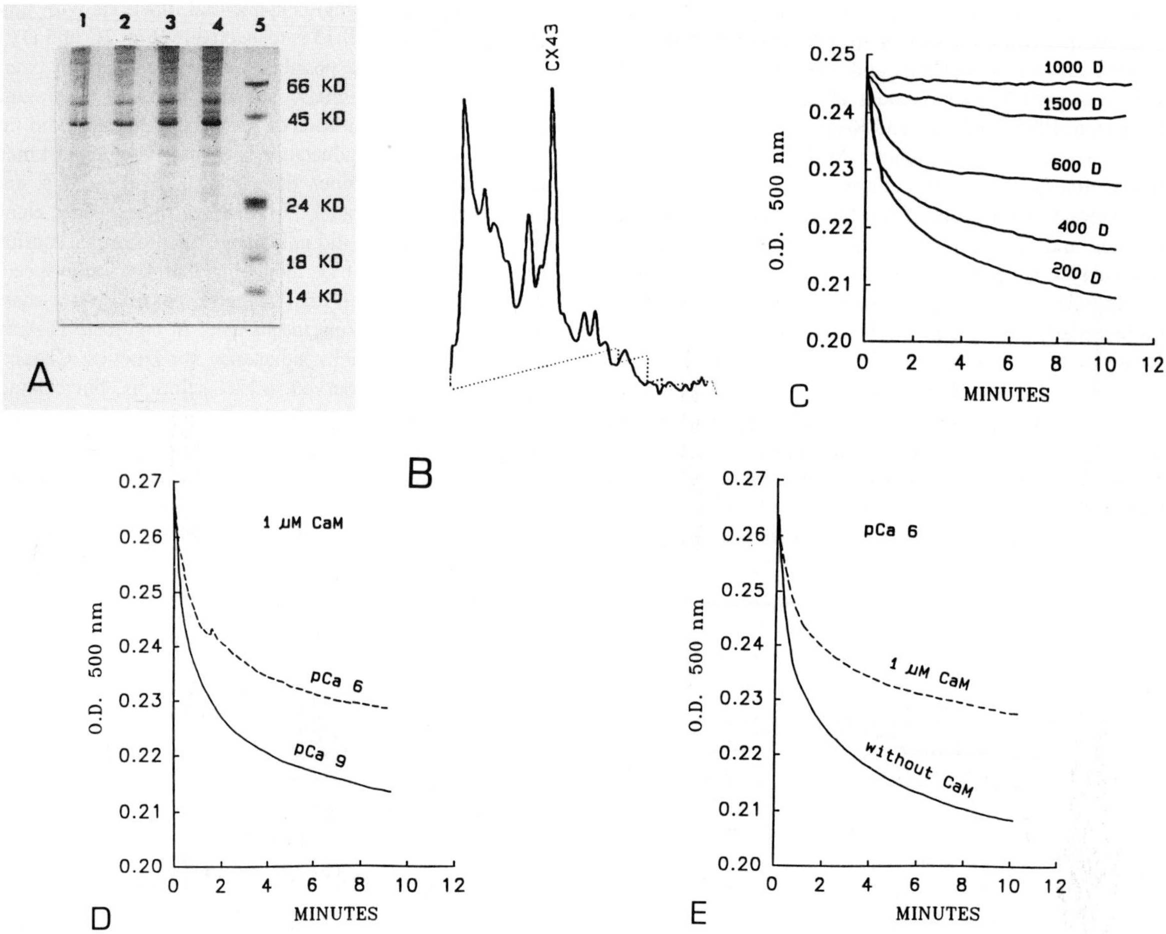

Figure 3. (A) SDS-PAGE of silver-stained proteins extracted from rat heart gap junctions (lines 1 through 4) and markers (line 5). (B) Densitometric scan of line 4 of the gel shown in (A). The trace shows a major peak of 43 kDa (Cx43), in addition to several small peaks and a large peak (first from left), probably a polymer. (C) Traces of changes in optical density (light scattering) in suspensions of liposomes reconstituted with heart gap-junction protein. The liposomes were loaded with a channel impermeant (Dextran T-10, MW = 10,000) and suspended into solutions of polyethyleneglycols (PEG) of different sizes (MW = 200, 400, 600, 1000, 1500), slightly hypertonic to T-10. Note that the liposomes swell in PEG 200–600 Da but do not in PEG 1000 and 1500 Da. This indicates that the reconstituted channels are permeable to molecules smaller than 1000 Da. (D) Reconstituted liposomes loaded with T-10 and CaM (1 µM) swell rapidly in sucrose solutions (pH 7.4) buffered to pCa 9.0, but much more slowly in the same solutions buffered to pCa 6.0 (E) The swelling rate of liposomes reconstituted with heart gap-junction protein and suspended in sucrose (pCa 6) is much reduced by the addition of 1 µM CaM to the sucrose solution. Addition of CaM to both sides of the liposome membrane further reduces swelling in Ca^{2+} solutions, supporting previous evidence for symmetrical orientation of the channels (Girsch and Peracchia, 1985a).

functional channels can re-form after channel protein purification under harsh conditions. This is also the first study that provides data on MIP26-channel kinetics derived from single-channel records. The single-channel conductance (120 pS) is smaller than the conductance steps (160 and 380 pS) reported by Ehring et al. (1990), possibly because of the presence of cholesterol in our bilayer. Recently, an increase in cholesterol has been shown to decrease the conductance of acetylcholine receptor channels (Lasalde and Zuazaga, 1991). Since the plasma membrane of bovine lens fibers is rich in cholesterol (cholesterol/phospholipid = 0.6–1.2; Zampighi et al., 1985; Malewicz et al., 1990), the conductance measured in our bilayers (cholesterol/phospholipid = 0.96) may be closer to that of in vivo MIP26 channels. Cholesterol may modulate channel function by affecting the microviscosity of the lipid bilayer (Lasalde and Zuazaga, 1991). In terms of permeability and electrical properties, the HPLC-purified MIP26 channels are similar to gap-junction channels, but whether they function as communicating channels is still unclear.

Are MIP26 channels involved in cell communication?

Although permeability, conductance and gating properties of MIP26 channels are similar to those of gap-junction channels, the involvement of these channels in cell-to-cell communication is still uncertain. MIP26 differs biochemically (Gorin et al., 1984) from connexins (Beyer et al., 1990) and conflicting immunological data on MIP26 localization in lens junctions have been reported (Bok et al., 1982; Fitzgerald et al., 1983; Paul and Goodenough, 1983; Sas et al., 1985; Vallon et al., 1985).

Recently, another protein (MP70) has been localized in thick lens junctions (Kistler et al., 1985; Gruijters et al., 1985; Zampighi et al., 1989). MP70 is similar to connexins at the N-terminus and thus is likely to be a lens fiber connexin (Kistler et al., 1988). However, MP70 is expressed only as deeply as 100 μm from the lens surface and, in spite of evidence for degradation products of MP70 (MP64 and MP38) in deeper fibers (Kistler and Bullivant, 1987), the biochemical nature of gap-junction channels of these fibers, the vast majority of lens fibers, is still unknown.

There is evidence that intracellularly injected monoclonal anti-MIP26 antibodies block dye diffusion between lentoid cells (Johnson et al., 1988), and for both CaM participation in the regulation of lens fiber communication (Girsch and Peracchia, 1985a; Gandolfi et al., 1990; Brewer and Dong, 1990) and CaM interaction with MIP26 (Girsch and Peracchia, 1985b, 1991; Louis et al., 1990). Taken together, these studies are consistent with the idea that MIP26 is involved in cell-to-cell communication. Our latest results confirm the absence of a direct effect of Ca^{2+} (up to 1 mM) on the channel gates.

Ehring et al. (1990) have proposed a new role for MIP26 channels in maintaining fluid balance and minimizing extracellular space. Based on evidence for MIP26 antibody binding to convex surfaces only of "wavy junctions" (Zampighi et al., 1989), this model assumes that MIP26 channels are hemichannels rather than junctional channels. However, these "wavy" membrane appositions are not typical lens fiber junctions; they are found almost exclusively among degenerating fiber cells of the lens nucleus and could very well result from asymmetrical apposition of single junctional membranes following junction membrane separation. On the other hand, there is strong evidence that MIP26 forms symmetrical junctions. Cortical lens fiber membranes with orthogonal and rhombic particle arrays, typical of MIP26 arrays (Dunia et al., 1987), have been shown to form junctions with precisely matched particle and pit arrays both in isolated fractions (Peracchia, 1978; Peracchia and Peracchia, 1980a, b) and in intact fiber cells (Bernardini and Peracchia, 1981).

Recently, a new connexin (Cx46, Chapter 11 and Chapter 24) and another non-connexin protein (MP20, Chapter 25) have been identified in lens fiber membranes. Interestingly, MP20 was also found to co-localize with both MIP26 and MP70 in thick lens junctions. This suggests that lens junctions may be composed of different communicating channels, whose expression could vary in different lens regions for special functional needs. This hypothesis is supported by freeze fracture data showing heterogeneity of particle size in cortical fiber junctions, and by the apparent inability of lens fiber junctions to form crystalline arrays in situ, as crystallization processes usually require molecular homogeneity.

Reconstitution of heart connexin into liposomes

Recently, we have studied permeability and gating of reconstituted heart gap-junction channels. The gap-junction protein (Fig. 3A and B), extracted with octylglucoside from pellets of isolated junctions (Manjunath and Page, 1986), was mixed with lecithin in octylglucoside. Cx43 was reconstituted into liposomes by dialysis, sonication and dehydration/rehydration. Channel (particle) reconstitution in liposomes was evaluated by freeze fracture, and channel permioselectivity was assayed with the spectrophotometric osmotic method using KCl, sucrose, or polyethylene glycols (PEG: 200, 400, 600, 1000 or 1500 Da) as probes.

Reconstituted liposomes swelled rapidly in KCl, sucrose, PEG-200 and PEG-400, but very slowly in PEG-600 and not at all in PEG-1000 and PEG-1500 (Fig. 3C), indicating that the channels are permeable to molecules smaller than 1 kDa. In contrast, control (protein-free) liposomes did not swell in any of the probe-containing solutions (data not shown). In preliminary experiments, the reconstituted channels were insensitive to Ca^{2+} alone but closed with Ca^{2+} in the presence of CaM (Fig. 3D and E). The participation of CaM in gating reconstituted heart channels is evidence for CaM binding sites in Cx43 (see below, Fig. 4) and confirms previous reports on the effectiveness of CaM-inhibitors in preventing uncoupling of cardiac cells (Wojtczak, 1985; Tuganowski et al., 1989).

CaM may participate directly in the regulation of other channels as well. Recently, evidence for a direct involvement of Ca^{2+}-CaM in channel gating has been obtained by patch-clamp experiments on Ca^{2+}-dependent Na^+ channels of *Paramecium* (Saimi and Ling, 1990). However, in this case, addition of CaM caused channel opening. The Ca^{2+}-CaM effect appeared to result from direct CaM binding to the channels as it did not require ATP.

CaM receptor sites in MIP26 and in connexins (Cx43, Cx32 and Cx38)

Although there is little doubt about Ca^{2+} participation in gap-junction regulation (Loewenstein, 1966; Rose and Loewenstein, 1976), it is still unclear whether Ca^{2+} acts directly on the channel gates. Channels reconstituted in vitro from either lens, liver or heart (Fig. 3E) gap-junction proteins are insensitive to Ca^{2+} alone (Girsch and Peracchia, 1985a; Zampighi et al., 1985; Young et al., 1987; Ehring et al., 1990; Harris, 1991), suggesting the existence of Ca^{2+} intermediates.

Johnston and Ramon (1981) proposed the existence of an uncoupling intermediate because neither Ca^{2+} nor H^+ uncoupled internally perfused crayfish axons. Based on

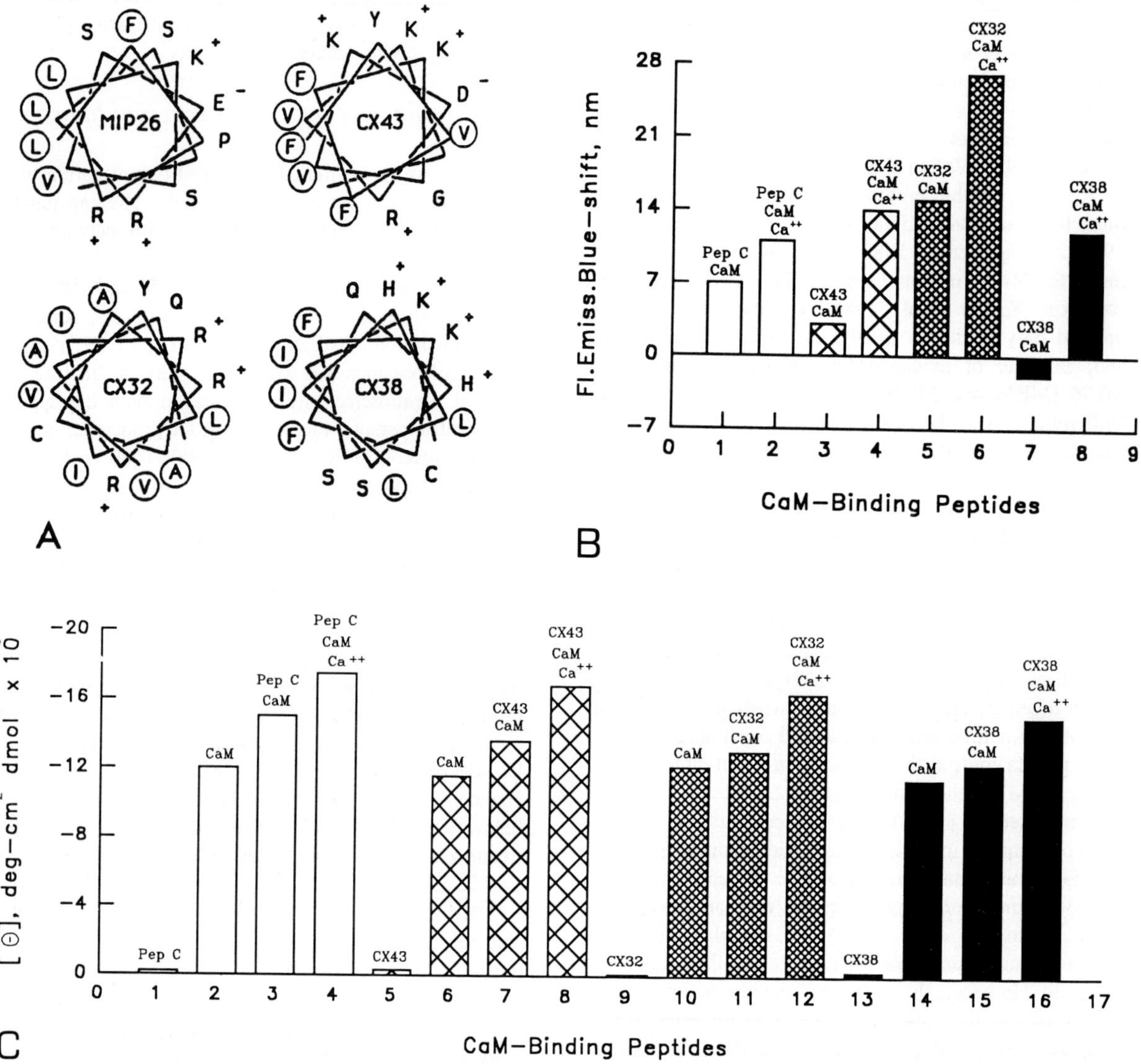

Figure 4. (A) Axial helical projections (helical wheels) of C-terminus CaM-binding sites of MIP26 (res. 223–235), Cx43 (229–241), Cx32 (209–224) and Cx38 (216–229). The hydrophobic residues are encircled and the charged residues are depicted with the appropriate sign. Note the amphiphilic, mostly basic, nature of the peptides. (B,C) Comparison among the four CaM:Peps complexes in terms of fluorescence emission blue-shift at 360 nm (B) and ellipticity (C), determined by CD spectroscopy. In the absence of Ca^{2+}, CaM induces a blue shift (in all but Cx38-Pep) and an increase in helicity, indicative of Ca^{2+}-independent interaction. In the presence of Ca^{2+}, greater blue shift and increase in helicity take place, reflecting further conformational changes in the CaM:Pep complexes.

data with CaM inhibitors, evidence for both CaM interaction with connexins and MIP26, and CaM participation in gating reconstituted channels, Peracchia (1988) suggested that a CaM-like protein is the intermediate.

Recently, potential CaM-binding sites were located in MIP26 and in the gap-junction proteins Cx43, Cx32 and Cx383 (Peracchia, 1988; Peracchia and Girsch, 1989; Girsch and Peracchia, 1991, 1992) (Fig. 4A) through a computer algorithm that predicts potential CaM binding sites by scanning protein sequences in terms of helical hydrophobic moment, average hydrophobicity and charge. Four 19-mer peptides: Pep C (MIP26), Cx43-Pep, Cx32-Pep and Cx38-Pep were synthesized and tested for CaM binding and conformational change with Ca^{2+}-CaM. The synthetic peptides matched the native peptides (Pep C, 223–242; Cx43-Pep, 227–245; Cx32-Pep, 208–226; Cx38-Pep, 215–233), except for the second residue (leucine or valine) that was substituted with tryptophan for monitoring fluorescence emission.

Spectrofluorometric titration shows that the four peptides bind CaM with 1:1 stoichiometry and K_d's of ~10 nM (Pep C), ~24 nM (Cx43-Pep), ~16 nM (Cx32-

Pep) and ~15 nM (Cx38-Pep). The peptides were not affected by Ca^{2+} or H^+ alone. In Ca^{2+}-free solutions containing CaM they underwent both a dramatic blue shift in tryptophan fluorescence emission (Fig. 4B, all except Cx38), indicative of strong hydrophobic interaction, and an increase in circular dichroism absorption in the α-helical region (Fig. 4C). Additional fluorescence blue shift (Fig. 4B) and α-helical content (Fig. 4C) occurred when Ca^{2+} was added to CaM-Peps solutions.

The data provide evidence for direct interaction between CaM and synthetic peptides matching C-terminal sequences of MIP26 and connexins of rat heart, liver, and *Xenopus* embryo. Our results extend earlier word showing the involvement of CaM-like proteins in gap-junction function (Peracchia, 1988).

Conclusions

Recent successes by several laboratories in reconstituting gap-junction proteins into liposomes and planar bilayers and in studying channel function under well controlled conditions testify to the potential of in vitro approaches. Much can be learned from reconstitution experiments about channel permioselectivity, channel gating and modulation of channel function by phosphorylation, different lipidic and ionic environments, and peripheral proteins, as well as about the effects of pH, ionic strength, temperature, membrane potential, etc. However, there are important drawbacks, e.g. possible changes in protein structure, resulting from proteolysis, dephosphorylation, oxidation, denaturation; altered connexon configurations, such as an abnormal number of connexins per connexon, unusual interactions among connexins, aggregation of connexins with opposite polarity; or the presence of hemichannels as opposed to whole channels (hemichannels may have unusual permeability and gating properties). In view of these and other possible artifacts, extreme caution should always be exercised in interpreting data from in vitro studies and the results should be continuously compared with data obtained in intact cells.

Acknowledgement

This work was supported by NIH grant GM20113.

References

Bernardini, G. and Peracchia, C. (1981) Gap-junction crystallization in lens fibers after an increase in cell calcium. Invest. Ophthalmol. Vis. Sci. 21, 291–299.

Beyer, E.C., Paul, D.L. and Goodenough, D.A. (1990) Connexin family of gap-junction proteins. J. Membr. Biol. 116, 187–194.

Bok, D., Dockstader, J. and Horwitz, J. (1982) Immunocytochemical localization of the lens main intrinsic polypeptide (MIP26) in communicating junctions. J. Cell Biol. 92, 213–220.

Brewer, G.J. (1991) Reconstitution of lens channels between two membranes. In: C. Peracchia (Ed.), Biophysics of Gap-Junction Channels, CRC Press, Boca Raton, FL, pp. 301–316.

Brewer, G.J., and Dong, R.G. (1990) Transjunctional lens channels reconstituted: regulation by Ca^{2+} and calmodulin. J. Cell Biol. 111, 65a.

Dunia, I., Manenti, S., Rousselet, A. and Benedetti, E.L. (1987) Electron microscopic observations of reconstituted proteoliposomes with the purified major intrinsic membrane protein of eye lens fibers. J. Cell Biol. 105, 1679–1689.

Ehring, G.R., Zampighi, G., Horwitz. J., Bok, D. and Hall, J.E. (1990) Properties of channels reconstituted from the major intrinsic protein of lens fiber membranes. J. Gen. Physiol. 96, 631–664.

Fitzgerald, P.G., Bok, D. and Horwitz, J. (1983) Immunocytochemical localization of the main intrinsic polypeptide (MIP) in ultrathin frozen sections of rat lens. J. Cell Biol. 97, 1491–1499.

Gandolfi, S.A., Duncan, G., Tomlinson, J. and Maraini, G. (1990) Mammalian lens inter-fiber resistance is modulated by calcium and calmodulin. Curr. Eye Res. 9, 533–541.

Girsch, S.J. and Peracchia, C. (1985a) Lens cell-to-cell channel protein: I. Self-assembly into liposomes and permeability regulation by calmodulin. J. Membr. Biol. 83, 217–225.

Girsch, S.J. and Peracchia, C. (1985b) Lens cell-to-cell channel protein: II. Conformational changes in the presence of calmodulin. J. Membr. Biol. 83, 227–233.

Girsch, S.J. and Peracchia, C. (1991) Calmodulin interacts with a C-terminus peptide from the lens membrane protein MIP26. Curr. Eye Res. 10, 839–849.

Girsch, S.J. and Peracchia, C. (1992) Calmodulin binding sites in connexins. Biophys. J. 61, A506.

Gooden, M.M., Rintoul, D.A., Takehana, M. and Takemoto, L. (1985) Major intrinsic polypeptide (MIP26) from lens membrane: reconstitution into vesicles and inhibition of channel forming activity by peptide antiserum. Biochem. Biophys. Res. Commun. 128, 993–999.

Gorin, M.B., Yancey, S.B., Cline, J., Revel, J-P. and Horwitz, J. (1984) The major intrinsic protein (MIP) of the bovine lens fiber membrane: characterization and structure based upon cDNA cloning. Cell 39, 49–59.

Gruijters, W.T.M., Kistler, J., Bullivant, S. and Goodenough, D.A. (1987) Immunolocalization of an MP70 in lens fiber 16–17 nm intercellular junctions. J. Cell Biol. 104, 565–572.

Harris, A.L. (1990) Connexin32 forms ion channels in single artificial membranes. In: C. Peracchia (Ed.), Biophysics of Gap-Junction Channels, CRC Press, Boca Raton, FL, pp. 373–389.

Johnson, R.G., Klukas, K.A., Tze-Hong, L. and Spray, D.C. (1988) Antibodies to MP26 are localized to lens junctions, alter intercellular permeability, and demonstrate increased expression during development. In: E.L. Hertzberg, and R.G. Johnson, (Eds.), Gap Junctions, Alan R. Liss, New York, pp. 81–98.

Johnston, M.F. and Ramon, F. (1981) Electrotonic coupling in internally perfused crayfish segmented axons. J. Physiol. 317, 509–518.

Kistler, J. and Bullivant, S. (1987) Protein processing in lens intercellular junctions: cleavage of MP70 to MP38. Invest. Ophthalmol. Vis. Sci. 28, 1687–1692.

Kistler, J., Kirkland, B. and Bullivant, S. (1985) Identification of a 70,000-D protein in lens membrane junctional domains. J. Cell Biol. 101, 28–35.

Kistler, J., Christie, D. and Bullivant, S. (1988) Homologies between gap-junction proteins in lens, heart and liver. Nature 331, 721–723.

Lasalde, J.A. and Zuazaga, C. (1991) Cholesterol enrichment decreases the conductance of nicotinic acetylcholine receptor channels in tissue-cultured chick muscle. Biophys. J. 59, 444a.

Lea, J.A. and Duncan, G. (1991) Lens cell communication – from the whole organ to single channels. In: C. Peracchia (Ed.), Biophysics of Gap-Junction Channels, CRC Press, Boca Raton, FL, pp. 353–371.

Loewenstein, W.R. (1966) Permeability of membrane junctions. Ann. N.Y. Acad. Sci. 137, 441–472.

Louis, C.F., Hogan, P., Visco, L. and Strasburg, G. (1990) Identity of the calmodulin-binding proteins in bovine lens plasma membrane. Exp. Eye Res. 50, 495–503.

Malewicz, B., Kumar, V.V., Johnson, R.G. and Baumann, W.J. (1990) Lipids in gap-junction assembly and function. Lipids 25, 419–427.

Manjunath, C.K. and Page, E. (1986) Rat heart gap junctions as disulfide-bonded connexon multimers: Their depolymerization and solubilization in deoxycholate. J. Membr. Biol. 90, 43–57.

Nikaido, H. and Rosenberg, E.Y. (1985) Functional reconstitution of lens gap-junction proteins into proteoliposomes. J. Membr. Biol. 85, 87–92.

Paul, D.L. and Goodenough, D.A. (1983) Preparation, characterization and localization of antisera against bovine MP26, an integral protein from the lens fiber plasma membrane. J. Cell Biol. 96, 625–632.

Peracchia, C. (1978) Calcium effects on gap-junction structure and cell coupling. Nature 271, 669–671.

Peracchia, C. (1984) Communicating junctions and calmodulin: inhibition of electrical uncoupling in *Xenopus* embryo by calmidazolium. J. Membr. Biol. 91, 49–58.

Peracchia, C. (1987) Calmodulin-like proteins and communicating junctions – electrical uncoupling of crayfish septate axons is inhibited by the calmodulin inhibitor W7 and is not affected by cyclic nucleotides. Pflügers Arch. 408, 379–385.

Peracchia, C. (1988) The calmodulin hypothesis for gap-junction regulation six years later. In: E.L. Hertzberg and R.G. Johnson (Eds.), Gap Junctions, Alan R. Liss, New York, pp. 267–282.

Peracchia, C. (1990a) The increase in gap-junction resistance with acidification in crayfish septate axons is closely related to changes in intracellular calcium but not hydrogen ion concentration. J. Membr. Biol. 113, 75–92.

Peracchia, C. (1990b) Effects of caffeine and ryanodine on low pH-induced changes in gap-junction conductance and calcium concentration in crayfish septate axons. J. Membr. Biol. 117, 79–89.

Peracchia, C., Bernardini, G. and Peracchia, L.L. (1983) Is calmodulin involved in the regulation of gap-junction permeability? Pflügers Arch. 399, 152–154.

Peracchia, C. and Girsch, S.J. (1985) An *in vitro* approach to cell coupling: Permeability and gating of gap-junction channels incorporated into liposomes. In: M.V.L. Bennett and D.C. Spray (Eds.), Gap Junctions, Cold Spring Harbor Laboratory, Cold Spring Harbor, NY, pp. 191–250.

Peracchia, C. and Girsch, S.J. (1989) Calmodulin site at the C-terminus of the putative lens gap-junction protein MIP26. Lens Eye Toxic. Res. 6, 613–621.

Peracchia, C. and Peracchia, L.L. (1980a) Gap-junction dynamics: reversible effects of divalent cations. J. Cell Biol. 87, 708–718.

Peracchia, C. and Peracchia, L.L.: (1980b) Gap-junction dynamics: reversible effects of hydrogen ions. J. Cell Biol. 87, 719–727.

Rose, B. and Loewenstein, W.R. (1976) Permeability of a cell junction and the local cytoplasmic free ionized calcium concentration: a study with aequorin. J. Membr. Biol. 28, 87–119.

Saimi, Y. and Ling, K-Y. (1990) Calmodulin activation of calcium-dependent sodium channels in excised membrane patches of *Paramecium*. Science 249, 1441–1444.

Sas, D.F., Sas, J., Johnson, K.R., Menko, A.S. and Johnson, R.G. (1985) Junctions between lens fiber cells are labeled with a monoclonal antibody shown to be specific for MP26. J. Cell Biol. 100, 216–225.

Shen, L., Shrager, P., Girsch, S.J., Donaldson, P.J. and Peracchia, C. (1992) Channel reconstitution in liposomes and planar bilayers with HPLC-purified MIP26 of bovine lens. J. Membr. Biol. 124, 21–32.

Tuganowski, W., Korczynska, I., Wasik, K. and Piatek, G. (1989) Effects of calmidazolium and dibutyryl cyclic AMP on the longitudinal internal resistance in sinus node strips. Pflügers Arch. 414, 351–353.

Vallon, O., Dunia, I., Favard-Sereno, C., Hoebeke, J. and Benedetti, E.L. (1985) MP26 in the bovine lens: a post-embedding immunocytochemical study. Biol. Cell. 53, 85–88.

Wojtczak, J.A. (1985) Electrical uncoupling induced by general anesthetics: a calcium-independent process? In: M.V.L. Bennett and D.C. Spray (Eds.), Gap Junctions, Cold Spring Harbor Laboratories, Cold Spring Harbor, NY, pp 167–175.

Young, D-L., Cohen, Z.A. and Gilula, N.B. (1987) Functional assembly of gap-junction conductance in lipid bilayers: demonstration that the major 27 kDa protein forms the junctional channel. Cell 48, 733–743.

Zampighi, G.A., Hall, J.E. and Kreman, M. (1985) Purified lens junctional protein forms channels in planar lipid films. Proc. Natl. Acad. Sci. USA 82, 8468–8472.

Zampighi, G., Hall, J.E., Ehring, G.R. and Simon, S.A. (1989) The structural organization and protein composition of lens fiber junctions. J. Cell Biol. 108, 2255–2275.

J.E. Hall, G.A. Zampighi and R.M. Davis (Eds.)
Progress in Cell Research, Vol. 3

CHAPTER 24

Expression of multiple connexins by cells of the cardiovascular system and lens

ERIC C. BEYER[a–c], H. LEE KANTER[b], DIANE M. RUP[a], EILEEN M. WESTPHALE[a], KAREN E. REED[a], DAVID M. LARSON[d] and JEFFREY E. SAFFITZ[b,e]

Departments of [a]Pediatrics, [b]Medicine, [e]Pathology and [c]Cell Biology, Washington University School of Medicine, St. Louis, MO 63110 and [d]Mallory Institute of Pathology, Boston University School of Medicine, Boston, MA 02118, USA

Introduction

Our recent studies test the hypothesis that intercellular communication may be regulated by the expression of multiple gap-junction proteins (connexins) with distinct physiological properties. Previous studies have demonstrated that hepatocytes express both Cx26 and Cx32 (Zhang and Nicholson, 1989; Traub et al., 1989). However, in the liver, the function of the channels is unknown, and there is little data for physiologically distinct channels. Our research has focused on the connexins expressed in the cardiovascular system and lens where biophysical studies have demonstrated multiple gap-junctional channels. Cell-to-cell coupling mediated by gap-junctional channels is crucial to the function of these tissues, facilitating electrical conduction in the heart and smooth muscle and metabolic cooperation in the lens (Dewey and Barr, 1962; Barr et al., 1965; Goodenough, 1979). The junctions may also mediate other processes, since work in other systems implicates intercellular communication in the coordination of hormonal responses, growth regulation, and embryogenesis and development (Loewenstein, 1987; Green, 1988; Guthrie and Gilula, 1988). Therefore, a multiplicity of gap-junction proteins might play many roles.

The connexin family of gap-junction proteins

Gap junctions are formed by members of a family of related subunit proteins, termed connexins. Connexin cDNAs or genomic sequences have now been cloned from a number of sources (reviewed by Beyer et al., 1990). All of these sequences predict polypeptides which share similar structures. Each of the connexins contains two major conserved regions with many identical amino acids; each protein also contains a region in the middle of the molecule and a region at the carboxyl end which show little homology between connexins.

A model has been developed to explain the structure and topography of the connexin polypeptides in the junctional membrane (Fig. 1). Hydropathy analysis suggests that each connexin has four major hydrophobic domains, which are proposed to be membrane-spanning regions. The model places the amino and carboxy termini of the

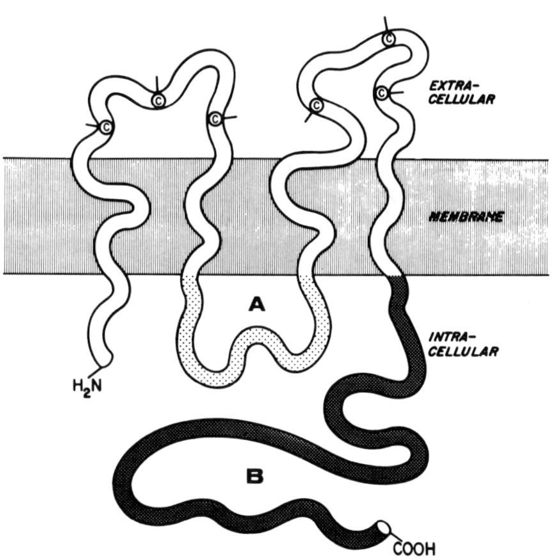

Figure 1. The structure and topology of connexins relative to the junctional plasma membrane. A model predicts that the connexins have two transmembrane spans and are oriented with their amino-and carboxy-termini on the cytoplasmic face of the junctional membrane. This model was developed based on hydropathy analysis and proteolysis, and immunocytochemical studies (reviewed by Beyer et al., 1990). The transmembrane and extracellular domains (including 6 invariant cysteines indicated by circled C's) are relatively conserved in all connexins. The cytoplasmic domains (A,B) differ dramatically among the connexins, both in sequence and length. (Reprinted from J. Membr. Biol. (1990) 116:188, with permission of the publisher.)

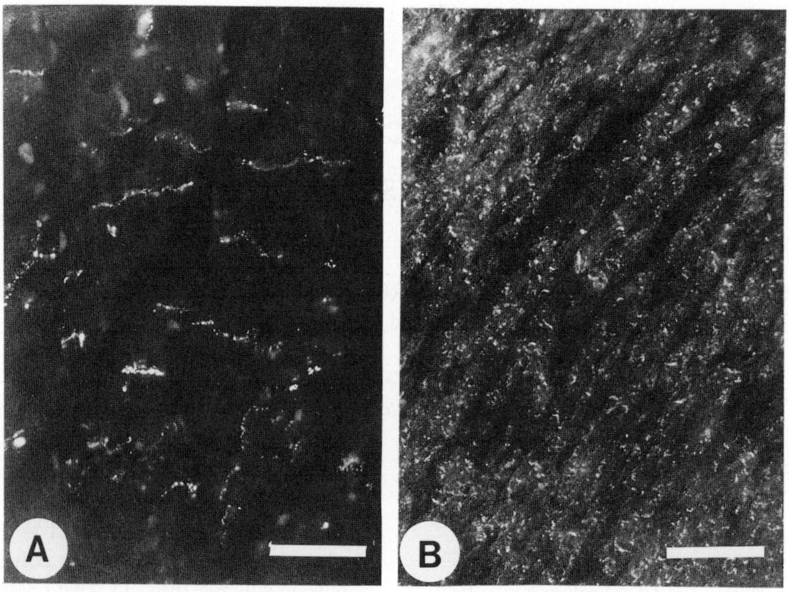

Figure 2. Immunohistochemical localization of heart connexins. (A) Cx43 in adult dog myocardium. Frozen sections were treated with a rabbit antiserum to amino acids 252–271 of rat Cx43 followed by a fluorescent secondary reagent; the antiserum specifically reacts with the intercalated disks, which contain the gap junctions. (B) Cx42 in chick embryo heart. Frozen sections of myocardium from a 19-day old chick embryo were similarly reacted with an antiserum directed against the corresponding region in chick Cx42, demonstrating a pattern consistent with cardiac gap junctions. Bars, 10 μm.

connexins on the cytoplasmic side of the junctional membrane. Sequence localization of Cx32 and Cx43 by proteolysis (Zimmer et al., 1987; Hertzberg et al., 1989) or immunocytochemistry (Goodenough et al., 1988; Milks et al., 1988; Beyer et al., 1989; Yancey et al., 1989) have confirmed many predictions of the model. The connexin model also predicts the locations of conserved and unique regions within the connexins (Fig. 1). The conserved regions correspond to the membrane-spanning and extracellular domains, while the cytoplasmic loop and carboxyl tail are divergent. It has been speculated that the unique cytoplasmic sequences might be gating domains, which confer different properties on channels composed of different connexins.

Multiple connexins expressed in heart

In the heart, physiological studies have identified multiple gap-junctional channels. Molecular biological studies have identified numerous connexins. These multiple gap-junction proteins may thus provide mechanisms for controlling communication.

The physiology of these channels has been examined using the double whole-cell patch-clamp technique in cardiac myocytes from chick embryos (Veenstra and DeHaan, 1986; Veenstra, 1991), neonatal rats (Burt and Spray, 1988; Rook et al., 1988), and adult guinea pigs (Rüdisüli and Weingart, 1989). While some differences are due to technical factors such as electrode-filling solutions, channels with differing unitary conductances have been observed. Veenstra and DeHaan (1986) and Rook et al. (1988) both recorded channels with two different conductances (60–80 pS and 160 pS, ~30 pS and ~60 pS respectively). While the junctions between neonatal or adult myocytes show only mild voltage dependence, Veenstra (1991) has detected a highly voltage-dependent channel, whose expression decreases with maturation in early chick embryos.

The first identified component of cardiac gap junctions was Cx43, which was cloned from a rat heart cDNA library (Beyer et al., 1987). The Cx43 polypeptide encoded by this cDNA corresponds to the major component of rodent cardiac gap junctions which had been previously purified (Manjunath and Page, 1985) and partially sequenced by Edman degradation (Nicholson et al., 1983; Manjunath et al., 1987). Antisera directed against synthetic peptides predicted from its sequence specifically react with mammalian cardiac gap junctions (Beyer et al., 1989; Luke et al., 1989; Yancey et al., 1989; El Aoumari et al., 1990), as shown in Fig. 2A.

We have investigated the connexins expressed by the chick embryo heart in an attempt to explain the multiple channels and have shown that Cx43 is not the only cardiac myocyte connexin. Musil et al. (1990) cloned chick Cx43 cDNA and demonstrated that this protein is highly conserved (more than 92% of amino acid residues are identical to those in rat), and that Cx43 mRNA is expressed in chick embryo heart, but at much lower levels than in lens. Interestingly, using an antipeptide antibody, El Aoumari et

```
Chk Cx43   MGDWSALGKLLDKVQAYSTAGGKVWLSVLFIFRILLLGTAVESAWGDEHVAFRCNTQQPG    60
Chk Cx42   *****F**EF*EE*HKH**VV*****T******M*V****AGPL****QSD*M*D*****    60
Chk Cx45   *-S**F*TR**EEIHNH**FV**I*****IV***V*TAVGG**IYY**QSK*V***E***    59

Chk Cx43   CENVCYDKSFPISHVRFWVLQIIFVSVPTLLYLAHVFY----------------------    98
Chk Cx42   ********A**********************T*S*V*MG*AMH--------------------TV   100
Chk Cx45   *******AFA*L*******F***L*A**SVM**GYAIHKIARMVEHSDVDRRFRSKSFST   119

Chk Cx43   --VMRKEEKLNKREEELKVVQNDGVNVDMHLKQIEIKKFKYGIEEHGKVKMRGGLLRTYI   156
Chk Cx42   RMEEKRKM*EAE**AQEMKNSG*TYYQQKCPVAEKTELSCWDESGGKIILR-*S**N**V   159
Chk Cx45   RWKQHRGLEEAEDDH*EDPMMYPEIELESERENK*QQPPAKAKHDGRRRIRED**M*I*V   179

Chk Cx43   ISILFKSVFEVAFLLIQWYIYGFSLSAIYTCERDPCPHRVDCFLSRPTEKTIFIVFMLVV   216
Chk Cx42   Y***IRTAM*I**IVG*YIL**IF*ETL*I*Q*A****P*N*YV******NV**I***A*   219
Chk Cx45   LQL*V*AT***G**IG**LL***EV*PVFV*S*K****KI***I*******LLI*YG*   239

Chk Cx43   SLVSLALNIIELFYVFFKGVKDRVKGKTDPYSHSGTMSPSKDCGSPKYAYYNGCSSPTAP   276
Chk Cx42   AVL**F*SLA**YHLGW*KA*E*CSRAYK*SPSTAPRRLESAPQ-----------VER*Q   268
Chk Cx45   *CMC*L**VM*MLHLG*GTIR*TLNN*RKELED***YNYPFTWNT*SAPPGYNIAVKPDQ   299

Chk Cx43   LSPMSPPGYKLVTGDRNNSSCRNY-NKQASEQNWANYSAEQNRMGQAGSTISNSHAQPFD   335
Chk Cx42   MYT-P**DFNQCLASP*GKFSPFIS**M**Q**T**FAT*--*VHSQEDAAGEGPFMKSS   325
Chk Cx45   MQYTELSNA*MAYKQNKANIAQEQQYGSNE*NIP*DLENLQREIKV*QERLDMAIQAYNN   359

Chk Cx43   FADEHQNTKKLASGHELQPLTIVDQRPS-SRASSRASSRPRPDDLEI                381
Chk Cx42   YMESPEVASE-CAAPA*-*ESYFNEKRRF*K******-KA*S***SV                369
Chk Cx45   QNNPG-----------SSSREKKSKAG*N-K-**-***KSGDGKNSVWI              394
```

Figure 3. Comparison of the deduced amino-acid sequences of three chick connexins. The sequences of Cx43, -42, and -45 were optimally aligned to match identical residues. Residues which match those in Cx43 are represented by asterisks. Dashes represent spaces added to optimize alignment. Homologous regions between connexins include the putative membrane-spanning regions (boxed) and extracellular domains. Proposed cytoplasmic domains in the middle and carboxyl end of the molecules show little similarity between connexins, except for a small serine-rich region.

al. (1990) were unable to detect Cx43 in chick embryo heart by immunoblotting. We have detected little, if any, Cx43 by immunofluorescence in the chick embryo myocytes.

Beyer (1990) used low-stringency hybridization to clone cDNAs for two additional chick connexins, Cx42 and -45. While their sequences differ from other known connexins, they contain the conserved-sequence features of other family members (Fig. 3). Interestingly, they also share similar serine-rich sequences near their carboxy termini, a region proposed to regulate Cx43 by phosphorylation. Cx45 is a unique connexin; it shares fewer residues than any other reported family members, and has a very long cytoplasmic loop. Cx45 expression is developmentally regulated; its mRNA is more abundant in hearts of early embryos than in more mature animals (Beyer, 1990).

Cx42, -43 and -45 are expressed in chick embryo heart (Fig. 4A). Indeed, Cx42 may be the major connexin expressed in chick embryo heart, based on the appearance of Northern blots given comparable exposures (Beyer and Westphale, unpublished data). We have recently prepared an antiserum directed against a unique peptide sequence in the tail of Cx42. That antiserum stains sections of chick embryo heart in a distribution consistent with that of myocyte gap junctions (Fig. 2B).

We have cloned and sequenced mammalian homologs of Cx42 and Cx45 using the polymerase chain reaction (PCR) and genomic cloning (Kanter et al., 1991). Both clones are quite similar to their avian counterparts. The canine homolog of Cx42 encodes a protein of only 40 kDa; therefore, it is identified as dog Cx40. High-stringency Northern blotting identifies specific Cx40, -43, and -45 mRNAs in total RNA prepared from mouse heart (Fig. 4B) or canine heart or isolated myocytes (Kanter et al., 1991). Preliminary immuno-localization studies using antisera directed against unique oligopeptides from the carboxy-terminal tails of Cx40, -43, and -45 show staining of

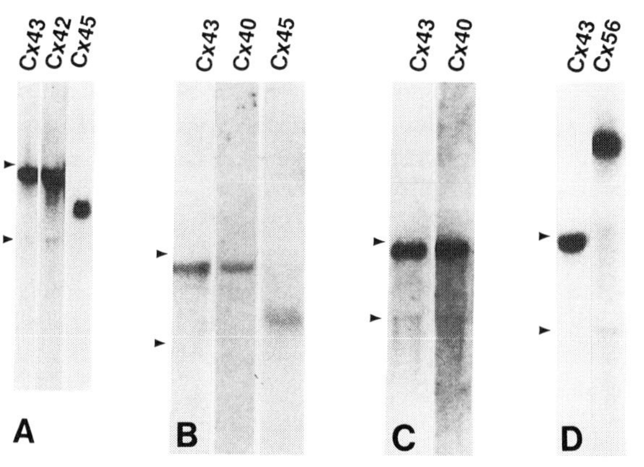

Figure 4. Northern blots of connexin mRNA expression. In each lane 10 μg of total RNA from various sources was loaded per lane, prepared for RNA blots (according to Beyer et al., 1987) and hybridized to ^{32}P-labeled connexin-specific DNA probes as indicated. (A) RNA from 9-day chick heart was hybridized with chick Cx43, -42, and -45. (B) RNA from adult mouse heart was hybridized with probes for rat Cx43 and canine Cx40 and -45. (C) RNA from cultured A7r5 cells was hybridized with rat Cx43 and dog Cx40 probes. (D) RNA from 19-day old chick embryo lens was hybridized with probes for chick Cx43 and -56.

myocyte gap junctions (Kanter, Saffitz, and Beyer; manuscript in preparation). Thus, adult canine ventricular myocytes may express three different connexins in gap junctions.

We have also been investigating the expression of connexins by vascular wall cells. Examination of RNA prepared from endothelial cells, smooth muscle cells, and pericytes demonstrated that all of these cells express Cx43 (Larson et al., 1990). However, Moore et al. (1991) found that A7r5 vascular smooth muscle cells express gap-junctional channels which have unique physiological properties compared to other Cx43-expressing cells (such as neonatal rat cardiac myocytes), i.e., their channels have different responses to second messengers and they contain channels with two distinct unitary conductances (~30 pS and ~90 pS). Neonatal myocytes and Cx43-transfected cells express 45–50 pS channels (Burt and Spray, 1988; Fishman et al., 1990). We have demonstrated that the A7r5 cells express Cx43 mRNA and immunoreactive Cx43 protein (Fig. 4C and Moore et al., 1991). Different post-translational modifications of the Cx43 protein may explain some of the physiological differences.

We also have evidence that these cells express a second, different connexin. The dog Cx40 probe hybridizes to A7r5 RNA (Fig. 4C). Indeed, we have isolated rat Cx40 sequences from a rat aorta cDNA library and by PCR analysis of rat genomic DNA. Cx40 mRNA is also expressed in dog coronary artery and aorta and in primary cultures of bovine and rat vascular smooth muscle cells. Future experiments will determine which channel properties are due to Cx43 and which to Cx40.

Lens connexins

Intercellular communication between the different cells of the eye lens is achieved by a network of gap junctions. This system includes several distinct gap junctions; epithelial and fiber junctions differ morphologically and in physiological sensitivity to cytoplasmic acidification (Goodenough, 1979; Schuetze and Goodenough, 1982; Miller and Goodenough, 1986; Menko et al., 1987). One of the component gap-junction proteins has been identified: Cx43 in lens epithelial cells (Beyer et al., 1989; Musil et al., 1990). Lens expression of Cx43 mRNA is demonstrated in Fig. 4D.

There has been substantial controversy about the composition of junctions between the lens fiber cells, but two connexin-like proteins are candidates, MP70 (Kistler et al., 1985) and Cx46 (Beyer et al., 1988; Goodenough et al., 1990). We have investigated lens gap junctions in the chick embryo system because of its advantages for developmental and tissue-culture studies (Menko et al., 1987). We have attempted to clone a homolog of Cx46 or MP70. Using PCR, we amplified a chicken genomic DNA sequence related to MP70 and rat Cx46 (Rup and Beyer, manuscript in preparation). Northern blots demonstrate that this sequence hybridizes to an abundant, ~8 kb mRNA in chick embryo lens (Fig. 4D), which is not detectable in RNA from any other tissue tested. Southern blots of chicken genomic DNA demonstrate that the amplified fragment corresponds to a single-copy gene which is distinct from those of any previously characterized chick connexins. This sequence was used to isolate a chick genomic clone which encodes a 56 kDa protein, chick Cx56. The sequences of its predicted extracellular and transmembrane domains are ~90% homologous to rat Cx46. However, the predicted cytoplasmic loop and tail of this novel chick connexin are substantially different and may endow the connexin with unique biophysical properties.

Summary and future directions; or multiple connexins, so what?

We have examined several systems where multiple physiologically distinct gap junctional channels have been identified. Our evidence suggests that within a single organ system (the cardiovascular system), within a single organ (the heart or lens), and within a single cell type (the A7r5 smooth muscle cell and the cardiac myocyte), multiple connexins are expressed. While we do not yet know their exact roles, it seems reasonable that these different gap-junction proteins may provide multiple mechanisms for controlling coupling. They may allow a single cell (e.g., a myocyte) to make diverse channels; or, they may allow two different cells in a single tissue (e.g., the lens epithelial and fiber cells) to make composite channels.

Clearly, more definitive research is needed. We are currently performing immunolocalization studies to determine whether a single cardiac myocyte expresses multiple connexins and to identify the lens cell type that expresses Cx56. Our functional-expression studies are designed to elucidate the biophysics of channels formed from these different connexins (see Chapter 13). A communication-deficient cell line has been stably transfected with constructs encoding different connexins. This system will allow us to determine properties of different connexins, including voltage dependence and unitary conductances. We should be able to answer such questions as: Is Cx45 the voltage-gated channel so prominent in early chick embryos? Which connexin forms the large (160 pS) channel in the chick heart? Do the unique cytoplasmic sequences indeed confer specific biophysical properties?

References

Barr, L., Dewey, M.M. and Berger, W. (1965) Propagation of action potentials and the structure of the nexus in cardiac muscle. J. Gen. Physiol. 48, 797–823.

Beyer, E.C. (1990) Molecular cloning and developmental expression of two chick embryo gap-junction proteins. J. Biol. Chem. 265, 14439–14443.

Beyer, E.C., Kistler, J., Paul, D.L. and Goodenough, D.A. (1989) Antisera directed against connexin43 peptides react with a 43-kD protein localized to gap junctions in myocardium and other tissues. J. Cell Biol. 108, 595–605.

Beyer, E.C., Goodenough, D.A. and Paul, D.L. (1988) The connexins, a family of related gap-junction proteins. In: E.L. Hertzberg and R.G. Johnson (Eds.) Gap Junctions. Alan R. Liss, New York, pp. 167–175

Beyer, E.C,, Goodenough, D.A. and Paul, D.L. (1990) Connexin family of gap-junction proteins. J. Membr. Biol. 116, 187–194.

Beyer, E.C,, Paul, D.L. and Goodenough, D.A. (1987) Connexin43, a protein from rat heart homologous to a gap-junction protein from liver. J. Cell Biol. 105, 2621–2629.

Burt, J.M. and Spray, D.C. (1988) Single-channel events and gating behavior of the cardiac gap-junction channel. Proc. Natl. Acad. Sci. USA 85, 3431–3434.

Dewey, M.M. and Barr, L. (1962) Intercellular connection between smooth muscle cells: the nexus. Science 137, 670–672.

El Aoumari, A., Fromaget, C., Dupont, E., Reggio, H., Durbec, P., Briand, J-P., Boller, K., Kreitman, B. and Gros, D. (1990) Conservation of a cytoplasmic carboxy-terminal domain of connexin43, a gap-junction protein, in mammalian heart and brain. J. Membr. Biol. 115, 229–240.

Fishman, G., Spray, D.C. and Leinwand, L.A. (1990) Molecular characterization and functional expression of the human cardiac gap-junction channel. J. Cell Biol. 111, 589–598.

Goodenough, D.A. (1979) Lens gap junctions: a structural hypothesis for nonregulated low-resistance intercellular pathways. Invest. Ophthalmol. Vis. Sci. 18, 1104–1122.

Goodenough, D.A., Paul, D.L. and Jesaitis, L. (1988) Topological distribution of two connexin32 antigenic sites in intact and split rodent hepatocyte gap junctions. J. Cell Biol. 107, 1817–1824.

Goodenough, D.A., Paul, D.L. and Takemoto L. (1990) Molecular cloning and expression of connexin46 (Cx46), a lens fiber junction protein distinct from MP70. J. Cell Biol. 111, 153A.

Green, C.R. (1988) Evidence mounts for the role of gap junctions during development. Bio. Essays 8, 7–10

Guthrie, S.C. and Gilula, N.B. (1989) Gap-junctional communication and development. Trends Neurosci. 12, 12–16.

Hertzberg, E.L., Disher, R.M., Tiller, A.A., Zhou, Y. and Cook, R.G. (1988) Topology of the M_r 27,000 liver gap-junction protein. J. Biol. Chem. 263, 19105–19111.

Kanter, H.L., Saffitz, J.E. and Beyer, E.C. (1991) Canine cardiac myocytes express multiple gap-junction proteins. Clin. Res. 39, 193A.

Kistler, J., Kirkland, B. and Bullivant, S. (1985) Identification of a 70,000-D protein in lens membrane junctional domains. J. Cell Biol. 101, 28–35.

Larson, D.M., Haudenschild, C.C. and Beyer, E.C. (1990) Gap-junction messenger RNA expression by vascular wall cells. Circ. Res. 66, 1074–1080.

Loewenstein, W.R. (1987) The cell-to-cell channel of gap junctions. Cell 48, 725–726.

Luke, R.A., Beyer, E.C., Hoyt, R.H. and Saffitz, J.E. (1989) Quantitative analysis of intercellular connections by immunohistochemistry of the cardiac gap-junction protein, connexin43. Circ. Res. 65, 1450–1457.

Manjunath, C.K. and Page, E. (1985) Cell biology and protein composition of cardiac gap junctions. Am. J. Physiol. 17, H783–H791.

Manjunath, C.K., Nicholson, B.J., Teplow, D., Hood, L., Page, E. and Revel J-P. (1987) The cardiac gap-junction protein (M_r 47,000) has a tissue-specific cytoplasmic domain of M_r 17,000 at its carboxy-terminus. Biochem. Biophys. Res. Commun. 142, 228–234

Menko, A.S., Klukas, K.A., Liu, T-F., Quade, B., Sas, D.F., Preus, D.M. and Johnson, R.G. (1987) Junctions between lens cells in differentiating cultures: structure, formation, intercellular permeability and junctional protein expression. Dev. Biol. 123, 307–320

Milks, L.C., Kumar, N.M., Houghten, R., Unwin, N. and Gilula, N.B. (1988) Topology of the 32-kDa liver gap-junction protein determined by site-directed antibody localizations. EMBO J. 7, 2967–2975.

Miller, T.M. and Goodenough, D.A. (1986) Evidence for two physiologically distinct gap junctions expressed by the chick lens epithelial cell. J. Cell Biol. 102, 194–199

Moore, L.K., Beyer, E.C. and Burt, J.M. (1991) Characterization of gap-junction channels in A7r5 vascular smooth muscle cells. Am. J. Physiol. 260 C975–C981.

Musil, L.S., Beyer, E.C. and Goodenough, D.A. (1990) Expression of the gap-junction protein connexin43 in embryonic chick lens: molecular cloning, ultrastructural localization, biosynthesis and post-translational phosphorylation. J. Membr. Biol. 116, 163–175.

Nicholson, B.J., Gros, D.B., Kent, S.B.H., Hood, L.E. and Revel J-P. (1985) The M_r 28,000 gap-junction proteins from rat heart and liver are different but related. J. Biol. Chem. 260, 6514–6517.

Rook, M.B., Jongsma, H.J. and van Ginneken, A.C. (1988) Properties of single gap-junctional channels between isolated neonatal rat heart cells. Am. J. Physiol. 255, H770–H782.

Rüdisüli, A. and Weingart, R. (1989) Electrical properties of gap-junction channels in guinea pig ventricular cell pairs revealed by exposure to heptanol. Pflügers Arch. 415, 12–21.

Schuetze, S.M. and Goodenough, D.A. (1982. Dye transfer between cells of the embryonic chick lens becomes less sensitive to CO_2 treatment with development. J. Cell Biol. 92, 694–705.

Traub, O., Look, J., Dermietzel, R., Brummer, F., Hulser, D. and Willecke, K. (1989) Comparative characterization of the 21-kDa and 26-kDa gap-junction proteins in murine liver and cultured hepatocytes. J. Cell Biol. 108, 1039–1051.

Veenstra, R.D. (1991) Developmental changes in regulation of embryonic chick heart gap junctions. J. Membr. Biol. 119, 253–265.

Veenstra, R.D. and DeHaan, R.L. (1986) Measurement of single-channel currents from cardiac gap junctions. Science 233, 972–974.

Yancey, S.B., John, S.A., Lal, R., Austin, B.J. and Revel, J-P. (1989) The 43-kDa polypeptide of heart gap junctions: immunolocalization, topology and functional domains J. Cell Biol. 108, 2241–2254.

Zhang, J-T. and Nicholson B. (1989) Sequence and tissue distribution of a second protein of hepatic gap junctions, Cx26, as deduced from its cDNA. J. Cell Biol. 109, 3391–3401.

Zimmer, D.B., Green, C.R., Evans, W.H. and Gilula, N.B. (1987) Topological analysis of the major protein in isolated intact rat liver gap junctions and gap junction-derived single membrane structures. J. Biol. Chem. 262, 7751–7763

J.E. Hall, G.A. Zampighi and R.M. Davis (Eds.)
Progress in Cell Research, Vol. 3

CHAPTER 25

The developmental expression and organization of membrane proteins of the mammalian lens

CHARLES F. LOUIS[a,b], MARK ARNESON[a], LISA JARVIS[a] and ERICA M. TENBROEK[b]

Departments of [a]Biochemistry and [b]Veterinary Biology, University of Minnesota, 1988 Fitch Avenue, Room 295m, St. Paul, MN 55108, USA

Introduction

Ultrastructural examination of mammalian lens fiber membranes by freeze-fracture electron microscopy has identified extensive regions of plaque-like clusters of intramembranous particles (Goodenough et al., 1980). These plaques comprise the 16–17 nm junctions between neighboring fiber cell plasma membranes, and are thought to resemble the gap junctions that have now been identified in many different tissues (Paul and Goodenough, 1983). That these junctions form communicating channels has been demonstrated in conductance experiments which indicated that lens fiber cells are electrically coupled throughout the lens (Cooper et al., 1991).

The major protein in bovine lens membranes, termed MIP or MP26, has a molecular weight of 28 kDa based on its cDNA sequence (Gorin et al., 1984) and was originally proposed to be the major component of lens fiber cell junctions (Goodenough, 1979). However, there is now considerable controversy as to whether it is indeed a junctional protein. While MP26 has been demonstrated to exhibit single-channel activity (Ehring et al., 1990), a recent study has proposed that these channels may serve to regulate osmolarity in the lens (Zampighi et al., 1989). Furthermore, immunoelectron microscopy has shown MP26 in both junctional and non-junctional domains; one study demonstrated that MP26 was, in fact, excluded from junctional domains (Fitzgerald et al., 1985; Sas et al., 1985).

The second most abundant protein in lens fiber cell membranes migrates as an 18 kDa protein on polyacrylamide gels, and has been termed MP20 based on its molecular weight of 19.6 kDa derived from its cDNA sequence (Louis et al., 1989; Gutekunst et al., 1990). This component, which has been identified in all regions of the lens, is a major substrate for both cAMP-dependent protein kinase and protein kinase C (Louis et al., 1985; Lampe et al., 1986; Hur and Louis, 1989). Although we have previously localized this protein to the 16–17 nm junctions of lens fiber cell membranes by immunoelectron microscopy (Louis et al., 1989), Voorter et al. (1989) were unable to confirm this observation using immunogold negative staining.

The first connexin-related protein to be identified in the mammalian lens fiber cell junctions was a 70 kDa lens membrane protein MP70 (Kistler et al., 1985). This protein has N-terminal homology to the well-characterized connexin family of gap-junction proteins (Kistler et al., 1988). The second connexin-related protein identified in lens fiber cell junctions, connexin46 (Cx46), has been localized to the same junctional plaques as MP70 (Paul et al., 1991).

To examine the possible role and organization of these membrane proteins in the mammalian lens, we have examined the major substrates for cAMP-dependent protein kinase and receptors for the Ca^{2+}-signaling protein calmodulin (CaM). In addition, we have developed a primary ovine lens culture system that should be of value in defining the role of these proteins *in vivo*. We have demonstrated that MP20 is stoichiometrically phosphorylated by cAMP-dependent protein kinase on a single serine residue (Galvan et al., 1989), and interacts with the Ca^{2+} receptor CaM (Louis et al., 1990). Using a novel immunological technique with antibodies to MP20 and Cx46, we now show that MP20 can be co-localized with Cx46 to the 16–17 nm lens junctions in a small region of the lens located 0.5 to 1.0 mm into the lens cortex equatorial region; MP20 is more diffusely distributed in other regions of the lens. We have developed an ovine culture system that contains differentiated fiber-like cells which exhibit cell-to-cell transfer of injected dye. Junctional plaques in this culture system stain positively for Cx46. MP20 is diffusely distributed over only the differentiated cells of the culture; this protein does not appear to be concentrated in junctional plaques.

Results and discussion

Lens membrane substrates for cAMP-dependent protein kinase; receptors for CaM.

MP26 and MP20 are the major membrane substrates for cAMP-dependent protein kinase in lens fiber cell membranes (Louis et al., 1985). We have demonstrated that, while the lens adenylate cyclase is restricted to the cortical regions of the lens, cAMP-dependent protein kinase is present in all regions of the mammalian lens (Hur and Louis, 1989). To examine the physiological relevance of the phosphorylation of MP26 and MP20, the stoichiometry of phosphorylation of these two proteins was estimated by phosphorylating them in intact membranes and subsequently purifying them by electroelution from polyacrylamide gels. The level of phosphorylation of MP26 was considerably less than stoichiometric (0.24 mole P/mole MP26) while the stoichiometry of MP20 phosphorylation was 1.3.

The stoichiometric phosphorylation of MP20 prompted us to examine whether a single amino acid was labeled. To identify the site of phosphorylation, MP20 was first phosphorylated in intact membranes, then purified by electroelution from polyacrylamide gels. The exhaustively dialyzed protein was digested with varying concentrations of the protease V8, and then electrophoretically analyzed on polyacrylamide gels (Galvan et al., 1989). This resulted in the production of a major 17 kDa-peptide (just visible below the 18-kDa MP20 band), and a number of poorly resolved peptides in the 14–15 kDa region. Coomassie Blue-stained bands of higher molecular mass than MP20 (notably the 28-kDa component) observed in the absence of protease (Fig. 1A), are the aggregated forms of MP20 described by Galvan et al. (1989). In the autoradiogram corresponding to this gel (Fig. 1B), it is apparent that the protease-derived major Coomassie Blue-stained 17-kDa component was not phosphorylated; the major phosphorylated component had a molecular mass of 15 kDa. However, following protease V8 digestion, 90% of the ^{32}P originally present in MP20 was associated with small peptides that were lost from the polyacrylamide gel during staining and destaining. Thus, the 17-kDa peptide had lost its phosphorylation site, indicating that the 90% of the counts associated with a small (< 1 kDa) phosphopeptide(s) derived from either the N- or C-terminus of MP20. This phosphopeptide was purified by gel chromatography and HPLC, when its sequence was determined: (Cys)-Arg-Arg-Leu-Ser-Thr-Pro-Arg. Phosphoamino-acid analysis of phosphorylated MP20 detected phosphoserine as the only phosphoamino acid, indicating that the serine in the phosphopeptide, which is located in a consensus sequence for cAMP-dependent protein kinase, is the single residue that is stoichiometrically phosphorylated by cAMP-dependent protein kinase in intact lens membranes. The sequence of this 969 Da peptide corresponds precisely to the 8 C-terminal amino acids of MP20 (Gutenkust et al., 1990).

CaM has been identified in the mammalian lens (Liu and Schwartz, 1978) and may regulate the channel activity of MP26 (Girsch and Peracchia, 1985), calcium pumping

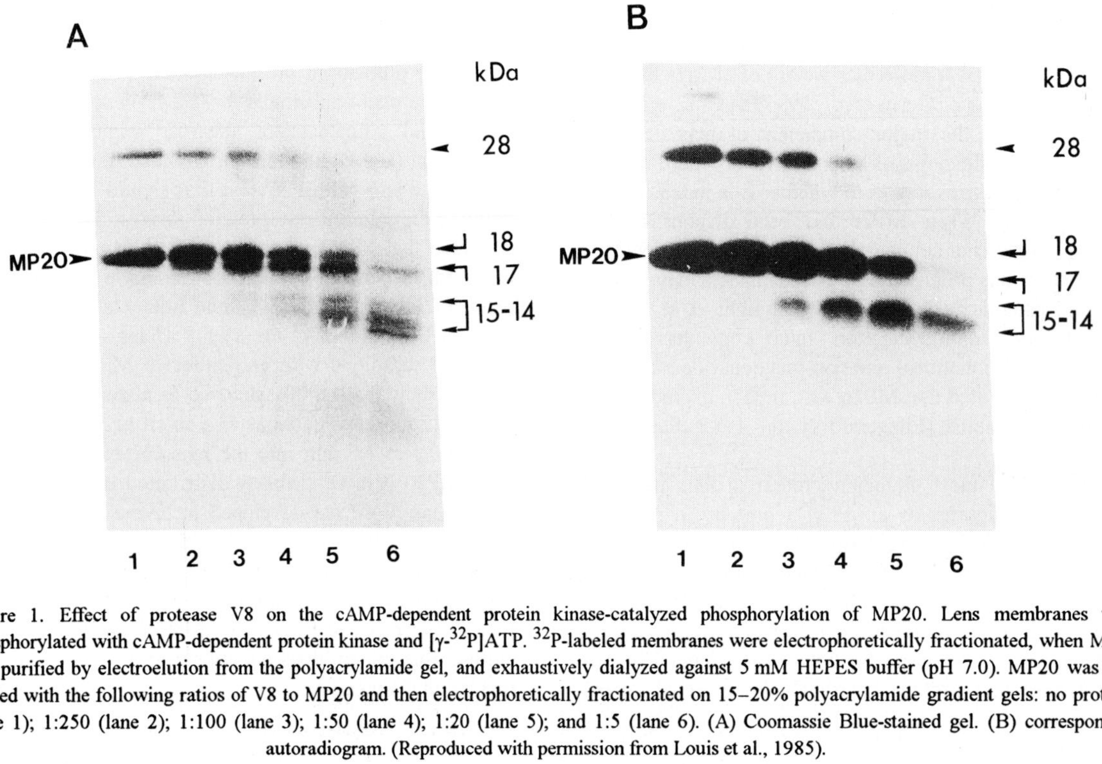

Figure 1. Effect of protease V8 on the cAMP-dependent protein kinase-catalyzed phosphorylation of MP20. Lens membranes were phosphorylated with cAMP-dependent protein kinase and [γ-^{32}P]ATP. ^{32}P-labeled membranes were electrophoretically fractionated, when MP20 was purified by electroelution from the polyacrylamide gel, and exhaustively dialyzed against 5 mM HEPES buffer (pH 7.0). MP20 was then treated with the following ratios of V8 to MP20 and then electrophoretically fractionated on 15–20% polyacrylamide gradient gels: no protease (lane 1); 1:250 (lane 2); 1:100 (lane 3); 1:50 (lane 4); 1:20 (lane 5); and 1:5 (lane 6). (A) Coomassie Blue-stained gel. (B) corresponding autoradiogram. (Reproduced with permission from Louis et al., 1985).

(Iwata, 1985), and adenylate cyclase activity (Louis et al., 1987). To identify the major lens membrane receptors for CaM, we have affinity labeled urea-washed membranes with the photoactivatable CaM affinity label benzophenone-[^{125}I]CaM (Fig. 2). Two major affinity-labeled components with molecular masses of 46 kDa and 36 kDa were identified, reflecting Ca^{2+}-dependent 1:1 complexes of CaM (17 kDa) with 29 kDa and 19 kDa lens-membrane components, respectively (Strasburg et al., 1988). Proteolytic digestion of these two proteins correlates precisely with the loss of the 46 and 36 kDa affinity-labeled components (Louis et al., 1990).

Thus, these various approaches demonstrate that MP20 is the major lens membrane substrate for a cAMP-dependent protein kinase that is stoichiometrically phosphorylated. Furthermore, this protein interacts with the Ca^{2+} receptor, CaM, indicating that MP20 may function to transduce Ca^{2+} signals in the mammalian lens.

Immunofluorescence labeling of lens fiber bundles with antibodies to MP20 and Cx46

Gruijters et al. (1987) developed a novel approach to identify the distribution of MP70 and MP26 in the mammalian lens fiber cell membrane, using immunofluorescent labeling in thin (0.1 to 0.2 mm thick) bundles of lens fiber cells. To examine the relationship between Cx46 and MP20 in the lens plasma membrane, we have immunolabeled fiber cell bundles with antibodies to Cx46 and MP20. We generated an IgM monoclonal antibody to Cx46, and a polyclonal IgG antibody to the C-terminal 8 amino acids of MP20; having two classes of antibody allowed us to double-label the fiber bundles with rhodamine and fluorescein-coupled second antibodies. Fiber bundles were selected from the outer 0.5 mm of the lens equatorial region (i.e. immediately below dividing epithelial cells), from approximately 0.5 mm, 0.5–1.0 and 2 mm into the lens equatorial region.

In the outer 0.5 mm bow region of the lens, there was excellent immunofluorescence labeling of junctional plaques with the Cx46 antibody (Fig. 3A). Some of the larger junctions appeared to have regions which excluded the Cx46 antibody, as noted previously by freeze-fracture electron microscopy of lens fiber cell membranes (Costello et al., 1989). In contrast, MP20 immunofluorescent labeling was very diffuse, with no specific labeling of the junctional plaques (Fig. 3B). Approximately 0.5 mm into the lens equatorial region, Cx46 could still be localized to junctional plaques on the broad side of the fiber cell membrane (data not shown). Although MP20 still had a predominantly disperse distribution, some of the MP20 appeared to be concentrated in plaques (data not shown). At a depth of 0.5 to 1.0 mm into the lens equatorial region, Cx46-containing plaques were again labeled on the broad sides of the fibers (Fig 4A). These plaques no longer seemed to contain regions which excluded the Cx46 antibody. MP20 was present in discrete plaques in this region (Fig. 4B) and frequently co-localized with Cx46 immunoreactivity (arrows in Fig. 4). While MP20 apparently clustered into foci in these plaques, these foci did not seem to be distributed like those of Cx46.

At a depth of 2 mm into the lens equatorial region, the Cx46 junctional plaques dispersed; very little immunofluorescent labeling of Cx46 was detectable (data not shown). Paul et al. (1991) also demonstrated that this protein could only be detected in the outer cortical regions of the lens. MP20 was no longer associated with junctional plaques, but had a disperse distribution and more intense labeling on the narrow, interdigitated sides (containing the ball-and-socket junctions) of the fiber cells (data not shown). Our data for the Cx46 antibody, like those of Gruijters et al. (1987) using an MP70 antibody, specifically localize both of these proteins to the outer cortical regions of the mammalian lens. In contrast, MP20, which, as we have shown, is present throughout the cortical and nuclear regions, appeared to be localized to junctional plaques in only a small fraction of lens fiber cells 0.5 to 1.0 mm into the lens equatorial region. Immunofluorescence labeling of lens cryosections also demon-

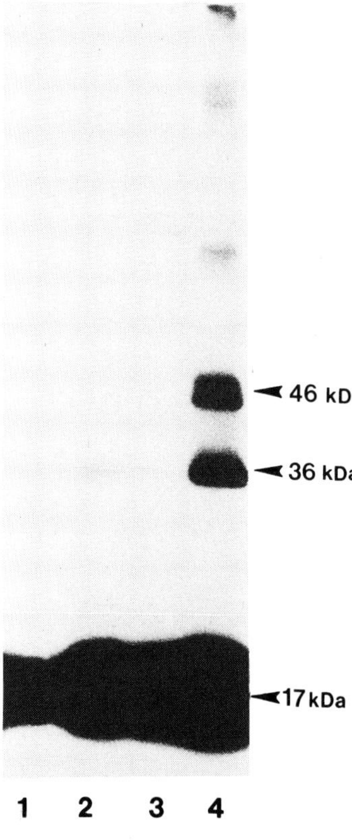

Figure 2. Identification of the [^{125}I]benzophenone-calmodulin affinity-labeled complexes in lens membranes. Autoradiogram of lens membranes affinity labeled with [^{125}I]benzophenone-calmodulin in the presence of 1 mM EGTA (lane 1), 1 mM EGTA + 10 mM $MgCl_2$ (lane 2), 0.1 mM $CaCl_2$ (lane 3), or 0.1 mM $CaCl_2$ + 1 mM $MgCl_2$ (lane 4) when samples were electrophoretically fractionated on 10–20% polyacrylamide gradient gels. (Reproduced with permission from Louis et al., 1990)

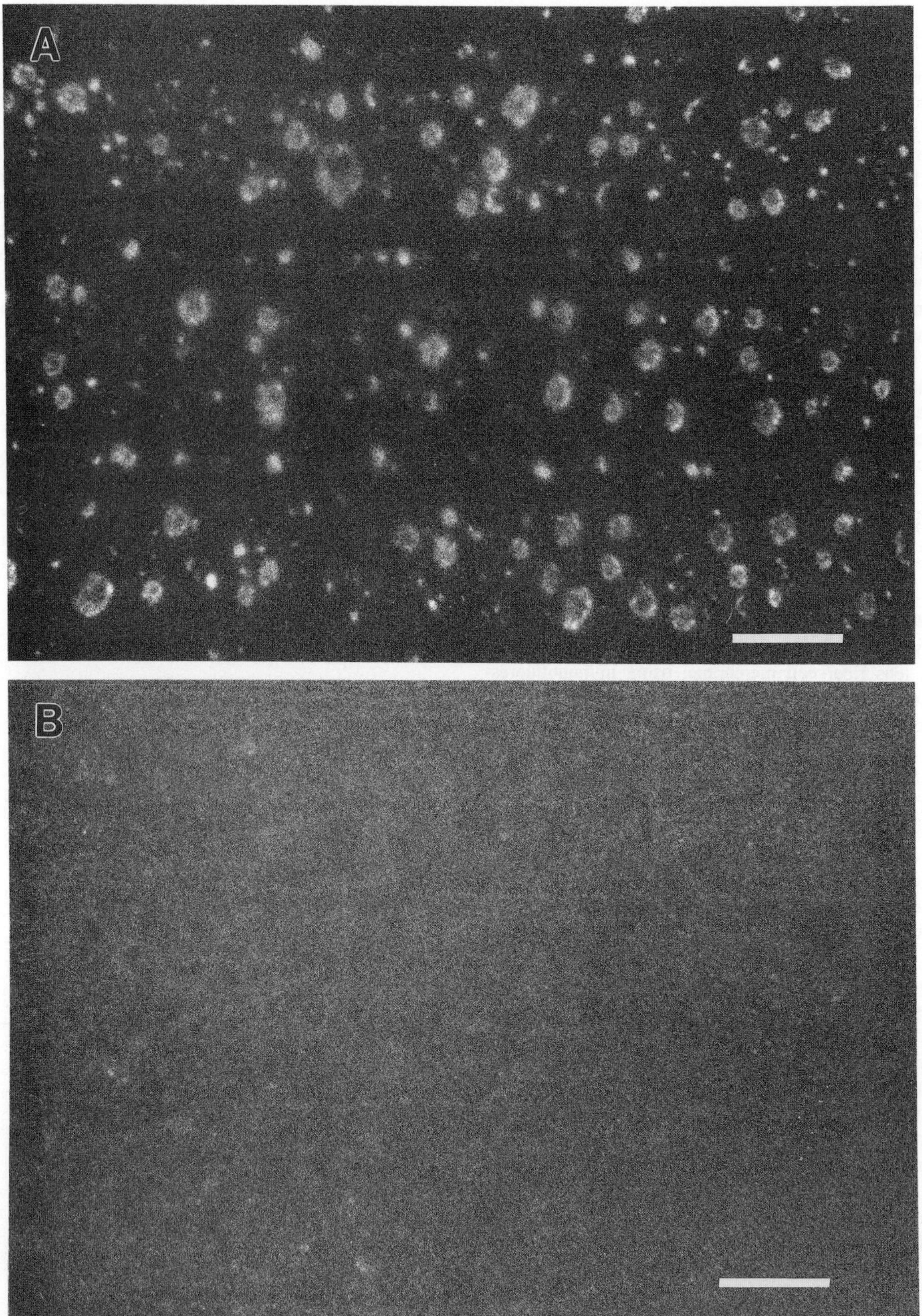

Figure 3. Immunofluorescent double labeling, using Cx46 and MP20 antibodies, of dissected lens fiber bundles derived from the outer 0.5 mm of the lens equatorial region. Lens fiber cell bundles were dissected from different regions of calf lenses, and immunolabeled essentially as described by Gruijters et al. (1987). (A) Cx46, detected with fluorescein-conjugated secondary antibody, is localized to discrete plaques on the fiber cell surface. (B) MP20, detected with rhodamine-conjugated secondary antibody, was dispersed on the fiber cell surface and did not specifically co-localize with Cx46. Bar = 20 μm.

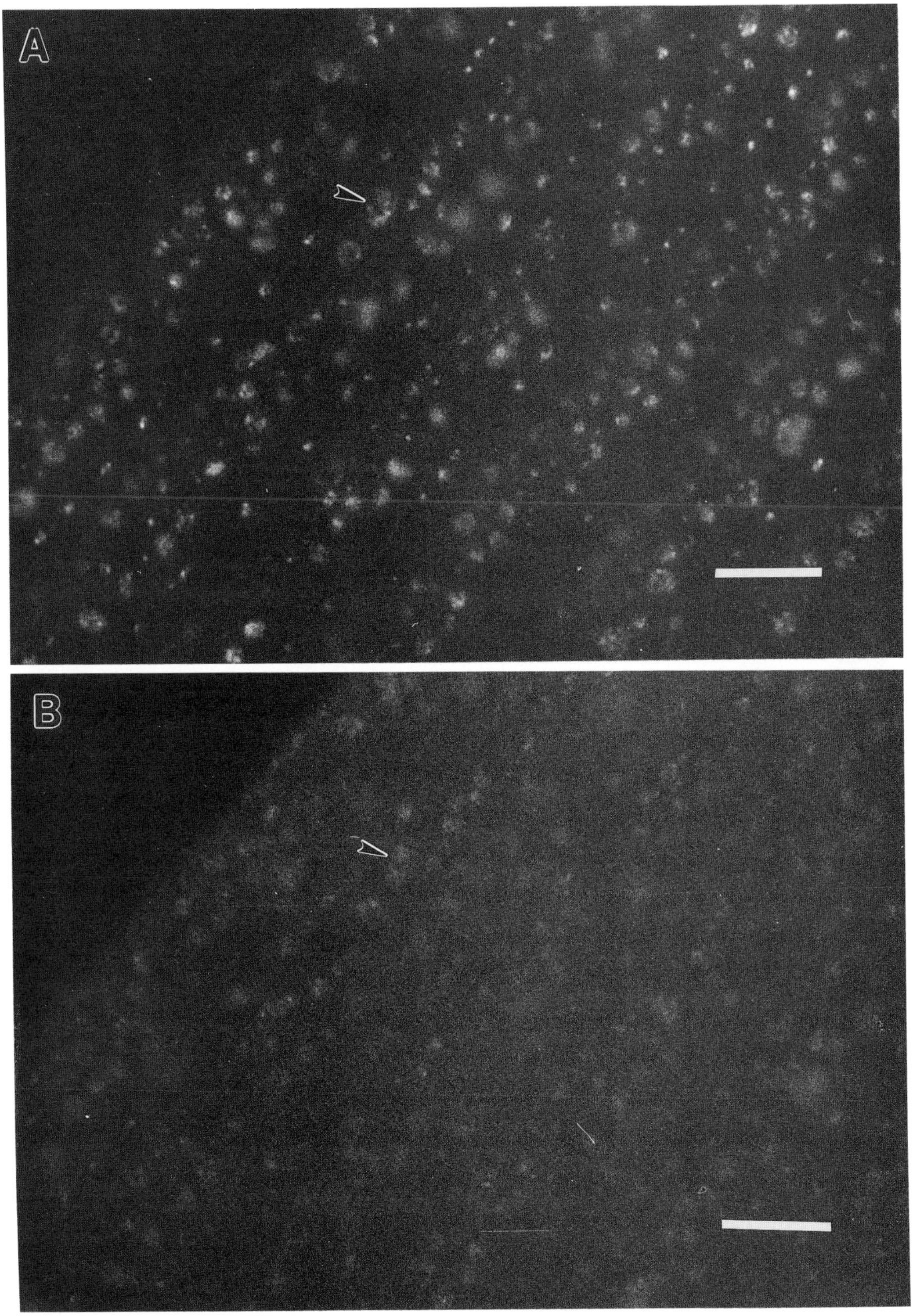

Figure 4. Immunofluorescent double labeling, with Cx46 and MP20 antibodies, of lens fiber bundles dissected from approximately 0.5 to 1.0 mm in the lens equatorial region. (A) Cx46 was localized in discrete plaques on the fiber-cell surface. (B) MP20 co-localized with Cx46 in these plaques (arrowhead). Bar = 20 μm.

strated a diffuse distribution of MP20 on the broad and narrow faces of fiber cells throughout the lens (TenBroek et al., 1992).

Immunogold electron microscopic localization of Cx46 and MP20 in lens membranes

Confirmatory studies were performed by immunolabeling purified lens fiber cell membranes with antibodies to both MP20 and Cx46 followed by a gold-conjugated secondary antibody. MP20 was clearly localized to both sides of the pentalaminar junctions; in addition, there was a small amount of labeling of single membranes with the MP20 antibody (Fig. 5A–C). In contrast, Cx46 labeling was confined to the pentalaminar junctions, confirming the restriction of this protein to the junctions (Paul et al., 1991) (Fig. 5D). The immunogold labeling data confirm the presence of MP20 in lens gap junctions, and further indicate that unlike Cx46, MP20 is also present in extra-junctional regions. The distribution of MP20 is remarkably similar to that reported for MP26 (Gruijters, 1989): Both can be localized to junctions in a small region of the developing lens. At later stages of lens fiber cell development, although labeling of ball and socket domains may be higher than that of the broad side of the fibers, both proteins are more uniformly distributed over the fiber cell membrane.

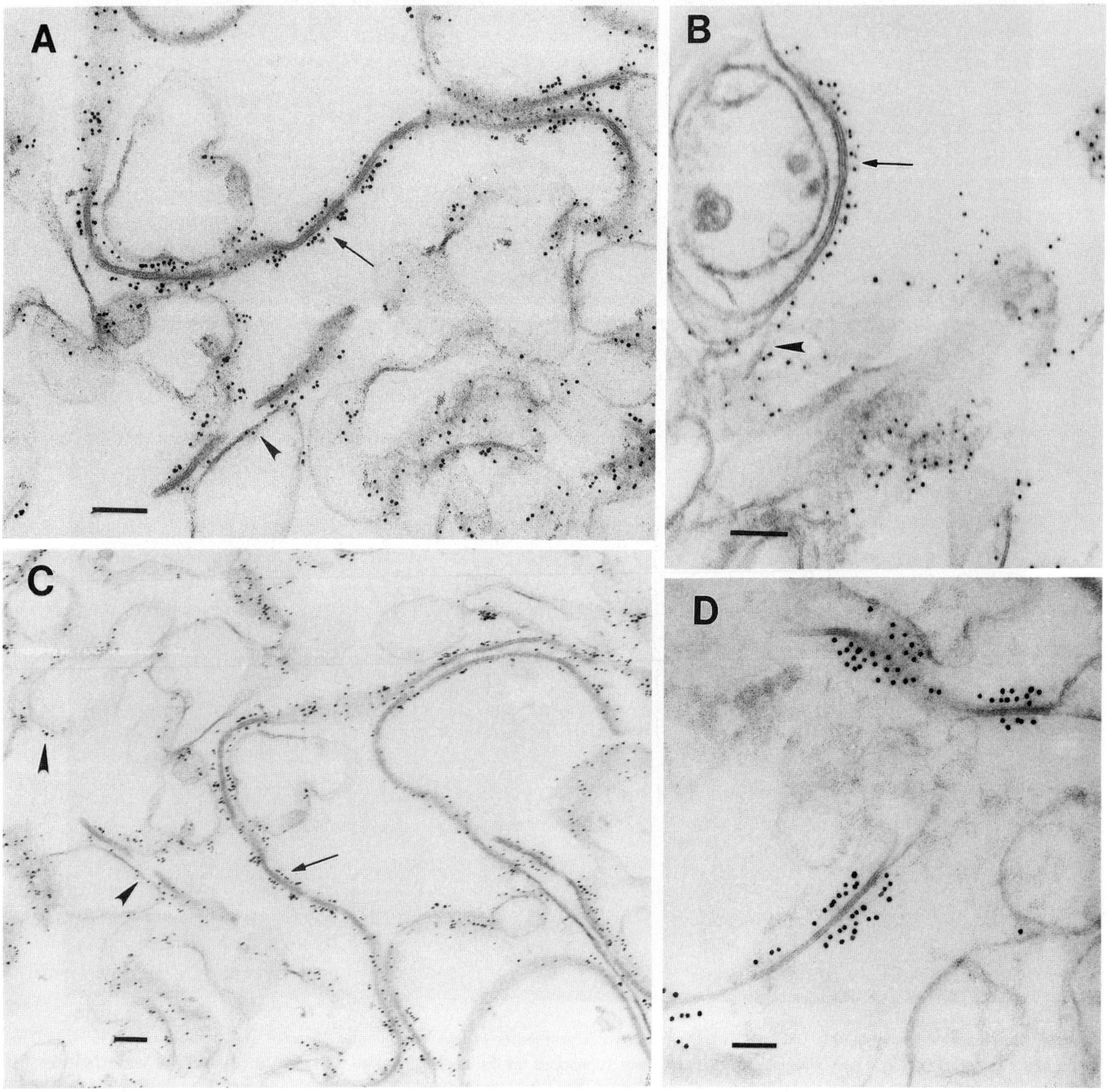

Figure 5. Immunogold electron microscopic localization of MP20 and Cx46 in lens fiber cell membranes. Urea-washed lens membranes were labeled with MP20 antibody (A–C), or Cx46 antibody (D), followed by gold-conjugated secondary antibody. Bar = 0.1 μm. While both antibodies label the pentalaminar gap junctions (arrows), there is some labeling of single membranes with the anti-MP20 antibody (arrowheads).

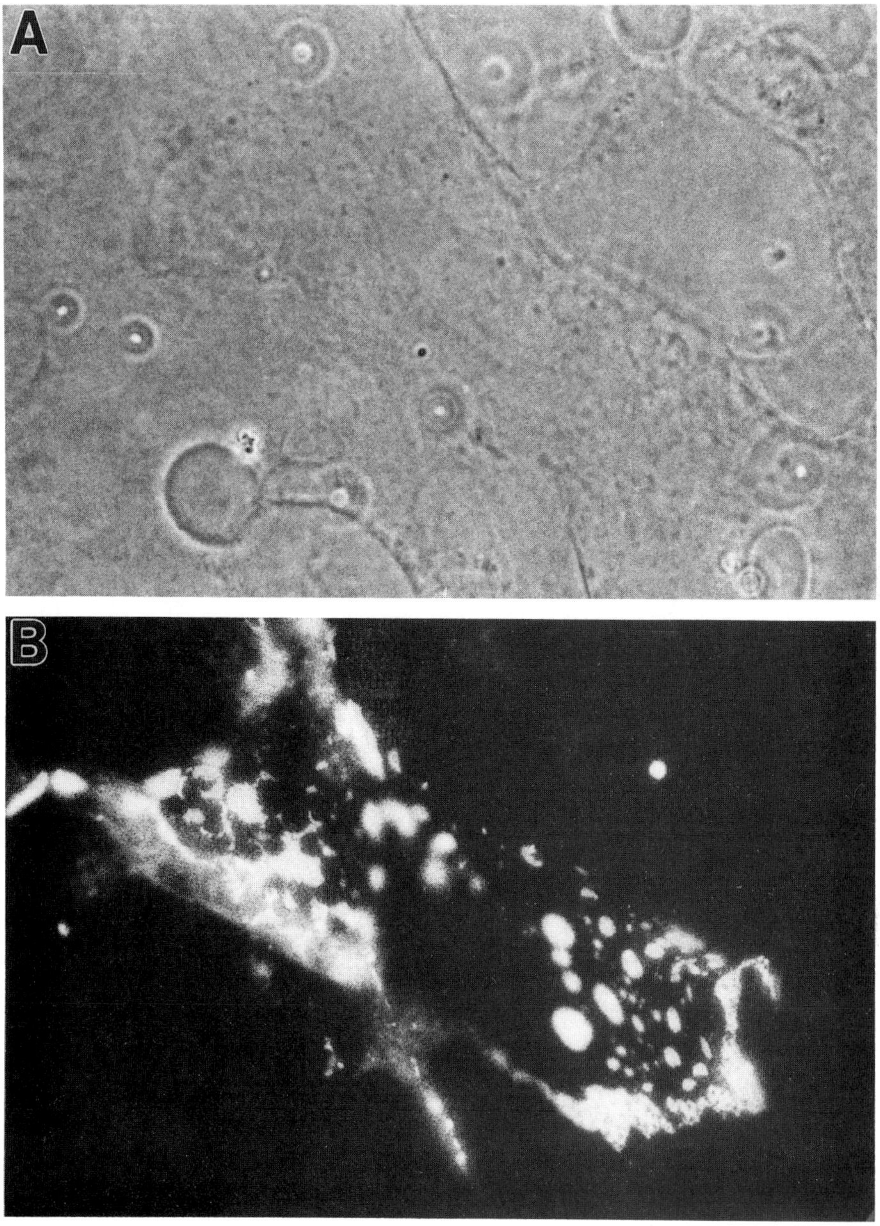

Figure 6. Immunofluorescent Cx46 antibody labeling of a differentiating cell from an ovine lens primary culture. Phase contrast micrograph (A) of a differentiating cell, 10 days in culture, labeled with Cx46 antibody (1:50), and detected with rhodamine-conjugated secondary antibody (1:250) (B) ×1400.

The distribution of lens membrane proteins in an ovine lens primary culture system

To identify the mechanisms regulating the developmental and structural organization of gap junctions in the mammalian lens, it would be highly advantageous to have a primary culture of differentiating lens cells. While differentiation of chick lens epithelial cells occurs in culture (Menko et al., 1984), many analogous proteins from mammalian lenses are absent or differ significantly in the chicken lens (Pitiagorsky, 1981). Although both human and rat lens cultures have been developed (Hamada et al., 1979; Creighton et al., 1981; Fitzgerald and Goodenough, 1986; Arita et al., 1990), no functional gap junctions have been documented (Fitzgerald and Goodenough, 1986). Therefore, our laboratory has focused on cultured ovine lenses; these contain a large number of differentiating lens cells and, unlike rat lens cultures (Fitzgerald and Goodenough, 1986), the differentiating regions exhibit cell-to-cell transfer of microinjected Lucifer Yellow. The transfer is inhibited by mM concentrations of n-octanol, indicating that functional junctions are present.

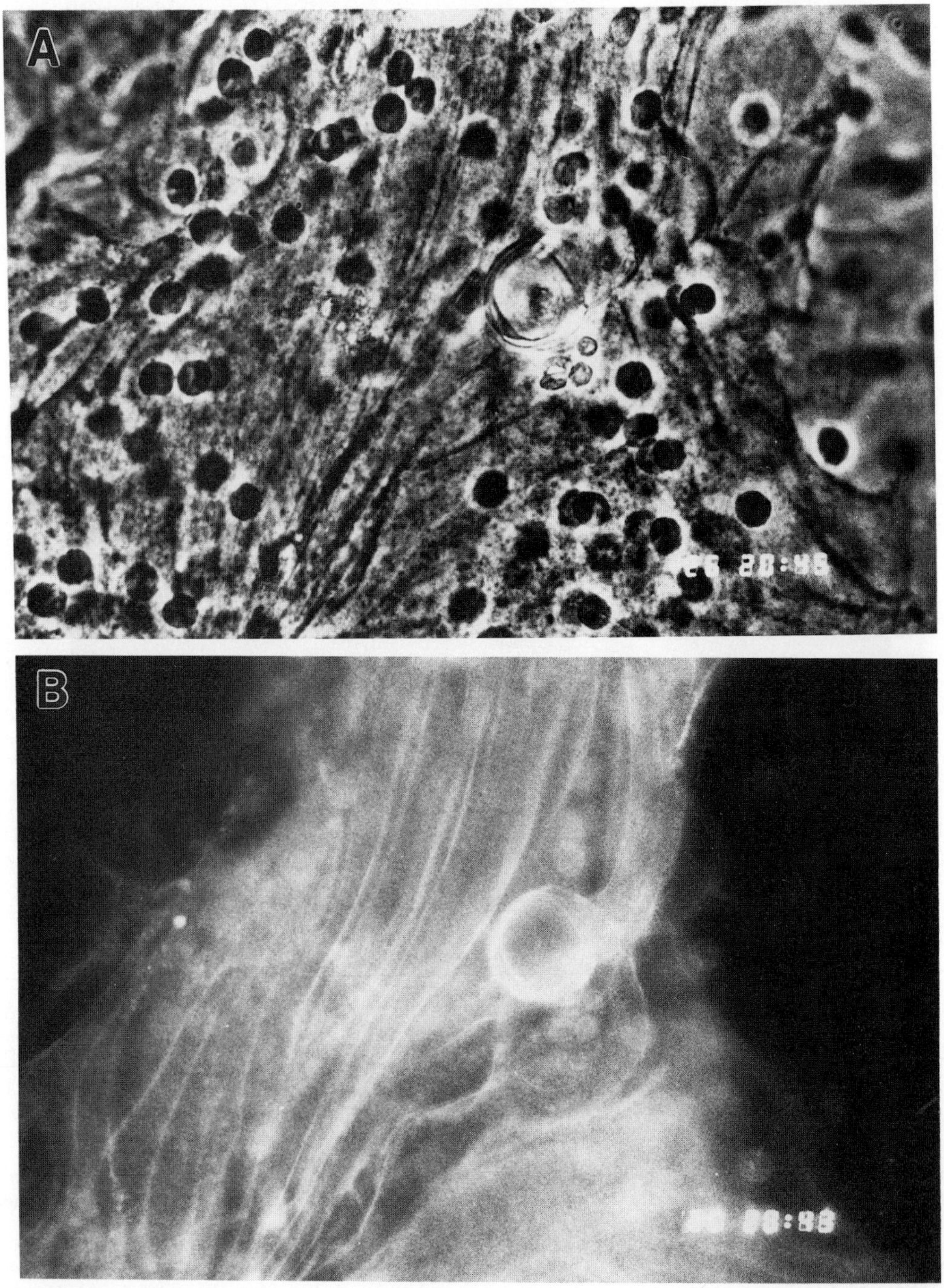

Figure 7. Immunofluorescent MP20 labeling of a differentiating region of ovine lens cultures. Phase contrast micrograph (A) of a region of elongated differentiating cells, 10 days in culture. MP20 antibody (1:100) was detected with a rhodamine-conjugated secondary antibody (1:250) (B) ×1400.

The differentiating lens cells of these cultures appear as large, clear cells overlying an epithelial cell layer (Fig. 6A). Fluorescent Cx46 antibody revealed both large, plaque-like and smaller, punctate labeling of fiber-like cells (Fig. 6B). The plaque-like labeling is remarkably similar to that seen in the fiber-cell-bundle immunolabeling in Figs. 3 and 4; there was essentially no labeling of the epithelial cell layer with this antibody (Fig. 6B).

Membranes in regions containing aggregated, differentiated cells displayed considerable fluorescence with the antibody to MP20 (Fig. 7). There was no labeling of the epithelial cell layer. The labeling with this antibody was very different from that using the antibody to Cx46: MP20 immunofluorescence was diffuse, lacking the discrete foci characteristic of junctional plaques. This supports data from immunolabeled fiber cell bundles, i.e., MP20 is restricted to junctional plaques during a small period of fiber cell differentiation; the bulk of MP20 in the lens is distributed throughout the lens fiber cell plasma membrane.

Our experiments using different antibody markers indicate that the differentiation of ovine lens cells in culture resembles the process of differentiation in the outer cortical region of the lens. The cells not only display conventional markers of differentiation (e.g. β-crystallin accumulation), but also have connexin-related proteins in junctional regions between cells. Thus, the ovine culture system provides easy access to mammalian cells which are analogous to lens fiber cells heretofore inaccessible. This system should enable closer examination of the factors regulating the developmental and structural organization of gap junctions in the mammalian lens.

Acknowledgements

We acknowledge the assistance and good humor of Dr. Terry Gruijters, who visited during a Minnesota winter to show us how to dissect and immunolabel lens fiber bundles. Research supported by NIH grant EY05684.

References

Arita, T., Lin, L., Susan, S.R. and Reddy, V.N. (1990) Enhancement of differentiation of human lens epithelium in tissue culture by changes in cell-substrate adhesion. Invest. Ophthalmol. Vis. Sci. 31, 2395–2404.

Beyer, E.C., Kistler, J., Paul, D.L. and Goodenough, D.A. (1989) Antisera directed against connexin 43 peptides react with a 43-kD protein localized to gap junctions in myocardium and other tissues. J. Cell. Biol. 108, 595–605.

Cooper, K., Mathias, R.T. and Rae, J.L. (1991). The physiology of lens junctions. In: C. Peracchia (Ed.), Biophysics of Gap-Junction Channels. CRC Press, Boca Raton, FL, pp. 57–74.

Costello, M.J., McIntosh, T.J. and Robertson, J.D. (1989) Distribution of gap junctions and square array junctions in the mammalian lens. Invest. Ophthalmol. Vis. Sci. 30, 975–989.

Creighton, M.O., Mousa, G.Y., Miller, G.G., Blair, D.G. and Trevithick, J.R. (1981) Differentiation of rat lens epithelial cells in tissue culture. IV. Some characteristics of the process, including possible *in vitro* models for pathogenic processes in catatactogenesis. Vis. Res. 21, 25–35.

Ehring, G.R., Zampighi, G., Horwitz, J., Bok, D. and Hall, J.E. (1990) Properties of channels reconstituted from the major intrinsic protein of lens fiber membranes. J. Gen. Physiol. 96, 631–664.

Fitzgerald, P.G., Bok, D. and Horwitz, J. (1985) The distribution of the main intrinsic membrane polypeptide in ocular lens. Current Eye Res. 4, 1203–1217.

Fitzgerald, P.G. and Goodenough, D.A. (1986) Rat lens cultures: MIP expression and domains of intercellular coupling. Invest. Ophthalmol. Vis. Sci. 27, 755–771.

Galvan, A., Lampe, P.D., Hur, K.-C., Howard, J., Eccleston, E., Arneson, M. and Louis, C.F. (1989) Structural organization of the lens fiber cell plasma membrane protein MP18. J. Biol. Chem. 264, 19974–19978.

Girsch, S.J. and Peracchia, C. (1985) Lens cell-to-cell channel protein: I. Self assembly into liposomes and permeability regulation by calmodulin. J. Membr. Biol. 83, 217–225.

Goodenough, D.A. (1979) Lens gap junctions: a structural hypothesis for non-regulated low-resistance intercellular pathways. Invest. Ophthalmol. Vis. Sci. 18, 1104–1122.

Goodenough, D.A., Dick, J.S.B. and Lyons, J.E. (1980) Lens metabolic cooperation: A study of mouse lens transport and permeability visualized with freeze-substitution autoradiography and electron microscopy. J. Cell. Biol. 86, 576–589.

Gorin, M.B., Yancey, S.B., Cline, J., Revel, J.-P. and Horwitz, J. (1984) The major intrinsic protein (MIP) of the bovine lens fiber cell membrane: characterization and structure based on cDNA cloning. Cell 39, 49–59.

Gruijters, W.T.M. (1989) A non-connexon protein (MIP) is involved in eye lens gap-junction formation. J. Cell. Sci. 93, 509–513.

Gruijters, W.T.M., Kistler, J. and Bullivant, S. (1987a) Formation, distribution and dissociation of intercellular junctions in the lens. J. Cell. Sci. 88, 351–359.

Gruijters, W.T.M., Kistler, J., Bullivant, S. and Goodenough, D.A. (1987b) Immunolocalization of MP70 in lens fiber 16–17 nm intercellular junctions. J. Cell. Biol. 104, 565–572.

Gutekunst, K.A., Rao, G.N. and Church, R.L. (1990) Molecular cloning and complete nucleotide sequence of the cDNA encoding a bovine lens intrinsic membrane protein (MP19). Curr. Eye Res. 9, 955–961.

Hur, K.C. and Louis, C.F. (1989) Regional distribution of the enzymes and substrates mediating the action of cAMP in the mammalian lens. Biochim. Biophys. Acta 1010, 56–63.

Hamada, Y., Watanabe, K., Aoyama, H. and Okada, T.S. (1979) Differentiation and dedifferentiation of rat lens epithelial cells in short- and long-term cultures. Dev., Growth Diff. 21, 205–220.

Iwata, S. (1985) Calcium pump and its modulator in the lens: a review. Curr. Eye Res. 4, 299–305.

Kistler, J., Kirkland, B. and Bullivant, S. (1985) Identification of a 70,000-kD protein in lens membrane junctional domains. J. Cell. Biol. 101, 28–35.

Kistler, J., Christie, D. and Bullivant, S. (1988) Homologies between gap-junction proteins in lens, heart and liver. Nature 331, 721–723.

Lampe, P.D., Bazzi, M.D., Nelsestuen, G.L. and Johnson, R. (1986) Phosphorylation of bovine lens membranes by protein kinase C. Eur. J. Biochem. 156, 351–357.

Liu, Y.P. and Schwartz, H.S. (1978) Protein activator of cyclic AMP phosphodiesterase and cyclic nucleotide phosphodiesterase in bovine retina and bovine lens. Biochim. Biophys. Acta 526, 186–193.

Louis, C.F., Johnson, R., Johnson, K. and Turnquist, J. (1985) Characterization of the bovine lens plasma membrane substrates for cAMP-dependent protein kinase. Eur. J. Biochem. 150, 279–286.

Louis, C.F., Mickelson, J.R., Turnquist, J., Hur, K.-C. and Johnson, R. (1987) Regulation of lens cyclic nucleotide metabolism by Ca^{2+} plus calmodulin. Invest. Ophthalmol. Vis. Sci. 28, 806–814.

Louis, C.F., Hur, K.-C., Galvan, A.C., TenBroek, E.M., Jarvis, L.J., Eccleston, E.D. and Howard, J.B. (1989) Identification of an

18,000-dalton protein in mammalian lens fiber cell membranes. J. Biol. Chem. 264, 19967–19973.

Louis, C.F., Hogan, P., Visco, L. and Strasburg, G. (1990) Identity of the calmodulin-binding proteins in bovine lens plasma membranes. Exp. Eye Res. 50, 495–503.

Menko, S.A., Klukas, K.A. and Johnson, R.G. (1984) Chick embryo lens cultures mimic differentiation in the lens. Dev. Biol. 103, 129–141.

Paul, D.L. and Goodenough, D.A. (1983) Preparation, characterization and localization of antisera against bovine MP26, an integral protein from lens fiber plasma membrane. J. Cell. Biol. 96, 625–632.

Paul, D.L., Ebihara, L., Takemoto, L.J., Swenson, K.I. and Goodenough, D.A. (1991) Connexin46, a novel lens gap-junction protein, induces voltage-gated currents in non-junctional plasma membrane of *Xenopus* oocytes. J. Cell Biol. 115, 1077–1089.

Piatigorsky, J. (1981) Lens differentiation in vertebrates: A review of cellular and molecular features. Differentiation 19, 134–153.

Sas, D.F., Sas, M.J., Johnson, K.R., Menko, A.S. and Johnson, R.G. (1985) Junctions between lens fiber cells are labeled with a monoclonal antibody shown to be specific for MP26. J. Cell Biol. 100, 261–225.

Strasburg, G.M., Hogan, M.H., Birmachu, W., Thomas, D.D. and Louis, C.F. (1988) Site-specific derivatives of wheat germ calmodulin. J. Biol. Chem. 263, 542–548.

TenBroek, E.M., Arneson, M., Jarvis, L. and Louis, C.F. (1992) The relationship between the fiber cell intrinsic membrane proteins MP20 and MP70 in the mammalian lens. J. Cell Sci. 103, 245–257.

Voorter, C.E.M., Kistler, J., Gruijters, W.T.M., Mulders, J.W.M., Christie, D. and de Jong, W.W. (1989) Distribution of MP17 in isolated lens fibre membranes. Curr. Eye Res. 8, 697–706.

Zampighi, G.A., Hall, J.E., Ehring, G.R. and Simon, S.A. (1989) The structural organization and protein composition of lens fiber junctions. J. Cell Biol. 108, 2255–2276.

J.E. Hall, G.A. Zampighi and R.M. Davis (Eds.)
Progress in Cell Research, Vol. 3

CHAPTER 26

Are cardiac gap junctions voltage sensitive?

H.J. JONGSMA, R. WILDERS, B.R. TAKENS-KWAK and M.B. ROOK

Department of Physiology, University of Amsterdam, Meibergdreef 15, 1105 AZ Amsterdam, The Netherlands

Introduction

Several authors (Spray et al., 1984; Metzger and Weingart, 1985; Noma and Tsuboi, 1987; Burt and Spray, 1988) have shown that gap junctions between pairs of cardiac myocytes behave like ohmic conductors, i.e., the intercellular current flow is directly proportional to the transcellular voltage difference. The overall junctional conductance between pairs of mammalian cardiomyocytes ranges from 4 to 4000 nS (White et al., 1985; Maurer and Weingart, 1987; Noma and Tsuboi, 1987). With a single gap-junctional channel conductance, γ_j, of around 50 pS (Burt and Spray, 1988; Rook et al., 1988; Rüdisüli and Weingart, 1989), this means that these gap junctions consist of several hundred to several thousand channels. On the other hand, gap junctions composed of relatively few channels exhibit pronounced voltage sensitivity: transcellular voltage steps larger than 30 to 50 mV induce junctional currents which diminish during the voltage step (Rook et al., 1988, 1989; Veenstra, 1990). The time constant of relaxation decreases with increasing transcellular voltage. Measurement of the amplitude of single-channel currents at different transjunctional voltages indicates that γ_j itself is voltage insensitive; this suggests that the voltage sensitivity of gap-junction channels resides in the gating mechanism (Jongsma and Rook, 1990).

To determine why large gap junctions do not exhibit voltage sensitivity, we constructed a simple model to calculate the electrical field sensed by gap junctions in relation to the number of channels comprised therein (Jongsma et al., 1991). The computations were kept manageable by invoking simplified gap-junction structures, thus keeping the model rotationally symmetrical. We were able to establish the electrical field distribution across gap junctions comprising various numbers of channels. It appeared that large gap junctions sense only part of the electrical field applied between two patch electrodes when the double whole-cell voltage-clamp configuration is used to assay the conductance properties of gap junctions (Neyton and Trautmann, 1985). The magnitude of the effect indeed depended on the number of channels, however, not as strongly as suggested earlier (Rook et al., 1988, 1992).

To resolve this difference between theory and experiment, we have extended our analysis to the full three-dimensional case, using a different method (Wilders and Jongsma, 1992). We also account for the effect of pipette series resistance on the apparent junctional conductance. We then model the behavior of ensembles of gap-junction channels as they open and close stochastically.

Methods

Calculation of transjunctional voltage difference

Both cells of a pair connected by a gap junction are modeled as right-cylindrical volumes of diameter ≈35 μm, length ≈35 μm (Fig. 1A). Cell A (cell B) is voltage clamped at membrane potential V_A (V_B) using a patch pipette. Without loss of generality, we assume $V_A \geq V_B$. Each gap-junction channel is modeled as a right-cylindrical volume, with a diameter of 2 nm and a length of 20 nm. The model gap junction is assumed to have a disc-like cross-section, with the channels arranged in a hexagonal grid with a grid spacing of 9 nm (Fig. 1B). Consequently, the relation between the diameter of the gap junction, d_{gj}, and the number of channels, N, is

$$N = 1 + \frac{(d_{gj} - 9)(d_{gj} + 9)}{108} \tag{1}$$

If the gap junction consists of one single pore, the model shows rotational symmetry. In this case, using cylindrical coordinates, Laplace's three-dimensional potential equation can be reduced to a two-dimensional equation that can be readily solved using numerical techniques (Jongsma et al., 1991). From Fig. 4A of Jongsma et al. (1991), which shows a plot of equipotential planes, it becomes clear that the current density at the mouth of the pore is approximately constant. According to a well-known result from electrostatics, the cytosolic access resistance $R_{cytosol,A}$, i.e. the resistance along the convergent current paths from the "bulk" cytosol of cell A to the narrow pore, is then given by

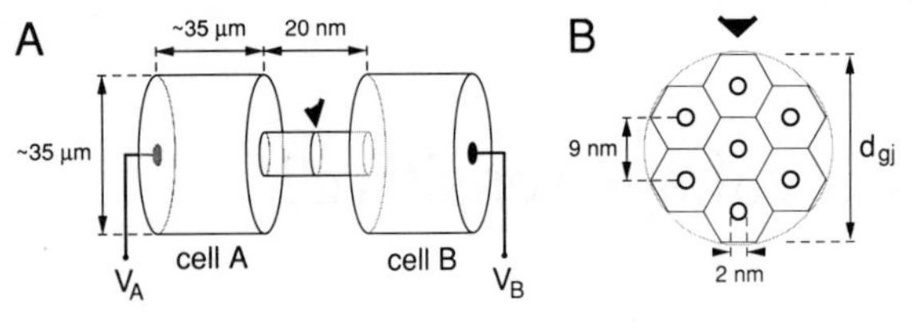

Figure 1. (A) Model of a cell pair connected by a gap junction. Cells and gap junction are modeled as right cylindrical volumes. V_A = membrane potential of cell A; V_B = membrane potential of cell B. (B) Cross-section of the model gap junction. The gap-junction channels, with a pore diameter of 2 nm, are arranged in a hexagonal grid with a grid spacing of 9 nm. d_{gj} = diameter of gap junction.

$$R_{cytosol,A} = \rho/\pi r_c \tag{2}$$

where ρ is the resistivity of the cytosolic medium and r_c is the radius of the pore. Assuming the conductive medium in cytosol and pore to be the same, pore resistance R_{pore} is given by

$$R_{pore} = \rho l_c/\pi r_c^{\,2} \tag{3}$$

where l_c is the length of the pore. Thus, the actual voltage difference across the pore, denoted by ΔV_j, is given by

$$\Delta V_j = \frac{R_{pore}}{R_{pore} + R_{cytosol,A} + R_{cytosol,B}} \cdot (V_A - V_B) \tag{4}$$

Combining Equations 2–4 yields

$$\Delta V_j = \frac{V_A - V_B}{1 + 2r_c / l_c} \tag{5}$$

According to Equation 5, in the single-pore case, the voltage difference across the pore is only 91% of $V_A - V_B$. Note that, also according to Equation 5, V_j does not depend on ρ. In a gap junction consisting of N channels ($N \geq 2$), the voltage difference across pore m, denoted by $\Delta V_{j,m}$, m = 1,2,…,N, is even a smaller fraction of $V_A - V_B$, due to the additional cytosolic voltage drops brought about by the remaining N – 1 channels. According to another well-known result from electrostatics, the cytosolic access resistance of channel n at the mouth of channel m, denoted by $R_{m,n}$, is given by

$$R_{m,n} = \rho/2\pi r_{m,n} \tag{6}$$

where $r_{m,n}$ is the spacing of channels m and n (Fig. 2). Denoting by $V_{m,A}$ ($V_{n,A}$) and $I_{j,m}$ ($I_{j,n}$) the potential at the mouth of pore m (n) in cell A and the strength of the junctional current through channel m (n) respectively, we have

$$V_{m,A} = V_A - I_{j,m} R_{cytosol,A} - \sum_{n \neq m} I_{j,n} R_{m,n} \tag{7}$$

and, in cell B,

$$V_{m,B} = V_B + I_{j,m} R_{cytosol,B} + \sum_{n \neq m} I_{j,n} R_{m,n} \tag{8}$$

Combining Equations 2, 3, 6, 7 and 8

$$\Delta V_{j,m} + \sum_{n \neq m} \frac{r_c^2}{l_c + 2r_c} \cdot \Delta V_{j,n} = \frac{V_A - V_B}{1 + 2r_c / l_c} \tag{9}$$

Equation 9 defines a set of N equations that can be readily solved to give the N unknown transjunctional voltage differences $\Delta V_{j,m}$, m = 1,2,…,N. Next $I_{j,m}$, m=1,2,…,N, can be calculated from

$$I_{j,m} = \gamma_j \Delta V_{j,m} \tag{10}$$

where γ_j is the single-channel conductance, which we have chosen to be 50 pS (see above). The series resistance of the microelectrodes, i.e. the access resistance from pipette interior to cell interior, generally ranges from 20 to 60 MΩ (Rook et al., 1988). Thus, if the junctional current is large, a considerable voltage drop across the pipettes accounts for a further decrease in ΔV_j. This voltage drop ΔV_s is given by

$$\Delta V_s = \left(\sum_{m=1}^{N} I_{j,m} \right) (R_{s,A} + R_{s,B}) \tag{11}$$

where $R_{s,A}$ ($R_{s,B}$) is the series resistance of the micro-electrode in cell A (B). Combining Equations 9–11, one can easily calculate, at a given configuration of gap-junction channels, $\Delta V_{j,m}$ and $I_{j,m}$, m = 1,2,…,N, once γ_j, $R_{s,A}$, $R_{s,B}$, and the applied voltage difference between the electrodes, denoted by ΔV_{app}, are known. In the above, the non-junctional membrane current of the cells has not been considered. Generally, due to the very high membrane resistance, typically 2 to 5 GΩ (Rook et al., 1988), this current is so small that its contribution to ΔV_s can be ignored.

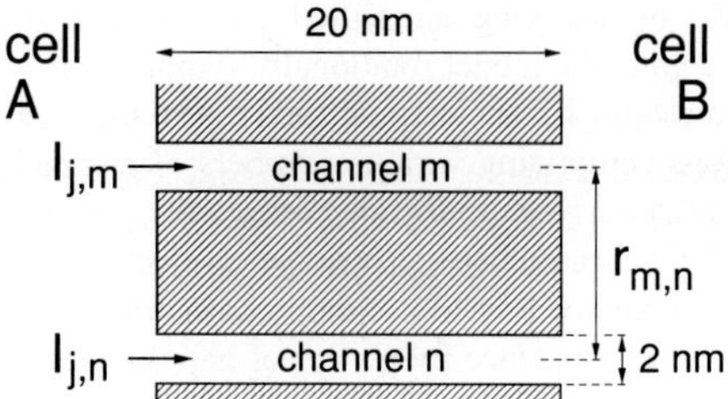

Figure 2. Longitudinal section through channels m and n of the model gap junction. $I_{j,m}$ = junctional current through channel m; $I_{j,n}$ = junctional current through channel n; $r_{m,n}$ = center-to-center spacing of channels m and n.

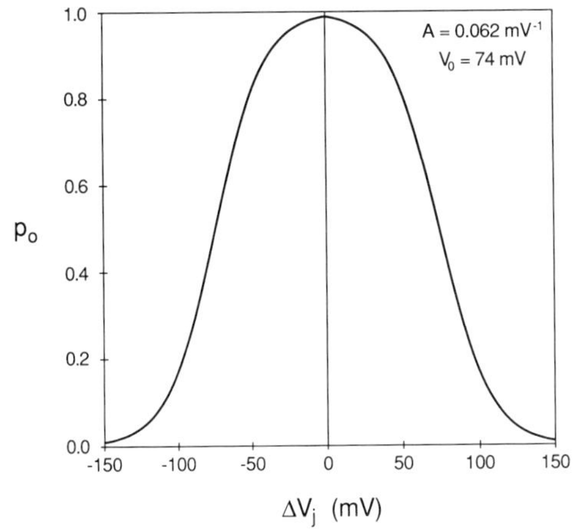

Figure 3. Relation between open probability (p_o) of an individual gap-junction channel and voltage difference across the channel (ΔV_j). Curve drawn according to Equation 14.

Gating of gap-junction channels

Following Harris et al. (1981), the model channel exists in either of two states, an open state with conductance γ_j and a closed state of zero conductance, and displays first-order kinetics, described by rate constants, α and β, that are functions of the transjunctional voltage difference

$$\alpha = \lambda \exp(-A_\alpha(\Delta V_j - V_0)) \tag{12}$$

and

$$\beta = \lambda \exp(A_\beta(\Delta V_j - V_0)) \tag{13}$$

where A_α and A_β are voltage sensitivities of α and β respectively, λ is the rate at which α equals β, and V_0 is the transjunctional voltage difference at which α equals β. Consequently, the open probability of the channel, denoted by p_o, is given by

$$p_o = \frac{1}{1 + \exp((A_\alpha + A_\beta)(\Delta V_j - V_0))} \tag{14}$$

The parameters λ, A_α A_β and V_0 of Equations 12–14 can be estimated from the open and closed times of a single gap-junctional channel measured by Rook et al. (1988), giving $\lambda = 2.5\ s^{-1}$, $A_\alpha = 0.041\ mV^{-1}$, $A_\beta = 0.021\ mV^{-1}$ and $V_0 = 74$ mV. The resulting open probability is depicted in Fig. 3. The rate constants α and β are the parameters of the negative exponential distributions of the single-channel open and closed times respectively (Colquhoun and Hawkes, 1983). Thus, a particular closed time, t_c, can be obtained by drawing a random number x from the uniform distribution on [0,1] and calculating t_c from

$$t_c = \frac{-\log x}{\alpha(\Delta V_j)} \tag{15}$$

(Clay and DeFelice, 1983). A particular open time, t_o, can be obtained in a similar way:

$$t_o = \frac{-\log x}{\beta(\Delta V_j))} \tag{16}$$

Junctional current during a voltage-clamp step

It is reasonable to assume that at $\Delta V_{app} = 0$ mV, and hence $\Delta V_j = 0$ mV, all channels are in the open state, because p_o is very close to 1 at $\Delta V_j = 0$ mV (Fig. 3). If, however, ΔV_{app} is stepped from 0 mV to some non-zero value, p_o becomes less than 1 and a stochastic process of channel closures, and subsequent openings, is initiated. This process is modeled as follows: Starting with all channels in the open state, $\Delta V_{j,m}$ and $I_{j,m}$, m = 1,2,...,N, are calculated from Equations 9–11. Next, Equation 15 is used to calculate the time at which the first channel closes, resulting in a change in the configuration of open gap-junction channels. Therefore, $\Delta V_{j,m}$ and $I_{j,m}$, m = 1,2,...,N, are recalculated, using the appropriate forms of Equations 9–11. Next, Equations 15 and 16 are used to calculate the time at which the next channel opening or closure takes place. These calculations are repeated until model time exceeds the duration of the voltage-clamp step.

Results and discussion

Figure 4 depicts the results of our model calculation for the static case, i.e. assuming that the junctional channels do not gate. As expected, it can be seen that the fraction of applied voltage actually present at the mouth of the channels decreases when the number of channels in a junction connecting adjacent cells increases. Three cases are presented: the upper trace shows the voltage sensed at the mouth of the channels in the ideal situation when the series resistance of the pipettes is zero; the lower two traces were obtained with pipette series resistances of 20 and 60 MΩ, respectively. These values comprise the range of pipette series resistances reported by Rook et al. (1988). It is apparent from these plots that the combined effect of cytosolic access resistance and pipette series resistance on the difference between the voltage applied and the voltage actually sensed by the junctional channels, is considerable, especially for large gap junctions.

It has been shown before that voltage dependence of mammalian gap-junction channels only occurs when the transjunctional voltage difference is larger than a certain "threshold" value. For liver gap junctions, a value of 12–20 mV was reported (Moreno et al., 1991), while for heart gap junctions, values between 30 mV (Veenstra, 1990) and 50 mV (Rook et al., 1988) were found. For an applied voltage difference of 100 mV, the 50 mV thresh-

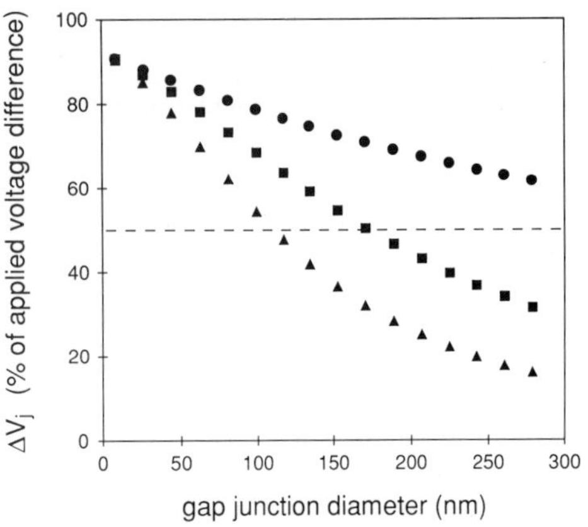

Figure 4. Relation between fraction of applied voltage difference actually sensed by the central pore of the model gap junction and gap-junction diameter, at pipette series resistances of 0 MΩ (circles), 20 MΩ (squares) and 60 MΩ (triangles). (See text.)

old is indicated by the interrupted line in Fig. 4. It can be seen that, depending on the series resistance of the pipettes, voltage dependence appears to be absent already in rather small junctions. In the examples given, gap junctions containing 130 or more channels would appear voltage independent when voltage clamping them using pipettes with a series resistance of 60 MΩ. For pipettes with a series resistance of 20 MΩ, a value that can easily be encountered even when series resistance is compensated, junctions containing 300 or more channels would appear voltage independent. Limiting the transjunctional voltage to lower values would make the detection of voltage dependence even more difficult. In reality, the situation is more complicated, because part of the junctional current is shunted through the membrane resistance to earth (Neyton and Trautmann, 1985; Rook et al., 1988). It is not surprising, therefore, that voltage dependence was reported to be absent in the rather large gap junctions between liver cells which have a high non-junctional membrane conductance (Spray et al., 1986). In a recent paper, the same laboratory reported that voltage dependence of liver gap-junction channels can be demonstrated when care is taken to avoid the problems of pipette series resistance (Moreno et al., 1991). Heart gap-junction channels are less steeply voltage dependent than liver gap-junction channels and have a higher "threshold" for voltage dependence to become apparent. According to Gourdie et al. (1991), the majority of gap junctions between atrial and ventricular cells in mammalian hearts have a diameter of 200 nm or more (although it should be kept in mind that these data obtained from fluorescently labeled antibody staining may suffer from a bias towards larger diameters due to the inability of the method to visualize smaller diameters) and thus contain 400 channels or more. Therefore, it is not surprising that voltage dependence has never been reported for gap junctions between pairs of cells isolated from mammalian hearts.

Many authors (see e.g. White et al., 1985; Spray et al., 1986; Maurer and Weigart, 1987) estimate the amount of coupling between cells, calculating total gap-junction conductance by dividing the junctional current measured by the voltage difference applied. Due to overestimation of the voltage actually dropping over the junction itself, the real junctional conductance is underestimated. The magnitude of this error strongly depends on the experimental conditions and on the protocol used to assess the junctional conductance. Take, for instance, the case of a 0.15 μm diameter junction consisting of 217 50 pS channels to which a transjunctional voltage difference of 100 mV is applied through pipettes with a series resistance of 60 MΩ each. According to Fig. 4, the voltage difference across the junction proper is about 38 mV. The junctional current measured will be 414 pA and total junctional conductance consequently amounts to 10.9 nS. When the junctional current measured is simply divided by the voltage difference applied, a junctional conductance of 4.14 nS will be obtained, corresponding to a junction containing only 83 channels. Even when the series resistance of the pipettes is known and corrected for, it is difficult to determine the number of channels in a junction from the measured conductance because the access resistance of the junctions depends on the number of channels they contain.

It is well known now that gap-junction channels are gated channels (Neyton and Trautmann, 1985; Burt and Spray, 1988; Rook et al., 1988; Veenstra and DeHaan, 1988) which exhibit voltage and time dependence. Therefore the static case considered above should be modified to account for the voltage and time dependence of the junctional channels in order to interpret experiments designed to measure junctional conductance and to assess its modulation by transjunctional voltage changes. To this end, we developed a simple model based on the Hodgkin-Huxley-like description of gap-junction channel kinetics by Harris et al. (1981), incorporating the stochastic properties of gap-junction channels. The voltage sensitivity of the open and closed times of the channels is determined by α and β (see Methods). Figure 5 compares such a calculation with an experimental record for a gap junction containing seven channels. The time-dependent relaxation of junctional current indicates the voltage sensitivity of the channels, which tend to close when a transjunctional voltage difference of 100 mV is applied. (According to Fig. 4, this amounts to 87 mV across the junction proper). The similarity between the experimental and the calculated record is quite striking. Due to the stochastic nature of channel opening and closure, the individual traces used to generate the average were widely different, however, in both the experimental and calculated individual traces. This same variability has been experimentally observed by Neyton and Trautmann (1985), Giaume et al., (1991) and Veenstra (1990). Only by averaging a sufficient number of traces does the similarity between experiment and model become apparent. Nonetheless, the instantaneous junctional current ($I_{j,0}$) measured at the beginning of the volt-

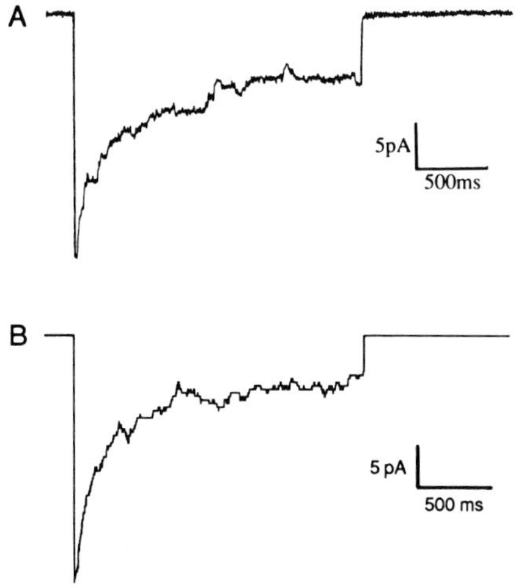

Figure 5. (A) Experimentally observed voltage dependence of a neonatal rat heart gap junction: average of 10 successive junctional current traces, at an applied voltage difference of 100 mV. (B) Voltage dependence of a model gap junction comprising 7 channels: average of 10 junctional current traces. Rate constants as explained in text; $\Delta V_{app} = 100$ mV; $R_{s,A} = R_{s,B} = 20$ MΩ.

age step and the quasi steady-state junctional current ($I_{j,\infty}$) measured at the end of a sufficiently long voltage step, i.e. ≥ 1–2 s, is the same at each voltage difference in each trace. Therefore, individual records can be used to obtain the normalized junctional conductance ($G_{j,\infty} = I_{j,\infty}/I_{j,0}$).

In Fig. 6, $G_{j,\infty}$ calculated from individual records from model experiments on gap junctions containing different numbers of channels was plotted against applied voltage difference. Figures 6A and 6B were obtained with pipette series resistances of 20 and 60 MΩ, respectively. As expected from Fig. 4, the plots for a one-channel junction are virtually identical in Fig. 6A and 6B because the difference in applied voltage and voltage "sensed" by the junction is almost the same for both pipette series resistances. With increasing numbers of channels in the gap junction, at higher applied voltages the downward deflection of the plots decreases because the actual voltage drop over the junctional resistance differs more and more from the applied voltage difference. The effect is stronger with larger pipette series resistances (cf. Fig. 6A and 6B): With pipette series resistances of 60 MΩ, the normalized conductance of a 0.135 μm diameter gap junction containing 169 channels would appear virtually constant in the range of voltage differences usually applied, leading to the erroneous conclusion that the junction has little or no voltage sensitivity. The apparent decrease of voltage sensitivity in larger junctions not only depends on the number of channels in the junction but also on A and V_0 of the Boltzmann distribution describing voltage dependence of the junctional channels. When V_0 decreases, the Boltzmann distribution shifts towards the $G_{j,\infty}$ axis (Fig. 6). As a result, in order for the voltage sensitivity to disappear due to the effects of series and access resistance, gap junctions with a lower V_0 must contain a larger number of channels. This probably explains why voltage sensitivity is easier to detect in large gap junctions from liver ($V_0 \approx$ 40 mV; Moreno et al., 1991) than in heart gap junctions of comparable size $V_0 \approx 70$ mV; Rook et al., 1988). Interventions which diminish p_o of gap-junction channels (e.g. uncoupling of gap junctions by heptanol or octanol (Burt and Spray, 1989; Takens-Kwak et al., 1992) would produce an inward shift of the Boltzmann distribution by decreasing V_0, thus voltage dependence would become apparent in gap junctions which are otherwise too large to show it. Recent experiments in our laboratory (Takens-Kwak et al., 1992) support this premise.

In research on voltage sensitivity of gap junctions, A and V_0 are often obtained by curve-fitting a two-state Boltzmann distribution to normalized conductance versus applied voltage plots. Veenstra (1991) concluded from experiments on embryonic chick heart cell pairs of different gestational ages that A decreased and V_0 increased with age so that gap-junction channel voltage sensitivity decreased during development. As it is well documented

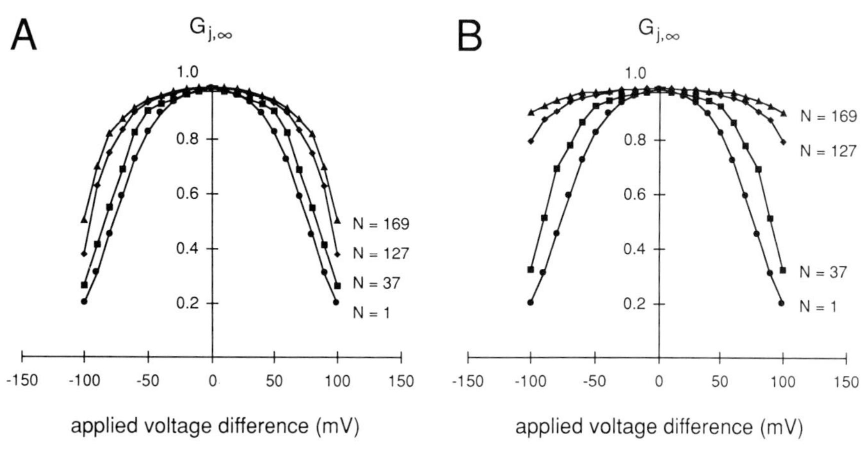

Figure 6. Relation between normalized steady-state conductance ($G_{j,\infty}$) of model gap junctions comprising different numbers of channels and applied voltage difference. (A) $R_{s,A} = R_{s,B} = 20$ MΩ. (B) $R_{s,A} = R_{s,B} = 60$ MΩ (see text).

(see, e.g. Gros et al., 1982) that heart gap junctions increase in size during development an alternative explanation could be that V_0 increased due to the larger contribution of the series- and access-resistance effect in older heart cell pairs. Thus, the apparent decrease of voltage sensitivity during development would not reflect a change in junctional conductance regulation but rather an increase in mean gap-junction size.

In summary, we have provided evidence to answer the question posed in title of this paper in the affirmative: cardiac gap junctions are moderately voltage sensitive. For a variety of reasons this voltage sensitivity cannot be observed in experiments on well-coupled heart cell pairs, mainly due to the presence of gap-junction channel access resistance and pipette-series resistance.

Acknowledgements

We are very grateful to Dr. A. van Oosterom for assistance with the electrical field distribution calculations. This work was supported in part by a grant from the Netherlands Organization for Scientific Research (NWO-SvB grant 810-406-151)

References

Burt, J.M. and Spray, D.C. (1988) Single-channel events and gating behavior of the cardiac gap-junction channel. Proc. Natl. Acad. Sci. USA 85, 3431–3434.

Burt, J.M. and Spray, D.C. (1989) Volatile anesthetics block intercellular communication between neonatal rat myocardial cells. Circ. Res. 65, 829–837.

Colquhoun, D. and Hawkes A.G. (1983) The principles of the stochastic interpretation of ion-channel mechanisms. In: B. Sakmann and E. Neher (Eds), Single Channel Recording, Plenum Press, New York, pp. 135–175.

DeFelice, L.J. and Clay, J.R. (1983) Membrane current and membrane potential from single-channel kinetics. In: B. Sakmann and E. Neher (Eds), Single Channel Recording, Plenum Press, New York, pp. 323–334.

Giaume, C., Fromaget, C., Abdelhakim el Aoumari, Cordier, J., Glowinski, J. and Gros, D. (1991) Gap junctions in cultured astrocytes: single-channel currents and characterization of channel-forming proteins. Neuron 6, 133–143.

Gourdie, R.G., Grenn, C.R. and Severs, N.J. (1991) Gap-junction distribution in adult mammalian myocardium revealed by an anti-peptide antibody and laser scanning confocal microscopy. J. Cell Sci. 99, 41–55.

Gros, D., Lee, I. and Challice, C.E. (1982) Formation and growth of myocardial gap junctions: *in vivo* and *in vitro* studies. In: L.N. Bouman and H.J. Jongsma (Eds.), Cardiac Rate and Rhythm, Nijhoff, Den Haag, pp. 243–264.

Harris, A.L., Spray, D.C. and Bennett, M.V.L. (1981) Kinetic properties of a voltage-dependent junctional conductance. J. Gen. Physiol. 77, 95–117.

Jongsma, H.J. and Rook, M.B. (1990) Cardiac gap junctions: gating properties of single channels. In: A.W. Robards and W.J. Lucas (Eds.), Parallels in Cell-to-Cell Junctions in Plants and Animals, Springer, Heidelberg, pp. 87–99.

Jongsma, H.J., Wilders, R., van Ginneken, A.C.G. and Rook, M.B. (1991) Modulatory effect of the transcellular electrical field on gap-junction conductance. In: C. Peracchia, (Ed.) Biophysics of Gap-Junction Channels. CRC Press, Boca Raton, FL. pp. 163–172.

Maurer, P. and Weingart, R. (1987) Cell pairs isolated from adult guinea pigs and rat hearts: effect of $[Ca^{2+}]_i$ on nexal membrane resistance. Pflügers Arch. 409, 394–402.

Metzger, P. and Weingart, R. (1985) Electrical current flow in cell pairs isolated from adult rat hearts. J. Physiol. 366, 177–195.

Moreno, A.P., Campos de Carvalho, A.C., Verselis, V., Eghbali, B. and Spray, D.C. (1991) Voltage-dependent gap-junction channels are formed by connexin32, the major gap-junction protein of rat liver. Biophys. J. 59, 920–925.

Neyton, J. and Trautmann, A. (1985) Single-channel currents of an intercellular junction. Nature 317, 331–335.

Noma, A. and Tsuboi, N. (1987) Dependence of junctional conductance on proton, calcium and magnesium ions in cardiac paired cells of guinea pig. J. Physiol. 382, 193–211.

Rook, M.B., Jongsma, H.J. and van Ginneken, A.C.G. (1988) Properties of single gap-junctional channels between isolated neonatal rat heart cells. Am. J. Physiol. 255, H770–H782.

Rook, M.B., Jongsma, H.J. and de Jonge, B. (1989) Single-channel currents of homo- and heterologous gap junctions between cardiac fibroblasts and myocytes. Pflügers Arch. 414, 95–98.

Rook, M.B., van Ginneken, A.C.G., de Jonge, B., el Aoumari, A., Gros, D. and Jongsma, H.J. (1992) Differences in gap junction channels between cardiac myocytes, fibroblasts, and heterologous pairs. Am. J. Physiol. 263, in press.

Rüdisüli, A. and Weingart, R. (1989) Electrical properties of gap-junction channels in guinea-pig ventricular cell pairs revealed by exposure to heptanol. Pflügers Arch. 415, 12–21.

Spray, D.C., White, R.L., Campos de Carvalho, A., Harris, A.L. and Bennett, M.V.L. (1984) Gating of gap-junction channels. Biophys. J. 45, 219–230.

Spray, D.C., Ginzberg, R.D., Morales, E.A., Gatmaitan, Z. and Arias, I.M. (1986) Electrophysiological properties of gap junctions between dissociated pairs of rat hepatocytes. J. Cell Biol. 103, 135–144.

Takens-Kwak, B.R., Jongsma, H.J., Rook, M.B. and van Ginneken, A.C.G. (1992) Mechanism of heptanol-induced uncoupling of cardiac gap junctions; a perforated-patch clamp study. Am. J. Physiol. 262, 1531–1538.

Veenstra, R.D. (1990) Voltage-dependent gating of gap-junction channels in embryonic chick ventricular cell pairs. Am. J. Physiol. 258, C662–C672.

Veenstra, R.D. (1991) Developmental changes in regulation of embryonic chick heart gap junctions. J. Membr. Biol. 119, 253–265.

Veenstra, R.D. and DeHaan, R.L. (1988) Cardiac gap-junction channel activity in embryonic chick ventricle cells. Am. J. Physiol. 254, H170–H180.

White, R.L., Spray, D.C., Campos de Carvalho, A.C., Wittenberg, B.A. and Bennett, M.V.L. (1985) Some electrical and pharmacological properties of gap junctions between adult ventricular myocytes. Am. J. Physiol. 249, C447–C455.

Wilders, R. and Jongsma, H.J. (1992) Limitations of the dual voltage clamp method in assaying conductance and kinetics of gap junction channels. Biophys. J. 63, 942–953.

J.E. Hall, G.A. Zampighi and R.M. Davis (Eds.)
Progress in Cell Research, Vol. 3

CHAPTER 27

Trypanosome infection decreases intercellular communication between cardiac myocytes

A.C. CAMPOS DE CARVALHO[a,*], H.B. TANOWITZ[b,c], M. WITTNER[b], R. DERMIETZEL[a,**] and D.C. SPRAY[a]

[a]*Department of Neuroscience,* [b]*Department of Pathology and* [c]*Department of Medicine, Albert Einstein College of Medicine, Bronx, NY 10461, USA*

Introduction

Gap junctions mediate direct intercellular current flow in heart and elsewhere, and disorders in cardiac gap junctions have been proposed to underlie various conduction disturbances. Chagas' disease, which is caused by the protozoan parasite, *Trypanosoma cruzi*, is the leading cause of cardiomyopathy in Latin America. Both acute and chronic Chagas' disease are characterized by cardiac conduction disorders, including ventricular arrhythmias. In order to assess the extent to which gap-junction abnormalities might contribute to aberrant current flow between *T. cruzi*-infected cardiac cells, we have assayed synchrony of contraction, intercellular communication, intracellular distribution of the cardiac gap-junction protein (Cx43) and abundance of Cx43 and its mRNA in neonatal rat cardiac myocytes cultured in the presence of *T. cruzi* (Tulahuen strain). Our results indicate that infection of heart cells by *T. cruzi* leads to the loss of gap junction-mediated intercellular communication, and thereby presumably contributes to the conduction disorders that are characteristic of Chagas' disease.

Gap junctions in heart mediate electrotonic current flow, thereby coordinating the spread of excitation and contraction throughout the organ (see Spray and Burt, 1990). Recent computer-modeling studies strongly implicate changes in electrical coupling as a contributory mechanism in re-entrant arrhythmias (e.g., Van Capelle and Durrer, 1980; Rudy and Quan, 1987; Quan and Rudy, 1990). Experimental studies indicate that conduction disorders may develop when junctional conductance is lowered by various manipulations (Spach et al., 1982, 1990; Delmar et al., 1987; Balke et al., 1988; Cole et al., 1988; Allessie et al., 1989). Moreover, there are indications that during aging and following myocardial infarction there is remodeling of gap-junction connections in ventricular muscle with an overall decrease in gap-junction expression (Wit and Rosen, 1983; Ursell et al., 1985; Spach and Dolber, 1986; Luke and Saffitz, 1991). Thus, changes in gap junction-mediated intercellular communication may contribute fundamentally to abnormal physiological functioning of cardiac tissue.

Cardiac arrhythmias are a common finding in Chagas' disease (Rosenbaum, 1965; Maguirre et al., 1987; dePaola et al., 1990; Higuchi et al., 1990), yet the underlying pathophysiological mechanisms are unclear (Morris et al., 1990). To investigate the role of gap junctions in arrhythmogenesis, ventricular myocytes were dissociated from neonatal rats and placed in tissue culture; one day later parasites were added to the cultures and the effects on junctional communication were assessed two days later by comparing infected cells with sister cultures maintained under identical conditions but not parasitized.

Results

Synchrony of contraction

Spontaneous contractile activity in cultures infected with *T. cruzi* was measured optically. Using an image-analysis system (IMAGE1/AT:Universal Imaging, Media, PA), brightness of 6–8 small areas (3–25 μm^2) of a field (100× magnification) of confluent cells was measured over time. Due to changes in refractility, contractions are detectable as changes in brightness, allowing evaluation of synchrony, rate and variability of intercontraction intervals. In control cultures, the intercontraction interval was essentially constant and contractions were synchronous throughout the field (Fig. 1, left). In infected cultures, the rate of spontaneous beating was lower than in controls [mean

*Permanent address: Laboratory of Excitable Membranes, Carlos Chagas Filho Institute of Biophysics, Federal University of Rio de Janeiro, Rio de Janeiro, Brazil.

**Permanent address: Department of Anatomy, University of Regensburg, Regensberg, Germany.

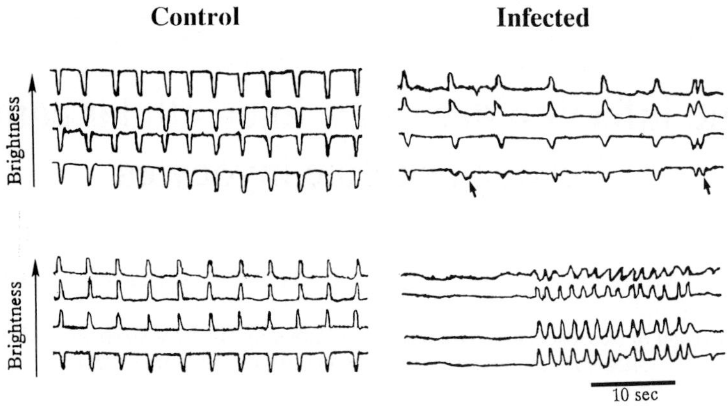

Figure 1. Beat rate and contraction synchrony are reduced in infected myocardial cells compared to uninfected cultures. Contraction of myocytes (at room temperature) in (usually 4–6) various areas of the visual field were monitored as pixel intensity within specified regions of interest, using "brightness over time" function in the IMAGE1/AT image-analysis package. Under control conditions (left panels), contractions of all cells within the field occurred synchronously and with a rapid beat rate. In infected cultures (right panels, each pair of recordings is from sister cultures, with or without added trypanosomes), spontaneous beat rate was significantly reduced. In addition, ectopic beats were commonly observed in the infected cultures, as well as contractions which did not spread into specific regions of the field (arrows). Modified from Campos de Carvalho et al. (1992).

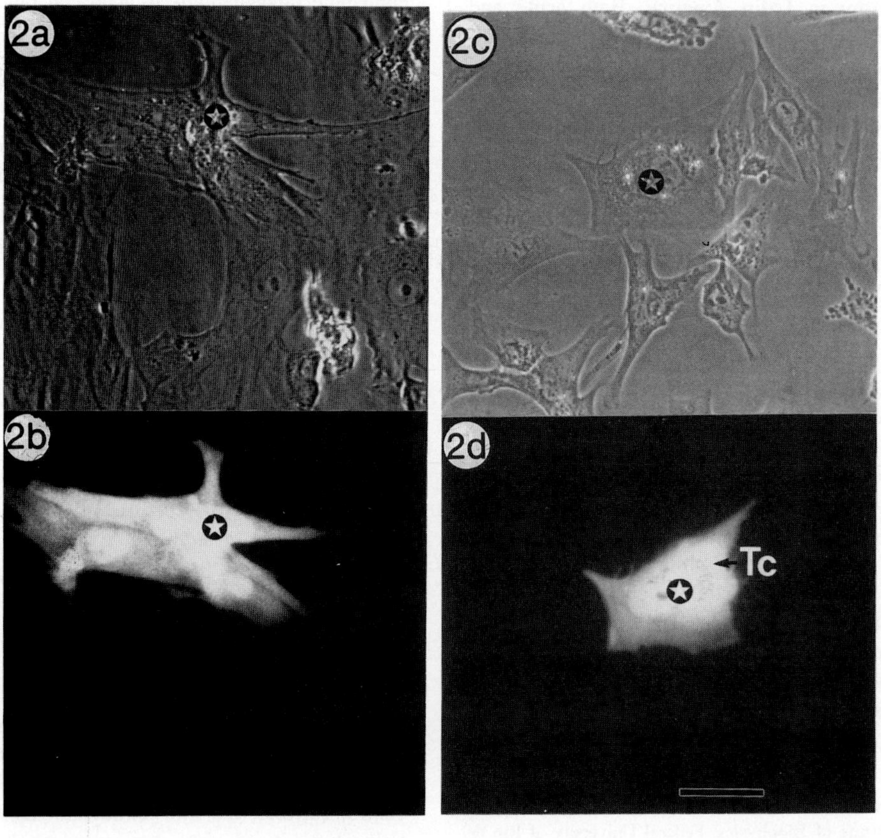

Figure 2. Dye coupling between cardiac myocytes in culture is normally extensive (a,b), and is reduced in cells infected with *T. cruzi* (c,d). Lucifer Yellow CH was injected into the cell in each field indicated by the asterisk. Within one min, Lucifer Yellow was visible in adjacent cells in the control cultures (fluorescence micrograph in b), but dye spread in the infected cells was much reduced (fluorescence micrograph in d). Note amastigotes in the Lucifer Yellow-injected cell, which appear as dark spots (arrow, d).

decrease 51 ± 9% (SE) in matched experiments on control and infected cells, n = 3], the intercontraction interval was variable, extra contractions were frequent, and rapid contractions separated by long periods of quiescence could occur (Fig. 1., right; preparations correspond to those on the left). Abnormal contraction patterns mimic cardiac arrhythmias developed during acute parasitic infection.

Dye and electrical coupling

In order to determine whether functional coupling was altered among the cells, Lucifer Yellow was injected into one cell of a cluster and the rapidity and extent of dye spread to neighbors was monitored. In uninfected dishes, the injected dye spread rapidly to second- and, often, third-order neighbors; in infected dishes, however, the dye was largely confined to the injected cell (Fig. 2). In order to quantify this apparent difference in junctional permeability, cells were cultured at very low density so that pairs were available for study at three days after plating (48 h after infection). Junctional conductance (g_j) was evaluated using the dual voltage-clamp technique with patch pipettes (White et al., 1985). In pairs where both cells were visibly infected, junctional conductance averaged 0.7 ± 0.5 nS (SE; n = 7 cell pairs, three experiments), compared to an average value of 24.6 ± 5.6 nS for matched controls (SE; n = 6, three experiments). These values are significantly different ($p < 0.01$). Similar strong coupling was seen between uninfected cells in infected dishes, indicating that only cells containing the parasite exhibit a reduction in junctional communication.

Immunocytochemistry

The reduction in dye coupling and junctional conductance between infected cells could be due to an alteration in gating mechanisms of the membrane's junctional channels or to decreased expression of the intercellular gap-junction protein. We evaluated whether gap-junction expression was altered in *T. cruzi*-infected myocardial cells using immunocytochemical techniques to localize Cx43 gap-junction protein within the infected cells. There were striking differences between controls and infected cells. In control cultures, typical gap-junction immunoreactivity was seen at virtually every interface between the cultured cells (Fig. 3a,b), but only rarely at interfaces between infected cells, although adjacent uninfected cells showed normal intercellular staining (Fig. 3c,d). The immuno-

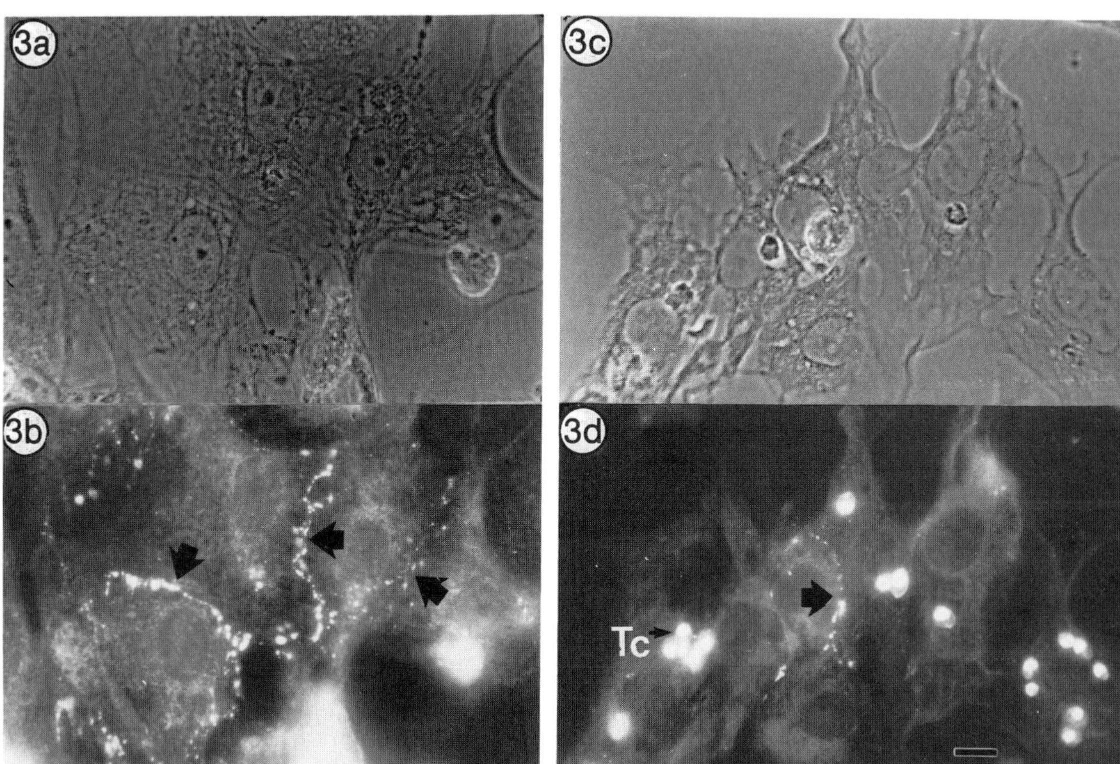

Figure 3. Localization of gap junctions in cardiac myocyte cultures. a,c, phase; b,d, fluorescence micrographs. Immunocytochemically localized gap junctions are prominent at interfaces between cells under control conditions (a,b), but are less extensive between infected cells (c,d). Note that cells adjacent to infected cells have relatively normal gap-junction staining (arrows in d), whereas punctate intercellular Cx43 immunoreactivity is almost entirely absent in infected cells. Modified from Campos de Carvalho et al. (1992).

reactivity localized to the parasites is not seen when affinity-purified primary antibody is used, indicating that Cx43 itself is not on the surface of the trypanosome (Campos de Carvalho et al., 1992).

Western and Northern blotting

Western blots (in collaboration with Dr. E.L. Hertzberg) using peptide-specific anti-Cx43 antibodies, revealed that the total Cx43 content of infected cultures was not detectably different from control cultures (not illustrated). Levels of Cx43 mRNA, detected by hybridization with a Cx43 cDNA probe, also showed no difference between infected and uninfected cultures (not illustrated). However, one should be extremely cautious in interpreting these results, since the degree of infection was highly variable and never exceeded 30–50% of the cultured cells.

Discussion and conclusions

Trypanosome infection reduces the strength of dye and electrical coupling between myocardial cells in culture. This reduction correlates well with disappearance of Cx43 immunoreactivity from junctional regions of plasma membrane. Our findings predict that *T. cruzi* infection should slow cardiac conduction and alter patterns of propagation through decreases in the strength of electrotonic coupling. It should be noted that, in the pathogenesis of Chagas' disease, two stages, quite distinct in onset, are evident. In the acute stage, the presence of the parasite may reduce the number of junctional channels at the surface, resulting in retarded conduction between otherwise functional cells. Because the parasites infect other Cx43-containing tissues during acute infection (Tafuri, 1971; Tanowitz et al., 1982), gap-junction dysfunction in these other tissues could contribute to the pathophysiology of the acute disease. In the chronic stage, when there may be few parasites, the occurrence of re-entrant arrhythmias might be accelerated due to a decreased "safety factor" for intercellular communication, because focal fibrosis, which results from infection-related myonecrosis, adds geometrically to the effects of other factors that tend to split the aging heart into slowly conducting units.

Acknowledgements

Supported by NIH grants NS16524, AI12770, AI26368, MH45654, GM30667, HL38449, NSF-CNPq Binational grant number INT8612496, CNPq grant 402655/91-5, a Grant-in-Aid from the New York Chapter of the American Heart Association, and a grant from the UNDP/World Bank/WHO Special Programme for Research and Training in Tropical Diseases (TDR ID 910101). We gratefully acknowledge the technical assistance of V. Braunstein, W. Rivera, and C. Roy. cDNA probes were provided by G.L. Fishman; antibodies specific for Cx43 were provided by E.L. Hertzberg and E.C. Beyer. We thank S. Morris and L. Weiss for informative and helpful discussions on many aspects of this work.

References

Allessie, M.A., Schalij, M.J., Kirchhof, C.J., Boersma, L., Huybers, M. and Hollen, J. (1989) Experimental electrophysiology and arrhythmogenicity. Anisotropy and ventricular tachycardia. Eur. Heart J. 10 (Suppl E.), 2–8.

Balke, C.W., Lesh, M.D., Spear, J.F., Kadish, A., Levine, J.H. and Moore, E.N. (1988) Effects of cellular uncoupling on conduction in anisotropic canine ventricular myocardium. Circ. Res. 63, 879–892.

Campos de Carvalho, A.C., Roy, C., Dermietzel, R., Hertzberg, E.L., Wittner, M., Tanowitz, H. and Spray, D.C. (1992) Gap junctions disappear between cardiac myocytes infected with *Trypanosoma cruzi*. Circ. Res. 70, 733–742.

Cole, W.C., Picone, J.B. and Sperelakis, N. (1988) Gap-junction uncoupling and discontinuous propagation in the heart. A comparison of experimental data with computer simulations. Biophys. J. 53, 809–918.

Delmar, M., Michaels, D.C., Johnson, T. and Jalife, J. (1987) Effects of increasing intercellular resistance on transverse and longitudinal propagation in sheep epicardial muscle. Circ. Res. 60, 780–785.

dePaola, A.A., Horowitz, L.N., Miyamoto, M.H., Pinheiro, R., Ferreira, D.F., Terzian, A.B., Cirenza, C., Guiguer N. Jr. and Portugal, O.P. (1990) Angiographic and electrophysiologic substrates of ventricular tachycardia in chronic Chagasic myocarditis. Am. J. Cardiol. 65, 360–363

Higuchi, M.L., DeMorais, C.F., Sambiase, N.V., Pereira-Barretto, A.C., Bellotti, G. and Pileggi, F. (1990) Histopathological criteria of myocarditis – a study based on normal heart, Chagasic heart and dilated cardiomyopathy. Jpn. Circ. J. 54, 391–400.

Luke, R.A. and Saffitz, J.E. (1991) Remodeling of ventricular conduction pathways in healed canine infarct border zones. J. Clin. Invest. 87, 1594–1602.

Maguire, J.H., Hoff, R. and Sherlock, I. (1987) Cardiac morbidity and mortality due to Chagas' disease: prospective electrocardiographic study of a Brazilian community. Circulation 75, 1140–1145.

Morris, S.A., Tanowitz, H.B., Wittner, M. and Bilezikian, J.P. (1990) Pathophysiological insights into the cardiomyopathy of Chagas' disease. Circulation 82, 1900–1909.

Quan, W. and Rudy, Y. (1990) Unidirectional block and reentry of cardiac excitation: a model study. Circ. Res. 66, 367–382.

Rosenbaum M.B. (1965) Chagas' disease. Prog. Cardiovasc. Dis. 7, 199–224.

Rudy, Y and Quan, W.L. (1987) A model of the effects of the discrete cellular structure on electrical propagation in cardiac tissue. Circ. Res. 61, 815–823.

Spach, M.S., Dolber, P.C. and Heidlage, J.F. (1990) Properties of discontinuous anisotropic propagation at a microscopic level. Ann. NY Acad, Sci. 591, 62–74.

Spach, M.S., Kootsey, J.M. and Sloan, J.D. (1982) Active modulation of electrical coupling between cardiac cells of the dog. A mechanism for transient and steady-state variations in conduction velocity. Circ. Res. 51, 347–362.

Spach, M.S. and Dolber, P.C. (1986) Relating extracellular potentials and their derivatives to anisotropic propagation at a microscopic level in human cardiac muscle. Evidence for electrical uncoupling of side-to-side fiber connections with increasing age. Circ. Res.. 58, 356–371.

Spray, D.C. and Burt, J.M. (1990) Structure-activity relations of the cardiac gap-junction channel. Am. J. Physiol. 258, C195–207.

Tafuri, W.L. (1971) Pathogenesis of lesions of the autonomic nervous system of the mouse in experimental acute Chagas' disease. Light and electron microscope studies. Am. J. Trop. Med. Hyg. 19, 405–417.

Tanowitz, H., Brosnan, C., Guastamacchio, D., Baron, G., Raventos-Suarez, C., Bornstein, M. and Wittner, M. (1982) Infection of organotypic cultures of spinal cord and dorsal root ganglia with *Trypanosoma cruzi*. Am. J. Trop. Med. Hyg. 31, 1090–1097.

Ursell, P.C., Gardner, P.I., Albala, A., Fenoglio, Jr., J.J. and Wit, A.L. (1985) Structural and electrophysiological changes in the epicardial border zone of canine myocardial infarcts during infarct healing. Circ. Res. 56, 436–451.

van Capelle, F.J. and Durrer, D. (1980) Computer simulation of arrhythmias in a network of coupled excitable elements. Circ. Res. 47, 454–466.

White, R.L., Spray, D.C., Campos de Carvalho, A.C., Wittenberg, B.A. and Bennett, M.V.L. (1985) Some physiological and pharmacological properties of cardiac myocytes dissociated from adult rat. Am. J. Physiol. 249, C447–461.

Wit, A.L. and Rosen, M.R. (1983) Pathophysiologic mechanisms of cardiac arrhythmias. Am. Heart J. 106, 798–811.

J.E. Hall, G.A. Zampighi and R.M. Davis (Eds.)
Progress in Cell Research, Vol. 3

CHAPTER 28

Rat pancreatic acinar cell coupling: comparison of extent and modulation in vitro and in vivo

M. CHANSON and P. MEDA

Department of Morphology, University of Geneva, 1211 Genève 4, Switzerland

Introduction

Because gap-junction channels provide cytoplasmic continuity between cells in contact (Hertzberg and Johnson, 1988) they could play a major role in tissue homeostasis (Loewenstein, 1981; Hertzberg and Johnson, 1988). However, in spite of considerable effort to assess the molecular and biophysical properties of gap junctions (Beyer et al., 1990; Bennett et al., 1991), the role of junctional communications between adult cells is still poorly understood, except in excitable tissues (Spray and Burt, 1990; Bennett et al., 1991).

The exocrine pancreas represents an interesting model to search for the function(s) of junctional coupling in a highly differentiated and non-excitable system. Indeed, pancreatic acinar cells uncouple during secretagogue stimulation (Iwatsuki and Petersen, 1978; Petersen, 1980). Conversely, secretion of amylase (the major enzyme produced by the exocrine pancreas) is increased by long-chain alcohols (Meda et al., 1986, 1987; Bruzzone et al., 1987) which uncouple acinar cells without altering the second messengers regulating the secretory machinery (Meda et al., 1986; Bruzzone et al., 1987; Chanson et al., 1989a, 1989b). Together, these observations have led to the suggestion that coupling of acinar cells is involved in the control of the secretion of the rat exocrine pancreas (Meda et al., 1988).

If cell coupling is of physiological relevance in the control of pancreatic function, acinar cell communications would be expected to be modulated by endogenous secretagogues in vivo. The purpose of this chapter is to summarize what we know about the extent and modulation of acinar cell coupling in vitro, and to compare these observations with those made in the intact pancreas.

Methods

For in vitro studies, acini were isolated by collagenase digestion (Bruzzone et al., 1985) from the pancreas of male SIVZ rats. To assess amylase secretion, the dispersed acini were incubated within glass vials in the absence (control) and presence of either 5 μM acetylcholine (ACh) or 10 μM carbamylcholine (CCh). At the end of 30 min incubation, aliquots of the acinar suspension were centrifuged and amylase was measured in both pellet (content) and supernatant (secretion). The amount of amylase released was then expressed as percentage of initial content (Bruzzone et al., 1985). To assess dye coupling, the dispersed acini were attached to Petri dishes coated with poly-L-lysine (M.W. 150,000–300,000), and individual cells were impaled with thin-tip high-resistance microelectrodes (Meda et al., 1986). Electrodes were filled with a 4% solution of Lucifer Yellow (LY) in 150 mM LiCl. The dye was then iontophoretically injected by applying negative square current pulses to the electrodes for 2–3 min. Extent of cell coupling was expressed as percent of the acinar area filled by Lucifer Yellow (Meda et al., 1986).

For in vivo studies, male SIVZ rats were anaesthetized with 37 mg/kg body weight of sodium pentobarbital. Release of endogenous ACh was elicited by electrical stimulation (4 mA, 1 ms, 12 Hz) of the neck portion of the left vagus nerve (Chanson et al., 1991). To assess amylase secretion, a catheter was inserted into the duodenal end of the common pancreatic bile duct. Samples of pancreatic juice were collected at 10 min intervals and assayed for amylase by the same method used in the in vitro studies. The amount of amylase released was expressed as U/10 min. To assess dye coupling, the spleen and the attached dorsal part of the pancreas were gently pulled out of the abdominal cavity and attached to a Sylgard-coated chamber by means of thin stainless-steel needles. LY was then injected into the exposed pancreas. Injections were performed as described above, and were judged successful only when a stable membrane potential could be recorded throughout the experiment. At the end of each injection, the exposed pancreas was immediately fixed in 4% paraformaldehyde. A small piece of tissue containing the labeled region was dissected from the fixed gland and processed for histology. Semi-thin sections were then cut serially throughout the injected area. Extent of cell coupling was evaluated by the proportion of acinar area stained by LY (Chanson et al., 1991).

Secretion and coupling under resting conditions

Isolated acini incubated for 30 min in the absence of secretagogues showed a modest basal release of amylase (Fig. 1). Under these resting conditions, injection of LY into one acinar cell resulted in the labeling of all neighboring cells comprising an acinus, indicating extensive cell coupling (Figs. 2 and 4). Communications between acinar cells could be ascribed to functional gap-junction channels since, on the one hand, they were abolished within 1 min, and in a reversible manner, by long chain alcohols (Meda et al., 1986, 1987; Chanson et al., 1989a), and, on the other hand, unitary gap-junctional currents were recorded from acinar cell pairs assessed by dual whole-cell recording patch clamp (Meda et al., 1991a). In the latter experiments, the conductance of most single channels was found to be about 110–130 pS, although smaller events were also detected, but less frequently. These values are consistent with those reported in other tissues which, as do the acinar cells of the exocrine pancreas (Dermietzel et al., 1987; Meda et al., 1991b), express Cx32 and Cx26 as the major gap-junction proteins (Neyton and Trautmann, 1985; Spray et al., 1986; Somogyi and Kolb, 1988; Eghbali et al., 1990).

In vivo, the basal release of amylase by acinar cells was also modest (Fig. 1) and stable throughout the 60 min duration of the experiments. Under these conditions, injection of LY also resulted in the extensive labeling of cells throughout each acinus (Figs. 3 and 4). Even though the fluorescent dye could be detected up to 60–100 μm distant, no diffusion of LY was ever observed from one acinus to another, nor between acinar cells and adjacent centroacinar and duct cells (Fig. 3).

Thus, in terms of cell coupling, resting pancreatic acini behaved as functional units both in vitro and in vivo.

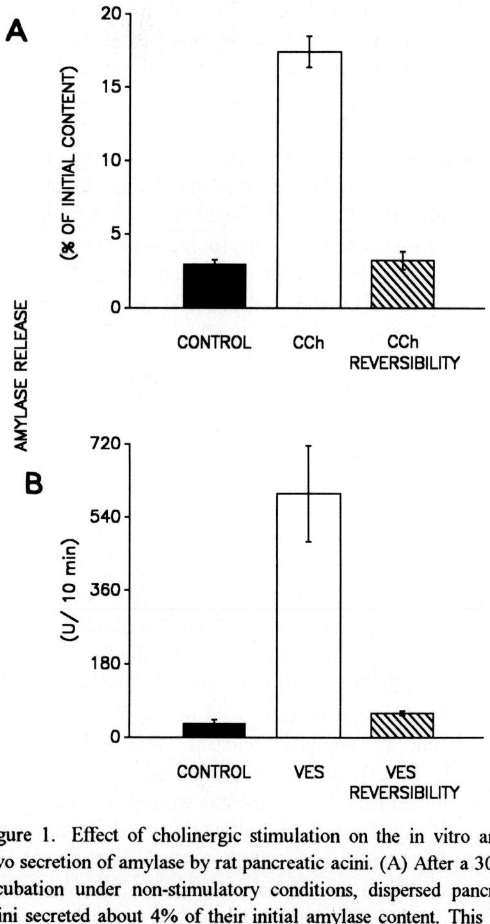

Figure 1. Effect of cholinergic stimulation on the in vitro and in vivo secretion of amylase by rat pancreatic acini. (A) After a 30 min incubation under non-stimulatory conditions, dispersed pancreatic acini secreted about 4% of their initial amylase content. This basal secretion increased 5–6 fold in the presence of 10^{-5} M CCh, and was reduced again to basal levels following return of the acini to a control medium devoid of secretagogues. (B) During infusion of 0.9% NaCl, the intact pancreas of anaesthetized rats released small levels of amylase. The output of this enzyme increased 16-fold during electrical stimulation of the vagal nerve (VES). Amylase secretion returned to control basal levels when the nerve stimulation was ended. Values are mean ± SEM of 4 separate experiments.

Secretion and coupling under stimulatory conditions

In the presence of either ACh or CCh, the release of amylase by isolated acini increased 4- to 5-fold (Fig. 1). This effect was fully and rapidly reversible after the acini were returned to a medium lacking secretagogues (Fig. 1). Under maximal stimulation by cholinergic secretagogues, acinar cells uncoupled within seconds (Fig. 2), as previously reported (Iwatsuki and Petersen, 1978; Petersen, 1980). We, however, have found that the extent of the ACh-induced uncoupling was variable (Meda et al., 1986); the average reduction in coupling was about 50%, as evaluated by LY diffusion (Figs. 2 and 4). A similar variable uncoupling was observed in pairs of acinar cells which were monitored by dual whole-cell patch-clamp recording during ACh superfusion (Meda et al., 1991a). Again, ACh reduced the macroscopic junctional conductance of acinar cells by an average of 50%.

In vivo stimulation of the vagus nerve enhanced amylase secretion by about 16-fold (Fig. 1), in a manner which was sustained throughout the 20–30 min stimulation. This increase was fully reversible, as indicated by the rapid return of amylase release to basal level after stimulation was stopped (Fig. 1). In vivo injection of LY during maximal amylase secretion (usually 20 min after starting the vagal nerve stimulation), consistently revealed a reduction in the extent of dye coupling (Fig. 3). However, as observed in vitro, the extent of this in vivo reduction was variable. On average, the area labeled by LY represented about half that of the acinar profiles (Fig. 4). This variable uncoupling is unlikely to result from the targeting of some, but

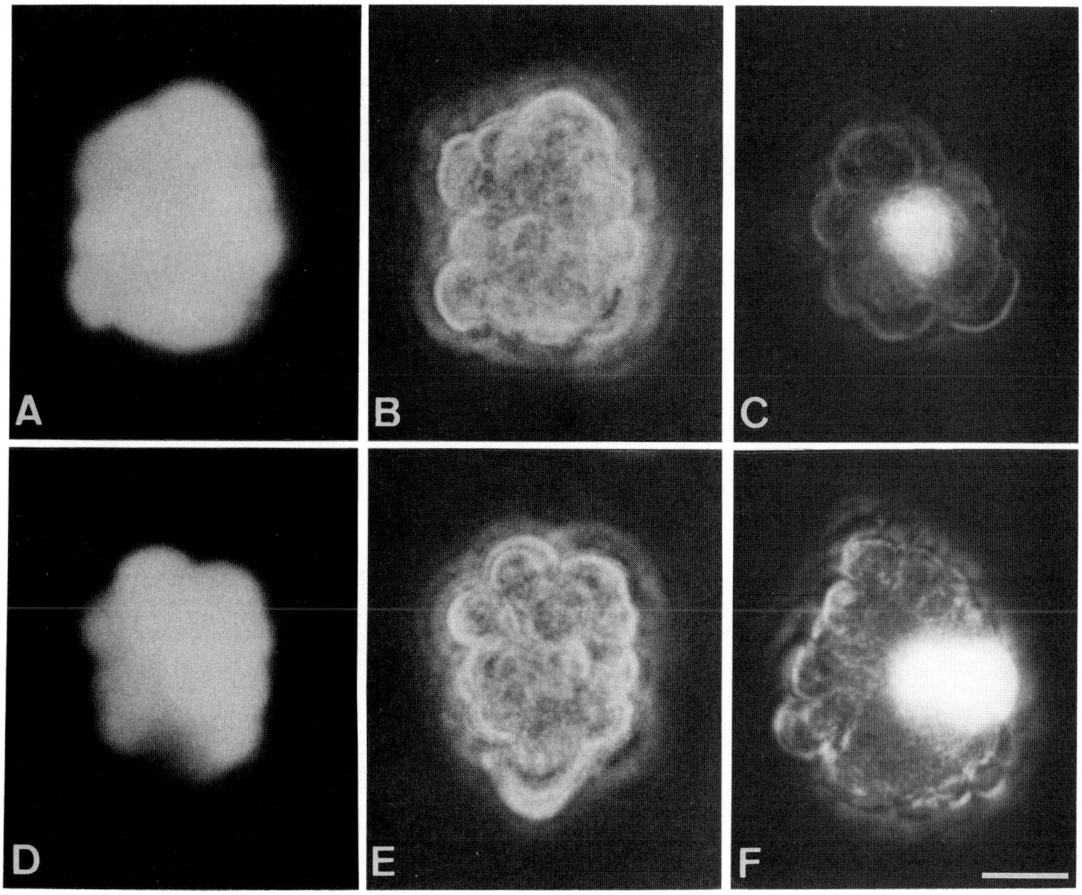

Figure 2. Dye coupling in isolated pancreatic acini injected with LY. Under control conditions (A and B), LY diffused throughout the entire acinus, indicating extensive coupling of acinar cells. Incubation of acini in the presence of 50 μM indomethacin did not affect this extensive coupling (D and E). Addition of 5 μM ACh restricted the intracellular diffusion of LY irrespective of whether the acini were incubated under control conditions (C) or in the presence of indomethacin (F), indicating that the ACh-induced gating of gap junctions did not require activation of the cyclo-oxygenase pathway. Bar = 20 μm.

not all, acinar cells by the endogenous ACh which had been released locally by intrapancreatic nerve terminals. Indeed, a variable extent of uncoupling was also observed in pancreas stimulated by vascular perfusion of exogenous ACh (Chanson et al., 1991). In the latter experiments, the 50% cell uncoupling evoked by ACh was prevented, as was the increase in amylase release, by pretreating the animals with atropine, indicating that the ACh effects were specifically related to the interaction of the secretagogue with muscarinic receptors present on the acinar cell membranes.

Thus, the increase of amylase release induced by physiological secretagogues is associated with the uncoupling of some acinar cells both in vitro and in vivo.

Mechanisms controlling secretion and cell coupling

The parallel changes of secretion and coupling observed in vitro and in vivo suggest that these two functions of acinar cells may be regulated by the same mechanism(s). Activation of muscarinic receptors by cholinergic secretagogues stimulates amylase release by triggering changes in several second messengers, including intracellular calcium, protein kinase C (PKC) and arachidonic acid metabolites (De Pont and Fleuren-Jakobs, 1984; Knight and Koh, 1984; Wooten and Wrenn, 1985; Sato et al., 1988; Sung et al., 1988). Increasing evidence indicates that these second messengers can also regulate the permeability and conductance of gap-junction channels in a variety of cell systems (Neyton and Trautmann, 1986; Sáez et al., 1990). However, their involvement in the control of rat pancreatic acinar cell coupling has not yet been verified experimentally. Rather, the following observations argue against a participation of these second messengers in modulating the junctional channels of acinar cells:

First, superfusion of acinar cells with ACh consistently reduced the junctional conductance by about 50% during dual whole-cell recording, even when the patch-clamp electrodes contained 11 mM EGTA. Under the strong Ca^{2+} buffering provided by the latter condition, ACh failed to activate the Ca^{2+}-dependent non-junctional currents which are normally detected using an electrode-filling solution devoid of EGTA (Meda et al., 1991a). Thus, a rise in cytosolic calcium does not appear to be a prerequisite for acinar cell uncoupling, confirming results for other types of rat exocrine cells (Neyton and Trautmann, 1986).

Second, 12-O-tetradecanoylphorbol-13-acetate (TPA), a tumor promoter which activates the PKC of pancreatic acinar cells within min, did not significantly affect the ionic and dye coupling of acinar cell pairs or dispersed acini, even after 45 min of incubation (Chanson et al., 1988, 1989b). TPA has since been reported not to affect the coupling of other cell types as well (Randriamampita et al., 1988; Spray and Burt, 1990). However, since high concentrations (100–250 μg/ml) of synthetic diacylglycerols (which also activate PKC) have been reported to evoke a progressive decline of coupling between the acinar cells of rat lacrimal glands (Randriamampita et al., 1988) and mouse pancreas (Somogyi and Kolb, 1990), we have tested these compounds on our preparations. At 50 μg/ml, a concentration which maximally activated PKC, 1-oleyl 2-acetylglycerol (OAG) and L-α-1,2-dioctanoyl glycerol (DiC_8) did not affect the extensive intercellular diffusion of LY that is normally observed within dispersed acini (Chanson et al., 1989b; Meda et al., 1991a).

Thus, our results argue against a role for PKC activation in the short-term regulation of gap-junction gating between pancreatic acinar cells. This implies that the rapid ACh-induced down regulation of acinar cell coupling is unlikely to be mediated by the modest activation of PKC usually achieved by the secretagogue (M. Chanson and R. Regazzi, unpublished data).

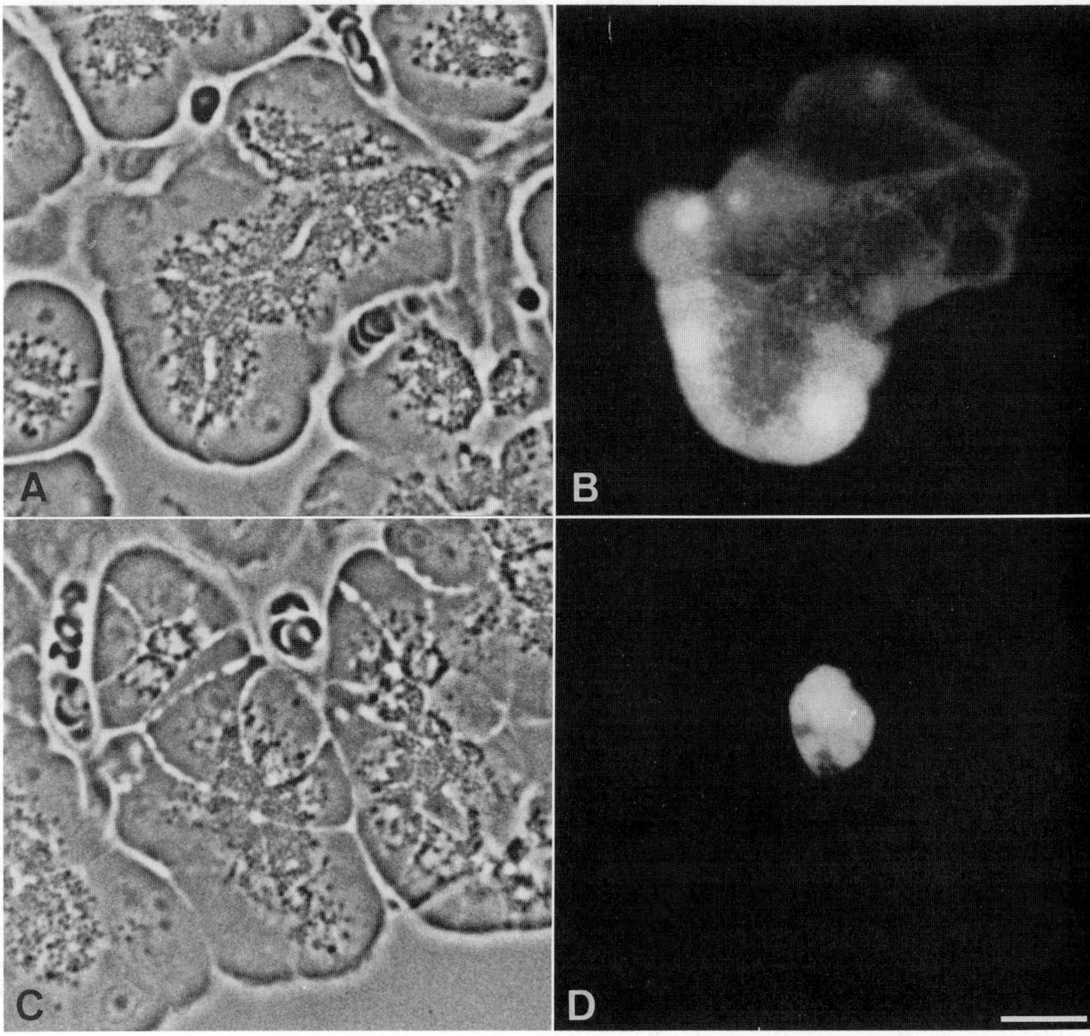

Figure 3. Modulation of dye diffusion in the intact pancreas of anaesthetized rats. Under resting conditions, LY usually diffused into all cells of each injected acinus (A and B), but not into nearby duct cells. After a 20 min electrical stimulation of the vagal nerve (C and D), the tracer diffusion was confined to fewer acinar cells. A and C are semi-thin sections observed under phase-contrast illumination. B and D are the very same sections seen under fluorescence illumination. Bar = 20 μm.

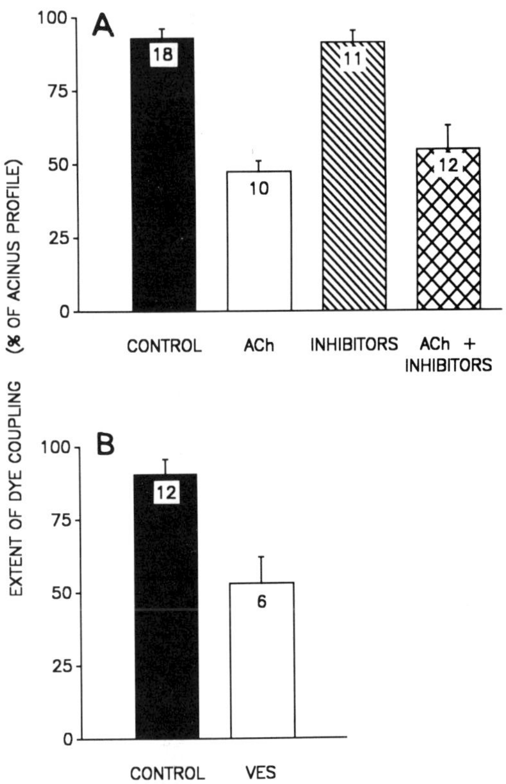

Figure 4. Effect of ACh on dye coupling of rat pancreatic acinar cells. (A) Under in vitro control conditions, most acinar cells were coupled, as indicated by the diffusion of LY throughout entire isolated acini. Inhibitors of protein kinase C, cyclo-oxygenases, lipoxygenases and phospholipase A_2 did not affect this extensive coupling. (The column labeled INHIBITORS shows the results of pooling the data obtained with different inhibitors). By contrast, LY diffusion was reduced by about 50% in acini stimulated by 5 μM ACh and incubated with or without inhibitors of arachidonic acid metabolism. (B) Coupling of acinar cells extended throughout virtually all injected acini in the intact pancreas of control anaesthetized rats. In this in vivo preparation, release of endogenous ACh by electrical stimulation of the vagal nerve (VES) also decreased the acinar cell coupling by half. Values are mean ± SEM of the number of injections indicated within columns.

Third, ACh causes the uncoupling of pancreatic acinar cells, even in the presence of inhibitors of arachidonic acid metabolism. Pancreatic secretagogues may trigger the production of arachidonic acid, either by activation of phospholipase A_2 or by a lipase-dependent hydrolysis of diacylglycerol. The newly formed arachidonic acid may then be metabolized in either the cyclo-oxygenase or the lipoxygenase pathway (Salmon, 1986). Therefore, we have tested a series of drugs which inhibit arachidonic acid metabolism at different levels. Neither polymyxin B (240 μg/ml), an inhibitor of PKC; 4-bromophenacylbromide (10–50 μM), an inhibitor of phospholipase A2; caffeic acid and nordihydroguaiaretic acid (5–10 μM), two inhibitors of lipoxygenases; nor indomethacin (10–50 μM), an inhibitor of cyclo-oxygenase, affected the basal coupling of pancreatic acinar cells or was able to prevent ACh-induced uncoupling (Figs. 2 and 4). Therefore, the arachidonic acid pathway is not obligatory for the gating of gap junctions that takes place during the physiological stimulation of acinar cells. However, high concentrations (50–100 μM) of arachidonic acid evoke cell uncoupling in different tissues (Giaume et al., 1989; Schmilinski-Fluri et al., 1990), including the rat exocrine pancreas (P. Meda, unpublished data), suggesting that, under certain conditions, the fatty acid may interact directly with the gap-junction proteins or their lipid environment (Sáez et al., 1986; Giaume et al., 1989; Schmilinski-Fluri et al., 1990). Precise measurements of arachidonic acid generated during ACh stimulation and a proven correlation between the time course of arachidonic acid production and cell uncoupling should help to validate this hypothesis.

The available data indicate that the parallel changes of coupling and secretion observed with pancreatic acinar cells are regulated by distinct intracellular mechanisms. This tentative conclusion stresses the independent role of junctional coupling in the control of pancreatic secretion. To date, the physiological mechanism(s) controlling gap-junction gating between pancreatic acinar cells is (are) still unknown. Increasing evidence from other systems suggests that covalent modification of connexins by phosphorylation may be crucial (Musil et al., 1990; Sáez et al., 1990; Spray et al., 1990). We have found that 10^{-6} M okadaic acid, an inhibitor of serine-threonine phosphatases (Cohen, 1989), prevents the uncoupling effect of ACh. Thus, in the presence of ACh and okadaic acid, LY diffused into 94.5 ± 3.8 % (n = 11) of the acinar area; charged ions are not secreted and protein secretion of acinar cells was also similar to that of controls. The coupling and secretion changes normally elicited by ACh were blocked by okadaic acid, but fully reversible after washout. Renewed exposure to ACh induced uncoupling of acinar cells, as expected; the diffusion of LY was restricted to 44.7 ± 12.6% (n = 6) of the acinar area, and amylase release increased normally. These preliminary observations suggest that phosphorylation of gap-junction connexins may be required to ensure a high level of junctional communication between acinar cells.

Conclusions

With a few exceptions (Petersen and Ueda, 1975; Berga, 1984), virtually all our knowledge about vertebrate cell coupling has been so far derived from in vitro studies. Because experimental conditions cannot reconstitute the native environment of an intact tissue, it is uncertain whether any function attributed to junctional communication in vitro is physiologically relevant. In vivo experiments evaluating the extent and modulation of junctional communication should help to resolve this issue. Our studies attempt to address this question, using the acinar cells of the exocrine pancreas as a model. This cell type is highly specialized for the biosynthesis, storage and release of a variety of digestive enzymes.

We have found that the pattern of junctional communication between acinar cells in vitro is consistent with that in vivo, in the intact pancreas. In both cases, non-stimulated acinar cells communicate extensively throughout each acinus, but not between different acini. Furthermore, during sustained stimulation by natural secretagogues, the junctional communication of acinar cells is reduced by half, both within the intact pancreas and in the less native environment of isolated acini and acinar cells. During both in vitro and in vivo stimulation, a tight relationship was observed between the increased secretion and the decreased coupling of acinar cells; this decrease varied from one cell to another. The reason for such a variability is still hypothetical. As are other secretory cells (Walker and Farquhar, 1980; Neill et al., 1983; Salomon and Meda, 1986), the acinar cells of the exocrine pancreas are functionally heterogeneous, as judging by the variable rate at which they release amylase in response to identical in vitro stimuli (Bosco et al., 1988). It is therefore possible that the variable effect of secretagogues on coupling correlates with their differential effect on the secretion of individual acinar cells. Based on the available data, the cells undergoing complete uncoupling would be expected to be those which also secrete more amylase. This question is now amenable to direct experimental testing by evaluating simultaneously the secretory output and the coupling of individual acinar cell pairs (Bosco et al., 1988).

Our in vivo observations provide significant evidence that junctional coupling is modulated in vivo within a highly differentiated and essentially non-proliferating system, under conditions which mimic closely those encountered under physiological conditions. Together with the evidence accumulated from several lines of in vitro work, these new observations support the view that cell-to-cell communications through gap-junction channels participate in the control of a physiological activity, namely the secretion of pancreatic enzymes. The challenge now is to elicidate the mechanisms of this control. These questions can now be addressed (Bosco et al., 1988; Petersen and Petersen, 1991) using simplified in vitro preparations which, as we have seen, provide data which are reasonably consistent with the in vivo observations.

Acknowledgments

We thank A. Charollais, L. Burkhardt, J.-P. Gerber, L. Iuliano, and P. Ruga for excellent technical assistance during these studies. This work was supported by grants from the Swiss National Science Foundation (32-34090-92), the Sir Jules Thorn Foundation and the Sandoz Foundation.

References

Bennett, M.V.L., Barrio, L.C., Bargiello, T.A., Spray, D.C., Hertzberg, E. and Sáez, J.C. (1991) Gap junctions: new tools, new answers, new questions. Neuron 6, 305–320.

Berga, S.E. (1984) Electrical potentials and cell-to-cell dye movement in mouse mammary gland during lactation. Am. J. Physiol. 247, C20–C25.

Beyer, E.C., Paul, D.L. and Goodenough, D.A. (1990) Connexin family of gap-junction proteins. J. Membr. Biol. 116, 187–194

Bosco, D., Chanson, M., Bruzzone, R. and Meda, P. (1988) Visualization of amylase secretion from individual pancreatic acini. Am. J. Physiol. 254, G664–G670.

Bruzzone, R., Halban, P.A., Gjinovci, A. and Trimble, E.R. (1985) A new, rapid method for preparation of dispersed pancreatic acini. Biochem. J. 226, 621–624.

Bruzzone, R., Trimble, E.R., Gjinovci, A., Traub, O., Willecke, K. and Meda, P. (1987) Regulation of pancreatic exocrine function: a role for cell-to-cell communication? Pancreas 2, 262–271.

Chanson, M., Bruzzone, R., Spray, D.C., Regazzi, R. and Meda, P. (1988) Cell uncoupling and protein kinase C: correlation in a cell line but not in a differentiated tissue. Am. J. Physiol. 255, C699–C704.

Chanson, M., Bruzzone, R., Bosco, D. and Meda, P. (1989a) Effects of n-alcohols on junctional coupling and amylase secretion of pancreatic acinar cells. J. Cell Physiol. 139, 147–156.

Chanson, M., Meda, P. and Bruzzone, R. (1989b) Increase in pancreatic exocrine secretion during uncoupling: evidence for a protein kinase C-independent effect. Exp. Cell Res. 182, 349–357.

Chanson, M., Orci, L. and Meda, P. (1991) Extent and modulation of junctional communication between pancreatic acinar cells *in vivo*. Am. J. Physiol. 261, G28–G36.

Cohen, P. (1989) The structure and regulation of protein phosphatases. Annu. Rev. Biochem. 58, 453–508.

De Pont, J.J. and Fleuren-Jakobs, A.M. (1984) Synergistic effect of A23187 and a phorbol ester on amylase secretion from rabbit pancreatic acini. FEBS Lett. 170, 64–68.

Dermietzel, R., Leibstein, A., Frixen, U., Janssen-Timmen, U., Traub, O. and Willecke, K. (1984) Gap junctions in several tissues share antigenic determinants with liver gap junctions. EMBO J. 3, 2261–2270.

Eghbali, B., Kessler, J.A. and Spray, D.C. (1990) Expression of gap-junction channels in communication-incompetent cells after stable transfection with cDNA encoding connexin 32. Proc. Natl. Acad. Sci. USA 87, 1328–1331.

Giaume, C., Randriamampita, C. and Trautmann, A. (1989) Arachidonic acid closes gap junction channels in rat lacrimal glands. Pflügers Arch. 413, 273–279.

Hertzberg, E.L. and Johnson, R.G., Eds. (1988) Gap Junctions, Alan Liss, New York.

Iwatsuki, N. and Petersen, O.H. (1978) Electrical coupling and uncoupling of exocrine acinar cells. J. Cell Biol. 79, 533–545.

Knight, D.E. and Koh, E. (1984) Ca^{2+} and cyclic nucleotide dependence of amylase release from isolated rat pancreatic acinar cells rendered permeable by intense electric fields. Cell Calcium 5, 401–418.

Loewenstein, W.R. (1981) Junctional intercellular communication: the cell-to-cell membrane channel. Physiol. Rev. 61, 829–913.

Meda, P., Bruzzone, R., Knodel, S. and Orci, L. (1986) Blockage of cell-to-cell communication within pancreatic acini is associated with increased basal release of amylase. J. Cell Biol. 103, 475–483.

Meda, P., Bruzzone, R., Chanson, M., Bosco, D. and Orci, L. (1987) Gap-junctional coupling modulates secretion of exocrine pancreas. Proc. Natl. Acad. Sci. (USA) 84, 4901–4904.

Meda, P., Bruzzone, R., Chanson, M. and Bosco, D. (1988) Junctional coupling and secretion of pancreatic acinar cells. In: E.L. Hertzberg and R.G. Johnson (Eds.), Gap Junctions, Alan R. Liss, New York, pp. 353–364.

Meda, P., Bosco, D. Giordano, E. and Chanson, M. (1991a) Junctional coupling modulation by secretagogues in two-cell pancreatic systems. In: C. Peracchia (Ed.), Biophysics of Gap-Junction Channels, CRC Press, Boca Raton, FL, pp. 191–208.

Meda, P., Chanson, M., Pepper, M., Giordano, E., Bosco, D., Traub, O., Willecke, K., El Aoumari, A., Gros, D., Beyer, E., Orci, L. and Spray, D.C. (1991b) *In vivo* modulation of connexin 43 gene expresssion and junctional coupling of pancreatic β-cells. Exp. Cell Res. 192, 469–480.

Musil, L.S., Cunningham, B.A., Edelman, G.M. and Goodenough, D.A. (1990) Differential phosphorylation of the gap-junction protein connexin 43 in junctional communication-incompetent and -deficient cell lines. J. Cell Biol. 111, 2077–2088.

Neill, J.D. and Frawley, L.S. (1983) Detection of hormone release from individual cells in mixed populations using a reverse hemolytic plaque assay. Endocrinology 112, 1135–1137.

Neyton, J. and Trautmann, A. (1985) Single-channel currents of an intercellular junction. Nature 317, 331–335.

Neyton, J. and Trautmann, A. (1986) Physiological modulation of gap-junction permeability. J. Exp. Biol. 124, 93–114.

Petersen, C.C.H. and Petersen, O.H. (1991) Receptor-activated cytoplasmic Ca^{2+} spikes in communicating clusters of pancreatic acinar cells. FEBS Lett. 284, 113–116.

Petersen, O.H. (1980) The Electrophysiology of Gland Cells. Academic Press, London.

Petersen O.H. and Ueda N., (1975). Pancreatic acinar cells: effect of acetylcholine, pancreozymin, gastrin and secretin on membrane potential and resistance *in vivo* and *in vitro*. J. Physiol. (London) 247, 461–471.

Randriamampita, C., Chanson, M. and Trautmann, A. (1988) Calcium- and secretagogue-induced conductances in isolated acinar cells. Pflügers Arch. 411, 53–57.

Sáez, J.C., Naim, A.C., Czernik, A.J., Spray, D.C., Hertzberg, E.L., Greengard, P. and Bennet, M.V.L. (1990) Phosphorylation of Cx32, a hepatocyte gap-junction protein, by cAMP-dependent protein kinase, protein kinase C and Ca^{2+}/calmodulin-dependent protein kinase II. Eur. J. Biochem. 194, 263–273.

Salmon, J.A. (1986) Inhibition of prostaglandin, thromboxane and leukotriene biosynthesis. Adv. Drug Res. 15, 111–167

Salomon, D. and Meda, P. (1986) Heterogeneity and contact-dependent regulation of hormone secretion by individual β cells. Exp. Cell Res. 162, 507–520.

Sato, S., Adachi, H., Noguchi, M., Honda, T., Oniki, S., Aoki, E. and Torizuka, K. (1988) Effect of AA861, a 5-lipoxygenase inhibitor, on amylase secretion from pancreatic acini. Biochim. Biophys. Acta 968, 1–8.

Schmilinsky-Fluri, G., Rüdisüli, A., Willi, M., Rohr, S. and Weingart, R. (1990) Effects of arachidonic acid on the gap junctions of neonatal rat heart cells. Pflügers Arch. 417, 149–156.

Somogyi, R. and Kolb, H.A. (1988) Cell-to-cell channel conductance during loss of gap-junctional coupling in pairs of pancreatic acinar and Chinese hamster ovary cells. Pflügers Arch. 412, 54–65.

Somogyi, R. and Kolb, H.A. (1990) A G-protein mediates secretagogue-induced gap-junctional channel closure in pancreatic acinar cells. FEBS Lett. 258, 216–218.

Spray, D.C. and Burt, J.M. (1990) Structure-activity relations of the cardiac gap junction channel. Am. J. Physiol. 258, C195–C205.

Spray, D.C., Ginzberg, R.D., Morales, E.A., Gatmaitin, Z. and Arias, I.M. (1986) Electrophysiological properties of gap junctions between dissociated pairs of rat hepatocytes. J. Cell Biol. 103, 135–144.

Sung, C.K., Hootman, S.R., Stuenkel, E., Kurolwa, C. and Williams, J.A. (1988) Down regulation of protein kinase C in guinea pig pancreatic acini: effects on secretion. Am. J. Physiol. 254, G242–G248.

Walker, A.M. and Farquhar, M.G. (1980) Preferential release of newly synthesized prolactin granules is the result of functional heterogeneity among mammotrophs. Endocrinology 107, 1095–1104.

Wooten, M.W. and Wrenn, R.W. (1985) Redistribution of phospholipid/calcium dependent protein kinase and altered phosphorylation of its soluble and particulate substrate proteins in phorbol ester-treated rat pancreatic acini. Cancer Res. 45, 3912–3917.

J.E. Hall, G.A. Zampighi and R.M. Davis (Eds.)
Progress in Cell Research, Vol. 3

CHAPTER 29

Delayed change in gap-junctional cell communication in the acinus of the rat submandibular gland after secretion of saliva

YOSHINOBU KANNO, YASUTO SASAKI, CHIKARA HIRONO and YOSHIKI SHIBA

Department of Physiology, Hiroshima University School of Dentistry, 2-3 Kasumi 1 Chome, Minami-ku, Hiroshima 734, Japan

Introduction

The acinar cells of the exocrine glands, such as the salivary glands, the lacrimal glands and the pancreas, communicate via a well-developed system of gap junctions (Kanno and Loewenstein, 1964; Hammer and Sheridan, 1978; Iwatsuki and Petersen, 1978; Meda et al., 1990). Intracellular substances with molecular weights of less than 1000 Da are able to pass between neighboring acinar cells through gap junctions. Transport of chloride ions and sodium ions from the basolateral to the luminal side is associated with the movement of water in the acinar cells of the salivary glands (Petersen, 1986). These ions seem to move between coupled neighboring cells.

The gap-junctional communication between cells is believed to play a role in the regulation of cell functions, such as secretion, via the intercellular movement of these ions; however, the exact role in such secretion remains to be determined. The acinar cells in the exocrine glands synthesize and store secretory substances which are released upon stimulation by secretagogues (Amsterdam et al., 1969). After secretion, the acinar cells enter a recovery phase during which they resynthesize the secretory substances and/or secretion-related substances.

If gap-junctional cell communication is closely related to the regulation of cell function, changes in cell function might be expected to induce changes in gap-junctional cell communication. Modulation of gap-junctional cell communication during the stimulation of secretion has been reported in many exocrine glands (Iwatsuki and Petersen, 1978; Meda et al., 1990), but fluctuations in gap-junctional cell communication associated with changes in cell function, such as those associated with the secretory cycle, have been not yet reported. Communication is also modulated by changes related to growth stimulation (Shiba et al., 1990).

In the present study, we investigated the relationship between changes in gap-junctional cell communication and changes in cell function during the secretory cycle.

Rapid inhibition of dye coupling in the acinar cells of the submandibular gland by secretagogues

Acinar cells isolated from rat submandibular glands by digestion with collagenase were strongly coupled, as visualized by use of the fluorescent dye, Lucifer Yellow CH (LY) (Sasaki et al., 1988). LY appeared in neighboring cells within 2 min after injection into an acinar cell. Dye coupling was found in 350 of 360 acini injected (97.2%). A high degree of coupling between acinar cells was also confirmed by an electrical-coupling method (Kanno et al., 1987).

Addition of acetylcholine (ACh) decreased the relative number of dye-coupled cells within 10 min. The suppressive effect of ACh was dose dependent at concentrations from 10^{-6} M to 10^{-4} M (75.0% at 10^{-6} M and 22.7% at 10^{-4} M). The reduction in numbers of dye-coupled cells was reversed upon the removal of ACh. Co-treatment of

TABLE I. Effects of secretagogues on dye coupling

	Percentage of dye-coupled cells
Control	97.2
10^{-4} M ACh	22.7
+10^{-5} M atropine	92.7
10^{-4} M adrenaline	75.3
+10^{-5} M phenoxybenzamine	92.0

Acini from rat submandibular glands were prepared by digestion with collagenase (Sasaki et al., 1988). The agonists and antagonists were added simultaneously or separately to the acini. The extent of intercellular communication was evaluated in isolated acini, which consisted of 5–7 acinar cells, by monitoring the transfer of dye after microinjection of a 10% solution of LY. If the dye diffused from the injected cell to one or more neighboring cells, the injected cell was scored as dye coupled. The percentage of dye-coupled cells was calculated from results obtained during the course of 25 min after addition of the drugs in each experiment.

TABLE II. Effects of calcium antagonists on ACh-induced inhibition of dye coupling

	Percentage of dye-coupled cells
Control	97.2
10^{-4} M ACh	22.7
10^{-5} M verapamil	92.3
5 mM Ni	80.0
ACh + verapamil	91.6
ACh + Ni	77.2
ACh + 10^{-4} M EGTA	82.9

The experimental conditions were the same as indicated in Table 1.

acinar cells with atropine blocked the ACh-induced inhibition of dye coupling (Table I). The addition of carbachol (CCh) suppressed dye coupling similarly, suggesting that the muscarinic action of ACh and CCh is responsible for the rapid inhibition of dye coupling.

Adrenaline also decreased the extent of dye coupling in isolated acini, but treatment with 10^{-4} M isoproterenol had no suppressive effects. The suppressive effect of adrenaline was antagonized by phenoxybenzamine (Table I). Thus, the action of α-adrenergic agonists induced a rapid inhibition of dye coupling in the acinar cells of the submandibular glands. The rapid inhibition of gap-junctional cell communication appears, therefore, to be related to the stimulation of salivary secretion by cholinergic and adrenergic agonists.

Mechanism of the rapid Ach-induced inhibition of dye coupling

Simultaneous treatment of acinar cells with H-7 and ACh blocked the rapid ACh-induced inhibition of dye coupling. Addition of HA1004 had no effect on the ACh-induced suppression of dye coupling. The suppressive effect of ACh was also blocked by verapamil, by nickel ions and by chelation of extracellular calcium ions with EGTA (Table II). Thus, the rapid inhibition of dye coupling by cholinergic agonists may result from an increased influx of calcium ions from the extracellular solution, as well as the activation of protein kinase C; these two factors have also been reported to inhibit gap-junctional cell communication in pancreatic acinar cells and cultured cells (Enomoto and Yamasaki, 1985; Yada et al., 1985).

Delayed inhibition of gap-junctional cell communication by ACh

The relative number of dye-coupled acinar cells isolated from submandibular glands 12 h after intraperitoneal application of carbachol (CCh) decreased, but returned to the control level when cells were isolated 24 h after this treatment (Table III). Intraperitoneal application of ACh and of pilocarpine also significantly decreased the number of dye-coupled acinar cells when isolated and examined 12 h later. Isoproterenol did not induce any delayed inhibition of dye coupling (Table III). Cholinergic agonists were able to increase the relative number of dye-uncoupled acinar cells (about 20% of total acinar cells) detected about 12 h after application.

Relationship between delayed inhibition of dye coupling and salivary secretion

Delayed inhibition of dye coupling, detectable 12 h after intraperitoneal application of 20 μg/kg, CCh was blocked by the simultaneous application of 100 μg/kg atropine. Intraperitoneal application of CCh stimulated salivary secretion for about 40 min, but secretion was completely suppressed by *simultaneous* application of atropine (100 μg/kg). However, the delayed inhibition of dye coupling by CCh was not suppressed if atropine was applied 60 min after application of CCh, i.e., after CCh-induced secretion stopped. The delayed increase in the number of dye-uncoupled acinar cells caused by cholinergic agonists is related to some as yet unknown change in cell function that occurs during the secretory cycle after saliva secretion.

Delayed inhibition of dye coupling by parasympathectomy

The number of dye-coupled acinar cells decreased when acinar cells were isolated 24h after parasympathectomy (Table IV). A sham operation did not influence dye coupling: The extent of dye coupling between acinar cells isolated from the non-denervated gland, which was on the opposite side from the denervated gland, was unchanged. Delayed inhibition of dye coupling was observed after parasympathectomy, as in the case of cholinergic agonists. Parasympathectomy resulted in stimulation of the secretion of saliva from denervated salivary glands 12 h later.

TABLE III. Delayed effects of intraperitoneal application of secretagogues on dye coupling

	Percentage of dye coupled cells
Control	96.2
CCh	75.3
ACh	85.2
Pilocarpine	84.4
Isoproterenol	97.7

The extent of the intercellular communication was examined by the transfer of LY in the acini isolated by treatment with collagenase 12 h after intraperitoneal application of saline (control), carbachol (CCh; 20 μg/kg body weight), ACh (20 μg/kg), pilocarpine (1.6 mg/kg) and isoproterenol (16 mg/kg).

TABLE IV. Effects of parasympathectomy on dye coupling

	Percentage of dye coupled cells
Control	96.2
Parasympathectomy	64.6
+ atropine 12 h later	93.9
+ atropine 24 h later	69.8

Parasympathectomy was carried out under a dissecting microscope by cutting the chordalingual nerve and then removing the nerves along the excretory duct. Atropine (100 μg/kg) was applied intraperitoneally 12 h after parasympathectomy (indicated as atropine 12 h later), or atropine (10^{-5} M) was applied extracellularly to acini isolated 24 h after parasympathectomy (indicated as atropine 24 h later). Dye coupling was measured in the acini isolated 24 h after parasympathectomy.

When atropine (100 μg/kg) was administered intraperitoneally 12 h after parasympathectomy, it suppressed the inhibition of dye coupling and stimulation of salivary secretion. When acinar cells isolated 24 h after parasympathectomy were treated with 10^{-5} M atropine in vitro, the denervation-induced inhibition of dye coupling was not blocked (Table IV).

Parasympathectomy is believed to stimulate the spontaneous release of acetylcholine from the terminals of injured nerve fibers near the salivary glands (Delfs and Emmelin, 1979; Thesleff, 1982). The rate of release of ACh may spontaneously increase around 12 h after parasympathectomy and may, in turn, stimulate salivary secretion. Dye coupling was inhibited in cells isolated 12 h after parasympathectomy-induced saliva secretion. Atropine blocked both the increase in saliva secretion and the induced uncoupling of acinar cells. Therefore, the release of ACh in vivo probably also induces secretion-related delayed inhibition of dye coupling.

Two types of inhibition of gap-junctional cell communication related to saliva secretion

Almost all acinar cells in the resting state are coupled with one another, and stimulation of saliva secretion induces two types of inhibition of dye coupling, namely, rapid inhibition and delayed inhibition, after the application of secretagogues. Rapid inhibition of dye coupling is observed only in the presence of secretagogues and may possibly occur as a result of the increased influx of calcium ions and the activation of protein kinase C. Rapid inhibition is induced by the low concentration of ACh (10^{-6} M) that induces prolonged secretion from perfused salivary glands (Nakahari et al., 1991). The inhibition of dye coupling may be related to saliva secretion itself. First, gap-junctional cell communication might affect the synchronous initiation of fluid secretion by mediating the transfer of a second messenger to neighboring cells. Inhibition of gap-junctional cell communication might increase the secretion of electrolytes from the basolateral to the luminal side by suppressing the passage of electrolytes into coupled neighboring acinar cells. However, a higher concentration of ACh (10^{-4} M) induces only a transient stimulation of salivary secretion (Nakahari et al., 1991). The inhibition of dye coupling by a high dose of a cholinergic agonist might be causally related to the inhibition of salivary secretion during continuous stimulation, by blocking the intercellular transfer of electrolytes and second messenger. High doses of ACh also induce a transient decrease in cell volume (Nakahari et al., 1991), so the inhibition of coupling might help to prevent the loss of intracellular electrolytes from uncoupled neighboring acinar cells.

The effects of cholinergic agonists on delayed inhibition of dye coupling is related to changes in cell function during the secretory cycle, but not to side effects from intraperitoneal application of cholinergic agonists, since parasympathectomy also induced secretion-related delayed inhibition of dye coupling. In the resting state, a few acinar cells are dye uncoupled. These might be in a phase of the secretory cycle that follows spontaneous secretion of saliva. Vigorous synchronous secretion of saliva induced by the intraperitoneal application of cholinergic agonists seems to increase the number of uncoupled acinar cells at some phase of the secretory cycle. The delayed inhibition of dye coupling might reflect changes in cell function during the secretory cycle.

Acknowledgements

The research was supported in part by Grants-in-Aid for Scientific Research (Nos. 62440022 and 03454430) from the Ministry of Education, Science and Culture, Japan.

References

Amsterdam, A., Ohad, I, and Schramm, M. (1969) Dynamic changes in the ultrastructure of the acinar cell of the rat parotid gland during the secretory cycle. J. Cell Biol. 41, 753–773.

Delfs, U. and Emmelin, N. (1979) Parasympathetic degeneration secretion of saliva in rats. Q. J. Exp. Physiol. 64, 109–117.

Enomoto, T. and Yamasaki, H. (1985) Rapid inhibition of intercellular communication between Balb/c 3T3 cells by diacylglycerol, a possible endogenous functional analogue of phorbol esters. Cancer Res. 45, 3706–3710.

Hammer, M.G. and Sheridan, J.D. (1978) Electrical coupling and dye transfer between acinar cells in rat salivary glands. J. Physiol. 275, 495–505.

Iwatsuki, N. and Petersen, O.H. (1978) Electrical coupling and uncoupling of exocrine acinar cells. J. Cell Biol. 79, 533–545.

Kanno, Y. and Loewenstein, W.R. (1964) Low-resistance coupling between gland cells. Some observations on intercellular contact membranes and intercellular space. Nature 201, 194–195.

Kanno, Y., Sasaki, Y. and Shiba, Y. (1987) Effects of denervation on cell-to-cell communication between acinar cells of rat submandibular salivary gland. Arch. Oral Biol. 32, 43–46.

Meda, P., Bosco, D., Giordano, E. and Chanson, M. (1990) Junctional coupling modulation by secretagogues in two-cell pancreatic systems. In: Peracchia, C. (Ed.), Biophysics of Gap-Junction Channels, CRC Press, Boston, pp. 191–208.

Nakahari, T., Murakami, M., Sasaki, Y., Kataoka, T., Imai, Y., Shiba, Y. and Kanno, Y. (1991) Dose effects of acetylcholine on the cell volume of rat mandibular salivary acini. Jpn. J. Physiol. 41, 153–168.

Petersen, O.H. (1986) Potassium channels and fluid secretion. News Physiol. Sci. 1, 92–95.

Sasaki, Y., Shiba, Y. and Kanno, Y (1988) Suppression of intercellular communication in acinar cells from rat submandibular gland by cholinergic and adrenergic agonists. Jpn. J. Physiol. 38, 531–543.

Shiba, Y., Sasaki, Y. and Kanno, Y (1990) Inhibition of gap-junctional intercellular communication and enhanced binding of fibronectin-coated latex beads by stimulation of DNA synthesis in quiescent 3T3-L1 cells. J. Cell Physiol. 145, 268–273.

Thesleff, P. (1982) Electrophysiological evidence of denervation activity in the parotid gland of the rat. Acta Physiol. Scand. 115, 97–101.

Yada, T., Rose. B. and Loewenstein, W.R. (1985) Diacylglycerol down regulates junctional membrane permeability. TMB-8 blocks this effect. J. Membr. Biol. 88, 217–232.

J.E. Hall, G.A. Zampighi and R.M. Davis (Eds.)
Progress in Cell Research, Vol. 3

CHAPTER 30

Gap junctions in human corpus cavernosum vascular smooth muscle: a test of functional significance

G.J. CHRIST[a], A.P. MORENO[b], C.M. GONDRÉ[a], C. ROY[b], A.C. CAMPOS DE CARVALHO[b,*], A. MELMAN[a] and D.C. SPRAY[b]

Departments of [a]Urology and [b]Neuroscience, Albert Einstein College of Medicine, 1300 Morris Park Avenue, Bronx, NY 10461, USA

Introduction

The past few years have witnessed an explosion of information, in a variety of tissues and cell types (see for example other chapters in Parts IV and V), concerning the physiological significance of the ever-increasing number of gap junctions that have been identified (see Part I). It is generally agreed that gap-junction channels interconnect the cytoplasm of adjacent cells in *most tissues* to provide a pathway for the intercellular transfer of current-carrying ions and second-messenger molecules (Bennett and Spray, 1985; Hertzberg and Johnson, 1988; Robards et al., 1990; Bennett et al., 1991). In fact, common intracellular second messengers such as Ca^{2+} and/or inositol 1,4,5-trisphosphate are known to traverse the junctional membrane in several cell types, such as hepatocytes (Sáez et al., 1989), epithelial cells (Sanderson et al., 1990), astrocytes (Cornell-Bell et al., 1990) and cultured kidney cells (Atkinson, 1989). As such, gap junctions have been hypothesized to play an important role in the synchronization and coordination of tissue responses.

However, establishing an obligatory role for gap junctions in tissue function, or dysfunction, has proved difficult; the strongest evidence available pertains only to the synchronization of contractions in ventricular myocardium (Spray and Burt, 1990; Campos de Carvalho et al., this volume), uterine smooth muscle at parturition (Cole and Garfield, 1985; Risek et al., 1990), and perhaps in coordinating secretory responses in pancreatic acinar cells as well (Chanson et al., 1991). Although a role for gap junctions in the physiology and pathophysiology of vascular function has been proposed (Segal and Duhling, 1986; Larson et al., 1990; Moore et al., 1991), the role of gap junctions in vascular wall cells remains controversial (Segal and Beny, 1991). The results of both clinical and experimental studies support a role for cell-to-cell coupling in coordinating tissue responses in the vascular smooth muscle of the human corpus cavernosum. In fact, the currently accepted theory of human penile erection maintains that erection and flaccidity are largely dependent on the synchronized relaxation and contraction, respectively, of the vascular smooth muscle cells in the corpus cavernosum (Saenz de Tejada, 1989; Krane et al., 1989; Heaton 1989; Christ et al., 1990; 1991a; 1991b).

Moreover, despite the sparse sympathetic innervation revealed by histological studies on corpus cavernosum (Melman et al., 1980), electromyographic studies indicate that electrical activity in the corpus cavernosum in vivo is well synchronized (Stief et al., 1991). In addition, the local injection of vanishingly small quantities of PGE_1 or VIP (e.g., mg amounts) under low-flow conditions (e.g., a blood flow of 8 ml/min per 100 g of tissue) elicits a rapid and robust erection. Furthermore, the spontaneous myotonic automaticity characteristic of corporal smooth muscle strips in vitro (Christ et al., 1990) is effectively inhibited when intercellular communication between smooth muscle cells is blocked by treatment with heptanol, a drug which reversibly closes gap-junction channels (unpublished observation).

The goal of these studies was to test the functional significance of the gap junctions previously identified in both tissue sections and cultured cells isolated from vascular smooth muscle of the human corpus cavernosum (Campos de Carvalho et al., 1990; Moreno et al., 1990). Our present results extend our previous observations and further support the hypothesis that the gap junctions present between human corpus cavernosum smooth muscle cells may play an important role in the initiation, maintenance and modulation of α_1-adrenergic contractions in human vascular smooth muscle.

---
*Permanent address: Carlos Chagas Filho Institute of Biophysics, Federal University of Rio de Janeiro, Brazil.

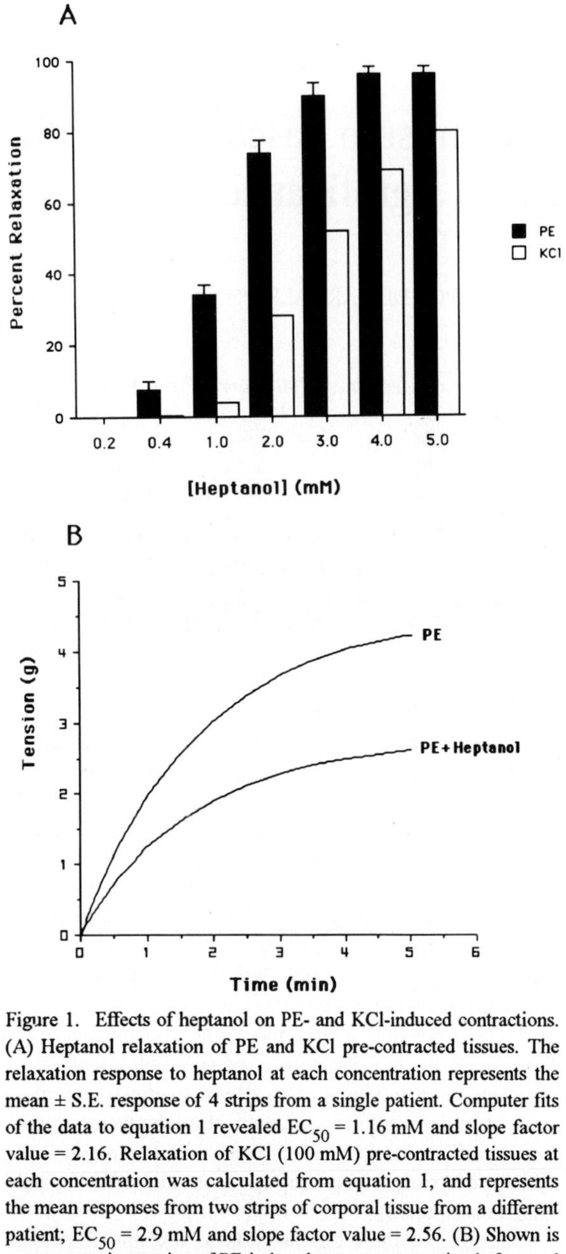

Figure 1. Effects of heptanol on PE- and KCl-induced contractions. (A) Heptanol relaxation of PE and KCl pre-contracted tissues. The relaxation response to heptanol at each concentration represents the mean ± S.E. response of 4 strips from a single patient. Computer fits of the data to equation 1 revealed EC_{50} = 1.16 mM and slope factor value = 2.16. Relaxation of KCl (100 mM) pre-contracted tissues at each concentration was calculated from equation 1, and represents the mean responses from two strips of corporal tissue from a different patient; EC_{50} = 2.9 mM and slope factor value = 2.56. (B) Shown is a representative tracing of PE-induced response generation before and after heptanol treatment (2 mM, 15 min) on a single strip of corporal tissue; data modified from Christ et al. (1991b)

Methods

Preparation and pretreatment of isolated tissues

Strips of corporal tissue were excised from impotent patients undergoing implantation of penile prostheses, and suspended between small surgical hooks in 20 ml organ chambers containing Krebs-Henseleit buffer, maintained at 32 ± 0.05°C and continuously gassed with 95% O_2 and 5% CO_2 to maintain pH = 7.5 ± 0.1 (Christ et al., 1990; 1991a; 1991b). Contractions were measured isometrically with a Grass Force-Displacement Transducer (Model FT-03), and recorded on a Grass Polygraph (Model 7D).

Steady-state protocol

Tissues were submaximally pre-contracted with PE (i.e., EC_{75}). After a stable response developed, a concentration response curve (CRC) to the cumulative addition of heptanol was assayed. CRC data were analyzed using a simple logistic equation of the following form:

$$E = E_{max}/(1 + (EC_{50}/[D])^n) \qquad (1)$$

where E is the observed effect in grams of tension, [D] is the molar concentration of drug, E_{max} is the calculated maximal effect, EC_{50} is [D] at 0.5 E_{max}, and n is the slope factor.

Kinetic protocol

After control kinetic responses were obtained to a submaximal concentration of phenylephrine (PE; i.e., EC_{75}) , tissues were pre-incubated for 15 min with 2 mM heptanol, and then contracted again with the same PE concentration.

Explant cell cultures

Corporal vascular smooth muscle cell cultures were prepared as previously described (Bhargava et al., 1991; Campos de Carvalho et al., 1991), and propagated for no more than four passages. Cellular homogeneity was verified by immunofluorescence staining using a monoclonal smooth muscle α-actin antibody.

Dye loading with Fura-2

Smooth muscle cell cultures were grown on coverslips and loaded with the acetoxymethylester form of Fura-2 (Fura-2 AM; 10–30 mM) for 60–90 min at room temperature to ensure optimal uptake and cytosolic distribution (Moore et al., 1990). Loaded cells were placed on a specially prepared chamber on a Zeiss Axiophot microscope in a 1 ml final volume. All experiments were carried out at room temperature (27°C) in a modified HEPES buffer of the following composition (mM): NaCl, 145; KCl, 5; $CaCl_2$, 2.5; $MgSO_4$, 1.2; glucose, 11; HEPES, 10. Nonperfusion conditions were used since the Ca^{2+} responses obtained under conditions of continuous perfusion were indistinguishable from those obtained under nonperfusion conditions (data not shown).

Optical measurement of cellular [Ca^{2+}]

Images were sequentially obtained from a manual gain intensified CCD camera (Quantex Corp.), digitized at excitation wavelengths of 340 and 380 nM using a filter wheel, with emission set above 480 nM by a dichroic mirror, and the resulting image was stored on an IBM compatible computer (Dell Computer Systems). Data were analyzed for intracellular Ca^{2+} levels using the Image 1AT/FL software package on a Dell Systems computer. Ratio imaging was used to obtain spatial maps of Ca^{2+} distribution in images of single cells as described fully elsewhere (Moore et al., 1990; Goldman et al., 1991).

Cellular studies

For PE stimulation, cells were preincubated with PE (100 μM) for 10–12 min in Ca^{2+}-free buffer containing 1 mM EGTA prior to the addition of 10 mM Ca^{2+}. In other experiments, cells were preincubated for 15 min in calcium-free buffer containing 1 mM EGTA and the calcium ionophore A23187 (1 mM), prior to re-addition of extracellular calcium. In all cases, drugs were added in

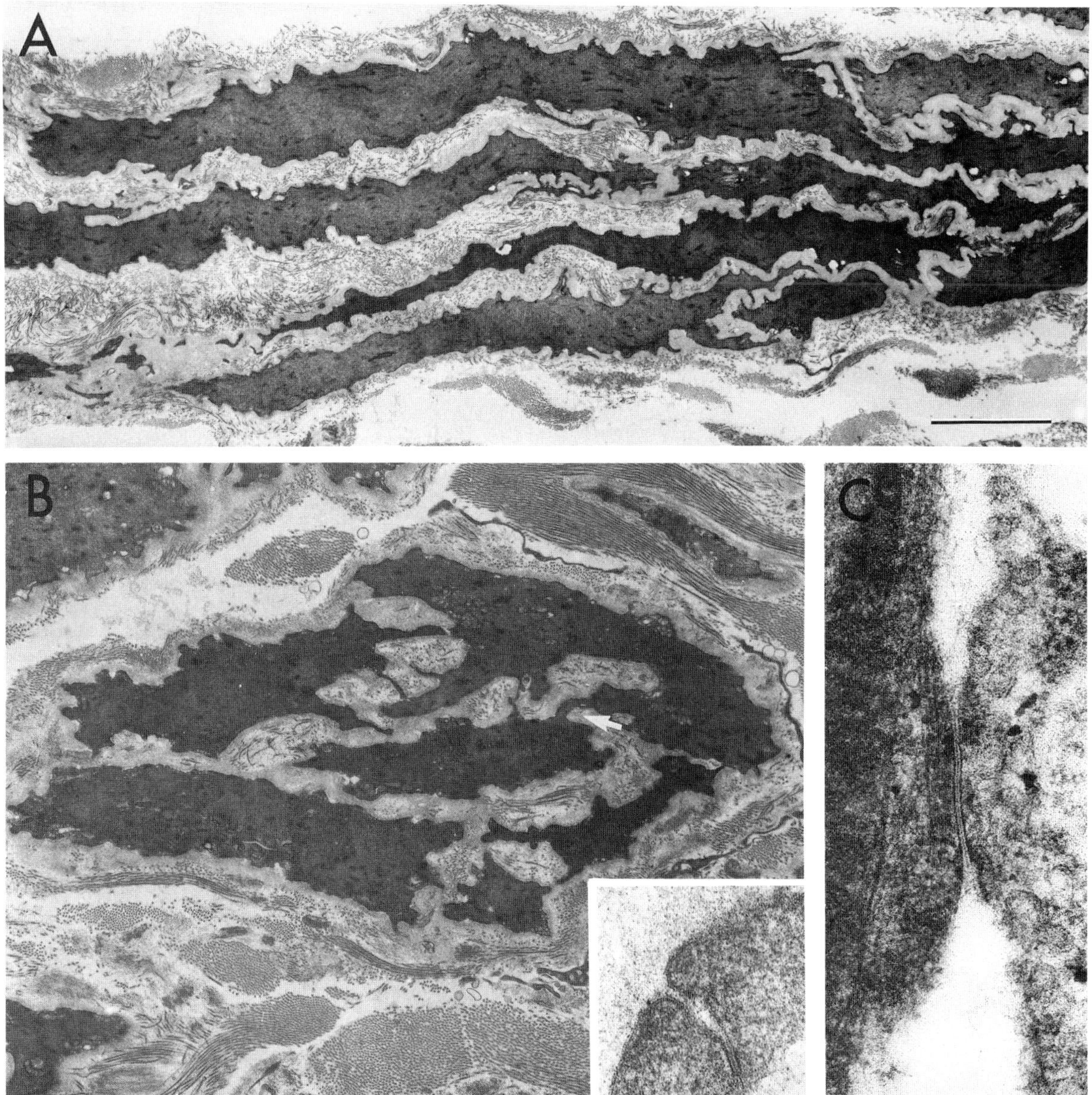

Figure 2. Ultrastructural aspects of corporal smooth muscle cells in situ (A,B) and in culture (C). A. Electron micrograph of corpus cavernosum, sectioned longitudinally to show the more electron-dense smooth muscle cells separated by lighter-staining connective tissue. Scale bar represents 5 μm. (B). Electron micrograph of corpus cavernosum showing a region of contact between smooth muscle cells (arrow). This region is enlarged in the inset to show the gap junction. Bars represent 5 mm in the main portion of B and 0.125 μm in the inset. (C) Electron micrograph of corporal smooth muscle cells in culture, showing a larger gap junction between the cells. Scale bar, 0.25 μm.

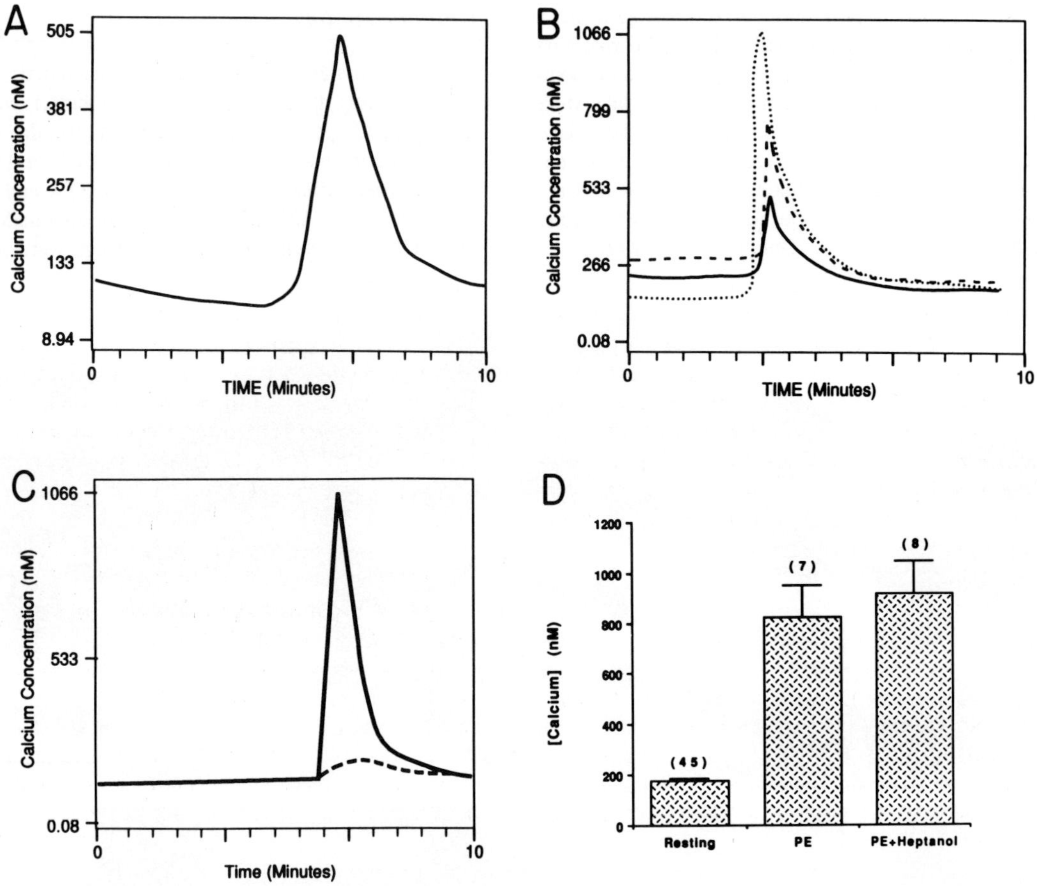

Figure 3. Ca^{2+} mobilization in cultured corporal smooth muscle cells. Panels A–C show representative tracings of the time course of changes in intracellular Ca^{2+} levels in response to various manipulations. (A) Changes in intracellular Ca^{2+} elicited by 100 μM PE. (B) Microinjection of Ca^{2+} into a single cell produced a significant increase in intracellular Ca^{2+} in the injected cell (dotted line), as well as significant, but temporally delayed, increases in intracellular Ca^{2+} in adjacent cells (solid and dashed lines). (C) Junctional transfer of intracellular Ca^{2+} is blocked by 2 mM heptanol; the solid line shows the increase in intracellular Ca^{2+} in the injected cell, and the dashed line shows the lack of change in intracellular Ca^{2+} in an adjacent, but coupled cell. (D) Calcium mobilization in response to 100 μM PE is not altered by 2 mM heptanol.

10–20 μl aliquots, placed directly over the cells being studied. For the microinjection experiments, microelectrodes (80 MΩ) were filled with 1 M Ca^{2+} or 2.5 mM IP_3 dissolved in 150 mM LiCl.

Results

Isolated tissue studies

Heptanol fully and reversibly relaxed both PE and KCl pre-contracted tissues in a concentration-dependent manner. A representative example of the dramatic differences in the magnitude of the heptanol-induced relaxation between PE and KCl pre-contracted tissues is shown in Fig. 1A. Moreover, Fig. 1B illustrates that preincubation of corporal tissue strips with 2 mM heptanol for 15 min prior to addition of PE (see Methods), significantly reduced the magnitude of the steady-state contraction by 40%, while having no effect on the time course of response generation.

Cellular studies

Electron microscopic examination of tissue sections excised from the corpus cavernosum in situ revealed bundles of smooth muscle cells embedded in a heavily collagenous extracellular matrix (Fig. 2A). Importantly, although difficult to detect, small gap junctions were present between the smooth muscle cells at areas of membrane apposition (Fig. 2B).

Optical measurement of intracellular Ca^{2+} in cultured corporal vascular smooth muscle cells

Resting calcium levels in cultured corporal vascular smooth muscle cells were typically in the range of

100–300 nM, with a mean ± SE of 176.9 ± 10.5 (n = 45 cells; see Fig. 3D); this value is in agreement with the results of several other studies on cultured vascular smooth muscle cells (Goldman, 1991; Papageorgiou and Morgan, 1991). Moreover, consistent with the known actions of PE in contracting isolated corporal tissue strips (Fig. 2), PE activation of the α_1-adrenergic receptor in cultured cells elicited a rapid, robust and transient increase in intracellular Ca^{2+} (Fig. 3A). In addition, Fig. 3B shows that microinjection of Ca^{2+} into a single corporal vascular smooth muscle cell resulted in an immediate rise in intracellular calcium levels in the injected cell, with a temporally delayed, but significant, increase in intracellular Ca^{2+} levels in contiguous cells. This intercellular transfer of Ca^{2+} was completely inhibited in the presence of 2 mM heptanol (Fig. 3C). Similar results were obtained subsequent to microinjection of IP_3 (data not shown). Importantly, 2 mM heptanol had no detectable effect on resting or PE-stimulated intracellular calcium levels (Fig. 3D).

Discussion

While gap junctions are generally recognized as playing a major role in intercellular communication, thereby integrating a diverse number of tissue responses (Bennett and Spray, 1985; Hertzberg and Johnson, 1988; Robards et al., 1990; Bennett et al., 1991), their role in modulating vascular smooth muscle tone remains controversial. However, there are now compelling data from both clinical and experimental studies which suggest that gap junctions are important modulators of vascular smooth muscle tone in the human corpus cavernosum. In fact, recent studies from our laboratory have demonstrated that the gap-junction protein Cx43 is ubiquitously distributed between corporal smooth muscle cells in culture and in situ (Campos de Carvalho et al., 1990; Moreno et al., 1990), and moreover, have suggested that Cx43 may play an integral role in pharmacomechanical coupling and syncytial tissue contraction during activation of the α_1-adrenergic receptor in isolated corporal strips (Fig. 1) (Christ et al., 1991b).

This report extends these previous observations, and provides direct evidence supporting a potentially important role for gap junctions in mediating the intercellular exchange of current-carrying ions and second messenger molecules between vascular smooth muscle cells (Fig. 3). That is, intracellular injection of both Ca^{2+} (Figs. 3B and C) and IP_3 (Christ et al., 1991c) was associated with an heptanol-sensitive increase in intracellular Ca^{2+} levels in adjacent cells. Since these studies were conducted in the absence of extracellular Ca^{2+} (see Methods), this presumably reflects intercellular diffusion of calcium ions through gap-junction channels formed between coupled corporal smooth muscle cells. Moreover, reversible disruption of intercellular communication by 2 mM heptanol significantly decreased the magnitude without altering the

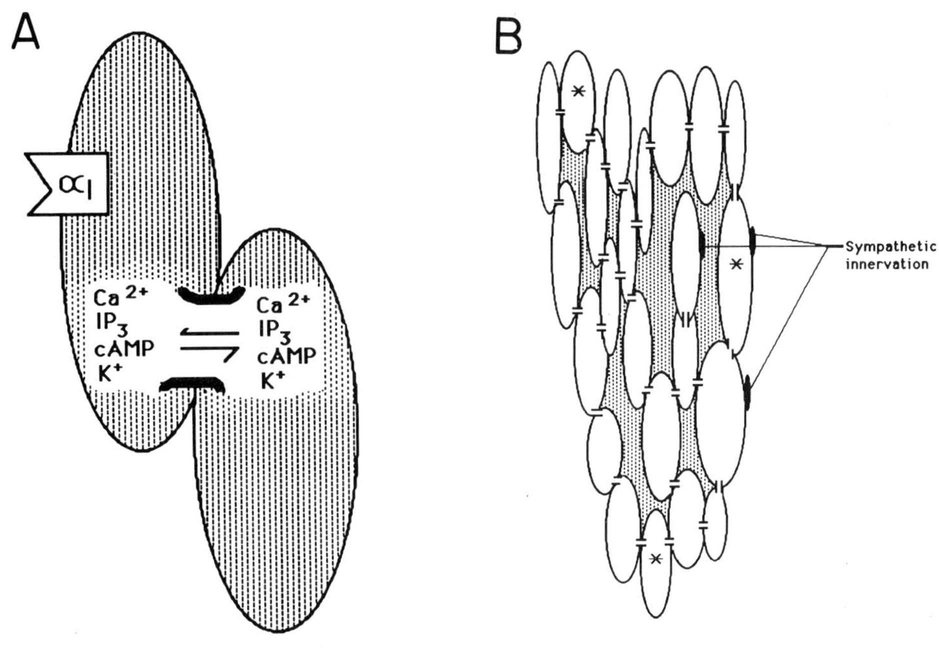

Figure 4. Schematic diagrams depicting the proposed role of gap junctions in coordinating α_1-adrenergic contractility in the corpus cavernosum. (A) Illustration of how second messengers known to be important modulators of smooth muscle contractility might be shared by diffusion through gap junctions between coupled cells. The stippled area here emphasizes the possible continuity of the intracellular environments in the two respective cells. (B) Depicted is a group of cavernosal smooth muscle cells connected by a network of small gap junctions that are a hallmark of this tissue; the asterisks emphasize how smooth muscle cells in distant areas of the corpora might be coupled with respect to both second messenger exchange and response generation, even if they were not directly activated by PE. The stippled area outlines the abundance of cavernous spaces in the corpora (compare with Fig. 1). For clarity and simplicity, endothelial cells are not shown.

time course of PE-induced contractions in isolated corporal strips (Fig. 1); thus, the overall effect of heptanol was to decrease the absolute rate of corporal contraction (i.e., in grams of tension/min) (Fig. 1). The implication of these findings is that intercellular communication among coupled cells may have important functional consequences on corporal tissue tone. The schematic diagram in Fig. 4 depicts the proposed role for gap junctions in corporal smooth muscle.

Our conclusion depends explicitly on the assumption that heptanol exerts relatively selective effects on gap junctions at concentrations of 2 mM (Christ et al., 1991b). In fact, several lines of experimental evidence support this assumption. First, recent evidence obtained from cells of different species suggests that heptanol, at the concentrations used here, has a more potent effect on gap junctions than on other ionic channels (Spear et al., 1990a; 1990b). Second, 2 mM heptanol has no effect on resting or stimulated intracellular calcium levels of corporal smooth muscle cells (Fig. 3D). Third, at heptanol concentrations that are apparently selective for gap junctions, relaxation of PE pre-contracted tissues is greater than 50% (Fig. 1A); furthermore, the lack of change in the time course of PE-induced contractions in the presence of heptanol (Fig. 1B) suggests that heptanol does not alter the contractility of individual smooth muscle cells, but rather decreases the number of recruited smooth muscle cells. Importantly, this effect is not mimicked by shorter-chain alcohols, such as ethanol, that have no effect on gap junctions (Christ et al. 1991b). Lastly, the effects of heptanol on both isolated tissues (Christ et al., 1991b) and cultured cells (Spray and Bennett, 1985) are readily reversible, even at concentrations as high as 7 mM.

Our data are consistent with the hypothesis that heptanol diminishes PE-induced contractility, at least in part, by decreasing junctional transfer of receptor-activated second messengers and/or current-carrying ions (Fig. 4A). The proposed mechanism of action would be expected to decrease the ability of smooth muscle cells directly activated by PE (or norepinephrine in vivo) to recruit contiguous cells not directly activated by PE, and thereby diminish syncytial tissue contraction (Fig. 4B). Thus, the heptanol-induced decrease in the rate (expressed as the developed tension, g, over time) and magnitude of PE-induced contractions depicted in Fig. 1 supports this mechanism of action. Moreover, syncytial tissue contraction could occur in vivo (see Fig. 2) following receptor activation of only a fraction of available cells (Fig. 4B).

In conclusion, these steady-state and kinetic studies on isolated tissues extend our previous observations on intact tissues and cultured cells (Campos de Carvalho et al., 1990; Moreno et al., 1990; Christ et al., 1991b; 1991c), are consistent with clinical observations in vivo and therefore provide compelling empirical evidence for the supposition that intercellular communication may play a significant role in the initiation, maintenance and modulation of α_1-adrenergic receptor-mediated contractions in human corporal vascular smooth muscle.

Acknowledgements

This work was supported in part by USPHS grants DK42027 and NS07521.

References

Atkinson, M.M. (1989) Calcium and inositol 1,4,5-trisphosphate traverse gap-junction channels. J. Cell Biol. 109,100a.

Bennett, M.V.L. and Spray, D.C. (Eds.) (1985) Gap Junctions. Cold Spring Harbor, New York.

Bennett, M.V.L., Bario, L., Bargiello, T.A., Spray, D.C., Hertzberg, E.L. and Sáez, J.C. (1991) Gap Junctions: New tools, new answers, new questions. Neuron. 6, 305–320.

Bhargava, G., Valcic, M. and Melman, A. (1990) Human corpora cavernosa smooth muscle cells in culture: Influence of catecholamines and prostaglandins on cAMP formation. Int. J. Impotence Res. 2, Suppl, 35–36.

Campos de Carvalho, A.C., Moreno, A.P., Christ, G.J., Bhargava, G., Melman, A. Roy, C., Hertzberg, E.L. and Spray, D.C. (1990) Junctional communication between vascular smooth muscle cells (human corpus cavernosum) in culture. Identity of the connexin type and unitary conductance events. Soc. Cell. Biol. 835 (suppl.), 153.

Chanson, M., Lelio, O. and Meda, P. (1991) Extent and modulation of junctional communication between pancreatic acinar cells in vivo. Am. J. Physiol. 261, G28–G36.

Christ, G.J., Maayani, S., Valcic, M. and Melman, A. (1990) Pharmacological studies of human erectile tissue: characteristics of spontaneous contractions and alterations in α-adrenoceptor responsiveness with age and disease in isolated tissues. Br. J. Pharmacol. 101,375–381.

Christ, G.J., Stone, B.S. and Melman, A. (1991a) Age-dependent alterations in the efficacy of phenylephrine-induced contractions in vascular smooth muscle isolated from the corpus cavernosum of impotent men. Can. J. Physiol. Pharmacol. 69, 909–913.

Christ, G.J., Moreno, A.P., Parker, M.E., Gondre, C.M., Valcic, M., Melman, A. and Spray, D.C. (1991b) Intercellular communication through gap junctions: A potential role in pharmacomechanical coupling and syncytial tissue contraction in vascular smooth muscle isolated from the human corpus cavernosum. Life Sci. 49, PL195–PL200.

Christ, G.J., Moreno, A.P., Valcic, M., Parker, M.E., Gondre, M.E., Melman, A. and Spray, D.C. (1991c) Role of gap junctions in contraction of human corpus cavernosum: Smooth muscle uncoupling by heptanol alters contractility without affecting Ca^{2+} mobilization. Proc., Fifth World Congress for Microcirc. p. 16.

Cole, W.C. and Garfield, R.E. (1985) Alterations in coupling in uterine smooth muscle. In: Gap Junctions, M.V.L. Bennett and D.C. Spray (Eds.), Cold Spring Harbor, New York. pp. 215–230

Cornell-Bell, A.H., Finkbeiner, S.M., Cooper, M.S. and Smith, S.J. (1990) Glutamate induces calcium waves in cultured astrocytes: Long range glial signalling. Science. 247, 470–473.

Cory, R.N., Osman, R., and Maayani, S. 1984. Kinetic characterization of the rabbit aorta contractile response to an alpha adrenergic agonist. J. Pharmacol. Exp. Ther. 230, 162–170.

Goldman, W.F. (1991) Spatial and temporal resolution of serotonin-induced changes in intracellular calcium in a cultured arterial smooth muscle cell line. Blood Vessels 28, 252–261.

Goldman, W.F., Bova S. and Blaustein, M.P. (1990) Measurement of intracellular Ca^{2+} in cultured arterial smooth muscle cells using Fura-2 and digital imaging microscopy. Cell Calcium. 11, 221–231.

Heaton, J.P.W. (1989) Synthetic nitrovasodilators are effective, in vitro, in relaxing penile tissue from impotent men: the findings and their implications. Can. J. Physiol. Pharmacol. 67, 78–81.

Hertzberg, E.L. and Johnson, R. (Eds.) (1988) Gap Junctions. Alan R. Liss, New York.

Krane, R.J., Goldstein I., and Saenz de Tejada, I. (1989) Medical progress: impotence. N. Engl. J. Med. 321, 1648–1658.

Larson, D.M., Haudenschild C.C. and Beyer, E.C. (1990) Gap-junction messenger RNA expression by vascular wall cells. Circ. Res. 66, 1074–1080.

Melman, A., Henry, D.P., Felten, D.L. and O'Connor, B.L. (1980) Effect of diabetes upon penile sympathetic nerves in impotent patients. South. Med. J. 73, 307.

Moore, E.D.W., Becker, P.L., Fogerty, K.E., Williams, D.A. and Fay, F.S. (1990) Ca^{2+} imaging in single living cells: Theoretical and practical issues. Cell Calcium. 11, 157–179.

Moore, L.K., Beyer, E.C. and Burt, J.M. (1991) Characterization of gap-junction channels in A7r5 vascular smooth muscle cells. Am. J. Physiol. 260, C975–C981.

Moreno, A.P., Campos de Carvalho, A.C., Christ, G.J., Melman, A., Hertzberg, E.L. and Spray, D.C. (1990) Gap junctions between human corpus cavernosum smooth muscle cells in primary culture: Electrophysiological and biochemical characteristics. Int. J. Impotence Res. 2, Suppl. 2, 55–56.

Papageorgiou, P. and Morgan, K.G. (1991) Increased Ca^{2+} signaling after α-adrenoceptor activation in vascular hypertrophy. Circ. Res. 68, 1080–1084.

Risek, B., Guthrie, S., Kumar, N. and Gilula, N.B. (1990) Modulation of gap-junction transcript and protein expression during pregnancy in the rat. J. Cell Biol. 110, 269–282.

Robards, A., Lucas, W., Jongsma, H., Pitts, J. and Spray, D.C. (Eds.). (1991) Plasmodesmata and Gap Junctions: Parallels in Evolution. Springer-Verlag, Berlin.

Sáez, J.C., Connor, J.A., Spray, D.C. and Bennett, M.V.L. (1989) Hepatocyte gap junctions are permeable to the second messenger, inositol 1,4,5-trisphosphate, and to calcium ions. Proc. Natl. Acad. Sci. USA 86, 2708–2712.

Saenz De Tejada, I. Goldstein, I., Azadzoi, K., Krane, R.J. and Cohen, R.A. (1989) Impaired neurogenic and endothelium-mediated relaxation of penile smooth muscle from diabetic men with impotence. N. Engl. J. Med. 320, 1025–1030.

Sanderson, M.J., Charles A.C., and Dirksen, E.R. (1990)Mechanical stimulation and intracellular communication increases intracellular Ca^{2+} in epithelial cells. Cell Regulation 1, 585–596.

Segal, S.S. and Duling, B.R. (1986) Contractions among microvessels coordinated by intercellular conduction. Science. 234, 868–870.

Segal, S.S. and Bny, J.-L. (1991) Acetylcholine hyperpolarizes arterioles without dye coupling between smooth muscle and endothelial cells. Proceedings, Fifth World Congress for Microcirculation. p. 99.

Spear, J.F., Balke, C.W., Lesh, M.D., Kadish, A.H., Levine J.H., and Moore, E.N. (1990a) Effect of cellular uncoupling by heptanol on conduction in infarcted myocardium. Circ. Res. 66, 202–217.

Spray, D.C. and Bennett, M.V. (1985) Physiology and pharmacology of gap junctions. Annu. Rev. Physiol. 47, 281–303.

Spear, J.F., Balke, C.W., Lesh, M.D., Kadish, A.H., Levine, J.H., and Moore, E.N. (1990b) Letters to the Editor. Circ. Res. 67, 1299–1300.

Spray, D.C., Moreno, A.P., Carvalho, A.C., Melman, A., and Christ, G.J. (1991) Junctional communication between corpus cavernosum smooth muscle cells. Proceedings, Fifth World Congress for Microcirculation. p. 104.

Spray, D.C. and Burt, J.M. (1990) Structure-activity relations of the cardiac gap-junction channel. Am. J. Physiol. 27, C195–C205.

Steif, C.G., Djamilian, M., Anton, P., De Reise, W., Allhoff, E.P. and Jonas, U. (1991) Single-potential analysis of cavernous electrical activity in impotent patients: A possible diagnostic method for autonomic cavernous dysfunction and cavernous smooth muscle degeneration. J. Urol. 146, 771–776.

J.E. Hall, G.A. Zampighi and R.M. Davis (Eds.)
Progress in Cell Research, Vol. 3

CHAPTER 31

Role of gap junctions in mesoderm induction in *Patella vulgata* (Mollusca, Gastropoda): a reinvestigation

PETER DAMEN and WIM J.A.G. DICTUS

Department of Experimental Zoology, University of Utrecht, 3584 CH Utrecht, The Netherlands

Introduction

Several lines of evidence indicate that gap-junctional communication is involved in early development (Warner et al., 1984; Caveney, 1985; Guthrie and Gilula, 1989; Serras et al., 1989). For example, in embryos of insects (Warner and Lawrence, 1982), mice (Kalimi and Lo, 1988), teleosts (Kimmel et al., 1984), ascidians (Serras and Van den Biggelaar, 1989) and molluscs (Serras et al., 1989), it has been found that gap-junctional communication becomes restricted to domains of cells with the same developmental potential. This phenomenon may reflect a developmental event related to the regionalisation of the embryo.

In embryos of the common limpet *Patella vulgata,* evidence for a role of gap junctions in development has been obtained by means of dye-coupling experiments. It has been demonstrated that during early development communication compartments are formed which coincide, spatially as well as temporally, with developmental compartments (Serras et al., 1989). Cells within a compartment have a high degree of coupling, whereas cells across compartmental boundaries have reduced coupling. During further development, the communication compartments become subdivided into smaller compartments which again coincide with developmental compartments (Serras et al., 1989). In the present study, the role of gap-junctional communication in a specific stage of early development of *Patella vulgata*, i.e. during mesoderm induction, was studied.

Patella vulgata is an equally cleaving mollusc. Up to the fifth cleavage (32-cell stage) the embryo is radially symmetrical and cleavages are synchronous. The sixth cleavage is asynchronous and the 64-cell stage is reached via a number of intermediate stages, i.e. the 40-, 52-, 60- and 63-cell stage. Immediately after fifth cleavage all four vegetally located macromeres stretch into the blastocoel towards the animally located first quartet micromeres. Subsequently, three macromeres retract and only one macromere establishes a firm contact with the first quartet micromeres. This macromere occupies a central position in the embryo and is induced to mesentoblast mother cell (3D), the stem cell of the mesoderm. It divides significantly later and much more unequally than the other three macromeres (Van den Biggelaar, 1977). The presumptive 3D cell can also be recognised by its increased amount of F-actin (Serras and Speksnijder, 1990) and, as a result of its (internal) central position, by its smaller external surface. In addition, the presumptive 3D can be recognised by the presence of two very small cells on either side. These cells are the result of the very unequal division of 3c and 3d, two cells that flank the presumptive 3D. These divisions are the first deviation from the spiral cleavage pattern and demarcate 3D, i.e. the dorsal quadrant. The two corresponding cells, 3a and 3b, do not form small cells towards a macromere (Van den Biggelaar, 1977).

Deletion experiments have shown that initially all four macromeres are equipotent and able to become 3D (Van den Biggelaar and Guerrier, 1979). Moreover, the results of these experiments demonstrated that the first quartet micromeres are causally involved in the induction of the stem cell of the mesoderm (Van den Biggelaar and Guerrier, 1979; Arnolds et al., 1983). Not only the presence of the first quartet micromeres, but also their contact with the presumptive 3D appeared to be necessary. Similar results were obtained in all other equally cleaving molluscs tested so far, i.e. in *Lymnaea* (Van den Biggelaar, 1976; Martindale et al., 1985) and *Haminoea* (Boring, 1989). Induction by a diffusible factor, supposedly secreted by the first quartet micromeres, is very unlikely to occur since contact between the presumptive 3D cell and the first quartet micromeres appeared to be necessary. Moreover, this diffusible factor would reach all four competent macromeres via the blastocoel and would subsequently induce all four macromeres instead of only one.

Dye-coupling experiments have been a means of analysing how the micromeres induce 3D (De Laat et al., 1980; Dorresteijn et al., 1983). De Laat et al. observed

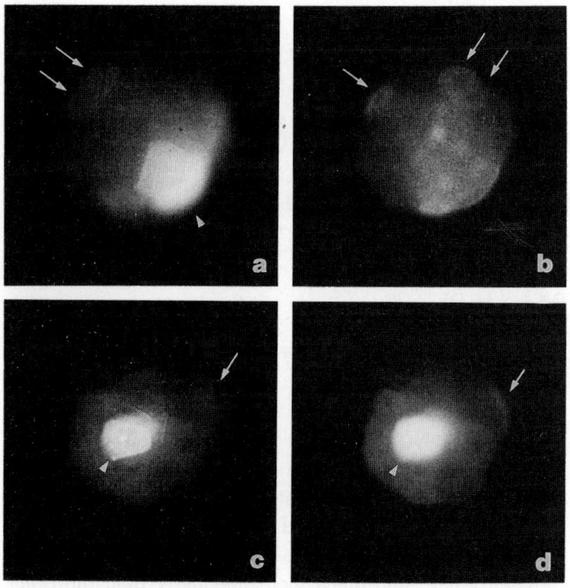

Figure 1. In vivo photographs of embryos injected in the presumptive stem cell of the mesoderm (3D cell). Direct dye transfer to first quartet micromeres (arrows) is visible. Arrowheads indicate the presumptive 3D cell. (a) Lateral view of the A-quadrant of a 52-cell embryo with direct coupling to four first quartet micromeres. Two of these cells are visible in this photograph ($1a^{121}$ and $1a^{122}$). (b) Latero-animal view of the C-quadrant of the same embryo. Direct coupling to corresponding cells in this quadrant ($1c^{121}$ and $1c^{122}$) is visible. One of the labelled cells of the A-quadrant ($1a^{121}$) is just visible. The presumptive 3D cell is located on the other side. (c,d) Vegetal view of another embryo (40-cell stage). The labelled first quartet micromere ($1d^{122}$) is at the periphery and clearly labelled. (c) Focus on the presumptive 3D cell. (d) Focus on the periphery.

dye-coupling between most animal micromeres and the presumptive stem cell of the mesoderm. In a later report, Dorresteijn and co-workers (1983) could not confirm this earlier observation. However, these results are not necessarily contradictory if the described dye-coupling only occurs during a limited phase of the interval between the fifth and the end of sixth cleavage. The results described in this report demonstrate that gap-junctional communication occurs transiently between the presumptive stem cell of the mesoderm (3D) and a limited number of first quartet micromeres. This indicates that gap junctions may be involved in mesoderm induction.

Results

Intracellular injections of Lucifer Yellow CH (MW 434.4 Da) were performed either by iontophoresis or high pressure at the desired stage of development. Most embryos were injected from approximately 40 min after fifth cleavage, i.e. when the presumptive 3D cell makes contact with the first quartet micromeres, up to the 63-cell stage, when mesoderm induction has taken place. The coupling pattern was studied *in vivo* with an epifluorescence excitation Argon laser and an epifluorescence microscope. In addition, the embryos were fixed and sectioned or processed for confocal laser scanning microscopy (CLSM).

Before the 32-cell stage, no spread of Lucifer Yellow (LY) was observed. Approximately 15 min after fifth cleavage, dye-coupling became apparent. After injection of a presumptive 3D cell, dye-spread to all first-order neighbouring cells was observed (Figs. 1c, 1d and 2a). No restrictions were visible. Spread to second-order neighbouring cells was frequently seen and depended on the amount of dye injected (Fig. 1c).

In a number of embryos, some micromeres of the upper two tiers of first quartet micromeres, or their daughter cells (depending on the cell stage), appeared to be labelled (Figs. 1–3). In most of these embryos, this labelling was restricted to cells of the second tier of first quartet micromeres. In contrast, the two lower tiers of first quartet micromeres were never labelled (Figs. 1–3). The first quartet micromeres are relatively transparent. When

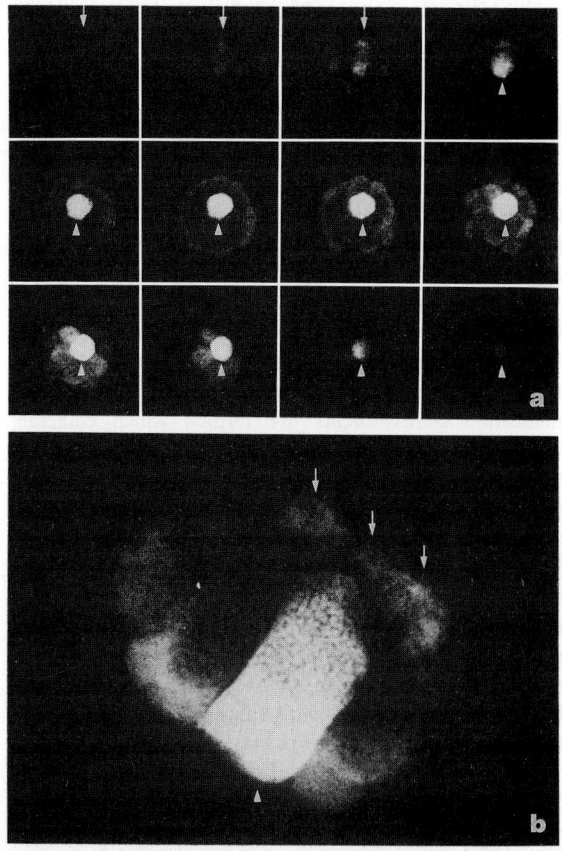

Figure 2. CLSM photographs. (a) Z-series from animal to vegetal of a 52-cell embryo. The first three optical sections show labelled first quartet micromeres ($1m^{111}$-, $1m^{112}$- and $1m^{12}$-cells) in the form of a cross (arrows). These cells contact the presumptive 3D macromere (centrally located, very fluorescent cell, indicated by arrowhead). Direct coupling is evident. (b) Lateral optical section showing the injected presumptive 3D (arrowhead) and direct coupling to first quartet micromeres (arrows).

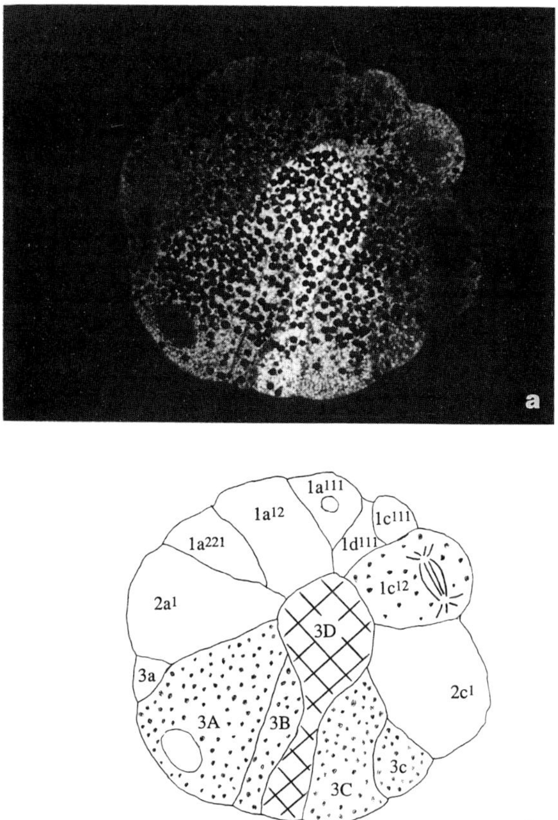

b

Figure 3. (a) Serial section (1 μm) of a 52-cell embryo injected in the presumptive 3D cell. Direct spread to a contacting first quartet micromere ($1c^{12}$) can be discerned. (b) Reconstruction. Cross-hatched area represents strong labelling of injected cell. Dense stippling indicates strong coupling, whereas light stippling indicates weak coupling to first quartet micromere.

the embryos were observed from the animal pole, these cells overlay the injected presumptive 3D, making it difficult to determine if these micromeres are actually labelled. Therefore, the embryos were rotated and examined carefully. Micromeres which are actually labelled will not lose their apparent fluorescence when their position relative to the presumptive 3D is changed (Figs. 1c and 1d). Labelling of the cells of the upper two tiers of first quartet micromeres was confirmed in sections and by CLSM (Figs. 2 and 3). In addition, both the CLSM preparations and the sectioned material showed that, in all embryos with dye-coupling between the first quartet micromeres and the presumptive 3D cell, the labelled cells were in direct contact with the presumptive 3D, which is located in the centre of the embryo (Figs. 2 and 3). The dye-spread which we have observed must have occurred via these internal cell contacts. Therefore, this dye-spread is referred to as direct dye-spread.

In the series of embryos injected during the beginning of the fifth and the end of sixth cleavage, the highest incidence of dye-spread occurred at the 52-cell stage. Four out of a total of 13 embryos (31%) showed labelling of one or more first quartet micromeres. At the 40-cell stage, direct dye-spread was observed in only 10% of the embryos (3 out of 31 embryos). Direct dye-spread to first quartet micromeres was never observed at the 32-cell and 63-cell stages.

Other dyes, Rhodamine B (MW 479 Da) and Cascade Blue (MW 596 Da), were used to study direct coupling. However, direct coupling could not be demonstrated with any of these dyes. Apparently, spread of Rhodamine B via the periphery of the embryo was as fast as or faster than direct spread. As a consequence, the whole embryo became homogeneously fluorescent. Injection of Cascade Blue resulted in only a very moderate spread. Only first-order neighbouring cells became faintly labelled. Dye-spread was not sufficient to study direct coupling.

In order to investigate whether dye-transfer can occur from the first quartet micromeres to the presumptive 3D cell, first quartet micromeres were injected with LY. Thirteen embryos were injected at various stages between the fifth and the end of sixth cleavage. In none of these embryos direct coupling to the presumptive 3D cell was observed.

Lymnaea stagnalis was used as an alternative to study direct coupling between the presumptive 3D cell and animal micromeres. *Lymnaea* embryos were injected with LY around the moment 3D was induced. This resulted in labelling of all, or nearly all cells of the embryo. This was observed both after injecting a presumptive 3D macromere or after injecting a first quartet micromere. As a result of this, no direct dye-spread to first quartet micromeres could be observed.

Discussion

Direct coupling between the presumptive stem cell of the mesoderm (3D) and one or more first quartet micromeres was observed transiently during the interval between the fifth and the end of sixth cleavage. In a number of 40-cell embryos (10%), direct coupling was observed. The highest incidence of direct coupling (31%) was found at the 52-cell stage. The results suggest that direct dye-coupling starts at the 40-cell stage. The low percentage of embryos showing direct dye-coupling observed at the 40-cell stage can be explained by the fact that these embryos were injected at the late 32-cell stage, i.e., before direct coupling starts. Therefore, direct dye-spread could only occur during a limited period of time. In contrast, the embryos observed at the 52-cell stage were injected at the 40-cell stage, after direct coupling had started. In these embryos, the whole period between start of injection and observation could be used to transfer dye to first quartet micromeres. Consequently, the incidence of direct dye-spread at the 52-cell stage is significantly higher than at the 40-cell stage.

We may consider the possibility that the appearance of direct coupling is not completely related to the 40-cell stage. In a given embryo, the appearance of direct coupling may be at the 40-cell stage, whereas in another embryo it may occur at the 52-cell stage. Variations in the moment the presumptive 3D cell contacts first quartet micromeres have been observed by Van den Biggelaar (1977). These variations may result, in turn, in small variations in the start of dye-coupling between the presumptive 3D cell and first quartet micromeres.

We only observed direct coupling in a limited number of embryos after injection of LY. If too much dye was injected, first quartet micromeres became labelled via the periphery of the embryo. In these embryos we did not know whether direct spread contributed to the labelling of first quartet micromeres. If, on the other hand, a small amount of dye was injected, direct dye-spread was never detected. In the embryos where we observed direct coupling, labelling of the micromeres was always very weak and just above the background level of autofluorescence. Apparently, the amount of LY that has to be injected in order to study direct coupling is critical.

The results suggest that only a small number of functional gap junctions are present between the presumptive 3D cell and first quartet micromeres. This is consistent with the results of Dorresteijn et al. (1982), who studied gap junctions in early *Patella* embryos by means of transmission electron microscopy. They did not find gap junctions between the presumptive 3D macromere and first quartet micromeres. However, due to the limitations of transmission electron microscopy, gap junctions might have been overlooked. In conclusion, the reasons discussed in this and the previous paragraphs may account for the relatively low percentage of embryos in which direct coupling can effectively be demonstrated, even though gap-junctional communication may actually exist in all embryos.

In most cases, direct coupling was limited to the second tier of first quartet micromeres (or their daughter cells). This result is consistent with the observation of Van den Biggelaar (1977) that second-tier cells make contact with the presumptive 3D cell before the first tier does and, in addition, appear to have a larger contact area with the presumptive 3D cell.

We never observed direct coupling from first quartet micromeres to the presumptive 3D cell, i.e., in the opposite direction. Probably, the amount of LY that can be injected in a relatively small first quartet micromere is insufficient to label the relatively large presumptive 3D cell.

The observations that the gap-junctional communication between the presumptive 3D cell and first quartet micromeres only occurs in a small time interval, and that this time interval starts at the 40-cell stage, may explain the conflicting results described in the literature (De Laat et al., 1980; Dorresteijn et al., 1983).

We used several other dyes to establish direct coupling, but none with success. Some dyes, e.g. Rhodamine B, diffuse very rapidly in our system. Therefore, dye-spread via the periphery of the embryo to first quartet micromeres is as fast as, or faster than, direct spread. Another dye, Cascade Blue, diffuses very slowly. Even first-order neighbouring cells were only faintly labelled. LY appears to be the most suitable dye to study direct coupling.

In all equally cleaving molluscs tested so far, contact with first quartet micromeres appears to be necessary to induce 3D (Van den Biggelaar, 1976; Martindale et al., 1985; Boring, 1989). In *Lymnaea stagnalis*, gap junctions have been observed between the presumptive 3D macromere and first quartet micromeres both by freeze-fracture electron microscopy and by transmission electron microscopy (Dorresteijn et al., 1981). Therefore, we also tried to study direct coupling between the presumptive 3D cell and first quartet micromeres in *Lymnaea stagnalis*. After injecting a presumptive 3D cell or a first quartet micromere with LY all cells of the embryo became labelled. In comparison with *Patella*, LY appeared to diffuse very rapidly in *Lymnaea*. In addition, in *Lymnaea*, a 24-cell embryo is produced at fifth cleavage because only the vegetal two tiers divide (Van den Biggelaar, 1976). At this stage, the mesodermal stem cell is induced. Because spread of LY in the cytoplasm is faster than spread through gap junctions, dye-spread via the periphery of the embryo to first quartet micromeres takes place much more easily in a 24-cell embryo than in a 32-cell embryo. For these reasons, *Lymnaea stagnalis*, and probably all other molluscs in which the stem cell of the mesoderm is induced at the 24-cell stage, is not suitable to study direct dye-coupling.

At this moment we know that dye-coupling exists between the presumptive 3D cell and first quartet micromeres. It has only been observed with specific cells, i.e. the upper two tiers of first quartet micromeres (or their daughter cells), and not with the other first quartet micromeres. These results support those of Van den Biggelaar and Guerrier (1979), which show the importance of the upper two tiers of first quartet micromeres for the induction of the 3D cell. After deletion of these cells, but with the lower two tiers of first quartet micromeres still present, 3D induction fails to occur. In conclusion, the direct gap-junctional communication between the presumptive 3D cell and the upper two tiers of first quartet micromeres (or their daughter cells) may be necessary for mesoderm induction. Currently, experiments are being performed to block gap-junctional communication in order to further investigate the role of gap-junctional communication in mesoderm induction in *Patella vulgata*.

Acknowledgements

The authors wish to thank Dr. F. Serras for initiating this research and Prof. Dr. J.A.M. van den Biggelaar for his constructive discussions. The photography department is thanked for excellent services. Last, but not least, we thank Mr. W.J. Hage for his assistance with the CLSM.

References

Arnolds, W.J.A., Van den Biggelaar, J.A.M. and Verdonk, N.H. (1983) Spatial aspects of cell interactions involved in the determination of dorsoventral polarity in equally cleaving gastropods and regulative abilities of their embryos, as studied by micromere deletions in *Lymnaea* and *Patella*. Wilhelm Roux's Arch. Dev. Biol. 192, 75–85.

Boring, L. (1989) Cell-cell interactions determine the dorsoventral axis in embryos of an equally cleaving opisthobranch mollusc. Dev. Biol. 136, 239–253.

Caveney, S. (1985) The role of gap junctions in development. Annu. Rev. Physiol. 47, 319–335.

De Laat, S.W., Tertoolen, L.G.J., Dorresteijn, A.W.C. and Van den Biggelaar, J.A.M. (1980) Intercellular communication patterns are involved in cell determination in early molluscan development. Nature 287, 546–548.

Dorresteijn, A.W.C., Van den Biggelaar, J.A.M., Bluemink, J.G. and Hage, W.J. (1981) Electron microscopical investigations of the intercellular contacts during the early cleavage stages of *Lymnaea stagnalis* (Mollusca, Gastropoda). Wilhelm Roux's Arch. Dev. Biol. 190, 215–220.

Dorresteijn, A.W.C., Bilinski, S.M., Van den Biggelaar, J.A.M. and Bluemink, J.G. (1982) The presence of gap junctions during early *Patella* embryogenesis: an electron microscopical study. Dev. Biol. 91, 397–401.

Dorresteijn, A.W.C., Wagemaker, H.A., de Laat, S.W. and Van den Biggelaar, J.A.M. (1983) Dye-coupling between blastomeres in early embryos of *Patella vulgata* (Mollusca, Gastropoda): its relevance for cell determination. Wilhelm Roux's Arch. Dev. Biol. 192, 262–269.

Guthrie, S.C. and Gilula, N.B. (1989) Gap-junctional communication and development. Trends Neurosci. 12, 12–15.

Kalimi, G.H. and Lo, C.W. (1988) Communication compartments in the gastrulating mouse embryo. J. Cell Biol. 107, 241–255.

Kimmel, C.B., Spray, D.C. and Bennet, M.V.L. (1984) Developmental uncoupling between blastoderm and yolk cell in the embryo of the teleost *Fundulus*. Dev. Biol. 102, 483–487.

Martindale, M.Q., Doe, C.Q. and Morrill, J.B. (1985) The role of animal-vegetal interaction with respect to the determination of dorsoventral polarity in the equal-cleaving spiralian, *Lymnaea palustris*. Wilhelm Roux's Arch. Dev. Biol. 194, 281–295.

Serras, F. and van den Biggelaar, J.A.M. (1989) Progressive restrictions in gap-junctional communication during development. In: A.W. Robards, H. Jongsma, W.J. Lucas, J. Pitts and D. Spray (Eds.), Parallels in Cell-to-Cell Junctions in Plants and Animals. Springer-Verlag, London, UK.

Serras, F., Damen, P., Dictus, W.J.A.G., Notenboom, R.G.E. and Van den Biggelaar, J.A.M. (1989) Communication compartments in the ectoderm of embryos of *Patella vulgata*. Wilhelm Roux's Arch. Dev. Biol. 198, 191–200.

Serras, F. and Speksnijder, J.E. (1990) F-actin is a marker of dorsal induction in early *Patella* embryos. Wilhelm Roux's Arch. Dev. Biol. 199, 246–250.

Van den Biggelaar, J.A.M. (1976) Development of dorsoventral polarity preceding the formation of the mesentoblast in *Lymnaea stagnalis*. Proc. K. Ned. Akad. Wet. Ser. C79, 112–126.

Van den Biggelaar, J.A.M. (1977) Development of dorsoventral polarity and mesentoblast determination in *Patella vulgata*. J. Morphol. 154, 157–186.

Van den Biggelaar, J.A.M. and Guerrier, P. (1979) Dorsoventral polarity and mesentoblast determination as concomitant results of cellular interactions in the mollusk *Patella vulgata*. Dev. Biol. 68, 462–471.

Warner, A.E. and Lawrence, P.A. (1982) Permeability of gap junctions at the segmental border in insect epidermis. Cell 28, 243–252.

Warner, A.E., Guthrie, S.C. and Gilula, N.B. (1984) Antibodies to gap-junctional protein selectively disrupt junctional communication in the early amphibian embryo. Nature 311, 127–131.

J.E. Hall, G.A. Zampighi and R.M. Davis (Eds.)
Progress in Cell Research, Vol. 3

CHAPTER 32

Gap-junction proteins and communication in human epidermis

D. SALOMON, E. MASGRAU, S. VISCHER, M. CHANSON[a], J.-H. SAURAT, D. SPRAY[b] and P. MEDA[a]

Clinic of Dermatology and [a]Department of Morphology, University of Geneva, Medical School, 1211 Geneva, Switzerland and [b]Department of Neuroscience, Albert Einstein College of Medicine, Bronx, NY 10461, USA

Introduction

Human epidermis is a self-renewing, multilayered tissue, comprising a proliferative and a differentiative compartment. The proliferative compartment is formed by the basal layer, where keratinocytes undergo mitosis. Once they move above this layer, keratinocytes lose their ability to divide and undergo a complex differentiation process, as they migrate throughout the suprabasal layers (the differentiative compartment). It is the coordinated balance between the proliferative and the differentiative compartments which ensures the turnover and preserves the characteristics of the epidermis. How keratinocytes maintain this proper equilibrium is presently unknown. Gap junction-mediated communications between keratinocytes could play a significant role in the renewal/differentiation process of the epidermis (Pitts et al., 1988). As proposed for other cell systems (Loewenstein, 1979; Hertzberg and Johnson, 1988; Pitts et al., 1988), such a role implies the existence of a cell subpopulation (stem cells, generator cells) able to produce one or more signal molecule(s) of less than 1000 Da. By diffusing through gap junctions into nearby coupled cells, such signals are expected to elicit graded changes in the division/differentiation program, as a function of their decreasing concentration with transfer from one cell to another (Loewenstein, 1979; Pitts et al., 1988; Hertzberg and Johnson, 1988). In this way, the division and specialization of individual epidermal cells may be controlled differentially, depending on the spatial location of each keratinocyte with respect to the cells generating signal molecule(s). In this model, an increase or a decrease in gap-junctional communication is also expected to affect keratinocyte division and/or differentiation, and, thus, to perturb epidermal homeostasis.

We first assessed the presence and extent of gap-junctional communication between keratinocytes of normal human epidermis, using both microinjection of Lucifer Yellow (LY) and dual patch-clamp electrophysiological monitoring. We then identified the gap-junction connexins expressed by human keratinocytes and quantitated their distribution under both control conditions and after treatment by all-*trans* retinoic acid. This drug, which markedly affects skin function (Fuchs, 1990; Saurat, 1991), has been reported to modulate the permeability and expression of gap junctions in different cell lines (Pitts et al., 1986; Rogers et al., 1990).

Junctional communication of human keratinocytes

Intercellular spread of electrotonic currents has been shown both within mechanically separated human epidermis (Van Heukelom et al., 1972) and in cultures of human keratinocytes (Cavoto and Flaxman, 1972). In vitro, keratinocytes were also found able to exchange LY (Kam et al., 1987; Salomon and Meda, 1987a), as well as tritium-labeled nucleotides (Hunter and Pitts, 1981; Pitts and Finbow, 1986). The latter metabolic coupling was seen not only between keratinocytes but also between these cells and melanocytes or fibroblasts (Hunter and Pitts, 1981; McKay and Taylor-Papadimitriou, 1982). Junctional communication between epidermal and dermal cells has also been observed with intact skin of newborn mice (Kam et al., 1986), an organ which differs markedly from adult human skin in terms of epidermal thickness, distribution of pilosebaceous organs, and absence of eccrine sweat glands.

To determine whether the gap junction-mediated transfer of molecules takes place non-selectively also between the different cell types which make up human skin, we have microinjected LY within individual keratinocytes of intact epidermis, which was sampled by keratome from healthy volunteers (Salomon et al, 1988a). We observed coupling between groups of 25–50 keratinocytes spanning all living layers of the epidermis (Fig. 1), indicating that gap junctions functionally interconnect relatively small groups of cells in various differentiation states. Quantitative analysis on serial sections of injected samples showed that keratinocyte coupling is more extensive in epidermal ridges than in suprapapillary plates, and, in both regions, is smaller after injection of a basal cell than after injection of a suprabasal cell (Fig. 1). These regional differences indicate that the ability of keratinocytes to intercommunicate is somehow dependent on their state of differentiation, as well as on their cellular and noncellular environment (Salomon et al., 1988a). We also showed that, contrasting with the situation within the epidermis, dermal coupling is extensive and usually involves more than 100 fibroblasts. In this study, however, no injection revealed coupling between keratinocytes and fibroblasts (Salomon et al., 1988a). Thus, while the extent and pattern of coupling between keratinocytes and between fibroblasts of human skin is not markedly dissimilar from that reported in newborn mouse skin (Kam et al., 1986), the two studies have revealed a major difference in the organization of coupling between keratinocytes and fibroblasts. This difference may be accounted for by the specific characteristics of mouse and human skin and/or may also result from differences in the tissue sampling and technical approaches used in the two studies.

To compare the pattern of dye coupling between keratinocytes of the undifferentiated (basal) and of the differentiated layers (suprabasal), we microinjected LY in pairs of keratinocytes which had been just isolated from normal human epidermis (Salomon et al., 1992). These experiments failed to reveal coupling in pairs of basal keratinocytes, which were identified by their diameter of less than 10 µm. By contrast, dye coupling was readily detected in 75% of the pairs of suprabasal keratinocytes, which were identified by their larger diameter (20–30 µm) (Fig. 1). Thus, in this fresh and highly simplified preparation there was a difference in the coupling of undifferentiated and differentiated keratinocytes which was reminiscent of that observed within intact epidermis.

To determine whether such a difference is preserved in vitro, we turned to cultures of cells dissociated from control human epidermis (Salomon et al., 1988b). We found that keratinocytes displaying a suprabasal phenotype were almost consistently coupled to neighboring cells, whereas keratinocytes with a basal phenotype were much less frequently and extensively coupled (Fig. 1). Thus, these in vitro observations are consistent with those made in vivo in showing that undifferentiated keratinocytes intercommunicate in a more restricted manner than differentiated ones.

Together, these results indicate that junctional communications do not occur to the same extent between all keratinocytes of the human epidermis and that this pattern is comparable in situ and in vitro. In vitro preparations may therefore be used to investigate aspects of keratinocyte communications which cannot be approached in vivo.

One such aspect is the biophysical characterization of keratinocyte junction channels. We have monitored keratinocyte pairs under dual whole-cell patch-clamp conditions, in order to assess the extent of their electrical coupling and to evaluate kinetically the conductance of their gap junctions. The analysis was restricted to pairs of suprabasal keratinocytes since, as shown by dye coupling, these cells are more frequently coupled than other epidermal cells. Under these conditions, direct current exchanges between contiguous keratinocytes were readily observed. These exchanges were attributed to ion flow through junctional channels, as they were blocked by heptanol, an alcohol which gates gap junctions in a variety of other systems (Johnston et al., 1980; Meda, et al., 1988). The mean junctional conductance at individual keratinocyte interfaces was about 9.3 nS. Assuming that 95 pS is the average conductance of a single channel (see below), the data indicate that adjacent keratinocytes are coupled by about 100 fully opened junctional channels. If 10% of these channels were open at any given time, the persistent coupling of two suprabasal keratinocytes would thus require 1000 channels per interface, a figure which is consistent with the number of connexons found on freeze-fracture replicas of keratinocyte interfaces.

The dual patch-clamp approach has also been instrumental in evaluating the unitary conductance of single channels coupling keratinocytes. Transitions of channels between the open and the closed state were readily resolved in pairs spontaneously undergoing partial uncoupling. The unitary currents measured under such conditions were demonstrably junctional in that they appeared simultaneously, but with opposite polarity, in the two paired cells and were gated by pharmacologic concentrations of either retinoic acid (see below) or heptanol, two drugs that also decreased macroscopic junctional conductance. Analysis of channel openings revealed a unitary conductance of about 95 pS and a median open time of about 1 s. If these properties are dependent on the type of connexin forming the channels (see next section), then keratinocyte channels may be somewhat different from those which have been previously described in other systems expressing connexin43 (Cx43) and Cx26. The elementary conductance of cells expressing Cx43 is about 50–60 pS (Veenstra et al., 1986; Spray and Burt, 1990) and that of cells expressing Cx26 is about 70 pS (Spray, unpublished). None of these conductances fits with those of keratinocytes. Until the reason for this apparent discrepancy is elucidated, a number of possibilities should be considered. First, it is possible that the same connexin may determine different conductances in different cell types. Second, the somewhat unusual unitary conductance of keratinocyte junctions may result from the simultaneous opening of two or more channels whose unitary conductance may add to each other. Third, as tissue and developmental variations in the electrophoretic mobility of Cx43, possibly due to the variable phosphorylation of this protein, have been documented (Filson et al., 1990; Musil et al., 1990; Swenson et al., 1990; Laird et al., 1991), it is possible that the junctional conductance of human keratinocytes reflects the incomplete phosphorylation of Cx43. Finally, keratinocyte conductance may be determined by a gap-junction protein other than Cx43 and Cx26 (see below). Indeed, several new molecules related to connexins have been identified by sequencing cDNA or genomic clones (Beyer et al., 1991; Haefliger and Paul, 1991; Willecke et al., 1991).

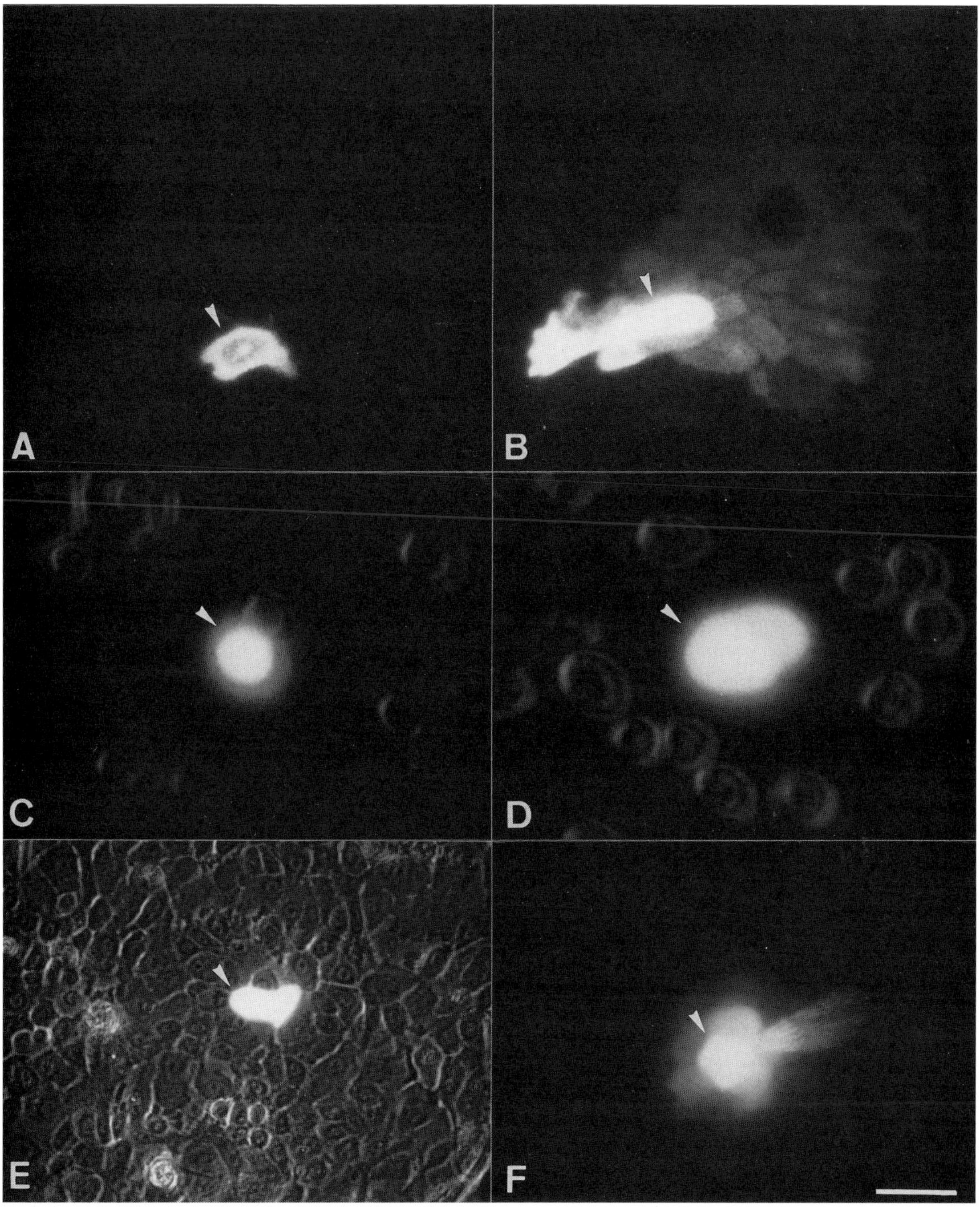

Figure 1. Dye coupling between human keratinocytes. (A,B) Microinjection of LY into individual keratinocytes (arrowheads) of intact epidermis, revealed variable levels of junctional communication depending on the location of the injected cell. Most of the injections performed in the basal epidermal layer did not result in a detectable transfer of LY between adjacent keratinocytes (A). By contrast, most injections performed in the suprabasal layers revealed the occurrence of extensive keratinocyte coupling, across the entire living portion of the epidermis (B). (C,D) An analogous difference was observed in pairs of keratinocytes freshly dissociated from human epidermis. Thus, cells showing the small size characteristic of basal keratinocytes were rarely coupled (C), whereas larger cells, with the more polygonal contour characteristic of suprabasal keratinocytes, were almost consistently coupled to each other (D). (E,F) These characteristics were not lost in vitro. Injection of cultured cells with a basal keratinocyte phenotype revealed, infrequently, a restricted coupling (E). By contrast, injection of cells with a suprabasal keratinocyte phenotype almost consistently revealed a much more extensive cell-to-cell transfer of the injected tracer (E). The bar represents 15 μm in A, C, D and E and 30 μm in B and F.

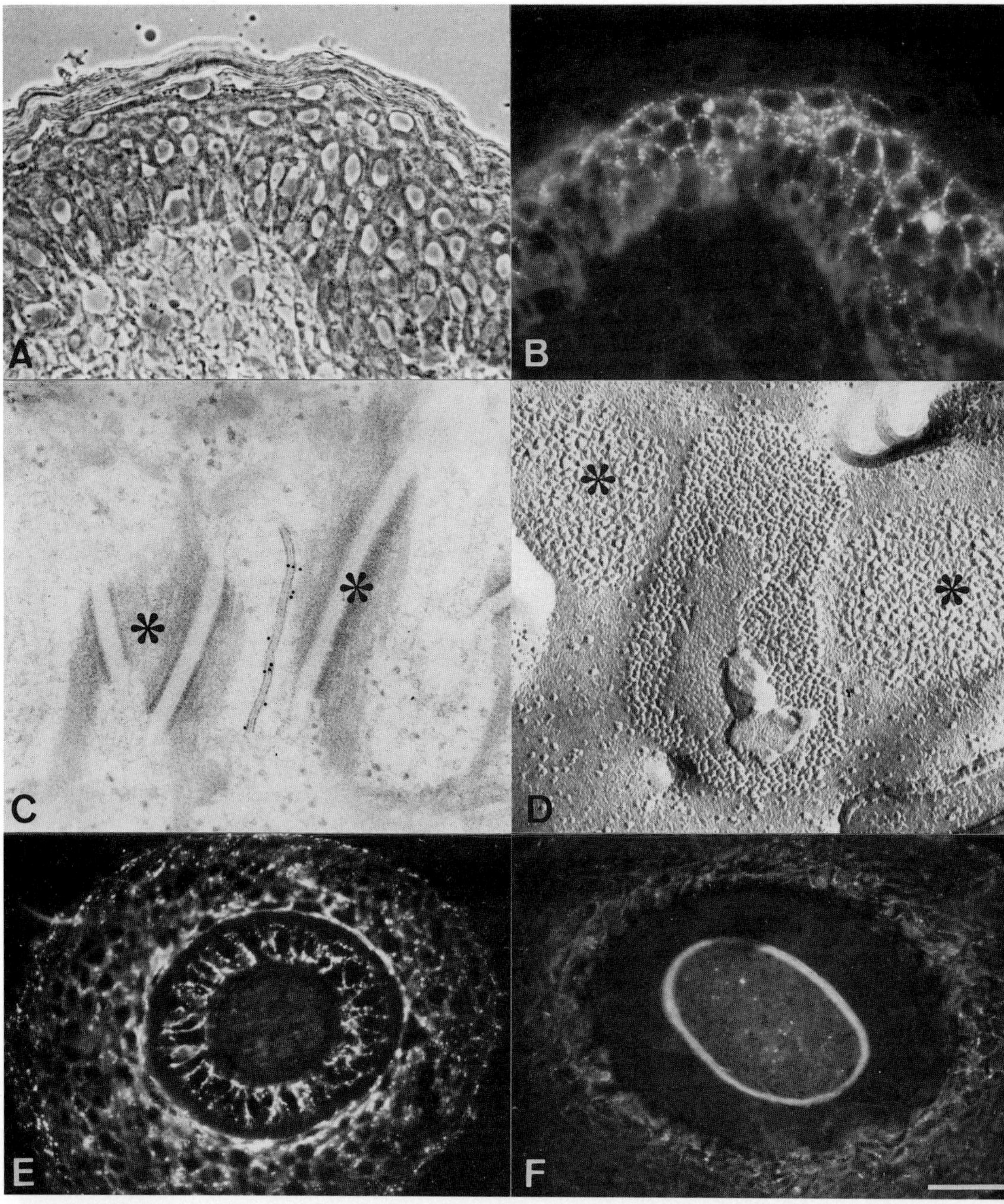

Figure 2. Connexins and gap junctions of human epidermis. (A) Phase-contrast appearance of a cryostat section from a neck biopsy, showing the characteristic organization of human epidermis and dermis. (B) Immunostaining of the same section with antibodies to Cx43 resulted in a punctate immunofluorescence labeling of the periphery of numerous keratinocytes, most of which were located in the spinous layers comprising the suprabasal living compartment of the epidermis. (C) At the electron microscopy level, the same serum labeled membrane appositions typical of gap junctions which, in this case, are decorated by 10 nm protein A gold particles. Other cell structures, including surrounding desmosomes (*), were not decorated at all. (D) The topographical distribution and size of the structures labeled by antibodies to Cx43 are consistent with the intraepidermal distribution and size of typical gap-junction plaques, as revealed by freeze fracture of spinous layer keratinocytes. (E,F) The different abundance and topographic distribution of various connexins within human epidermis is readily apparent on two cross-sections of hair follicles. Antibodies to Cx43 immunostained most keratinocytes heavily in both the external and the internal hair follicular sheets (E). By contrast, antibodies to Cx26 immunostained only some keratinocytes much more sparsely, most of which were located in the inner layers of the external hair follicular sheet (E). In both cases, no specific labeling was observed over the fully keratinized hair shaft. The bar represents 40 μm in A and B, 120 nm in C and D, and 60 μm in E and F.

Although it remains to be assessed whether these proteins are actually expressed to any significant level in human skin, this possibility is suggested by the presence of specific coding messages in both fetal and adult mouse skin (Beyer et al., 1991; Willecke et al., 1991).

Connexins expressed in human epidermis

Electron microscopy has shown that human keratinocytes are connected by gap junctions (Wolff and Schreiner, 1968; Hashimoto, 1971; Breathnach et al., 1972; Caputo and Peluchetti, 1977). The number and size of these structures vary across the epidermis (Breathnach et al., 1972; Caputo and Peluchetti, 1977). Few and small gap junctions are observed between the keratinocytes of the basal layer, whereas numerous and larger gap junctions are detected between the more differentiated keratinocytes of the spinous layers (Fig. 2). Gap junctions are less abundant in the stratum granulosum, and are no longer detected in the stratum corneum (Caputo and Peluchetti, 1977), the most superficial region of the epidermis, which is formed by fully keratinized and anucleate keratinocytes.

The regional heterogeneity in the distribution of ultrastructurally detectable gap junctions is consistent with the pattern of epidermis immunolabeling by antibodies to Cx43. This labeling, which by electron microscopy was restricted to gap junctions (Fig. 2), was seen, at the light microscopy level, as fluorescent spots at the periphery of keratinocytes (Fig. 2). Analysis of sections made perpendicular to the epidermal surface showed that these spots were rather sparse in the basal layer, much more extensive throughout the spinous layers and virtually non-existent in the upper keratinized layers of the epidermis which are formed by dead keratinocytes (Fig. 2). Thus, the distribution of Cx43 immunostaining was consistent with that of gap-junction plaques and with the heterogenous extent of coupling observed in different regions of the human epidermis (Salomon et al., 1988a). Antibodies to Cx43 also labeled epidermal adnexae with the following distribution: hair bulb (++), outer and inner root hair sheaths (+++), basal layer of the sebaceous gland (+), differentiated layers of the sebaceous gland (++++). Western blot analysis confirmed the presence of Cx43 in human epidermis by revealing that sera against Cx43 specifically immunostained a band at approximately 43 kDa, which migrated similarly to Cx43 of rat heart muscle.

A similar analysis, using antibodies to Cx26, showed that this protein is also expressed in human skin, although at lower levels than Cx43 (Fig. 2), and is located predominantly in hair follicles and ducts of eccrine sweat glands. Within the epidermis, a faint immunolabeling for this protein is observed throughout the differentiated layers, mostly toward the skin surface. No labeling of any skin structure was ever observed using antibodies to Cx32.

Thus, as in neonatal rat epidermis (Risek and Gilula, 1991), Cx43 and Cx26 are co-localized in certain layers of human epidermis, although most keratinocytes appear to be primarily interlinked by Cx43.

The expression of specific connexin messages in human skin was evaluated by hybridizing total epidermis RNA with cDNA clones for Cx26 (Zhang et al., 1989), Cx32 (Paul et al., 1986) and Cx43 (Beyer et al., 1987). Cx43 and Cx26 mRNAs but not Cx32 mRNA were detected in human epidermis. Northern blot analysis showed that the message for Cx43 was predominant, thus confirming that Cx43 is, by far, the major keratinocyte connexin. It remains to be tested whether human keratinocytes also express other members of the connexin family, such as the recently identified Cx30.9, Cx31.1 and Cx37 whose specific messages are abundant in mouse tail skin (Beyer et al., 1991; Chapter 5).

Effects of retinoic acid on keratinocyte coupling and Cx43 expression

Natural and synthetic retinoids are of particular interest in skin biology because of their complex effects on the growth, differentiation and development of keratinocytes (Sherman, 1986; Fuchs, 1990, Saurat, 1991). Some of these effects may involve, if not require, changes in keratinocyte gap junctions and coupling. Indeed, all-*trans* retinoic acid (RA) has been reported to increase gap junctions in human epidermis (Prutkin, 1975; Elias and Friend, 1976; Elias et al., 1980), and the synthetic retinoid tetrahydrotetramethylnaphthalenylpropenylbenzoic acid (TTNPB) has been shown to stimulate gap-junctional communication and Cx43 expression in a fibroblast cell line (Rogers et al., 1990). Furthermore, retinol, RA and TTNPB affect coupling of different skin-unrelated cell types, in a concentration-dependent manner (Pitts et al., 1986; Mehta et al., 1986,1989; Hossain et al., 1989).

We have also found that the effects of natural RA on the coupling of human keratinocytes varies with its concentration. 10^{-4} M RA abolished coupling of freshly prepared keratinocyte pairs in a rapid and irreversible way, whereas 10^{-6} M RA did not significantly modify LY transfer or junctional conductance. Thus, as in other systems, the junctional communication of primary keratinocytes is blocked by pharmacological concentrations of RA but not by concentrations which, although lower, are still 100–1000 times higher than physiologically relevant levels (Vahlquist, 1982). This consideration, together with the irreversibility of the RA-induced uncoupling, raises serious concerns about the possibility of keratinocyte coupling regulation by endogenous retinoids. We cannot, therefore, rule out the idea that the uncoupling effect observed with pharmacological levels of RA may reflect the cellular toxicity of the compound rather than its specific gating of junctional channels. The lack of effect of 10^{-6} M RA in our system is at variance with other reports (Hossain et al., 1989; Mehta et al., 1989). This discrepancy may be explained by the different time courses assessed in our (minutes) and previous studies (days). Since RA is a ligand for a class of DNA-binding proteins which are homologous to the nuclear receptors for vitamin D3, thyroid and steroid hormones (Giguere et al., 1987; Petkovich et al., 1987), retinoids may directly regulate the expression of different proteins. Whether a chronic exposure to RA stimulates the expression of connexins, thus leading to changes in epidermal junctional communication, remains to be assessed.

To determine the in vivo effects of RA on coupling, we have studied LY transfer in the intact epidermis of normal volunteers who were either untreated or who applied RA (0.1% in 95% ethanol) for 4 days, under occlusion. Two patients who were treated orally with TTNPB (35 μg/day) were also studied. Both topical and systemic retinoid treatments significantly increased ($p < 0.0025$) the levels of cytosolic RA-binding proteins in the skin samples which were obtained for microinjection, indicating that the retinoids had actually targeted the epidermis under study. However, no major change in either the organization of the tissue or the extent of dye coupling was detected (Salomon et al., 1988a). Again, these observations indicate that treatments which achieve concentrations of circulating RA mimicking endogenous intraepidermal levels, are ineffective in uncoupling human keratinocytes.

To assess whether the expression of Cx43 is altered after chronic topical treatment with RA, we used a protocol which markedly changes the differentiation pattern of human epidermis (Plewig and Braun-Falco, 1975). Seven volunteers (including two of us) applied 0.1% RA twice a day, for three months, on one side of the neck. The solvent used to dilute RA was similarly applied on the other side. Under these conditions, the RA treatment induced a 50% increase in epidermal thickness and quantitative analysis showed that the immunofluorescent spots decorated by antibodies against Cx43 increased significantly ($p < 0.001$) throughout the living layers of the RA-treated epidermis. The labeling for Cx32 (which is nil in control skin) was not affected by the RA treatment. These data are consistent with the view that a sustained change in the proliferation/differentiation program of human epidermis is associated with a parallel change in the amount of the protein which forms the bulk of keratinocyte gap junctions.

Preliminary studies on the lesional epidermis of a patient suffering from lamellar ichthyosis, a rare congenital skin disease characterized by a dramatic increase in keratinocyte turnover (Frost et al., 1966), revealed an extent of keratinocyte coupling which was significantly increased over control levels. This increase was markedly reduced after a few weeks of systemic treatment with TTNPB, a synthetic retinoid which also greatly improved the clinical appearance of the patient's epidermis (Salomon et al., 1987b). These preliminary findings suggest that keratinocyte coupling may be altered in a disease which perturbs epidermis homeostasis, and that retinoid treatment can correct in parallel, at least partially, the abnormal coupling of keratinocytes and the disturbed skin phenotype.

Together, these observations highlight the complexity of the relationship between retinoids, gap junctions and epidermal homeostasis.

Conclusions

At this stage, our studies have provided basic information on the pattern of coupling between keratinocytes of human skin. These studies have shown the feasibility of studying this coupling in intact epidermis, as well as in primary cultures or even cell pairs derived from it, and have not revealed major discrepancies between the observations made using different preparations. The possibility of characterizing keratinocyte coupling in intact epidermis opens the way to studies aimed at investigating the biological function(s) of skin gap junctions and their possible alterations in a variety of common skin diseases. The availability of simplified in vitro systems will permit analysis of the modulation of keratinocyte communications under conditions which cannot be met in vivo, and to further characterize the biochemical and biophysical characteristics of the channels which mediate these communications.

Acknowledgments

We thank L. Burkhardt, A. Charollais, J.-P. Gerber, L. Iuliano, and P. Ruga for excellent technical assistance. This work was supported by grants from the Swiss National Science Foundation (32-34090.92 and 32-30211.90), the Sir Jules Thorn Foundation and the Sandoz Foundation.

References

Beyer, E.C., Kanter, H.L., Rup, D.M., Westphale, E.M. Reed, K.E., Larson, D.M. and Saffitz, J.E. (1993) Expression of multiple connexins by cells of the cardiovascular system and lens. In: J.E. Hall, G.A. Zampighi and R.M. Davis, (Eds.) Progress in Cell Research, Vol. 3. Elsevier, Amsterdam, pp. 171–175.

Beyer, E.C., Paul, D.L. and Goodenough, D.A. (1987) Connexin43: a protein from rat heart homologous to a gap junction protein from liver. J. Cell Biol. 105, 2621–2629.

Breathnach, A.S., Stolinski, C. and Gross, M. (1972) Ultrastructure of foetal and postnatal human skin as revealed by the freeze-fracture replication technique. Micron 3, 287–304.

Caputo, R. and Peluchetti, D. (1977) The junctions of normal human epidermis. A freeze-fracture study. J. Ultrastruct. Res. 61, 44–61.

Cavoto, F.V. and Flaxman, B.A. (1972) Communication between normal human epidermal cells in vitro. J. Invest. Dermatol. 59, 370–374.

Elias, P.M. and Friend, D.S. (1976) Vitamin-A-induced mucous metaplasia. An in vitro system for modulating tight and gap-junction differentiation. J. Cell Biol. 68, 173–188.

Elias, P.M., Grayson, S., Caldwell, T.M. and McNutt, N.S. (1980) Gap junction proliferation in retinoic acid-treated human basal cell carcinoma. Lab. Invest. 42, 469–474.

Filson, A.J., Azarnia, R., Beyer, E.C., Loewenstein, W.R. and Brugge, J.S. (1990) Tyrosine phosphorylation of a gap junction protein correlates with inhibition of cell-to-cell communication. Cell Growth Diff. 1, 661–668.

Frost, P., Weinstein, G.D. and Van Scott, E.J. (1966) The ichthyosiform dermatoses II. Autoradiographic studies of epidermal proliferation. J. Invest. Dermatol. 47, 561.

Fuchs, E. (1990) Epidermal differentiation: The bare essentials. J. Cell Biol. 111, 2807–2814.

Giguere, V., Ong, E., Segui, P. and Evans, R.M. (1987) Identification of a receptor for the morphogen retinoic acid. Nature 330, 624–629.

Hashimoto, K. (1971) Intercellular spaces of the human epidermis as demonstrated with lanthanum. J. Invest. Dermatol. 57, 17–31.

Hossain, Z.H., Wilkens, L.R., Mehta, P.P., Loewenstein, W.R. and Bertram, J.S. (1989) Enhancement of gap junctional communication by retinoids correlates with their ability to inhibit neoplastic transformation. Carcinogenesis 10, 1743–1748.

Hunter, K. and Pitts, J.D. (1981) Non-selective junctional communication between some different mammalian cell types in primary culture. J. Cell. Sci. 49, 163–175.

Johnston, M.F., Simons, M.F. and Ramon, F. (1980) Interactions of anaesthetics with electrical synapses. Nature 286, 498–500.

Kam, E., Melville and Pitts, J.D. (1986) Patterns of junctional communication in skin. J. Invest. Dermatol. 87, 748–753.

Kam, E., Watt, F.M., Pitts, J.D. (1987) Patterns of junctional communication in skin: studies on cultured keratinocytes. Exp. Cell Res. 173, 431–438.

Laird, D.W., Puranam, K.L. and Revel, J.P. (1991) Turnover and phosphorylation dynamics of connexin43 gap junction protein in cultured cardiac myocytes. Biochem. J. 273, 67–72.

Loewenstein, W.R. (1979) Junctional intercellular communication and the control of growth. Biochim. Biophys. Acta 560, 1–65.

McKay, I. and Taylor-Papadimitriou, J. (1982) The nonselective junctional communication phenotype of normal and transformed human epidermal keratinocytes in vitro. Exp. Cell Res. 141, 171–180.

Meda, P., Bruzzone, R., Chanson, M. and Bosco, D. (1988) Junctional coupling and secretion of pancreatic acinar cells. In: E.L. Hertzberg and R.G. Johnson (Eds.), Gap Junctions, Alan R. Liss, New York, pp. 353–364.

Mehta, P.P., Bertram, J.S. and Loewenstein, W.R. (1989) The actions of retinoids on cellular growth correlate with their actions on gap-junctional communication. J. Cell Biol. 108, 1053–1065.

Musil, L.S., Cunningham, B.A., Edelman, G.M. and Goodenough, D.A. (1990) Differential phosphorylation of the gap junction protein connexin43 in junctional communication-competent and -deficient cell lines. J. Cell Biol. 111, 2077–2088.

Paul, D.L. (1986) Molecular cloning of cDNA for rat liver gap junction protein. J. Cell Biol. 103, 123–134.

Paul, D.L., Bruzzone, R. and Haefliger, J.-A. (1993) Identification of novel connexins by reduced-stringency hybridization and PCR amplification using degenerate primers. In: J.E. Hall, G.A. Zampighi and R.M. Davis (Eds.), Progress in Cell Research, Vol. 3. Elsevier, Amsterdam, pp. 15–20.

Petkovich, M., Brand, N.J., Krust, A. and Chambon, P. (1987) A human retinoic acid receptor which belongs to the family of nuclear receptors. Nature 330, 444–450.

Plewig, G. and Braun-Falco, O. (1975) Kinetics of epidermis and adnexae following vitamin A acid in the human skin. Acta Dermatovener., suppl., 74, 87–98.

Pitts, J.D., Hamilton, A.E., Kam E., Burk, R.R. and Murphy, J.P. (1986) Retinoic acid inhibits junctional communication between animal cells. Carcinogenesis 7, 1003–1010.

Pitts, J.D., Finbow, M.E. and Kam, E. (1988) Junctional communication and cellular differentiation. Br. J. Cancer 58, 52–57.

Prutkin, L. (1975) Mucous metaplasia and gap junctions in the vitamin A acid treated skin tumor, keratoacanthoma. Cancer Res. 35, 364–369.

Risek, B. and Gilula, N.B. (1991) Spatiotemporal expression of three gap-junction gene products involved in fetomaternal communication during rat pregnancy. Development 113, 165–181.

Rogers, M., Berestecky, J.M., Hossain, M.Z., Guo, H., Kadle, R., Nicholson, B.J. and Bertram, J.S. (1990) Retinoid-enhanced gap junctional communication is achieved by increased levels of connexin 43 mRNA and protein. Mol. Carcinogen. 3, 335–343.

Salomon, D. and Meda, P. (1987a) Junctional communication in human skin. International Gap Junction Conference, Asilomar, California, July 1987. Abstract booklet, p. 29.

Salomon, D., Meda P. and Saurat, J.H. (1987b) Gap-junctional cell-to-cell communication: quantitative analysis in normal and pathological human epidermis. J. Invest. Dermatol. (abstr.), 89, 316.

Salomon, D., Saurat, J.H. and Meda, P. (1988a) Cell-to-cell communication within intact human skin. J. Clin. Invest. 82, 248–254.

Salomon, D., Spray, D., Chanson, M. and Saurat, J.H. (1988b) In vitro cell-to-cell communication of human keratinocytes, different technical approaches. J. Invest. Dermatol. (abstr.) 91, 383.

Salomon, D., Chanson, M., Vischer, S., Masgrau, E., Vozzi, C., Saurat, J.-H., Spray, D.C. and Meda, P. (1992) Gap junctional communication of primary human keratinocytes: characterization by dual voltage clamp and dye transfer. Exp. Cell Res. 201, 452–461.

Saurat, J.-H. (1991) Retinoids: 10 Years On. J.-H. Saurat (Ed.), Karger, Basel, pp. 1–356.

Sherman, M.I. (1986) Retinoids and Cell Differentiation. M.I. Sherman (Ed.), CRC Press, Boca Raton, FL, p. 178.

Swenson, K.I., Piwnica-Worms, H., McNamee, H. and Paul, D.L. (1990) Tyrosine phosphorylation of the gap junction protein connexin43 is required for the pp60v-*src*-induced inhibition of communication. Cell Reg. 1, 989–1002.

Vahlquist, A. (1982) Vitamin A in human skin: I. Detection and identification of retinoids in normal epidermis. J. Invest. Dermatol. , 79, 89–93.

Van Heukelom, J.S., Slaaf, D.W. and Van Der Leun, J.C. (1972) Cell communication in the basal cells of the human epidermis. Biophys. J., 12, 1266–1284.

Veenstra, R.D. and DeHaan, R.L. (1986) Measurement of single-channel currents from cardiac gap junctions. Science, 233, 972–974.

Willecke, K., Hennemann, H., Dahl, E., Jungbluth, S., Nicholson, B. and Grzeschnik, K.H. (1991) Molecular genetics of connexin genes. International Meeting on Gap Junctions, Asilomar, 1991. Abstract booklet, p. 180.

Wolff, K., Schreiner, E. (1968) An electron microscopic study on the extraneous coat of keratinocytes and the intercellular space of the epidermis. J. Invest. Dermatol., 51, 418–430.

Zhang, J.T. and Nicholson, B.J. (1989) Sequence and tissue distribution of a second protein of hepatic gap junctions, Cx26, as deduced from its cDNA. J. Cell Biol., 109, 3391–3401.

J.E. Hall, G.A. Zampighi and R.M. Davis (Eds.)
Progress in Cell Research, Vol. 3

CHAPTER 33

Expression patterns of α_1 and β_2 gap-junction gene products during rat skin and hair development

BORIS RISEK, F. GEORGE KLIER and NORTON B. GILULA

Department of Cell Biology, The Scripps Research Institute, La Jolla, CA 92037, USA

Introduction

The mammalian skin represents the largest organ of the body, consisting of an epithelium (epidermis) which is separated by a basement membrane from the underlying connective tissue matrix (dermis). The epidermis consists of continually renewing cells (keratinocytes) organized into several layers with specific structural properties and functional activities. The different cell layers represent various stages of epidermal differentiation (keratinization), a complex, controlled series of events regulated by both extrinsic and intrinsic factors. In the mature epidermis, only cells of the basal layer proliferate and give rise to the superficial, differentiated layers.

Based on morphological and biochemical criteria, Dale and Holbrook (1987) defined four stages of human epidermal development which are also applicable for rat epidermis. These stages consist of: (1) embryonic period (from approximately 12 to 14 days of gestation; E12–E14); (2) epidermal stratification (E14–E16); (3) follicular keratinization (E16–E18); and (4) period of interfollicular keratinization (E18–E20). The embryonic epidermis develops from a single layer of ectodermal cells which becomes double layered during an early embryonic stage (E12). The inner layer (embryonic basal layer) gives rise to the epidermis and its appendages (hair follicles and sebaceous glands), while the outer layer, the periderm, surrounds the developing epidermis until birth. The periderm has been analysed ultrastructurally in the rat (Bonneville, 1968) and human embryonic and fetal skin (Holbrook and Odland, 1975). Periderm cells undergo mitosis and contain microvilli exposed to the amniotic cavity. Since skin develops in a fluid environment during embryonic and fetal life, it is likely that there is an interaction between the amniotic fluid and total skin prior to keratinization. It has been suggested that the periderm might play a role in material transport between the developing skin and amniotic fluid (Breathnach and Wyllie, 1965; Bonneville, 1968; Holbrook and Odland, 1975). Although direct evidence for the peridermal functions has not been documented thus far, there is evidence that the fetal epidermis contributes to material transport during human (Parmley and Seeds, 1970) and rat embryogenesis (Stern et al. 1971). However, the potential role of the amniotic fluid in influencing skin development has not yet been determined. At about E14, the epidermis stratifies into three layers by formation of an intermediate layer between the basal and peridermal cells. Following epidermal stratification, hair germs start to develop from epidermal basal cells at approximately E16. The last stage of epidermal development, interfollicular keratinization, is accomplished by formation of the stratum corneum, a process which occurs in the rat at about E20.

In spite of well-documented morphological changes during embryonic and fetal skin development, little is known about biochemical events that accompany the differentiation of the undifferentiated epithelium into a highly organized, multilayered squamous epidermis. Thus far, keratins are the best described cellular markers whose expression coincides with the course of epidermal terminal differentiation (Fuchs and Green, 1980; Banks-Schlegel, 1982; Kopan and Fuchs, 1989).

Our knowledge about the contribution of gap junctions (GJs) to skin development and function is very limited, and it is primarily based upon ultrastructural investigations. GJs have been described at some fetal stages of human development (Breathnach, 1971) and in the adult human epidermis (Caputo and Peluchetti, 1977). In addition, de novo GJ formation has been reported during the wound-healing process in rat and rabbit skin (Gabbiani et al., 1976). During this process, epithelial GJ formation was accompanied by a reduction of desmosomes in proliferating epithelium. More recently, the presence of epidermal and dermal GJs has been demonstrated in situ by injections of LY into skin isolated from the newborn mouse (Kam et al., 1986; Pitts et al., 1987) and from the adult human (Salomon et al., 1988). By measuring intercellular dye transfer, the authors concluded that the epidermis is divided into many small communication compartments. However, since these studies were reported prior to the discovery of the gap-junction multigene family (for recent reviews, see Beyer et al., 1990; and Bennett et al., 1991) it is not known which members of the GJ multigene family are utilized during skin development.

The hair follicle is a highly compartmentalized structure consisting of concentrically arranged cylinders of different cell types. A detailed description of developmental stages of mouse hair follicles combined with a histochemical analysis of various follicular compartments has been provided by Hardy (1952). The cells of the most centrally located hair shaft (composed of medulla, cortex and cuticle) and the inner root sheath are derived from the pluripotential matrix cells which are located at the common base (hair bulb region) of these compartments. The highly mitotic matrix cells surround the dermal papilla, a cluster of dermal fibroblasts, which appear to be crucial for hair follicle morphogenesis (Jahoda et al., 1984). The mammalian hair growth cycle (Butcher, 1934) represents a continuous process of proliferation and differentiation which is subdivided into three distinct phases (Chase, 1954; Kligman, 1959): (a) active growth (anagen); (b) breakdown with regression of the hair follicle (catagen); and (c) resting period (telogen). The sebaceous gland is an outgrowth of the hair follicle formed by the cells of the external root sheath, and both together constitute the pilosebaceous unit (Parnell, 1949; Chase, 1954). The sebaceous gland cells undergo differentiation from the periphery (undifferentiated cells) towards the center of the gland. This process reflects the life cycle of sebaceous cells and is characterized by the production, segregation and accumulation of lipids resulting in enlarged cells that fragment and disperse their sebum into the pilosebaceous duct (Montagna, 1974). These properties indicate a biological similarity with the epidermis, since in both systems the cell populations are organized into three compartments, each with its own functional and structural characteristics (Tosti, 1974): (a) a population of proliferative stem cells at the gland periphery which restores lost cells; (b) an expanding compartment, where differentiation occurs; and (c) a terminal compartment, where the cells complete their maturation and are eliminated. Desmosomes (Bell, 1974) were identified in an ultrastructural analysis of sebaceous glands.

In a previous report (Risek and Gilula, 1991), GJ expression was analysed within the rat implantation chamber at different stages of gestation. High levels of α_1 (Cx43) and β_2 (Cx26) GJ proteins were detected in the developing fetal epidermis. For these reasons, we have studied skin as a well-defined developmental system for understanding the expression of GJ gene products during

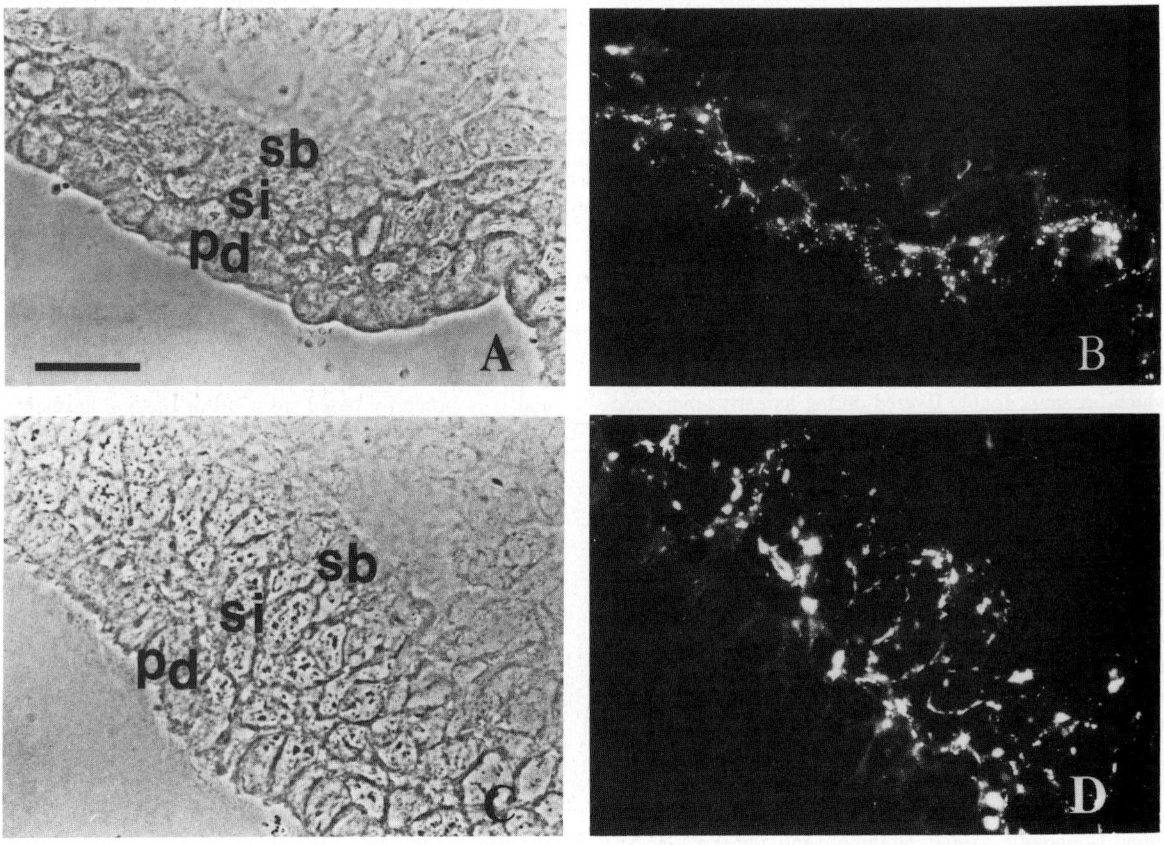

Figure 1. Immunolocalization of α_1 and β_2 GJ antigens in fetal rat epidermis at E16. (A,C) Phase-contrast micrographs of frozen ventral skin incubated with α_1S (B) and β_2J (D) peptide antibodies. The developing epidermis consists of three distinct layers: stratum basale (sb), stratum intermedium (si) and the outermost layer, periderm (pd). (B) Localization of α_1 in periderm, in stratum intermedium and stratum basale. (D) Localization of β_2 protein in basal and intermedium layer. β_2 protein is not detectable in periderm. Scale bar, 20 μm.

the course of epidermal morphogenesis, proliferation and terminal differentiation of keratinocytes. In addition, we extended the analysis on epidermal appendages, the hair follicle and sebaceous gland. The expression of α_1 and β_2 GJ gene products was determined by immunohistochemistry, electron microscopy, immunoblot and Northern blot analysis using affinity-purified peptide antibodies (α_1S and β_2J) and cDNA probes as described by Risek et al. (1990).

Results and discussion

The expression of α_1 and β_2 GJ proteins coincided with the differentiation of the epidermis during the course of embryonic and fetal rat development. Both α_1 and β_2 proteins were co-expressed in the single-layered undifferentiated epithelium at E12, and they were also present in the periderm following the formation of this stratum at about E13/14. The period of epidermal stratification (formation of stratum intermedium at approximately E14) was characterized by high expression levels of both α_1 and β_2 connexins in this rather undifferentiated suprabasal layer. However, differential expression was observed at E16, coinciding with the onset of follicular keratinization (Fig. 1). α_1 was expressed at high levels in the periderm and, to

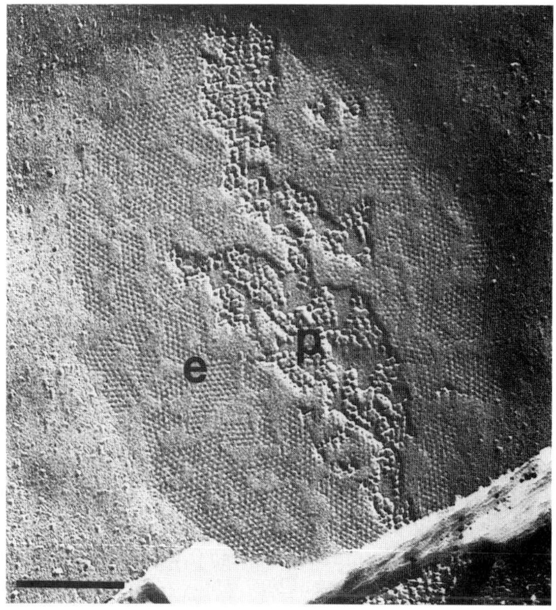

Figure 2. Freeze-fracture image of a gap junction in fetal rat epidermis at E16. Large gap junction in the stratum intermedium consists of interjoined islands of 40–80 particles each. P-fracture face (p) with complementary pits on the E-fracture face (e). Scale bar, 200 nm.

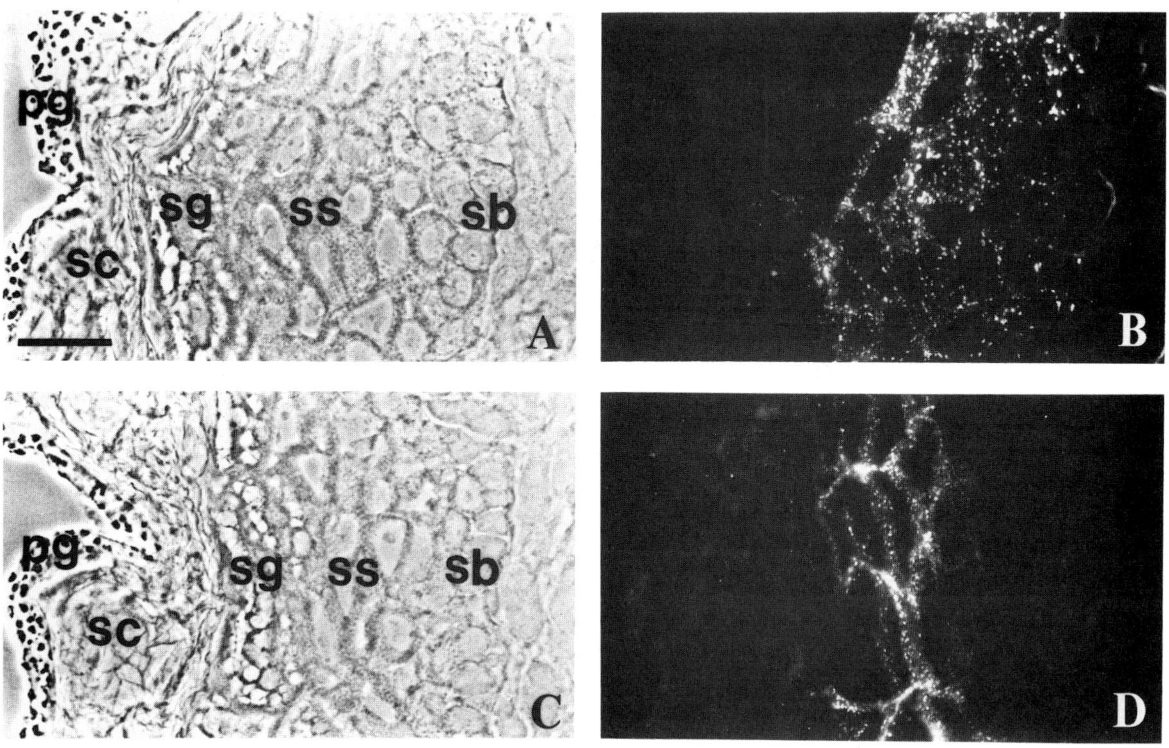

Figure 3. Immunolocalization of α_1 and β_2 GJ antigens in fetal rat epidermis at E21. (A, C) Phase-contrast micrographs of frozen dorsal skin incubated with α_1S (B) and β_2J (D) peptide antibodies. The epidermis is fully differentiated one day before birth and consists of stratum basale (sb), spinosum (ss), granulosum (sg) and stratum corneum (sc). Periderm, containing granules (pg) forms the outermost layer. (B) Localization of α_1 protein in the basal and spinous layer. (D) Localization of β_2 protein in the more differentiated cells of the spinous layer and in the stratum granulosum. Scale bar, 20 μm.

a lesser extent, in the intermedium and basal layer. In contrast, β_2 was not present in the periderm of ventral skin, but was expressed at high levels in stratum intermedium and stratum basale. The freeze-fracture analysis of E16 epidermis revealed large GJs which comprised a latticework of angular islands (Fig. 2). The average size of islands consisted of about 60 particles which were incompletely segregated by particle-free regions. Interestingly, these freeze-fracture characteristics were very similar to GJs observed in embryonal chick epidermis following vitamin-A treatment (Elias and Friend, 1976). Differential expression of epidermal α_1 and β_2 GJ proteins was even more prominent at later stages of fetal development, coinciding with epidermal differentiation and maturation. At E21, α_1 connexin was expressed at high levels in the spinous layer, and to a lesser extent in the basal and granular strata (Fig. 3). β_2 expression was confined to the differentiated cells of the spinous and granular layers. GJ antigen was not detected between the terminally differentiated cells of the cornified layer, and it was also absent in the degenerating periderm. Thus, while α_1 and β_2 GJ proteins were co-expressed in a specific manner relative to the epidermal maturation at early and middle stages of development, in the fully differentiated epidermis the expression of α_1 and β_2 coincided with the program of terminal keratinocyte differentiation. The high abundance of α_1 antigen throughout the spinous layer, as well as at the interface between the spinous and granular layers, potentially reflects a high level of intercellular coupling between the keratinocytes of these two differentiated strata. Furthermore, it appears that the intercellular coupling may be increased between the spinous and granular layers by a contribution from β_2 GJs. One might speculate that this extensive, but differential expression of α_1 and β_2 GJs in the fully differentiated fetal epidermis may be responsible for creating large communication compartments within one single epidermal layer, as well as between two adjacent strata consisting of several subcompartments.

α_1 and β_2 GJ expression was analysed immunohistochemically during the morphogenesis of hair follicles and during different stages of hair growth cycle. Both α_1 and β_2 connexins were expressed in the hair cone of elongated hair follicles and in the outer root sheath in the skin of newborn rats. α_1 was also detected between dermal fibroblasts, but was not detectable in dermal papillae, a structure composed of dermal fibroblasts. α_1 was detectable in this hair compartment in 10-day-old animals (d + 10 stage). α_1 and β_2 GJ expression increased in different follicular compartments with progressing growth and differentiation of follicular cell layers. In 15-day-old rats, when hair follicles attained their maximal size (late anagen of the first growth cycle, d + 15), α_1 was co-expressed with β_2 protein in the outer root sheath, in partially keratinized layers of the inner root sheath, as well as in the cortex and in matrix cells of the hair bulb. A cell-type specific expression was observed for α_1 in dermal papillae and for β_2 in medulla of the hair shaft. The resting period of hair follicles during the first growth cycle in 25-day-old rats

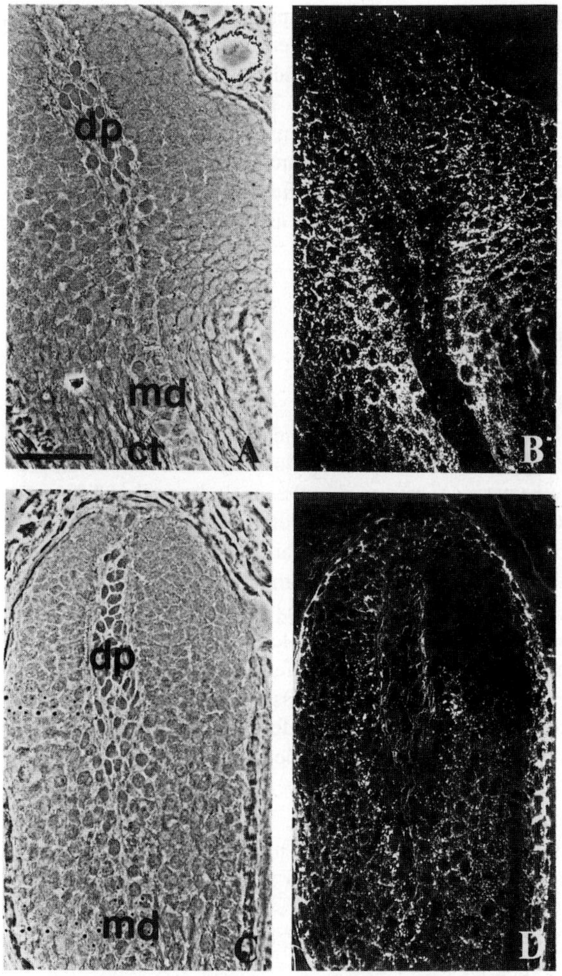

Figure 4. Immunolocalization of α_1 and β_2 GJ antigens in bulbar regions of growing hair follicles during the second hair growth cycle. (A, C) Phase-contrast micrographs of frozen dorsal skin from 40-day-old rats incubated with α_1S (B) and β_2J (D) peptide antibodies. Longitudinal section through the follicular hair bulb at anagen, illustrating the dermal papillae (dp), cortex (ct) and medulla (md) of the hair shaft. (B) Localization of α_1 protein; note the absence of α_1 in the medulla. (D) Localization of β_2 protein; note the absence of β_2 in dermal papillae. Scale bar, 50 μm.

(d + 25, telogen) was characterized by low levels of α_1 protein in the remaining follicular compartments, and by the absence of β_2 GJ protein. Following re-entry of hair follicles into the growing period of the second growth cycle (anagen at d + 40 was analysed), α_1 and β_2 GJs had essentially the same expression pattern as observed during the active growth of the first cycle (Fig. 4). Analysis of GJ expression during the resting period of the second growth cycle (telogen at d + 55 was analysed) resembled the expression pattern of the resting stage from the first growth cycle. The dynamic changes of α_1 and β_2 expression during different stages of the hair growth cycle, monitored by immunohistochemistry, was confirmed at the protein and transcript levels by using immunoblot and Northern blot analysis.

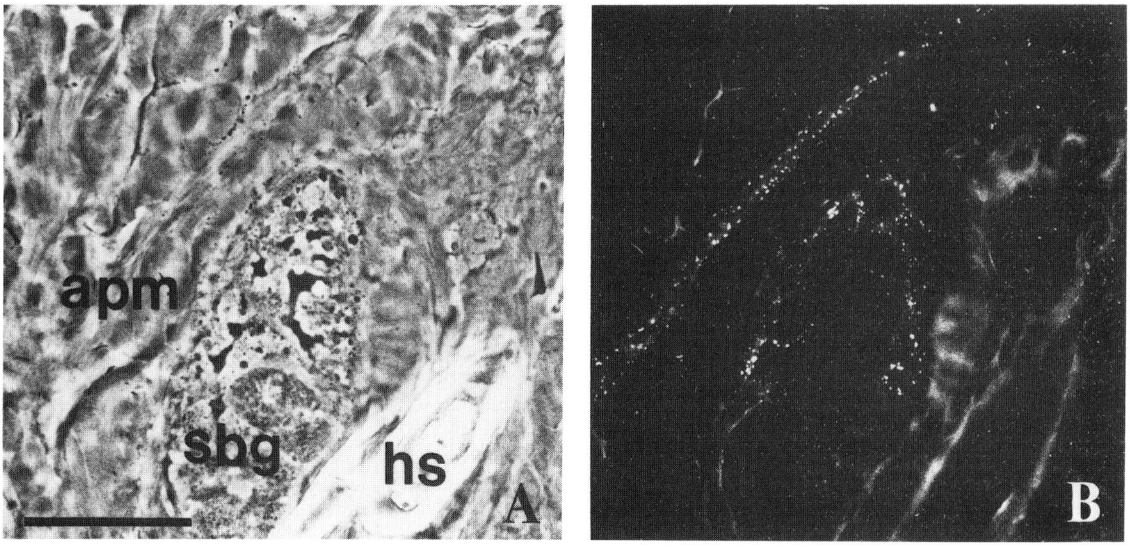

Figure 5. Immunolocalization of α_1 GJ antigen in the arrector pili muscle and in the sebaceous gland. (A) Phase-contrast micrograph of frozen dorsal skin of a 40-day old rat illustrating the pilosebaceous apparatus consisting of a hair follicle (hs, keratinized hair shaft), sebaceous gland (sbg) and arrector pili muscle (apm). (B) Localization of α_1 protein in the smooth muscle fibres of the arrector pili muscle and in the undifferentiated, peripheral cells of sebaceous gland. Scale bar, 20 μm.

Following the onset of sebaceous gland formation in newborn rats, α_1 GJs were detected at the periphery of these structures in 6-day-old rats (mid-anagen; d + 6). Concomitant with α_1 detection in the sebaceous gland, α_1 was also expressed in the arrector pili muscle (Fig. 5). The arrector pili muscle is formed at about the same stage by a grouping of smooth muscle fibres (Gibbs, 1941), and it represents the third part of a functional pilosebaceous apparatus. The contraction of this muscle causes the perpendicular orientation of hair, and probably squeezing of the sebaceous glands (Gibbs, 1941). The α_1 GJ protein localized in the arrector pili muscle was constitutively expressed at relatively high levels, independent of the hair cycle or animal age.

It is worthwhile to compare the constitutive α_1 expression in the arrector smooth muscle with the rapidly inducible α_1 expression in the smooth muscle of uterine myometrium at term (Risek et al., 1990). In both systems, the α_1 GJs appear to provide low-resistance pathways between muscle cells, thus allowing a rapid synchronization of action potentials, resulting in well-coordinated contractions. Conversely, the absence or low abundance of myometrial GJs in the non-pregnant uterus reflects the quiescent or inactive state of the uterine wall. Note that β_2 GJ protein was not detected in sebaceous glands or in arrector pili muscle. The expression of α_1 connexin in the undifferentiated, proliferating cells of the sebaceous gland, as well as in the undifferentiated, proliferating cells in the basal layer of the differentiated epidermis, may provide further evidence for a biological similarity between these two compartments, as suggested by Tosti (1974). α_1 expression increased with the overall growth of sebaceous glands with no detectable differences found at different stages of the hair growth cycle.

Conclusions

The results of this study indicate that there is an intimate relationship between morphology and GJ protein expression during both the ontogenesis and terminal differentiation of rat epidermis. Since this cell-type specific expression coincides with epidermal differentiation and maturation, a correlation of biochemistry and morphology may provide insights into understanding the cellular functions of certain members of the GJ multigene family. The biology of GJs represents, at present, a challenging area where many basic questions remain to be answered. Some of these questions address the biological significance of the diversity of the GJ multigene family, while others focus on the structure/function relationships.

Finally, the control mechanisms regulating a spatiotemporal, cell-type specific (co)-expression of different GJ genes or their products are completely unknown. Organ and cell culture systems using skin explants and isolated hair follicles may provide an approach for studying some aspects of GJ-mediated cell-cell communication in the broader context of hair biology. On the other hand, the cyclicity and dynamic properties of follicular hair growth may provide an experimental model that will be helpful for understanding the control of gap-junction expression at the molecular and cellular level.

References

Banks-Schlegel, S.P. (1982) Keratin alterations during embryonic epidermal differentiation: a presage of adult epidermal maturation. J. Cell Biol. 93, 551–559.

Bell, M. (1974) A comparative study of the ultrastructure of the sebaceous glands of man and other primates. J. Invest. Dermatol. 62, 132–143.

Bennett, M.V.L., Barrio, L.C., Bargiello, T.A., Spray, D.C., Hertzberg, E. and Sáez, J.C. (1991) Gap junctions: new tools, new answers, new questions. Neuron 6, 305–320.

Beyer, E.C., Paul, D.L. and Goodenough, D.A. (1990) Connexin family of gap junction proteins. J. Membr. Biol. 116, 187–194.

Bonneville, M.A. 1968. Observations on epidermal differentiation in the fetal rat. Am. J. Anat. 123, 147–164.

Breathnach, A.S. and Wyllie, L.M. (1965) Fine structure of cells forming the surface layer of the epidermis in human fetuses at fourteen and twelve weeks. J. Invest. Dermatol. 45, 179–189.

Breathnach, A.S. (1971) Embryology of human skin. A review of ultrastructural studies. J. Invest. Dermatol. 57, 133–143.

Butcher, E.O. (1934) The hair cycles in the albino rat. Anat. Rec. 61, 5–14.

Caputo, R. and Peluchetti, D. (1977) The junctions of normal human epidermis. A freeze-fracture study. J. Ultrastruct. Res. 61, 44–61.

Chase, H.B. (1954) Growth of the hair. Physiol. Rev. 34, 113–126.

Dale, B.A. and Holbrook, K.A. (1987) Developmental expression of human epidermal keratins and fillagrin. Curr. Top. Dev. Biol. 22, 127–155.

Elias, P.M. and Friend, D.S. (1976) Vitamin-A induced mucous metaplasia. An *in vitro* system for modulating tight and gap-junction differentiation. J. Cell Biol. 68, 173–188.

Fuchs, E. and Green, (1980) Changes in keratin gene expression during terminal differentiation of the keratinocyte. Cell 19, 1033–1042.

Gabbiani, G., Chaponnier, C. and Huttner, I. (1976) Cytoplasmic filaments and gap junctions in epithelial cells and myofibroblasts during wound healing. J. Cell Biol. 76, 561–568.

Gibbs, H.F. (1941) A study of the post-natal development of the skin and hair of the mouse. Anat. Rec. 80, 61–82.

Hardy, M.H. (1952) The histochemistry of hair follicles in the mouse. Am. J. Anat. 90, 285–335.

Holbrook, K.A. and Odland, G.F. (1975) The fine structure of developing human epidermis: light, scanning and transmission electron microscopy of the periderm. J. Invest. Dermatol. 65, 16–38.

Jahoda, C.A.B., Horne, K.A. and Oliver, R.F. (1984) Induction of hair follicle growth by implantation of cultured dermal papilla cells. Nature 311, 560–562.

Kam, E., Melville, L. and Pitts, J.D. (1986) Patterns of junctional communication in skin. J. Invest. Dermatol. 87, 748–753.

Kligman, A.M. (1959) The human hair cycle. J. Invest. Dermatol. 33, 307–316.

Kopan, R. and Fuchs, E. (1989) A new look into an old problem: keratins as tools to investigate determination, morphogenesis, and differentiation in skin. Genes Develop. 3, 1–15.

Montagna, W. (1974) An introduction to sebaceous glands. J. Invest. Dermatol. 62, 120–123.

Parmley, T.H. and Seeds, A.E. (1970) Fetal skin permeability to isotopic water (THO) in early pregnancy. Am. J. Obstet. Gynecol. 108, 128–131.

Parnell, J.P. (1949) Postnatal development and functional histology of the sebaceous glands in the rat. Am. J. Anat. 85, 41–71.

Pitts, J., Kam, E., Melville, L. and Watt, F.M. (1987) Patterns of junctional communication in animal tissues. Ciba Foundn. Symp. 125, 140–153.

Risek, B., Guthrie, S., Kumar, N. and Gilula, N.B. (1990) Modulation of gap-junction transcript and protein expression during pregnancy in the rat. J. Cell Biol. 110, 269–282.

Risek, B. and Gilula, N.B. (1991) Spatiotemporal expression of three gap-junction gene products involved in fetomaternal communication during rat pregnancy. Development 113, 165–181.

Salomon, D., Saurat, J.-H. and Meda, P. (1988) Cell-to-cell communication within intact human skin. J. Clin. Invest. 82, 248–254.

Stern, I.B., Dayton, L. and Duecy, J. (1971) The uptake of tritiated thymidine by the dorsal epidermis of the fetal and newborn rat. Anat. Rec. 170, 225–234.

Tosti, A. (1974) A comparison of the histodynamics of sebaceous glands and epidermis in man: a microanatomic and morphometric study. J. Invest. Dermatol. 62, 147–152.

J.E. Hall, G.A. Zampighi and R.M. Davis (Eds.)
Progress in Cell Research, Vol. 3

CHAPTER 34

Double whole-cell patch-clamp of gap junctions in insect epidermal cell pairs: single channel conductance, voltage dependence, and spontaneous uncoupling

DENNIS CHURCHILL and STANLEY CAVENEY

Department of Zoology, University of Western Ontario, London, Ontario, Canada N6A 5B7

Introduction

The cells of most tissues in multicellular organisms are connected cytosol-to-cytosol via a lattice of integral membrane protein channels called gap junctions that are permeable to ions and small metabolites (MW < 2000, in arthropods). These may function in tissue homeostasis, in mediating hormonal responses, and in coordinating growth and development in tissues (reviewed in Loewenstein, 1981; Caveney 1985; Guthrie and Gilula, 1989: Bennett et al., 1991).

The insect integument, a simple monolayer of epidermal cells producing an exoskeleton, is a useful model to study the role of intercellular communication in growth control and pattern formation. The gap junctions of the segmented epidermis have been shown (1) to regulate junctional communication in a predictable pattern during the moult cycle; (2) to increase permeability to inorganic ions, but not to fluorescent tracers on epidermal exposure to the developmental hormone 20-hydroxyecdysone or to L-glutamate in vitro; and (3) to establish developmental compartments between segments through the presence of strips of cells (found at segment borders) having reduced junctional permeability to organic molecules.

Intercellular resistance to the movement of ions among cells of the intact segment changes predictably in phase with each moult cycle associated with beetle larva growth (Caveney, 1978). One of the key hormones controlling the moult cycle in vivo, 20-hydroxyecdysone (20-HE), increases junctional permeability in vitro (Caveney and Blennerhassett, 1980). Because this occurs in the presence of inhibitors of protein synthesis, and without an increase in the size of gap-junction plaques or particles, a change in some property of existing gap-junction channels rather than new channel synthesis or recruitment is suspected (Caveney et al., 1980).

Unlike junctional permeability to inorganic ions, fluorescent tracer permeability does not vary within the moult cycle, even though it does increase as the larval insect grows in size (Caveney and Safranyos, 1990). Experimentally, a similar response can be elicited *in vitro*. A powerful Na^+-dependent L-glutamate uptake mechanism operates in the epidermis (McLean and Caveney, 1991). 100 μM L-glutamate in the bath decreases intercellular resistance 50% in 2 h without changing dye permeability (Caveney, unpublished result). In addition, dye passage is restricted by a strip of border cells at the compartment boundary between segments of a developing insect epidermis while ionic coupling is unaffected (Warner and Lawrence, 1982; Blennerhassett and Caveney, 1984).

In the epidermis, the mechanisms by which ionic conductance is up-regulated and permeability to ions and small molecules is regulated discoordinately are not understood. By examining the properties of single gap-junctional channels using the recently developed double whole-cell variant of the patch-clamp technique (Neyton and Trautmann, 1985; Veenstra and DeHaan, 1986), we hope to gain insight into these mechanisms. We present here preliminary results obtained using patch-clamp techniques to investigate spontaneous uncoupling, voltage dependence and single-channel conductance of gap-junction channels in isolated epidermal cells pairs.

Materials and methods

Cell preparation

Dissection was similar to that described in detail elsewhere (Caveney and Blennerhassett, 1980). Segments of integument (4 mm^2) were dissected into bath saline from late pre-pupal larvae of the yellow mealworm *Tenebrio molitor* following apolysis, and were incubated in bath saline with elastase and collagenase for a few min. The digestion was terminated by dilution (bath saline + 2.5 mM EGTA) and was followed by gently pipetting the tissue up and down with a large bore Pasteur pipette. Cells were washed by centrifugation at 100 × g for 5 min, resuspended in bath saline, plated on glass coverslips which were fixed in a plexiglass perfusion chamber, and used within 8 h. Isolation and all experiments were carried out at 24–28°C.

Salines

Bath saline used for isolation and patch-clamp experiments (mM): 105 Na^+, 128 Cl^-, 43 K^+, 1.4 Ca^{2+}, 1 Mg^{2+}, 98 sucrose, 20 PIPES, approx. 400 mOsM, pH 6.7.

Pipette solution (mM): 165 K^+, 116 Cl^-, 10 Na^+, 2 Mg^{2+}, 11 EGTA, 80 sucrose, 20 PIPES, approx. 400 mOsM, pH 6.7.

Data recording

Single and double whole-cell patch-clamp recording methods have been described in detail elsewhere (Hamill et al., 1981; Neyton and Trautmann, 1985; Veenstra and DeHaan, 1986 and 1988; Rook et al., 1988). Junctional currents were measured in electrically coupled cell pairs using 5–8 MΩ fire-polished patch pipettes filled with pipette solution and attached to independent EPC-7 (List, Germany) patch-clamp type voltage-clamp amplifiers. The electrical seal that formed between the plasma membrane and borosilicate glass of the pipette was typically >10 GΩ. Junctional conductance (G_J) was calculated by dividing the junctional current (I_j) by the transjunctional potential (V_j). In these experiments, the apparent G_J was not corrected for series resistance (R_S) errors since only relative G_J were of concern in spontaneous uncoupling experiments. Errors due to R_S were negligible (error < 10%) during measurement of single-channel currents.

On-line stimulation and storage of data was achieved using an AT computer, a TL-1 DMA interface and pClamp software (Axon Instruments). Data were also stored to VCR tape and chart recorder. Prior to digitization, data were filtered at 0.2 × f_S (sample frequency) using a low-pass Bessel filter. Data analysis and display were performed off-line using pClamp, Axotape (Axon Instruments), Supercalc (Computer Associates) and Inplot (Graphpad) analysis and graphics software.

Results and discussion

Single whole-cell recording

Cells 10–15 μm in diameter were isolated. Whole-cell voltage clamp of single cells using control bath and pipette solutions indicated that the non-junctional membrane does not have voltage-gated channel activity and responds in ohmic fashion to 1000 ms voltage jumps (± 100 mV) applied from a holding potential of –40 mV. The current-voltage relationships (I–V) in all cells studied were linear, and time-dependent gating was not observed. RS and V_m typically ranged from 20–40 MΩ and –10 to –60 mV, respectively. The high input resistance [14.1 ± 2.6 GΩ (n = 29, ± SEM)] of these cells reduces the signal-to-noise ratio significantly, which is important for recording single gap-junction channel activity.

Spontaneous uncoupling of cell pairs and effect of ATP

The initial macroscopic junctional conductance detected in 41 pairs was 9.0 ± 1.2 nS (± SEM; V_m of both cells was clamped at –50 mV and depolarizing 20 mV pulses were applied to one of the cells). When control pipette solution was used (Fig. 1A), cell pairs spontaneously uncoupled following rupture of the patches in the two cells. (Rupture of both patches was the point at which recording of junctional could begin.) As the junctional currents dropped, openings and closings of single gap-junction channels became detectable. The average half-time to uncoupling for 12 pairs was 7 to 8 min (Fig. 1B).

When 5 mM ATP was included in the pipette solution, spontaneous uncoupling was inhibited (Fig. 1B,). Cells remained coupled up to 60 min. Single gap-junction channel events have never been observed at the end of experiments with ATP. These data suggest that ATP is important in maintaining channels in the open state. We are presently investigating whether this occurs through a phosphorylation-dependent or -independent mechanism using non-hydrolyzable analogues of ATP. The role of ATP and phosphorylation in inhibiting spontaneous uncoupling has been demonstrated in several vertebrate cell preparations (e.g. Somogyi and Kolb, 1988; Cooper et al., 1989; Sugiura et al., 1990) and also seems to be important in stabilizing insect gap-junction channels.

Single-channel properties

Typical open/closed channel events are shown in Fig. 2. The average properties of the main conductance state observed were: single-channel conductance (g_j) 288 ± 2 pS (168 events, 14 cell pairs, ± SEM); transition time from open to closed state 21 ± 0.8 ms (190 events, 14 cell pairs ± SEM); and open times 0.06 to 11 s.

The dominant conductance of insect gap-junctional channels is larger than that of vertebrate channels, which range from 30 to 188 pS (Neyton and Trautmann, 1986; Burt and Spray, 1988; Rook et al., 1988; Somogyi and Kolb, 1988; Veenstra and DeHaan, 1988), although 200–300 pS channels have recently been found in embryonic chicken lens (Miller et al., 1990). Single-channel conductances of 30–40 pS and 100 pS were found in the earthworm septate junction (Brink and Fan, 1989). Previous experiments using tracer molecules have demonstrated the apparent limiting pore size in *Chironomus* salivary gland junctions to be 2–3 nm (Flagg-Newton et al., 1979; Schwarzmann et al., 1981; Zimmerman and Rose, 1985), in contrast to 1.6 to 2 nm in junctions of cultured mammalian cells (Flagg-Newton et al., 1979; Schwarzmann et al., 1981). In *Tenebrio*, techniques similar to those used by Zimmerman and Rose estimate pore size to be 2.4–3.4 nm (Safranyos, 1985). Assuming the gap-junction channel to be a right cylindrical pore, the resistance of the channel can be calculated using the equation described in Burt and Spray (1988): $r_j = 2\rho/4r + \rho(l/\rho r2)$, where r is the pore radius, l is channel length (15–20 nm), and ρ (Ω-cm) is the resistivity of the pore fluid (100–200 Ω-cm). If all parameters except channel radius are kept constant (i.e., l = 17 nm and ρ = 150 Ω-cm) then single-channel conductance is calculated ($1/r_j$) for r = 1 nm or 1.5 nm to be 112 pS and 243 pS, respectively, values comparable with typical experimental values.

Subconductances

Subconductance states of the major conductance level were found (Fig. 2) to occur primarily in the range of 50–100 pS, although we have observed infrequent subconductances as large as 238 pS. Following the subconductance criteria outlined in Fox (1987), these smaller conductances would appear to be subconductances of the 288 pS channel, and not separate channel events. In nearly all cases, they interconverted directly with the 288 pS conductance level (Fig. 2), they occurred infrequently and were always observed in the presence of the major conductance level and, finally, the main state did not appear to result from the superimposition of conductances of independent channels. We never observed isolated channel events of smaller conductance which could sum to give the main conductance.

The observation of four equal subconductances adding up to the fully opened state has been reported for channels made from purified lens junctional protein, each channel being composed of 4 subunits (Zampighi et al., 1985). Statistically we do not have enough data (a minimum of 200 events are needed to describe the simplest distribution; Colquhoun and Sigworth, 1983) to determine if we have a channel with a single subconductance, many non-equal subconductances, or, several subconductances that

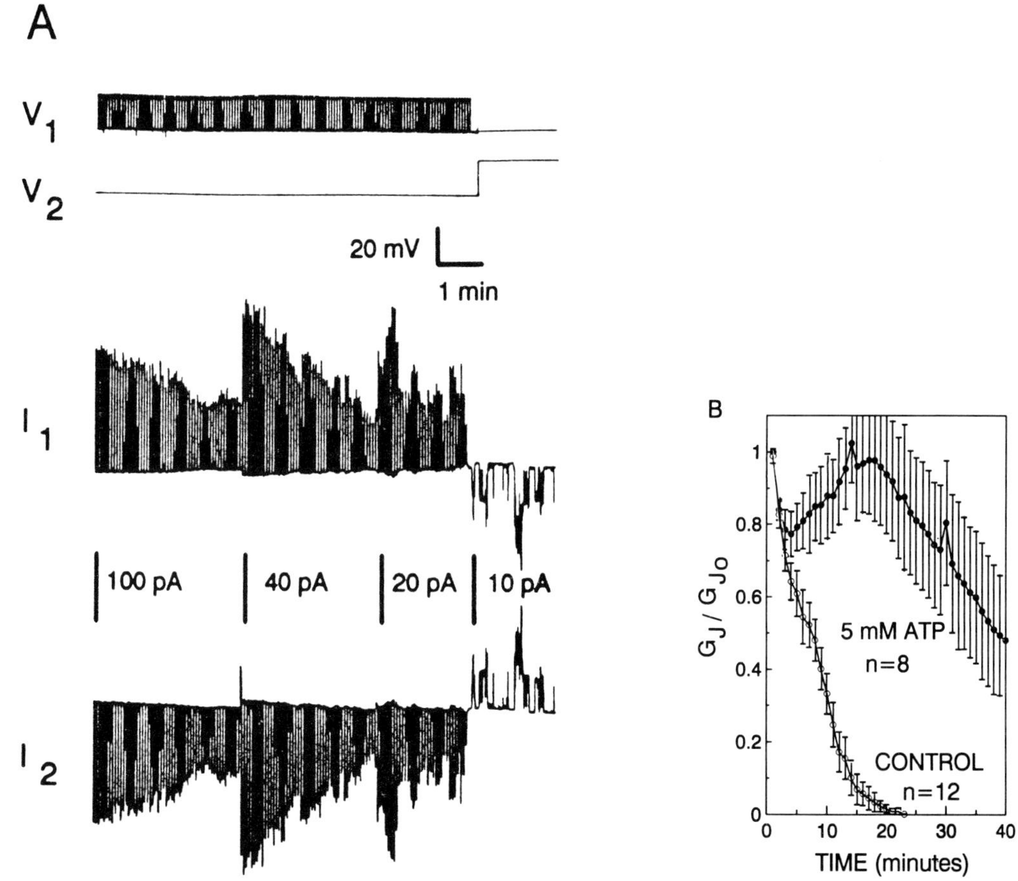

Figure 1. (A) In order to measure changes in G_J with time, each cell of an isolated pair was voltage clamped at −50 mV, while one of the cells received 1 s depolarizing 20 mV voltage pulses at 3 s intervals. Raw data on a compressed time scale are shown in A (f_S = 15 Hz). The data represent the final 8 min of recording from this cell. Four gain changes are shown (scale bars shown between the current traces). At the very end of the recording, currents through single gap-junction channels are observed. (cell 90d04c1/4) (B) The inset in B shows the average of 12 uncoupling pairs with an initial conductance > 10 nS. G_J is normalized to G_{Jo}. All cell pairs studied with control pipette solution (α) spontaneously uncoupled and, on average, uncoupling occurred with a half-time of 7.6 min. 5 mM ATP significantly inhibited spontaneous uncoupling (●).

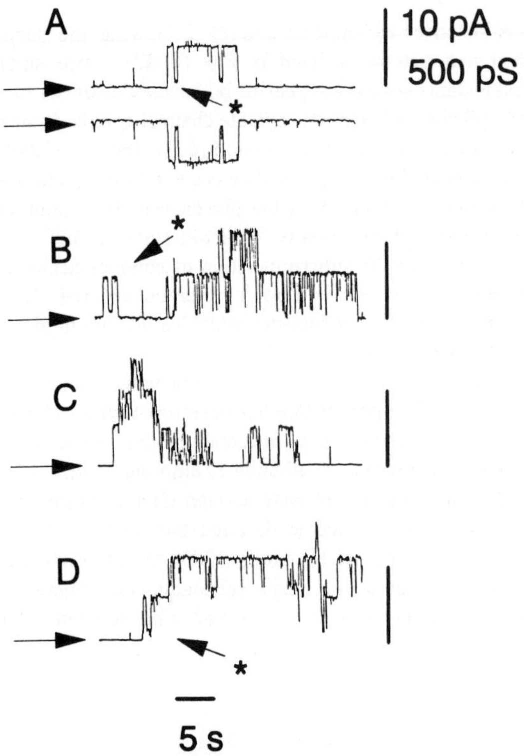

Figure 2. Examples of single-channel events are shown in (a)–(d). In order to measure single-channel properties in this system, only clear, single-level, single-channel events from the fully open to fully closed states were measured. V_j was usually held constant at 20 mV. Data were obtained from 4 cell pairs (f_S = 300 Hz). The horizontal arrows denote the fully closed state. Traces from both cells are shown in (a). Only one trace is shown for the other examples. In all examples V_j = ±20 mV. Stable subconductances of 30–80 pS that last for tens of milliseconds are shown at *→.

depend on the number of protein subunits contributing to the channel.

The possibility of subconductance states in the gap junctions of patch-clamped vertebrate cells has been suggested (Neyton and Trautmann, 1985; Veenstra and DeHaan, 1986, 1988; Somogyi and Kolb, 1988). It remains unclear whether the conductance levels observed represent subconductances of the main state or the activity of different populations of channels, as the observations made may not meet the above criteria. (Whether an attempt was made to apply them is uncertain.)

V_m dependence of junctional conductance (inside-outside voltage)

Junctional currents were measured in cell pairs in response to a 20 mV transjunctional voltage jump applied over a range (20 mV to –70 mV) of membrane potentials (Fig. 3). The currents were very small or undetectable when the cells were held positive to 0 mV, and increased in sigmoidal fashion as the cells were hyperpolarized. These data show that macroscopic conductance of beetle gap junctions is strongly dependent on the holding potential of the cells (i.e. inside-outside voltage dependence), as reported for other insect junctions (Obaid et al., 1983; Verselis and Bargiello, 1990). This voltage sensitivity differs from that of certain vertebrate junctions, which are predominantly sensitive to transjunctional voltage (reviewed by Spray, 1990a,b). The membrane potential at the estimated half-maximal conductance (V_0) in these insect cells lies between –30 and –60 mV. Because V_0 lies within the normal range of membrane potential in the intact epidermis (Caveney and Blennerhassett, 1980), V_m may play a role in controlling junctional conductance in this tissue.

V_j dependence of junctional currents

Beetle epidermal cell pairs respond to transjunctional voltage in a manner analogous to that reported in

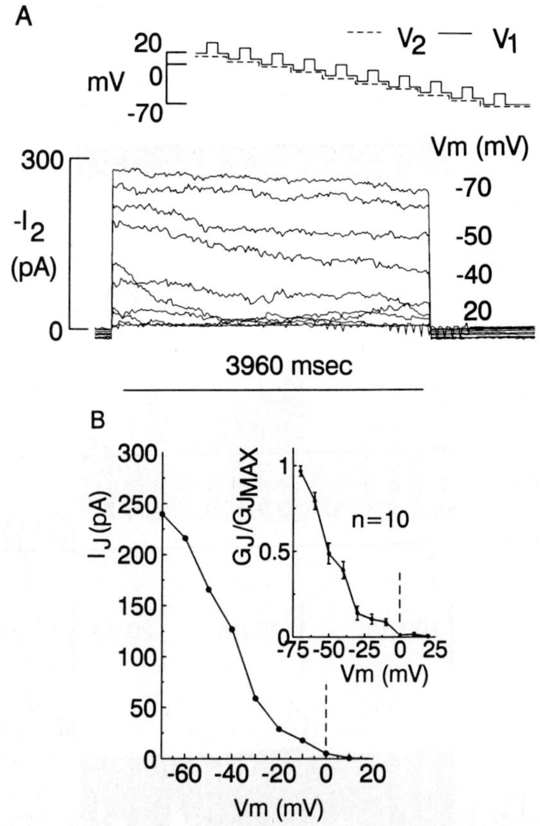

Figure 3. (A) Junctional currents were measured in cell pairs in response to 20 mV voltage jumps applied from a range of membrane potentials (protocol as diagrammed). The currents recorded in cell 2 are shown inverted (cell 90D13c15; f_S = 500 Hz). The pipette solution does not contain ATP. (B) The I–V curve for the junctional currents ($-\Delta I_2$) is shown. Each point represents the average current during each jump and is graphed against the common holding potential (V_m) of the two cells. The inset shows this same curve converted to conductance and normalized to maximal conductance.

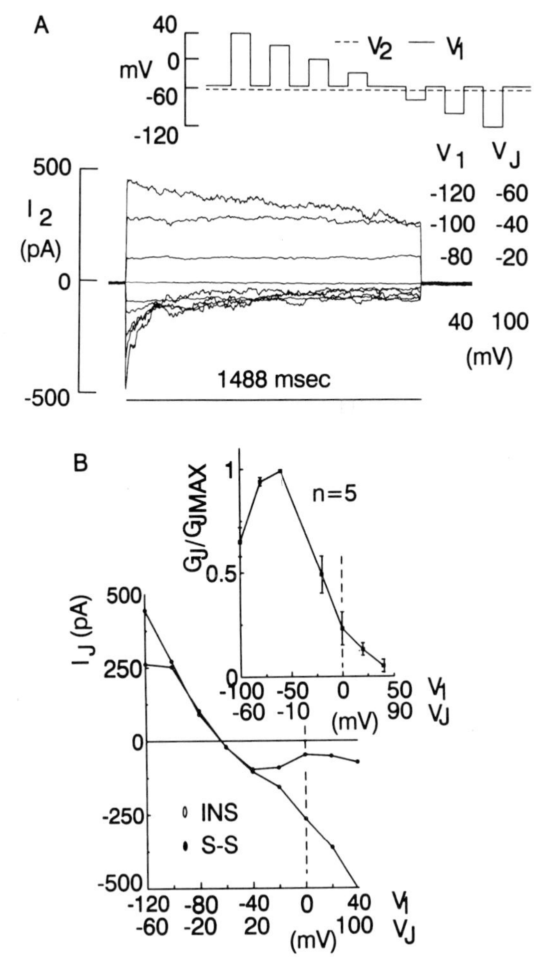

Figure 4. (A) Each cell of a well-coupled pair was voltage clamped at –60 mV and one cell of the pair (cell 1) received a range of 1488 ms voltage jumps. The currents recorded in cell 2 are shown (cell 90d20c17; f_S = 500 Hz). At the left of the curves, the voltage of the jump as well as the corresponding transjunctional voltage is displayed. (B) Both the instantaneous and steady-state I–V curves are shown. Steady-state current is determined by averaging the points in the last 150 ms of the jump. The inset displays the average steady-state conductance normalized to maximal conductance vs. V_1 from 5 cell pairs, each with the membrane potential clamped at –40 mV.

Drosophila salivary gland pairs (Verselis and Bargiello, 1990). The instantaneous I–V curve is linear, indicating that the dependence of single-channel conductance on V_j is linear (i.e., the gap-junction channel itself does not rectify). The steady-state curve, calculated from the average current during the last 150 ms of each current trace, is more complex (Fig. 4). As seen also in the G_j/G_{JMAX} vs. V_m curve (Fig. 3B, inset), conductance approaches zero as cell 2 is depolarized, and increases as it is hyperpolarized. This suggests that V_m is the major determinant of G_j. However, the decrease of G_J when V_1 is hyperpolarized greater than –60 mV, and the time-dependent inactivation of I_j at large transjunctional voltage suggests that junctional conductance is sensitive to both V_j and V_m.

Since these cells are so well coupled in the intact tissue, with an estimated 90,000 active channels per cell, it is possible that this V_j dependence is an artifact of the relatively poorly coupled nature of isolated cell pairs. For example, it has been shown that the junctions of poorly coupled cardiac cells have a stronger voltage dependency than well-coupled cells. It is claimed that each channel between poorly coupled cells senses a greater proportion of the transmembrane electric field (Jongsma et al. 1990). For this reason we think that, unlike V_m dependence, V_j dependence is probably not physiologically relevant in the normally tightly coupled epidermis.

Voltage dependence of single-channel activity

We have been able to demonstrate at the single-channel level in poorly coupled cells that channel activity increases with hyperpolarization of one or both cells of the pair. The experiments were performed by either hyperpolarizing one cell of a voltage-clamped, poorly coupled pair while the other was clamped at 0 mV or, to test V_m dependence alone, hyperpolarizing both cells while maintaining a constant V_j of 20 mV. Channel activity increased more when both cells were hyperpolarized and less when only one cell was hyperpolarized. It appears that the single-channel conductance, at least in poorly coupled cells, remains constant, and only the number (and likely open times) of active channels increases with hyperpolarization. These data support the macroscopic data and confirm that V_m dependence is the predominant voltage-dependent gating mechanism controlling the activity of gap-junction channels in these cells.

Conclusions

Our primary goal in examining double whole-cell patch-clamped epidermal cells has been to gain insight into the junctional mechanism(s) accounting for the separate control of junctional permeability to inorganic ions and to small organic molecules. To explain selective changes in junctional permeability, either different classes of channels with different permeability properties, or a single channel with different substate permeabilities may be required. Each channel or substate in such models is presumably subject to differential regulation.

In these preliminary results, only one channel was observed, as defined by its single-channel conductance and ability to assume subconductance states. One might conclude that to be the only class of channel present in the epidermis. It is possible, however, that the observed channel class is the only one (of perhaps several) whose activity is not lost during the procedure used to isolate cell pairs. Similarly, since the cells uncouple spontaneously, it is possible that other classes of channels exist that shut down earlier during the uncoupling process. In either case, the method as used here may not be able to detect all the channels present, or even reveal channel behavior typical of cells in normally well-coupled tissue.

The existence of a channel with subconductance states suggests the possibility that the fully open main state is permeable to small organic molecules and the substates are not. In order to validate such an hypothesis, we must demonstrate whether the observed substates are a component of normal channel behaviour, or a property of poorly coupled cells alone. One would also have to test whether treatments that affect junctional permeability in the intact epidermis (exposure to 20-hydroxyecdysone or L-glutamate?) alter the frequency at which different channel states are detected.

From the calculations above, there seems to be a correlation between the size of molecule capable of passing through a channel pore and the main channel conductance (e.g. vertebrate vs. arthropod junctional channels). Similarly, the permeability of a channel may change because it shifts from its main state to a substate conductance level. Even if this were the case, it is not easy to demonstrate as it may not be possible to correlate dye coupling and the pattern of single-channel activity. In order to detect the movement of fluorescent tracers from cell to cell, many channels may have to be open to allow enough tracer to pass at a detectable level, a condition not suitable for observing single-channel activity with patch-clamp techniques. For this reason it may be difficult to assign to any given channel state a conductance and permeability characteristic and, therefore, the possibility that channels may have high single-channel conductances and low permeability to small organic molecules exists (i.e. the channel pore and/or its substates may not be right cylindrical in shape and instead locally constricted). It is worth noting that if both the channel main state and its substate were right cylindrical pores the conductance of each alone (288 pS and 50–100, respectively) would be associated with a channel predicted to pass typical tracer molecules.

It may be that until cDNAs for arthropod gap-junction channels are identified and reconstitution experiments performed with purified protein, many of the questions we are interested in may not be adequately addressed.

Acknowledgements

This work was supported by NSERC of Canada grant No. A6797.

References

Blennerhassett, M.G. and Caveney, S. (1984) Separation of developmental compartments by a cell type with reduced junctional permeability. Nature 309, 361–364.

Brink, P.R. and Shih-fang, P. (1989) Patch clamp recordings from membranes which contain gap-junction channels. Biophys. J. 56, 579–593.

Burt, J.M. and Spray, D.C. (1988) Single-channel events and gating behaviour of the cardiac gap-junction channel. Proc. Natl Acad. Sci. USA 85, 3431–3434.

Caveney, S. (1978) Intercellular communication in insect development is hormonally controlled. Science 199, 192–195.

Caveney, S., Berdan, R.C. and McLean, S. (1980) Cell-to-cell ionic communication stimulated by 20HE occurs in the absence of protein synthesis and gap-junction growth. J. Insect. Physiol. 26, 557–567.

Caveney, S. and Blennerhassett, M.G. (1980) Elevation of ionic conductance between insect epidermal cells by β-ecdysone *in vitro*. J. Insect. Physiol. 26, 13–25.

Caveney, S. (1985) The role of gap junctions in development. Annu. Rev. Physiol. 47, 319–335.

Caveney, S. and Safranyos, R.G.A. (1989) Developmental physiology of insect epidermal gap junctions. In: N. Sperelakis and W.C. Cole (Eds.) Cell Interactions and Gap Junctions, Vol. 1. CRC Press, Boca Raton, FL.

Colquhoun, D. and Sigworth, F.J.. (1983) Fitting and statistical analysis of single-channel records. In: B. Sakmann and E. Neher (Eds.) Single-Channel Recording. Plenum Press, New York, p. 192.

Cooper, K.E., Rae, J.L. and Gates, P. (1989) Membrane and junctional properties of dissociated frog lens epithelial cells. J. Membr. Biol. 111, 215–227.

Flagg-Newton, J., Simpson, I. and Loewenstein, W.R. (1979) Permeability of the cell-to-cell membrane channels in mammalian cell junction. Science 205, 404–407.

Fox, J.A. (1987) Ion channel subconductance states. J. Membr. Biol. 97, 1–8.

Guthrie, S.C. and Gilula, N.B. (1989) Gap-junctional communication and development. Trends Neurol. Sci. 12, 12–16.

Hamill, O.P., Marty, A., Neher, E., Sakmann, B. and Sigworth, F.J. (1981) Improved patch-clamp techniques for high-resolution current recording from cell and cell-free membrane patches. Pflügers Arch. 391, 85–100.

Jongsma, H.J., Wilders, R., van Ginneken, A.C.G. and Rook, M.B. (1990) Modulatory effect of the transcellular electrical field on gap-junction conductance. In: C. Perracchia (Ed.) Biophysics of Gap Junctions. CRC Press, Boca Raton, FL.

Loewenstein, W.R. (1981) Junctional intercellular communication: the cell-to-cell membrane channel. Physiol. Rev. 61, 829–913

McLean, H. and Caveney, S. (1991) Na^+/L-glutamate co-transport in the epidermis of the beetle *Tenebrio molitor*: kinetics of uptake. J. Comp. Physiol. B, submitted.

Miller, A.G., Hall, J.E. and Zampighi, G.A. (1990) Cell-to-cell permeability in dissociated embryonic chicken lenses. Biophys. J. 57, 245a.

Neyton, J. and Trautmann, A. (1985) Single-channel currents of an intercellular junction. Nature 317, 331–335.

Neyton, J. and Trautmann, A. (1986) Acetylcholine modulation of the conductance of intercellular junctions between rat lacrimal cells. J. Physiol. 377, 283–295.

Obaid, A.L., Socolar, S.J. and Rose, B. (1983) Cell-to-cell channels with two independently regulated gates in series: analysis of junctional channel modulation by membrane potential, calcium and pH. J. Membr. Biol. 73, 69.

Rook, M.B., Jongsma, H.J. and van Ginneken, C.G. (1988) Properties of single gap-junctional channels between isolated neonatal rat heart cells. Am. J. Physiol. 255, H770–H782.

Safranyos, R.G.A. (1985) Diffusion via cell-to-cell membrane channels. PhD thesis, University of Western Ontario, London, Canada.

Schwarzmann, H., Wiegandt, H., Rose, B., Zimmermann, A., Ben-Haim, D. and Loewenstein, W.R. (1981) Diameter of cell-to-cell junctional channels as probed with neutral molecules. Science 213, 551–553.

Somogyi, R. and Kolb. H.A. (1988) Cell-to-cell channel conductance during loss of gap-junctional coupling in pairs of pancreatic acinar and Chinese hamster ovary cells. Pflügers Arch. 412, 54–65.

Spray, D.C. (1990a) Electrophysiological properties of gap-junction channels. In: A.W. Robards, H. Jongsma, W.J. Lucas, J. Pitts and D. Spray (Eds.), Parallels in Cell-to-Cell Junctions in Plants and Animals. Springer-Verlag, Berlin.

Spray, D.C. (1990b) Transjunctional voltage dependence of gap-junction channels. In: C. Perracchia (Ed.), Biophysics of Gap Junctions. CRC Press, Boca Raton, FL.

Sugiura, H., Toyama, J., Tsuboi, N., Kamiya, K. and Kodama, I. (1990) ATP directly affects junctional conductance between paired ventricular myocytes isolated from guinea pig heart. Circ. Res. 66, 1095–1102.

Veenstra, R.D. and DeHaan, R.L. (1986) Measurement of single-channel currents from cardiac gap junctions. Science 233, 972–974.

Veenstra, R.D. and DeHaan, R.L. (1988) Cardiac gap-junction channel activity in embryonic chick ventricle cells. Am. J. Physiol. 254, H170–180.

Verselis, V.K. and Bargiello, T.A. (1990) Dual voltage control in a *Drosophila* gap-junction channel. In: C. Perracchia (Ed.), Biophysics of Gap Junctions. CRC Press, Boca Raton, FL.

Warner, A.E. and Lawrence, P.A. (1982) Permeability of gap junctions at the segmental border in insect epidermis. Cell 28, 243.

Zampighi, G., Hall, J. and Kremen, M. (1985) Purified lens junctional protein forms channels in planar lipid films. Proc. Natl. Acad Sci. USA 82, 8468–8472.

Zimmerman, A.L. and Rose, B. (1985) Permeability properties of cell-to-cell channels: Kinetics of fluorescent tracer diffusion through a cell junction. J. Membr. Biol. 84, 269–283.

J.E. Hall, G.A. Zampighi and R.M. Davis (Eds.)
Progress in Cell Research, Vol. 3

CHAPTER 35

Insect cell pairs: electrical properties of intercellular junctions

ROBERT WEINGART[a], FELIKSAS F. BUKAUSKAS[b] and CHRISTOPH KEMPF[c]

[a]Department of Physiology, University of Berne, Bühlplatz 5, 3012 Berne, Switzerland, [b]Kaunas Medical Academy, Mickeviciaus 9, 233000 Kaunas, Lithuania and [c]Department of Biochemistry and Central Laboratory, Blood Transfusion Service, Swiss Red Cross, University of Berne, Freistr. 3, 3012 Berne, Switzerland

Introduction

Salivary glands of insect larvae have gap junctions whose properties differ from those of mammalian tissues (for review, see Weingart, 1987). For example, diffusion studies have shown that insect gap junctions have pores about twice as large as mammalian gap junctions (Zimmerman and Rose, 1985). And, electrical measurements have demonstrated that gap junctions of insects are regulated by both the junctional and non-junctional membrane potential (Obaid et al., 1983; Verselis et al., 1991). These properties render insect tissues extremely interesting for physiologists and biochemists.

This work was initiated to explore the biophysical mechanisms underlying the electrical properties of insect gap junctions. Experiments were performed using the double voltage-clamp approach in conjunction with patch pipettes (see, e.g., Weingart, 1986). It allows examination of macroscopic and microscopic currents in the same preparation. The classical preparation, the salivary glands of *Chironomus* larvae, was inappropriate for this study because of large cells (diameter: 80–150 μM) and membrane coating. Therefore, we used an established cell line derived from *Aedes albopictus* larvae (Igarashi, 1978). When grown at low density ($(0.2–1) \times 10^6$ cells/ml), cell pairs formed spontaneously in abundance.

To carry out experiments, cell pairs were visually selected under the microscope. Initially, a major problem was that the cell pairs exhibited different electrical properties. Eventually, functional tests enabled us to distinguish between intercellular contacts containing (1) cytoplasmic bridges, (2) gap junctions, (3) cytoplasmic bridges *and* gap junctions, or (4) no conducting path (Bukauskas et al., 1991, 1992).

This paper summarizes the basic properties of cytoplasmic bridges and gap junctions.

Cytoplasmic communications

Influence of junctional membrane potential on g_j

To determine the relationship between the junctional conductance and the transjunctional voltage gradient, both cells of a cell pair were clamped to the same holding potential (V_h). Subsequently, test pulses of constant duration (5–15 s), variable amplitude and either polarity were applied to cell 1 (V_1) while the voltage of cell 2 was maintained ($V_2 = V_h$). This gave rise to junctional voltage gradients (V_j) and transjunctional current flow (I_j). Individual values of junctional conductance were determined ($g_j = I_j/V_j$) and plotted vs. V_j. Over the voltage range examined (V_j = –100 mV to +100 mV), g_j remained constant. This finding is compatible with the presence of cytoplasmic bridges between the cells. On average, g_j of cytoplasmic bridges was 19.6 ± 3.1 nS (n = 13).

Influence of non-junctional membrane potential on g_j

Cell pairs with putative cytoplasmic bridges were also used to examine the effect of the non-junctional membrane potential (V_m) on g_j. The following pulse protocol was adopted.

Starting from a common V_h, identical conditioning pulses were applied concomitantly to cell 1 and cell 2, maintaining V_j at 0 mV. Small test pulses were then superimposed on cell 1 (–12.5 mV; 1 s; 0.2 Hz) to assess g_j. Values of g_j were estimated and plotted vs. V_m. It turned out that g_j remained constant (voltage range explored: –142 mV to +58 mV). This behavior is different from that seen in insect gap junctions (Obaid et al.; see also below).

Modulation of g_j

Cell pairs with a g_j insensitive to V_j and V_m were also characterized pharmacologically. For this purpose, they were exposed to 3 mM heptanol, a dose sufficient to fully uncouple gap junctions (see below). In these preparations, however, this intervention had no effect on g_j, even after prolonged exposures. This finding is compatible with the existence of cytoplasmic connections between adjacent cells.

Intercellular diffusion

Diffusion studies were also performed on cell pairs whose g_j obeyed the criteria for cytoplasmic bridges. Experimentally, one cell of a cell pair was injected with fluorescent tracer molecules of different size. Redistribution of the intracellular tracer was monitored under the microscope. Preliminary studies revealed cell-to-cell diffusion of FITC-dextran (molecular weight = 4400 Da), suggesting an involvement of cytoplasmic bridges rather than gap junctions. Previous studies established a cut-off limit of 2500 Da for permeation across insect gap junctions (Schwarzmann et al., 1981).

Conclusions

We propose that junctional membranes insensitive to V_j, V_m, and heptanol, consist of cytoplasmic bridges rather than gap junctions. This view is supported by the observation that these junctions allow intercellular diffusion of FITC-dextran. Cytoplasmic bridges have been observed between dividing cells in culture (see Karasiewicz, 1981). During cell division, nuclear division or mitosis is followed by cytoplasmic division or cytokinesis, a process which leads to the formation of a narrow bridge, the midbody. The midbody may persist for some time until it breaks, leaving two separate daughter cells. Cytoplasmic bridges are relevant for the interpretation of studies on cell-to-cell coupling performed on cultured cells. Criteria commonly used to identify gap junctions, e.g. intercellular current flow or tracer diffusion, may not be sufficient. The existence of more than one type of intercellular junctions requires more sophisticated experimental tests.

Gap junctions

Dependence of g_j on junctional membrane potential

The relationship between g_j of gap junctions and the transjunctional voltage gradient was assessed utilizing an asymmetrical pulse protocol. Initially, the membrane potential of cell 1 and cell 2 was set to the same voltage, usually −40 mV. Thereafter, voltage pulses were administered to cell 1 every 30 s. The pulses lasted 15 s and varied in amplitude and polarity. As shown in Fig. 1A, a V_j pulse of −25 mV led to an I_j signal with slowly growing amplitude; a V_j pulse of −100 mV gave rise to a I_j signal with a pronounced time-dependent decay. The complete analysis of this experiment is illustrated in Fig. 1B. The amplitudes of I_j were determined at the beginning and end of a voltage-clamp pulse (I_j(inst) and I_j(steady state), respectively), the ratios I_j(inst)/V_j and I_j(steady state)/V_j calculated and plotted vs. V_j. The graph of g_j(inst) revealed a horizontal line (○), indicating that, immediately after establishment of a voltage gradient, g_j was insensi-

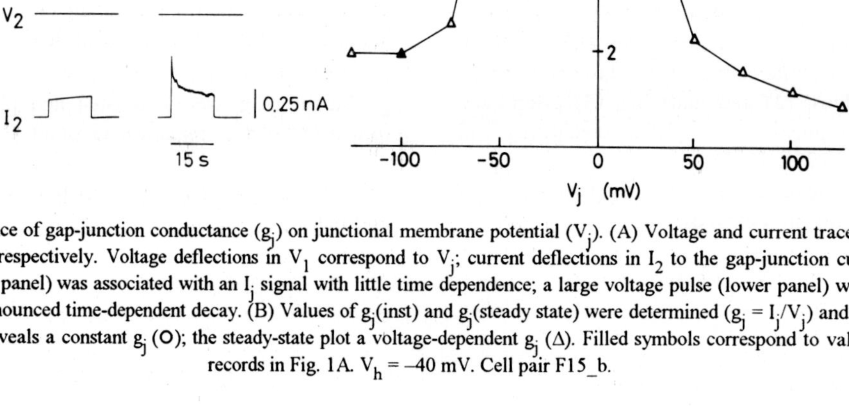

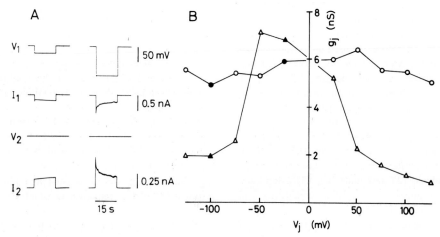

Figure 1. Dependence of gap-junction conductance (g_j) on junctional membrane potential (V_j). (A) Voltage and current traces of cell 1 (V_1, I_1) and cell 2 (V_2, I_2), respectively. Voltage deflections in V_1 correspond to V_j; current deflections in I_2 to the gap-junction current (I_j). A small voltage pulse (upper panel) was associated with an I_j signal with little time dependence; a large voltage pulse (lower panel) was accompanied by an I_j signal with pronounced time-dependent decay. (B) Values of g_j(inst) and g_j(steady state) were determined ($g_j = I_j/V_j$) and plotted vs. V_j. The instantaneous plot reveals a constant g_j (○); the steady-state plot a voltage-dependent g_j (Δ). Filled symbols correspond to values extracted from records in Fig. 1A. V_h = −40 mV. Cell pair F15_b.

tive to V_j. The graph of g_j(steady state) yielded an asymmetrical bell-shaped curve (Δ). Depolarization gave rise to a decrease in g_j. It declined rapidly for V_j values from 0 mV to +50 mV. Above +50 mV, g_j decreased slowly without reaching a steady level. Hyperpolarization led to a biphasic alteration in g_j. Small changes in V_j provoked an increase in g_j, culminating at V_j = –50 mV. Between V_j = –50 mV and –75 mV, g_j decreased steeply; near –100 mV it leveled off. Two features of g_j(steady state) are of particular interest: (1) it is not maximal at V_h = –40 mV; (2) it reaches a steady-state level at negative voltages, but not at positive ones. This suggests that g_j is not controlled by V_j alone, but also by the non-junctional membrane potential, V_m. Dependency of g_j on V_j is

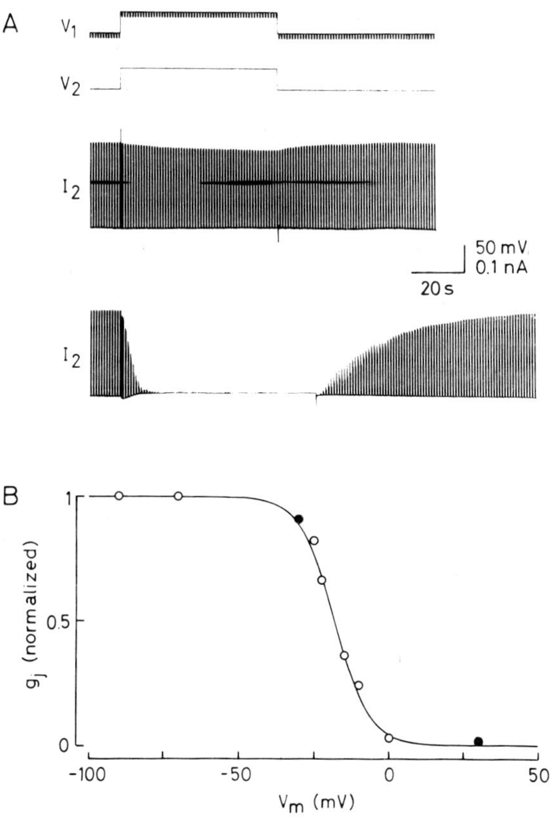

Figure 2. Dependence of gap-junction conductance (g_j) on non-junctional membrane potential (V_m). (A) An identical conditioning pulse was administered simultaneously to cell 1 (V_1) and cell 2 (V_2), while test pulses (–12.5 mV; 500 ms; 1 Hz) were applied repetitively to cell 1 (V_1) The latter gave rise to voltage gradients (V_j; downward deflections in V_1) and current flow across the gap junction (I_j; upward deflections in I_2). Depolarization to –30 mV produced a small decrease in I_j (upper I_2 signal), depolarization to +30 mV led to a large decrease in I_j (lower I_2 signal). (B) Values of g_j(steady state) were determined at the end of a step change in V_m, normalized and plotted vs. V_m. The plot shows a decrease in g_j for values of V_m positive to –40 mV. Curve: best fit of function $1 + \exp[(V - V_h)/k]^{-1}$ to data. V_h = –18 mV; k = 6. Filled symbols correspond to values obtained from records shown in Fig. 2A. V_h = –70 mV. Cell pair F88_d.

a widespread phenomenon. It has been observed in a number of different tissues (for review, see Brink, 1991).

Dependence of g_j on non-junctional membrane potential

The influence of the non-junctional membrane potential (V_m) on the gap-junction conductance was explored utilizing a double-pulse protocol. Starting from a common V_h (–70 mV), identical conditioning pulses were applied concomitantly to cell 1 and cell 2. Small test pulses were then superimposed on cell 1 (–12.5 mV; 1 s; 0.2 Hz) to determine g_j. The conditioning pulses lasted 10 to 30 s, long enough for g_j to reach a steady state. As indicated in Fig. 2A, depolarization of V_m to –30 mV led to a small decrease in I_j (upper I_2 trace), and depolarization to +30 mV to a pronounced decrease (lower I_2 trace). Upon return to V_j, g_j recovered completely. The analysis of these records and others from the same experiment is illustrated in Fig. 2B. Values of g_j(steady state) were determined from I_j prevailing at the end of the conditioning pulses and plotted vs. V_m. As shown, g_j(steady state) remained constant when V_m was changed from –100 mV to –40 mV. However, it decreased when V_m was depolarized towards positive voltages. This suggests that g_j is sensitive to V_m. On average, half-maximal decrease in g_j(steady state) was found at V_m = +21 ± 5 mV (n = 11). In cell pairs with putative gap junctions, the average g_j was 5.0 ± 1.2 nS (n = 41). To avoid interference from V_m, these measurements were performed at a V_h of –50 to –70 mV. Dependence of g_j on V_m is unusual for gap junctions. So far, it has been observed only in few tissues, i.e. squid blastomeres (Spray et al., 1984), salivary glands of *Chironomus* (Rose, 1970; Socolar and Politoff, 1971; Obaid et al., 1983) and *Drosophila* larvae (Verselis et al., 1991).

Modulation of g_j by heptanol

Cell pairs whose g_j was sensitive to V_j and V_m were also characterized pharmacologically. For this purpose they were treated with heptanol, an alkanol previously shown to impair gap junctions reversibly (see, e.g. Rüdisüli and Weingart, 1989). Exposure to 3 mM heptanol, a supramaximal dose, gave rise to complete uncoupling. This finding is consistent with the idea that these preparations possess gap junctions.

Single gap-junction channels

Cell pairs were exposed to 3 mM heptanol to reduce the amplitude of I_j and hence visualize putative single gap-junction channel events. Shortly before complete uncoupling and early during recovery from uncoupling, I_j signals revealed discrete current steps arising from random opening and closing of gap-junction channels. An example is illustrated in Fig. 3A. Application of a transjunctional voltage gradient (V_j) was associated with an I_j signal with repetitive single-channel activity. Fig. 3B shows the amplitude histogram of the current record. The amplitude of the peaks corresponds to the dwell time at a particular current level, distances between the peaks to the size of the current levels. The first peak corresponds to the background fluctuations when V_j was zero ($peak_{background}$), the following peaks to the open channel fluctuations when V_j was −50 mV ($peak_{channel}$). The distances between ($peak_{channel}$) and ($peak_{background}$) are multiples of the unitary current, i_j. The analysis yielded an i_j of 7.4 pA, corresponding to a single-channel conductance, γ_j, of 148 pS. Analyzing 80 records from 5 different cell pairs, we obtained γ_j values ranging from 100 to 250 pS. The large scatter indicates that the data were collected at different levels of V_m (see below), and in the presence or absence of 3 mM heptanol. Our values of γ_j are larger than those found in earthworm, 100 pS (Brink and Fan, 1989) and vertebrates (50–165 pS; for review, see Brink, 1991). Diffusion studies established a cut-off limit for channel permeation of 2.45 kDa in insect salivary glands (for review, see Weingart, 1987) and 0.86 kDa in mammalian heart muscle (for review, see Imanaga, 1987), and thus predict that γ_j is larger in insects than in mammals.

Dependence of single-channel activity on V_j and V_m

The conductance of the gap junction may be expressed as $g_j = N \times p \times \gamma_j$ (N = number of operational channels, p = probability of a channel to be open, γ_j = single-channel conductance). Hence, a change in g_j may arise from a modification of N, p, or γ_j. Analysis of I_j signals with single-channel activity may help to distinguish between these possibilities. This rationale was adopted to explore the mechanisms underlying V_j- and V_m -dependent gating of g_j (Bukauskas and Weingart, 1993). For this purpose, we used cell pairs with weak intercellular coupling, i.e. $g_j < 1.5$ nS, equivalent to the simultaneous operation of few channels only. To explore the mechanism of V_j-dependent gating, V_j pulses of variable amplitude and either polarity were administered to one of the cells and the single-channel activity analyzed. Preliminary experiments of this kind revealed that V_j primarily affects p rather than γ_j. This suggests that V_j may interfere directly with the channel kinetics.

To study the mechanism of V_m-dependent gating, we used the protocol with conditioning pulses and superimposed test pulses (see Fig. 2). Figure 4 illustrates the response of a gap junction to a depolarizing conditioning pulse. The panels in Fig. 4A show selected I_j records. Figs. 4B and 4C depict the graphs of γ_j and the maximal number of channels simultaneously open (n), respectively, plotted vs. time. At V_m = −75 mV, γ_j was 175–225 pS and n was 3–4. After depolarization to −15 mV, both parameters changed dramatically. After a short delay, γ_j and n started to decrease. During the conditioning pulse, γ_j declined to 70–125 pS and n to 1–2. After repolarization, γ_j

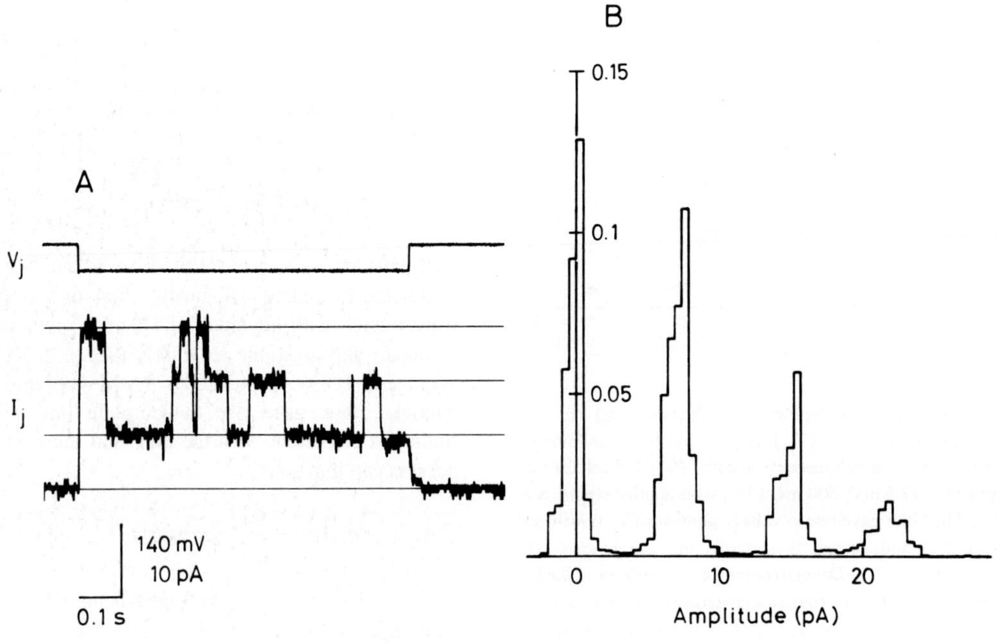

Figure 3. Electrical activity of gap-junction channels visualized by exposure to 3 mM heptanol. (A) Transjunctional voltage gradient (V_j) and associated gap-junction current (I_j). Dashed lines refer to discrete current levels extracted from the analysis. B) Amplitude histogram of current signal, plotting the normalized frequency of occurrences vs. amplitude of I_j. Binwidth = 0.3 pA. The inter-peak intervals revealed a single-channel conductance, c_j, of 148 pS. V_h = −46 mV. Cell pair F17_b.

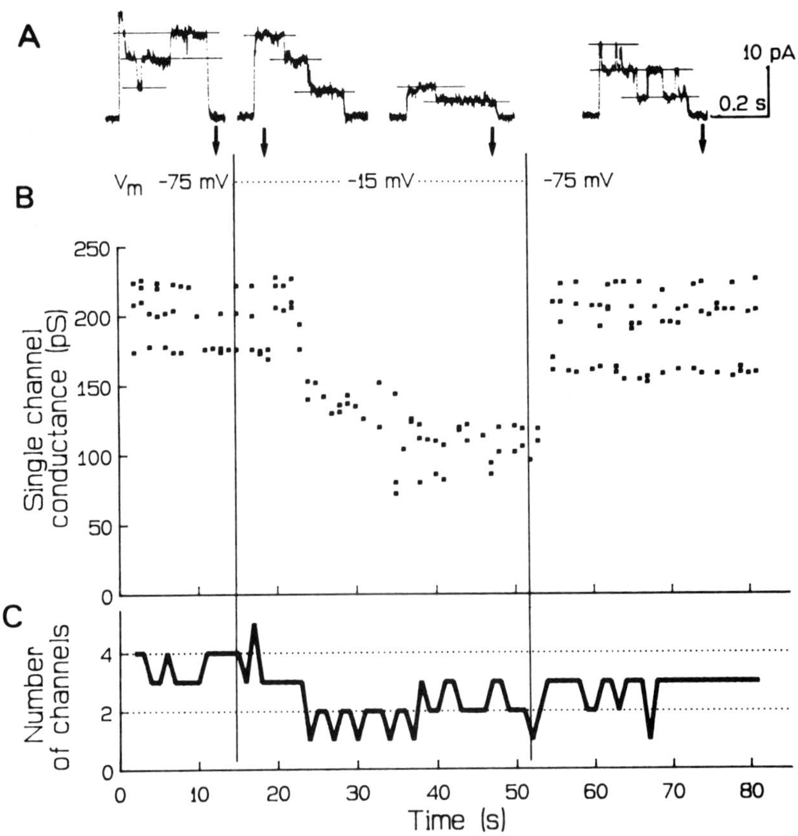

Figure 4. Dependence of γ_j on non-junctional membrane potential (V_m). Test pulses (–25 mV; 300 ms; 1 Hz) were administered repetitively to one cell, superimposed on a conditioning pulse (+60 mV; 37 s) delivered to both cells concomitantly. (A) Selected I_j signals recorded during control (V_m = –75 mV), early and late after administration of conditioning pulses (V_m = –15 mV), and late during re-control (V_m = –75 mV). (B) Plot of single-channel conductance (γ_j) versus time. (C) Plot of maximal number of channels simultaneously open (n) versus time. Arrows indicate the incidence of recording of I_j signals exemplified in A. Cell pair F106_d.

and n recovered quickly, returning to 155–225 pS and 3, respectively. Hyperpolarizing conditioning pulses had no effect on γ_j or n. These findings demonstrate that γ_j is voltage dependent; it was large at negative V_m, and small at positive V_m. We postulate a V_m-dependent gate, possibly located between the two connexons forming a channel. V_m may control this gate via voltage gradient between the intercellular cleft and the channel lumen.

Conclusions

We suggest that junctional membranes sensitive to V_j, V_m, and heptanol possess gap junctions. This interpretation is consistent with previous studies on insect salivary glands (for review, see Weingart, 1987; Verselis et al., 1991). It is also compatible with the finding that these junctions may exhibit I_j signals with discrete current steps attributable to single channels. Based on our experiments, we propose that V_j and V_m act via separate mechanisms. Macroscopic current measurements revealed that V_j- and V_m-dependent regulation of g_j are different with respect to the voltage range and degree of uncoupling. Microscopic current measurements indicate that V_j-dependent gating involves channel kinetics, while V_m-dependent gating involve both γ_j and channel kinetics.

Finally, the question arises whether or not cytoplasmic bridges exist in other types of dividing cells as well, and hence represent widespread structures. In this case, caution is due when interpreting cell-coupling studies. Simple criteria used to identify gap junctions, e.g. the presence of intercellular current flow or tracer diffusion, may not be sufficient. The existence of different types of intercellular structures requires more sophisticated experimental tests.

Acknowledgements

We appreciate the financial support by the Swiss National Science Foundation (grant 31.25732.88).

References

Brink, P.R. (1991) Patch-clamp studies of gap junctions. In: C. Peracchia (Ed.) Biophysics of Gap Junction Channels, CRC Press, Boca Raton, pp. 29–42.

Brink, P.R. and Fan, S. (1989) Patch-clamp recordings from membranes which contain gap-junction channels. Biophys. J. 56, 579–593.

Bukauskas, F., Kempf, C. and Weingart, R. (1991) Electrical properties of gap junctions studied in pairs of insect cells. Experientia 47, A19.

Bukauskas, F. and Weingart, R. (1993) Voltage-dependent conductance of gap-junction single channels, in preparation.

Bukauskas, F., Kempf, C. and Weingart, R. (1992) Electrical coupling between cells of the insect *Aedes albopictus*. J. Physiol., (London) 448, 321–337.

Igarashi, A. (1978) Isolation of a Singh's *Aedes albopictus* cell clone sensitive to Dengue and Chikungunya viruses. Gen. Virol. 40, 531–544.

Imanaga, I. (1987) Cell-to-cell coupling studied by diffusional methods in myocardial cells. Experientia 43, 1080–1083.

Karasiewicz, J. (1981) Electron microscopic studies of cytokinesis in metazoan cells. In: A. M. Zimmerman and A. Forer (Eds.) Mitosis/Cytokinesis. Academic Press, New York, pp. 419–436

Obaid, A.L., Socolar, S.J. and Rose, B. (1983) Cell-to-cell channels with two independently regulated gates in series: Analysis of junctional conductance modulation by membrane potential, calcium, and pH. J. Membr. Biol. 73, 69–89.

Rose, B. (1970) Junctional membrane permeability: Restoration by repolarizing current. Science 169, 607–609.

Rüdisüli, A. and Weingart, R. (1989) Electrical properties of gap-junction channels in guinea-pig ventricular cell pairs revealed by exposure to heptanol. Pflügers Arch. 415, 12–21.

Schwarzmann, G., Wiegandt, H., Rose, B., Zimmerman, A.L., Ben-Haim, D. and Loewenstein, W.R. (1981) Diameter of the cell-to-cell channels as probed with neutral molecules. Science 213, 551–553.

Socolar, S.J. and Politoff, A.L. (1971) Uncoupling cell junctions of a glandular epithelium by depolarizing current. Science 172, 492–494.

Spray, D.C., White, R.L., Campos De Carvalho, A.C., Harris, A.L. and Bennett, M.V.L. (1984) Gating of gap-junction channels. Biophys. J. 45, 219–230.

Verselis, V.K., Bennett, M.V.L. and Bargiello, T.A. (1991) A voltage-dependent gap junction in *Drosophila melanogaster*. Biophys. J. 59, 114–126.

Weingart, R. (1986) Electrical properties of the nexal membrane studied in rat ventricular cell pairs. J. Physiol. (London) 370, 267–284.

Weingart, R. (1987) Cell-to-cell communication in salivary glands. In: W.C. De Mello (Ed.) Cell-to-Cell Communication. Plenum, New York, pp. 269–297.

Zimmerman, A.L. and Rose, B. (1985) Permeability properties of cell-to-cell channels: Kinetics of fluorescent tracer diffusion through a cell junction. J. Membr. Biol. 84, 269–283.

Part VI. Regulation and Biochemistry

J.E. Hall, G.A. Zampighi and R.M. Davis (Eds.)
Progress in Cell Research, Vol. 3

CHAPTER 36

Phosphorylation, intracellular transport, and assembly into gap junctions of connexin43

LINDA S. MUSIL and DANIEL A. GOODENOUGH

Department of Anatomy and Cellular Biology, Harvard Medical School, Boston, MA 02115, USA

Introduction

The assembly of gap junctions is a complex and poorly understood process. As inferred from structural analysis of isolated gap junction preparations (reviewed by Caspar et al., 1988), at least three assembly steps are involved. First, six connex*in* monomers must oligomerize into a connex*on* (hemichannel) in an as yet undefined cellular compartment. Next, a connexon in the plasma membrane of one cell must join in mirror symmetry with a connexon in an opposing cell membrane to form an intercellular channel. Finally, these channels become clustered at cell–cell interfaces into very high density clusters (~10^4 channels/μm^2 of membrane) referred to as gap junctional plaques or maculae. Although both pharmacological (Ito et al., 1974) and, more recently, mutational (Dahl et al., 1991) perturbation of gap junction formation has been reported, in all cases only the endpoint of gap junction assembly, production of junctional plaques capable of mediating intercellular coupling, was assessed, providing little direct information concerning the assembly process itself. A major challenge in the gap junction field is to determine how the various steps in junctional plaque assembly are regulated, any one of which could, in principle, serve as an important control point in gap junction-mediated cell–cell communication. The structural features of the connexin molecule that are required for these interactions have also not been established.

The goal of our studies has been to characterize the assembly intermediates involved in junction formation and to determine how progression from one assembly step to the next is regulated. We have focused our work on Cx43 (Beyer et al., 1987), a widely distributed vertebrate connexin which is believed (among other potential functions) to play a role in the synchronization of myocardial and uterine contractions and to be a component of certain electrical synapses in the brain. The general experimental approach has been to metabolically radiolabel and immunoprecipitate Cx43 from various cell types using affinity-purified anti-Cx43 antibodies directed against a unique region of the cytoplasmic tail domain of Cx43 (residues 252–271). Figure 1A, lanes 1 and 2 summarize our results obtained with cell types that are well coupled and form easily visualized gap junctional plaques, in this case normal rat kidney (NRK) cells. When NRK cell cultures are metabolically labeled with [^{35}S]-methionine for 5 h, lysed, and immunoprecipitated with anti-Cx43 (252–271) antibodies, three labeled Cx43-related bands (M_r = 42 kDa, 44 kDa, and 46 kDa) are recovered (lane 1). The two slower-migrating forms of Cx43, but not the 42 kDa species, can also be metabolically labeled with [^{32}P]-orthophosphate (lane 2). Phosphorylation occurs exclusively on serine residues under basal conditions (Musil et al., 1990b). Pulse-chase analysis (not shown) reveals that the 42-kDa band represents newly synthesized Cx43, which we refer to as Cx43-NP (**n**ot **p**hosphorylated). With increasing chase times, Cx43-NP is converted first to the 44-kDa and then to the 46-kDa phosphorylated species, which we have designated Cx43-P_1 and Cx43-P_2, respectively. Dephosphorylation experiments demonstrate that this post-translational shift in the apparent molecular weight of Cx43 is due exclusively to the addition of phosphate onto the Cx43 molecule (Musil et al., 1990a, 1990b). Serine phosphorylation of Cx43 to species with multiple electrophoretic mobilities has also been reported by others in a variety of communication-competent cell types (Crow et al., 1990; Laird et al., 1991; Filson et al., 1990; Swenson et al., 1990).

The first clue that serine phosphorylation of Cx43 may be of functional significance came from examination of the biosynthesis of Cx43 in cell lines that are severely deficient (S180 cells) (Furshpan and Potter, 1968; Mege et al., 1988) or completely lacking (L929 cells) (Larson et al., 1990) in morphologically or physiologically recognizable gap junctions (communication-deficient cells). Immunoprecipitation with anti-Cx43 (252–271) antibodies demonstrated that both of these cell lines constitutively synthesize Cx43, which is as metabolically stable as it is in NRK cells. However, neither S180 (Fig. 1B, lanes 1 and 2) nor L929 cells (not shown) significantly process Cx43-NP to the P_2 form. Metabolic labeling of other cellular proteins with [^{32}P] appears to be normal, suggest-

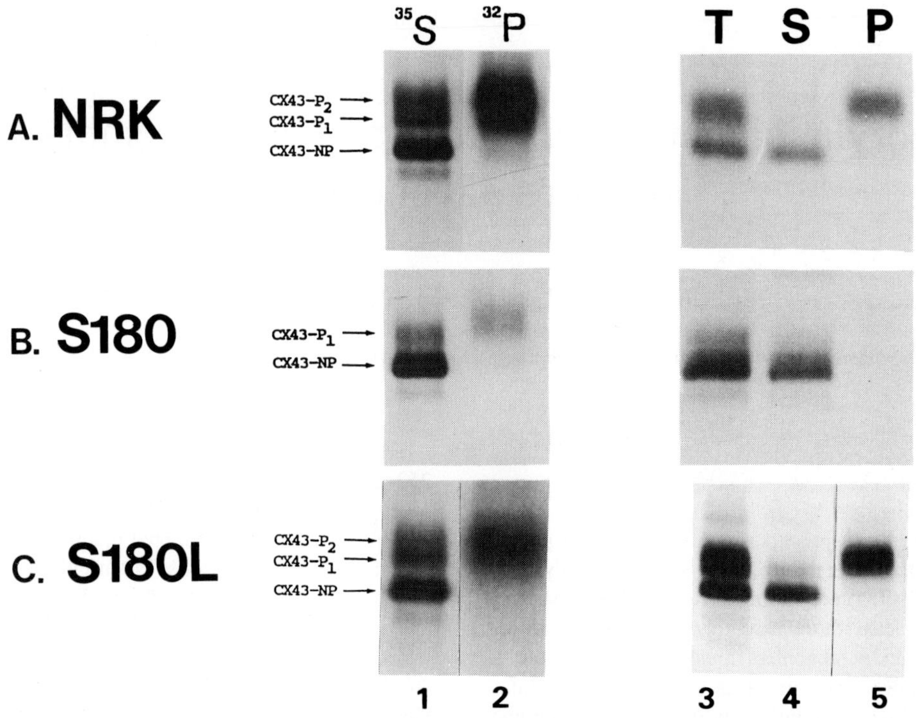

Figure 1. Solubility of phosphorylated and non-phosphorylated forms of Cx43 in Triton X-100. Confluent cultures of communication-competent NRK (panel A) or S180L (panel C) cells or communication-deficient S180 cells (panel B) were metabolically labeled with either [^{35}S]met (lanes 1, 3, 4, and 5) or [^{32}P]O_4 (lane 2) for 5 h. In lanes 1 and 2, the cells were immediately lysed and immunoprecipitated with affinity-purified anti-Cx43 (252–271) antibodies. Lanes 3–5 are from a separate experiment in which cells were homogenized and incubated with 1% Triton X-100 in PBS for 30 min at 4°C. One-half of the Triton-treated lysate was reserved at 4°C (= total cellular lysate) whereas the remainder was subjected to centrifugation at 100,000 × g for 50 min. Cx43 was then immunoprecipitated from equal amounts of the total cellular lysate (T, lane 3), Triton-soluble supernatant (S, lane 4), or Triton-insoluble pellet (P, lane 5) fractions.

ing a specific defect in Cx43 phosphorylation in these cells (Musil et al., 1990b).

S180 and L929 cells lack detectable cell–cell adhesion molecules (CAMs) (Mege et al., 1988, Nagafuchi et al., 1987). Mege et al. (1988) have demonstrated that transfection of S180 cells with a cDNA encoding L-CAM corrects the intercellular adhesion defect in these cells and results in the formation of large, functional gap junctional plaques. We have determined that these L-CAM-expressing S180 cells (S180L cells) phosphorylate Cx43 to the P_2 form in a manner indistinguishable from NRK cells (Fig. 1C, lanes 1 and 2; see also Musil et al., 1990b). Thus, conversion of S180 cells to a communication-competent phenotype results in processing of Cx43 to the P_2 form. Conversely, disruption of gap junctional communication in either S180L cells (with anti-L-CAM Fab fragments) or NRK cells (with the uncoupling agent heptanol) abolishes processing of Cx43 to the P_2 form (Musil et al., 1990b).

The studies summarized above have established a strong correlation between the ability of cells to phosphorylate Cx43 to the P_2 form and to produce functional gap junctional plaques. Conversion to the P_2 form could, in theory, be involved in such processes as Cx43 oligomerization, intracellular transport, or gating of the junctional channel. In order to evaluate these possibilities, we have developed biochemical assays for transport of Cx43 to the plasma membrane and for subsequent assembly of Cx43 into gap junctional plaques, and have used these techniques to further examine the relationship between Cx43 phosphorylation, gap junction assembly, and cell–cell communication.

Assembly of Cx43 into gap junctional plaques

Gap junctions are notoriously insoluble structures, with their relative resistance to the ionic detergent sarcosine forming the basis for several standard gap junction isolation procedures (Goodenough and Stoeckenius, 1972; Kensler and Goodenough, 1980; Manjunath et al., 1984). Since detergent insolubility is rare among normal (i.e., non-mutant) nascent membrane proteins, it seemed likely that connexins are synthesized in a detergent-soluble state and acquire resistance to such agents post-translationally. To determine when in the lifetime of the Cx43 molecule detergent insolubility is attained, NRK cells were metabolically labeled with [^{35}S]met for 5 h, homogenized by repeated passage through a 21 g needle, and the resulting cell lysates incubated under isotonic conditions with 1%

Triton X-100 for 30 min at 4°C. One-half of the lysate served as the total Cx43 sample (Fig. 1A, lane 3) whereas the remainder was subjected to centrifugation at 100,000 × g for 50 min to separate the Triton-soluble from the -insoluble material. Immunoprecipitation of Cx43 from the resulting supernatant fraction revealed that 80–90% of [^{35}S]met-Cx43-NP was Triton soluble (Fig. 1A, lane 4). In contrast, Cx43-P_2 (and most Cx43-P_1) was recovered in the insoluble pellet fraction (Fig. 1A, lane 5). Similar results were obtained with all concentrations of Triton tested (up to 3.0%) that were above the critical micellar concentration, or if the solubilization time was extended from 30 min to 24 h. Pulse-chase analysis (not shown) revealed that newly synthesized [^{35}S]met-Cx43-NP was totally Triton soluble and acquired Triton insolubility as it became phosphorylated. These results indicate a strong correlation between processing of Cx43 to the P_2 form and the acquisition of insolubility in Triton X-100. Whether these two events normally occur in rapid succession or simultaneously is not, however, known.

The relationship between the phosphorylation state of Cx43 and its solubility in Triton X-100 held true for other cell types as well. For example, 80–90% of the Cx43-NP in communication-competent S180L cells was Triton-soluble whereas Cx43-P_2 was insoluble (Fig. 1C, lanes 3–5). Similar results were obtained with Cx43 from whole embryonic chick lenses, indicating that differential detergent solubility of phosphorylated and non-phosphorylated forms of Cx43 was not confined to tissue culture cells (not shown). Furthermore, communication-deficient S180 (Fig. 1B, lanes 3–5) and L929 cells (Musil and Goodenough, 1991), which do not process Cx43 to the P_2 form, also lack detectable Triton-insoluble Cx43. This last observation rules out the possibility that Cx43 acquires resistance to Triton over time regardless of whether it becomes phosphorylated or not.

The Triton insolubility of Cx43-P_2 was exploited to determine the intracellular distribution of the various forms of Cx43 using an in situ extraction procedure (Fig. 2). Monolayer cultures of NRK cells were extracted by addition of 1% Triton-containing buffer directly to the tissue culture dish. After a 30 min incubation, the cultures were carefully rinsed to remove solubilized cellular components, and the Cx43 remaining with the extracted monolayer was analyzed either biochemically by immunoprecipitation or morphologically by immunofluorescence using the anti-Cx43 antibodies. As shown in the inset, ~80% of metabolically labeled cellular [^{35}S]met-Cx43-NP was removed by this extraction procedure; This is comparable to the fraction of Cx43-NP that was solubilized by Triton from homogenized NRK cell lysates (see Fig. 1A, lanes 3–5). In contrast, Triton-insoluble Cx43-P_2 was quantitatively recovered with the extracted monolayer. Immunofluorescent localization of Cx43 in NRK monolayers mock-extracted in the absence of Triton showed typical punctate gap junctional staining as well as specific intracellular staining of the Golgi region, as expected for this cell type (Musil et al., 1990b) (Fig. 2A). In situ extraction of these cells with Triton abolished the Golgi-like signal but had no discernible effect on the punctate intercellular staining (Fig. 2B), which thin section electron microscopy confirmed represented actual gap junctional plaques (not shown). Taken together, these results indicate that Triton-insoluble Cx43, the overwhelming majority of which is Cx43-P_2, is localized primarily in cell-surface gap junctional plaques. Conversely, Triton-soluble Cx43-NP appears to be predominantly intracellular, with the caveat that diffusely distributed Cx43 would not be detectable by morphological means.

Transport of Cx43 to the cell surface

Although the results depicted in Fig. 2 demonstrated that Cx43-P_2 accumulated in the plasma membrane in junctional plaques, the cellular location for phosphorylation of Cx43-NP to the P_2 form was not established. Two possibilities existed: (1) Cx43 was converted to Cx43-P_2 inside the cell and was then rapidly transported to the cell surface, in which case phosphorylation might serve as an intracellular transport signal targeting Cx43 to the plasma membrane, or (2) processing to the P_2 form occurred after arrival of Cx43 on the cell surface. To distinguish between these two scenarios, we developed a biochemical assay to selectively label and monitor Cx43 on the plasma membrane, based on the technique of cell-surface biotinylation (Le Bivic et al., 1989 and 1990b). Briefly, cell monolayers were metabolically labeled with [^{35}S]met and then incubated at 4°C with the membrane-impermeant protein biotinylating reagent NHS-LC-biotin, which in intact cells reacts covalently with primary amine groups (mainly on lysine residues) located in the extracellular domains of plasma-membrane proteins (Sargiacomo et al., 1989). After 30 min, the reaction was quenched with excess glycine, the cells lysed in SDS, and Cx43 immunoprecipitated using our standard protocols. A fraction of the immunoprecipitated Cx43 was reserved as a sample of total cellular Cx43, and the remainder subjected to a second round of precipitation with avidin-agarose to selectively recover only cell-surface biotinylated Cx43 molecules. Both the total cellular Cx43 and the cell-surface biotinylated Cx43 were then analyzed on SDS-PAGE.

When NRK cells were labeled with [^{35}S]-met for 5 h and then subjected to this procedure, two distinct biotinylated [^{35}S]met-Cx43 species were obtained (Fig. 3A, lane 3). The higher mobility form co-migrated with total cellular Cx43-NP (Fig. 3A, lane 2) and was quantitatively solubilized by 1% Triton X-100 (Fig. 3B, lane 2). The other biotin-labeled band corresponded to phosphorylated Cx43 and was resistant to Triton (Fig. 3B, lane 3). We refer to the latter species as simply Cx43-P since Cx43-P_1 was not well resolved from Cx43-P_2 after the double-precipitation procedure. Control experiments demonstrated that recovery of either Cx43-NP or Cx43-P was completely dependent on both protein biotinylation (Fig. 3A, lane 4) and on specific immunoprecipitation of Cx43 (not shown), as expected. In addition, virtually no surface-labeled [^{35}S]met-Cx43 was detectable when NRK cells were biotinylated immediately after a 30 min pulse with [^{35}S]-met (Fig. 4A, lane marked with an open arrow) whereas high levels of biotinylated [^{35}S]met-Cx43 were recovered if the cells were chased for an hour prior to cell-surface labeling (Fig. 4A, lane 1). Thus, newly synthesized Cx43 molecules still present in intracellular compartments were not accessible to the biotinylating reagent, indicating that biotinylation was indeed confined to the cell surface.

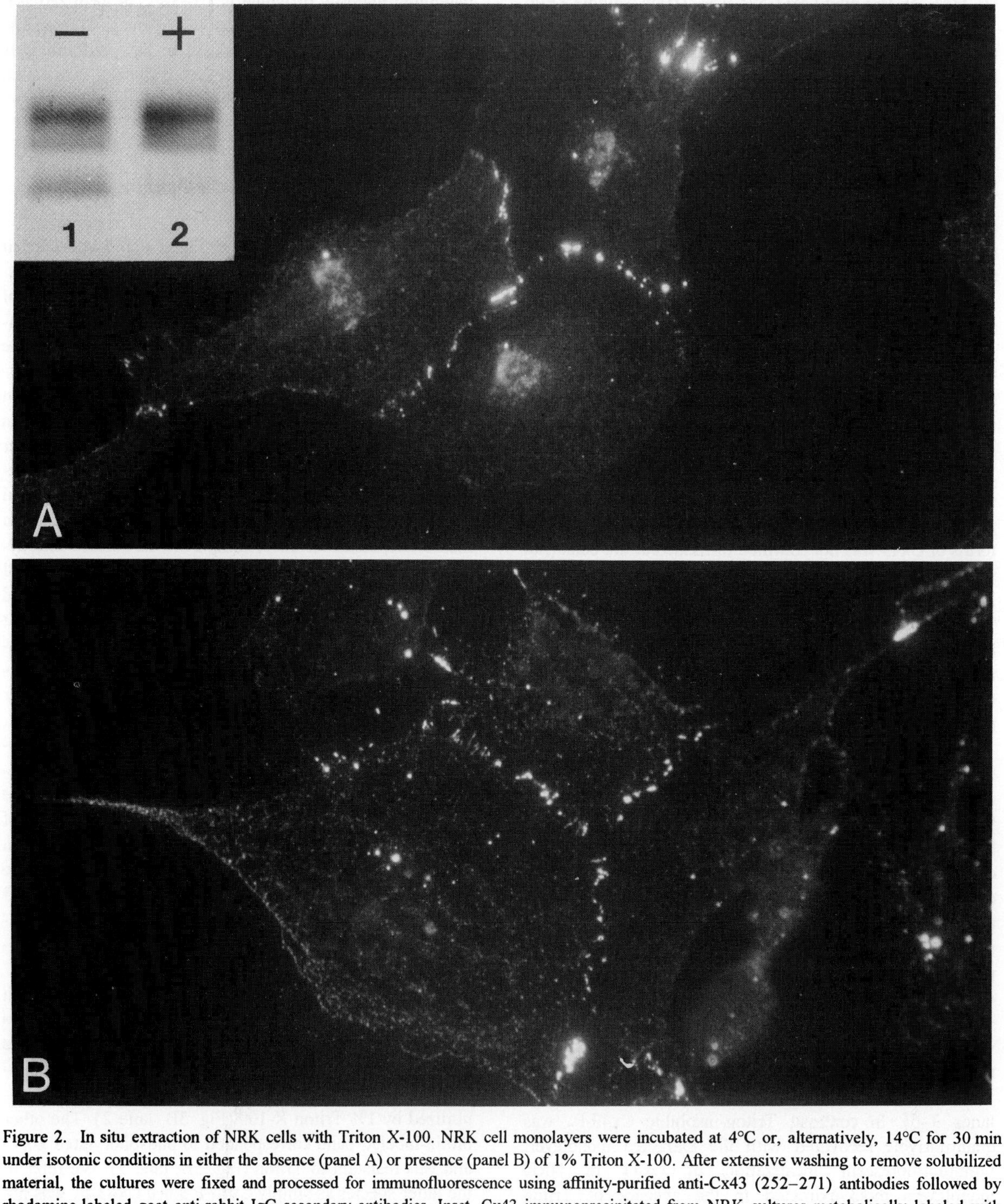

Figure 2. In situ extraction of NRK cells with Triton X-100. NRK cell monolayers were incubated at 4°C or, alternatively, 14°C for 30 min under isotonic conditions in either the absence (panel A) or presence (panel B) of 1% Triton X-100. After extensive washing to remove solubilized material, the cultures were fixed and processed for immunofluorescence using affinity-purified anti-Cx43 (252–271) antibodies followed by rhodamine-labeled goat anti-rabbit IgG secondary antibodies. Inset, Cx43 immunoprecipitated from NRK cultures metabolically labeled with [^{35}S]met for 5 h and then extracted either with (lane 2) or without (lane 1) 1% Triton at 14°C, exactly as described for unlabeled cells. Note that [^{35}S]met-Cx43-P_2 is resistant to Triton solubilization and appears to be localized to brightly staining maculae at intercellular interfaces.

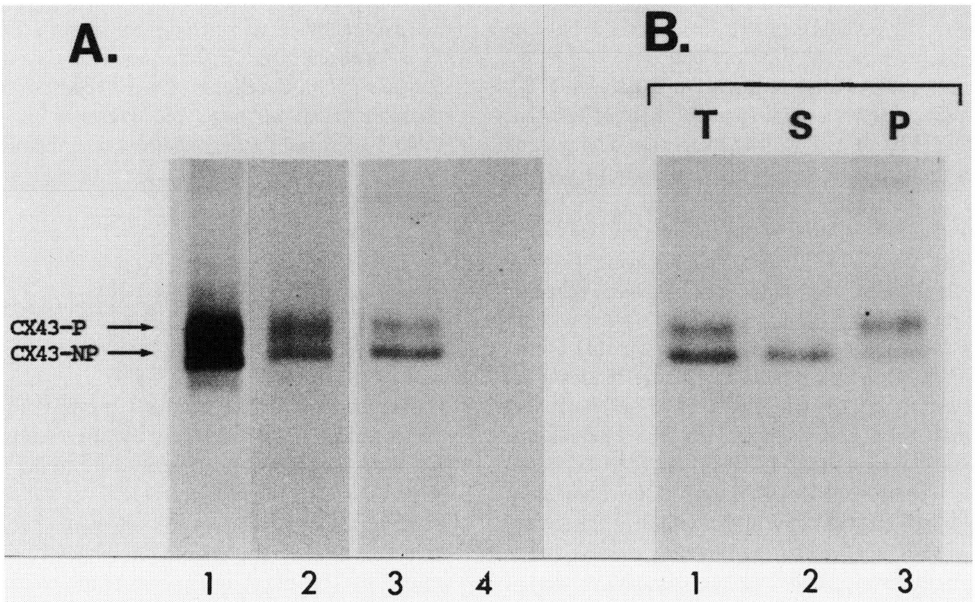

Figure 3. Cell surface biotinylation of Cx43 in intact NRK cells. Confluent 60 mm NRK cell monolayers were labeled with [^{35}S]met at 37°C for 5 h and subjected to cell-surface biotinylation at 4°C. After termination of the reaction with excess glycine, the cells were either lysed directly into SDS (Panel A) or solubilized in 1% Triton X-100 in PBS for 30 min at 4°C prior to centrifugation at 100,000 × g for 50 min (Panel B). Panel A: lane 1, total cellular [^{35}S]met-Cx43 immunoprecipitated from 1/15 of a biotinylated cell culture. Lane 2, shorter exposure of lane 1. Lane 3: biotinylated Cx43 recovered from the remaining 14/15 of the culture by sequential precipitation with affinity purified anti-Cx43 (252–271) antibodies followed by avidin-agarose. Same exposure as lane 1. Lane 4: same as lane 3, except that NHS-LC-biotin was quenched with 15 mM glycine prior to the biotinylation reaction. Panel B: Cell-surface biotinylated Cx43 recovered from equal amounts of the total cellular lysate (lane 1), Triton-soluble supernatant (lane 2), or Triton-insoluble pellet (lane 3) fractions.

The presence of Cx43-NP on the plasma membrane suggested that phosphorylation to the P_1 and P_2 forms occurred after arrival of Cx43 on the cell surface. This possibility was confirmed by an experiment shown in Fig. 4. NRK cells were pulsed with [^{35}S]met for 30 min and then chased for 1 h at 37°C to allow transport of newly synthesized Cx43 to the plasma membrane. The cultures were then biotinylated at 4°C, after which the reaction was quenched with excess glycine and the cells incubated at 37°C for 0–3 h prior to cell lysis and analysis of biotin-labeled Cx43. Cell-surface biotinylation has been shown not to interfere with subsequent processing of plasma-membrane proteins in other systems (Matter et al., 1990; Le Bivic et al., 1990a and 1990b), and so it was not unreasonable to expect that Cx43 phosphorylation and assembly would proceed as usual after biotinylation. In the absence of a chase period after biotinylation, almost all (80–90%) of the surface-labeled [^{35}S]met-Cx43 recovered was in the Cx43-NP form (Fig. 4A, lane 1) and was soluble in Triton X-100 (Fig. 4B, lane 2). When cell cultures were chased for 1 h after biotinylation, the amount of biotin-conjugated [^{35}S]met-Cx43 decreased ~40%, reflecting the 2–2.5 h half-life of total Cx43 (Musil et al., 1990b) (Fig. 4A, lane 2). Of the surface-labeled [^{35}S]met-Cx43 remaining, ~50% had been converted to the Cx43-P form. By 3 h of chase, all of the biotinylated [^{35}S]met-Cx43 still detectable was in the Cx43-P form (Fig. 4A, lane 4), which was Triton-insoluble (Fig. 4B, lane 6) and therefore likely to be part of a gap junctional plaque. These observations demonstrate that cell-surface Cx43-NP serves as a precursor to Cx43-P. Transport of Cx43 to the plasma membrane in the NP form appears to be part of the normal pathway leading to Cx43 phosphorylation and gap junctional plaque formation in communication-competent cells.

Is Cx43 also transported to the surface of cells which do not phosphorylate it to the P_2 form? To answer this question, monolayers of communication-deficient S180 or L929 cells were metabolically labeled for 5 h with [^{35}S]-met and then biotinylated at 4°C as described for NRK cells (Fig. 5). A biotinylated protein that co-migrated with total cellular Cx43-NP was recovered from both cell types (lane 2), the amount of which (relative to total cellular [^{35}S]met-Cx43) ranged from 0.75 to 1.5 times as much as in NRK cells. Control experiments identical to those conducted with communication-competent cells confirmed that this band represented Cx43-NP on the cell surface (not shown). Since S180 and L929 cells lack gap junctions, these results provide biochemical proof for extra-junctional channel precursors in the plasma membrane, the existence of which has long been hypothesized but has been very difficult to demonstrate directly.

A drawback of the cell-surface biotinylation technique is that it is not quantitative. Published reports have estimated that the efficiency with which plasma membrane proteins become biotinylated varies from ~9%–50%, depending on the particular protein examined (Le Bivic et al., 1989; Matter et al., 1990). Since only ~1% of the total cellular Cx43 radiolabeled during a 5-h pulse with [^{35}S]-met was recovered with avidin-agarose after cell-surface biotinylation (Figs. 3 and 5), it is likely that biotinylation of Cx43 is inefficient as well. This appears to be particularly true of junctional plaque-associated (i.e., Triton-insoluble) Cx43-P, which is proportionately less well biotinylated than Cx43-NP (see Fig. 3A, lane 1 vs. 3; 0.4% of total cellular [^{35}S]met-Cx43-P was recovered after cell surface biotinylation vs. 0.95% of total cellular [^{35}S]met-Cx43-NP). Although the reason for this under-representation of phosphorylated Cx43 after biotinylation is unknown, it is possible that Cx43-P in junctional plaques may be a poor substrate for the biotinylation reagent due to restricted diffusion of NHS-LC-biotin (molecular weight = 556 Da; 2.24 nm) in the 2 nm intra-junctional "gap" or to changes in the tertiary or quaternary structure of Cx43 that render the protein's reactive lysine residues inaccessible after plaque formation (for a complete discussion, see Musil and Goodenough, 1991). In light of these quantitative limitations, we cannot calculate the $t_{1/2}$ of transport of Cx43 to the plasma membrane nor directly determine the fraction of total Cx43 present on the cell surface at a given time. Our demonstration that phosphorylation of Cx43 to the P_2 form occurs after transport to the plasma membrane and that this species accumulates in cell-surface plaques does, however, allow us to conclude that most (if not all) Cx43-P_2 is localized to the plasma membrane. Conversely, the majority of intracellular Cx43 appears to be non-phosphorylated, although the presence of internalized Cx43-P_2 destined for degradation cannot be ruled out.

Discussion

Potential role of Cx43 phosphorylation in gap junction formation and function

On the basis of our current and previous (Musil et al., 1990a and 1990b) data, we can eliminate certain potential functions for phosphorylation of Cx43 to the P_1 or the P_2 form. First, the presence of Cx43-NP on the plasma membrane of both communication-competent (Fig. 3) and -deficient (Fig. 5) cells rules out an obligatory role for processing to either the P_1 or P_2 form in the transport of Cx43 to the plasma membrane. Second, phosphorylation does not appreciably affect the metabolic stability of Cx43: the $t_{1/2}$ of degradation of Cx43 is equivalent in cells that process Cx43 to both the P_1 and P_2 forms (NRK cells) and those that do not (L929 cells) (Musil et al., 1990b). Lastly, phosphorylation of Cx43 is not exclusively associated with actively communicating cells. Blocking junctional permeability in NRK cells by brief exposure to

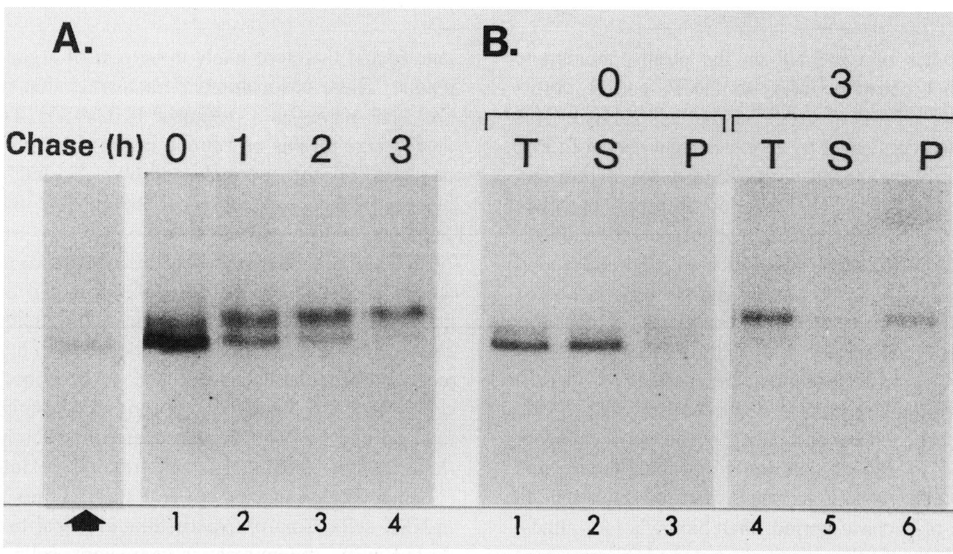

Figure 4. Processing of cell surface Cx43-NP to Triton-insoluble Cx43-P. NRK cell cultures were incubated for 30 min with [^{35}S]met to metabolically label newly synthesized intracellular Cx43. The cells were then chased in the presence of an excess of unlabeled met for 1 h at 37°C to permit transport of [^{35}S]met-Cx43 to the plasma membrane. The labeled monolayers were cell-surface biotinylated at 4°C, after which the reaction was quenched with excess glycine and the cells chased at 37°C for 0–3 h to follow the fate of biotinylated Cx43. Panel A, lanes 1–4: at the end of the specified chase period, the cells were lysed directly in SDS, and biotinylated Cx43 was recovered by sequential precipitation with anti-Cx43 antibodies followed by avidin-agarose. Panel B: cells were subjected to the standard Triton solubility assay (1% Triton X-100 in PBS, 30 min, 4°C) either immediately after biotinylation (lanes 1–3) or after a 3 h chase (lanes 4–6). Cell-surface biotinylated Cx43 was recovered from equal amounts of the total cellular lysate (lanes marked T), Triton-soluble supernatant (lanes marked S), or Triton-insoluble pellet (lanes marked P) fractions. Panel A, lane marked with an open arrow: cell surface biotinylated Cx43 recovered immediately after the 30-min pulse with [^{35}S]-met (no chase), demonstrating the inability of the biotinylation reagent to react with newly synthesized intracellular Cx43.

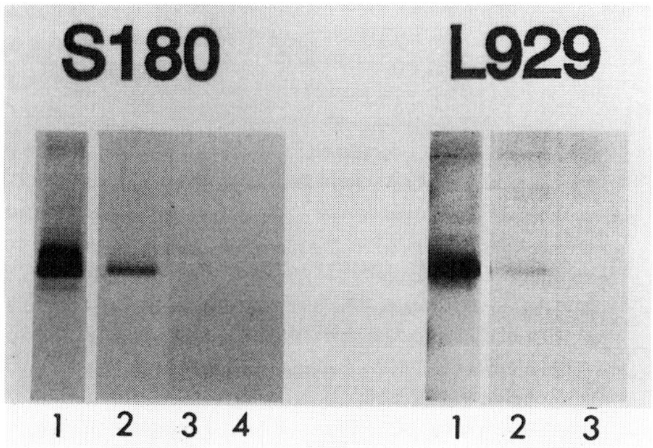

Figure 5. Expression of Cx43 on the surface of communication-deficient S180 and L929 cells. Monolayers (60 mm) of S180 (left, lanes 1–4) or L929 (right, lanes 1–3) cells were metabolically labeled at 37°C with [^{35}S]met for 5 h prior to cell-surface biotinylation at 4°C. The cultures were lysed in SDS and total cellular Cx43 immunoprecipitated, a fraction of which was then incubated with avidin-agarose to selectively recover biotinylated Cx43 molecules. Lane 1: total cellular Cx43 immunoprecipitated from 1/15 of a biotinylated cell culture. Lane 2: biotinylated Cx43 recovered from the remaining 14/15 of the corresponding cell culture. Lane 3: same as lane 2, except that NHS-LC-biotin was quenched with 15 mM glycine prior to the biotinylation reaction. Lane 4: same as lane 2, except that Cx43 immunoprecipitation was conducted in the presence of an excess of competing Cx43(252–271) peptide.

heptanol or to 100% CO_2 did not result in detectable dephosphorylation of Cx43-P_2 nor in its conversion to a Triton-soluble form (Musil and Goodenough, 1991). Given that Triton-insoluble Cx43-P_2 is largely assembled into gap junctional plaques and is therefore likely to be involved in cell-cell communication, this lack of obvious modulation upon uncoupling suggests that dephosphorylation of Cx43-P_2 is not the mechanism whereby the gap junctional channel is closed. In the absence of specific knowledge concerning the properties or the amount of junctional Cx43 that actually mediates intercellular coupling in these cells we cannot, however, discount the possibility that reversible phosphorylation events (perhaps ones that do not result in a shift in the electrophoretic mobility of Cx43) are involved in some aspect of channel gating.

What functional role, then, could phosphorylation of Cx43 to the P_2 form play? Under all conditions examined, terminal phosphorylation of Cx43 was tightly linked to acquisition of Triton insolubility and therefore to accumulation of Cx43 into gap junctional plaques. Although the exact kinetic relationship between these three events is unclear, phosphorylation to the Cx43-P_2 form could thus potentially be involved in gap junctional plaque establishment and/or maintenance. One possibility is that phosphorylation of Cx43 may serve to promote association of Cx43 monomers or connexons into higher-order homo-oligomers, reminiscent of the role of phosphorylation in the self-assembly of the SV40 large T antigen (Montenarh and Muller, 1987). In this case, phosphorylation could be permissive for the *cis* association of connexins within the plane of the membrane and/or the *trans* association necessary to assemble the intercellular channel.

Alternatively, phosphorylation to the Cx43-P_2 form may be necessary for complex formation between Cx43 and other, as yet unknown, accessory proteins involved in gap junctional plaque establishment or stabilization. Creation of novel binding sites as a result of protein phosphorylation has been well established for several plasma membrane receptors (reviewed by Ullrich and Schlessinger, 1990). In either case, it is conceivable that phosphorylation of Cx43 to the P_2 form is not necessary for the formation of individual cell–cell channels but is involved only in their subsequent assembly into, and/or

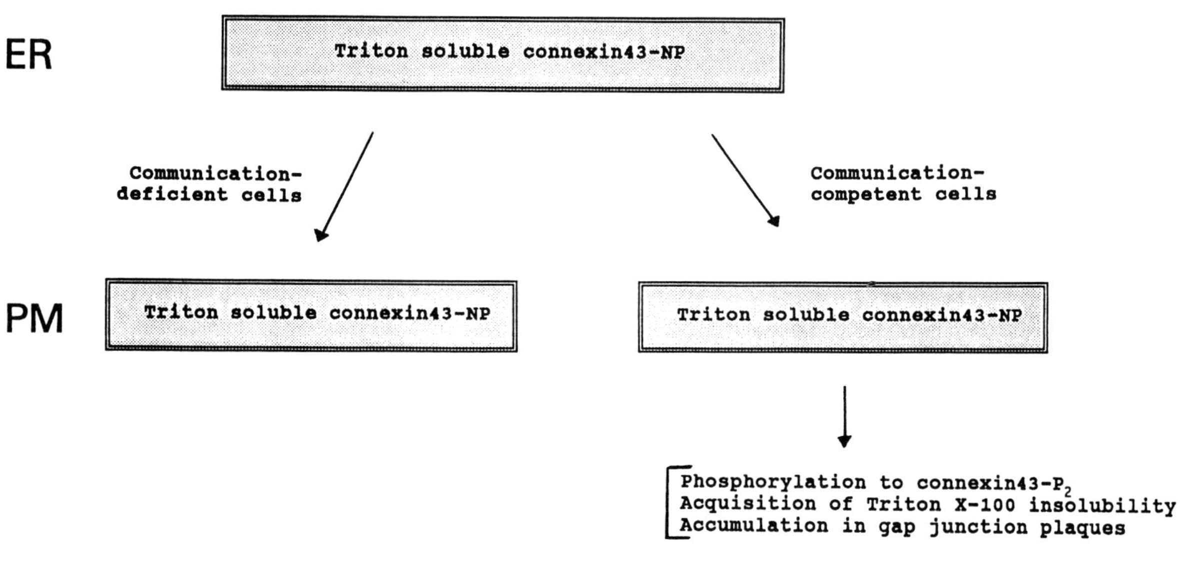

Figure 6. Summary of Cx43 post-translational processing and assembly into gap junctional plaques.

maintenance in, large, high-density gap junctional plaques. Determination of the role of Cx43 phosphorylation will require careful evaluation of the assembly state of phosphorylation-deficient site-directed Cx43 mutants in addition to assessment of their ability to mediate cell–cell communication.

Conclusions

Our results support the following model of gap junctional plaque assembly (Fig. 6). First, Cx43 is translated in both communication-competent and -deficient cell lines in the Triton-soluble Cx43-NP form. This species is then transported to the plasma membrane, independent of whether or not it will be incorporated into a gap junction. Since both S180 and L929 cells transport Cx43 to the plasma membrane despite their lack of cell–cell adhesion molecules, specific intercellular association cannot play an obligatory role in the cell-surface expression of Cx43. Only in communication-competent cells is plasma-membrane Cx43-NP then phosphorylated to the P_2 form in a process that is temporally associated with acquisition of Triton insolubility and with accumulation of Cx43 into gap junctional plaques. Current studies are directed towards determining the cause-and-effect relationship between these events and their role in the establishment and maintenance of cell–cell communication.

References

Beyer, E.C., Paul, D.L. and Goodenough, D.A. (1987) Connexin43: a protein from rat heart homologous to a gap junction protein from liver. J. Cell Biol. 105, 2621–2629.

Caspar, D.L.D., Sosinsky, G.E., Tibbitts, T.T., Phillips, W.C. and Goodenough, D.A. (1988) Gap junction structure. In: E.L. Hertzberg and R.G. Johnson (Eds.), Gap Junctions, Alan R. Liss, New York, pp. 117–133.

Crow, D.S., Beyer, E.C., Paul, D.L., Kobe, S.S. and Lau, A.F. (1990) Phosphorylation of connexin43 gap junction protein in uninfected and Rous sarcoma virus–transformed mammalian fibroblasts. Mol. Cell Biol 10, 1754–1763.

Dahl, G., Levine, E., Rabadan-Diehl, C. and Werner, R. (1991) Cell/cell channel formation involves disulfide exchange. Eur. J. Biochem. 197, 141–144.

Filson, A.J., Azarnia, R., Beyer, E.C., Loewenstein, W.R. and Brugge, J.S. (1990) Tyrosine phosphorylation of a gap junction protein correlates with inhibition of cell-to-cell communication. Cell Growth Diff. 1, 661–668.

Furshpan, E.J. and Potter, D.D. (1968) Low-resistance junction between cells in embryos and tissue culture. In: A.A. Moscona and A. Monroy (Eds.), Current Topics in Developmental Biology (vol. 3), Academic Press, New York, pp. 95–127.

Goodenough, D.A. and Stoeckenius, W. (1972) The isolation of mouse hepatocyte gap junctions. Preliminary chemical characterization and x-ray diffraction. J. Cell Biol. 54, 646–656.

Ito, S., Sato, E. and Loewenstein, W.R. (1974) Studies on the formation of a permeable cell membrane junction. I. Coupling under various conditions of membrane contact. Effects of colchicine, cytochalasin B, dinitrophenol. J. Membr. Biol. 19, 305–337.

Kensler, R.W. and Goodenough, D.A. (1980) Isolation of mouse myocardial gap junctions. J. Cell Biol. 86, 755–764.

Laird, D.W., Puranam, K.L. and Revel, J.P. (1991) Turnover and phosphorylation dynamics of connexin43 gap junction protein in cultured cardiac myocytes. Biochem. J. 273, 67–72.

Larson, D.M., Haudenschild, C.C. and Beyer, E.C. (1990) Gap junction messenger RNA expression by vascular wall cells. Circ. Res. 66, 1074–1080.

Le Bivic, A., Real, F.X. and Rodriguez-Boulan, E. (1989) Vectorial targeting of apical and basolateral plasma membrane proteins in a human adenocarcinoma epithelial cell line. Proc. Natl. Acad. Sci. USA. 86, 9313–9317.

Le Bivic, A., Quaroni, A., Nichols, B. and Rodriguez-Boulan, E. (1990a) Biogenetic pathways of plasma membrane proteins in Caco-2, a human intestinal epithelial cell line. J. Cell Biol. 111, 1351–1362.

Le Bivic, A., Sambuy, Y., Mostov, K. and Rodriguez-Boulan., E. (1990b) Vectorial targeting of an endogenous apical membrane sialoglycoprotein and uvomorulin in MDCK cells. J. Cell Biol. 110, 1533–1539.

Manjunath, C.K., Goings, G.E. and Page, E. (1984) Cytoplasmic surface and intramembrane components of rat heart gap junctional proteins. Am. J. Physiol. 246, H865–H875.

Matter, K., Brauchbar, M., Bucher, K. and Hauri., H.-P. (1990) Sorting of endogenous plasma membrane proteins occurs from two sites in cultured human intestinal epithelial cells (Caco-2). Cell 60, 429–437.

Mege, R.M., Matsuzaki, F., Gallin, W.J., Goldberg, J.I., Cunningham, B.A. and Edelman, G.M. (1988) Construction of epithelioid sheets by transfection of mouse sarcoma cells with cDNAs for chicken cell adhesion molecules. Proc. Natl. Acad. Sci. USA. 85, 7274–7278.

Montenarh, M. and Muller, D. (1987) The phosphorylation at Thr 124 of simian virus 40 large T antigen is crucial for its oligomerization. FEBS Lett. 221, 199–204.

Musil, L.S., Beyer, E.C. and Goodenough, D.A. (1990a) Expression of the gap junction protein connexin43 in embryonic chick lens: molecular cloning, ultrastructural localization and post-translational phosphorylation. J. Membr. Biol. 116, 163–175.

Musil, L.S., Cunningham, B.A., Edelman, G.M. and Goodenough, D.A. (1990b) Differential phosphorylation of the gap junction protein connexin43 in junctional communication-competent and -deficient cell lines. J. Cell. Biol. 111, 2077–2088.

Musil, L.S. and Goodenough, D.A. (1991) Biochemical analysis of connexin43 intracellular transport, phosphorylation and assembly into gap junctional plaques. J. Cell Biol. 115, 1357–1374.

Nagafuchi, A., Shirayoshi, Y., Okazaki, K., Yasuda, K. and Takeichi., M. (1987) Transformation of cell adhesion properties by exogenously introduced E-cadherin cDNA. Nature 329, 341–343.

Sargiacomo, M., Lisanti, M., Graeve, L., Le Bivic, A. and Rodriguez-Boulan., E. (1989) Integral and peripheral protein composition of the apical and basolateral membrane domains in MDCK cells. J. Membr. Biol. 107, 277–286.

Swenson, K.I., Piwnica-Worms, H., McNamee, H. and Paul, D.L. (1990) Tyrosine phosphorylation of the gap junction protein connexin43 is required for the pp60v-*src*-induced inhibition of communication. Cell Regulation 1, 989–1002.

Ullrich, A. and Schlessinger, J. (1990) Signal transduction by receptors with tyrosine kinase activity. Cell 61, 203–212.

J.E. Hall, G.A. Zampighi and R.M. Davis (Eds.)
Progress in Cell Research, Vol. 3

CHAPTER 37

Identification of intermediate forms of connexin43 in rat cardiac myocytes

DALE W. LAIRD, KASTURI L. PURANAM* and JEAN-PAUL REVEL

Division of Biology, California Institute of Technology, Pasadena, CA 91125, USA

Introduction

Gap junctions are composed of a family of homologous proteins known as connexins (Cx). Gap-junction communication occurs when connexins assemble into hexameric arrangements (connexons) and pair with connexons from a neighboring cell. Connexons undergo further steps of assembly as they cluster into aggregates or plaques (see Johnson et al., 1989 for review). It has become clear that at least two connexin proteins (Cx32 and Cx43) are phosphorylated during maturation and assembly (Sáez et al., 1986; Takeda et al., 1987; 1989; Sáez et al., 1990; Crow et al., 1990; Laird and Revel, 1990; Musil et al., 1990a; b; Laird et al., 1991). In our studies on rat neonatal cardiac myocytes, the nonphosphorylated Cx43 protein at 40 kDa is phosphorylated to 42- and 44-kDa forms (Laird et al., 1991). Likewise, a nonphosphorylated (NP) and two phosphorylated forms of Cx43 (designated P_1 and P_2) were identified in NRK cells (Musil et al., 1990b). In S180 cells, the assembly of Cx43 into gap-junction plaques only occurs when the cells are transfected with a cDNA encoding L-CAM and this, in turn, is coincidental with Cx43 phosphorylation (P_2 form) and communication competence (Musil et al., 1990b). The maturation and assembly of Cx43 may also involve other post-translational intermediates. For example, Cx43 in leptomeningeal cells was shown to be fatty acylated (Hertzberg et al., 1989).

We have attempted to elucidate the intracellular maturation of Cx43 in rat cardiac myocytes by perturbing its translocation with the ionophore monensin. In addition, we have identified early forms of Cx43 in reassociating cells that were in the initial phases of recovering from enzymatic dissociation and assembling new gap junctions.

*Present address: Howard Hughes Medical Institute, Duke University Medical Center, Durham, NC 27710, USA.

Results and discussion

Cardiomyocyte cultures

We have used primary cultures of rat neonatal cardiac myocytes (Laird and Revel, 1990) in our studies of Cx43 maturation and translocation. Cardiomyocytes began beating within 24 h of plating and developed synchronous beating patterns after 1–2 days in culture. We used 2-day-old cultures for the majority of our experiments. An antibody generated against residues 360–382 of Cx43 (CT-360) stained gap junctions at sites of cell-cell apposition (Fig. 1A). Close examination of this confocal image also shows a light staining of intracellular processes and some intensely labeled sites (see arrowheads) that may represent Cx43 in the endoplasmic reticulum and/or Golgi apparatus.

As a morphological marker of the Golgi apparatus, cardiomyocytes were stained with an antibody raised against an intrinsic membrane Golgi protein (MG-160, Gonatas et al., 1989; Croul et al., 1990). A perinuclear immunofluorescent-staining pattern, typical of the Golgi apparatus, was observed with this anti-MG-160 antibody (Fig. 1B).

Identification of the phosphorylated forms of Cx43 in control and monensin-treated cardiomyocytes

Cx43 was immunoprecipitated from metabolically labeled cells by using an antibody (AT-2) to the amino terminal tip of Cx43 (Yancey et al., 1989; Kadle et al., 1991) or CT-360 (Laird et al., 1991). Cardiomyocytes were labeled with either [^{35}S]-Met for 2 h or [^{32}P]-P_i for 5 h, lysed and immunoprecipitated with anti-Cx43 antibodies. The forms of Cx43 were separated by SDS-PAGE and analyzed by fluorography or autoradiography. Both the AT-2 and CT-360 antibodies immunoprecipitated major [^{35}S]-Met labeled proteins at 40 and 42 kDa (Fig. 2, lanes a and c) as well as a protein at 45 kDa. A large fraction of the band at

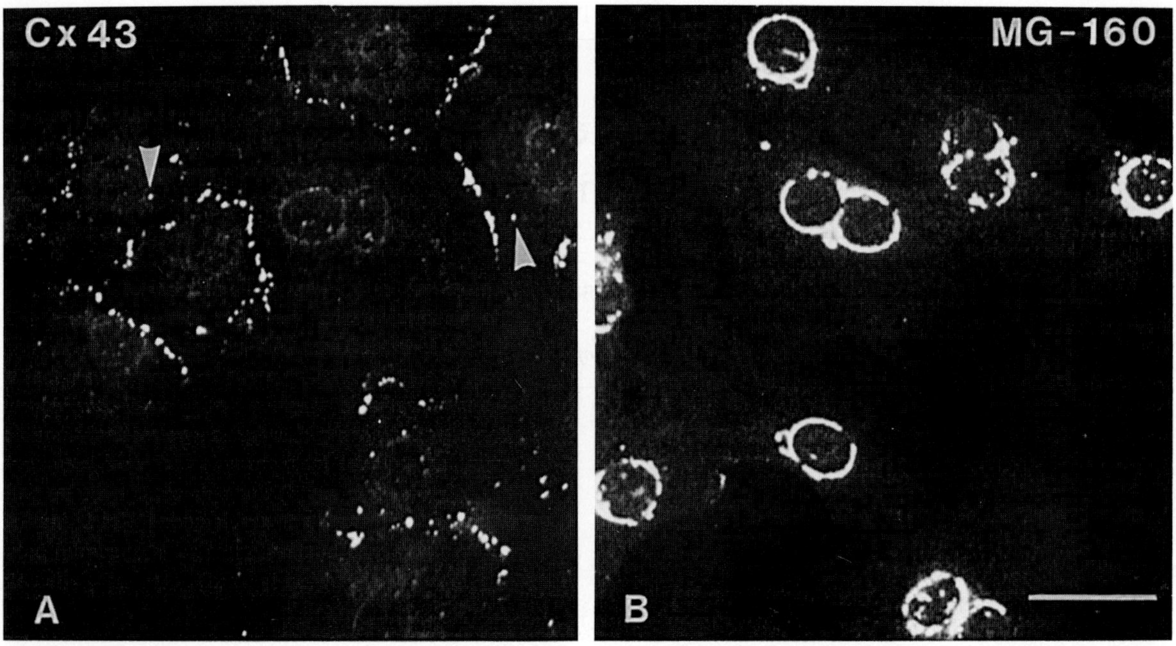

Figure 1. Localization of Cx43 and the resident Golgi protein, MG-160, in primary cultures of rat cardiac myocytes. Cardiomyocytes were treated with CT-360 (A) or anti-MG-160 (B) antibody followed by goat anti-rabbit antibody conjugated to fluorescein. Cells were viewed by a Zeiss LSM laser confocal microscope. Arrowheads indicate possible intracellular pools of Cx43. Bar = 25 μm.

45 kDa was due to the nonspecific immunoprecipitation of actin (Laird et al., 1991). The AT-2 and CT-360 antibodies were differentially effective in immunoprecipitating Cx43 from cell lysates as the AT-2 antibody was more reactive with the detergent-solubilized 40-kDa form of Cx43. Two [^{32}P]-P$_i$-labeled proteins were immunoprecipitated by both antibodies at 42 and 44 kDa (Fig. 2, lanes b and d) demonstrating that there were at least two phosphorylated forms of Cx43 in cardiac myocytes. Treatment of [^{35}S]-Met-labeled cell lysates with alkaline phosphatase prior to anti-Cx43 immunoprecipitation eliminated the 42-kDa protein, confirming that this was a phosphorylated form of Cx43 (Fig. 2, lane e) as previously shown (Laird et al., 1991). Likewise, the [^{32}P]-P$_i$-labeled proteins at 42 and 44 kDa were eliminated when treated with alkaline phosphatase (Fig. 2, lane f). These results indicate that in cultured cardiac myocytes, the parent form of Cx43 at 40 kDa is phosphorylated to 42 and 44-kDa proteins.

When cardiomyocytes were labeled with [^{35}S]-Met in the presence of the ionophore monensin, equally intense protein bands at 40 and 41 kDa were immunoprecipitated from the cell lysates by anti-Cx43 antibodies (Fig. 2, lane g). The almost complete absence of the band at 42 kDa suggested that at least one phosphorylation step was inhibited in the presence of monensin. Incubation of cell lysates from monensin-treated cells with alkaline phosphatase resulted in the loss of the band at 41 kDa, with an increase in the major band at 40 kDa (Fig. 2, lane h), thus suggesting that the 41-kDa protein was a phosphorylated form of Cx43. This phosphorylation step must occur prior to movement to the *trans*-Golgi, the site where monensin blocks protein translocation (Tartakoff and Vassalli, 1978; Tartakoff, 1983; Mollenhauer et al., 1990).

Monensin-treated cells were labeled with anti-MG-160 and CT-360 antibodies in order to determine if monensin were altering the morphology of the Golgi apparatus and to determine if there were any intracellular accumulation of Cx43 (Fig. 3). Immunofluorescent labeling showed that Cx43 had accumulated as distinct vesicles within the cytoplasm of monensin-treated cells and the number of gap-junction plaques was significantly reduced (Fig. 3A). Immunofluorescent images of MG-160 staining clearly showed that the normal perinuclear arrangement of the Golgi apparatus (Fig. 1B) was lost, and instead, the Golgi appears as a series of swollen membrane vesicles (Fig. 3B). Double-labeling experiments in conjunction with confocal microscopy showed that many swollen Golgi membrane vesicles contained Cx43 (unpublished results). These results suggest that the 40- and 41-kDa forms of Cx43 accumulate in the swollen Golgi membrane vesicles. The inability of newly synthesized Cx43 to translocate beyond the site of monensin blockage resulted in a gradual decrease in the ability of cells to transfer Lucifer Yellow (LY) intercellularly. Dye coupling was observed in 74% of the cells pairs tested (n = 27) when treated with monensin for 4 to 7 h. However, longer periods of time (7–15 h; n = 94) resulted in only 45% of the cell pairs being coupled. The loss of functional coupling between paired cardiomyocytes was correlated with the reduction in gap-junction plaques as determined by immunofluorescence. The gradual loss of gap-junction plaques in the presence of monensin may be due to turnover (Fallon and

Goodenough, 1981; Yancey et al., 1981; Traub et al., 1987; Musil et al., 1990b; Laird et al., 1991), monensin-induced clearing or toxicity of the drug when used for long periods of time (9–10 h).

Immature forms of Cx43 in reassociating cardiomyocytes

Another approach we used to study the maturation of Cx43 in cardiomyocytes was to identify the newly synthesized forms of the protein in enzymatically dissociated cardiomyocytes that were in the process of reassociating. Cardiomyocytes from one-day-old neonatal rats were dissociated with pancreatin and allowed to recover in culture medium. After only 2 h of recovery, the cells were labeled with [^{35}S]-Met for 2 h, lysed and immunoprecipitated with anti-Cx43 antibodies (Laird et al., 1991). Under these conditions, only bands at 40 and 41 kDa were found in reassociating cells (Fig. 2, lane i). The results of [^{32}P]-P_i incorporation into the 41-kDa protein were inconclusive, as the time required to label the cells (2 h of recovery + 5 h labeling) resulted in further maturation of Cx43 and detection of the more highly phosphorylated forms of the protein (unpublished results). Likewise, attempts to dephosphorylate this 41-kDa form of the protein under the same conditions used to dephosphorylate the 41-kDa protein found in monensin-treated cells were unsuccessful. Consequently, the 41-kDa form of Cx43 trapped in monensin-treated cells was designated "41s kDa" and the 41-kDa form identified in reassociating cells "41r kDa" to reflect their sensitivity to alkaline phosphatase (s = sensitive; r = resistant). The inability to dephosphorylate the 41r kDa protein may be due to the possibility that the protein is phosphorylated at sites that are inaccessible to alkaline phosphatase. Alternatively, it is possible that the 41r kDa protein in reassociating cells is not the result of a post-translational modification, but rather may represent a homologous protein (Harfst et al., 1990). However, this latter explanation seems unlikely since the 41r kDa form of the protein was immunoprecipitated with antibodies to both the amino- and carboxy-termini of Cx43. Moreover, cross-reactivity with the gap-junction protein, Cx46, has been ruled out, as the CT-360 antibody does not label lens fibers (Barbara Yancey; personal communication) where Cx46 is abundant (Beyer et al., 1988).

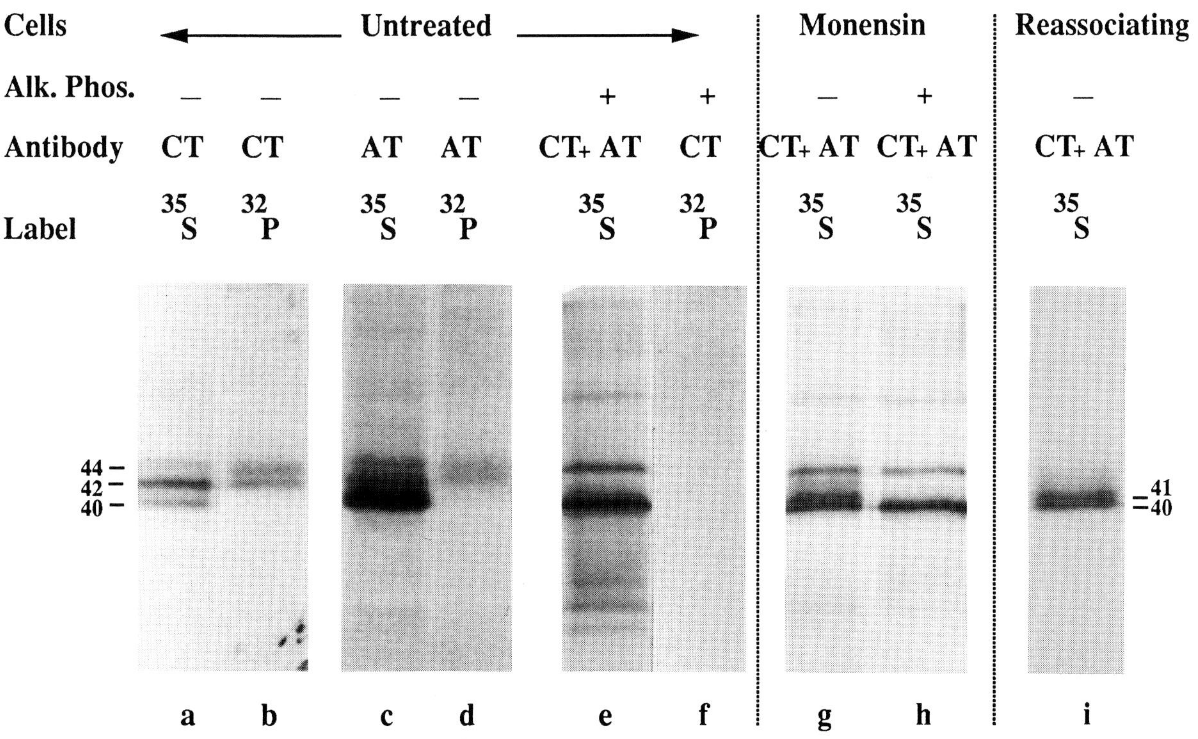

Figure 2. Metabolic labeling of Cx43 in rat cardiac myocytes. Untreated (lanes a–f) or monensin-treated (10 μM for 3 h; lanes g, h) cardiomyocytes were labeled with [^{35}S]-Met or [^{32}P]-P_i for 2 or 5 h, respectively (Laird et al., 1991). Cells were washed, lysed, and in some cases treated with alkaline phosphatase (Alk. Phos.), prior to being immunoprecipitated with CT-360 (CT) and/or AT-2 (AT) anti-Cx43 site-directed antibodies. In one scenario, cells that were enzymatically dissociated and in the process of reassociating were labeled with [^{35}S]-Met for 2 h, washed, lysed and immunoprecipitated with anti-Cx43 antibodies (lane i). All immunoprecipitates were subjected to SDS-PAGE and analysis as previously described (Laird et al., 1991). Molecular weights are in thousands.

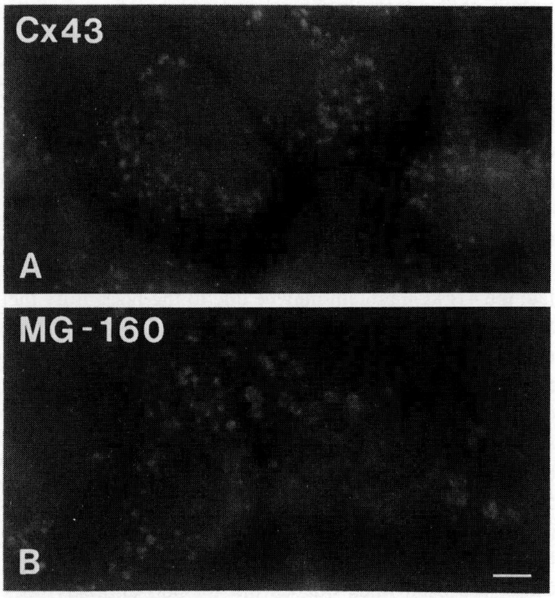

Figure 3. Effect of monensin on the localization of Cx43 and MG-160. Cardiac myocytes were treated with 10 μM monensin for 9–10 h prior to fixation in 1% paraformaldehyde and permeabilization in 1% Triton X-100. Cells were labeled with CT-360 (A) or MG-160 (B) antibody followed by goat anti-rabbit antibody conjugated to fluorescein. Cells were viewed and photographed under a Zeiss microscope equipped with epifluorescence. Bar = 10 μm

At early stages (< 5 h) of reassociation, the majority of the cardiomyocytes were unpaired. In addition, of the cells that had formed visible cell-cell contacts, only 20% (n = 50) were capable of transferring Lucifer Yellow, indicating that very few cells had assembled functional gap-junction channels. Confirmation that Cx43 had not been assembled into gap junctions was provided by immunofluorescent labeling with the CT-360 anti-Cx43 antibody: punctate and diffuse staining was found but not at sites of cell-cell contact (Fig. 4A). The locations of Cx43 could represent residual protein that is destined for degradation (cell surface or annular profiles) or pools of newly synthesized protein. When reassociating cells were stained with the anti-MG-160 antibody, a punctate perinuclear staining pattern was observed (Fig. 4B). To date, we have not determined if any Cx43 is localized to the Golgi apparatus at this stage of cell reassociation.

Conclusions

We have used two approaches to identify the immature forms of Cx43 in cardiomyocytes. The first involved using the drug monensin to inhibit the translocation of Cx43 within the Golgi apparatus. In the presence of monensin, Cx43 accumulated in Golgi membrane vesicles, and newly synthesized Cx43 was incompletely processed resulting in 40- and 41s-kDa forms of the protein. The 41s-kDa form of the protein was sensitive to alkaline phosphatase, suggesting that this represents a phosphorylated form of the protein and that phosphorylation occurs prior to translocation of Cx43 beyond the Golgi apparatus. A logical extension of these observations would suggest that further phosphorylation of Cx43 to more highly phosphorylated forms must occur after the site of monensin blockage.

In the second approach, newly synthesized forms of Cx43 in reassociating cells were isolated prior to their translocation to the cell surface and assembly into functional gap-junction channels. In these preparations of reassociating cardiac myocytes, Cx43 biosynthesis began within 4 h of enzymatic dissociation, with the formation of 40- and 41r-kDa forms of the protein. The resistance of the 41r-kDa protein to alkaline phosphatase makes this protein biochemically distinct from the 41s-kDa protein identified in monensin-treated cells. The origin or post-translational modification responsible for generating the 41r-kDa protein will require further analysis.

Following the nomenclature assigned by Musil et al. (1990b), the 42- and 44-kDa phosphorylated forms of Cx43 identified in cardiac myocytes could be designated

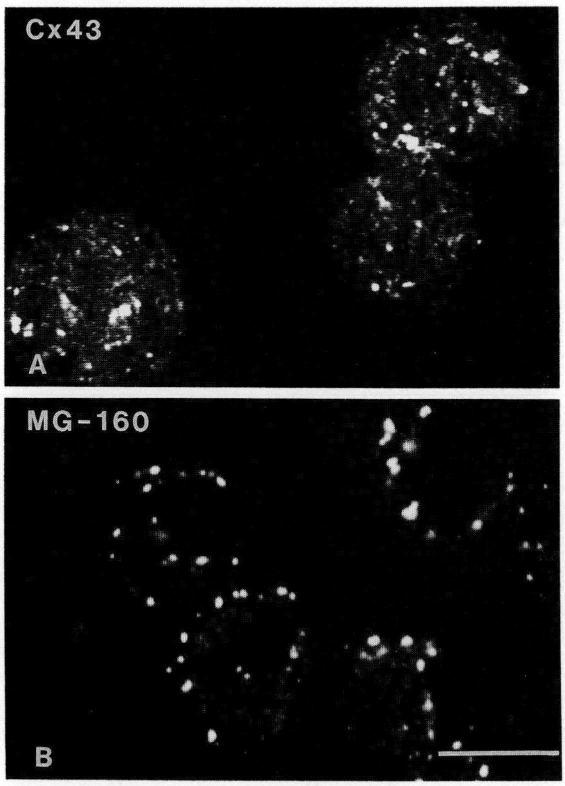

Figure 4. Localization of Cx43 and MG-160 in reassociating cardiac myocytes. Freshly dissociated cells were allowed to recover for 2–3 h in culture medium prior to being absorbed to poly-L-lysine-coated coverslips, fixed in 1% paraformaldehyde and permeabilized with 1% Triton X-100. The cells were either labeled with CT-360 (A) or anti-MG-160 (B) antibody followed by goat anti-rabbit IgG conjugated to fluorescein. Samples were viewed on a Zeiss LSM confocal microscope. Bar = 10 μm

P_1 and P_2, respectively. In addition, we have identified a 41s kDa phosphorylated form of Cx43 in monensin-treated cardiomyocytes. Our results raise the question as to whether a transient 41s-kDa form of Cx43 is found in untreated cardiac myocytes. Upon close examination of several fluorographs and autoradiograms, low amounts of a 41s-kDa protein were occasionally immunoprecipitated from untreated cells labeled with [^{35}S]-Met or [^{32}P]-P_i; however, an abundance of the 40-kDa parent form of the protein or weak signals, respectively, made clear conclusions difficult. Crow et al. (1990) reported the presence of a 44-kDa band in vole fibroblasts that may be equivalent to the 41s-kDa band that we observe in cardiomyocytes. We suggest that in coupled cardiac myocytes, the first phosphorylation events occur prior to Cx43 translocation beyond the Golgi apparatus. Once initially phosphorylated, the protein is rapidly processed, resulting in relatively low quantities of the 41s-kDa form compared to all forms of Cx43. Confirmation that the 41s-kDa form of Cx43 resides only within the Golgi apparatus may require subcellular fractionation and organelle isolation.

In summary, we are beginning to understand the pathway of Cx43 translocation and the post-translational modification(s) associated with each subcellular compartment. It is possible, if not likely, that each compartment and modification to Cx43 plays key roles in directing gap-junction assembly and/or function.

Acknowledgements

We are deeply indebted to Dr. Nicholas K. Gonatas for his generous gift of the anti-MG-160 antibody. We would also like to thank Bruce J. Nicholson and S. Barbara Yancey for their initial work in generating and characterizing the AT-2 antibody. Finally, we thank Jan H. Hoh for his comments on the manuscript and Lakshmi Bugga for her technical assistance.

References

Beyer, E.C., Goodenough, D.A. and Paul, D.L. (1988) The connexins: a family of related gap-junction proteins. In: E.L. Hertzberg and R.G. Johnson (Eds.) Modern Cell Biology: Gap Junctions. Vol. 7. Alan R. Liss, New York. pp. 167–175.

Croul, S., Mezitis, S.G.E., Stieber, A., Chen, Y., Gonatas, J.O., Goud, B. and Gonatas, N.K. (1990) Immunocytochemical visualization of the Golgi apparatus in several species, including human, and tissues with an antiserum against MG-160, a sialoglycoprotein of rat Golgi apparatus. J. Histochem. Cytochem. 38, 957–963.

Crow, D.S., Beyer, E.C., Paul, D.L., Kobe, S.S. and Lau, A.F. (1990) Phosphorylation of connexin43 gap-junction protein in uninfected and Rous sarcoma virus-transformed mammalian fibroblasts. Mol. Cell. Biol. 10, 1754–1763.

Fallon, R.F. and Goodenough, D.A. (1981) Five hour half-life of mouse liver gap-junction protein. J. Cell Biol. 90, 521–526.

Gonatas, J.O., Mezitis, S.G.E., Stieber, A., Fleischer, B. and Gonatas, N.K.. (1989) MG-160: A novel sialoglycoprotein of the medial cisternae of the Golgi apparatus. J. Biol. Chem. 264, 646–653.

Harfst, E., Severs, N.J. and Green, C.R. (1990) Cardiac myocyte gap junctions: evidence for a major connexon protein with an apparent relative molecular mass of 70 000. J. Cell Sci. 96, 591–604.

Hertzberg, E.L., Corpina, R., Roy, C., Dougherty, M.J. and Kessler, J.A. (1989) Analysis of the 43 kDa heart gap-junction protein in primary cultures of rat leptomeningeal cells. J. Cell Biol. 109, 47a.

Johnson, R.G., Meyer, R.A. and Lampe, P.D. (1989) Gap-junction formation: A "Self-Assembly" model involving membrane domains of lipid and protein. In: N. Sperelakis and W. C. Cole (Eds.). Cell Interactions and Gap Junctions, Vol. 1, CRC Press, Boca Raton, FL, pp. 159–179.

Kadle, R., Zhang, J.T. and Nicholson, B.J. (1991) Tissue-specific distribution of differentially phosphorylated forms of Cx43. Mol. Cell. Biol. 11, 363–369.

Laird, D.W. and Revel, J.-P. (1990) Biochemical and immunochemical analysis of the arrangement of connexin43 in rat heart gap-junction membranes. J. Cell Sci. 97, 109–117.

Laird, D.W., Puranam, K.L. and Revel, J.-P. (1991) Turnover and phosphorylation dynamics of connexin43 gap-junction protein in cultured cardiac myocytes. Biochem. J. 273, 67–72.

Mollenhauer, H.H., Morre, D.J. and Rowe, L.D. (1990) Alteration of intracellular traffic by monensin: mechanism, specificity and relationship to toxicity. Biochim. Biophys. Acta. 1031, 225–246.

Musil, L.S., Beyer, E.C. and Goodenough, D.A. (1990a) Expression of the gap-junction protein connexin43 in embryonic chick lens: molecular cloning, ultrastructural localization and post-translational phosphorylation. J. Membr. Biol. 116, 163–175.

Musil, L.S., Cunningham, B.A., Edelman, G.M. and Goodenough, D.A. (1990b) Differential phosphorylation of the gap-junction protein connexin43 in junctional communication-competent and -deficient cell lines. J. Cell Biol. 111, 2077–2088.

Sáez, J.C., Nairn, A.C., Czernik, A.J., Spray, D.C., Hertzberg, E.L., Greengard, P. and Bennett, M.V.L. (1990) Phosphorylation of connexin 32, a hepatocyte gap-junction protein, by cAMP-dependent protein kinase, protein kinase C and Ca^{2+}/calmodulin-dependent protein kinase II. Eur. J. Biochem. 192, 263–273.

Sáez, J.C., Spray, D.C., Nairn, A.C., Hertzberg, E., Greengard, P. and Bennett, M.V.L. (1986) cAMP increases junctional conductance and stimulates phosphorylation of the 27-kDa principal gap-junction polypeptide. Proc. Natl. Acad. Sci. USA 83, 2473–2477.

Takeda, A., Hashimoto, E., Yamamura, H. and Shimazu, T. (1987) Phosphorylation of liver gap-junction protein by protein kinase C. FEBS Lett. 210, 169–172.

Takeda, A., Saheki, S., Shimazu, T. and Takeuchi, N. (1989) Phosphorylation of the 27-kDa gap-junction protein by protein kinase C *in vitro* and in rat hepatocytes. J. Biochem. (Tokyo) 106, 723–727.

Tartakoff, A. (1983) Perturbation of vesicular traffic with the carboxylic ionophore monensin. Cell 32, 1026–1028.

Tartakoff, A. and Vassalli, P. (1978) Comparative studies of intracellular transport of secretory proteins. J. Cell Biol. 79, 694–707.

Traub, O., Look, J., Paul, D. and Willecke, K. (1987) Cyclic adenosine monophosphate stimulates biosynthesis and phosphorylation of the 26 kDa gap-junction protein in cultured

mouse hepatocytes. Eur. J. Cell Biol. 43, 48–54.
Yancey, S.B., John, S.A., Lal, R., Austin, B.J. and Revel, J.-P. (1989) The 43-kD polypeptide of heart gap junctions: immunolocalization, topology, and functional domains. J. Cell Biol. 108, 2241–2254.
Yancey, S.B., Nicholson, B.J. and Revel, J.-P. (1981) The dynamic state of liver gap junctions. J. Supramol. Struct. Cell. Biochem. 16, 221–232.

J.E. Hall, G.A. Zampighi and R.M. Davis (Eds.)
Progress in Cell Research, Vol. 3

CHAPTER 38

Regulation of gap junctions by cell contact and phosphorylation in MDCK cells

VIVIANA M. BERTHOUD[a], MARY LEE S. LEDBETTER[b], ELLIOT L. HERTZBERG[a] and JUAN C. SÁEZ[a]

[a]*Department of Neuroscience, Albert Einstein College of Medicine, 1300 Morris Park Ave., Bronx, NY 10461 and* [b]*Department of Biology, College of the Holy Cross, Worcester, MA 01610, USA*

Introduction

The MDCK cell line has been extensively used to study regulation of certain components of the junctional cell complex, such as tight junctions, desmosomes and *adhaerens* junctions (Kartenbeck et al., 1982; Pasdar and Nelson, 1988; Stevenson et al., 1988). Gap junctions, other components of the junctional complex present between MDCK cells, have not been extensively studied. It is known that MDCK cells are functionally coupled (Imhof et al., 1983; Cereijido et al., 1984; Giaume et al., 1986; Ledbetter et al., 1986), and gap junctions have been morphologically identified (Cereijido et al., 1984). Although a 26-kDa phosphoprotein has been immunoprecipitated from MDCK cells with an anti-Cx32 antibody (Traub et al., 1987), the molecular components of these junctions have not been further characterized.

Regulation of gap-junctional intercellular communication by various factors has been studied in different tissues and cell types. Among factors which regulate gap junctions are cell density (Loewenstein, 1981) and second messenger pathways (Sáez et al., 1990; Bennett et al., 1991). The nature of the effects depends (at least) on the cell type and member(s) of the connexin (Cx) family expressed (Sáez et al., 1990; Bennett et al., 1991). These effects include changes in Cx mRNA stability, gap-junctional area, the phosphorylation state of Cx, and gap-junctional permeability (Sáez et al., 1990).

Cx32 and Cx43, but not Cx26, are phosphoproteins (Sáez et al., 1990; Bennett et al., 1991). The role of phosphorylation is still not clearly understood. Recently, a correlation between expression of a cell-adhesion molecule (L-CAM) and formation of functional gap-junction channels has been reported (Mege et al., 1988). This process was associated with phosphorylation of Cx43 and its insertion into the plasma membrane (Musil et al., 1990).

Cell adhesion between MDCK cells is highly Ca^{2+}-dependent, and tight junctions, desmosomes and *adhaerens* junctions are all destabilized by lowering the extracellular Ca^{2+} concentration (Cereijido et al., 1978; Kartenbeck et al., 1982). Under these conditions, gap junctions between MDCK cells might be similarly affected. Indeed incubation of MDCK cells with an antibody that recognizes a Ca^{2+}-dependent cell-adhesion protein decreases the incidence of dye coupling (Imhof et al., 1983). Furthermore, in other cell, lines expression of adhesion molecules such as L-CAM and E-cadherin favors the formation of gap junctions (Mege et al., 1988; Musil et al., 1990; Jongen et al., 1991).

In this study, regulation of levels and state of phosphorylation of Cx43 by cell density, extracellular Ca^{2+} concentration and protein kinase pathways in MDCK cells is reported.

Methods

MDCK cells (Madin and Darby, 1958) obtained from ATCC were cultured in DMEM supplemented with 5% fetal calf serum, 50 units/ml penicillin G sodium and 5 μg/ml streptomycin sulfate. For experiments in Ca^{2+}-free medium, all components except Ca^{2+} were present. Distribution of Cxs was studied by indirect immunofluorescence using previously characterized anti-Cx32 and anti-Cx43 antibodies (Stevenson et al., 1988; Yamamoto et al., 1990). Immunoblots were performed following the method described by Towbin et al. (1979). Dye coupling was measured as the number of interfaces permeable to Lucifer Yellow (LY).

Results

Regulation of dye coupling by cell density

While Cereijido et al. (1984) found reduced dye coupling in confluent cultures of MDCK cells, Giaume et al. (1986)

found subconfluent and confluent cultures about equally well coupled electrically. In view of these differences, we decided to evaluate the incidence and extent of dye coupling in our cultures.

MDCK cells in subconfluent cultures showed extensive dye coupling, with an incidence of 85–90% (Fig. 1C). Cells in confluent cultures, however, showed a low incidence of dye coupling (5–10%). In the cases where dye coupling was observed, the dye spread only to one or two adjacent cells. Figure 1D shows an example of the absence of dye coupling in confluent cultures.

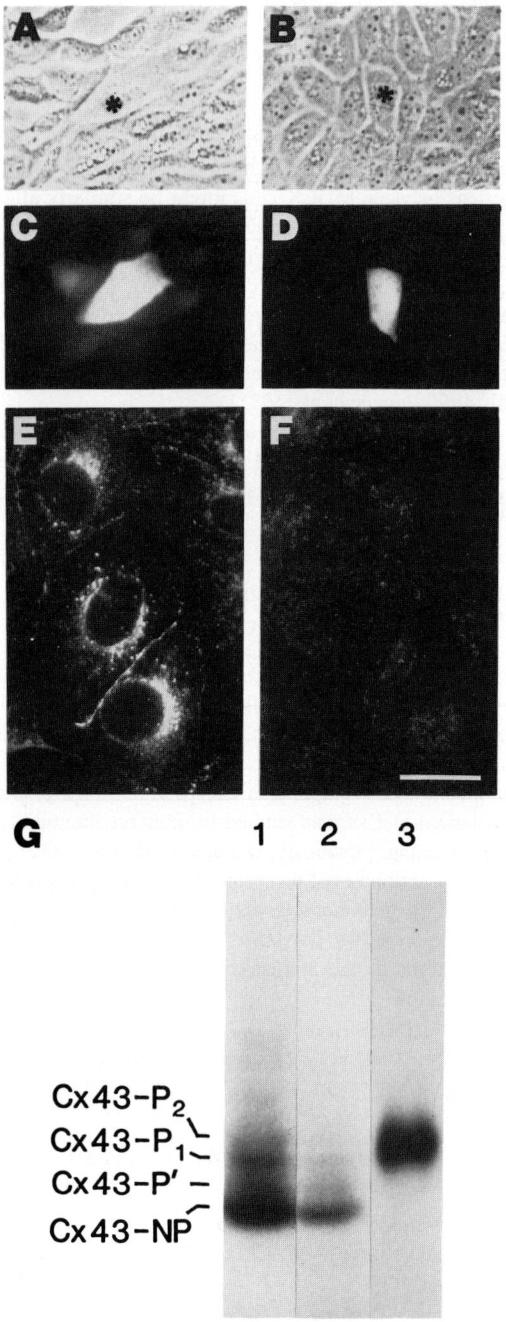

MDCK cells express Cx43

Since coupling was dependent on cell density, we first used specific antibodies to identify a molecular component of gap junctions and then determined whether the difference in the incidence of dye coupling observed between subconfluent and confluent cultures was associated with changes in the amount of the molecular component present.

Immunofluorescence and Western blot analyses indicated that MDCK cells expressed Cx43 (Fig. 1) but not Cx32 (not illustrated). In subconfluent cultures, immunofluorescence demonstrated punctate immunoreactivity with anti-Cx43 antibodies at appositional membranes as well as a more diffuse labeling in the cytoplasm (Fig. 1E). In confluent cultures, immunoreactivity was lower than in subconfluent cultures and was localized preferentially in the cytoplasm (Fig. 1F). Even though levels of Cx43 were much lower in confluent than in subconfluent cultures, the major band detected by immunoblotting at both cell densities was the rapidly migrating form of Cx43 (Fig. 1G, lanes 1 and 2). This form represents the protein in its unphosphorylated state (not illustrated).

Regulation of Cx43 by extracellular Ca^{2+}

As mentioned above, expression of Ca^{2+}-dependent cell-adhesion molecules favors formation of functional gap junctions (Mege et al., 1988; Jongen et al., 1991); and expression of L-CAM or E-cadherin correlates with insertion of Cx43 into the plasma membrane (Musil et al., 1990; Jongen et al., 1991). Since, in MDCK cells, several components of the junctional complex are highly dependent on the extracellular Ca^{2+} concentration, the effect of low extracellular Ca^{2+} on the distribution and levels of Cx43 was assessed.

When normal culture medium was replaced with Ca^{2+}-free medium, MDCK cells lost their typical morphology, becoming almost round and losing most of their intercellular contacts (Fig. 2B, C and D). The effect became more pronounced the longer the cells remained in Ca^{2+}-free medium. The loss of intercellular contact was associated

Figure 1. Effect of cell density on dye coupling and abundance of Cx43. (A) and (B) Phase-contrast micrographs of cultures shown in (C) and (D). (C) Dye coupling in a subconfluent culture of MDCK cells. (D) Absence of dye coupling in a confluent culture of MDCK cells. (E) and (F) Distribution of anti-Cx43 antibodies in subconfluent and confluent cultures of MDCK cells, respectively. The cell injected with LY is indicated with an (*) in (A) and (B). Calibration bar: A–D, 39 μm; E and F, 20 μm. (G) Immunoblot of Cx43 in cell lysates (50 μg protein/lane) of subconfluent (lane 1) and confluent (lane 2) cultures of MDCK cells, and in whole rat heart homogenate (lane 3). The positions of the different forms of Cx43 are indicated. The dephosphorylated form of Cx43 has been named Cx43-NP and the phosphorylated forms $Cx43\text{-}P_1$ and $Cx43\text{-}P_2$ (Musil et al., 1990). A fourth band with an electrophoretic mobility between that of Cx43-NP and $Cx43\text{-}P_1$ has been labeled as Cx43-P'.

with a decrease in gap-junctional intercellular communication assayed by dye coupling and in Cx43 immunoreactivity at appositional membranes. By 15 min, the majority of the cells showed no labeling at appositional membranes (Fig. 2C). When the Ca^{2+} concentration of the medium was restored to normal levels, Cx43 immunoreactivity reappeared at membrane appositions. The kinetics of this process depended on the length of time the cells remained in Ca^{2+}-free medium. Cells that were incubated for 15 min in Ca^{2+}-free medium showed labeling at appositional membranes 1 h after 2 mM Ca^{2+} was added (Fig. 2E), while cells that remained 3 hr in Ca^{2+}-free medium showed labeling at appositional membranes 3 h after Ca^{2+} was restored (Fig. 2H). However, no changes were detected in the levels or state of phosphorylation of Cx43 under any of these conditions (Fig. 2I).

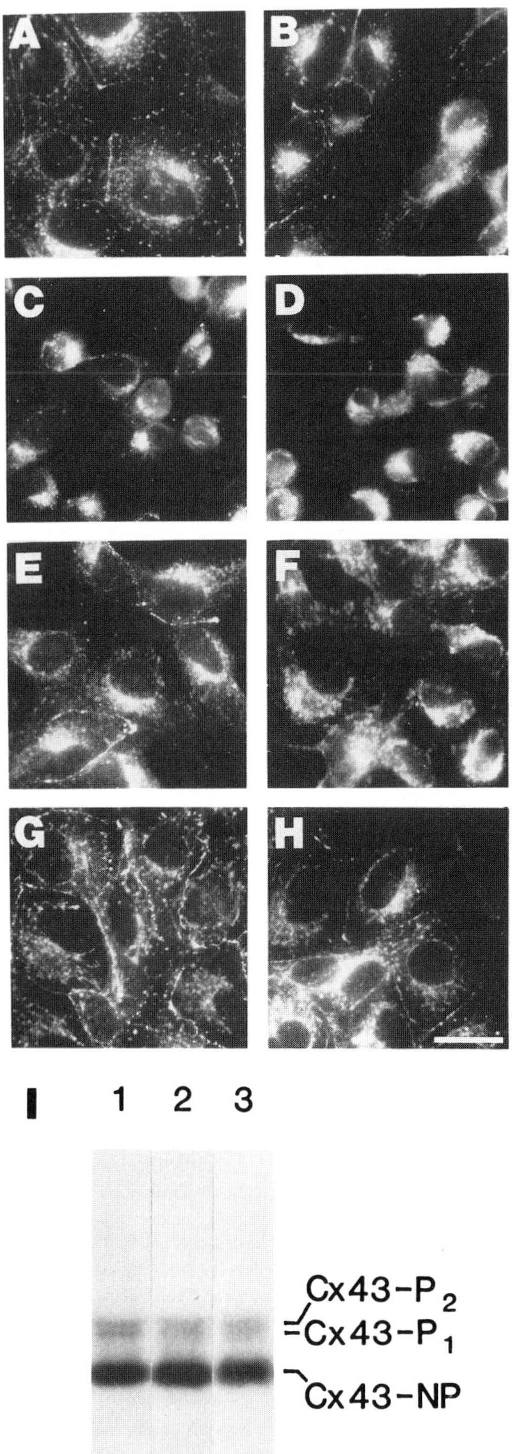

Figure 2. Effect of extracellular Ca^{2+} concentration on Cx43 in MDCK cells. Photomicrographs show the immunolocalization of Cx43 in cultures maintained (A) in medium containing 2 mM Ca^{2+}, (B) in Ca^{2+}-free medium for 10 min, (C) 15 min or (D) 3h, in Ca^{2+}-free medium for 15 min and then cultured in medium containing 2 mM Ca^{2+} for (E) 1 h or (G) 3 h, in Ca^{2+}-free medium for 3 h and then cultured in medium containing 2 mM Ca^{2+} for (F) 1 h or (H) 3 h. Calibration bar: 20 μm. (I) Immunoblot of Cx43 in total homogenates of MDCK cells (50 μg protein/lane) from cultures maintained in medium containing 2 mM Ca^{2+} (lane 1), Ca^{2+}-free medium for 3 h (lane 2), and Ca^{2+}-free medium for 3 h and then cultured in medium containing 2 mM Ca^{2+} for 3 h (lane 3). The different forms of Cx43 are indicated as in Fig. 1.

Regulation of the state of phosphorylation of Cx43 and of gap-junctional permeability

Since Cx43 is a phosphoprotein (Sáez et al., 1990; Bennett et al., 1991), changes in its phosphorylation state might be altered by changing the activity of protein kinase(s). Different phosphorylated forms of Cx43 show different electrophoretic mobilities on SDS-PAGE (Crow et al., 1990; Musil et al., 1990; Laird et al., 1991); thus, changes in its state of phosphorylation can be followed by immunoblotting (Crow et al., 1990). Cultures of MDCK cells were treated with compounds known to activate cAMP- or cGMP-dependent protein kinase, or protein kinase C (PKC).

Neither 8-Br-cAMP (0.5 mM) nor 8-Br-cGMP (0.5 mM) changed the immunoblot pattern of Cx43 after treatment for 1 h (not shown). However, treatment with TPA, a PKC activator (<200 nM) for 5–15 min increased the relative amounts of the phosphorylated forms at the expense of the dephosphorylated form (Fig. 3A, lane 2). Maximal response was observed after treatment with 100 nM TPA. Pretreatment with staurosporine, a protein kinase inhibitor (Kase et al., 1989), prevented the effect of TPA (Fig. 3A, lane 3). This effect was specific for TPA since 4-α-phorbol, an inactive analog of TPA (not illustrated) and 0.007% dimethylsulfoxide (DMSO; Fig. 3B, lane 2), used as solvent for both TPA and 4-α-phorbol, had no effect. TPA-induced phosphorylation of Cx43 occurred in the absence of serum in the incubation medium (Fig. 3B, lane 3).

MDCK cells were treated with TPA in Ca^{2+}-free medium in order to ascertain the requirement for extracellular Ca^{2+} ions in the phosphorylation of Cx43. Under

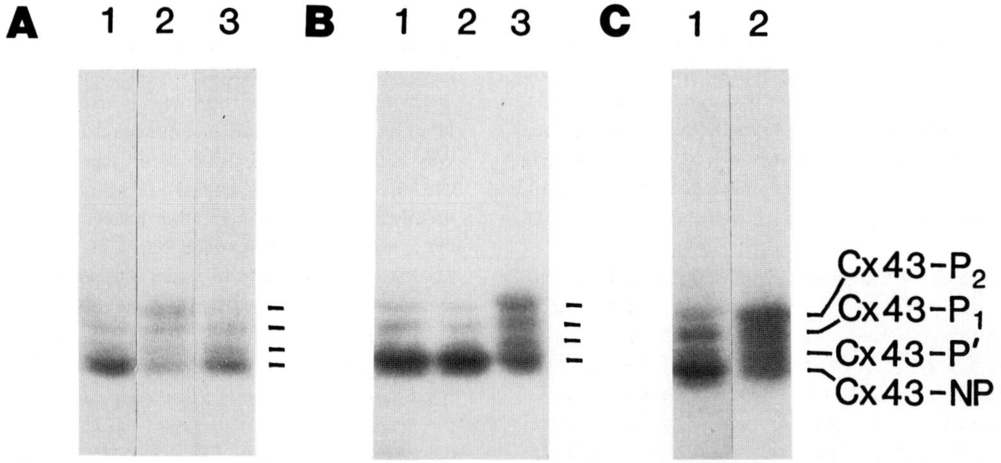

Figure 3. Regulation of the state of phosphorylation of Cx43 in MDCK cells. (A) Immunoblot of Cx43 in total homogenates of MDCK cells in cultures under control conditions (lane 1), in cultures treated with 200 nM TPA for 15 min (lane 2) and in cultures treated with 600 nM staurosporine for 20 min and then treated with 200 nM TPA for 15 min in the presence of the same concentration of staurosporine (lane 3). (B) Immunoblot of Cx43 in total homogenates of MDCK cells from cultures maintained in serum-free medium for 24 h (lane 1), and then treated with DMSO for 1 h (lane 2) or with 200 nM TPA for 1 h (lane 3). (C) Immunoblot of Cx43 in total homogenates of MDCK cells under control conditions (lane 1) and in cells kept in Ca^{2+}-free medium for 30 min and then treated with 200 nM TPA for 15 min in Ca^{2+}-free medium (lane 2); 50 μg protein per lane. The different forms of Cx43 are indicated as in Fig. 1. In A and B, marks indicate the positions of the different forms of Cx43 as in C.

this condition, the response to TPA was not affected (Fig. 3C, lane 2).

The cellular distribution of Cx43 was also unaffected by treatment with 8-Br-cAMP (0.5 mM) or 8-Br-cGMP (0.5 mM) for 1 h (not illustrated). Treatment with TPA (200 nM) for 15 min did not produce evident changes in the distribution of Cx43, although the cytoplasmic labeling acquired a more vesicle-like appearance (Fig. 4B). After longer periods of exposure to TPA, immunolabeling at appositional membranes gradually decreased, and cytoplasmic labeling increased along with changes in cell morphology (Fig. 4C, D).

Since, in many other systems, TPA reduces gap-junctional intercellular communication (Sáez et al., 1990; Bennett et al., 1991), it was of interest to study the effect of TPA on the functional state of gap junctions between MDCK cells. TPA (200 nM) induced a drastic reduction in dye coupling 10–15 min after its addition (Fig. 4F).

Discussion

In this study, we have been able to decrease gap-junctional intercellular communication in MDCK cells by three treatments, each affecting Cx43 in different ways. The transition of MDCK cell cultures from the subconfluent to confluent state was accompanied by a decrease in the total amount of Cx43 and its cellular distribution, without significant changes in its state of phosphorylation. Disruption of cell contacts by withdrawal of Ca^{2+} from the medium affects neither the amount nor the phosphorylation of Cx43, only its distribution. Treatment of cells with TPA enhanced the phosphorylation of Cx43 within 15 min, without altering either its amount or distribution.

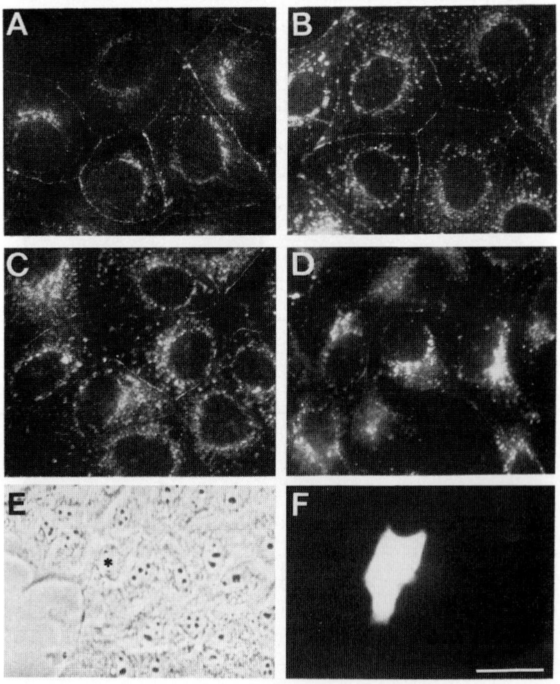

Figure 4. Effect of TPA on the distribution of Cx43 and dye coupling in MDCK cells. Photomicrographs show the immunolocalization of Cx43 in cultures (A) under control conditions or treated with TPA for (B) 15 min, (C) 30 min or (D) 1h. (E) Phase-contrast micrograph of cells shown in (F). (F) Representative example of the reduction in the extent of dye coupling between MDCK cells in subconfluent cultures observed after treatment with 200 nM TPA for 15 min at 37°C. The cell injected with LY is indicated by an (*) in (E). Calibration bar: A–D, 20 μm; E and F, 30 μm.

The decreases seen in dye coupling and in levels of Cx43 immunoreactivity in cultures of MDCK cells when they passed from a subconfluent to a confluent state are in agreement with reports correlating a decrease in dye coupling with a reduction in morphologically identifiable gap junctions as a function of the age of MDCK monolayers (Cereijido et al., 1984). Since MDCK cells did not contain detectable amounts of Cx32, and levels and distribution of Cx43 correlated with the presence of dye coupling, it is likely that this Cx is a major molecular component responsible for dye coupling between MDCK cells.

It is known that low extracellular Ca^{2+} decreases intercellular contact by destabilizing several components of the junctional complex (Cereijido et al., 1978; Kartenbeck et al., 1982). Low extracellular Ca^{2+} was also found to destabilize gap junctions in MDCK cells because labeling at appositional membranes disappeared shortly after incubation of the cells in a Ca^{2+}-free medium. The kinetics of gap junction reappearance after restoring extracellular Ca^{2+} to 2 mM suggests that the treatment induced internalization and degradation of Cx43 previously present at appositional membranes.

In the present study, Cx43 levels were not affected by a reduction in the extracellular Ca^{2+} concentration, suggesting that neither synthesis nor degradation of Cx43 was affected under these conditions. In support, the same amount of metabolically radiolabeled Cx43 was found in cells incubated in the presence or absence of anti-chicken L-CAM Fab fragments, a treatment that separated the cells (Musil et al., 1990), suggesting that during the periods of time studied, synthesis or degradation of Cx43 is not affected by changes in Ca^{2+}-dependent cell adhesion. The present results also agree with a recent report showing that Ca^{2+}-dependent formation of gap junctions occurs without changes in levels of Cx43 mRNA, and that the ability to establish functional gap-junctional communication is resistant to cycloheximide (Jongen et al., 1991). We also saw no change in the state of phosphorylation of the Cx43 protein during treatment with Ca^{2+}-free medium. This finding is contrary to the prediction of Musil et al. (1990), i.e., that phosphorylation of Cx43 depended on Ca^{2+}-mediated cell adhesion.

Treatment of cells with membrane-permeant derivatives of cyclic nucleotides did not affect the state of phosphorylation of Cx43, suggesting that, in subconfluent cultures of MDCK cells, Cx43 is not a target for the corresponding cyclic nucleotide-dependent protein-kinase pathways. Phosphorylation was increased, however, by TPA, a compound known to bind to PKC (Niedel et al., 1983) and activate it (Castagna et al., 1982). The TPA effect was not a permissive effect of a serum component, but rather a direct effect on PKC, since it was unaffected by excluding serum from the medium. These results suggest that the state of phosphorylation of Cx43 can be increased upon activation of PKC in MDCK cells. It remains to be determined if PKC phosphorylates Cx43 directly or through a PKC-activated protein kinase. The cellular compartment where the TPA-induced phosphorylation of Cx43 takes place in MDCK cells is so far unknown.

It has been proposed that phosphorylation of Cx43 induces its insertion and/or assembly into the plasma membrane (Musil et al., 1990). In MDCK cells, treatment with TPA that led to an increase in the relative amounts of the phosphorylated forms of Cx43 did not significantly change its cellular distribution within the first 15 min. Thus, the state of phosphorylation of Cx43 probably is not correlated with a specific cellular location. Moreover, considering that morphological changes and decreased immunoreactivity at appositional membranes were observed upon longer treatment with TPA, when the phosphorylated forms of Cx43 were still predominant, the changes observed in the distribution of Cx43 are more likely to be a consequence of a reduced intercellular contact rather than the phosphorylation state of Cx.

Although in other cell types it has been proposed that Ca^{2+}-dependent cell adhesion induces phosphorylation of Cx43 (Musil et al., 1990), in MDCK cells Ca^{2+}-dependent cell contact could be drastically altered with no hindrance to the TPA-induced phosphorylation of Cx43. Phosphorylation of Cx43 in MDCK cells by a PKC-dependent pathway thus appears to be independent of extracellular Ca^{2+}. Since phosphorylation in MDCK cells is associated with a reduction in gap-junctional permeability, while in other systems, phosphorylation is associated with functional coupling (Mege et al., 1988; Musil et al., 1990), suggests that different protein kinases, acting on different sites of phosphorylation, might operate in the two systems. Alternatively, TPA may trigger gating mechanisms other than phosphorylation, such as release of arachidonic acid (Coyne et al., 1990; Sáez et al., 1990), which could account for the functional uncoupling observed in MDCK cells.

Acknowledgements

Viviana Berthoud would like to thank the Weintraub Foundation, for their generosity and support. This work was supported by NIH grants DK-41368 (to J.C.S.), DK-40140 (to M.L.S.L.), GM-30667 (to E.L.H.) and NS-07512 (to M.V.L.B.).

References

Bennett, M.V.L., Bargiello, T.A., Barrio, L., Spray, D.C., Hertzberg, E.L. and Sáez, J.C. (1991) Gap junctions: new tools, new answers, new questions. Neuron 6, 305–320.

Castagna, M., Takai, Y., Kaibuchi, K., Sano, K., Kikkawa, U. and Nishizuka, Y. (1982) Direct activation of calcium-activated, phospholipid-dependent protein kinase by tumor-promoting phorbol esters. J. Biol. Chem. 257, 7847–7851.

Cereijido, M., Robbins, E.S., Dolan, W.J., Rotunno, C.A. and Sabatini, D.D. (1978) Polarized monolayers formed by epithelial cells on a permeable and translucent support. J. Cell Biol. 77, 853–880.

Cereijido, M., Robbins, E., Sabatini, D.D. and Stefani, E. (1984) Cell-to-cell communication in monolayers of epithelioid cells (MDCK) as a function of the age of the monolayer. J. Membr. Biol. 81, 41–48.

Coyne, D.W., Mordhorst, M. and Morrison, A.R. (1990) Regulation of eicosanoid biosynthesis by phorbol ester in Madin Darby canine kidney cells. Am. J. Physiol. 259, F698–F703.

Crow, D.S., Beyer, E.C., Paul, D.L., Kobe, S.S. and Lau, A.F. (1990) Phosphorylation of connexin43 gap-junction protein in uninfected and Rous sarcoma virus-transformed mammalian fibroblasts. Mol. Cell. Biol. 10, 1754–1763.

Giaume, C., Sahuquillo, C., Louvard, D. and Korn, H. (1986) Evidence for ionic coupling between MDCK cells at non-confluent and confluent stages of culture. Biol. Cell 57, 33–38.

Imhof, B.A., Vollmers, P.H., Goodman, S.L. and Birchmeier, W. (1983) Cell-cell interaction and polarity of epithelial cells: specific perturbation using a monoclonal antibody. Cell 35, 667–675.

Jongen, W.M.F., Fitzgerald, D.J., Asamoto, M., Piccoli, C., Slaga, T.J., Gros, D., Takeichi, M. and Yamasaki, H. (1991) Regulation of connexin 43-mediated gap-junctional intercellular communication by Ca^{2+} in mouse epidermal cells is controlled by E-cadherin. J. Cell Biol. 114, 545–555.

Kartenbeck, J., Schmid, E., Franke, W.W. and Geiger, B. (1982) Different modes of internalization of proteins associated with adhaerens junctions and desmosomes: experimental separation of lateral contacts induced endocytosis of desmosomal plaque material. EMBO J. 1, 725–732.

Kase, H., Iwahashi, K., Nakanishi, S., Matsuda, Y., Yamada, K., Takahashi, M., Murakata, C., Sato, A. and Kaneko, M. (1987) K-252 compounds, novel and potent inhibitors of protein kinase C and nucleotide-dependent protein kinases. Biochem. Biophys. Res. Commun. 142, 436–440.

Laird, D.W., Puranam, K.L. and Revel, J.-P. (1991) Turnover and phosphorylation dynamics of connexin43 gap-junction protein in cultured cardiac myocytes. Biochem. J. 273, 67–72.

Ledbetter, M.L.S., Young, G.J. and Wright, E.R. (1986) Cooperation between epithelial cells demonstrated by potassium transfer. Am. J. Physiol. 250, C306–C313.

Loewenstein, W. R. (1981) Junctional intercellular communication: the cell-to-cell membrane channel. Physiol. Rev. 61, 829–913.

Madin, S.H. and Darby, N.B. (1958) As catalogued in: American Type Collection Catalog of Strains. 2, 574–576.

Mege, R.M., Matsuzaki, F., Gallin, W.F., Golber, J.I., Cunningham, B.A. and Edelman, G.M. (1988) Construction of epithelioid sheets by transfection of mouse sarcoma cells with cDNAs for chicken cell-adhesion molecules. Proc. Natl. Acad. Sci. USA 85, 7274–7278.

Musil, L.S., Cunningham, B.A., Edelman, G.M. and Goodenough, D.A. (1990) Differential phosphorylation of the gap-junction protein connexin43 in junctional communication-competent and -deficient cell lines. J. Cell Biol. 111, 2077–2088.

Niedel, J.E., Kuhn, L.J. and Vandenbark, G.R. (1983) Phorbol diester receptor copurifies with protein kinase C. Proc. Natl. Acad. Sci. USA 80, 36–40.

Pasdar, M. and Nelson, W.J. (1988) Kinetics of desmosome assembly in Madin-Darby canine kidney epithelial cells: temporal and spatial regulation of desmoplakin organization and stabilization upon cell-cell contact. I. Biochemical analysis. J. Cell Biol. 106, 677–685.

Sáez, J.C., Spray, D.C. and Hertzberg, E.L. (1990) Gap junctions: biochemical properties and functional regulation under physiological and toxicological conditions. *In Vitro* Toxicol. 3, 69–86.

Stevenson, B.R., Anderson, J.M., Goodenough, D.A. and Mooseker, M.S. (1988) Tight junction structure and ZO-1 content are identical in two strains of Madin-Darby canine kidney cells which differ in transepithelial resistance. J. Cell Biol. 107, 2401–2408.

Traub, O., Look, J., Paul, D. and Willecke, K. (1987) Cyclic adenosine monophosphate stimulates biosynthesis and phosphorylation of the 26 kDa gap-junction protein in cultured mouse hepatocytes. Eur. J. Cell Biol. 43, 48–54.

Towbin, H., Staehelin, T. and Gordon, J. (1979) Electrophoretic transfer of proteins from polyacrylamide gels into nitrocellulose sheets: procedure and some applications. Proc. Natl. Acad. Sci. USA 76, 4350–4354.

Yamamoto, T., Ochalski, A., Hertzberg, E.L. and Nagy, J.I. (1990) LM and EM immunolocalization of the gap-junctional protein connexin 43 in rat brain. Brain Res. 508, 313–319.

J.E. Hall, G.A. Zampighi and R.M. Davis (Eds.)
Progress in Cell Research, Vol. 3

CHAPTER 39

Rat connexin43: regulation by phosphorylation in heart

JUAN C. SÁEZ[a], ANGUS C. NAIRN[b], ANDREW J. CZERNIK[b], DAVID C. SPRAY[a] and ELLIOT L. HERTZBERG[a,c]

[a]Department of Neuroscience, Albert Einstein College of Medicine, 1300 Morris Park Ave., Bronx, NY 10461, [b]Laboratory of Molecular and Cellular Neuroscience, The Rockefeller University, New York, NY 10021-6399 and [c]Department of Anatomy and Structural Biology, Albert Einstein College of Medicine, Bronx, NY 10461, USA

Introduction

Cardiac gap junctions are essential for the propagation of electrical activity between myocytes, enabling coordination of the contractile behavior of the myocardium (Forbes and Sperelakis, 1985). Furthermore, gap junctions in this tissue mediate the intercellular exchange of second messengers and small molecules, allowing metabolic cooperativity (Tsien and Weingard, 1976; Lawrence et al., 1978).

At least two gap-junctional proteins, connexins (Cxs), are found in the mammalian myocardium, Cx43 and Cx46 (Beyer et al., 1987; Beyer et al., 1988), the best characterized being Cx43. It is an abundant constituent of myocyte gap junctions in ventricle and atrium (Beyer et al., 1987; Yancey et al., 1989). The prediction that Cx43 has four transmembrane domains and that both the C- and N-terminal regions are localized in the cytoplasm (Beyer et al., 1987) has received empirical support using sequence-specific antibodies and site-specific proteases (Yancey et al., 1989; Laird and Revel, 1990).

Several physiological studies have shown that second messenger pathways affect the functional state of heart gap junctions. Hence, it has been proposed that the effect could be mediated by protein kinases, including cAMP- or cGMP-dependent protein kinase or protein kinase C (PKC), which would directly phosphorylate the gap-junctional protein(s) (Burt and Spray, 1988; De Mello, 1988). It has been clearly established that Cx43 is a phosphoprotein. In immunoblots of samples of heart ventricle or cultured neonatal cardiac myocytes subjected to SDS-PAGE, Cx43 is detected as a prominent doublet of about 43–45 kDa and a minor band of 41 kDa (Musil et al., 1990; Kadle et al., 1991; Laird and Revel, 1991; Lau et al., 1991). The two bands of slowest electrophoretic mobility incorporate $^{32}P[P_i]$ and, upon incubation with alkaline phosphatase, lose the radioactivity and their electrophoretic mobility is shifted to that of the 41-kDa band, indicating that the latter is the unphosphorylated form and the former are phosphorylated forms of Cx43 (Laird and Revel, 1991; Lau et al., 1991). Furthermore, Cx43 has been shown to be phosphorylated and rapidly dephosphorylated in mammalian fibroblasts (Crow et al., 1990). Although Cx43 contains several putative phosphorylation consensus sequences, it is unknown which sites are in fact phosphorylated and by which protein kinase(s). In addition, although it has been possible to dephosphorylate Cx43 in vitro (Musil et al., 1990; Laird et al., 1991; Lau et al., 1991) the identity of the phosphatase(s) responsible for this reaction in vivo is unknown.

In this study, using several compounds that alter the activity of protein kinases or protein phosphatases, we obtained evidence that suggests a role for PKC and phosphatases I and/or 2A in phosphorylation and dephosphorylation of Cx43, respectively. Furthermore, a synthetic peptide, whose sequence corresponds to a C-terminal region of Cx43, was phosphorylated by PKC at two identified seryl residues. We also found a direct correlation between the level of ^{32}P-incorporation into Cx43 and the extent of coupling between myocytes.

Results

Regulation of the state of phosphorylation of Cx43 in primary cultures of cardiac myocytes

Cx43 is a phosphoprotein, and at least two phosphorylated forms can be detected by immunoprecipitation from $[^{32}P]P_i$ labeled cells or by immunoblotting (Crow et al., 1990; Musil et al., 1990; Kadle et al., 1991; Laird et al., 1991; Lau et al., 1991). In immunoblots of total cell lysates of primary cultures of rat neonatal cardiac myocytes, three Cx43-immunoreactive bands were detected (Fig. 1). A doublet between 43–45 kDa was prominent and a minor band at 41 kDa was observed (Fig. 1, I, lane 1). From cultures labeled with $^{32}P[P_i]$, two phosphoproteins were immunoprecipitated with anti-Cx43 anti-

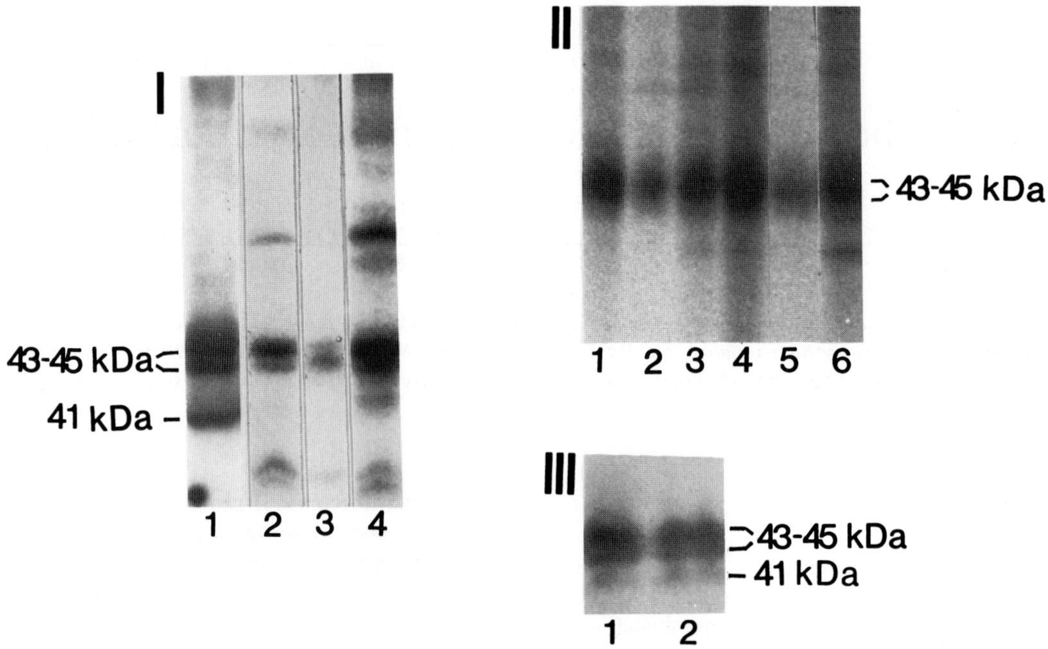

Figure 1. Regulation of the state of phosphorylation of Cx43. Proteins (50 μg/lane) of cell lysates of neonatal cardiocytes were separated by SDS-PAGE using 8% (14 × 11 cm; I) or 10% (8 × 6 cm; II and III) acrylamide gels and analyzed by immunoblotting (I) with an anti-Cx43-[346–360] antibody and ^{125}I-protein A. Different forms of Cx43 were resolved by immunoblotting (I, lane 1). Confluent cultures of cardiocytes were labeled with ^{32}Pi (0.5 mCi/60 mm plate) for 2 h. Under control conditions, both bands of the 43–45 kDa doublet were radiolabeled with ^{32}Pi, while the 41-kDa band was unlabeled (I, lane 2). Panel II: Effect of different compounds that alter the activity of protein kinases or protein phosphatases on ^{32}P-incorporation into Cx43. The compounds were added during the last 30 min of a 2 h labeling period. Cx43 was immunoprecipitated and samples were run in a 8% (I) or 10% (II) acrylamide SDS-PAGE of 14 × 11 cm. Treatment with 300 nM ST reduced (I, lane 3) and, with 60 nM OA, increased (I, lane 4) the incorporation of ^{32}P into Cx43. In another experiment, the effect of ST (II, lane 2) was reversed by a 20 min treatment with 600 nM (II, lane 3) or 900 nM TPA (II, lane 4) and by 2 μM ionophore A23187 (II, lane 6) but not by 4-α-phorbol (II, lane 5). The pattern of bands detected by immunoblotting in control (III, lane 1) and after treatment with 300 nM ST (III, lane 2) for 30 min in phosphate-free medium.

bodies that had the same electrophoretic mobilities as those of the doublet detected by immunoblot (Fig. 1, I, lane 2). No incorporation of $^{32}P_i$ into the 41-kDa band was detected. ^{32}P-incorporation into Cx43 was reduced by staurosporine (ST, 300 nM), a protein-kinase inhibitor (Kase et al., 1987) and was increased by okadaic acid (OA, 60 nM), an inhibitor of phosphatases types 1 and 2A (Ishihara et al., 1989) (Fig. 1, I). When SDS-PAGE was performed using 8% acrylamide gels to better resolve the bands, it was observed that the effects were more pronounced on the band of lowest electrophoretic mobility (Fig. 1, I lanes 3 and 4). In cells first treated with ST, phosphorylation of Cx43 recovered substantially, in a dose-dependent manner, after subsequent treatment with 12-O-tetradecanoylphorbol-13-acetate (TPA) but not after treatment with its inactive analogue (Fig. 1, II lanes 3, 4 and 5). Treatment with the ionophore A23187 (2 μM) also enhanced ^{32}P-incorporation into Cx43 (Fig. 1, II lane 6), but treatment with 0.5 mM 8Br-cAMP or 8Br-cGMP for 20 min either had no effect or, occasionally, reduced incorporation (not shown). The relative ratio of the 43-45/41-kDa bands of paired cultures was determined by immunoblotting. Neither the levels nor the ratio 43–45/41-kDa bands was affected by ST, TPA or OA. The result obtained with ST in shown in Fig. 1, III. This result was unexpected since ST clearly reduced ^{32}P-incorporation into Cx43 in parallel experiments.

Phosphorylation of synthetic peptides and identification of the sites of phosphorylation

Seven peptides (Table I, a–g), corresponding to sequences in the C-terminal region and the cytoplasmic loop (Fig. 2A) of the deduced primary structure of rat Cx43, were

TABLE I. Synthetic peptides used as substrates for protein kinases.

	Cx43 peptides	Residues
a	(C)L K V A Q T D G V N V	[113–123]
b	H G K V K M R G G L R T Y C	[142–155]
c	K G R S D P Y H A T T G P L S P S K D C	[241–260]
d	L S P S K D C G S P K Y A	[254–266]
e	G C S S P T A P L S P M S P P G	[270–285]
f	P P G Y K L V T G D R N N S S C	[283–298]
g	D Q R P S S R A S S R A S S R P	[360–375]
g'	D Q R P S S R A **A** S R A **A** S R **F**	[360–375]*

synthesized. The peptides were assayed as substrates for protein kinase C (PKC; from bovine brain), cAMP- and cGMP-dependent protein kinases (from bovine heart and lung, respectively) and Ca^{2+}/calmodulin-dependent protein kinase II (from rat brain). Only peptide Cx43-[360–375] was phosphorylated, and only by PKC (Fig. 2B and C). Aliquots of the reaction mixture were taken at different times and peptides were separated by HPLC using a C18 reverse-phase column and a gradient of acetonitrile in 0.1% TFA (Fig. 2B). Peptides were detected by absorbance at 214 nm and the amount of peptide was quantified by integration of the area under the curve. Standardization was carried out by measuring the area under the curve obtained with known amounts of dephosphopeptide. Radioactivity in each fraction was measured by Cerenkov counting. The specific activity of each peak was determined by dividing total radioactivity in each peak by the relative amount of the peptide. At time zero, only peak I was detected, corresponding to the unphosphorylated peptide (Fig. 2B). Upon phosphorylation for various times, three ^{32}P-labeled phosphopeptides (peaks II–IV), with distinct retention times, increased (Fig. 2B), while peak I decreased. The stoichiometry of phosphorylation for the peak II and peak III phosphopeptide was 1 mole of phosphate/mole of peptide, and for the peak IV phosphopeptide was 2 mol of phosphate/mole of peptide.

To establish the exact location of the phosphorylated residues, samples of peak II, III and IV phosphopeptides

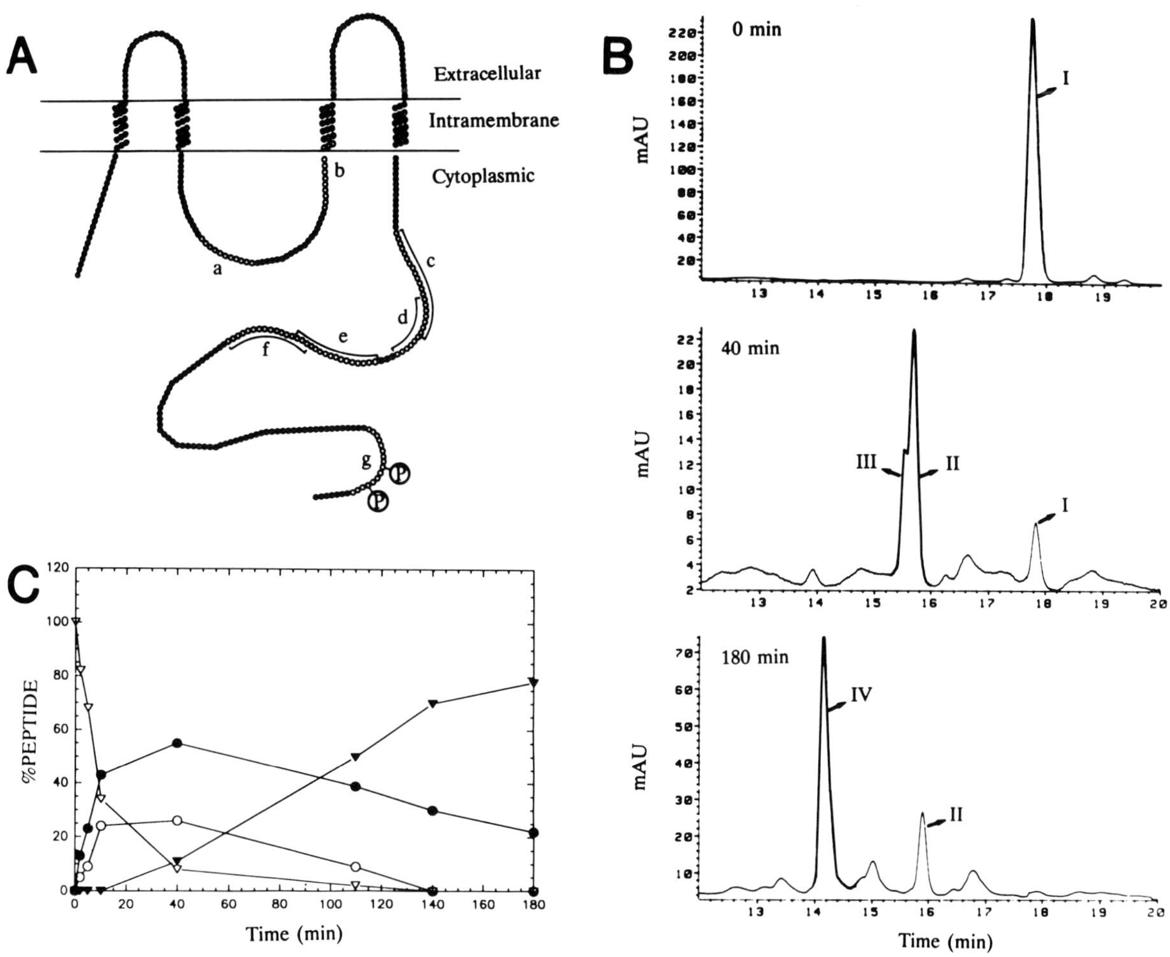

Figure 2. Phosphorylation of a synthetic peptide of Cx43 by PKC. (A) Molecular topology of Cx43 inserted in the plasma membrane (adapted from models proposed by Beyer et al., 1987 and Yancey et al., 1989). The regions of Cx43 that correspond to seven synthetic peptides assayed as substrates for different protein kinases are represented by (α). The region of overlap between two synthetic peptides (c and d; e and f) is indicated by a parallel trace to the molecule. Aliquots of the phosphorylation reaction mixture of the peptide Cx43-[360–375] by PKC were removed at different times and separated by HPLC. (B) HPLC peptide analyses of aliquots removed at times 0, 45 min and 100 min are illustrated. Peak I, unphosphorylated peptide; peak II, phosphorylated peptide at serine 368; peak III, phosphorylated peptide at serine 372 and peak IV, phosphorylated peptide at serines 368 and 372. (C) The time course for the disappearance of the unphosphorylated Cx43-[360–375] (▽) and the appearance of its phosphorylated forms at Ser-368 (●), Ser-372 (O) or at both Ser-368 and -372 (▼) are shown.

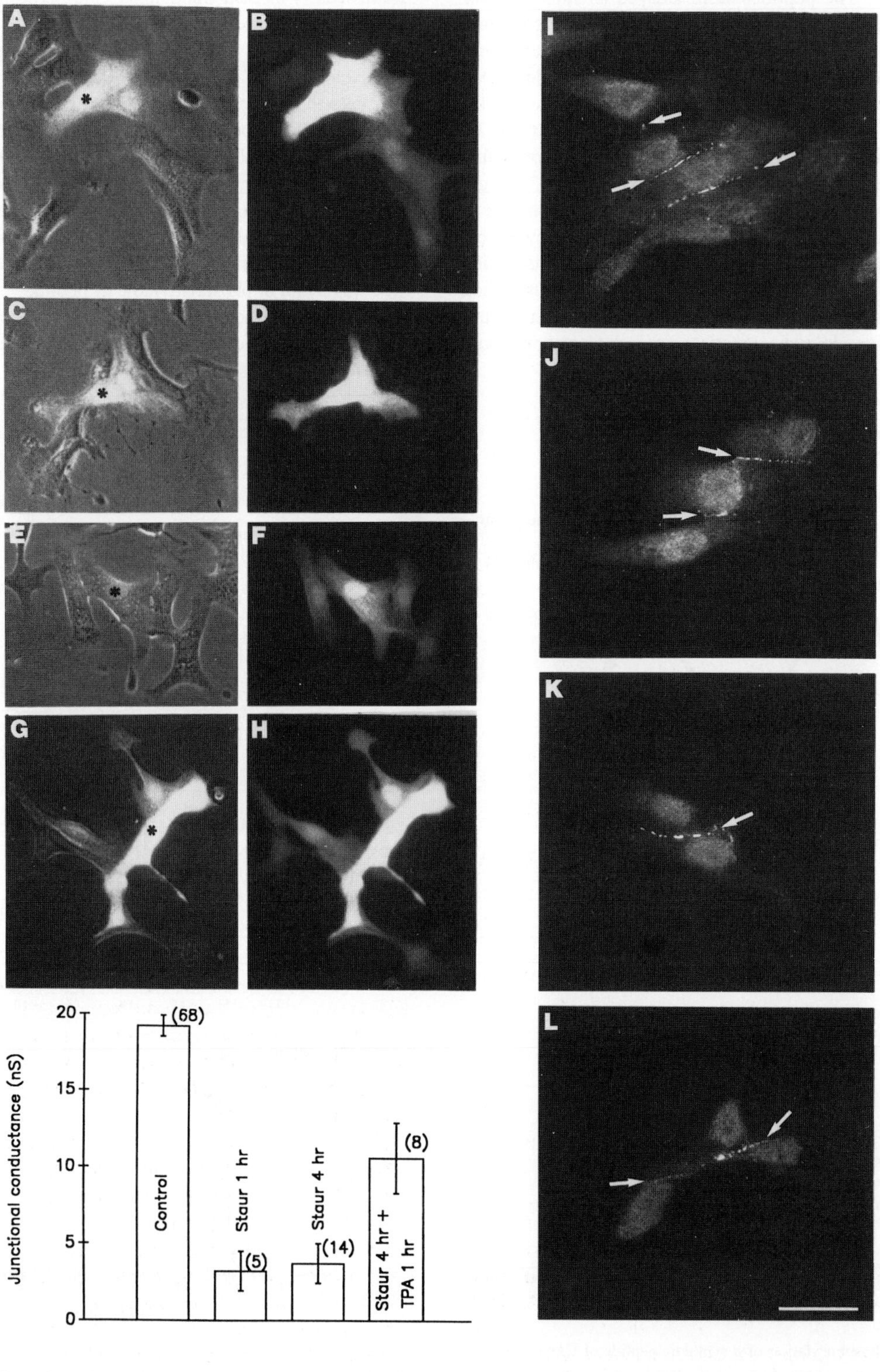

Figure 3. Effect of ST, OA and TPA on dye and electrical coupling and distribution of Cx43 in cardiocytes. A, C, E and G are phase-contrast photomicrographs of fluorescence fields shown in B, D, F and H, respectively. Photomicrographs show dye coupling between cardiocytes in culture under (B) control conditions or after treatment with (D) 300 nM ST for 4 h, (F) 300 nM ST and then 600 nM TPA for 1 h in the presence of ST or (H) 60 nM OA for 4 h. Graph shows junctional conductance between cardiocyte pairs under control conditions (Control) or after treatment with 300 nM ST for 1 h (Staur 1 h) or 4 h (Staur 4 h), or 300 nM ST for 4 h plus 600 nM TPA as above (Staur 4 h + TPA 1 h). Photomicrographs show the distribution of Cx43 studied by indirect immunofluorescence in (I) control cultures, and after treatment with (J) 300 nM ST, (K) 600 nM TPA or (L) 60 nM OA. Calibration bar: A–H, 32 μm; I–L, 20 μm.

(Fig. 2B) were re-chromatographed, derivatized with ethanethiol, and subjected to automated Edman degradation. Treatment with ethanethiol results in the conversion of phosphoserine to S-ethyl-cysteine (S-Et-Cys), while seryl residues are unmodified. The appearance of the phenylthiohydantoin (PTH) derivative of the S-Et-Cys in the amino-acid analysis of each Edman cycle indicated the position of the original phosphoseryl residue. For the peak II phosphopeptide, PTH-S-Et-Cys was detected at cycle 9; for the peak III phosphopeptide, PTH-S-Et-Cys was detected at cycle 13; for the peak IV phosphopeptide, PTH-S-Et-Cys was detected at cycles 9 and 13. Thus, the time of appearance of the derivatives indicated that residues corresponding to serine-368 and serine-372 of Cx43 were phosphorylated (Fig. 2A). However, the rate of phosphorylation at the two sites appeared to differ. If Ser-372 was phosphorylated first, rapid phosphorylation of Ser-368 occurred; however, the inverse sequence of phosphorylation was slower (Fig. 2C).

In agreement with the identification of the sites of phosphorylation in peptide 360–375, a synthetic peptide Cx43-[360–375]* containing alanine residues at the positions corresponding to residues 368 and 372 (Table I, g') was not phosphorylated by PKC.

Kinetics of phosphorylation of synthetic peptides by protein kinase C

The rate of phosphorylation of the synthetic peptide Cx43-[360–375] by PKC was compared to the rate of phosphorylation of another peptide, Cx32-[228–239], corresponding to the sequence in the cytoplasmic domain of Cx32 which contains a phosphorylation site for PKC and cAMP-dependent protein kinase (Sáez et al., 1990). PKC phosphorylated Cx43-[360–375] with a V_{max} of 7 μmol $min^{-1} mg^{-1}$ and a K_m of 140 μM and Cx32-[228–239] with a V_{max} of 2 μmol min^{-1} mg^{-1} and a K_m of 50 μM.

Regulation of permeability and conductance of gap junctions by agents that affect phosphorylation-dependent pathways

The functional state of gap-junctional communication was estimated by dye coupling (Lucifer Yellow [LY]) and dual whole-cell voltage-clamp techniques.

Treatment with 300 nM ST for 1 or 4 h at 37°C induced a substantial reduction of the spread of LY and electrical coupling between myocytes (Fig. 3). The effect was partially reversed upon the addition of 600 nM TPA in the presence of ST (Fig. 3). Furthermore, treatment with 60 nM OA did not reduce the extent or incidence of dye coupling (Fig. 3).

The cellular localization of Cx43 was studied by indirect immunofluorescence using anti-Cx43-[346–360] as primary antibody and FITC-conjugated anti-rabbit IgG as secondary antibody. Cardiac myocytes cultured on glass coverslips were treated with different compounds and then fixed with 70% ethanol at −20°C for 20 min. In cells under control conditions, labeling was localized at appositional membranes and also in the cytoplasm (Fig. 3I). After treatment for 20 min (not shown) or for as long as 4 h with 300 nM ST, 600 nM TPA or 60 nM OA the extent and the distribution of labeling were comparable to controls (Fig. 3J, K and L).

Discussion

Physiological studies with cultured cells have suggested that direct phosphorylation of cardiac connexins could be responsible for the changes in permeability and conductance of gap junctions observed after treatment with agents that activate specific protein kinases (Burt and Spray, 1988; De Mello, 1988; 1991; Spray and Burt, 1990). It has been established that Cx43, a major cardiac gap-junction protein, is a phosphoprotein (Crow et al., 1990; Musil et al., 1990; Laird et al., 1991; Lau et al. 1991) and that phosphorylated forms of Cx43 localize to gap junctions (Laird et al., 1991). Amino acid analyses have revealed that, in cardiocytes, both bands (43–45 kDa) of the Cx43 doublet are phosphorylated on seryl residues (Laird et al., 1991; Lau et al., 1991), and that the band with the lowest electrophoretic mobility can be phosphorylated, although to a minor extent, on threonyl residues (Lau et al., 1991). Results obtained in this study suggest that PKC phosphorylates Cx43. Although the state of phosphorylation was not affected by TPA, ST reduced ^{32}P-incorporation into Cx43 and its effect was reversed by TPA in a dose-dependent manner. Since PKC is the main binding site for TPA (Niedel et al., 1983) and the effect of ST was reversed upon addition of TPA, it is suggested that Cx43 can be phosphorylated by PKC. Furthermore, a synthetic peptide, Cx43-[360–375], was phosphorylated by PKC at sites corresponding to seryl residues 368 and 372 of Cx43. Phosphorylation was stoichiometric and kinetic properties of the reaction were similar to those of peptide Cx32-[228–239], derived from the sequence of Cx32, which has been shown to be a good substrate for PKC (Sáez et al., 1990).

A role for cAMP- and cGMP-dependent protein kinases has been proposed for regulating the functional state of gap junctions between cardiocytes (Burt and Spray, 1988; De Mello, 1988). Under the conditions used in our current experiments, the state of phosphorylation of Cx43 was not affected by agents which stimulate the activity of cAMP- or cGMP-dependent protein kinase. ^{32}P-incorporation into Cx43 was not significantly altered by treatment of cells with membrane-permeant derivatives of cAMP or cGMP, either alone or after pretreatment with ST. However, the lack of effect might be due to the possibility that, under basal conditions, the protein is already maximally phosphorylated, or to an insufficient time of exposure to these agents. Moreover, cyclic nucleotides might be unable to reverse the effect of ST due to persistent inhibition of the kinases. The lack of cAMP- or cGMP-mediated phosphorylation of Cx43 is consistent with the absence of consensus sequences for phosphorylation by their respective protein kinases in both rat (Beyer et al., 1987) and human (Fishman et al., 1990) Cx43. In addition, none of the synthetic peptides was phosphorylated by cAMP- or cGMP-dependent kinases. This cannot, however, exclude every possibility, since the peptides tested did not include all the seryl and threonyl residues contained within the cytoplasmic domains of Cx43, and higher-order structural features of this region might be important to confirm substrate specificity. Although a brief report has appeared indicating that the major protein of isolated dog heart gap junctions could be phosphorylated in vitro by cAMP-dependent protein kinase (Pressler and Hathaway, 1987), evidence that this reaction occurs in vivo has not yet been reported.

The sequence surrounding both serines 368 and 372 conforms to the consensus sequences for phosphorylation by PKC (Kennelly and Krebs, 1991) and these sites are conserved in the sequence of both the rat and the human isoforms (Beyer et al., 1987; Fishman et al., 1990). In contrast to the phosphorylation site identified in Cx32, which can be phosphorylated by both PKC and cAMP-dependent protein kinase (Sáez et al., 1990), the sites of phosphorylation identified in Cx43-[360–375] are not located within a consensus sequence for cAMP-dependent protein kinase (Kennelly and Krebs, 1991). Thus, if the phosphorylation sites identified in Cx43-[360–375] are physiologically relevant, this would support the hypothesis that alternative phosphorylation of the C-terminal region of connexin isoforms may account for the differential regulation of their functional properties (Sáez et al., 1990).

The protein kinase $p34^{cdc2}$, which is involved in cell cycle control, phosphorylates seryl or threonyl residues located in the consensus sequence S/T-P-X-R/K (Kennelly and Krebs, 1991). Interestingly, in the C-terminal domain of rat Cx43, adjacent to the plasma membrane, serine 255 is found in a consensus sequence of phosphorylation for this kinase which is also present in the synthetic peptide Cx43-[254–266]. The latter was not phosphorylated by any of the protein kinases studied, and the possibility that it could serve as a substrate for $p34^{cdc2}$ kinases is currently being studied.

The predominance of the phosphorylated forms of Cx43 in cardiocytes (Kadle et al., 1991; Laird et al., 1991; Lau et al., 1991), confirmed in this work, could be the result of a high protein kinase or a low protein-phosphatase activity. In neonatal rat heart myocytes, the turnover rate of phosphate groups on Cx43 is the same as the half-life of the protein, suggesting that once the protein is phosphorylated it is not dephosphorylated during its lifespan (Laird and Revel, 1991). In contrast, similar experiments carried out in fibroblasts indicated that Cx43 was subject to rapid cycles of dephosphorylation and phosphorylation (Crow et al., 1990). Analogously, we found that OA increased ^{32}P-incorporation into Cx43, suggesting that those phosphatases might be involved in dephosphorylation of Cx43 in cardiac myocytes. Moreover, the inhibition of protein-kinase activity with ST caused a drastic reduction in ^{32}P-incorporation into Cx43, and indicates that at least some sites in Cx43 are being persistently phosphorylated. Thus, it appears likely that the predominance of the phosphorylated forms of Cx43 that have been observed in cardiocytes is the result of steady-state conditions which favor a higher ratio of kinase-to-phosphatase activity towards Cx43 as substrate. The inverse situation has been found in MDCK cells, where the dephosphorylated form of Cx43 is predominant (Chapter 38).

The changes in ^{32}P-incorporation induced by ST were neither associated with increased levels of the dephosphorylated form of Cx43 detected in immunoblots nor with changes in cellular distribution of the protein. Thus, the residues whose state of phosphorylation were affected by the ST treatment did not appear to be responsible for the different electrophoretic mobilities of the bands detected by immunoblotting.

It has been proposed that phosphorylation of connexins by cyclic nucleotide-dependent protein kinases might regulate gating properties (Sáez et al., 1986; Burt and Spray, 1988; De Mello, 1988). In our current studies, the inability of cAMP or cGMP analogues to alter the phosphorylation state of Cx43 in cardiocytes was not consistent with the changes in macroscopic g_j previously observed in the presence of these agents (Burt and Spray, 1988; De Mello, 1988). It remains to be determined whether changes in the phosphorylation state of another Cx (e.g. Cx46; Beyer et al., 1988) could explain the macroscopic changes in junctional conductance induced by cyclic nucleotides in cardiocytes.

The involvement of PKC in regulating the functional state of gap junctions has been reported, although the findings remain controversial. We found a direct correlation between changes in the state of phosphorylation of Cx43 obtained with PKC activators and inhibitors and changes in dye and electrical coupling between cardiocytes. These results were consistent with data previously reported, where treatment with a tumor-promoting phorbol ester that activates PKC increases g_j (Spray and Burt, 1990). In contrast, a recent report indicated that inhibitors of PKC increase g_j in adult cardiocytes, suggesting that activation of PKC reduces g_j (De Mello, 1991). The use of cardiocytes at different stages of development might partially account for the discrepancy in physiological effects observed in these studies.

In another system, phosphorylation of Cx43 correlated with its insertion into the plasma membrane (Musil et al., 1990). In cardiocytes, we have observed that changes in the phosphorylation state of Cx43 occurred without a significant change in cellular distribution of the molecule. Therefore, it seems reasonable to suggest that Cx43, once translocated to the plasma membrane, remains subject to both phosphorylation and dephosphorylation. In support of this idea, treatment of Cx43 transfectants with agents that affect protein phosphorylation or dephosphorylation affect the unitary junctional conductance (Chapter 18). Thus, although differential phosphorylation of Cx43 could be involved in its insertion into the plasma membrane, it is also probable that phosphorylation plays a role in gap-junctional gating.

Acknowledgements

This work was supported in part by National Institute of Health DK-41368-02 (to J.C.S.), HL-38449 (to D.C.S.), MH-40899 (to A.C.N.) and GM-30667 (to E.L.H.) and a Grant-in-Aid from the New York Chapter of the American Heart Association (to D.C.S.). E.L.H. is a recipient of a Career Scientist Award from the Irma T. Hirschl Trust. We also want to thank Mr. Richard Corpina for his technical assistance.

References

Beyer, E.C., Paul, D.L. and Goodenough, D.A. (1987) Connexin43: a protein from rat heart homologous to a gap-junction protein from liver. J. Cell Biol. 105, 2621–2629.

Beyer, E.C., Goodenough, D.A. and Paul, D.L. (1988) The connexins, a family of related gap-junction proteins. In: E.L. Hertzberg and R.G. Johnson (Eds.), Modern Cell Biology: Gap Junctions. Vol. 7, Alan R. Liss, New York, pp. 167–175.

Burt, J.M. and Spray, D.C. (1988) Inotropic agents modulate gap-junctional conductance between cardiac myocytes. Am. J. Physiol. 254, H1206–H1210.

Crow, D.S., Beyer, E.C., Paul, D.L., Kobe, S.S. and Lau, A.F. (1990) Phosphorylation of connexin43 gap-junction protein in uninfected and Rous sarcoma virus-transformed mammalian fibroblasts. Mol. Cell. Biol. 10, 1754–1763.

De Mello, W.C. (1988) Increase in junctional conductance caused by isoproterenol in heart cell pairs is suppressed by cAMP-dependent protein-kinase inhibitor. Biochem. Biophys. Res. Commun. 154, 509–514.

De Mello, W.C. (1991) Effect of vasopressin and protein kinase C inhibitors on junctional conductance in isolated heart cell pairs. Cell Biol. Intern. Rep. 15, 467–478.

Fishman, G.I., Spray, D.C. and Leinwand, L.A. (1990) Molecular characterization and functional expression of the human cardiac gap-junction channel. J. Cell Biol. 111, 589–598.

Forbes, M.S. and Sperelakis, N. (1985) Intercalated discs of mammalian heart: a review of structure and function. Tissue and Cell 17, 605–648.

Ishihara, H., Martin, B.L., Brautigan, D.L., Karaki, H., Ozaki, H., Kato, Y., Fusetani, N., Watabe, S., Hashimoto, K., Uemura, D. and Hartshorne, D.J. (1989) Calyculin A and okadaic acid: inhibitors of protein phosphatase activity. Biochem. Biophys. Res. Commun. 159: 871–877.

Kadle, R., Zhang, J.T. and Nicholson, B.J. (1991) Tissue-specific distribution of differentially phosphorylated forms of Cx 43. Mol. Cell. Biol. 11: 363–369.

Kase, H., Iwahashi, K., Nakanishi, S., Matsuda, Y., Yamada, K., Takahashi, M., Murakata, C., Sato, A. and Kaneko, M. (1987) K-252 compounds, novel and potent inhibitors of protein kinase C and nucleotide-dependent protein kinases. Biochem. Biophys. Res. Commun. 142: 436–440.

Kennelly, P.J. and Krebs, E.G. (1991) Consensus sequences as substrate-specificity determinants for protein kinases and protein phosphatases. J. Biol. Chem. 266: 15555–15558.

Laird, D.W., Puranam, K.L. and Revel, J.-P. (1991) Turnover and phosphorylation of connexin43 gap-junction protein in cultured cardiac myocytes. Biochem. J. 273, 67–72.

Laird, D.W. and Revel, J.-P. (1990) Biochemical and immunochemical analysis of the arrangement of connexin43 in rat heart gap-junction membranes. J. Cell Sci. 97, 109–117.

Lau, F.A., Hatch-Pigott, V. and Crow, D.S. (1991) Evidence that heart connexin43 is a phosphoprotein. J. Mol. Cell Cardiol. 23, 659–663.

Lawrence, T.S., Beers, W.H. and Gilula, N.B. (1978) Transmission of hormonal stimulation by cell-to-cell communication. Nature 272, 501–506.

Musil, L.S., Cunningham, B.C., Edelman, G.M. and Goodenough, D.A. (1990) Differential phosphorylation of the gap-junction protein connexin43 in junctional communication-competent and -deficient cell lines. J. Cell Biol. 111, 2077–2088.

Niedel, J.E., Kuhn, L.J. and Vandenbark, G.R. (1983) Phorbol diester receptor copurifies with protein kinase C. Proc. Natl. Acad. Sci. USA 80, 36–40.

Pressler, M.L. and Hathway, D.R. (1987) Phosphorylation of purified dog heart gap junction by cAMP-dependent. Circulation 76, suppl. IV: (abstract) 18.

Sáez, J.C., Spray, D.C., Nairn, D.C., Hertzberg, E.L., Greengard, P. and Bennett, M.V.L. (1986) cAMP increases junctional conductance and stimulates phosphorylation of the 27-kDa principal gap-junction polypeptide. Proc. Natl. Acad. Sci. USA 83, 2473–2477.

Sáez, J.C., Nairn, A.C., Czernik, A.J., Spray, D.C., Hertzberg, E.L., Greengard, P. and Bennett, M.V.L. (1990) Phosphorylation of connexin32, a hepatocyte gap-junction protein, by cAMP-dependent protein kinase, protein kinase C and Ca^{2+}/calmodulin-dependent protein kinase II. Eur. J. Biochem. 192, 263–273.

Spray, D.C and Burt, J.M. (1990) Structure-activity relations of the cardiac gap-junction channel. Am. J. Physiol. 258, C195–C205.

Tsien, R.W. (1977) Cyclic AMP and contractile activity in heart. In: P. Greengard and G.A. Robison (Eds.). Advances in Cyclic Nucleotide Research. Raven Press, New York, vol. 8, pp. 363–420.

Yancey, S.B., John, S.A., Lal, R., Austin, B.J. and Revel, J.-P. (1989) The 43-kDa polypeptide of heart gap junctions: immunolocalization (I), topology (II), and functional domains (III). J. Cell Biol. 108, 2241–2254.

J.E. Hall, G.A. Zampighi and R.M. Davis (Eds.)
Progress in Cell Research, Vol. 3

CHAPTER 40

Gap junction assembly: the external domains in the connexins fulfill an essential function

ROSS G. JOHNSON and RITA A. MEYER

Department of Genetics and Cell Biology, University of Minnesota, St. Paul, MN 55108, USA

Introduction to gap junction assembly

As two cells come together and establish contact, a sequence of events can be triggered which culminates in the formation of a gap junction. As a result, within minutes each cell becomes capable of sharing small molecules and modifying the activities of the neighboring cell. Although junction assembly appears to be a relatively simple process, we have only a limited understanding of the events that lead to this widespread form of cell communication (Johnson et al., 1989).

We have developed a cell culture system, involving reaggregated Novikoff hepatoma cells, which allows us to address a variety of questions about the gap junction assembly process. We have learned that the process is cell-contact dependent (Preus et al., 1981a), occurs in the absence of protein synthesis (Epstein et al., 1977), does not require more than 5% of the normal ATP levels (Epstein et al., 1977) and even occurs at temperatures as low as 4°C (Li et al., 1984). In addition, cytoskeletal elements do not appear to anchor or direct the movement of junctional precursors (Preus et al., 1985). Consequently, gap junction assembly has a number of features which are exhibited by spontaneous assembly systems. The process is obviously more complex by virtue of not occurring in solution and by possibly involving adhesion factors (Musil et al., 1990; Meyer et al., 1992). However, we have suggested that gap junctions "self-assemble" within the environment of the two apposed lipid bilayers (Johnson et al., 1989).

Gap junction assembly appears to be a highly conserved process, as evidenced by the formation of these junctions between cells of different types and different species (Epstein and Gilula, 1977). More recently, the sequencing of numerous cDNAs which code for distinct proteins has established the concept of a family of gap junction proteins, the "connexins", and has emphasized the homology of external connexin domains, in particular (Kumar and Gilula, 1986; Paul, 1986; Beyer et al., 1987; Nicholson and Zhang, 1989). For example, there is an intriguing conservation of cysteine residues in the external loops of all connexins sequenced to date (Beyer et al., 1990). Presumably, these sequences facing the external side of the junctional membrane are involved in the assembly of gap junctions. Since this process is conserved, we would anticipate that findings in the Novikoff system will be applicable to our understanding of gap junction assembly in all vertebrate cells. Our strategy then is to capitalize on the possibilities of the Novikoff system, utilizing molecular approaches to connexins as well. We describe below our early studies on Novikoff connexins.

A complete understanding of gap junction assembly will require a variety of experimental approaches, including work on gap junction reconstitution (Lampe et al., 1991; Stauffer et al., 1991) and studies dealing with the expression of mutagenized connexins (Swenson et al., 1990). In addition, investigations of a cellular nature will still be required to examine certain aspects of junction assembly. For example, consider the recruitment of gap junction precursors to the actual site of assembly. This is a critical aspect of assembly which could have a major impact on the extent of junctional formation and the subsequent degree of communication. Recruitment could also be subject to a number of biological control mechanisms (Meyer et al., 1992). Thus, we emphasize the need for multi-faceted approaches to gap junction assembly.

Characterizing connexins in Novikoff cells

With a number of valuable probes now available for connexins, a variety of experimental possibilities exist. In order to pursue these studies in a knowledgeable manner, it is necessary to identify the connexin(s) expressed in a particular system. Therefore, to explore the possible expression of different connexins in Novikoff cells, a detailed Western blot analysis was performed with a set of different antibodies (generously provided by a number of investigators). Five different antibodies for Cx43 all reacted in a specific manner when tested against Novikoff

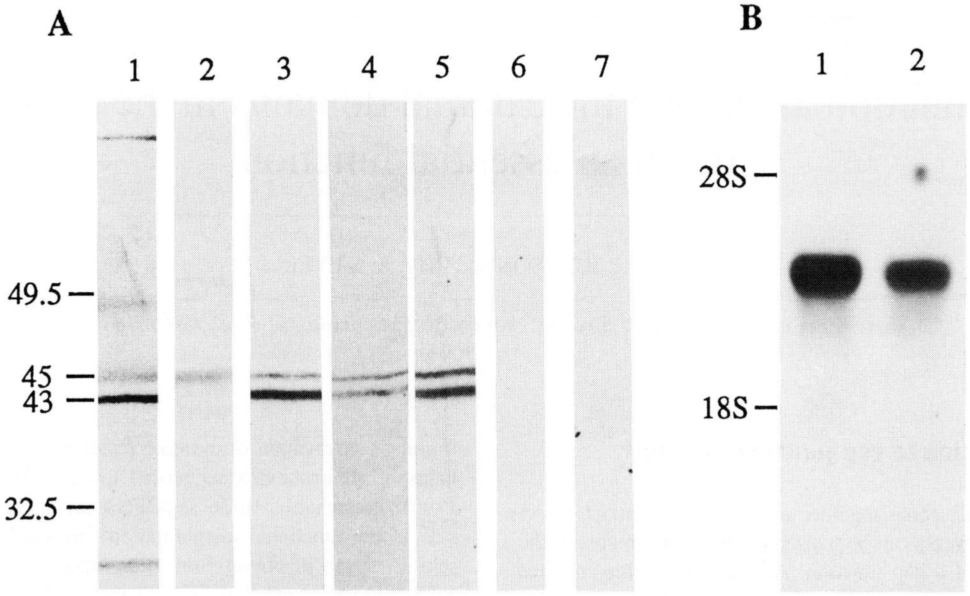

Figure 1. An analysis of connexin proteins and RNA found in Novikoff hepatoma cells. (A) This Western blot utilizes several specific rabbit antibodies for Cx43 peptides to label bands at ~43 kDa in plasma membrane fractions. See the legend to Table I for details on the peptide sequences. Lane 1 was reacted with an antibody for the N-terminus (AT-2); lane 2 with EL46; lane 3 with EL 186; lane 4 with an antibody for the second external loop in Cx32; lane 5 with an antibody for the C-terminus (CT360); lane 6 with the pre-immune antibodies for the Cx32 immunization; and lane 7 with a non-immune antibody. (B) In this Northern blot, a cDNA probe for rat Cx43 was used under conditions of high stringency to detect an RNA of ~3 kb from Novikoff cells (lane 1). Note that rat heart RNA contains a co-migrating species of ~3 kb (lane 2), as reported previously with the identification of Cx43 (Beyer et al., 1987).

cells or isolated plasma membrane fractions from these cells (see, e.g., Fig. 1A), while all but one of the Cx32 probes were negative (Fig. 1A). This exception is considered below.

A series of Northern blots was also evaluated, since with this method one can probe for homology throughout a molecule and not restrict one's approach to a single antigenic determinant. In this case, Novikoff RNA was reacted with cDNA probes for Cx26, Cx32 and Cx43. The Cx43 probe consistently gave a strong signal under conditions of high stringency, demonstrating the presence of a related RNA of ~3 kb (Fig. 1B). The only other significant reaction was a weak band at ~3 kb using the Cx32 probe at reduced stringency (data not shown). These results, coupled with the Western blot data on Novikoff proteins, strongly indicate that Cx43 or another closely related connexin species is expressed in these cells. Ongoing studies will determine: (a) whether this is actually Cx43 and (b) whether other connexins are also expressed in this system.

Since the Cx43 antibodies reacted specifically on Westerns, we proceeded to ask whether junctional structures could also be labeled with these probes in a specific manner. Immunofluorescence studies revealed a limited amount of punctate fluorescence at cell surfaces, along with significant levels of cytoplasmic staining (data not shown). The former was thought to represent gap junctions, while the latter could reflect sizeable stores of junctional protein and help explain the assembly of Novikoff gap junctions in the presence of protein-synthesis inhibitors (Epstein et al., 1977).

The immunological labeling of gap junctions at the ultrastructural level would be a more demanding means of asking whether Cx43 (or a closely related species) is actually a component of junctions linking Novikoff cells. Therefore, peptide-specific antibodies for the C-terminus of Cx43 were used to label isolated junctions in a crude membrane sample, followed by indirect labeling with gold-labeled antibodies (Fig. 2b). The Novikoff gap junctions were decorated with an impressive number of gold particles, while non-junctional membranes were essentially free of gold. In addition, non-immune antibodies failed to label the junctions (Fig. 2a). Thus, the specific binding of Cx43 antibodies to the junctions indicates that these membrane specializations contain Cx43-related proteins.

Additional support for the presence of these proteins in Novikoff gap junctions comes from work with "freeze-fracture replica immuno-labeling" or FRIL techniques (Gruijters et al., 1987). This provides for the labeling of gap junctions after the membranes have been fractured and the aggregates of junctional particles have been revealed, enabling one to definitively identify junctional regions. Cx43 antibodies were also found to label Novikoff gap junctions using this approach (Fig. 3). Since probes for the C-terminus of Cx43 were again used, we would expect the cytoplasmic side of the membrane to be

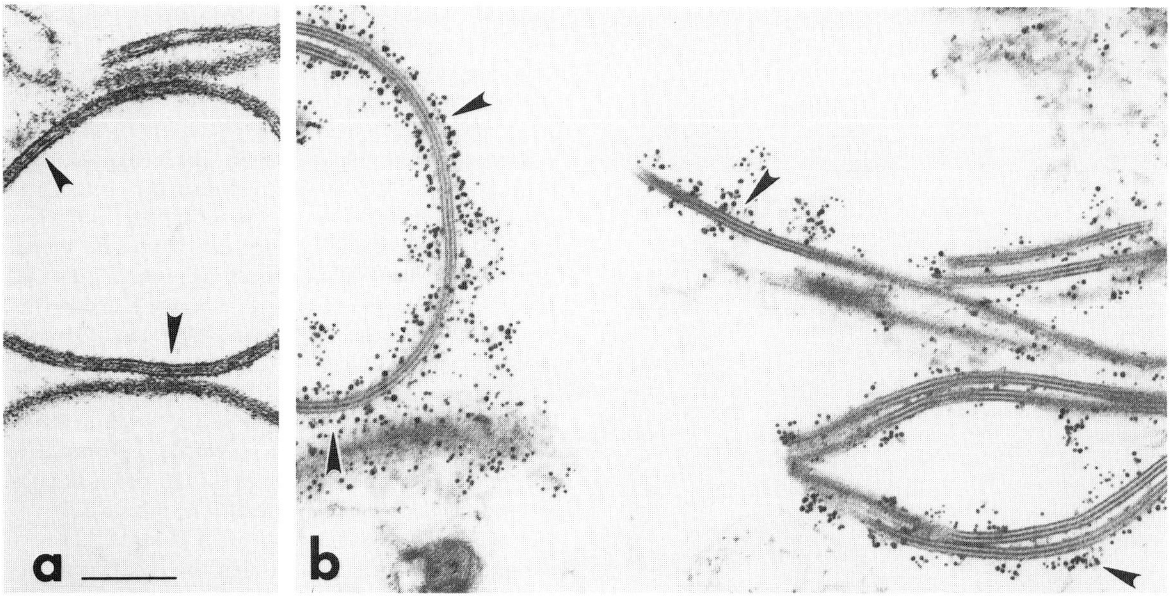

Figure 2. Indirect immuno-gold labeling of isolated gap junctions from Novikoff cells using rabbit antibodies for a peptide corresponding to the C-terminus of Cx43. In panel b, note the binding of large numbers of gold particles on both sides (left arrowhead) of the gap junction membranes. The specific nature of this binding is emphasized by the lack of label on non-junctional membranes and, in panel a, by the virtual absence of gold labeling (arrowheads) in the pre-immune samples. Bar is 200 nm. This figure is reproduced with the permission of the Journal of Cell Biology (Meyer et al., 1992).

labeled. The exposed junctional particles on the P-face were gold-labeled, with the binding actually occurring beneath the membrane after fracturing and replication. In addition, junctional pits were labeled with gold and were more obvious. The labeling of pits is thought to involve an unaltered junctional membrane (from cell 1), that binds antibody and is found beneath the fractured junctional membrane in which the pits are visualized (from cell 2).

TABLE I. Antibody effects on cell-to-cell dye transfer

Fab fragments used	Fab concentration (μg/ml)	Cell pairs without dye transfer after 4 min		
		No. of pairs	%	Total no. of injections
AT-2	250	2	5	42
CT-360	120	1	9	11
	250	3	5	55
MHC	250	4	9	42
Non-immune serum	30	0	0	12
	60	0	0	8
	120	0	0	12
	250	0	0	13
MIP	125	1	8	12
	250	1	3	31
Pre-Cx32 164–189	250	1	5	19
EL-46	30	0	0	15
	60	37	88	42
EL-186	30	52	88	59
Cx32 164–189	60	0	0	8
	120	0	0	13
	250	60	91	66
Total				460

*AT-2, polyclonal antibody specific for the N-terminus of Cx43; CT-360, affinity purified polyclonal antibody specific for the C-terminus of Cx43; MHC, monoclonal to the rat major histocompatibility complex; MIP, polyclonal antibody to the lens main intrinsic protein; EL-46, affinity purified polyclonal to the first extracellular domain of Cx43, residues 46–76; EL-186, affinity purified polyclonal antibody to the second extracellular domain of Cx43, residues 186–206; Cx32-164–189, polyclonal antibody to the second extracellular domain of Cx32, not affinity purified.

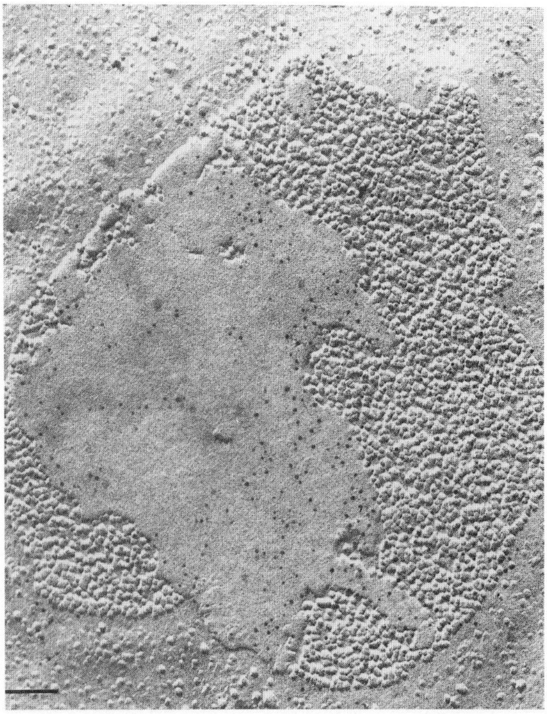

Figure 3. Labeling of a Novikoff gap junction with Cx43 antibodies (specific for the C-terminus) using the freeze-fracture replica immuno-label (FRIL) method. Note the presence of gold particles on both the particulate (P-face) and the pitted (E-face) aspects of the fractured gap junction membrane. The specificity of the binding is illustrated by the presence of very few gold particles on the non-junctional plasma membrane and by the low levels of binding in the pre-immune controls (data not shown). The interpretation of binding with respect to fracture planes is discussed in the text. Bar is 100 nm.

These findings with the FRIL methods reinforce the idea that Cx43-related proteins are present in Novikoff gap junctions. In the future, it will be of interest to carry out FRIL studies on reaggregating Novikoff cells to determine whether the 9–11 nm particles in developing gap junctions, i.e. in "formation plaques" (Johnson et al., 1989), are clearly junctional precursors.

Inhibiting gap junction assembly with antibodies

Since there is evidence that Cx43 is a transmembrane protein (Laird and Revel, 1990), it is logical to suggest that the external domains serve as surface receptors for an initial binding and subsequent "docking" of connexin-formed channels during gap junction assembly. At least some of the antibodies with specificity for these domains would be expected to interfere with connexin interactions during cell reaggregation and junction assembly. This could effectively perturb the gap junction assembly process. To test this hypothesis, Novikoff cells were dissociated with EDTA for 30 min, recovered for 90 min in a shaker-incubator (to eliminate previously existing junctions) and then reaggregated for 60 min by "settling out" in a petri dish (Meyer et al., 1992). During this last 60 min period, the cells were exposed to microgram quantities of Fab fragments of affinity-purified antibodies for Cx43 peptides (Laird and Revel, 1990). Three experimental antibodies were examined, two for external domains on Cx43 and one for the second external domain on Cx32 (Table I). This is the only Cx32 antibody that reacted in Western blots with Novikoff plasma membranes, labeling a 43 kDa band. This is not too surprising given the sequence similarities between connexins 32 and 43 in the second outside loop (approximately 70% homology). Fab fragments for a number of control antibodies were also examined. These included pre-immune antibodies, Cx43 antibodies which bind to the N- and C-termini on the cytoplasmic side of the membrane, antibodies for the rat histocompatibility complex and unrelated antibodies (Table I).

The antibody-treated cells were first evaluated by means of intracellular dye injection; the experiments were monitored on videotape with the use of a image-intensification camera (Fig. 4). In the controls, approximately 90% of the cells transferred dye after the 60-min reaggregations (Table I). In sharp contrast, less than 10% of the cells displayed dye coupling when treated with the appropriate concentration of Fab fragments for external connexin domains (Table I). Although the effect was concentration dependent, dye transfer was markedly inhibited with as little as 30 μg per ml of the affinity-purified Fab fragments. The clear concentration-dependent inhibition of junction formation may reflect the cooperative nature of this assembly process. While the average dye transfer time for Novikoff cell pairs is approximately 30 s in control preparations (Meyer et al., 1991), cells scored as negatives after antibody treatment showed no detectable transfer 4 min after injection. It should be noted that this study involved over 400 injections of cell pairs (Meyer et al., 1992). With the large difference in the control and experimental samples, we conclude that the antibodies for the external domains on Cx43 are altering gap junctions in a highly specific manner.

The most likely interpretation of these data was that the external domain antibodies were, indeed, inhibiting the assembly of gap junctions. However, there could have been an unexpected effect on the gating of junctional channels. Therefore, we next examined the antibody-treated cells with freeze-fracture methods and transmission EM. Cells were fixed with glutaraldehyde immediately after dye injection studies and processed for freeze fracture, in order to provide for the closest possible structure-function correlation. In cases where different antibody concentrations were examined and only a higher concentration had an effect on dye transfer (e.g., with the Cx32 antibody), then only the higher concentration was used. After treatment with Fab fragments for external connexin domains, no gap junctions were observed when 85 cell interfaces were examined (Meyer et al., 1992). In the control antibody preparations noted above, analysis of

134 interfaces revealed junctions on 67 interfaces. Thus, this structure-function approach clearly demonstrated that fewer cells were linked by gap junctions in the presence of the antibodies for the external domains in the connexins.

We interpret this reduced level of junctions as a specific effect on the gap junction assembly process for two reasons. First, we have reported the virtual absence of gap junctions at the beginning of the Novikoff reaggregation period (Preus et al., 1981b); this was confirmed by our evaluation of such samples in the present study. Second, we have found that junctions reappear over a time course of minutes, as monitored with both electrophysiological and ultrastructural methods (Johnson et al., 1974). The substantial presence of dye-permeable junctions in the control samples from this study (approximately 90% positive cell pairs) further emphasizes that we are dealing with extensive gap junction assembly during the course of cell reaggregation. The antibody treatments then dramatically reduced the extent of gap junction assembly as revealed by both the dye injections and freeze-fracture EM.

Implications of the antibody effects on assembly

Given the effect of antibodies on gap junction assembly, what then can we learn about this process? First, let us consider the formation plaque, which represents, to date, the earliest detectable stage in the process of gap junction assembly (Johnson et al., 1974; Preus et al., 1981b). This structure appears to involve both a specialization *within* the membrane (e.g., an accumulation of 9–11 nm membrane particles) and a linkage/association *between* apposed plasma membranes (since similar structures, though not necessarily mirror images, are matched in the two membranes). Our failure to detect any formation plaques in the presence of the connexin antibodies would suggest that connexin interactions are directly or indirectly required for the development of both of these features. We should note, however, that the present connexin antibody data do not allow us to distinguish between connexin–connexin interactions and connexin interactions with other molecules.

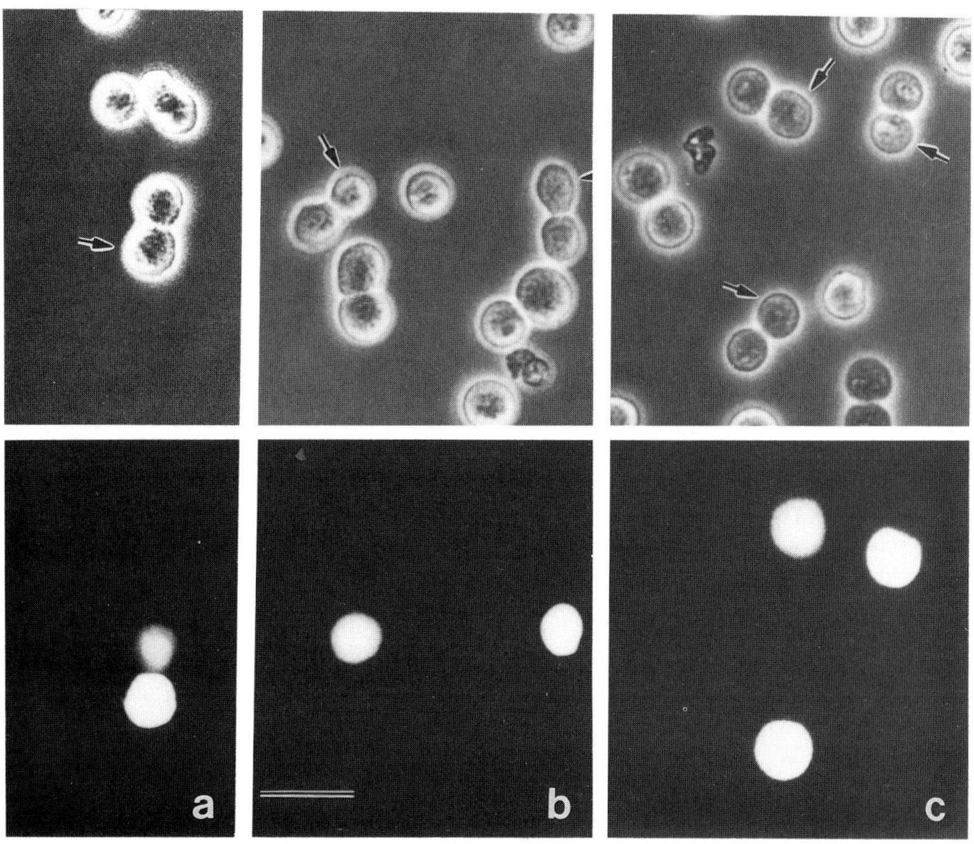

Figure 4. Results of dye injection studies following the reaggregation of Novikoff cells in the presence of antibodies for external domains on connexins. The micrographs are paired, with the phase contrast image presented above and the corresponding fluorescence image below. Antibodies for the N-terminus served as controls and did not block the transfer of dye (A). In contrast, dye transfer was blocked in five cell pairs treated with antibodies to the first external domain (anti-EL 46–76 in B) and the second external domain (anti-EL 186–206 in C). Bar is 20 nm.

We have suggested that the recruitment of 9–11 nm particles to the formation plaque is based on the lateral diffusion of these putative gap junction precursors (Johnson et al., 1989). Furthermore, there appears to be a mechanism for targeting these particles to the plaque and/or for retaining these particles within the plaque. Since the binding of Fab fragments would not be expected to dramatically alter the diffusion of the connexin complex within the bilayer (de Bradander et al., 1991), we suggest that the formation plaque is not initiated when connexin interactions are inhibited, or that an alteration has occurred in the retention of particles within this specialized membrane domain. For example, connexin–connexin binding *between* membranes could be a prerequisite for either of these aspects.

One obvious alternative to the lateral diffusion model involves exocytosis: the delivery and incorporation of a "pre-assembled" formation plaques into the plasma membrane. The present findings argue against such a mechanism. The presence of connexin antibodies in the culture medium would not be expected to alter the process of exocytosis, and yet no formation plaques were observed. Thus, we believe the exocytosis model provides an unsatisfactory alternative.

With the possibility of lateral diffusion playing an important role in gap junction assembly, a variety of questions quickly come to mind. Are all connexins which are present in the plasma membrane, found as hexamers? What is the mobility of the connexins within the bilayer? What factors influence the lifetime of connexins in the non-junctional membrane? For example, is protein phosphorylation involved? How is a formation plaque initiated? Are there factors which serve to stabilize the formation plaque? Do membrane lipids play a key role in formation plaque function? Are there critical interactions between connexin hexamers within the same formation plaque membrane? Are any non-connexin proteins involved in the actual process of gap junction assembly?

The effects of the connexin antibodies on junction assembly, as reported here, illustrate that we have identified methods to address the functions of specific connexin domains. Hopefully, these types of experiments will provide us with access to this array of questions dealing with the mechanisms of gap junction assembly.

Acknowledgements

This work was supported by NIH Grant CA-28548. For generously providing the peptide-specific antibodies, the authors thank Dale Laird, Barbara Yancey and Jean-Paul Revel for the Cx43 probes (Yancey et al., 1989; Laird and Revel, 1990) and Dan Goodenough (Goodenough et al., 1988) for the Cx32 probes. We also express our appreciation to David Paul and Eric Beyer for providing us with their cDNAs for Cx32 (Paul, 1986) and Cx43 (Beyer et al., 1987) and to Bruce Nicholson for his assistance in probing Novikoff RNA in a Northern blot with his Cx26 cDNA (Nicholson and Zhang, 1989).

References

Beyer, E.C., Paul, D.L. and Goodenough, D.A. (1987) Connexin43: A protein from rat heart homologous to a gap junction protein from liver. J. Cell Biol. 105, 2621–2629.

Beyer, E.C., Paul, D.L. and Goodenough, D.A. (1990) Connexin family of gap junction proteins. J. Membr. Biol. 116, 187–194.

de Bradander, M., Nuydens, R., Ishihara, A., Holifield, B., Jacobson, K. and Geerts, H. (1991) Lateral diffusion and retrograde movements of individual cell surface components on single motile cells observed with nanovoid microscopy. J. Cell Biol. 112, 111–124.

Epstein, M.L. and Gilula, N.B. (1977) A study of communication specificity between cells in culture. J. Cell Biol. 75, 769–787.

Epstein, M.L., Sheridan, J.D. and Johnson, R.G. (1977) Formation of low-resistance junctions *in vitro* in the absence of protein synthesis and ATP production. Exp. Cell Res. 104, 25–30.

Goodenough, D.A., Paul, D.L. and Jesaitis, L. (1988) Topological distribution of two connexin32 antigenic sites in intact and split rodent junctions. J. Cell Biol. 107, 1817–1824.

Gruijters, W.T.M., Kistler, J., Bullivant, S. and Goodenough, D.A. (1987) Immunolocalization of MP70 in lens fiber 16–17 nm intercellular junctions. J. Cell. Biol. 104, 565–572.

Johnson, R., Hammer, M., Sheridan, J. and Revel, J.P. (1974) Gap junction formation between reaggregating Novikoff hepatoma cells. Proc. Natl. Acad. Sci. USA 71, 4536–4540.

Johnson, R.G., Meyer, R.A. and Lampe, P.D. (1989). Gap junction formation: A self-assembly model involving membrane domains of lipid and protein. In: N. Sperelakis and W.C. Cole (Eds.), Cell Interactions and Gap Junctions. CRC Press, Boca Raton, pp. 159–179.

Kumar, N.M. and Gilula, N.B. (1986) Cloning and characterization of human rat liver cDNAs coding for a gap junction protein. J. Cell Biol. 103, 767–776.

Laird, D.W. and Revel, J.P. (1990) Biochemical and immunochemical analysis of the arrangement of connexin43 in rat heart gap junction membranes. J. Cell Sci. 97, 109–117.

Lampe, P.D., Kistler, J., Hefti, A., Bond, J., Muller, S., Johnson, R. G. and Engel, A. (1991) *In vitro* assembly of gap junctions. J. Struct. Biol. 107, 281–290.

Li, X.-R., Quade, B., Preus, D., Sheridan, J.D. and Johnson, R.G. (1984) Lack of filipin labeling and reduced particle aggregation at low temperatures. J. Cell Biol. 99, 344a.

Meyer, R.A., Lampe, P.D., Malewicz, B., Baumann, W. and Johnson, R.G. (1991) Low-density lipoprotein and apolipoprotein B enhance gap junction assembly and cell-cell communication. Exp. Cell Res. 196, 72–81.

Meyer, R.A., Laird, D.W., Revel, J.-P. and Johnson, R.G. (1992) Inhibition of gap junction and adherens junction assembly by connexin and A-CAM antibodies. J. Cell Biol. 119, 179–189.

Musil, L.S., Cunningham, B.A., Edelman, G.M. and Goodenough, D.A. (1990) Differential phosphorylation of gap junction protein connexin43 in junctional communication-competent and -deficient cell lines. J. Cell Biol. 111, 2077–2088.

Nicholson, B.J. and Zhang, J.-T. (1989) Sequence and tissue distribution of a second protein of hepatic gap junctions, Cx26, as deduced from its cDNA. J. Cell Biol. 109, 3391–3401.

Paul, D.L. (1986) Molecular cloning of cDNA for rat liver gap-junction protein. J. Cell Biol. 103, 123–134.

Preus, D., Johnson, R. and Sheridan, J. (1981a) Gap junctions between Novikoff hepatoma cells following dissociation and recovery in the absence of cell contact. J. Ultrastruct. Res. 77, 248–262.

Preus, D., Johnson, R., Sheridan, J. and Meyer, R. (1981b) Analysis of gap junctions and formation plaques between reaggregating Novikoff hepatoma cells. J. Ultrastruct. Res. 77, 263–276.

Preus, D., Li, X.-R., Sheridan, J.D., Balson, J., Wagner, M. and Johnson, R.G. (1985) Gap junction formation in the presence of colchicine and cytochalsin B. J. Cell Biol. 101, 179a.

Stauffer, K.A., Kumar, N.M., Gilula, N.B. and Unwin, N. (1991) Isolation and purification of gap junction channels. J. Cell Biol. 115, 141–150.

Swenson, K.I., Piwnica-Worms, H., McNamee, H. and Paul, D.L. (1990) Tyrosine phosphorylation of the gap junction protein connexin43 is required for pp60*src*-induced inhibition of communication. Cell Regul. 1, 989–1002.

Yancey, S.B., John, S.A., Lal, R., Austin, B.J. and Revel, J.P. (1989) The 43-kD polypeptide of heart gap junctions: Immunolocalization (I), topology (II), and functional domains (III). J. Cell Biol. 108, 2241–2254.

J.E. Hall, G.A. Zampighi and R.M. Davis (Eds.)
Progress in Cell Research, Vol. 3

CHAPTER 41

Characterization of rat gene regulatory elements

SHUANG BAI[a], DAVID C. SPRAY[b] and ROBERT D. BURK[c,d]

Marion Bessin Liver Research Center, Departments of [a]Medicine, [b]Neuroscience, [c]Pediatrics and [d]Microbiology and Immunology, Albert Einstein College of Medicine, 1300 Morris Park Avenue, Bronx, NY 10461, USA

Introduction

The biochemical components of gap junctions are connexins, which in each cell form hexameric assemblies that extend across the cell membrane to provide low-resistance channels between the cells. The isolation of junctional proteins allowed the cloning and sequencing of the cDNA encoding Cx32 (Kumar and Gilula, 1986; Paul, 1986). In addition, low-stringency hybridization of cDNA and genomic libraries with various connexin probes, and polymerase chain reaction (PCR) amplification of assorted genomes with conserved amplimers has identified a large number of homologous and/or related connexin genes (see Part I). Moreover, the cloning and characterization of genomic clones containing connexin coding regions has revealed a unique and perplexing structural organization. The connexin genes appear to contain the entire open reading frame on a single exon which is preceded by a large intron and a small noncoding upstream exon (Miller et al., 1988; Fishman et al., 1991).

Sequence-specific antibodies and cDNA probes for the various connexins have demonstrated that expression of individual connexins displays an overlapping tissue distribution, such that individual cells can express more than one type of gap-junction protein (e.g., Traub et al., 1989; Dermietzel et al., 1990; Spray et al., 1991). In addition to the tissue-specific distribution of connexins, expression has been shown to vary developmentally, both within the same cell types and for different cell types within a tissue (e.g., Dermietzel et al., 1989; Gimlich et al., 1990; Fishman et al., 1991), and in response to hormonal challenge for one connexin type in one tissue but not in others (Risek et al., 1990). Moreover, a growing number of reports indicate that expression of one connexin can be inhibited, while expression of another connexin may increase during tumorigenesis (Dermietzel et al., 1987; Fitzgerald et al., 1989; Neveu et al., 1990). The tissue specificity of connexin expression, and its apparent modulation by physiological and pathological stimuli, raise the issue of how transcription of this class of genes is regulated. In the initial description of the structure of the Cx32 gene, Werner and colleagues (Miller et al., 1988) suggested sequences that were candidates for possible regulatory function by several agents. Dr. Werner's laboratory kindly provided their rat genomic Cx32 clone, and we here report our characterization of the Cx32 gene regulatory elements. These findings include evidence for multiple transcription-initiation sites, localization of the native promoter through the use of the luciferase reporter, and the identification of DNase-hypersensitive sites corresponding to regions of the gene with negative enhancer activity in cultured hepatoma cell lines.

Materials and methods

Cell culture and transient expression assay

HuH-7, a highly differentiated human hepatoma cell line, was maintained in Dulbecco's modified Eagle minimal essential medium supplemented with 10% fetal bovine serum (GIBCO), 100 units/ml penicillin and 100 µg/ml streptomycin. Transient gene expression assays were performed by DNA transfection with supercoiled plasmid DNA, banded twice on a CsCl gradient using a calcium-phosphate precipitation technique (Okagama and Chen, 1987), followed by either luciferase assay or chloramphenicol acetyltransferase (CAT) assay. Briefly, cells were seeded on 35 mm plates 24 h prior to transfection. The DNA mixture for transfection included 2 µg of the test plasmid DNA and 0.1 µg of pXGH5 DNA, a human growth hormone expression vector which served as an internal control to monitor transfection efficiency (Nichols Diagnostic). Each transfection experiment was performed in duplicate. Cell extracts were prepared 48 h after transfection, either by lysing cells in 25 mM glycylglycine, 15 mM $MgSO_4$, 4 mM EGTA, 15 mM KH_2PO_4, and 1 mM DTT for determination of luciferase activity or by three cycles of freeze/thawing in 0.25 M Tris–HCl (pH 7.8) for the determination of CAT activity. Luciferase activity was determined as previously described (Brasier et al., 1989). Light output was measured for 20 s at room temperature in a monolight 2010 Luminometer (Analytical Luminescence Laboratory, San Diego, CA). Each luciferase assay was performed in duplicate. Luciferase activity was expressed as relative light units/μl cell extract/ml hGH.

CAT assays were performed as described by Gorman et al. (1982). Protein concentrations were determined by the method of Bradford (1976). Acetylated and non-acetylated forms of [^{14}C]chloramphenicol were separated by thin-layer chromatography. The percentage of acetylation of [^{14}C]chloramphenicol was determined by counting the regions of the TLC plates identified by autoradiography.

Plasmid constructions and preparation

The 8.1 kb *Eco*RI fragment containing the 5' region, first exon, and most of the intron of the rat Cx32 gene was isolated from Charon 4A (a kind gift from Dr. Rudolf Werner, University of Miami, College of Medicine) and subcloned into pGEM-3Z (Promega) to give pGC32-8.1 (see Fig. 2). The 2.2 kb 5' region [*Eco*RI to map position (mp) –33; mp1 corresponds to A in the the first ATG in the cDNA] was cloned from pGC32-8.1 after polymerase chain reaction (PCR) amplification using one oligonucleotide primer, 5' GCGGTACCTGGTTGCAACTG-CTTT 3', complementary to mp –33 to –52 which contained a *Kpn*I site and a second oligonucleotide primer corresponding to the pGEM-3Z polylinker sequence. The PCR product was isolated, digested with *Kpn*I, and ligated in both orientations to the unique *Kpn*I site in p19LUC (van Zonneveld et al., 1988) which contains a promoterless luciferase gene cassette, yielding recombinants p19C32-2.2-LUC and p19C32-2.2*-LUC (in reverse orientation). A set of luciferase vectors containing varying 5' deletions of the genomic region upstream of the first exon was constructed either by restriction enzyme digestion or by PCR amplification. For construction of p19C32-760/HBK-LUC, the *Hind*III/*Bgl*II fragment was isolated from pGC32-8.1 and cloned into p19C32-760-LUC. Additional PCR-generated deletions were constructed as shown in Fig. 2b. These new constructs were called p19C32-325-LUC (mp –358 to –33), p19C32-146-LUC (mp –179 to –33), p19C32-101-LUC (mp –134 to –33), and p19C32-79-LUC (mp –112 to –33) corresponding to the map position of the primers used in the PCR reactions (see Fig. 2). Three genomic DNA fragments from the rat Cx32 gene locus, called P (1.7 kb *Eco*RI/*Pst*I fragment), H (1.9 kb *Hind*III fragment), and K (1.4 kb *Kpn*I fragment) elements containing DNase hypersensitive (HS) regions identified in this study (shown in Fig. 3b) were isolated and inserted 3' to pCAT-P (Promega), an enhancerless CAT-expression vector driven by the SV40 early promoter. The new constructs were called pCAT-P, pCAT-H and pCAT-K which contained the P, H, and the K element, respectively (see Fig. 4).

Primer-extension assay

Total rat liver RNA was isolated by the method of Chirgwin et al. (1979). The primer extension assay was performed according to the method described by Sambrook et al. (1989). Briefly, 2 ng of a ^{32}P end-labeled oligonucleotide primer, 5' TGTAGAATGCCGAATCACG-CCACTGAGCAA 3', complementary to the published rat Cx32 cDNA sequence (mp 25 to 54), was added to 30 μl of hybridization buffer (40 mM PIPES (pH 6.4), 1 mM EDTA, 400 mM NaCl, 80% formamide) containing 30 μg of total liver RNA. After annealing at either 42°C or 37°C overnight, the nucleic acids were precipitated with ethanol

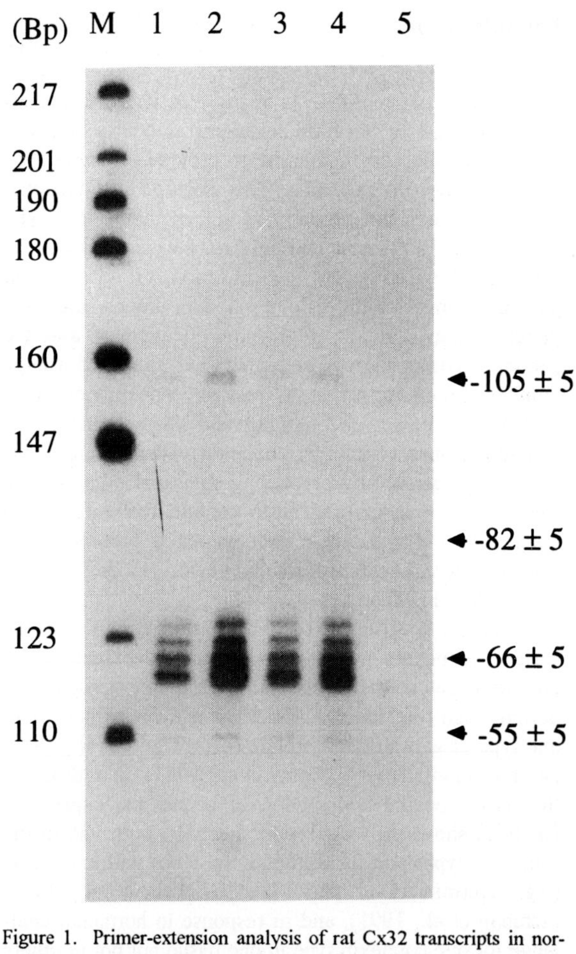

Figure 1. Primer-extension analysis of rat Cx32 transcripts in normal rat liver. 30 μg of rat liver RNA was annealed, and cDNA synthesized with an oligonucleotide primer as described in Materials and Methods. Samples in lanes 1 and 2 were annealed at 37°C and extended at 37°C and 42°C, respectively. Samples in lanes 3 and 4 were annealed at 42°C and extended at 37°C and 42°C, respectively. The extended cDNA products were analyzed on a 6% polyacrylamide sequencing gel, dried and exposed overnight to film at –70°C. The arrows on the right side of panel indicate the 5' ends of the Cx32 transcripts and the numbers correspond to their locations, upstream of the first ATG in the cDNA. Lane M contains *Msp*I-digested pBR322 DNA fragments end-labeled with ^{32}P, and the number of base pairs in each fragment is indicated at the left side of the figure.

and dissolved in 20 µl of reverse transcriptase buffer (50 mM Tris, pH 7.6, 60 mM KCl, 10 mM $MgCl_2$, 1 mM of each dNTP, 1 mM DTT and 1 unit/µl of RNase inhibitor). 30 units of AMV reverse transcriptase (United States Biochemical Corporation) was then added and the mixture incubated for 2 h followed by the addition of 1 µl of 0.5 M EDTA and 1 µl of 10 mg/ml RNase. The primer-extended cDNA products were denatured and analyzed on a 6% polyacrylamide sequencing gel.

Isolation of rat liver nuclei, DNase digestion and Southern blot hybridization

Nuclei were isolated as previously described (El-Ghor and Burk, 1988). Briefly, fresh isolated rat liver was minced and washed 3 times in an ice-cold buffer of 150 mM NaCl, 15 mM Na citrate, 10 mM Tris (pH 7.4) and centrifuged for 15 s. The cell pellet was resuspended in RSB buffer (10 mM NaCl, 3 mM $MgCl_2$ 10 mM Tris, pH 7.4) supplemented with 0.5% NP-40 and 1 mM PMSF and homogenized. The homogenate was filtered through two layers of cheesecloth and centrifuged for 10 min. The nuclei pellet was washed twice with, and resuspended in, RSB buffer. For DNase HS site analysis, aliquots of nuclei (approximately $1-2 \times 10^6$) were incubated at 37°C for 5, 10 and 20 min, respectively allowing endogenous DNase cleavage. The reactions were terminated by the addition of stop buffer (1% SDS, 600 mM NaCl, 20 mM EDTA, 400 µg proteinase K/ml, 20 mM Tris, pH 7.5), phenol/chloroform extracted, and ethanol precipitated. The purified DNA was then digested with the indicated restriction enzymes (New England Biolabs) according to manufacturer's recommendation. As a control, rat liver genomic DNA was purified in the presence of 25 mM EDTA and analyzed in parallel. Digested DNAs were electrophoresed on 1.0% agarose gels, transferred to genescreen membranes (NEN Research Products) and hybridized with a Cx32 cDNA probe, labeled with ^{32}P to a specific activity > 10^8 cpm/µg.

Results

Characterization of Cx32 mRNA start sites in rat liver

To identify the rat Cx32 mRNA 5' end(s), a single-stranded oligonucleotide primer, corresponding to mp 25 to 54 of the rat Cx32 cDNA, was synthesized and used for primer-extension analysis of rat liver RNA. As shown in Fig. 1, the major extended cDNA products were approximately 120 ± 5 bps. Less abundant cDNA products migrated around 158 ± 5 bps, 135 ± 5 bps and 110 ± 5 bps. Primer-extension assays, using either 37°C or 42°C for annealing, and cDNA extension gave similar results. Thus, the major transcription start sites correspond to mp –66 ± 5 nucleotides, and the minor start sites to mp –105, –82, and –55 ± 5 nucleotides. These results demonstrate that the rat Cx32 gene has multiple transcription-initiation sites localized upstream of the first exon.

Genetic analysis of rat Cx32 upstream regulatory regions

To determine the region(s) necessary for the transcriptional activity of the rat Cx32 gene, a series of deletions starting upstream of the first exon and extending toward the mRNA initiation sites, were generated and cloned 5' to a luciferase reporter gene in p19LUC (see Fig. 2). These recombinants were verified by restriction-enzyme and sequence analysis and tested for promoter activity by transfection into HuH-7 cells. The structure of the various deletions and their relative promoter activities are shown in Fig. 2b. The plasmid construct p19C32-2.2-LUC contained approximately 2.2 kb of the genomic region upstream of the rat Cx32 first exon; it had approximately 100-fold greater luciferase activity than p19LUC after normalization for transfection efficiency. In contrast, the 2.2 kb fragment, inserted in the reverse orientation, resulted in only a 10-fold increase in luciferase activity. Deletion to the *Bam*HI site, p19C32-760-LUC, increased the luciferase activity to about 400-fold over p19LUC, or 4 times the activity of p19C32-2.2-LUC. These results suggest that a negative regulatory element was present in the region between the *Eco*RI and *Bam*HI sites. The silencer activity of the *Eco*RI/*Bam*HI fragment was confirmed using additional constructs (see below). Removal of the region between 594 and 358 decreased the promoter activity to the level produced by p19C32-2.2-LUC. Further deletion of the promoter region to mp –179 had minimal effect on the level of luciferase gene expression. However, deletion to mp –134 and mp –112 essentially abolished luciferase activity, indicating the presence of a basal promoter in the region between mp –179 to –112.

Identification of DNase HS sites within the rat Cx32 gene locus

Recent studies indicate that the majority of DNase HS regions within chromatin represent alterations in the nucleosome structure of DNA. In fact, many of these DNase HS regions have been localized within the regulatory regions of a variety of genes (Montandon et al., 1982; Jongstra et al., 1984). Therefore, identification of DNase HS sites can be used to identify regulatory elements 5', 3' and within the exons or introns of a given gene.

To determine the chromatin structure in and around the rat Cx32 gene, nuclei were isolated from rat liver and digested by endogenous DNase activity. The nuclear DNA was then purified and analyzed by restriction enzyme digestion and Southern blot hybridization with a ^{32}P-labeled Cx32 cDNA probe. The location of the DNase HS cleavage sites was determined by the indirect end-labeling method and/or restriction-enzyme mapping (Wu et al., 1979). After DNase digestion, the purified nuclear DNA isolated from rat liver contained new hybridizing fragments compared to the rat liver genomic DNA isolated in the presence of EDTA, a DNase inhibitor.

As shown in Fig. 3a, genomic DNA contains two *Kpn*I hybridizing fragments, 11.5 kb and 1.3 kb, and one *Hind*III hybridizing fragment, 6.6 kb. After allowing the endogenous DNase to digest the DNA in intact nuclei, two new hybridizing fragments, 8.4 kb and 5.4 kb, could be seen (indicated by arrows on the left side of blot) after *Kpn*I digestion, and one new band, 3.9 kb, (indicated by arrows on the right side of blot) after *Hind*III digestion. As expected, increased time of incubation resulted in degradation of the DNA, as can be seen by the disappearance of the 11.5 kb *Kpn*I and 6.6 kb *Hind*III fragments.

Restriction-enzyme mapping revealed that these DNase HS regions were localized 1.2 kb upstream of the first exon, about 2.0 kb downstream of the first exon and approximately 0.5 kb downstream of the Cx32 ORF (Fig. 3b).

Functional analysis of DNAse HS regions

To investigate the possible function of the regions containing the DNase HS sites in Cx32 gene expression, we isolated and cloned each of the DNA fragments containing DNase HS sites 3' to a CAT reporter gene driven by the SV40 early promoter which lacked an enhancer. The function of each element was determined after transfection into HuH-7 cells and the results are shown in Fig. 4. The CAT activities of the vectors containing the P and H fragments were approximately 25% of the control pCAT-P vector, whereas insertion of the K fragment did not cause any significant change in CAT activity. To determine whether the P and H regions containing DNase HS sites also act as silencer elements with the native rat Cx32 gene promoter complex, the H element was cloned 5' to the rat Cx32 promoter sequence in p19C32-760-LUC to yield construct p19C32-760/HBK-LUC. The P fragment was already present in the p19C32-2.2-LUC. Both fragments resulted in a decrease in luciferase activity in comparison to the p19C32-760-LUC (see Fig. 2b). These results support the observation that these two DNase HS regions

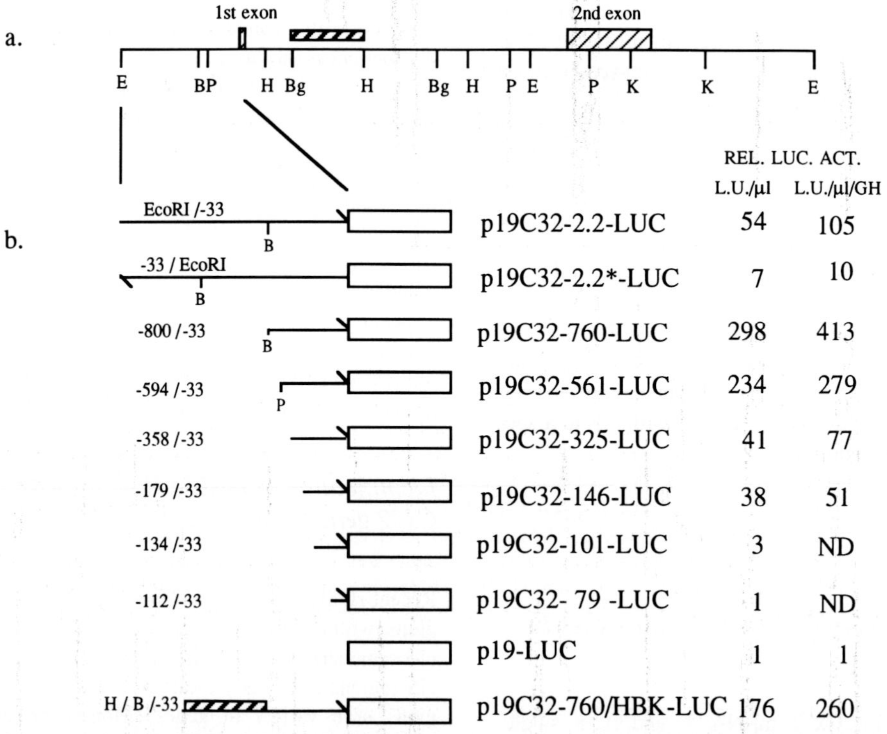

Figure 2. Deletion analysis of the rat Cx32 upstream genomic region. (a) A partial restriction map of the cloned rat DNA locus containing the Cx32 exons, intron and flanking sequences. (b) Diagram of DNA fragments, corresponding to the genomic region upstream of the splice donor site in the first exon and including the mRNA start sites, cloned into the p19LUC vector which contains a promoter/enhancerless luciferase reporter gene (open box). The indicated recombinants were transfected into HuH-7 cells and the luciferase activity was determined from 2 parallel transfections. Mean values are expressed relative to p19LUC. Relative luciferase activity (REL.LUC.ACT.) is expressed as light units (L.U.) per µl cell extract or light units per µl extract normalized for transfection efficiency by a co-transfected vector expressing human growth hormone. The numbers (i.e. –33 to –800 correspond to sequences upstream of the ATG initiation codon (+1 to +3). Restriction endonuclease abbreviations: E, *Eco*RI; B, *Bam*HI; P, *Pst*I; H, *Hind*III; Bg, *Bgl*II; K, *Kpn*I. The exons are as indicated and the box with bold hatch marks defines a *Bgl*II/*Hind*III fragment containing a DNase HS site (see Fig. 3).

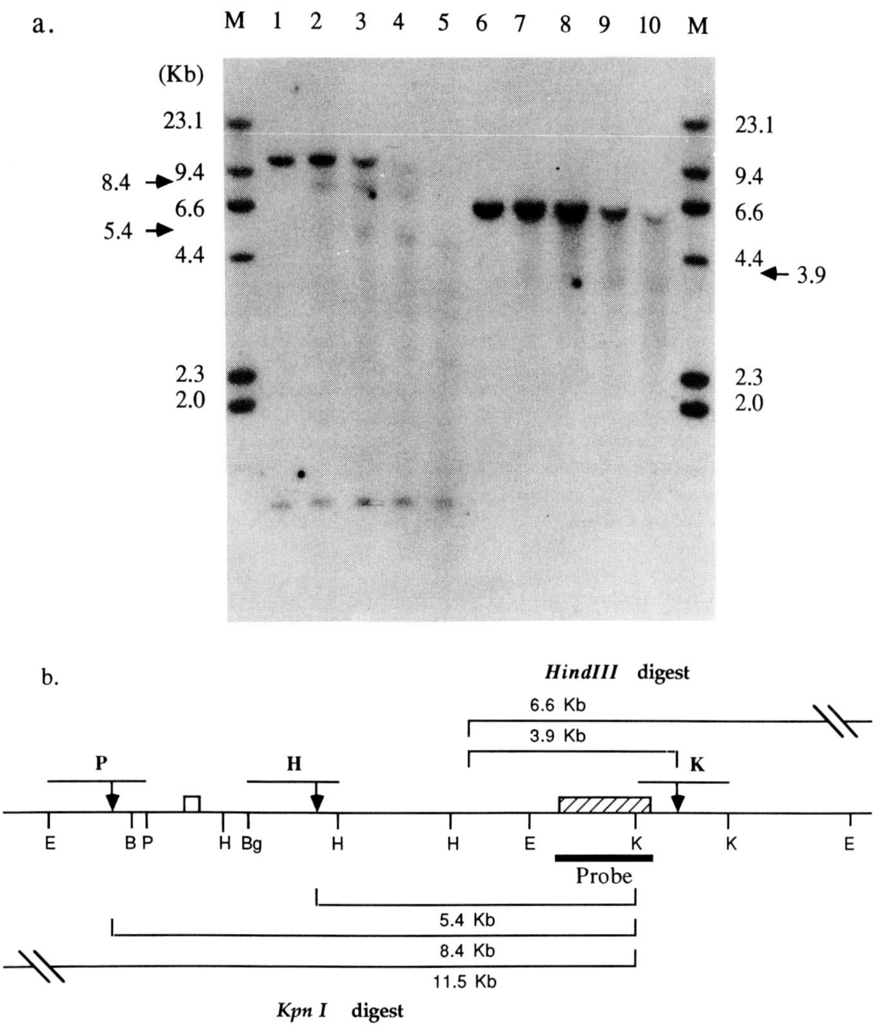

Figure 3. Detection of DNase HS sites in the rat Cx32 genomic locus in adult rat liver. (a) Nuclei were isolated from rat liver and incubated with endogenous DNase activity at 0°C for 20 min (lanes 2 and 7); or 37°C for 5 min (lanes 3 and 8), 10 min (lanes 4 and 9), or 20 min (lanes 5 and 10); genomic DNA was then isolated. As a control, rat liver DNA was isolated in the presence of 25 mM EDTA, a DNase inhibitor (lanes 1 and 6). DNAs were digested by either *Kpn*I (lanes 1–5) or *Hind*III (lanes 6–10) and subjected to Southern blot hybridization using a radiolabeled rat Cx32 probe indicated in panel b. The arrows in the left margin indicate the position of two new fragments, 8.4 kb and 5.4 kb, generated by digestion with DNase and *Kpn*I. The arrow in the right margin indicates the position of a new fragment, 3.9 kb, generated by digestion with DNase and HindIII. Lane M contains ^{32}P end-labeled λ HindIII fragments as size markers. (b) Location of DNase HS sites within the rat Cx32 locus. The Cx32 first and second exons are shown as an open and hatched box, respectively. The arrows correspond to the localization of the DNase HS sites and correspond to the size in kb indicated either above or below. The letters P, H, and K above a solid line correspond to cloned DNA fragments containing each DNase HS site region. The letters below the line are restriction endonuclease abbreviations described in the legend of Fig. 2.

act as repressor elements with both heterologous and native promoters when analyzed in HuH-7 cells.

Discussion

We identified and characterized the promoter complex, the basal promoter and distant regulatory elements of the rat Cx32 gene using transient expression assays in HuH-7 cells and DNase-HS-site analysis. To identify the promoter region, it was critical to document where the native rat Cx32 mRNA transcript(s) initiate, since this gene has a small, untranslated first exon many thousands of base pairs from the coding region. Primer-extension analysis of rat liver RNA using an oligonucleotide within the coding region of the second exon demonstrated multiple mRNA start sites clustered in a 60 bp region, 50–110 bp upstream of the first exon. This result corresponds well to the transcription start sites previously mapped by Miller et al., (1988) and is consistent with the size of the Cx32 transcript identified by Northern blot analysis.

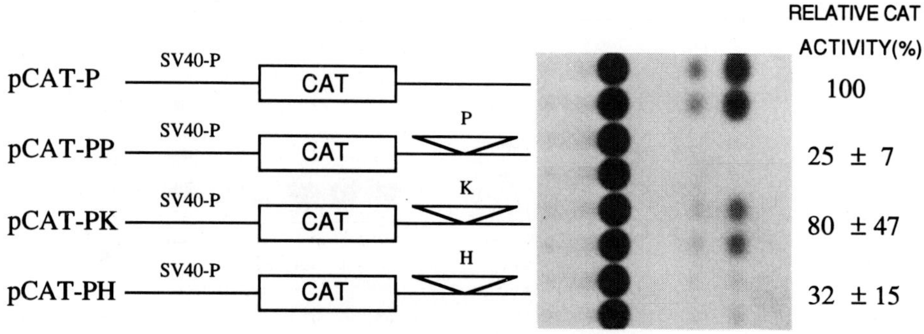

Figure 4. Functional analysis of cloned DNA fragments containing DNase HS regions. P, K, and H fragments, described in Fig. 3b, were inserted 3' to the CAT gene in the vector, pCAT-P, which contains the SV40 early promoter but no enhancer. The vectors pCAT-PP, pCAT-PK, and pCAT-PH contain fragments P, K, and H, respectively as indicated. The various plasmids were transfected into HuH-7 cells and the CAT activity was determined in cell extracts as described by Gorman et al. (1982). The CAT activity is presented as percent acetylation of ^{14}C-labeled chloramphenicol in cell extracts determined by TLC and liquid scintillation spectroscopy of acetylated and nonacetylated bands identified from autoradiography of the TLC plate. The relative CAT activity (%) was expressed in relation to the activity of pCAT-P (100%).

To analyze the promoter activity of the genomic region upstream of the first exon, we used PCR amplification to clone a 2.2 kb fragment lacking the splice donor site, but including the mRNA transcription initiation region, into a luciferase reporter vector. Transient transfection experiments were carried out in the highly differentiated human hepatoma cell line HuH-7, which was shown to express the 1.6 kb Cx32 transcript, but at a reduced level compared to adult rat liver (data not shown). A promoter complex was identified in a 561 bp region upstream from the first exon by its ability to direct maximal luciferase activity. Further deletion of this region led to a significant drop in luciferase activity suggesting the presence of an enhancer function. A basal promoter activity was identified in a 70 bp region immediately upstream of the mRNA start sites described in this report. Therefore, we suggest the promoter complex of the Cx32 gene contains an enhancer element(s) and a basal promoter.

The promoter complex region of the rat Cx32 gene differs from the corresponding regions of most other eukaryotic genes, by lacking a "TATA box". However, it does resemble the promoter regions of a group of constitutively expressed housekeeping genes, including human 3-phosphoglycerase kinase (Singer-Sam et al., 1984), mouse and human hypoxanine phosphoribosyltransferase (Metton et al., 1984; Kim et al., 1986), human epidermal growth factor (Ishii et al., 1985), and human insulin receptor gene (Araki et al., 1987). The promoters for this group of genes all have multiple transcription start sites and multiple "GC" boxes highly homologous to the consensus SP1 binding site (GGGCGG) (Kadonaga et al., 1986). In fact, multiple "GC" boxes are clustered in the Cx32 basal promoter region identified by us, whereas no promoter activity was detected after deletion of these elements. Therefore, we propose that Cx32 gene promoter may be responsive to, or regulated by, the SP1 factor. The nature of the promoters of the other connexin genes remains to be investigated.

A number of studies indicate that DNase HS regions within chromatin represent alterations in the nucleosome structure of the DNA in that region (Jongstra et al., 1984; Reeves, 1984; Jackson and Felsenfeld, 1985; Elgin, 1988; Gross and Garrod, 1988). The majority of DNase HS regions have been associated with active genes, and studies mapping their location indicate they can be 5', 3', or even within an intron of a gene. Interestingly, these HS regions have been localized within regulatory sequences of a variety of cellular genes including, for example, chicken globin genes (Weintraub et al., 1981), human μ immunoglobulin heavy-chain genes (Mills et al., 1983), and rat albumin genes (Nahon et al., 1987), as well as viral regulatory regions such as the enhancer of hepatitis B virus (El-Ghor et al., 1989) and the SV40 enhancer (Jongstra et al., 1984). More recent studies demonstrate that DNase I HS regions contain specific protein-binding sites (Gross, 1988; Jackson, 1985).

Here we provide the first evidence for the presence of at least 3 DNase HS sites in the rat Cx32 gene locus. Transient expression assays of DNA fragments containing these DNase HS sites showed that two, the P and H regions, inhibited CAT activity arising from the SV40 promoter in HuH-7 cells. The negative regulatory activity of the P and H fragments was confirmed with additional constructs containing the luciferase reporter gene directed by the Cx32 promoter complex. Interestingly, recent data show that Cx32 gene expression was significantly decreased during tumorigenesis and liver regeneration (Fitzgerald et al., 1989; Neveu et al., 1990). One explanation for the silencer activity of the DNase HS regions in hepatoma cells might be the reduced activity of factors regulating Cx32 in cancers.

In conclusion, both positive and negative regulatory elements possibly are involved in transcriptional regulation of the rat Cx32 gene. Whether or not the decreased expression of Cx32 in tumorigenesis and liver regeneration is mediated by the regulatory elements present in

DNase HS regions is currently under study as is the characterization of the cellular proteins which bind to the identified regulatory regions.

Acknowledgement

This work was supported by the NIH Grant #DK- 41918. The authors would like to thank Anna Caponigro for world-class preparation of the manuscript, Kurt Gustin for critically reading the manuscript, and Allan Wolkoff for continued support and encouragement.

References

Beyer, E.C., Paul, D.L. and Goodenough, D.A. (1987) A protein from rat heart homologous to a gap-junction protein from liver. J. Cell Biol. 105, 2621–2629.

Bradford, M.M. (1976) A rapid and sensitive method for the quantitation of microgram quantities of protein utilizing the principle of protein-dye binding. Anal. Biochem. 72, 248–254.

Brasier, A.R., Tate, J.E. and Habener, J.F. (1989) Optimized use of the firefly luciferase assay as a reporter gene in mammalian cell lines. Bio. Techniques 7, 1116–1122.

Chen, C. and Okayama, H. (1987) High efficiency transformation of mammalian cells by plasmid DNA. Mol. Cell. Biol. 7, 2745–2752.

Chirgwin, J.M., Przybyla, A.E., MacDonald, R.J. and Rutter, W.J. (1979) Isolation of biologically active ribonucleic acid from sources enriched in ribonuclease. Biochemistry 18, 5294–5299.

Dermietzel, R., T. K. Hwang and D. C. Spray. (1990) The gap-junction family, structure, function and chemistry. Anat. Embryol. 182, 517–528.

Dermietzel, R., Traub, O., Hwang, T.K., Beyer, E.C., Bennett, M.V.L., Spray, D.C. and Willecke, K. (1989) Differential expression of three gap-junction proteins in developing and mature brain tissues. Proc. Natl. Acad. Sci. USA 86, 10148–10152.

Dermietzel, R., Yancey, S.B., Traub, O., Willecke, K. and Revel, J.-P. (1987) Simultaneous light and electron microscope observation of immunolabeled liver 27 KD gap-junction protein on ultra-thin cryosections. J. Cell Biol. 105, 1925–1934.

El-Ghor, M.A.A. and Burk, R.D. (1989) DNase I hypersensitive site maps to the HBV enhancer. Virology 172, 478–88.

Fishman, G.I., Hertzberg, E.L., Spray, D.C. and Leinwand, L.A. (1991) Expression of connexin43 in the developing rat heart. Circ. Res. 68, 782–787.

Fitzgerald, D.J., Mesnil, M., Oyamada, M., Tsuda, H., Ito, N. and Yamasaki, H. (1989) Changes in gap-junction protein (connexin 32) gene expression during rat liver carcinogenesis. J. Cell Biochem. 41, 97–102.

Gimlich, R.L., Kumar, N.M. and Gilula, N.B. (1990). Differential regulation of the level of three gap-junction mRNAs in *Xenopus* embryos. J. Cell Biol. 110, 597–605.

Gorman, C., Moffat, L. and Howard, N. (1982) Recombinant genomes which express chloramphenicol acetyltransferase in mammalian cells. Mol. Cell. Biol. 2, 1044–1051.

Jongstra, J., Reudelhuber, T.L., Oudet, P., Benoist, C., Chae, C., Jeltsch, J., Mathis, D.J. and Chambon, P. (1984) Induction of altered chromatin structures by simian virus 40 enhancer and promoter elements. Nature 307, 708–714.

Kadonaga, J.T., Jones, K.A. and Tijian, R. (1986) Promoter-specific activation of RNA polymerase II transcription by Sp1. Trends Biochem. Sci. 11, 20–23.

Kumar, N. and Gilula, N.B. (1986) Cloning and characterization of human and rat liver cDNAs coding for a gap-junction protein. J. Cell Biol. 103, 767–776.

Miller, T., Dahl, G. and Werner, R. (1988) Structure of a gap-junction gene: connexin32. Biosci. Rep. 8, 455–464.

Montandon, P.E., Montando, F. and Fan, H. (1982) Methylation state and DNase I sensitivity of chromatin containing Moloney murine leukemia virus DNA in exogenously infected mouse cells. J. Virol. 44, 475–486.

Neveu, M.J., Hully, J.R., Paul, D.L. and Pitot, HC. (1990) Reversible alteration in the expression of the gap-junctional protein connexin 32 during tumor promotion in rat liver and its role during cell proliferation. Cancer Commun. 2, 21–31.

Paul, D.L. (1986). Molecular cloning of cDNA for rat liver gap-junction protein. J. Cell Biol. 103, 123–134.

Risek, B., Guthrie, S., Kumar, N.M. and Gilula, N.B. (1990) Modulation of gap-junction transcript and protein expression during pregnancy in rat. J. Cell Biol. 110, 269–282.

Sambrook, J., Fritsch, E.F. and Maniatis, T. (1989) Molecular Cloning. A Laboratory Manual, 2nd edition. Cold Spring Harbor Laboratory Press, Cold Spring Harbor, NY.

Spray, D.C., Chanson, M., Moreno, A.P., Dermietzel, R. and Meda, P. (1991) Distinctive gap-junction channel types connect WB cells, a clonal cell line derived from rat liver. Am. J. Physiol. 260, C513–C527.

Traub, O., Look, J., Dermietzel, R., Brummer, F., Hulser, D. and Willecke, K. (1989). Comparative characterization of the 21-kD and 26-kD gap-junction proteins in murine liver and cultured hepatocytes. J. Cell Biol. 108, 1039–1051.

van Zonneveld, A.J., Curriden, S.A. and Loskutoff, D.J. (1988) Type 1 plasminogen activator inhibitor gene: functional analysis and glucocorticoid regulation of its promoter. Proc. Natl. Acad. Sci. USA 85, 5525–5529.

Willecke, K., Jungbluth, S., Dahl, E., Hennemann, H., Heynkes, R. and Grzeschik, K.-H. (1990) Six genes of the human connexin gene family coding for gap-junctional proteins are assigned to four different human chromosomes. Eur. J. Cell Biol. 53, 275–280.

Wu, C., Bingham, P.M., Livak, K.J., Holmgren, R. and Elgin, S.C.R. (1979) The chromatin structure of specific genes. Evidence for higher order domains of defined DNA sequences. Cell 16:797–806.

Part VII. Cancer

J.E. Hall, G.A. Zampighi and R.M. Davis (Eds.)
Progress in Cell Research, Vol. 3

CHAPTER 42

Retinoids and carotenoids upregulate gap-junctional communication: correlation with enhanced growth control and cancer prevention

MOHAMMAD Z. HOSSAIN, LI-XIN ZHANG and JOHN S. BERTRAM

Molecular Oncology Unit, Cancer Research Center of Hawaii, University of Hawaii, Honolulu, HI 96813, USA

Introduction: junctional communication and cancer

Gap junctions are membrane specializations found in most metazoan tissues and have been characterized as direct channels connecting adjacent cells. The involvement of gap junctions in intercellular communication was reported by Loewenstein's group who showed that communication competence in cells goes hand in hand with the presence of gap junctions (Azarnia et al., 1974). These hydrophilic channels have an inner diameter of 16–20 Å and allow the passive diffusion of small molecules and ions up to about 1000 Da in size. Although considerable progress has been made in our understanding of the structure, permeability and biochemistry of gap junctions, much less conclusive evidence has been reported for the biological roles played by the junctions. A notable exception is the heart, where gap junctions are involved in the propagation of electrical impulses (De Mello, 1977). It has been speculated that junctional communication is also involved in the exchange of nutrients and in tissue homeostasis (reviewed by Loewenstein, 1979 and Caveney, 1985). One important function of gap junctions is considered to be the regulation of growth and development. We and others have proposed that gap junctions serve as conduits for growth regulatory signals from one cell to another (Loewenstein, 1979; Mehta et al., 1986; Hossain et al., 1989). The role of gap junctions in tissue development has been documented in several studies (Warner et al., 1984; Fraser et al., 1987; Allen et al., 1990).

The growth regulatory function of gap-junctional communication may play a crucial role in cancer. When the behavior of normal and cancerous fibroblasts is compared, there is an essential difference: whereas normal cells grow and interact with surrounding cells to produce a growth-inhibited monolayer, cancerous cells grow autonomously to produce multilayered cell masses. The association of junctional communication with neoplasia was first suggested by Loewenstein and Kanno (1966) who demonstrated that cancerous cells did not communicate; their normal counterparts were capable of cell communication. Since then, numerous studies have demonstrated that loss of junctional communication is a common feature in carcinogenesis (for review see Klaunig and Ruch, 1990 and Yamasaki, 1990). The proposed role of junctional communication was further strengthened when several classes of tumor promoters, and growth factors with tumor promoting activity, were shown to inhibit communication (Enomoto and Yamasaki, 1985; Hamel et al., 1988; Ruch and Klaunig, 1988; Madhukar et al., 1989). Furthermore, several oncogenes inhibited intercellular communication (Atkinson et al., 1981; Bignami et al., 1988; El-Fouley et al., 1989). Conversely, transfection of communication-deficient transformed cells with a gap-junction protein, Cx43, resulted in increased junctional communication *and* reduced cell proliferation (Zhu et al., 1991). These results strongly suggest that loss of junctional communication is a factor in the development of neoplasia.

The demonstration that the incidence of tumorigenesis can be diminished by elevated communication would provide a corollary to this statement. This has been shown: neoplastic cells can undergo reversible growth arrest when in junctional communication with surrounding normal cells (Mehta et al., 1986). Yamasaki's group also demonstrated a very similar phenomenon (Yamasaki and Katoh, 1988). All available evidence points to a role for intercellular communication in the transfer of growth regulatory signals from normal cells, thus causing the growth arrest of neoplastic cells.

Cancer preventive agents: role of induced junctional communication

In this chapter, we discuss the involvement of junctional communication in the prevention of cancer in greater

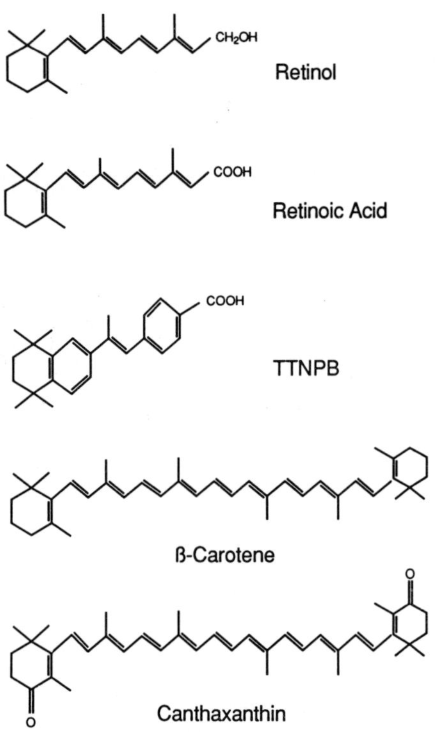

Figure 1. Structures of retinoids and carotenoids tested.

detail. For many years we have studied the mechanism whereby retinoids (natural and synthetic compounds with vitamin-A activity) and carotenoids (plant pigments found in vegetables and fruits) act as cancer chemopreventive agents. The anti-neoplastic effects of retinoids have been demonstrated in animal models (reviewed by Moon, 1989) and in in vitro cell-culture systems (Merriman and Bertram, 1979; Bertram, 1980). In clinical trials, retinoids delay neoplasia or cause reversion of preneoplastic pathology at several anatomic sites (Alfthan et al., 1983; Hong et al., 1990). For carotenoids, epidemiological studies revealed that individuals with a high dietary intake of β-carotene have a reduced cancer risk at different anatomic sites (Bertram et al., 1987; Connett et al., 1989). Because, β-carotene has the greatest provitamin A activity, it has been suggested to exert its effect by being converted to vitamin A (Stich et al., 1984). Other studies revealed that not only β-carotene but also canthaxanthin, a carotenoid without any known provitamin A activity in mammals, inhibited carcinogen-induced neoplastic transformation in animal model systems and in vitro cell cultures (Mathews-Roth, 1982; Pung et al., 1988; Schwartz and Shklar, 1988). These studies indicate carotenoids have a direct antitumor action.

In our in vitro carcinogenesis and chemoprevention studies, we have utilized C3H/10T1/2, a mouse fibroblast cell line. This sub-tetraploid cell line of mesenchymal origin was developed in the laboratory of the late Charles Heidelberger (Reznikoff et al., 1973). It has a very low spontaneous transformation rate and exhibits a high degree of post-confluence inhibition of cell division. The latter property is vital for the chemoprevention assay, since a major criterion of the transformed phenotype in fibroblasts is a loss of "contact inhibition" and subsequent growth as a dense, multilayered focus. This cell line reliably reproduces the results obtained from experimental animal models of chemical- and radiation-induced carcinogenesis (Bertram, 1985).

We have examined a variety of retinoids and carotenoids for their in vitro chemopreventive activity (Bertram, 1980; Bertram et al., 1991). For the purpose of clarity, only a selected few of these compounds will be discussed here (Fig. 1). We have compared three retinoids, i.e., retinol (the principal retinoid in blood), retinoic acid (the metabolite of retinol believed to be responsible for normal growth and differentiation) and the highly stable synthetic benzoic acid analog of retinoic acid, tetrahydrotetramethylnaphthalenyl-propenyl benzoic acid (TTNPB). The carotenoids, β-carotene and canthaxanthin were also studied.

Retinoids and carotenoids prevent in vitro carcinogenesis

In carcinogen-treated 10T1/2 cultures, formation of transformed foci was inhibited when non-toxic concentrations of these three retinoids were added 7 days after carcinogen removal (i.e., in the post-initiation phase of carcinogenesis) (Hossain et al., 1989). Among the tested retinoids, TTNPB was the most potent: at 10^{-9} M it com-

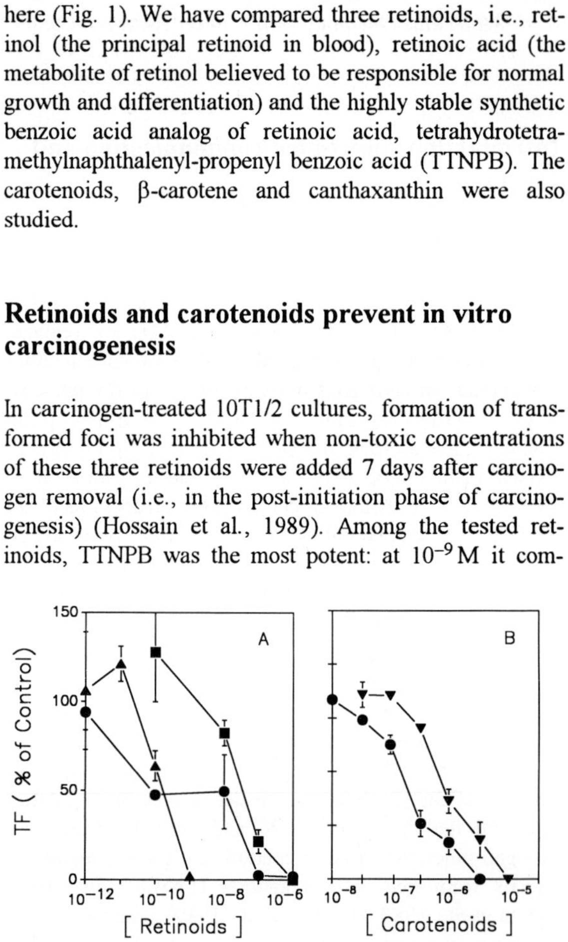

Figure 2. Inhibition of 3-methylcholanthrene induced transformation by retinoids (A) and carotenoids (B). 10T1/2 cultures were initiated with 3-methylcholanthrene, then received retinoids or carotenoids 7 days after the removal of carcinogen. The drug treatments were given every 3 days (retinoids) or every 7 days (carotenoids) with weekly medium change for the remaining 35-day duration of the experiment. Results are expressed as a % of the transformation (TF) observed in cultures exposed to methylcholanthrene, then to the solvent or control beadlets used to deliver the drugs. (Fig. 2A is reproduced from Hossain et al., 1989 with permission. Fig. 2B is derived from Pung et al., 1988 with permission). Symbols: (A) retinol (●), retinoic acid (■), TTNPB (▲). (B) β-Carotene (▼), canthaxanthin (●).

pletely inhibited 3-methylcholanthrene-induced transformation and retained strong activity at 10^{-10} M (Fig. 2A). Both retinol and retinoic acid exhibited a similar degree of inhibition at 10^{-6} M and, in addition, caused a dose-dependent decrease in cell saturation density at confluence over similar concentration ranges (Mordan and Bertram, 1983). Cell saturation density is considered to be a measure of cellular growth control. Both of these effects of the retinoids are reversible (Merriman and Bertram, 1979; Mordan and Bertram, 1983). These findings demonstrated that retinoid action on cell transformation and cell saturation density were not due to cytotoxicity.

Both β-carotene and canthaxanthin also induced a dose-dependent inhibition of transformed focus formation when administered using a similar protocol (Fig. 2B). These compounds, however, required higher concentrations (10^{-5} M) than did retinoids for their antineoplastic action. Like the retinoids, carotenoids were reversibly active in the post-initiation phase of carcinogenesis (Pung et al., 1988). The activity of canthaxanthin suggested that the chemopreventive action of carotenoids was not dependent on their pro-vitamin A activity (Pung et al., 1988), an observation later substantiated in studies with structurally diverse carotenoids (Bertram et al., 1991).

Induction of junctional communication by retinoids and carotenoids

The similarity in the biological effects of retinoids and carotenoids in inhibiting neoplastic transformation in 10T1/2 cells, and their structural similarities (Fig. 1), suggest that these compounds share similar molecular mechanisms of action. Because of the relationship between junctional communication and cancer prevention,

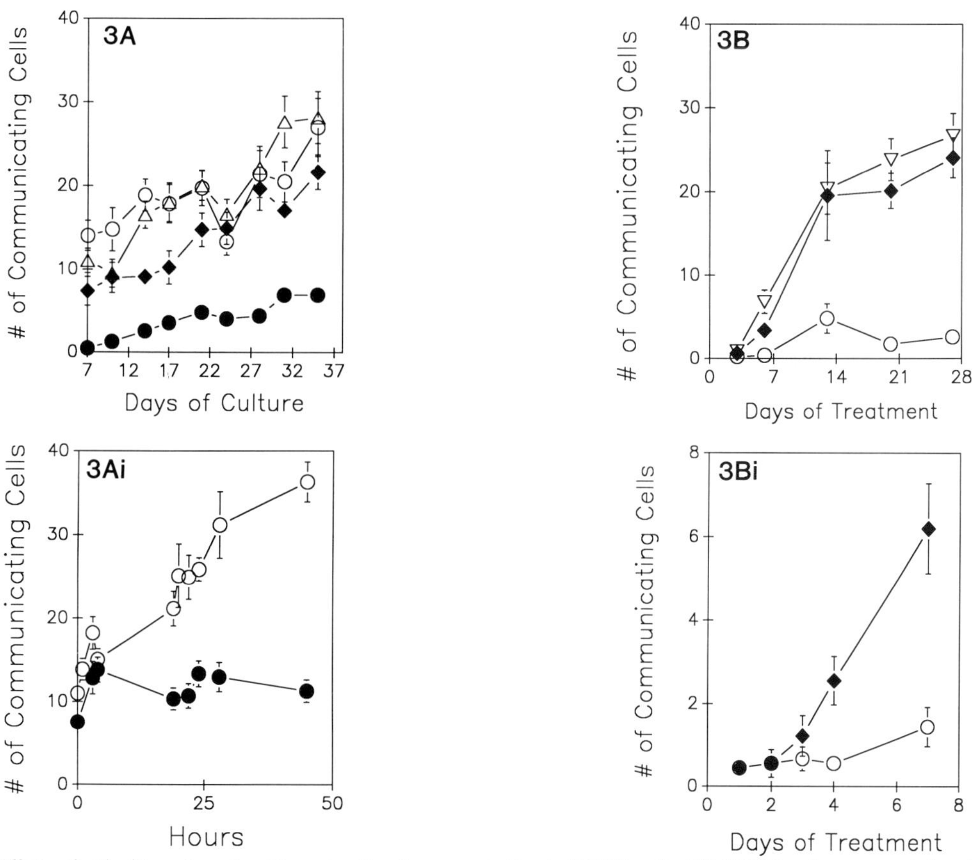

Figure 3. Effects of retinoids and carotenoids on junctional communication in 10T1/2 cells. (A) 10T1/2 cells were seeded and treated with retinoids on day 1 post-seeding and every 3 days thereafter. Junctional transfer was indexed by the number of cells showing transfer of LY within 10 min of injection into a test cell ("number of communicating cells"). Cultures were probed at intervals shown. Data points are the means ± SE of at least 10 microinjection trials performed in each of two dishes. ●, acetone control 0.2%; Δ, retinol 10^{-6} M; ♦ , retinoic acid 10^{-6} M; O, TTNPB 10^{-9} M. (Ai) Expanded timescale: TTNPB 10^{-9} M, a separate group of cultures were treated and probed as above. Symbols as in (A). Results represent the mean ± SE of at least 20 microinjections in two dishes each (from Hossain et al., 1989 with permission). (B) Carotenoid effects on gap-junctional communication in 10T1/2 cells. The cells were seeded at 1000 cells/60 mm culture dish and treated with carotenoids dissolved in tetrahydrofuran (THF), or with 0.5% THF as control, on day 7 when confluent and weekly thereafter. Data points are the means ± SE of 15 microinjections performed in two dishes each. O, solvent control; ∇, β-carotene 10^{-5} M; ♦, canthaxanthin 10^{-5} M. (Bi) expanded time scale, symbols as main graph (from Zhang et al., 1991 with permission).

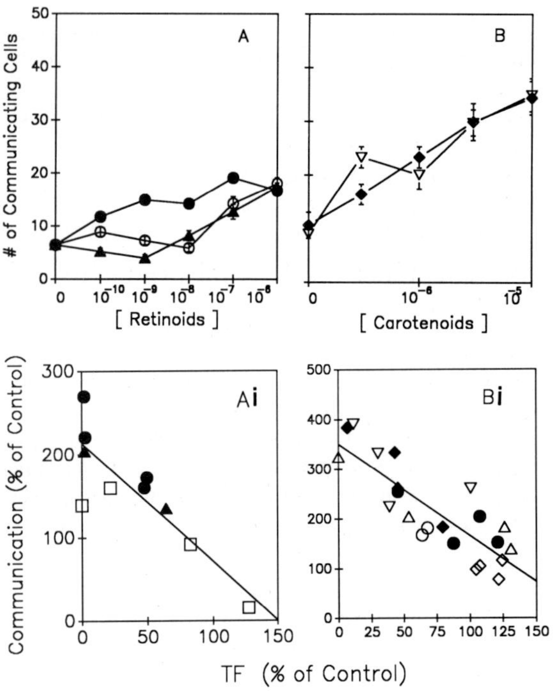

Figure 4. Dose response for induction of junctional communication by retinoids (A) and carotenoids (B). (A) Cultures were seeded and treated as in Fig. 3A. They were probed for junctional communication 2 days after reaching confluence. Data points represent means ± SE of two experiments, each involving about 20 microinjections in two separate cultures. Retinol, O; retinoic acid, ▲; TTNPB, ●. Dose-response relationships were statistically significant for TTNPB: P = 0.002; retinol: P = 0.03. We excluded the zero-dose treatment data for retinoic acid, since 10^{-10} M retinoic acid inhibited communication. The resulting dose-response relationship was statistically significant: P = 0.0003. (Ai) Correlation between transformation and junctional communication. Symbols as in (A). The Pearson correlation coefficient was –0.86, indicating a strong negative association between the two events. This was highly statistically significant (P = 0.001) (from Hossain et al., 1989 with permission). (B) Dose response for induction of junctional communication by carotenoids. Cells were seeded and treated as in Fig. 3B and probed after two weeks of treatment. β-carotene, ▼; canthaxanthin, ●. (Bi) Correlation between inhibition of carcinogen-induced neoplastic transformation and induction of gap-junctional communication by various carotenoids and α-tocopherol. ∇, β-carotene; ●, α-carotene; ♦, canthaxanthin; ■, lutein; Δ, lycopene; ◊, m-bixin; O, α-tocopherol. Pearson correlation coefficient –0.823; P = 0.005 (from Zhang et al., 1991 with permission).

the involvement of gap-junctional communication in retinoid and carotenoid action was examined.

Concentrations of retinoids which were equipotent in transformation assays were added to cultures of 10T1/2 cells and junctional communication was assayed over the time course of a typical transformation experiment. As seen in Fig. 3A, junctional communication was markedly increased by all three retinoids and this increase was sustained over the 35-day period of observation. The increases produced by the various retinoids were of comparable magnitude, but required a 1000-fold higher concentration of retinol and retinoic acid than that of TTNPB. Enhancement of junctional communication induced by TTNPB (10^{-9} M) required several hours; no effect was detectable before 18 h of treatment. At this time, communication was significantly different ($p < 0.001$) from acetone-treated controls and continued to increase thereafter (Fig. 3Ai). As is typical of 10T1/2 cells, junctional communication increased as cultures aged; this was observed in both treated and control cultures. As with effects on neoplastic transformation and cell saturation density, this action was found to be reversible (Mehta et al., 1989). At their maximally effective concentrations (10^{-5} M), carotenoids also markedly elevated junctional communication to a level comparable to the effect of retinoids (Zhang et al., 1991). As seen in Fig. 3B, both β-carotene and canthxanthene increased junctional communication in a time-dependent manner. A notable difference, however, was that whereas retinoid-induced communication occurred within 18 h of treatment (Fig 3Ai), carotenoids required 4 days to produce similar effects (Fig 3Bi).

Both TTNPB and retinol enhanced junctional communication in 10T1/2 cells in a dose-dependent manner (Fig. 4A). Retinoic acid however caused a biphasic response. It inhibited communication at the lowest concentration (10^{-10} M) but at higher concentrations (> 10^{-8} M) induced communication. The inhibitory action of retinoic acid at 10^{-10} M was particularly interesting, since at this concentration, it enhanced cell transformation (Fig 2A). The relationship between junctional communication and neoplastic transformation is demonstrated in Fig. 4Ai. Analysis of the data showed a strong negative correlation between the two parameters (Pearson correlation coefficient of –0.86) which was highly statistically significant ($p < 0.001$). Thus, treatments that enhanced junctional communication inhibited neoplastic transformation; conversely, inhibition of intercellular communication increased cell transformation. The ability of diverse retinoids to decrease cell saturation density at confluence was also negatively correlated with the induction of gap-

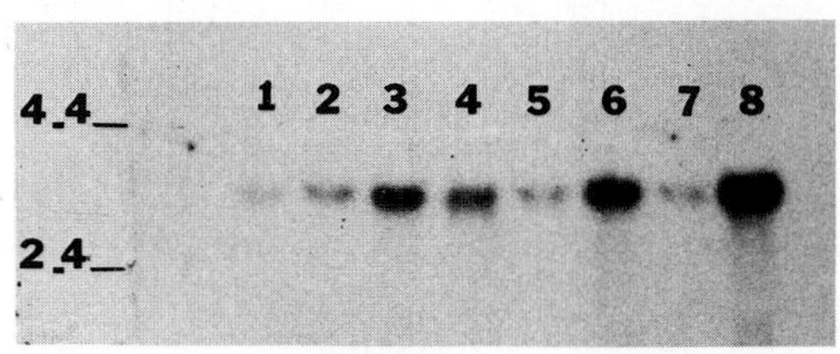

Figure 5. Time course of TTNPB-induced increase in Cx43 mRNA. Confluent 10T1/2 cells were stimulated with 10^{-8} M TTNPB for the times indicated. 10 μg of total RNA was loaded per lane, electrophoresed, and transferred to nitrocellulose by blotting. The blot was hybridized at 42°C with ^{32}P-labeled Cx43 cDNA and stringently washed at 65°C. Lane 1: acetone, 6 h; lane 2: TTNPB, 6 h; lane 3: TTNPB, 12 h; lane 4: TTNPB, 24 h; lane 5: acetone, 48 h; lane 6: TTNPB, 48 h; lane 7: acetone: 72 h; lane 8: TTNPB, 72 h. Positions of RNA standards (kb) are shown. (from Rogers et al., 1990 with permission).

junctional communication (Mehta et al., 1989). These linkages between neoplastic transformation, growth control and junctional communication suggest a functional relationship between these parameters. Similar dose-response curves were obtained for the carotenoids, when tested in the junctional communication assay (Fig. 4B). As in the transformation assay (Fig. 2B), β-carotene and canthaxanthin required higher concentrations (≥ 10^{-6} M) than those of retinoids to elevate junctional communication in 10T1/2 cells; these two responses were strongly correlated (Fig. 4Bi).

Molecular mechanism of retinoid and carotenoid action: induction of Cx43 gene expression

Enhancement of Cx43 mRNA levels and up-regulation of Cx43 and its phosphorylation

To detect Cx43, we have generated a rabbit polyclonal antibody against a 15-mer polypeptide corresponding to the C-terminal region (position 368–382) of the predicted amino acid sequence of Cx43 (Beyer et al., 1987). This region is believed to be in the cytoplasmic tail of the protein; a computer search revealed no homology with other known proteins. The specificity of the antibody was confirmed by its ability to stain Cx43 in the intercalated disc of rat heart (data not shown). Analysis of total cellular proteins by sodium dodecylsulfate polyacrylamide gel electrophoresis (SDS-PAGE) under reducing conditions, followed by immunoblotting with rabbit antiserum (described above), showed that TTNPB treatment resulted in a major increase in immunoreactive protein in the appropriate 43–45 kDa region of the gel (Fig. 6A). However, when cellular lysates were not reduced with β-mercaptoethanol, higher M_r protein bands (70, 107 and 126 kDa) were detected (Rogers et al., 1990). This finding suggests extensive dimerization and tetramerization of Cx43 molecules through intermolecular disulfide bonds, as reported by Manjunath and Page (1986). Time-course studies showed increased immunolabeling of Cx43 within 6 h of TTNPB (10^{-8} M) treatment and this labeling increased progressively over the 96 h duration of the experiment (Fig 6A). This time course correlates well with our previous observations on Cx43 message induction (Fig. 5). TTNPB increased the levels of Cx43 in a dose-dependent

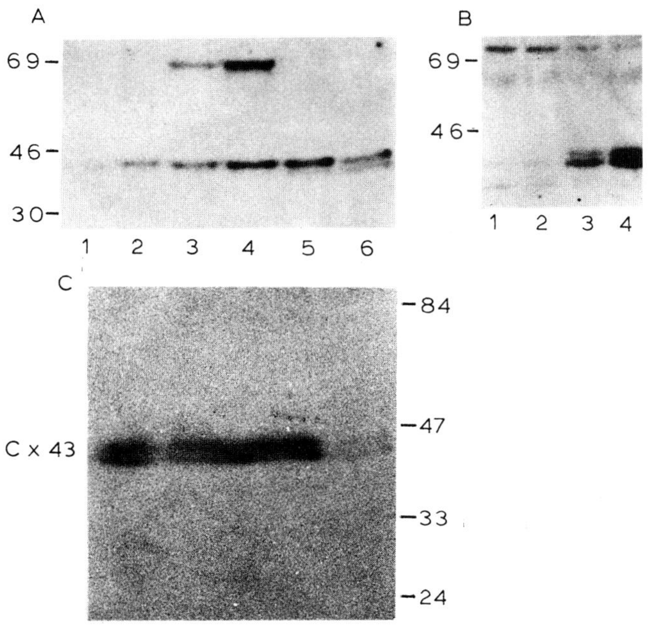

Figure 6. Up-regulation of Cx43 by TTNPB and carotenoids. (A) Time course of TTNPB-induced increase in Cx43. Confluent cultures of 10T1/2 cells were treated with TTNPB (10^{-8} M) for the times indicated. Cell lysates were reduced prior to electrophoresis, equal protein amounts were loaded to each lane, and Western blots performed. Cx43 protein bands were detected using a rabbit polyclonal antibody. Lane 1: 0 h solvent control; lane 2: 6 h TTNPB; lane 3: 24 h TTNPB; lane 4: 48 h TTNPB; lane 5: 96 h TTNPB; lane 6: 96 h solvent control. The bands in the 70-kDa region (lanes 3 and 4) are believed to represent dimers of Cx43. Molecular-weight markers are expressed in kDa. (From Rogers et al., 1990 with permission). (B) Dose response for Cx43 induction by TTNPB. Confluent cultures were treated with TTNPB or acetone control for 96 h. Levels of Cx43 were examined following the above-mentioned protocol. Lane 1, acetone control; Lane 2, TTNPB 10^{-10} M; Lane 3, TTNPB 10^{-9} M; Lane 4, TTNPB 10^{-8} M. (From Rogers et al., 1990 with permission). (C) Induction of Cx43 by carotenoids. 10T1/2 cultures were treated with β-carotene or canthaxanthin (10^{-5} M) for 7 days and levels of Cx43 determined as above. For comparison, cell lysates from 10T1/2 cultures that had been treated with TTNPB for 3 days were loaded in the same gel. Lane 1: TTNPB (10^{-8} M); lane 2: β-carotene (10^{-5} M); lane 3: canthaxanthin (10^{-5} M); lane 4: solvent control (0.5% tetrahydrofuran).

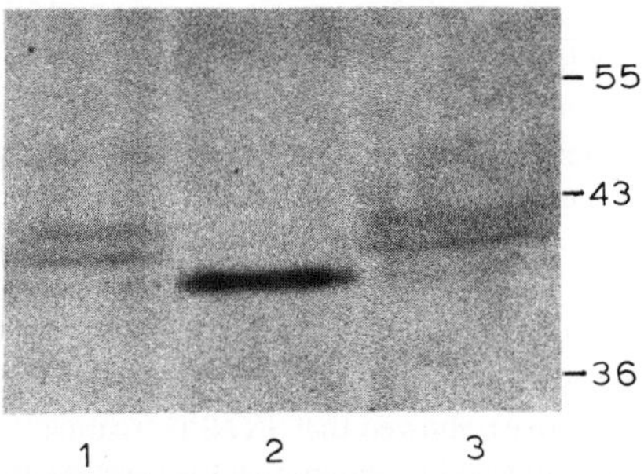

Figure 7. Effects of phosphatase digestion on Cx43. Lysates from cells treated with TTNPB (10^{-8} M) for 96 h were immunoprecipitated with N-terminal Cx43 antibody (Yancey et al., 1989) and proteins incubated with or without alkaline phosphatase. In some samples, phosphatase action was inhibited with 1 mM vanadate. Resulting digests were electrophoresed and immunoblotted. Lane 1: TTNPB treated; lane 2: TTNPB treated + alkaline phosphatase; lane 3: TTNPB treated + alkaline phosphatase + 1 mM vanadate. Molecular-weight markers of 18, 24, 36, 43, 55 and 95 kDa were used to calibrate this gel. Calculated molecular masses of immunoreactive bands based on semilog plots of these markers were about 43–45 kDa for the lower and upper bands. Based on these calculations, the 43 kDa marker shown ran anomalously slowly on this gel. (From Rogers et al., 1990 with permission).

manner (Fig. 6B). Retinol and retinoic acid also caused similar elevations in Cx43 protein levels, but required higher concentrations (10^{-6} M) (Hossain and Bertram, unpublished data). Carotenoid treatment also produced a comparable response. Both β-carotene and canthaxanthin at their maximally effective concentrations (10^{-5} M) increased Cx43 levels (Fig. 6C). Carotenoids required longer time periods (7 days) than TTNPB (3 days) for producing equivalent magnitudes of Cx43 induction. The induction of Cx43 levels by β-carotene and canthaxanthin is dose-dependent; this response correlates well with that of carotenoid-induced junctional communication (Zhang et al., 1992).

It is important to note that treatment with retinoids and carotenoids produced increases in the immunoreactive protein bands of apparent molecular masses 43 kDa and 45 kDa (Fig. 6). Since Cx43 can be phosphorylated, resulting in altered electrophoretic mobility (Musil et al., 1990; Crow et al., 1990), we were curious about the phosphorylation state of the higher M_r protein band seen after TTNPB treatment. As presented in Fig. 7, when retinoid-treated 10T1/2 cell lysates were immunoprecipitated with Cx43 antibody and then treated with alkaline phosphatase prior to loading on a SDS-polyacrylamide gel, the high M_r protein bands disappeared and the intensity of Cx43 band increased proportionately. Using vanadate to block phosphatase action inhibited this band shift. These results clearly demonstrate that in 10T1/2 cells, retinoids elevate the gene expression of Cx43; this protein is then post-translationally modified by phosphorylation. Cx43 phosphorylation at Ser/Thr residues is required for junctional competence (Musil et al., 1990). Although the mechanistic significance of Cx43 phosphorylation in 10T1/2 cells is presently unknown, we speculate that it is necessary for the proper assembly of gap junctions in the membrane (Rogers et al., 1990).

Retinoids and carotenoids increase the assembly of gap-junctional plaques

To be functional, connexins must localize to the plasma membrane, organize into hemi-connexons, dock with

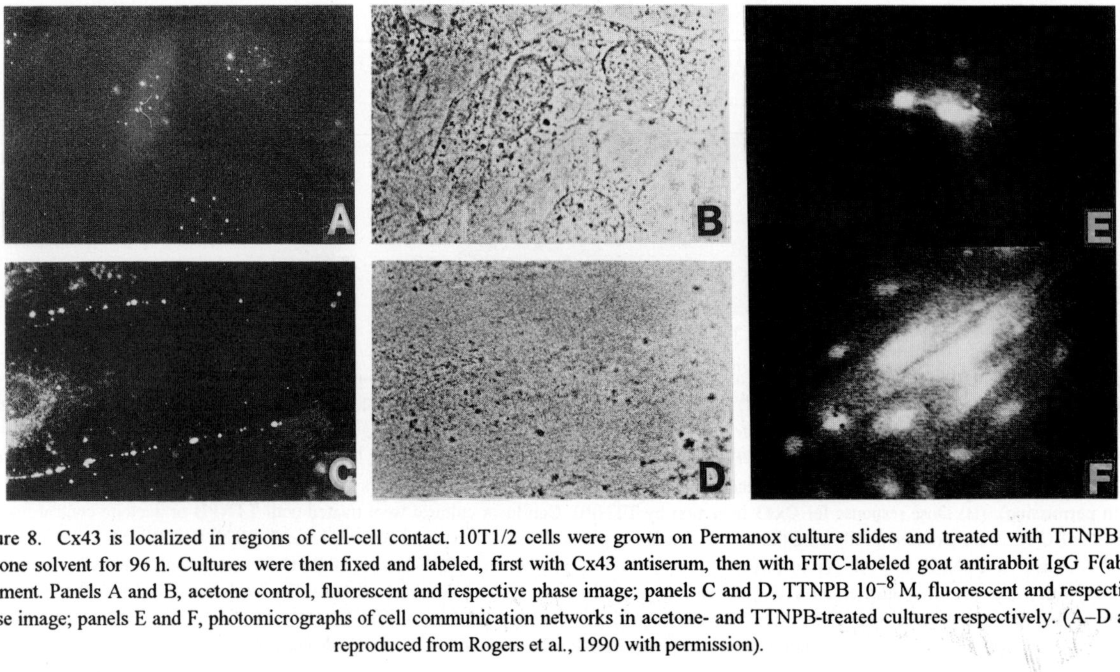

Figure 8. Cx43 is localized in regions of cell-cell contact. 10T1/2 cells were grown on Permanox culture slides and treated with TTNPB or acetone solvent for 96 h. Cultures were then fixed and labeled, first with Cx43 antiserum, then with FITC-labeled goat antirabbit IgG $F(ab')_2$ fragment. Panels A and B, acetone control, fluorescent and respective phase image; panels C and D, TTNPB 10^{-8} M, fluorescent and respective phase image; panels E and F, photomicrographs of cell communication networks in acetone- and TTNPB-treated cultures respectively. (A–D are reproduced from Rogers et al., 1990 with permission).

hemi-connexons of adjacent cells to form a connexon and aggregate into plaques containing many connexons. To localize Cx43, cells were fixed in situ and subjected to indirect immunofluorescence microscopy using the Cx43-antibody described above for Western blot experiments. A dramatic increase in fluorescent plaques was seen in regions of cell/cell contact in retinoid-treated cells in comparison with solvent-treated controls (Fig 8A–D). Microinjection of identically treated cells with Lucifer Yellow showed enhanced junctional communication in retinoid-treated cells (Fig. 8E and F). The increased levels of Cx43 became localized in regions of the cell membrane where gap junctions would be expected to form and were associated with an increased ability of the cells to communicate. Comparable results were seen after carotenoid treatment (Zhang et al., 1992).

Conclusions

The correlation between gap-junctional communication and growth control supports the hypothesis first proposed by Loewenstein (1979) that gap junctions serve as conduits for growth regulatory signals. The correlation between induced communication and suppression of neoplastic transformation by two important classes of chemopreventive agents active in the post-initiation phase of carcinogenesis strongly suggests that junctionally transferred signals can inhibit the transformation of carcinogen-initiated cells. Enhanced junctional communication by retinoids and carotenoids could allow increased transfer of these signals, resulting into the suppression of neoplastic transformation (Mordan and Bertram, 1983; Hossain et al., 1989; Zhang et al., 1992).

As we have pointed out (Mehta et al., 1986), these data do not define these signals as growth inhibitory (being transferred from normal to initiated cells) or stimulatory (being lost from initiated cells to surrounding normal cells). However, by analogy with tumor suppressor genes, which must be deleted or mutated in order for neoplasia to develop, the concept of transfer of anti-proliferative signals is attractive. Dr. R. Sagar presented evidence at this meeting that deleting a connexin gene may have similar consequences to deleting a tumor suppressor gene. These signals would be expected to pass readily through gap junctions. Therefore, they would be electrically charged (so as to limit membrane diffusion), hydrophilic, and have low protein binding. Due to the dimensions of the channel pore, the size limit of these molecules or ions should be about 1000 Da. Such signals should also be capable of rapid generation and efficient destruction or sequestration. Cyclic AMP, inositol trisphosphate and free Ca^{2+} have all been shown to traverse gap junctions (Fletcher et al., 1987; Sáez et al., 1989) and would satisfy these criteria.

The molecular mechanism of retinoid or carotenoid action on Cx43 gene expression is yet to be determined. Recently, nuclear retinoic acid receptors (RARs) have been discovered (Giguere et al., 1987; Petkovitch et al., 1987); the retinoid-RAR complex interacts with a genomic retinoic acid responsive element (RARE) sequence usually located 5'-upstream to the target gene to regulate its expression. The presence of RARE sequences has been reported in several genes regulated by retinoids (Vasios et al., 1989; deThe et al., 1990). It remains to be elucidated whether a similar mechanism regulates Cx43 gene expression. How carotenoids modulate Cx43 gene expression is still obscure, since no such receptors for carotenoids have yet been described. However, the striking similarity in the biological and cellular actions of these compounds suggests a similar pathway. We have proposed that this apparently identical mode of action is not a coincidence but, instead, is evidence for a low rate of conversion of carotenoids into active retinoids (Zhang et al., 1991). Although we have been unable to show such conversion in 10T1/2 cells (Rundhaug et al., 1988), it may be possible that, as a consequence of oxidative damage, occasional cleavage of carotenoid molecules results in the formation of retinoid-like molecules. These metabolites, even at a low concentrations, would be sufficient to produce a significant biological effect since RAR molecules have been reported to be activated by sub-nanomolar concentrations of retinoids (Zelent et al., 1989). We are currently investigating the involvement of RARs in Cx43 expression and examining whether carotenoids regulate other retinoid-inducible genes.

Note added in proof

Recent studies have shown that the retinoid-inducible gene RAR-β is not inducible by canthaxanthin in 10T½ or in F9 cells (Zhang et al., 1992). These results indicate that this carotenoid does not activate retinoic acid receptors.

Acknowledgement

This work has been supported by Grants CA 39947 from the National Institutes of Health and BC 686 from the American Cancer Society.

References

Alfthan, O., Tarkkanen, J., Grohn, P., Heinonen, E., Pyrhonen, S. and Saila, K. (1983) Tigason (etretinate) in prevention of recurrence of superficial bladder cancer. Eur. Urol. 9, 6–9.

Allen, F., Tickle, C. and Warner, A. (1990) The role of gap junctions in patterning of the chick limb bud. Development 108, 623–634.

Atkinson, M.M., Menko, A.S., Johnson, R.G., Sheppard, J.R. and Sheridan, J.D. (1981) Rapid and reversible reduction of junctional permeability in cells infected with a temperature-sensitive mutant of avian sarcoma virus. J. Cell Biol. 91, 573–578.

Azarnia, R., Larsen, W.L. and Loewenstein, W.R. (1974) The membrane junctions in communicating and non-communicating cells, their hybrids and segregants. Proc. Natl. Acad. Sci. USA 71, 880–884.

Bertram, J.S. (1980) Structure-activity relationships among various retinoids and their ability to inhibit neoplastic transformation and to increase cell adhesion in the C3H/10T1/2 Cl 8 cell line. Cancer Res. 40, 3141–3146.

Bertram, J.S. (1985) Neoplastic transformation in cell cultures: *In vitro/in vivo* correlations. In T. Kakunaga and H. Yamasaki (Eds.), Transformation Assay of Established Cell Lines: Mechanisms and Applications, IARC Sci. Publ. # 67, Lyon, pp. 77–91.

Bertram, J.S., Kolonel, L.N. and Meyskens, F.L (1987) Rationale and strategies for chemoprevention of cancer in humans. Cancer Res. 47, 3012–3031.

Bertram, J.S., Pung, A., Churley, M., Kappock, T.J., Wilkens, L.R. and Cooney, R.V. (1991) Diverse carotenoids protect from chemically induced neoplastic transformation. Carcinogenesis 12, 671–678.

Beyer, E.C., Paul, D.L. and Goodenough, D.A. (1987) Connexin43: A protein from rat heart homologous to a gap-junction protein from liver. J. Cell Biol. 105, 2621–2629.

Bignami, M., Rosa, S., Falcone, G., Tato, F., Katoh, F. and Yamasaki, H. (1988) Specific viral oncogenes cause differential effects on cell-to-cell communication, relevant to the suppression of the transformed phenotype by normal cells. Mol. Carcinogenesis 1, 67–75.

Caveney, S. (1985) The role of gap junctions in development. Annu. Rev. Physiol. 47, 319–335.

Connett, J.E., Kuller, L.H., Kjelsberg, M.O., Polk, B.F., Collins, G., Rider, A. and Hulley, S.B. (1989) Relationship between carotenoids and cancer: The multiple risk factor intervention trial (MRFIT) study. Cancer 64, 126–134.

Crow, D.S., Beyer, E.C., Paul, D.L., Kobe, S.S. and Lau, A.F. (1990) Phosphorylation of connexin43 gap-junction protein in uninfected and RSV-transformed mammalian fibroblasts. Mol. Cell Biol. 10, 1754–1763.

De Mello, W.C. (1977) Intercellular communication in heart muscle. In W.C. De Mello (Ed.), Perspectives in Cardiovascular Research, Plenum, New York, pp. 87–125.

deThe, H., Vivanco-Ruiz, M.D.M., Tiollais, P., Stunnenberg, H. and Dejean, A. (1990) Identification of a retinoic acid response element in the retinoic acid receptor β gene. Nature 343, 177–180.

El-Fouly, M.H., Trosko, J.E., Chang, C.C. and Warren, S.T. (1989) Potential role of the human Ha-*ras* oncogene in the inhibition of gap-junctional intercellular communication. Mol. Carcinogenesis 2, 131–135.

Enomoto, T. and Yamasaki, H. (1985) Phorbol ester mediated inhibition of intercellular communication in BALB/c 3T3 cells: Relationship to enhancement of cell transformation. Cancer Res. 45, 2681–2688.

Fletcher, W.H., Byus, C.V. and Walsh, D.A. (1987) Receptor-mediated action without receptor occupancy: A function for cell-cell communication in ovarian follicles. Adv. Exp. Med. Biol. 219, 299–323.

Fraser, S.E., Green, C.R., Bode, H.R. and Gilula, N.B. (1987) Selective disruption of gap-junctional communication interferes with a patterning process in *Hydra*. Science 237, 49–55.

Giguere, V., Ong, E.S., Segui, P. and Evans, R.M. (1987) Identification of a receptor for the morphogen retinoic acid. Nature 330, 624–629.

Hamel, E., Katoh, F., Mueller, G., Birchmeier, W. and Yamasaki. H. (1988). Transforming growth factor β as a potent tumor promoter in two-stage BALB/c 3T3 cell transformation. Cancer Res. 48, 2832–2836.

Hong, W.K., Lippman, S.M., Itri, L.M., Karp, D.D., Lee, J.S., Byers, R.M., Schantz, S.P., Kramer, A.M., Lotan, R., Peters, L.J., Dimery, I.W., Brown, B.W. and Goeppert, H. (1990) Prevention of secondary tumors with isotretinonin in squamous-cell carcinoma of the head and neck. New Engl. J. Med. 323, 795–801.

Hossain, M.Z., Wilkens, L.R., Mehta, P.P., Loewenstein, W.R. and Bertram, J.S. (1989) Enhancement of gap-junctional communication by retinoids correlates with their ability to inhibit neoplastic transformation. Carcinogenesis 10, 1743–1748.

Klaunig, J.E. and Ruch, R.J. (1990) Role of intercellular communication in carcinogenesis. Lab. Invest. 62, 135–145.

Loewenstein, W.R. (1979) Junctional communication and the control of growth. Biochim. Biophys. Acta 560, 1–65.

Loewenstein, W.R. and Kanno Y. (1966). Intercellular communication and the control of tissue growth. Lack of communication between cancer cells. Nature 209, 1248–1249.

Madhukar, B.V., Oh, S.Y., Chang, C.C., Wade, M. and Trosko, J.E. (1989) Altered regulation of intercelllular communication by epidermal growth factor, transforming growth factor-β and peptide hormones in normal human keratinocytes. Carcinogenesis 10, 13–20.

Manjunath, C. and Page, E. (1986) Rat heart gap junctions as disulfide-bonded connexon multimers: Their depolymerization and solubilization in deoxycholate. J. Membr. Biol. 90, 43–57.

Mathews-Roth, M.M. (1982) Antitumor activity of β-carotene, canthaxanthin and phytoene. Oncology 39, 33–37.

Mehta, P.P., Bertram, J.S. and Loewenstein, W.R. (1986) Growth inhibition of transformed cells correlates with their junctional communication with normal cells. Cell 44, 187–196.

Mehta, P.P., Bertram, J.S. and Loewenstein, W.R. (1989) The actions of retinoids on cellular growth correlates with their actions on gap-junctional communication. J. Cell Biol. 108, 1053–1065.

Merriman, R.L. and Bertram, J.S. (1979) Reversible inhibition by retinoids of 3-methylcholanthrene-induced neoplastic transformation in C3H/10T1/2 clone 8 cells. Cancer Res. 39, 1661–1666.

Moon, R.C. (1989) Comparative aspects of carotenoids and retinoids as chemopreventive agents for cancer. J. Nutr. 119, 127–134.

Mordan, L.J. and Bertram, J.S. (1983) Retinoid effects on cell-cell interactions and growth characteristics of normal and carcinogen-treated C3H/10T1/2 cells. Cancer Res. 43, 567–571.

Musil, L.S., Cunningham, B.A., Edelman, G.M. and Goodenough, D.A. (1990) Differential phosphorylation of the gap-junction protein connexin43 in junctional communication-competent and -deficient cell lines. J. Cell Biol. 111, 2077–2088.

Petkovich, M., Brand, N.J., Krust, A. and Chambon, P. (1987) A human retinoic acid receptor which belongs to the family of nuclear receptors. Nature 330, 444–450.

Pung, A., Rundhaug, J.E., Yoshizawa, C.N. and Bertram, J.S. (1988) β-Carotene and canthaxanthin inhibit chemically and physically induced neoplastic transformation. Carcinogenesis 9, 1533–1539.

Reznikoff, C.A., Brankow, D.W. and Heidelberger, C. (1973) Establishment and characterization of a cloned line of C3H mouse embryo cells sensitive to postconfluence inhibition of division. Cancer Res. 33, 3231–3238.

Rogers, M., Berestecky, J., Hossain, M.Z., Guo, H., Nicholson, B.J.,

Kadle, R. and Bertram, J.S. (1990) Retinoid-enhanced gap-junctional communication is achieved by increased levels of connexin43 mRNA and protein. Mol. Carcinogenesis, 3, 335–343.

Ruch, R.J. and Klaunig, J.E. (1988) Kinetics of phenobarbital inhibition of intercellular communication in mouse hepatocytes. Cancer Res. 48, 2519–2523.

Rundhaug, J.E., Pung, A., Read, C.M. and Bertram, J.S. (1988) Uptake and metabolism of β-carotene and retinal by C3H/10T1/2 cells. Carcinogenesis 9, 1541–1545.

Sáez, J.C., Connor, J.A., Spray, D.C. and Bennett, M.V.L. (1989) Hepatocyte gap junctions are permeable to the second messenger, inositol 1,4,5-trisphosphate and to calcium ions. Proc. Natl. Acad. Sci. USA 86, 2708–2712.

Schwartz, J. and Shklar, G. (1988) Regression of experimental oral carcinomas by local injection of β-carotene and canthaxanthin. Nutr. Cancer 11, 35–40.

Stich, H.F., Stich, W., Rosin, M.P. and Vallejera, M.O. (1984) Use of the micronucleus test to monitor the effect of vitamin A, β carotene and canthaxanthin on the buccal mucosa of betel nut/tobacco chewers. Int. J. Cancer 34, 745–750.

Vasios, G.W., Gold, J.D., Petkovich, M., Chambon, P. and Gudas, L.J. (1989) A retinoic acid-responsive element is present in the 5' flanking region of the laminin β1 gene. Proc. Natl. Acad. Sci. USA 86, 9099–9103.

Warner, A.E., Guthrie, S.C. and Gilula, N.B. (1984) Antibodies to gap-junction protein selectively disrupt communication in the early amphibian embryo. Nature 311, 127–131.

Yamasaki, H. (1990) Gap-junctional intercellular communication and carcinogenesis. Carcinogenesis 11, 1051–1058.

Yamasaki, H. and Katoh, F. (1988) Further evidence for the involvement of gap-junctional intercellular communication in induction and maintenance of transformed foci in Balb/c 3T3 cells. Cancer Res. 45, 2681–2688.

Yancey, S.B., John, S.A., Lal, R., Austin, B.J. and Revel, J.-P. (1989) The 43 kD polypeptide of heart gap junctions: Immunolocalization, topology and functional domains. J. Cell Biol. 108, 2241–2254.

Zelent, A., Krust, A., Petkovitch, M., Kastner, P. and Chambon, P. (1989) Cloning of murine α and β retinoic acid receptors and a novel receptor γ predominantly expressed in skin. Nature 339, 714–717.

Zhang, L.-X., Cooney, R.V. and Bertram, J.S. (1991) Carotenoids enhance gap-junctional communication and inhibit lipid peroxidation in C3H/10T1/2 cells: Relationship to their cancer chemopreventive action. Carcinogenesis 12, 2109–2114.

Zhang, L.-X., Cooney, R.V. and Bertram, J.S. (1992) Carotenoids up-regulate connexin43 gene expression independent of their provitamin A or antioxidant properties. Cancer Res. 52, 5707–5712.

Zhu, D., Caveney, S., Kidder, G.M. and Naus, C.C.G. (1991) Transfection of C6 glioma cells with connexin 43 cDNA: Analysis of expression, intercellular coupling, and cell proliferation. Proc. Natl. Acad. Sci. USA 88, 1883–1887.

J.E. Hall, G.A. Zampighi and R.M. Davis (Eds.)
Progress in Cell Research, Vol. 3

CHAPTER 43

Gap-junctional communication alterations at various regulatory levels of connexin expression and function during animal and human carcinogenesis

M. MESNIL, M. OYAMADA*, D.J. FITZGERALD**, W.M.F. JONGEN***, V. KRUTOVSKIKH and H. YAMASAKI

Unit of Multistage Carcinogenesis, International Agency for Research on Cancer, 150, cours Albert Thomas, 69372 Lyon Cédex 08, France

Introduction

The loss of homeostasis which characterizes tumor growth is probably the result of a deregulated intercellular communication capacity of the cancer cells. The deregulation of gap-junctional intercellular communication (GJIC) was observed in in vivo and in vitro carcinogenesis models. Our studies using gap-junction protein (connexin) molecular probes indicate that the loss of GJIC during carcinogenesis and in tumors involves aberrations at multiple levels of regulatory mechanisms of connexins. Tumor promoters, such as phorbol esters and phenobarbital, are known to inhibit GJIC. Our recent work with mouse epidermal cells suggests that TPA inhibits GJIC without affecting the amount of Cx43 mRNA, but that it acts at a post-translational level. This is in contrast to our previous in vivo observation that the liver-tumor promoter, phenobarbital, decreases Cx32 mRNA level in the rat liver. These results suggest that different types of tumor promoters inhibit GJIC at different regulatory levels. Similarly, analysis of GJIC and connexin expression of cells at different stages of carcinogenesis suggests multiple mechanisms of GJIC deregulation. For example, we found a progressive decrease of GJIC in a series of rat liver epithelial cell lines and mouse epidermal cell lines which paralleled the progressively malignant phenotypes. However, the level of Cx43 gene expression was similar in each series of cell lines. On the other hand, GJIC loss in primary pre-neoplastic foci and tumors induced in rat liver was associated with a decreased level of Cx32 mRNA and protein, as measured by immunostaining. In surgically removed primary human hepatocellular carcinomas (HCC), however, the level of Cx32 mRNA was not changed. Instead, the Cx43 mRNA level was significantly increased in all HCC samples. Such aberrant expression of Cx43 in HCC apparently had not altered their GJIC, since direct measurements of GJIC by a dye-transfer assay slices of freshly isolated human livers suggests similar levels of GJIC in HCC and their surrounding normal tissues.

A clear selective lack of GJIC between transformed and non-transformed cells has been observed in cultured cells. We have now confirmed such a selective lack of GJIC between preneoplastic foci and surrounding cells in rat liver in vivo. Our results from connexin gene transfection experiments suggest that selective GJIC is not due to different connexin species being expressed between transformed cells and their non-transformed counterparts, but may be due to a lack of cell-cell recognition between transformed and nontransformed cells. Our recent results, and those from others, indirectly support such a hypothesis. E-cadherin gene transfection into mouse epidermal cells indicates that this cell-adhesion molecule is a prerequisite for Cx43 molecules to form functional GJIC. These results suggest that during carcinogenesis GJIC regulation mechanisms are vulnerable to various factors.

Studies on various models of experimental carcinogenesis have provided clear evidence that cancer is an evolutionary process that can be divided into distinctive stages (Berenblum, 1941). This evolutionary process seems to correspond to an accumulation of genetic disorders, as has been observed in human colon cancer (Fearon and Vogelstein, 1990). However, the mechanisms governing tumor growth are still unclear, and the critical point of this process is to elucidate how an initiated cell may escape from the control of normal surrounding cells. Since disruption of homeostasis is the principal characteristic of tumor growth, it is important to investigate how cell-cell interactions are regulated during carcinogenesis. Among

*Present address: Department of Pathology, Sapporo Medical College, Sapporo 060, Japan.
**Present address: Public and Environmental Health Division, South Australia Health Commission, Adelaide SA, Australia.
***Present address: Agrotechnological Research Institute ATO, Wageningen, The Netherlands.

the various candidates of cell-cell interactions responsible for homeostasis is GJIC, which permits direct contact between the cytoplasms of adjacent cells. The passage of small cytoplasmic compounds, including second messengers, through gap junctions (Sáez et al., 1989) was suggested to be a pathway for growth regulators (Loewenstein, 1979). Furthermore, experimental work performed on different cell systems has indicated that GJIC may indeed play a critical role at the different stages of carcinogenesis (Yamasaki and Mesnil, 1987; Yamasaki, 1990).

The discovery that tumor-promoting agents inhibit GJIC during the tumor promotion phase of carcinogenesis in various cell types suggested that such inhibition would facilitate the escape of initiated cells from the control of normal surrounding cells (Murray and Fitzgerald, 1979; Yotti et al., 1979). Hence, the selective lack of GJIC between normal and transformed cell populations, often observed in in vitro models (Enomoto and Yamasaki, 1984), would be, next to "initiation", a second necessary event leading to tumor progression.

This hypothesis was mainly formulated from in vitro studies performed on rodent fibroblastic cell cultures. Since more than 90% of human tumors have an epithelial origin, we thought it would be important to study the variations of GJIC in models more adapted to the human cancer situation, i.e. epithelial cell cultures, when in vitro studies are required. In addition, taking the advantage of available molecular probes, we employed animal models of cancer progression (rat liver carcinogenesis) and multistage carcinogenesis (mouse skin) in order to avoid artefacts due to cell-culture that may change considerably the communication capacity of cells (Eldridge et al., 1989). Finally, to complete this approach, we studied GJIC directly in human liver tumors. These different studies provide information on the regulation of GJIC at the various stages of carcinogenesis.

Gap-junctional communication and tumor promotion

Tumor promotion is a critical phase of carcinogenesis that permits the selective growth of initiated cells as a tumor (Börzsönyi et al., 1984). This stage, originally described by Berenblum (1941) from experimental animal carcinogenesis, is reversible, and tumor promoters are thought to act through non-genotoxic mechanisms (Börzsönyi et al., 1984). Tumor-promoting agents are numerous and diverse, often with non-related chemical properties. It is important to emphasize, however, that many types, though not all of them, inhibit GJIC (Fitzgerald and Yamasaki, 1990). This capacity seems to be tissue-specific and realized through different pathways. For instance, the phorbol ester TPA, a potent tumor-promoting agent in mouse skin carcinogenesis, strongly inhibits GJIC in a non-transformed rat liver epithelial cell line, IAR 20. This effect, demonstrated by dye-transfer assay, is transient and disappears a few hours after the treatment (Mesnil et al., 1986). Immunocytologic analysis shows that inhibition of dye transfer is correlated with the disappearance of Cx43 from the cell membranes. Since the amounts of not only mRNA coding for Cx43, but also the protein as revealed by Western blotting, are not changed, TPA probably decreases GJIC of IAR 20 cells through post-translational mechanisms (Asamoto et al., 1991). Phenobarbital is a commonly used tumor-promoting agent of rat hepatocarcinogenesis (Pitot and Sirica, 1980). In rat liver, phenobarbital induces a significant decrease in the number of gap junctions within 4 weeks based on immunofluorescence studies of the major liver gap-junction protein, Cx32, (Krutovskikh et al., 1991); this might be due to a decreased amount of the corresponding mRNA (Mesnil et al., 1988). The effect of phenobarbital on the mRNA coding for Cx32 is tissue specific, since no variation was detected in other organs such as stomach and kidney, and thus is correlated with the specific tumor-promoting action of this compound on the liver. A similar effect of phenobarbital has been observed in primary cultures of rat hepatocytes, both on GJIC and expression of Cx32 mRNA (M. Mesnil, unpublished results). The above results clearly indicate that the inhibition of GJIC by TPA and phenobarbital involves different mechanisms. Whatever their various properties, the tumor promoters, acting in a tissue-dependent manner, appear to inhibit GJIC of target tissues; therefore this property was suggested to be used as an endpoint for developing an in vitro screening test for tumor promoters (Zeilmaker and Yamasaki, 1986).

Gap-junctional communication and cell transformation

Carcinogenesis is an evolutionary process which, in humans, takes many decades. The histological characterizations of tumors can be subdivided into successive grades, and cells from these different stages can be isolated and studied. A few such studies have suggested that a decreased communication capacity is correlated with a more transformed phenotype. For example, we found a good correlation among a panel of rat liver epithelial cells: the more the cells are transformed the less they are able to communicate through gap junctions (Mesnil et al., 1986; Mesnil and Yamasaki, 1988). A similar result was obtained with mouse skin cell lines isolated from different stages of carcinogenesis (Klann et al., 1989).

In order to make our work as directly relevant to human disease as possible, we decided to compare the GJIC capacity of various human carcinoma cell lines (mesothelioma, liver and skin carcinomas) with their corresponding non-transformed primary cells or immortalized counterparts. In every case, we observed that human carcinoma cells communicate less extensively than their normal counterparts. In some cases, immortalization was sufficient to induce this decreased communication capacity, as seen in normal human liver cells transformed by SV40. These results seemed to confirm our observations on animal cell cultures.

In order to examine whether the decreased GJIC capacity also occurs in vivo, we developed a method to measure the communication in tissues freshly removed from animals or humans (Krutovskikh et al., 1991). This assay involves microinjection of a gap-junction-permeable fluorescent dye directly into the cells of a piece of tissue and, after fixation, estimation of the number of communicating cells. Using this technique in rat liver carcinogenesis experiments, we observed a decreased GJIC in primary preneoplastic foci and tumors compared to normal surrounding hepatocytes; this is consistent with an earlier finding of reduced levels of RNA coding for the major liver gap-junction protein, Cx32, in rat hepatocellular carcinomas and hyperplastic nodules (Fitzgerald et al., 1989).

However, when the in vivo microinjection assay was performed on human liver containing carcinomas, no GJIC difference was observed between tumor cells and the normal surrounding tissue (V. Krutovskikh, unpublished observations). This observation was confirmed at the mRNA level (Cx32). Interestingly, in human liver carcinomas, there is an appearance of Cx43 mRNA that is not detected in normal liver tissues (Oyamada et al., 1990). The presence of excess endothelial cells due to angiogenesis could explain the presence of this Cx43 mRNA (Larson et al., 1990). Thus, the decreased communication capacity that we previously thought to be a general phenomenon of transformed epithelial cells is not observed in human liver carcinomas. This discrepancy emphasizes the importance of studying more human tumors, both at the functional and molecular levels. Furthermore, such a result directly raises the question of the extrapolation of animal experimental carcinogenesis results to human situations.

Selective lack of communication and tumor progression

It has been suggested that tumor growth could be the consequence of a lack of communication of tumor cells with the normal surrounding tissue (Enomoto and Yamasaki, 1984). This idea derived principally from in vitro experimental transformation of fibroblasts that gives rise to transformed foci in situ. In these circumstances, the lack of communication between transformed and normal cells is so well established that it is possible to kill selectively the transformed cells (Yamasaki and Katoh, 1988a). The importance of selective lack of communication in tumor growth was emphasized by the fact that, after treatments which led to reversion of the transformed phenotype, communication between the two cell populations was established (Mehta et al., 1986; Yamasaki and Katoh, 1988b). The selective lack of communication between normal and transformed cells was also seen with epithelial cells (Mesnil and Yamasaki, 1988).

Using our in vivo dye-transfer assay, we found that the selective intercellular communication can also be seen be-

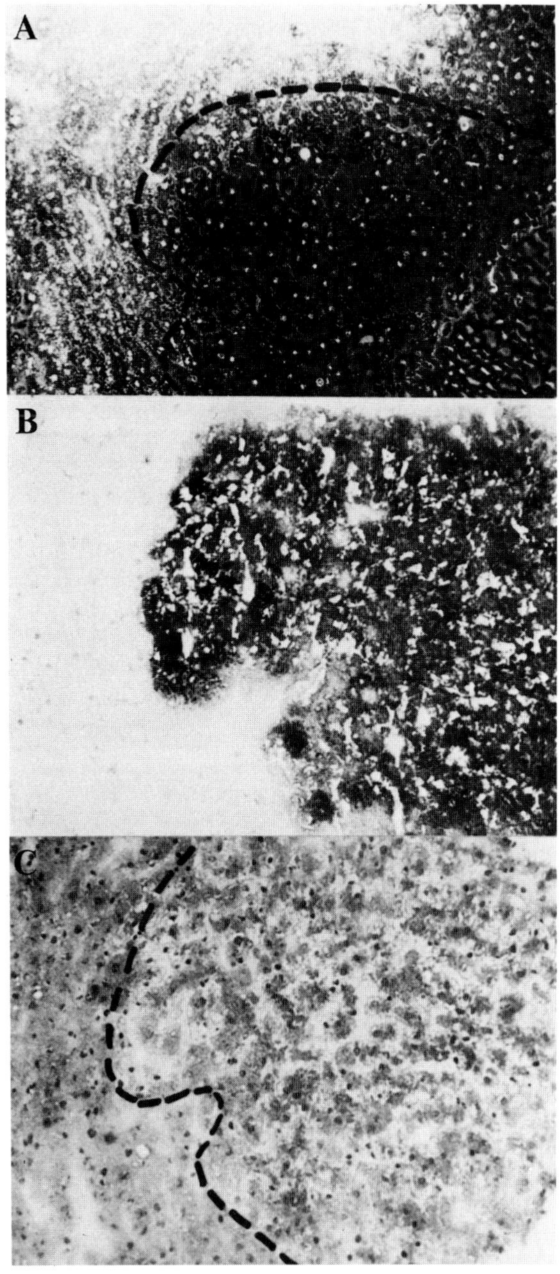

Figure 1. Selective lack of gap-junctional communication (dye transfer) between a pre-neoplastic focus and the surrounding hepatocytes of a Fischer-344 rat treated by a modified Solt-Farber carcinogenic diet (Krutovskikh et al., 1991). (A) LY was microinjected in the hepatocytes bordering the preneoplastic lesion (*, microinjected cells) that was characterized (B) by a positive anti-GST-P serum reaction and (C) hematoxylin-eosin staining. Note the absence of dye transfer between the two cell populations.

tween tumor and normal tissue in rat liver (Krutovskikh et al., 1991) (Fig. 1). It is interesting to note that some small tumor foci in rat liver are able to communicate with the normal surrounding tissue. This might be a sign of phenotypic reversion, as is often observed in chemically induced hepatocarcinogenesis, or might be due to an insufficiently high level of transformation. We still do not know whether a selective lack of communication also occurs in human cancers in vivo, probably because we use human liver samples and, interestingly, most liver tumors are encapsulated in an extracellular matrix-like structure that would not permit extended GJIC between the carcinoma and the normal surrounding tissues to be established.

Several hypotheses may explain the selective lack of GJIC between normal and transformed cells. For instance, a considerably decreased communication capacity of the

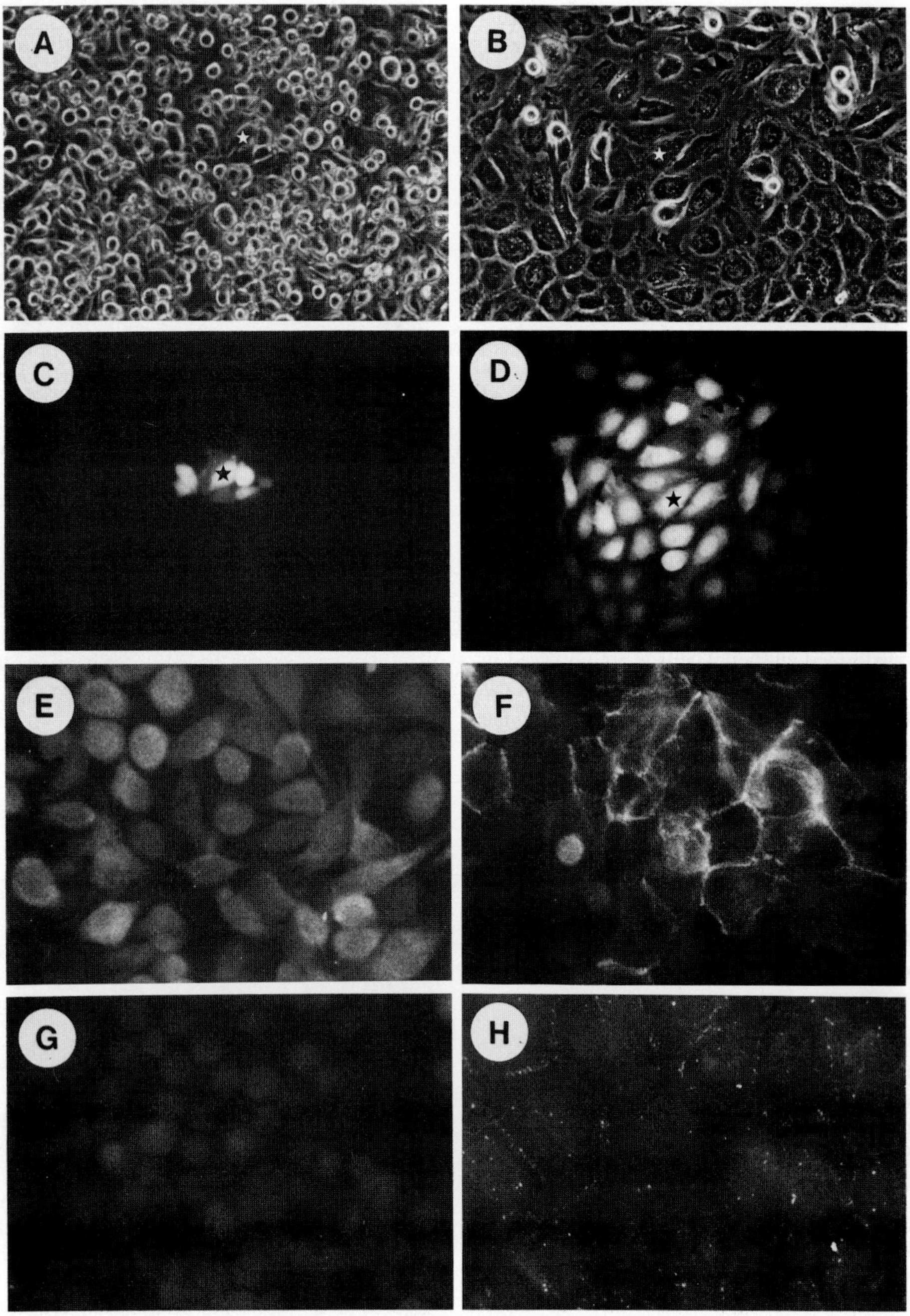

Figure 2. Morphology, gap junctional communication and immunostaining of E-cadherin and Cx43 in mouse epidermal cells before (left) and after (right) transfection of the E-cadherin gene. A and B, phase contrast micrographs; C and D, gap junctional communication estimated by LY microinjection (*, microinjected cells); E and F, immunocytochemical staining of E-cadherin; G and H, immunocytochemical staining of Cx43.

transformed cells compared to the normal counterparts would be sufficient, as may happen in rat liver foci (Krutovskikh et al., 1991). There are also cases in which the two cell populations have a high GJIC capacity (Yamasaki et al., 1987). In these cases, two possibilities must be considered: First, if the populations express different connexin types, we might conclude that a common gap-junctional channel cannot be made because of the lack of recognition between the two different connexin types. Indeed, we have observed the selective lack of communication in co-cultures of rat hepatocytes (expressing Cx32) and liver epithelial cells (expressing Cx43) (Mesnil et al., 1987; and unpublished results). However, if two apposing cell populations express a similar connexin type, another phenomenon may occur: In co-cultured IAR 20 and IAR 6-1 cells—rat liver epithelial cells that show a selective lack of communication—the gap junctions which are mainly present among the well-communicating IAR 20 cells are not present in the membranes of those cells which are in contact with the transformed IAR 6-1 cells (M. Oyamada, unpublished results). The selectivity of the distribution of the connexins which form gap junctions seems to be, in this particular case, the result of some alteration in a specific cell-cell recognition process.

Cell-cell recognition and connexin regulation

Previous studies suggested that cell-adhesion molecules (CAMs), especially the Ca^{2+}-dependent cell-adhesion molecules (cadherins), may facilitate the formation of gap junctions between cells. This was clearly demonstrated by transfecting an L-CAM cDNA into non-communicating mouse sarcoma cells. After transfection, these cells took on an epithelial shape, a non-tumorigenic behaviour, and became communication competent through gap junctions (Mege et al., 1988). These results not only suggest that cell-adhesion molecules are important for the establishment of gap junctions between cells, but also that they may play a role in carcinogenesis.

We have observed that mouse epidermal cells isolated from skin papillomas and carcinomas had GJIC capacity, depending on their level of transformation (Klann et al., 1989). We also observed that most of these cell lines had a higher GJIC capacity when Ca^{2+} concentrations were increased in the culture media; this increased capacity is not correlated with any change in the amount of Cx43 mRNA. However, one papilloma cell line, P3/22, was poorly communicating, even in high Ca^{2+} conditions. This was intriguing since these cells had an amount of Cx43 mRNA similar to that of cell lines which respond to calcium. Further investigations showed that, in contrast to other cell lines, P3/22 cells were unable to express detectable levels of mRNA for the epithelial form of cadherin, E-cadherin. In order to test directly whether E-cadherin is involved in the Ca^{2+}-dependent regulation of GJIC of this cell line, we transfected an E-cadherin expression vector into P3/22 cells and obtained clones expressing high levels of E-cadherin mRNA. All the transfectants expressed E-cadherin molecules, as well as functional gap junctions in a calcium-dependent manner at cell-cell contact areas (Fig. 2). These results suggest that Ca^{2+}-dependent regulation of GJIC in mouse epidermal cells is directly controlled by a calcium-dependent cell-adhesion molecule, E-cadherin (Jongen et al., 1991).

The fact that cell-adhesion molecules directly regulate the function of connexin molecules suggests that malfunctioning of certain cell-adhesion molecules is involved in carcinogenesis. Several lines of evidence support this hypothesis: (1) Transfection of cadherin genes to tumorigenic human and animal cell lines suppressed tumorigenicity or invasive behavior (Mege et al., 1988; Chen and Obrink, 1991; Frixen et al., 1991; Vleminckx et al., 1991). (2) There is a correlation between expression of cadherins and invasiveness of human tumors (Eidelman et al., 1989; Shimoyama et al., 1989; Shimoyama and Hirohashi, 1991a, b; Shiozaki et al., 1991). (3) One of the identified human colorectal tumor suppressor genes, DCC, has sequence homology to N-CAM (Fearon et al., 1990). Taken together, we propose that connexins and cell-adhesion molecules form a family of tumor suppressor genes (Yamasaki, 1990).

References

Asamoto, M., Oyamada, M., El Aoumari, A., Gros, D. and Yamasaki, H. (1991) Molecular mechanisms of TPA-mediated inhibition of gap-junctional intercellular communication. Evidence for action on the assembly or function but not the expression of connexin 43 in rat liver epithelial cells. Mol. Carcinogen. 4, 322–327.

Berenblum, I. (1941) The mechanism of carcinogenesis: A study of the significance of co-carcinogenic action and related phenomena. Cancer Res. 1, 807–814.

Börzsönyi, M., Day, N.E., Lapis, K. and Yamasaki, H. (Eds.) (1984) Models, Mechanisms and Etiology of Tumour Promotion. IARC Scientific Publications No. 56, International Agency for Research on Cancer, Lyon, France.

Chen, W. and Obrink, B. (1991) Cell-cell contacts mediated by E-cadherin (uvomorulin) restrict invasive behavior of L-cells. J. Cell Biol. 114, 319–327.

Eidelman, S., Damsky, C.H., Wheelock, M.J. and Damjanov, I. (1989) Expression of the cell-cell-adhesion glycoprotein cell-CAM 120/80 in normal human tissues and tumors. Am. J. Pathol. 135, 101–110.

Eldrige, S.R., Martens, T.W., Sattler, C.A. and Gould, M.N. (1989) Association of decreased intercellular communication with the immortal but not the tumorigenic phenotype in human mammary epithelial cells. Cancer Res. 49, 4326–4331.

Enomoto, T. and Yamasaki, H. (1984) Lack of intercellular communication between chemically transformed and surrounding non-transformed BALB/c 3T3 cells. Cancer Res. 44, 5200–5203.

Fearon, E.R. and Vogelstein, B. (1990) A genetic model for colorectal tumorigenesis. Cell 61, 759–767.

Fearon, E.R., Cho, K.R., Nigro, J.M., Kern, S.E., Simons, J.W.,

Ruppert, J.M., Hamilton, S.R., Preisinger, A.C., Thomas, G., Kinzler, K.W. and Vogelstein, B. (1990) Identification of a chromosome 18q gene that is altered in colorectal cancers. Science 247, 49–56.

Fitzgerald, D.J., Mesnil, M., Oyamada, M., Tsuda, H., Ito, N. and Yamasaki, H. (1989) Changes in gap-junction protein (connexin 32) gene expression during rat liver carcinogenesis. J. Cell Biochem. 41, 97–102.

Fitzgerald, D.J. and Yamasaki, H. (1990) Tumor promotion: models and assay systems. Teratogen., Carcinogen. Mutagen. 10, 89–102.

Frixen, U.H., Behrens, J., Sachs, M., Eberle, G., Voss, B., Warda, A., Löchner, D. and Birchmeier, W. (1991) E-Cadherin-mediated cell-cell-adhesion prevents invasiveness of human carcinoma cells. J. Cell Biol. 113, 173–186.

Jongen, W.M.F., Fitzgerald, D.J., Asamoto, M., Piccoli, C., Slaga, T.J., Gros, D., Takeichi, M. and Yamasaki, H. (1991) Regulation of connexin 43-mediated gap-junction intercellular communication by Ca^{2+} in mouse epidermal cells is controlled by E-cadherin. J. Cell Biol. 114, 545–555.

Klann, R.C., Fitzgerald, D.J., Piccoli, C., Slaga, T.J. and Yamasaki, H. (1989) Gap-junctional intercellular communication in epidermal cell lines from selected stages of SENCAR mouse skin carcinogenesis. Cancer Res. 49, 699–705.

Krutovskikh, V.A., Oyamada, M. and Yamasaki, H. (1991) Sequential changes of gap-junctional intercellular communications during multistage rat liver carcinogenesis: Direct measurement of communication *in vivo*. Carcinogenesis 12, 1701–1706.

Larson, D.M., Haudenschild, C.C. and Beyer, E.C. (1990) Gap-junction messenger RNA expression by vascular wall cells. Circ. Res. 66, 1074–1080.

Loewenstein, W.R. (1979) Junctional intercellular communication and the control of growth. Biochim. Biophys. Acta 560, 1–65.

Mege, R.M., Matsuzaki, F., Gallin, W.J., Goldberg, J.I., Cunningham, B.A. and Edelman, G.M. (1988) Construction of epithelioid sheets by transfection of mouse sarcoma cells with cDNAs for chicken cell-adhesion molecules. Proc. Natl. Acad. Sci. USA 85, 7274–7278.

Mehta, P.P., Bertram, J.S. and Loewenstein, W.R. (1986) Growth inhibition of transformed cells correlates with their junctional communication with normal cells. Cell 44, 187–196.

Mesnil, M. and Yamasaki, H. (1988) Selective gap-junctional communication capacity of transformed and non-transformed rat liver epithelial cell lines. Carcinogenesis 9, 1499–1502.

Mesnil, M., Fraslin, J.M., Piccoli, C., Yamasaki, H. and Guguen-Guillouzo, C. (1987) Cell contact but not junctional communication (dye coupling) with biliary epithelial cells is required for hepatocytes to maintain differentiated functions. Exp. Cell Res. 173, 524–533.

Mesnil, M., Montesano, R. and Yamasaki, H. (1986) Intercellular communication of transformed and non-transformed rat liver epithelial cells. Exp. Cell Res. 165, 391–402.

Mesnil, M., Fitzgerald, D.J. and Yamasaki, H. (1988) Phenobarbital specifically reduces gap-junction protein mRNA level in rat liver. Mol. Carcinogen. 1, 79–81.

Murray, A.W. and Fitzgerald, D.J. (1979) Tumor promoters inhibit metabolic cooperation in co-cultures of epidermal and 3T3 cells. Biochem. Biophys. Res. Commun. 91, 395–401.

Oyamada, M., Krutovskikh, V.A., Mesnil, M., Partensky, C., Berger, F. and Yamasaki, H. (1990) Aberrant expression of gap-junction gene in primary human hepatocellular carcinomas: Increased expression of cardiac-type gap-junction gene connexin 43. Mol. Carcinogen. 3, 273–278.

Pitot, H.C. and Sirica, A.E. (1980) The stages of initiation and promotion in hepatocarcinogenesis. Biochim. Biophys. Acta 605, 191–212.

Sáez, J.C., Connor, J.A., Spray, D.C. and Bennett, M.V.L. (1989) Hepatocyte gap junctions are permeable to the second messengers, inositol 1,4,5-trisphosphate, and to calcium ions. Proc. Natl. Acad. Sci. USA 86, 2708–2712.

Shimoyama, Y. and Hirohashi, S. (1991a) Expression of E- and P-cadherin in gastric carcinomas. Cancer Res. 51, 2185–2192.

Shimoyama, Y. and Hirohashi, S. (1991b) Cadherin intercellular adhesion molecule in hepatocellular carcinomas; loss of E-cadherin expression in an undifferentiated carcinoma. Cancer Lett. 57, 131–175.

Shimoyama, Y., Hirohashi, S., Hirano, S., Noguchi, M., Shimosato, Y., Takeichi, M. and Abe, O. (1989) Cadherin cell-adhesion molecules in human epithelial tissues and carcinomas. Cancer Res. 49, 2128–2133.

Shiozaki, H., Tahara, H., Oka, H., Miyata, M., Kobayashi, K., Tamara, S., Iihara, K., Doki, Y., Hirano, S., Takeichi, M. and Mori, T. (1991) Expression of immunoreactive E-cadherin adhesion molecules in human cancers. Am. J. Pathol. 139, 17–23.

Vleminckx, K., Vakaet, L. Jr, Mareel, M., Fiers, W. and Van Roy, F. (1991) Genetic manipulation of E-cadherin expression by epithelial tumor cells reveals an invasion suppressor role. Cell 66, 107–119.

Yamasaki, H. (1990) Gap-junctional intercellular communication and carcinogenesis. Carcinogenesis 11, 1051–1058.

Yamasaki, H. and Katoh, F. (1988a) Novel method for selective killing of transformed rodent cells through intercellular communication, with possible therapeutic applications. Cancer Res. 48, 3203–3207.

Yamasaki, H. and Katoh, F. (1988b) Further evidence for the involvement of gap-junctional intercellular communication in induction and maintenance of transformed foci in Balb/c 3T3 cells. Cancer Res. 48, 3490–3495.

Yamasaki, H. and Mesnil, M. (1987) Cellular communication in cell transformation. In: H.A. Milman and E. Elmore (Eds.), Advances in Modern Environmental Toxicology, vol. XIV, Biochemical Mechanisms and Regulation of Intercellular Communication, Princeton Scientific Publishing, Princeton, NJ, pp. 181–207.

Yamasaki, H., Hollstein, M., Mesnil, M., Martel, N. and Aguelon, A.M. (1987) Selective lack of intercellular communication between transformed and non-transformed cells as a common property of chemical and oncogene transformation of Balb/c 3T3 cells. Cancer Res. 47, 5658–5664.

Yotti, L.P., Chang, C.C. and Trosko, J.E. (1979) Elimination of metabolic cooperation in Chinese hamster cells by a tumor promoter. Science 206, 1089–1091.

Zeilmaker, M.J. and Yamasaki, H. (1986) Inhibition of junctional intercellular communication as a possible short-term test to detect tumor-promoting agents: Results with nine chemicals tested by dye-transfer assay in Chinese hamster V79 cells. Cancer Res. 46, 6180–6186.

J.E. Hall, G.A. Zampighi and R.M. Davis (Eds.)
Progress in Cell Research, Vol. 3

CHAPTER 44

Gap junctions and tumorigenesis: transfection of communication-deficient tumor cells with connexin32 retards growth in vivo

B. EGHBALI[a], J.A. KESSLER[a,b], L.M. REID[c], C. ROY[a] and D.C. SPRAY[a]

Departments of [a]Neuroscience, [b]Neurology and [c]Molecular Pharmacology, Albert Einstein College of Medicine, 1410 Pelham Parkway South, Bronx, NY 10461, USA

Summary

Exchange of ions and small molecules through gap-junction channels is believed to be critical for normal tissue growth and development. As a test for a role of gap junction-mediated intercellular communication in control of abnormal cell growth, we have compared growth rates of communication-deficient human tumor cells (SKHep1) with those of clones stably transfected with cDNA encoding the rat liver gap-junction protein, Cx32. In culture, rates of proliferation for parental and transfected clones were similar. However, when sizes of tumors in vivo formed by these cell lines were compared, growth rates for two well-coupled clones were significantly lower than for communication-deficient or poorly coupled clones. This study demonstrates that the growth rate of these cells as tumors is negatively correlated with strength of intercellular communication.

Intercellular communication through gap junctions allows the exchange of nutrients, ions and regulatory molecules between adjacent cells. This type of intercellular communication has long been considered crucial in providing signals for normal cell proliferation and development (Furshpan et al., 1965; Loewenstein, 1979; Bennett et al., 1981; Pitts and Finbow, 1986), and disruption of gap-junctional communication has been postulated to underlie tumorigenesis and developmental defects (e.g. Loewenstein, 1979; Welsch and Stedman, 1984; Guthrie and Gilula, 1989). In support of this hypothesized role, gap-junction expression and function are reduced in some tumors relative to levels in host tissue (but are seldom altogether absent) (Weinstein and Pauli, 1986), expression of certain oncogenes is associated with reduced coupling (Atkinson and Sheridan, 1985; Azarnia et al., 1988) and phorbol esters and other tumor promoters decrease junctional conductance in some cell types (but do not generally uncouple completely) (e.g., Yotti et al., 1979; Kanno, 1985; Trosko et al., 1988; Yamasaki, 1988).

In order to explicitly test the hypothesis that gap-junction disappearance is causally linked to carcinogenesis, it would be desirable to determine the effects on growth rate of either inserting gap junctions into cells that do not have them or deleting junctions from cells in which junctions are normally present. We have performed the first type of experiment, using as an experimental preparation the SKHep1 cell line, which is highly metastatic, rapidly growing and communication deficient (Fogh et al., 1977; Doerr et al., 1989; Eghbali et al., 1990). These studies revealed that expression of gap-junctional communication in this cell type had no effect on growth rate of the cells in culture, but markedly reduced the growth of these cells as tumors (Eghbali et al., 1991b).

Materials and methods

SKHep1 cells were co-transfected with vectors containing full-length coding sequence for Cx32 and/or the selectable marker *neo* using the $CaPO_4$-precipitation technique. Colonies were doubly selected: by antibiotic resistance and, subsequently, on the basis of Lucifer Yellow transfer (Eghbali et al., 1990). In this study, we have compared the growth of five colonies: Tr2 and Tr3, which were transfected only with *neo*; Tr1 and Tr4, which were transfected with both Cx32 and *neo* and were well coupled; and Tr5, which was transfected with both genes but was only poorly coupled.

Growth rate was determined in 60 mm tissue culture dishes after seeding from confluent cultures at a density of 5×10^4 cells/dish on day 0. On subsequent days, the concentration of cells was determined either by counting the number of cells/unit area of the dish or by counting the number of cells/unit volume (using a hemocytometer) after cells were dissociated using mild trypsinization. For each time point in each experiment, two or three dishes of cells were counted and mean values from repeated experiments were compared to determine means and variance.

Results and discussion

Growth rate in culture

We found no differences in rates of growth in the various clones, although at 11 days in culture the number of cells was slightly higher in the parental line than for the transfectants (Fig. 1). This result differs from those obtained in a similar type of study in which transfection of C6 glioma cells with Cx43 was reported to reduce the growth rate of colonies (Zhu et al., 1991). In that study, cell growth was evaluated for only 4 days, variability was not evaluated, a different connexin was transfected, and it was possible that the endogenous expression of gap junctions in that system affected the results. The small effect on cell number that we detected here, compared to the parental line, agrees with previous studies in which *neo* gene expression has been shown to depress cell proliferation (e.g., Cheng et al., 1989).

Differential adhesiveness (which results from transfection of pc12 cells with Cx26: Eghbali et al., 1991a) might contribute to a difference in the number of cells counted after the trypsinization treatment. Therefore, we compared cell counts obtained in culture dishes during the phase of rapid proliferative growth (days 3–6). These measures of growth rates were also statistically indistinguishable from one colony to another (Eghbali et al., 1991b). We conclude from both sets of data that the presence of gap junctions did not alter the growth rates of the cells in culture.

Tumor studies

Although assessment of contact inhibition in culture has been used by others to evaluate whether coupled cells grow less rapidly (Loewenstein, 1979; Nicolson et al., 1988; Mehta et al., 1986), it is two-dimensional and may not adequately represent what happens in a tumor, where cells contact heterologous neighbors and the normal complement of extracellular matrix and hormones is present (see Reid et al., 1988). In order to test directly the effects of transfection on tumor growth in vivo, the SKHep1 cells expressing only *neo* and those expressing *neo* plus Cx32 were injected into the backs of nude mice (Shouval et al., 1983; Doerr et al., 1989).

In the first experiment, clonal populations of Tr2 and Tr1 cells were compared: tumor sizes (both weight and dimensions) were measured at four and ten weeks after injection ($Tr1_1$ and $Tr2_1$ in Fig. 2). At both time points, tumors formed from Cx32-transfected cells were significantly smaller ($p < 0.01$). Moreover, the growth rate (evaluated as the change in size from 4 to 10 weeks) was strikingly lower for tumors containing Cx32 than for the controls (Fig. 2).

We evaluated additional clones in order to verify that this retardation in growth of the coupled cells was due to the expression of gap junctions, rather than to disrupted expression of a growth-promoting gene (or even enhanced expression of a growth inhibitor) caused by transcriptional interference of the incorporated construct DNA. In these experiments, the growth rate was determined from measurements of tumor dimensions each week (beginning at the third week after injection), and tumors were harvested at the end of the ten-week period. Growth was again found to be less rapid for the well-coupled cells expressing high levels of Cx32 (compare Tr4 with Tr2 and Tr3, Fig. 2). Growth rates of the independent communication-deficient clones Tr2 and Tr3 were not significantly different from one another ($p > 0.05$).

For the experiments described above, tumor growth was assessed after bilateral injection of the cells. We also compared growth rates of unilateral tumors in Tr4 and Tr3; these rates were significantly different from one another ($p < 0.05$), but not from growth rates of bilateral tumors of the same clones shown in Fig. 2.

We also evaluated the growth rate of tumors formed by Tr5, a very weakly coupled Cx32 transfectant that expresses little Cx32 mRNA (only 8% of that of Tr1); weak LY transfer is generally detected in these cells, but only to one or two neighbors and only to a small degree (Eghbali et al., 1991b). Growth of the poorly coupled transfectant, Tr5, was significantly ($p < 0.05$) more rapid than of either the communication-deficient *neo* transfectants or the well-coupled clones (Fig. 2, Tr5, last columns).

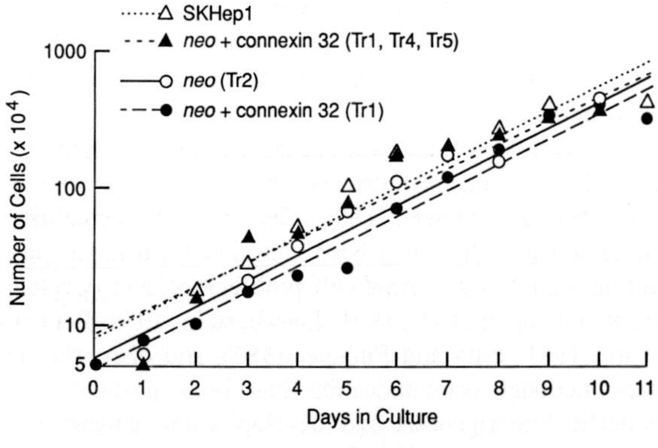

Figure 1. Growth rates of the parental SKHep1 cell line and transfectants in tissue culture. This graph is composed of seven experiments comparing parental SKHep1 cells with three clones transfected with *neo* + Cx32 (the data were not significantly different for Tr1, Tr4 and Tr5, which are combined), and a single experiment run in triplicate comparing a colony transfected with *neo* alone (Tr2) with a colony transfected with *neo* + Cx32 (Tr1). Standard errors about the means are smaller than the symbols in many cases; lines for each group (see legend in inset) indicate the semi-logarithmic best fits to the means over the entire experiment (days 0–11); for each, regression coefficient > 0.95. Neither slopes nor intercepts were significantly different between groups. Modified slightly from Eghbali et al., 1991b.

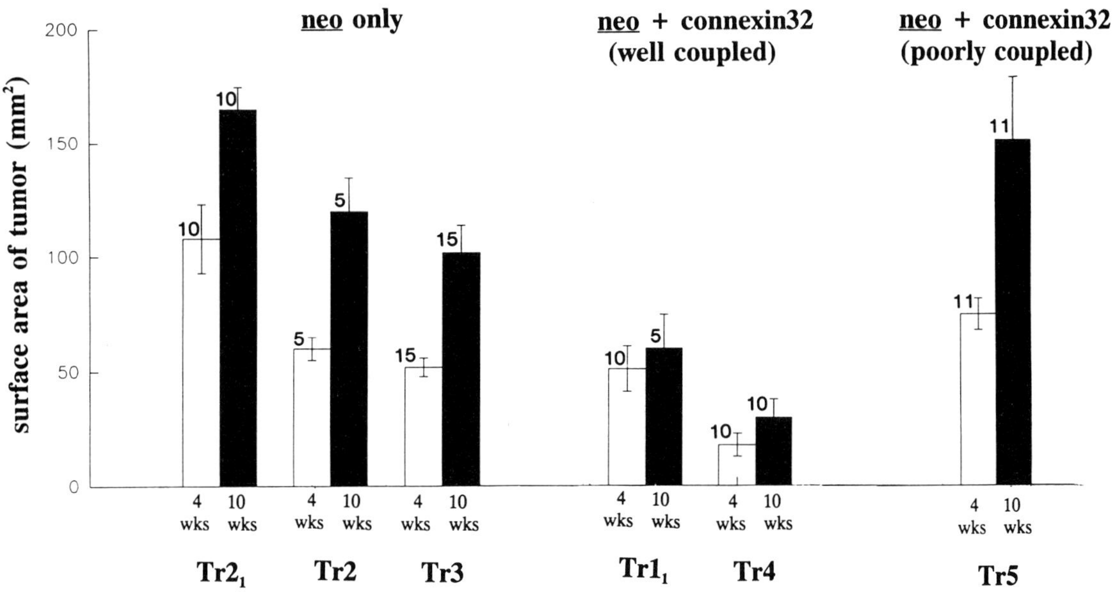

Figure 2. Growth rates of tumors comprised of *neo*- and *neo*- + Cx32-transfected SKHep1 cells in nude mice. Groups of animals (numbers above bars in histogram indicate the number of animals in each group) were injected subcutaneously with 10^7 cells/0.1 ml (in RPMI medium with 10% fetal calf serum) bilaterally into the subcutaneous space of the back. Four and ten weeks later, largest and smallest linear dimensions of all tumors (which were generally oval in shape) were measured transcutaneously with calipers; area computations represent sums of the products of these measurements for each tumor. Clones are indicated beneath the bars in the histogram; subscripts refer to histograms obtained from the first experiments. Open bars represent the areas of tumors at 4 weeks after injection; solid bars represent areas at 10 weeks. Note that in *neo* transfectants and in weakly coupled *neo* + Cx32 transfectants, sizes of tumors increased significantly between weeks 4 and 10, whereas in the well-coupled *neo* + Cx32 tumors, the increase in size over this period was not significant.

These results indicate that tumors formed of well-coupled Cx32 transfectants cells grew less rapidly than those with little or no Cx32. In order to determine whether the coupled cells would form tumors that would continue to grow, albeit at a slower rate, we examined a small group of animals in which well-coupled transfectants (Tr1) were introduced unilaterally into the backs of nude mice at a lower concentration. Tumors were very small or not detectable at ten weeks in any of these animals, but the sizes of the tumors expressed in two of the five animals, measured at 18 weeks, were as large as those present at ten weeks in the *neo* controls (Eghbali et al., 1991b). Thus, transfection with Cx32 retarded growth rate but did not prevent the tumors from eventually attaining a large size.

In summary, well-coupled cells formed tumors that grew more slowly than the poorly coupled transfectants, even though the presence of functional gap junctions did not detectably alter growth rates of the cells in tissue culture. The difference in behavior of cells grown in culture and as tumors is consistent with previous reports in which expression of a number of genes by tumor cells in culture correlated poorly with expression of the same genes by the tumors in vivo (Reid et al., 1988).

The parental SKHep1 cell line is not entirely devoid of gap junction-mediated intercellular communication, but the incidence and strength of coupling are quite low; only about 15% of cell pairs are coupled at all, and junctional conductance in the coupled pairs is 0.6 nS or less; LY transfer has never been detected in the parental line or in the *neo* transfectants, presumably due to the low extent of endogenous coupling. The high metastatic potential of the parental cell line (Fogh et al., 1977; Doerr et al., 1989) and the high rates of growth of the tumors in which Cx32 was absent ($Tr2_1$, Tr2 and Tr3) suggest that a threshold of intercellular communication may be necessary for retardation of tumor growth. [That the poorly coupled transfectant (Tr5) had the highest growth rate of all may have been due to transcriptional interference with a growth inhibitor and could also reflect the existence of a threshold degree of coupling for retardation of tumor growth.] Current views regarding the involvement of gap junctions in carcinogenesis (Loewenstein, 1979; Trosko et al., 1988; Yamasaki, 1988) consider that interactions with surrounding cells serve to restore the mutant cells to the non-malignant phenotype. Weak coupling might be inadequate for this purpose, which would also explain the frequent observation of at least a low incidence of gap junctions within rapidly dividing tumors (Weinstein and Pauli, 1986; Sheridan, 1987).

Clearly, extent of cell coupling is not the only determinant of growth rate. In various tissues, differentiated cells behave differently; in transformed cells the extent of coupling can vary widely among tumorigenic cell types. Nevertheless, this study clearly shows that in a single cell line which lacks detectable gap junctions, the growth rate of tumors can be modified substantially by expression of intercellular communication.

Acknowledgements

Supported in part by NIH grants DK41918, NS16524 and HL37449 (to DCS), NS20778 and NS20013 (to JAK), and NS07512 (PI: MVL Bennett, subprojects to DCS and JAK), a Grant-in-Aid from the New York Chapter of the American Heart Association (to DCS) and grants from the American Cancer Society (BC439 and BC675 to LMR).

References

Atkinson, M.M. and Sheridan, J.D. (1985) Reduced junctional permeability in cells transformed by different viral oncogenes. In: M.V.L. Bennett and D.C. Spray (Eds.), Gap Junctions. Cold Spring Harbor Laboratory, Cold Spring Harbor, NY, pp. 205–213.

Azarnia, R., Reddy, S, Kmiecik, T.E., Shalloway, D. and Loewenstein, W.R. (1988) The cellular *src* gene: regulation of communication and growth. In: E.L. Hertzberg and R.G. Johnson (Eds.), Gap Junctions. Alan R. Liss, New York, pp. 423–433.

Bennett, M.V.L., Spray, D.C. and Harris, A.L. (1981) Gap junctions in development. Am. Zool. 21, 413–427.

Cheng, C.Y., Ryan, R.F., Vo, T.P. and Hornsby, P.J. (1989) Cellular senescence involves stochastic processes causing loss of expression of differentiated function genes: Transfection with SV40 as a means for dissociating effects of senescence on growth and on differentiated function gene expression. Exp. Cell Res. 180, 49–62.

Doerr, R., Zvibel, I., Chiuten, D., D'Olimpio, J. and Reid, L.M. (1989) Clonal growth of tumors on tissue-specific biomatrices and correlation with organ site specificity of metastases. Cancer Res. 49, 384–392.

Eghbali, B., Kessler, J.A. and Spray, D.C. (1990) Expression of gap-junction channels in a communication-incompetent cell line after transfection with connexin32 cDNA. Proc. Natl. Acad. Sci. USA 87, 1328–1331.

Eghbali, B., Chalazonitis, A., Dermietzel, R., Kalberg, C., Chiu, F.C., Kessler, J.A., and Spray, D.C. (1991a) Transfection of pc12 cells with connexin26 (cx26) cDNA alters phenotype and response to NGF. J. Cell Biol. 115, 191a (abstr).

Eghbali, B., Kessler, J.A., Reid, L.M., Roy, C. and Spray, D.C. (1991b) Involvement of gap junctions in tumorigenesis: Transfection of hepatoma cells with connexin32 cDNA retards growth *in vivo*. Proc. Natl. Acad. Sci. USA 88, 10701–10705.

Fogh, J., Wright, W.C., and Loveless, J.D. (1977) Absence of HeLa cell contamination in 169 cell lines derived from human tumors. J. Natl. Cancer Inst. 68, 507–517.

Furshpan, E.J. and Potter, D.D. (1968) Low-resistance junctions between cells in embryos and tissue culture. Curr. Top. Dev. Biol. 3, 95–127.

Guthrie, S.C. and Gilula, N.B. (1989) Gap-junctional communication and development. Trends Neurosci 12, 12–16.

Kanno, Y. (1985) Modulation of cell communication and carcinogenesis. Jpn. J. Physiol. 35, 693–707.

Loewenstein, W.R. (1979) Junctional intercellular communication and the control of growth. Biochim Biophys Acta 56, 1–65.

Mehta, P.P., Bertram, J.S. and Loewenstein, W.R. (1986) Growth inhibition of transformed cells correlates with their junctional communication with normal cells. Cell 44, 187–196.

Nicolson, G., Dulski, K.M. and Trosko, J.E. (1988) Loss of intercellular junctional communication correlates with metastatic potential in mammary adenocarcinoma cells. Proc. Natl. Acad. Sci. USA 85, 473–476.

Pitts, J. and Finbow, M.E. (1986) The gap junction. J. Cell Sci., Suppl. 4, 239–266.

Reid, L.M., Abreu, S.L. and Montgomery, K. (1988) Extracellular matrix and hormonal regulation of synthesis and abundance of messenger RNAs in cultured liver cells. In: I.M. Arias, W.B. Jakoby, H. Popper, D. Schachter and D.A. Shafritz (Eds.), The Liver: Biology and Pathobiology, 2nd edition. Raven Press, New York, pp. 717–737.

Sheridan, J.D. (1987) Cell communication and growth. In: W. DeMello (Ed.), Cell-to-Cell Communication. Plenum, New York, pp. 187–222.

Shouval, D., Rager-Zisman, B., Quan, P., Shafritz, D.A., Bloom, B.R. and Reid, L.M. (1983) Role in mice of interferon and natural killer cells in inhibiting human hepatocellular carcinoma cells infected with hepatitis B virus. J. Clin. Invest. 72, 707–717.

Trosko, J.E., Chang, C.C., Madhukar, B.V., Oh, S.Y., Bombick, D., and El-Fouly, M.H. (1988) Modulation of gap-junction intercellular communication by tumor promoting chemicals, oncogenes and growth factors during carcinogenesis. In: E.L. Hertzberg and R.G. Johnson (Eds.) Gap Junctions. Alan R. Liss, New York, pp. 435–448.

Weinstein, R.S. and Pauli, B.U. (1986) Cell junctions and the behavior of cancer. In: Junctional Complexes in Epithelial Cells (Ciba Foundation Symposium 125). Wiley, Chichester, UK, pp. 240-260.

Welsch, F. and Stedman, D.B. (1984) Inhibition of metabolic cooperation between Chinese hamster V79 cells by structurally diverse teratogens. Teratogen. Carcinogen. Mutagen. 4, 285–301.

Yamasaki, H. (1988) The role of gap-junctional intercellular communication in malignant cell transformation. In: E. Hertzberg and R.G. Johnson (Eds.), Gap Junctions. Alan R. Liss, New York, pp. 449–465.

Yotti, L.P., Chang, C.C., and Trosko, J.P. (1979) Elimination of metabolic cooperation in Chinese hamster cells by a tumor promoter. Science 206, 1089–1091.

Zhu, D., Caveney, S., Kidder, G.M. and Naus, C.G. (1991) Transfection of C6 glioma cells with connexin43 cDNA: Analysis of expression, intercellular coupling and cell proliferation. Proc. Natl. Acad. Sci. USA 88, 1883–1887.

J.E. Hall, G.A. Zampighi and R.M. Davis (Eds.)
Progress in Cell Research, Vol. 3

CHAPTER 45

Suppression of gap-junction gene expression by growth factors and TPA in human epidermal keratinocytes in vitro

EMMANUEL DUPONT, BURRA V. MADHUKAR and JAMES E. TROSKO

Department of Pediatrics/Human Development, Michigan State University, East Lansing, MI 48824, USA

Introduction

Gap junction protein-mediated intercellular communication (GJIC) has been implicated to regulate fundamental biological processes in many types of cells and tissues. A role for inhibited communication has also been inferred in carcinogenesis, based upon the observation that many cancer cells have lost the ability to communicate with their normal neighbors via gap junctions, and that many tumor-promoting agents inhibit GJIC both in vitro and in vivo. Several mechanisms have been postulated to be involved in the regulation of gap-junction permeability, some involved in the up-regulation of GJIC, while others (e.g., Ca_i^{2+}, pH, free radicals, protein kinase C) in the down-regulation of GJIC (see reviews by Loewenstein, 1990; Trosko et al., 1990a, b). Since tumor promoters, which block cell-cell communication and promote growth of initiated cells, are in essence acting as growth factors, it is conceivable that growth factors which promote proliferation of normal cells can also block cell-cell communication, albeit transiently. In order to study the regulation of gap-junctional intercellular communication, we have developed a primary cell culture model of normal human epidermal keratinocytes (NHEK) (Madhukar et al., 1989). Using this model, we have shown that, in a serum-free medium epidermal growth factor (EGF), transforming growth factor-β (TGF-β), and a tumor promoter, 12-O-tetradecanoyl-13-phorbol acetate (TPA), down-regulated GJIC of these cells. Furthermore, we have observed that, in these cells, EGF sustained the proliferative potential, whereas transforming TGF-β and TPA both inhibited DNA synthesis and induced differentiation response and down-regulated GJIC. These observations have led us to consider a possible role for GJIC in both cell proliferation and differentiation. In light of the recent reports that multiple gap-junction proteins are expressed by many cell types in culture, we extended our investigations of gap junctions in NHEK cells to determine the mechanisms of GJIC regulation by these agents. Here we report that EGF-, TGF-β- or TPA-induced down-modulation of GJIC involves transcriptional regulation of gap-junction genes, and that each of these agents has specific effects on the regulation of the two connexin proteins, Cx43 and Cx26.

Materials and methods

Cell culture

Primary cultures of normal human epidermal keratinocytes (NHEK) derived from neonatal foreskin were obtained as proliferating or frozen secondary cultures from Clonetics Corporation, San Diego, CA. They were subcultured and used between passages 3 to 5, after which they showed extensive differentiation and decreased proliferation. When cells were maintained in keratinocyte basal medium (KBM) for the various experiments, they were first seeded in complete medium for at least 24 h and then shifted to KBM. The cell-culture conditions have been described in detail (Madhukar et al., 1989). Briefly, NHEK were grown in KGM (keratinocyte growth medium) containing 10 ng/ml EGF, 5 μg/ml insulin, 0.5 μg/ml hydrocortisone and 10 μg/ml transferrin (transferrin replaces the whole bovine pituitary extract of the original formula) plus 50 μg/ml gentamicin in a humidified 5% CO_2 incubator at 37°C until they were 70–80% confluent.

For treatment by growth factors or chemicals, the KGM was removed, cells were washed with phosphate-buffered saline without divalent cations (Mg–Ca free PBS). The cells were thereafter treated with EGF 10 ng/ml, TGF 1 ng/ml, TPA 10 or 100 ng/ml, or 10% fetal bovine serum in KBM for different times. Control experiments were carried out by incubating the cells in KGM or KBM for the same amount of time.

Chemicals and biochemicals

EGF, transferrin, insulin from bovine pancreas and hydrocortisone were obtained from Sigma Chemical Co., St. Louis, MO. Human TGF-β was obtained from R and D

Systems, Inc., Minneapolis, MN. 12-O-tetradecanoyl-phorbol-13-acetate (TPA) was purchased from LC Services Corporation, Woburn, MA. All the test compounds were used as 1000- or 200-fold stock solutions and were added directly to the cultures. All other chemicals and biochemicals were high-purity analytical grade or molecular biology grade. All solutions for RNA preparation were rendered RNase free with diethylpyrocarbonate (DEPC) or were made in DEPC-treated water followed by sterilization by autoclaving or filtration.

Assay for cell-cell communication

GJIC of NHEK was measured using a dye-transfer technique, as described previously (Oh et al., 1988; Madhukar et al., 1989). In brief, monolayers of NHEK (80% confluent) under different treatment conditions were rinsed in PBS; 2 ml of PBS containing 0.05% Lucifer Yellow (LY) was added. The dye uptake was facilitated by scraping the monolayer with a sharp surgical knife. The transient membrane rupture caused by scraping allowed the dye to be taken up by the ruptured cells (primary loaded cells). After 3 min, the dye solution was removed and the monolayers were washed three times with PBS followed by fixing in 10% formaldehyde in PBS. Photographs of the dye-loaded cells were taken under a Nikon epifluorescent microscope. The relative gap-junctional communication competence under different treatments was determined by the extent of dye distribution from the primary dye-loaded cells into the neighboring contacting cells.

Northern blot analysis

Total RNA was extracted from monolayers according to Chomczynski and Sacchi (1989). Twenty micrograms of total RNA was electrophoresed on formaldehyde-containing gel and capillary-blotted on nylon membrane (Hybond-N Amersham, exactly as described by Sambrook et al. (1989). After baking at 80°C for 2 h, the membrane was exposed to shortwave UV light for 4–5 min in order to crosslink RNAs.

cDNA inserts for Cx26 (Zhang and Nicholson, 1989), Cx43 (Beyer et al., 1987) and chicken β-actin (PvuII fragment of $p\alpha_1$, 0.8 kb, Cleveland et al., 1980) were used for hybridization of the RNA blots. The inserts were separated from the vectors after digestion with the appropriate restriction enzymes (*Eco*RI for Cx26 and 43, *Eco*RI and *Hind*III for β-actin) and electrophoresis in low-melting agarose gel. The cDNA fragments were sliced from the gel and purified using phenol and chloroform extraction as described by Sambrook et al. (1989). The purified DNA inserts were labeled using ^{32}P α-dCTP (Dupont NEN 3000 Ci/mmol.) and a random primer labeling kit supplied by Boehringer Mannheim Biochemicals exactly as described by the supplier (specific activities were always $> 1.5 \times 10^9$ dpm/μg).

Following pre-hybridization for 30 min at 65°C in 10 ml of 5× SSC, 5× Denhardt reagent, 0.5% SDS, 250 μg/ml sonicated and heat-denatured herring sperm DNA and 5% dextran sulfate, the membranes were hybridized by adding the ^{32}P-labeled probes to the pre-hybridization solution and incubating for 15 h (overnight) at 65°C. The hybridized membranes were washed twice in 2× SSC plus 0.1% SDS for 15 min at room temperature, 2× in 0.5% SSC plus 0.1% SDS for 30 min at 65°C, and membranes were exposed to a Kodak x-omat x-ray film at –70°C for 12–24 h. The films were developed in an automatic x-ray film processor.

Membrane dehybridizations were carried out in a solution containing 50% Formamide, 2× SSC, 0.1% SDS at 75°C for 1 h followed by a brief rinse in 2× SSC. The dehybridized membranes were used for hybridization with a different probe as described above.

Densitometry

Quantitation of the levels of RNA expressed under different conditions was achieved by densitometric scanning of the autoradiographs with a Gibson Response Spectrophotometer equipped with autoradiogram holder and scanning accessory.

Results

Gap-junctional intercellular communication of NHEK

The results of LY transfer in human keratinocytes further confirmed our previous report (Madhukar et al., 1989) on the regulation of GJIC by the growth factors, EGF and TGF-β. Data presented in Fig. 1A–F clearly indicate that when NHEK were grown in the completely defined growth medium, KGM, containing EGF, insulin, hydrocortisone and bovine pituitary extract or transferrin, the communication among these cells was minimal, or none at all, as measured by the limited spread of LY into the monolayer (2–3 rows). A 24-h incubation of the cells in the basal medium devoid of these proliferation-promoting factors significantly increased the level of communication among these cells as indicated by the extensive appearance of the dye (10–12 rows) (Fig. 1C). However, when the incubation was done in the presence of EGF (10 ng/ml) or TGF-β (1 ng/ml), the down regulation of dye transfer continued, suggesting that the presence of EGF in the complete growth medium might be responsible for the lack of GJIC of keratinocytes in KGM. The tumor-promoting agent, TPA, also completely abolished dye transfer in NHEK following 1 h or 24 h of exposure (Fig. 1E,F).

Connexin genes expressed by NHEK

In order to determine whether the down regulation of GJIC of NHEK in KGM, or in the presence of growth fac-

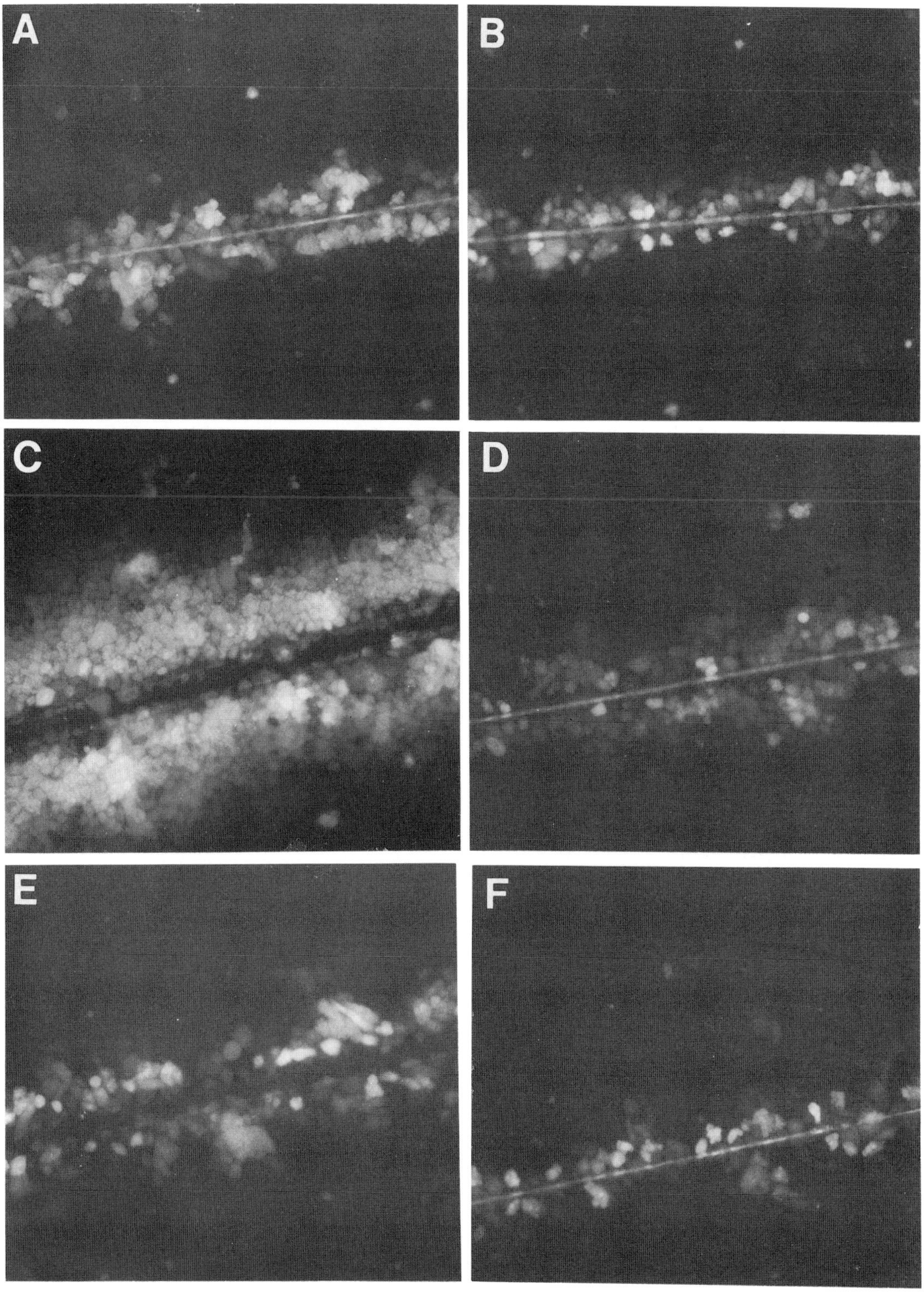

Figure 1. Gap-junctional intercellular communication (GJIC) in NHEK cultured in a serum-free medium, with or without growth factors. GJIC was measured by a dye-transfer technique as described. LY distribution in: (A) KGM; (B) EGF, 10 ng/ml in KBM; (C) KBM; (D) TGF-β, 1 ng/ml in KBM; (E) TPA, 10 ng/ml, 1 h; (F) TPA, 10 ng/ml, 24 h. In B–D, GJIC was assayed 24 h after growing the cells in KBM without or with the test agents.

tors, was due to an altered regulation of connexin (Cx) gene expression, we analyzed the RNA isolated from these cells for the expression of Cx26, Cx32 and Cx43 genes by Northern blot hybridization. The results indicated that NHEK express both Cx26 and Cx43 mRNA (Fig. 2A,B). We failed to detect any messenger for Cx32 under the same conditions. The molecular sizes of these two connexins were the same as those for rat heart Cx43 (3.0 kb) and rat liver Cx26 (2.5 kb).

Modulation of connexin transcription by growth factors

After determining the types of connexin gene expressed by NHEK, we examined whether EGF, TGF-β or other growth factors, as well as the tumor promoter, TPA, or 10% FBS influences the expression of these connexin genes to down-regulate GJIC. The data on the changes in the levels of Cx43 or Cx26 mRNA under different growth conditions of NHEK are presented in Figs. 2 and 3. Following culture of NHEK in the complete growth medium, KGM, they were shifted to growth factor-free medium, KBM, for 12 or 24 h with or without added EGF or TGF-β. There was a 50–70% increase in the level of Cx43 over that of cells grown in KGM when the cells were incubated in KBM for 24 h (Fig. 2A).

On the other hand, in cells incubated in KBM plus EGF (10 ng/ml) for 12 or 24 h, the level of Cx43 expression was comparable to that of KGM (18–20% of KBM levels, Figs. 2A and 3). Cx43 expression was not appreciably decreased when NHEK were incubated in KBM in the presence of TGF-β (1 ng/ml) for 24 h (Figs. 2A and

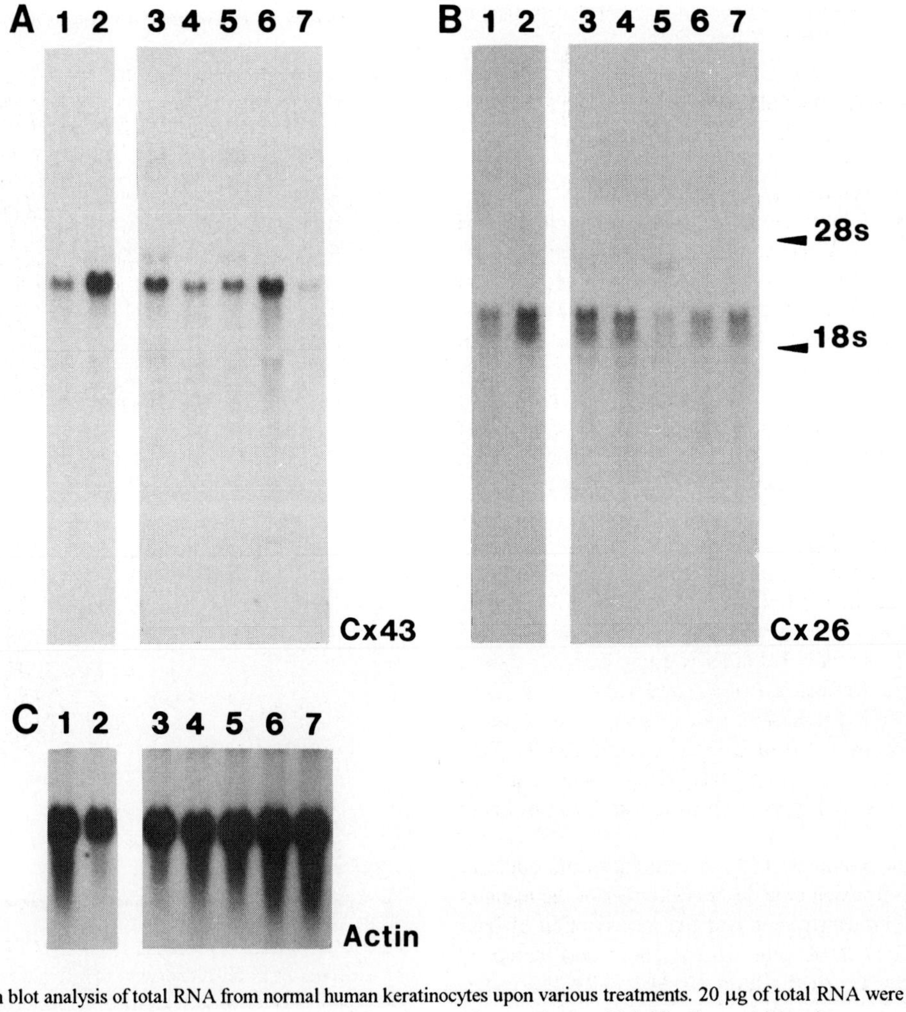

Figure 2. Northern blot analysis of total RNA from normal human keratinocytes upon various treatments. 20 μg of total RNA were fractioned by electrophoresis in formaldehyde-denaturing agarose gel and transferred onto nylon membranes. Hybridizations at high stringency were carried out successively with the following random primer labeled probes: Panel A: Clone G2 insert for Cx43; Panel B: Clone Cx26-1 insert for Cx26; Panel C: Clone pAI insert for β-actin. In each panel the lanes 1–7 are as follows: lane 1, KGM for an additional 24 h; lane 2, KBM for 24 h; lane 3, KBM + EGF 10 ng/ml for 12 h; lane 4, KBM + EGF 10 ng/ml for 24 h; lane 5, KBM + EGF 10 ng/ml and TGF-β 1 ng/ml for 24 h; lane 6, KBM + TGF-β 1 ng/ml for 24 h; lane 7, KBM + 10% FCS for 24 h. Arrowhead on the right side indicates the position of the 28S and 18S rRNA. All probes react with a single messenger at 3 kb for Cx43, 2.5 kb for Cx26, and 2 kb for β-actin.

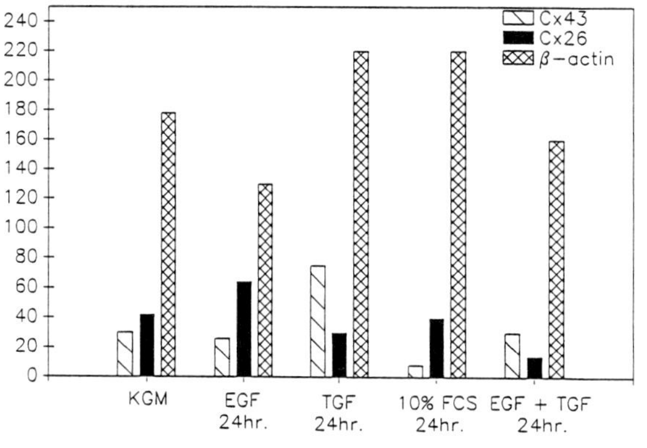

Figure 3. Densitometric analysis of the autoradiograph of the Northern blots hybridized with Cx43, Cx26 or β-actin probes as described under Fig. 2. The mRNA levels in each case are expressed as % of the mRNA level of these genes in cells grown in KBM. The level of their expression in keratinocytes maintained in KBM for 24 h is considered 100%.

3). When the cells were incubated in KBM with EGF and TGF-β together, the reduction in the level of Cx43 mRNA was comparable to that of EGF alone, suggesting no additive or synergistic action. The most significant reduction (90%) in Cx43 message levels was seen after 24 h in KBM with 10% FBS (Figs. 2A and 3). The changes in the Cx43 levels of NHEK following growth factor treatment or incubation in KBM were not due to the amount of RNA hybridized, as indicated by subsequent hybridization of the same membrane with a cDNA probe for β-actin (Figs. 2C and 3).

The changes in the expression of Cx26 of NHEK were markedly different from those of Cx43 under similar treatment conditions. Data in Figs. 2B and 3 show that replacing KGM with KBM significantly increased the Cx26 expression. However, unlike in the case of Cx43, incubation of the cells in KBM containing EGF for 12 h did not change the level of Cx26. After 24 h in KBM plus EGF, there was a 30% reduction in the level of Cx26 mRNA compared with that of KBM (Fig. 3). In contrast to the insignificant effect of EGF on Cx26 expression, incubation with TGF-β in KBM for 24 h decreased the level of Cx26 by 70% over that of KBM (Figs. 2B and 3). EGF and TGF-β, when added together, showed synergistic interaction in decreasing the expression of Cx26 by more than 85%.

The tumor promoter, TPA, in contrast to the effect of EGF or TGF-β which only decreased either of the connexin levels, significantly inhibited the transcription of both Cx26 and Cx43 RNA (Fig. 4A,B). There was, however, an initial increase in the level of Cx26 mRNA between 3–6 h after treatment with 10 ng/ml of TPA; this declined to 35% of the level in KBM by 12 h. At the end of 18, 24, and 48 h post-treatment, the expression of both Cx26 and Cx43 RNAs was completely abolished. A similar abolition of the expression of the two connexin mRNAs was observed at 100 ng/ml TPA treatment. There was, however, no increase in the level of Cx26 message at 3 and 6 h post-treatment. Furthermore, connexin mRNA showed extensive degradation at the high dose of TPA treatment as indicated by the smeary appearance of the hybridized mRNA in the autoradiographs.

Discussion

In the present investigation we have demonstrated that, in primary cultures of NHEK, gap-junction gene expression was suppressed during active proliferation of cells in growth medium containing growth factors such as EGF. Upon switching the cells to a basal medium lacking these factors, specifically EGF, the cells showed enhanced communication, as measured by dye transfer, concomitant

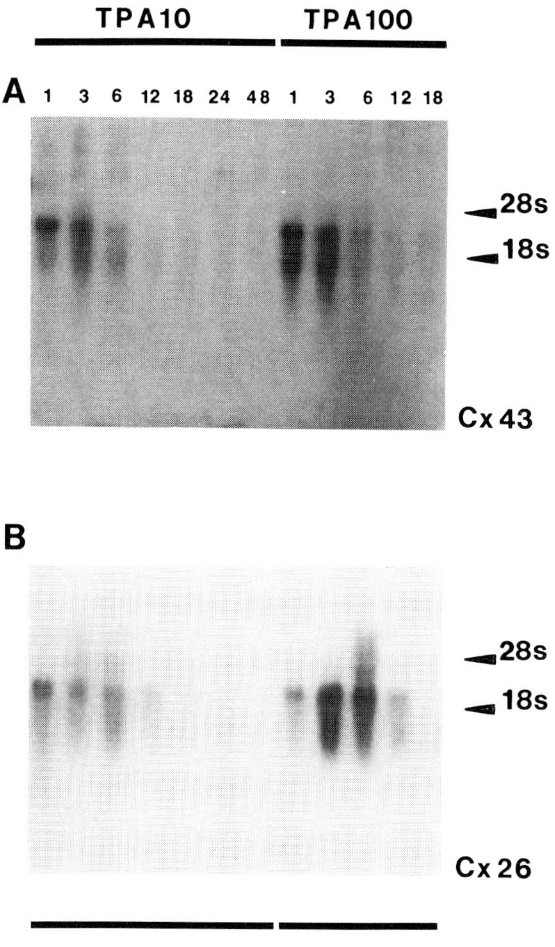

Figure 4. Northern blot analysis of total RNA from normal human keratinocytes after TPA treatment. NHEK were treated with 10 or 100 ng/ml TPA for indicated period of time (numbers above each lane). 20 μg of total RNA were fractionated in formaldehyde-containing agarose gel, capillary transferred on nylon membranes and hybridized at high stringency successively with Cx43 and Cx26 radiolabeled probes. Note that at longer incubation times RNA becomes increasingly smeary, indicating some degradation of the messenger.

with a marked increase in the expression of two gap-junction genes, Cx43 and Cx26. These observations suggest that suppressed GJ gene expression and the resultant down regulation of gap-junctional communication might be involved in cell proliferation. Other workers have previously reported that GJIC was markedly decreased in hepatocytes after partial hepatectomy, as well as during cell division, strengthening our contention that mitogenic activity and mitogenic agents, such as EGF or serum, can down-regulate GJIC (Meyer et al., 1981; Dermietzel et al., 1987). Although the exact mechanisms or factors involved in cell proliferation are not yet clearly understood, it is reasonable to postulate that down regulation of gap junctions can facilitate the buildup to threshold levels of the mitogenic signals required for cell division, which otherwise might not reach threshold levels.

While transcriptional regulation of gap-junction genes by EGF could be considered to be involved in the proliferation of NHEK, in the current study we have observed that TGF-β, which inhibits proliferation of NHEK and induces differentiation, also down-regulated gap junctions, as well as the expression of Cx26, but did not significantly affect the levels of Cx43 mRNA. TGF-β has been shown in various cell culture systems, particularly the epithelial cells, to inhibit DNA synthesis and induce reversible differentiation (Coffey et al., 1987, Coffey et al., 1988; Choi and Fuchs, 1990; Matsumoto et al., 1990; Stainano-Caico et al., 1990; Sporn and Roberts, 1991; Vollberg et al., 1991). On the other hand, it is a mitogen for several types of mesenchymal cells (see reviews by Moses et al., 1990; Sporn and Roberts, 1990). The differential effects of EGF and TGF-β on the expression of Cx26 and Cx43 thus suggest distinct roles for these two types of gap junctions in cell proliferation and differentiation.

We have also demonstrated that the potent skin-tumor promoter, TPA, down-regulated gap-junction gene expression, resulting in decreased GJIC. In primary NHEK cultures, TPA inhibited DNA synthesis and induced differentiation. In addition, TPA exerts its pleiotropic biochemical effects by activating the calcium- and phospholipid-dependent protein kinase C (PKC) (Neidel et al., 1983; Ashendal, 1985; Castagna, 1987). This kinase has increasingly been recognized to play a central role in cell proliferation, tumor promotion, as well as differentiation (reviewed in Nishizuka, 1984). While it is still not clear whether specific isozymes of PKC have distinct roles in cellular proliferation or differentiation, such specificity has been implicated in human promyelocytic leukemia (HL-60) cells where TPA induces differentiation (Trayner and Clemens, 1990; Zwelling et al., 1990; Shimizu et al., 1991; see also review by Kiss, 1990). TPA can exert a dual effect on keratinocyte growth and differentiation through transcriptional regulation of gap-junction gene expression. This dual effect should facilitate isolating the differentiation-committed cells from those which are proliferatively active. It has in fact been suggested that TPA can exert this dual effect by transcriptional regulation of the expression of *both* TGF-α and TGF-β (Akhurst et al., 1988; Pittelkow, 1989). Since TGF-α is a proliferative growth factor for NHEK while TGF-β is a growth inhibitor/differentiation inducer, the effect of TPA in down-regulating both Cx26 and Cx43 genes may be related to its induction of TGF-α and TGF-β expression. Unlike EGF or TGF-β, which affected the transcription of either Cx43 or Cx26 genes, TPA blocked the transcription of both the genes, suggesting its dual role as a growth promoter as well as an inducer of differentiation.

In the present study, we have also demonstrated that addition of 10% FBS to NHEK in KBM induced a marked down regulation in the levels of both Cx26 and Cx43, similar to the effect of TPA. Serum addition also induces a dramatic change in the morphology of NHEK within a few hours, followed by inhibition of proliferation and induction of terminal differentiation. It is conceivable, though speculative, that serum induces the expression of a set of genes which, in turn, inhibit the expression of connexin genes. It should be interesting to study which factors in serum affect gap junctions.

In summary, we have demonstrated that growth factors and TPA regulate gap-junction protein expression of NHEK at the level of transcription. This regulation may possibly play a role in the proliferative and differentiation responses evoked by these agents. It is unclear at the present time whether such regulation of gap junctions by these agents is restricted to only NHEK or is a general response in other primary epithelial cells. Therefore, further investigations on the regulation of gap-junction proteins by growth factors in other primary cell culture systems to determine the generality of their effect on connexin proteins are clearly necessary.

Acknowledgments

Research on which this chapter was based was supported by grants from the U.S. Air Force Office for Scientific Research (AFOSR-89-0325) and the NIEHS (1P42ES-04911).

References

Akhurst, R.J., Fee, F. and Balmain, A. (1988) Localized production of TGF-β mRNA in tumor promoter-stimulated mouse epidermis. Nature 331, 363–365.

Ashendel, C.L. (1985) The phorbol ester receptor: A phospholipid-regulated protein kinase. Biochim. Biophys. Acta 822, 219–242.

Bertram, J.S. (1990) Role of gap-junctional cell-cell communication in the control of proliferation and neoplastic transformation. Radiat. Res. 123, 252–256.

Beyer, E.C., Paul, D.L. and Goodenough, D.A. (1987) Connexin 43: A protein from rat heart homologous to a gap-junction protein from liver. J. Cell Biol. 105, 2621–2629.

Castagna, M. (1987) Phorbol esters as signal transducers and tumor promoters. Biol. Cell 59, 3–14.

Choi, Y. and Fuchs, E. (1990) TGF-β and retinoic acid: Regulators of growth and modifiers of differentiation in human epidermal cells. Cell Reg. 1, 791–809.

Chomczynski, P. and Sacchi, N. (1987) Single-step method of RNA isolation by acid guanidinium thiocyanate-phenol-chloroform extraction. Anal. Biochem. 162, 156–159.

Cleveland, D.W., Lopata, M.A., MacDonald, R.J., Cowan, N.J., Rutter, W.J. and Kirschner, M.W. (1980) Number and evolutionary conservation of α- and β-tubulin and cytoplasmic β- and γ-actin genes using specific cloned cDNA probes. Cell 20, 95–105.

Coffey, R.J., Derynck, R., Wilcox, J.N., Bringman, T.S., Goustin, T.S., Moses, H.L. and Pittelkow, M.R. (1987) Production and autoinduction of transforming growth factor-α in human keratinocytes. Nature 328, 817–820.

Coffey, R.J., Bascom, C.C., Sipes, N.J., Graves-Deal, R., Weissman, B.E. and Moses, H.L. (1988) Selective inhibition of growth-related gene expression in murine keratinocytes by transforming growth factor-β. Mol. Cell. Biol. 8, 3088–3093.

Dermietzel, R., Yancey, S.B., Traub, O., Willecke, K. and Revel, J.P. (1987) Major loss of the 28 kD protein of gap junctions in proliferating hepatocytes. J. Cell Biol. 105, 1925–1934.

Dermietzel, R., Hwang, T.K. and Spray, D.S. (1990) The gap-junction family: structure, function and chemistry. Anat. Embryol. 182, 517–528.

Eckert, R.L. and Rorke, E.A. (1989) Molecular biology of keratinocyte differentiation. Environ. Health Perspect. 80, 109–116.

Edwards, D.R. and Heath, J.K. (1991) Regulation of transcription by transforming growth factor-β. In: P. Cohen and J.G. Foulkes (Eds.), The Hormonal Control and Regulation of Gene Transcription, Elsevier, Amsterdam, pp. 333–347.

Flagg-Newton, J.L., Dahl, G. and Loewenstein, W.R. (1981) Cell junction and cyclic AMP: Up-regulation of junctional membrane permeability and junctional membrane particles by administration of cyclic nucleotide or phosphodiesterase inhibitor. J. Membr. Biol. 63, 105–121.

Fuchs, E. (1990) Epidermal differentiation: The bare essentials. J. Cell Biol. 111, 2807–2814.

Kam, E., Watt, F.M. and Pitts, J.D. (1987) Patterns of junctional communication in skin: Studies on cultured keratinocytes. Exp. Cell Res. 173, 431–438.

Kiss, Z. (1990) Effects of phorbol ester on phospholipid metabolism. Prog. Lipid Res. 29, 141–166.

Loewenstein, W.R. (1990) Regulation of cell-to-cell communication by phosphorylation. Biochem. Soc. Symp. 50, 43–58.

Madhukar, B.V., Oh, S.Y., Chang, C.C., Wade, M. and Trosko, J.E. (1989) Altered regulation of intercellular communication by epidermal growth factor, transforming growth factor-β and peptide hormones in normal human keratinocytes. Carcinogenesis 10, 13–20.

Matsumoto, K., Hashimoto, K., Hashiro, M., Yoshimasa, H. and Yoshikawa, K. (1990) Modulation of growth and differentiation in normal human keratinocytes by transforming growth factor-β. J. Cell Physiol. 145, 95–101.

Meyer, D.J., Yancey, S.B. and Revel, J.P. (1981) Intercellular communication in normal and regenerating rat liver: A quantitative analysis. J. Cell Biol. 91, 505–523.

Neidel, J.E., Kuhn, L.J. and Vanderbank, G.R. (1983) Phorbol diester receptor co-purifies with protein kinase C. Proc. Natl. Acad. Sci. USA 80, 36–40.

Nishizuka, Y. (1984) The role of protein kinase C in cell surface signal transduction and tumor promotion. Nature 308, 693–698.

Oh, S.Y., Madhukar, B.V. and Trosko, J.E. Inhibition of gap-junctional blockage by palmitoyl carnitine and TMB-8 in a rat liver epithelial cell line. Carcinogenesis 9, 135–139.

Pittelkow, M.R., Lindquist, B.P., Abraham, R.T., Graves-Deal, R., Derynck, R. and Coffey, R.J. (1989) Induction of transforming growth factor-α expression in human keratinocytes by phorbol esters. J. Biol. Chem. 264, 5164-5171.

Sambrook, J., Fritsch, E.F. and Maniatis, T. (1989) Molecular Cloning: A Laboratory Manual. Cold Spring Harbor Laboratory Press, Cold Spring Harbor, NY.

Shimizu, N., Ohta, M., Fujiwara, C., Sagara, J., Mochizuki, N., Oda, T. and Utiyama, H. (1991) Expression of a novel immediate early gene during 12-O-tetradecanoylphorbol-13-acetate-induced macrophagic differentiation of HL-60 cells. J. Biol. Chem. 266, 12157–12161.

Sporn, M.B. and Roberts, A.B. (1991) Minireview: Interactions of retinoids and transforming growth factor-β in regulation of cell differentiation and proliferation. Mol. Endocrinol. 5, 3–7.

Staiano-Coico, L., Khandke, L., Krane, J.F., Sharf, S., Gottlieb, A.B., Krueger, J.G., Heim, L., Rigas, B. and Higgins, P.J. (1990) TGF-α and TGF-β expression during sodium-N-butyrate-induced differentiation of human keratinocytes — evidence for subpopulation-specific up regulation of TGF-β messenger RNA in suprabasal cells. Exp. Cell Res. 191, 286–291.

Trayner, I.D. and Clemens, M.J. (1990) Phorbol ester-induced macrophage-like differentiation of human promyelocytic leukemia (HL-60) cells occurs independently of transferrin availability. Cancer Res. 50, 7221–7225.

Trosko, J.E., Chang, C.C. and Madhukar, B.V. (1990a) Modulation of intercellular communication during radiation and chemical carcinogenesis. Radiat. Res. 123, 241–251.

Trosko, J.E., Chang, C.C. and Madhukar, B.V. (1990b) Cell-cell communication: Relationship of stem cells to the carcinogenic process. In: D.E. Stevenson, J.A. Popp, J.M. Ward, R.M. McClain, T.J. Slaga and H.C. Pitot (Eds.), Mouse Liver Carcinogenesis: Mechanisms and Species Comparisons, Alan R. Liss, New York, pp. 259–276.

Vollberg, T.M., George, M.D. and Jetten, A.M. (1991) Induction of extracellular matrix gene expression in normal human keratinocytes by transforming growth factor-β is altered by cellular differentiation. Exp. Cell Res. 193, 93–100.

Watt, F. (1988) Keratinocyte cultures: An experimental model for studying how proliferation and terminal differentiation are coordinated in the epidermis. J. Cell Sci. 90, 253–259.

Wilke, M.S., Hsu, B.M., Wille, J.J., Pittelkow, M.R. and Scott, R.E. (1988) Biological mechanisms for the regulation of normal human keratinocyte proliferation and differentiation. Am. J. Pathol. 131, 171–181.

Yuspa, S.H., Ben, T., Hennings, H. and Lichti, W. (1982) Divergent responses in epidermal basal cells exposed to the tumor promoter 12-O-tetradecanoylphorbol-13-acetate. Cancer Res. 42, 2344–2349.

Zhang, J.T. and Nicholson, B.J. (1989) Sequence and tissue distribution of a second protein of hepatic gap junctions, Cx26, as deduced from its cDNA. J. Cell Biol. 109, 3391–3401.

Zwelling, L.A., Hinds, M., Chan, D., Altschuler, E., Mayes, J. and Zipf, T.F. (1990) Phorbol ester effects on topoisomerase II activity and gene expression in HL-60 human leukemia cells with different proclivities toward monocytoid differentiation. Cancer Res. 50, 7116–7122.

Index